PRINCIPLES OF COMMUNICATIONS
Systems, Modulation, and Noise

Fourth Edition

R. E. Z I E M E R
University of Colorado at Colorado Springs

W. H. T R A N T E R
University of Missouri–Rolla

JOHN WILEY & SONS, INC.
New York • Chichester • Brisbane • Toronto • Singapore

Cover designer: DesignHeads, Boston

Cover image: Andrea Way, Tidepool, 1988.
Ink and pencil on paper, 36 x 51½"
Private collection
Courtesy Brody's Gallery, Washington DC

Printed in the U.S.A.

Library of Congress Catalog Card Number: 94–76558

ISBN: 0-471-12496-6

10 9 8 7 6

CONTENTS

6 NOISE IN MODULATION SYSTEMS   394

7 BINARY DATA TRANSMISSION   453

8 ADVANCED DATA COMMUNICATIONS TOPICS   529

9 OPTIMUM RECEIVERS AND SIGNAL-SPACE CONCEPTS   603

10 INFORMATION THEORY AND CODING   667

APPENDIX A: PHYSICAL NOISE SOURCES AND NOISE CALCULATIONS IN COMMUNICATION SYSTEMS   747

APPENDIX B: JOINTLY GAUSSIAN RANDOM VARIABLES   773

PREFACE

In the fourth edition of this book we have maintained the objectives of the previous three editions. For the past 20 years we have attempted to provide in a single book a relatively complete treatment on the principles of communications suitable for seniors and beginning graduate students in electrical engineering. In addition to topics in communications, sections have been included on signals, systems, probability, and random processes, so that all of the necessary background is included within the pages of a single book. This has worked well for undergraduates, for new graduate students who may have forgotten some of the fundamentals, and for the working engineer who may use the book as a reference or who may be taking a course after hours.

A significant feature of the fourth edition of *Principles of Communications* is the inclusion of a number of "computer exercises" at the end of each chapter. These exercises are designed to make use of a computer in order to illustrate the principles involved. They follow the end-of-chapter problems in a section marked by a computer symbol. There are a number of motivations for the inclusion of these problems. The computer is now accepted as a valuable tool in engineering education and is becoming integrated into engineering curricula. Computer usage allows students to work more complex problems and to gain important insights through the working of these problems. While these problems can all be worked by writing programs in FORTRAN, Pascal, or C, we strongly recommend the use of scientific problem-solving programs such as MATLAB®, Mathcad® and Mathematica® to name only a few. Through the use of these tools, the programming time for the student is significantly minimized. In addition, the graphing support provided by these tools allows the student to plot waveforms and performance characteristics with ease. We also feel that it is important for students to be familiar with these tools and to use them in their day-to-day problem solving efforts.

The past twenty years have seen many new applications for communication systems and a number of new technologies have been developed for the implementation of these systems. While there is always a strong desire to include a wide variety of new applications and technologies in a new edition of a book, we strongly feel that a first course in communications serves the student best if the emphasis is placed on fundamentals. We feel that applications and technologies, which often have short lifetimes, are best treated in subsequent courses after students have mastered the basic theory and analysis techniques. We have, however, been sensitive to new techniques that are fundamental in nature and have added material as appropriate. A new section on trellis-coded modulation is an example. Other new sections are discussed in later paragraphs. In addition, a number of new problems and in-chapter examples have been added. Several less-useful sections have been deleted, resulting in a fourth edition that is actually a few pages shorter than the previous edition.

In order to compare the fourth edition of this book with the previous edition, we briefly consider the changes chapter by chapter. In Chapter 1, the tables have been updated and a brief discussion on atmospheric characteristics has been added. Chapter 2, which is essentially a review of signal and system theory, has been tightened by deleting several examples that seemed superfluous and a brief discussion of the decimation-in time FFT algorithm has been added.

Chapter 3, which is devoted to basic modulation methods, has been significantly expanded. A number of examples have been added which illustrate the effect of modulation in the time domain. Students usually find it much easier to visualize the effect of modulation in the frequency domain than in the time domain and therefore a number of waveforms corresponding to different modulation techniques have been added. In addition, the effect of interference in linear and nonlinear modulation has been combined into a single section. This section has also been expanded to include a discussion on spike noise resulting from origin encirclement in FM systems due to signal-tone interference. An understanding of this simple deterministic process will significantly help the student understand the origin encirclement phenomenon in the presence of additive Gaussian noise in Chapter 5 and thresholding in FM systems, studied in Chapter 6.

Chapters 4 and 5, which deal with probability and stochastic processes, have changed little from previous editions of this book. The most significant change in Chapter 4 is the use of the probability Q-function rather than the error function, $\text{erf}(x)$, to represent probabilities of events in a Gaussian environment. While this really represents only a notational change, it recognizes that the probability Q-function is now used extensively in communications engineering. In edition four, a table has been added that gives the most common probability density functions along with the means and variances of the random variables described by the density functions. Chapter 5 has been shortened slightly, and a short section has been added illustrating the determination of the system equivalent-noise bandwidth from the impulse response.

In Chapter 6, which treats the noise performance of various modulation schemes, the section on threshold extension has been rewritten so that the results are more general. The thresholding characteristics of an FM discriminator with a sinusoidal message signal are presented as a special case in an example problem.

Binary digital data transmission is the subject of Chapter 7. A new section has been added to this chapter dealing with fading channels. This is an important subject in today's world because of the rapidly expanding use of mobile communication systems. Chapter 8 treats more advanced topics in data communication systems including M-ary systems, bandwidth efficiency, synchronization, spread-spectrum systems and satellite links.

Chapter 9, which deals with optimum receivers and signal-space concepts is little changed from the previous edition. The chapter has been tightened, however, by deleting the topics of Weiner filtering and nonlinear estimation.

Chapter 10 provides the student with a brief introduction to the subjects of information theory and coding. As mentioned previously, an addition to this chapter is a brief discussion of the important subject of trellis-coded modulation, a very useful technique for the implementation of bandwidth and power efficient communication systems.

We have used this text for a two-semester course sequence for a number of years. The first course is a first-semester senior-level one covering Chapters 1, 2, 3, and Appendix A. The second course is a one-semester senior-level one covering Chapters 4 and 5 and selected topics from the remainder of the book. Some good students who have had a previous course on signal and system theory have successfully taken the second course without the first. A previous course on signal and system theory also allows the material in Chapter 2 to be reviewed rather than covered in detail. Hence, more material from Chapter 3 can be covered; for example, the coverage of the phase-lock loop can be expanded. Similarly, a previous course on probability allows the material in Chapters 4 and 5 to be reviewed, thereby allowing more topics on communications to be covered.

Although the text was designed primarily for a two-semester sequence, we included sufficient material to allow considerable flexibility in structuring many different course sequences based on the text. A solutions manual is available from the publisher. This manual contains complete solutions to all odd-numbered problems and answers to all even-numbered problems. Partial solutions are given to even-numbered problems in those cases in which we felt that some extra guidance would be helpful. In addition, complete solutions to all computer exercises are given in the solutions manual.

We wish to thank the many persons who have contributed to the development of this textbook and who have suggested improvements for the fourth edition. We especially thank our colleagues and students at the University of

Colorado at Colorado Springs and at the University of Missouri-Rolla for their comments and suggestions. We also express our thanks to the many colleagues who have offered suggestions to us by correspondence or verbally at technical conferences. The industries and agencies that have supported our research deserve special mention since, by working on these projects, we have expanded our knowledge and insight significantly. These include the National Aeronautics and Space Administration, the Office of Naval Research, the National Science Foundation, GE Aerospace, Motorola, Inc., Emerson Electric Company, Battelle Memorial Institute, and the McDonnell Douglas Corporation.

We also thank the reviewers of all editions of this book. The reviewers for the fourth edition deserve special mention. They were:

Derald Cummings, Penn State
Bruce Black, Rose-Hulman Institute of Technology
Steven Bibyk, Ohio State
Michael Fitz, Purdue
Laurence Hassebrook, Kentucky
Donley Winger, California Polytechnic State University (San Luis Obispo)
John Kieffer, Minnesota
Toby Berger, Cornell
B.V.K. Vijaya Kumar, Carnegie Mellon
John Komo, Clemson
Dan Schonfeld, Illinois (Chicago)
H.K. Hwang, California State Polytechnic University (Pomona)
Rao Yarlagadda, Oklahoma State

Dr. Ronald Iltis, from the University of California at Santa Barbara, deserves special mention. He provided a very helpful checkout review by reading the entire manuscript in page-proof form. All reviewers have contributed significantly to the fourth edition of this book. All remaining shortcomings are solely the responsibility of the authors.

Our students have provided continuing encouragement for our writing projects, and it is to them that we dedicate this book.

Finally, our wives, Sandy and Judy, deserve much more than a simple thanks for the patience that they have shown throughout our seemingly endless writing projects.

R. E. Z.
W. H. T.

PRINCIPLES OF COMMUNICATIONS

INTRODUCTION 1

As we approach the close of the twentieth century, we live in a world in which electrical communication is so commonplace that we pick up our cordless telephones without a second thought. Yet the importance of such communication in today's world is so crucial that we cannot imagine modern society without it. We are in an era of change, which some people refer to as the "information age," much like the era—more than 100 years ago— when the world underwent drastic changes because of the industrial revolution. The prosperity and continued development of modern nations will depend more on the originating and disseminating of *information,* rather than of manufactured goods. Examples illustrating that the value of information is often more than that of goods and services are provided by the fact that many portable computers are equipped with modems so that persons traveling on business are linked with their offices; by the number of motorists using cellular telephones in metropolitan areas; and by the fact that more and more mail transactions are now accomplished through computer, or electronic, mail.*

This book is concerned with the theory of systems for the conveyance of information. A *system* is a combination of circuits and devices put together to accomplish a desired result, such as the transmission of intelligence from one point to another. A characteristic of communication systems is the presence of uncertainty. This uncertainty is due in part to the inevitable presence in any system of unwanted signal perturbations, broadly referred to as *noise,* and in part to the unpredictable nature of information itself. Systems analysis in the presence of such uncertainty requires the use of probabilistic techniques.

Noise has been an ever-present problem since the early days of electrical communication, but it was not until the 1940s that probabilistic systems-analysis procedures were used to analyze and optimize communication sys-

* For more on this fascinating subject, see the *IEEE Communications Magazine,* Vol. 30, Oct. 1992 (special issue on "Realizing Global Communications").

tems operating in its presence [Wiener 1949; Rice 1944, 1945].* It is also somewhat surprising that the unpredictable nature of information was not widely recognized until the publication of Claude Shannon's mathematical theory of communications [Shannon 1948] in the late 1940s. This work was the beginning of the science of information theory, a topic that will be considered in some detail later.

A better appreciation of the accelerating pace at which electrical communication is developing can perhaps be gained by the historical outline of selected communications-related events given in Table 1.1.

With this brief introduction and history, we now look in more detail at the various components that make up a typical communication system.

1.1   THE BLOCK DIAGRAM OF A COMMUNICATION SYSTEM

Figure 1.1 shows a commonly used model for a communication system. Although it suggests a system for communication between two remotely located points, this block diagram is also applicable to remote sensing systems, such as radar or sonar, in which the system input and output may be located at the same site. Regardless of the particular application and configuration, all information transmission systems invariably involve three major subsystems—a transmitter, the channel, and a receiver. In this book we will usually be thinking in terms of systems for transfer of information between remotely located points. It is emphasized, however, that the techniques of systems analysis developed are not limited to such systems.

We will now discuss in more detail each functional element shown in Figure 1.1.

Input Transducer   The wide variety of possible sources of information results in many different forms for messages. Regardless of their exact form, however, messages may be categorized as *analog* or *digital*. The former may be modeled as functions of a continuous-time variable (for example, pressure, temperature, speech, music), whereas the latter consist of discrete symbols (for example, written text). Almost invariably, the message produced by a source must be converted by a transducer to a form suitable for the particular type of communication system employed. For example, in electrical communications, speech waves are converted by a microphone to voltage variations. Such a converted message is referred to as the *message signal*. In this book, therefore, a *signal* can be interpreted as the variation of a quantity, often a voltage or current, with time.

---

* References in brackets [   ] refer to Historical References in the Bibliography.

TABLE 1.1   A Brief History of Communications

Time Period	Year	Event
80 years	1826	Ohm's law
	1838	Samuel F. B. Morse demonstrates telegraph
	1864	James C. Maxwell predicts electromagnetic radiation
	1876	Alexander Graham Bell patents the telephone
	1887	Heinrich Hertz verifies Maxwell's theory
	1897	Marconi patents a complete wireless telegraph system
40 years	1904	Fleming invents the diode
	1906	Lee De Forest invents the triode amplifier
	1915	Bell System completes a transcontinental telephone line
	1918	B. H. Armstrong perfects the superheterodyne radio receiver
	1920	J. R. Carson applies sampling to communications
	1937	Alec Reeves conceives pulse-code modulation (PCM)
	1938	Television broadcasting begins
20 years	World War II	Radar and microwave systems are developed; statistical methods are applied to signal extraction problems
	1948	The transistor is invented; Claude Shannon's "A Mathematical Theory of Communications" is published
	1950	Time-division multiplexing is applied to telephony
	1956	First transoceanic telephone cable
10 years	1960 to 1970	Laser demonstrated by Maiman (1960) First communication satellite, Telstar I, launched (1962) Live television coverage of moon exploration; experimental PCM systems; experimental laser communications; integrated circuits; digital signal processing; color TV; time-shared computing
10 years	1970 to 1980	Commercial relay satellite communications (voice and digital); gigabit signaling rates; large-scale integration; integrated circuit realization of communications circuits; intercontinental computer communication nets; low-loss light fibers; off-the-shelf optical communication systems; packet-switched digital data systems; interplanetary grand tour launched (1977)— communications accomplished from Jupiter, Saturn, Uranus, and Neptune (encounter in August, 1989); microprocessor, computed tomography, and supercomputers developed
	1980	Satellite "switchboards in the sky"; mobile, cellular telephone systems; multifunction digital displays; 2 gigasample/s digital oscilloscopes; desktop publishing systems; programmable digital signal processors; digitally tuned receivers with autoscan;

(continued)

TABLE 1.1 *(continued)*

Time Period	Year	Event
10 years	to	cryptography on a chip; single-chip digital encoders and decoders; infrared data/control links; compact disk audio players; 200,000-word optical storage media; Ethernet developed; Bell system disbands, allowing competing long-haul telephone services;
	1990	digital signal processors developed
	1990's	Global positioning system (GPS) completed; high-definition television, very small antenna aperture satellites (VSATs), global satellite–based cellular telephones and integrated services digital network (ISDN) developed; personal communications systems come of age

Transmitter   The purpose of the transmitter is to couple the message to the channel. Although it is not uncommon to find the input transducer directly coupled to the transmission medium, as for example in some intercom systems, it is often necessary to *modulate* a carrier wave with the signal from the input transducer. *Modulation* is the systematic variation of some attribute of the carrier, such as amplitude, phase, or frequency, in accordance with a function of the message signal. There are several reasons for using a carrier and modulating it. Important ones are (1) for ease of radiation, (2) to reduce noise and interference, (3) for channel assignment, (4) for multiplexing or transmission of several messages over a single channel, and (5) to overcome equipment limitations. Several of these reasons are self-explanatory; others, such as the second, will become more meaningful later.

In addition to modulation, other primary functions performed by the

FIGURE 1.1   Block diagram of a communication system

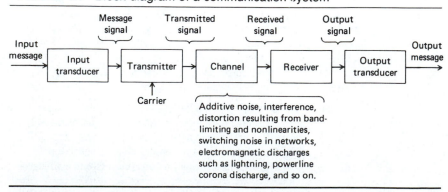

transmitter are filtering, amplification, and coupling the modulated signal to the channel (for example, through an antenna or other appropriate device).

Channel   The channel can have many different forms; the most familiar, perhaps, is the channel that exists between the transmitting antenna of a commercial radio station and the receiving antenna of a radio. In this channel, the transmitted signal propagates through the atmosphere, or free space, to the receiving antenna. However, it is not uncommon to find the transmitter hard-wired to the receiver, as in most local telephone systems. This channel is vastly different from the radio example. However, all channels have one thing in common: the signal undergoes degradation from transmitter to receiver. Although this degradation may occur at any point of the communication system block diagram, it is customarily associated with the channel alone. This degradation often results from noise and other undesired signals or interference but also may include other distortion effects as well, such as fading signal levels, multiple transmission paths, and filtering. More about these unwanted perturbations will be presented shortly.

Receiver   The receiver's function is to extract the desired message from the received signal at the channel output and to convert it to a form suitable for the output transducer. Although amplification may be one of the first operations performed by the receiver, especially in radio communications, where the received signal may be extremely weak, the main function of the receiver is to *demodulate* the received signal. Often it is desired that the receiver output be a scaled, possibly delayed, version of the message signal at the modulator input, although in some cases a more general function of the input message is desired. However, as a result of the presence of noise and distortion, this operation is less than ideal. Ways of approaching the ideal case of perfect recovery will be discussed as we proceed.

Output Transducer   The output transducer completes the communication system. This device converts the electric signal at its input into the form desired by the system user. Perhaps the most common output transducer is a loudspeaker. However, there are many other examples, such as tape recorders, personal computers, meters, and cathode-ray tubes, to name only a few.

1.2   CHANNEL CHARACTERISTICS

Noise Sources

Noise in a communication system can be classified into two broad categories, depending on its source. Noise generated by components within a communication system, such as resistors, electron tubes, and solid-state active devices is referred to as *internal noise*. The second category, *external noise*, results from sources outside a communication system, including atmospheric, man-made, and extraterrestrial sources.

Atmospheric noise results primarily from spurious radio waves generated by the natural electrical discharges within the atmosphere associated with thunderstorms. It is commonly referred to as *static* or *spherics*. Below about 100 MHz, the field strength of such radio waves is inversely proportional to frequency. Atmospheric noise is characterized in the time domain by large-amplitude, short-duration bursts and is one of the prime examples of noise referred to as *impulsive*. Because of its inverse dependence on frequency, atmospheric noise affects commercial AM broadcast radio, which occupies the frequency range from 540 kHz to 1.6 MHz, more than it affects television and FM radio, which operate in frequency bands above 50 MHz.

Man-made noise sources include high-voltage powerline corona discharge, commutator-generated noise in electrical motors, automobile and aircraft ignition noise, and switching-gear noise. Ignition noise and switching noise, like atmospheric noise, are impulsive in character. Impulse noise is the predominant type of noise in switched wireline channels, such as telephone channels. For applications such as voice transmission, impulse noise is only an irritation factor; however, it can be a serious source of error in applications involving transmission of digital data.

Yet another important source of man-made noise is radio-frequency transmitters other than the one of interest. Noise due to interfering transmitters is commonly referred to as *radio-frequency interference* (RFI). RFI is particularly troublesome in situations in which a receiving antenna is subject to a high-density transmitter environment, as in mobile communications in a large city.

Extraterrestrial noise sources include our sun and other hot heavenly bodies, such as stars. Owing to its high temperature (6000°C) and relatively close proximity to the earth, the sun is an intense, but fortunately localized source of radio energy that extends over a broad frequency spectrum. Similarly, the stars are sources of wideband radio energy. Although much more distant and hence less intense than the sun, nevertheless they are collectively an important source of noise because of their vast numbers. Radio stars such as quasars and pulsars are also intense sources of radio energy. Considered a signal source by radio astronomers, such stars are viewed as another noise source by communications engineers. The frequency range of solar and cosmic noise extends from a few megahertz to a few gigahertz.

Another source of interference in communication systems is multiple transmission paths. These can result from reflection off buildings, the earth, airplanes, and ships or from refraction by stratifications in the transmission medium. If the scattering mechanism results in numerous reflected components, the received multipath signal is noiselike and is termed *diffuse*. If the multipath signal component is composed of only one or two strong reflected rays, it is termed *specular*. Finally, signal degradation in a communication system can occur because of random changes in attenuation within the transmission medium. Such signal perturbations are referred to as *fading*, although it should be noted that specular multipath also results in fading due to the constructive and destructive interference of the received multiple signals.

Internal noise results from the random motion of charge carriers in electronic components. It can be of three general types: the first is referred to as *thermal noise*, which is caused by the random motion of free electrons in a conductor or semiconductor excited by thermal agitation; the second is called *shot noise* and is caused by the random arrival of discrete charge carriers in such devices as thermionic tubes or semiconductor junction devices; the third, known as *flicker noise*, is produced in semiconductors by a mechanism not well understood and is more severe the lower the frequency. The first two types of noise sources are modeled analytically in Appendix A, and examples of system characterization using these models are given there.

Types of Transmission Channels

There are many types of transmission channels. We will discuss the characteristics, advantages, and disadvantages of three common types: electromagnetic-wave propagation channels, guided electromagnetic-wave channels, and optical channels. The characteristics of all three may be explained on the basis of electromagnetic-wave propagation phenomena. However, the characteristics and applications of each are different enough to warrant our considering them separately.

Electromagnetic-Wave Propagation Channels   The possibility of the propagation of electromagnetic waves was predicted in 1864 by James Clerk Maxwell (1831–1879), a Scottish mathematician who based his theory on the experimental work of Michael Faraday. Heinrich Hertz (1857–1894), a German physicist, carried out experiments between 1886 and 1888 using a rapidly oscillating spark to produce electromagnetic waves, thereby experimentally proving Maxwell's predictions. Therefore, by the latter part of the nineteenth century, the physical basis for many modern inventions utilizing electromagnetic-wave propagation—such as radio, television, and radar—was already established.

The basic physical principle involved is the coupling of electromagnetic

energy into a propagation medium, which can be free space or the atmosphere, by means of a radiation element referred to as an *antenna*. Many
different propagation modes are possible, depending on the physical configuration of the antenna and the characteristics of the propagation medium.
The simplest case—which never occurs in practice—is propagation from a
point source in a medium that is infinite in extent. The propagating wave
fronts (surfaces of constant phase) in this case would be concentric spheres.
Such a model might be used for the propagation of electromagnetic energy
from a distant spacecraft to earth. Another idealized model, which approximates the propagation of radio waves from a commercial broadcast antenna,
is that of a conducting line perpendicular to an infinite conducting plane.
These and other idealized cases have been analyzed in books on electromagnetic theory. Our purpose is not to summarize all the idealized models,
but to point out basic aspects of propagation phenomena in practical
channels.

Except for the case of propagation between two spacecraft in outer space,
the intermediate medium between transmitter and receiver is never well
approximated by free space. Depending on the distance involved and the
frequency of the radiated waveform, a terrestrial communication link may
depend on line-of-sight, ground-wave, or ionospheric skip-wave propagation
(see Fig. 1.2). Table 1.2 lists frequency bands from 3 kHz to 10^7 GHz, along
with typical uses in each band. Note that the frequency bands are given in
decades; the VHF band has 10 times as much frequency space as the HF
band. Table 1.3 shows some bands of particular interest.

General application allocations are arrived at by international agreement.
The present system of frequency allocations is administered by the International Telecommunications Union (ITU), which is responsible for the periodic convening of Administrative Radio Conferences on a regional (RARC)
or a worldwide (WARC) basis.* In the United States, the Federal Communications Commission (FCC) awards specific applications within a band as
well as licenses for their use. For example, in 1974, the FCC decided to
alleviate congestion of land mobile radio in the 150 and 450 MHz bands by
allocating 115 MHz of bandwidth in the 806 to 947 MHz band to mobile
radio. The reasons for the move to UHF were twofold: (1) frequency space
was available; and (2) a new concept in mobile radio, called *cellular mobile
radio,* in which the geographic area of coverage is divided into cells with a
base station in each cell and handoff between cells, had emerged.†

* See A. F. Inglis, *Electronic Communications Handbook,* New York: McGraw-Hill (1988), Chapter
3.
† See John Oetting, "Cellular Mobile Radio—An Emerging Technology." *IEEE Communications Magazine,* Vol. 21, pp. 10–15, Nov. 1983. Also see *IEEE Communications Magazine,* Vol. 24,
Feb. 1986 (the whole issue is devoted to mobile radio communications).

FIGURE 1.2   The various propagation modes for electromagnetic waves (LOS stands for line of sight)

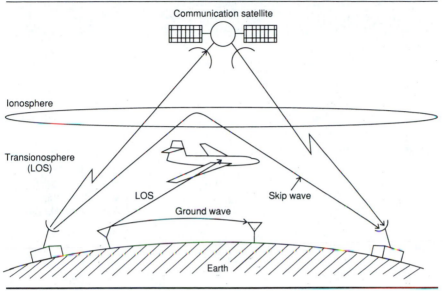

WARC-92, the most recent World Administrative Radio Conference, was attended by 127 out of 166 member countries. It made the following changes:

1. Added 790 kHz of extra spectrum to the shortwave bands—200 kHz below 10 MHz and 590 kHz between 11 and 19 MHz
2. Made an allocation between 1.70 and 2.69 GHz for future public land-mobile telecommunication systems
3. Allocated the 1.605 to 1.675 and 1.800 to 1.805 GHz bands worldwide for public telephone service to and from commercial aircraft
4. Allocated additional spectrum for low-earth-orbit satellite communications, mostly below 1 GHz

At lower frequencies, or long wavelengths, propagating radio waves tend to follow the earth's surface. At higher frequencies, or short wavelengths, radio waves propagate in straight lines. Another phenomenon that occurs at lower frequencies is reflection (or refraction) of radio waves by the ionosphere (a series of layers of charged particles at altitudes between 30 and 250 miles above the earth's surface). Thus, for frequencies below about 100 MHz, it is possible to have skip-wave propagation. At night, when lower ionospheric layers disappear due to less ionization from the sun (the E, F_1, and F_2 layers coalesce into one layer—the F layer), longer skip-wave prop-

TABLE 1.2   Frequency Bands with Typical Uses

Frequency Band[a]	Name	Microwave Band (GHz)	Letter Designations		Typical Uses
3–30 kHz	Very low frequency (VLF)				Long-range navigation; sonar
30–300 kHz	Low frequency (LF)				Navigational aids; radio beacons
300–3000 kHz	Medium frequency (MF)				Maritime radio; direction finding; distress and calling; Coast Guard comm.; commercial AM radio
3–30 MHz	High frequency (HF)				Search and rescue; aircraft comm. with ships; telegraph, telephone, and facsimile; ship-to-coast
30–300 MHz	Very high frequency (VHF)		P R E V I O U S	C U R R E N T	VHF television channels; FM radio; land transportation; private aircraft; air traffic control; taxi cab; police; navigational aids
0.3–3 GHz	Ultrahigh frequency (UHF)	0.5–1.0	VHF	C	UHF television channels; radiosonde; nav. aids; surveillance radar; satellite comm.; radio altimeters; microwave links; airborne radar; approach radar; weather radar; common carrier land mobile
		1.0–2.0	L	D	
		2.0–3.0	S	E	
		3.0–4.0	S	F	
		4.0–6.0	C	G	
3–30 GHz	Superhigh frequency (SHF)	6.0–8.0	C	H	
		8.0–10.0	X	I	
		10.0–12.4	X	J	
		12.4–18.0	Ku	J	
		18.0–20.0	K	J	
		20.0–26.5	K	K	
		26.5–40.0	Ka	K	
30–300 GHz	Extremely high frequency (EHF)				Railroad service; radar landing systems; experimental

(continued)

TABLE 1.2 (*continued*)

Frequency Band[a]	Name	Microwave Band (GHz)	Letter Designations	Typical Uses
300 GHz–3 THz	(0.1–1 mm)			Experimental
43–430 THz	Infrared (7–0.7 μm)			Optical comm. systems
430–750 THz	Visible light (0.7–0.4 μm)			Optical comm. systems
750–3000 THz	Ultraviolet light (0.4–0.1 μm)			Optical comm. systems

[a] Abbreviations: kHz = kilohertz = $\times 10^3$ hertz; MHz = megahertz = $\times 10^6$ hertz; GHz = gigahertz = $\times 10^9$ hertz; THz = terahertz = $\times 10^{12}$ hertz; μm = micrometers = 10^{-6} meters.

agation occurs as a result of reflection from the higher, single reflecting layer of the ionosphere.

Above about 300 MHz, propagation of radio waves is by line of sight, because the ionosphere will not bend radio waves in this frequency region sufficiently to reflect them back to the earth. At still higher frequencies, say above 1 or 2 GHz, atmospheric gases (mainly oxygen), water vapor, and precipitation absorb and scatter radio waves. This phenomenon manifests itself as attenuation of the received signal, with the attenuation generally being more severe the higher the frequency (there are resonance regions for absorption by gases that peak at certain frequencies). Figure 1.3 shows

TABLE 1.3 Selected Frequency Bands for Public Use and Military Communications

Use	Frequency
Omega navigation	10–14 kHz
Worldwide submarine communication	30 kHz
Loran C navigation	100 kHz
Standard (AM) broadcast	540–1600 kHz
FM broadcast	88–108 MHz
Television: Channels 2–4	54–72 MHz
Channels 5–6	76–88 MHz
Channels 7–13	174–216 MHz
Channels 14–83	420–890 MHz

FIGURE 1.3 Specific attenuation for atmospheric gases and rain. (a)
Specific attenuation due to oxygen and water vapor (concentration of 7.5
g/m^3). (b) Specific attenuation due to rainfall at rates of 10, 50, and 100 mm/h.

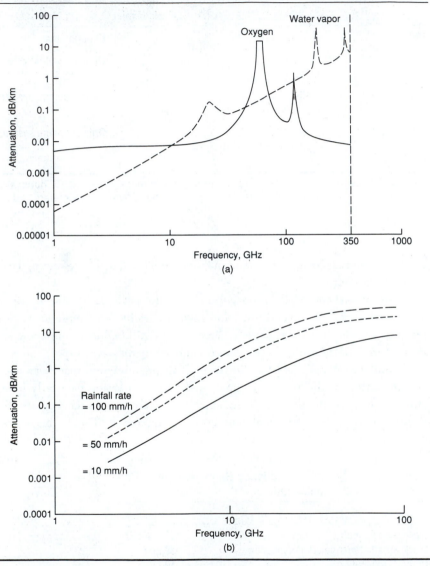

specific attenuation curves as a function of frequency* for oxygen, water
vapor, and rain [recall that 1 decibel (dB) is ten times the logarithm to the
base 10 of a power ratio]. One must account for the possible attenuation by

* Data from Louis J. Ippolito, Jr., *Radiowave Propagation in Satellite Communications*. New York,
Van Nostrand Reinhold, 1986, Chapters 3 and 4.

such atmospheric constituents in the design of microwave links, which are used in transcontinental telephone links and ground-to-satellite communications links.

At about 23 GHz, the first absorption resonance due to water vapor occurs, and at about 62 GHz a second one occurs due to oxygen absorption. These frequencies should be avoided in transmission of desired signals through the atmosphere, or undue power will be expended (one might, for example, use 62 GHz as a signal for cross-linking between two satellites, where atmospheric absorption is no problem, and thereby prevent an enemy on the ground from listening in). Another absorption frequency for oxygen occurs at 120 GHz, and two other absorption frequencies for water vapor occur at 180 and 350 GHz.

Communication at millimeter-wave frequencies (that is, at 30 GHz and higher) is becoming more important now that there is so much congestion at lower frequencies (the Advanced Technology Satellite, to be launched in the mid-1990s, employs an uplink frequency band around 20 GHz and a downlink frequency band at about 30 GHz). Communication at millimeter-wave frequencies is becoming more feasible because of technological advances in components and systems.

Somewhere above 1 THz (1000 GHz), the propagation of radio waves becomes optical in character. At a wavelength of 10 μm (0.00001 m), the carbon dioxide laser provides a source of coherent radiation, and visible-light lasers (for example, helium-neon) radiate in the wavelength region of 1 μm and shorter. Terrestrial communications systems employing such frequencies experience considerable attenuation on cloudy days, and laser communications over terrestrial links are restricted to optical fibers for the most part. Analyses have been carried out for the employment of laser communications cross-links between satellites, but there are as yet no optical satellite communications links actually flying.

Guided Electromagnetic-Wave Channels   Perhaps the most extensive example of guided electromagnetic-wave channels is the part of the long-distance telephone network that uses wire lines.* Communication between persons a continent apart was first achieved by means of voice frequency transmission (below 10,000 Hz) over open wire. Quality of transmission was rather poor. By 1952, use of the types of modulation known as *double-sideband* and *single-sideband* on high-frequency carriers was established. Communication over predominantly multipair and coaxial-cable lines produced transmission of

---

* For a summary of guided transmission systems as applied to telephone systems, see F. T. Andrews, Jr., "Communications Technology: 25 Years in Retrospect. Part III, Guided Transmission Systems, 1952–1973." *IEEE Communications Society Magazine*, Vol. 16, pp. 4–10, Jan. 1978.

much better quality. With the completion of the first trans-Atlantic cable in 1956, intercontinental telephone communication was no longer dependent on high-frequency radio, and the quality of intercontinental telephone service improved significantly.

Bandwidths on coaxial-cable links are a few megahertz. The need for greater bandwidth initiated the development of millimeter-wave waveguide transmission systems. However, with the development of low-loss optical fibers, efforts to improve millimeter-wave systems to achieve greater bandwidth have ceased. The development of optical fibers, in fact, has made the concept of a wired city—wherein digital data and video can be piped to any residence or business within a city—a reality. Modern coaxial-cable systems can carry only 13,000 voice channels per cable, but optical links are capable of carrying several times this number (the limiting factor being the current driver for the light source).

Optical Links   The use of optical links was, until recently, limited to short and intermediate distances. With the installation of trans-Pacific and trans-Atlantic optical cables in 1988 and early 1989, this is no longer true.* The technological breakthroughs that have made possible the use of light waves for communication are the development of small coherent light sources (semiconductor lasers), low-loss (approximately 1 dB/km) optical fibers or waveguides, and low-noise detectors.†

A typical fiber-optic communication system has a light source, which may be either a light-emitting diode or a semiconductor laser, in which the intensity of the light is varied by the message source. The output of this modulator is the input to a light-conducting fiber (analogous to the coaxial cable of a standard telephone channel). The receiver, or light sensor, typically consists of a photodiode. In a photodiode, an average current flows that is proportional to the optical power of the incident light. However, the exact number of charge carriers (that is, electrons) is random. The output of the detector is the sum of the average current—which is proportional to the

---

* See Inglis, op. cit., Chapter 8.

† For an overview of optical communication systems, see S. D. Personick, "Optical Fibers: A New Transmission Medium," *IEEE Communications Society Magazine,* Vol. 13, pp. 20–24, Jan. 1975. Also see S. D. Personick, "Fiber Optic Communications: A Technology Coming of Age," *IEEE Communications Society Magazine,* Vol. 16, pp. 12–20, Mar. 1978. For an overview of future trends in terrestrial optical fiber communications systems see P. Kaiser, J. Midwinter, and S. Shimada, "Status and Future Trends in Terrestrial Optical Fiber Systems in North America," *IEEE Communications Magazine,* Vol. 25, pp. 8–21, Oct. 1987. For an overview on the use of signal-processing methods to improve optical communications, see J. H. Winters, R. D. Gitlin, and S. Kasturia, "Reducing the Effects of Transmission Impairments in Digital Fiber Optic Systems," *IEEE Communications Magazine,* Vol. 31, pp. 68–76, June 1993.

modulation—and a noise component. This noise component differs from the thermal noise generated by the receiver electronics in that it is "bursty" in character. It is referred to as *shot noise,* in analogy to the noise made by shot hitting a metal plate. Another source of degradation is the dispersion of the optical fiber itself. For example, pulse-type signals sent to the fiber are observed as "smeared out" at the receiver. Losses also occur as a result of the connections between cable pieces and between cable and system components.

Finally, it should be mentioned that optical communications can take place through free space.*

1.3 SUMMARY OF SYSTEMS-ANALYSIS TECHNIQUES

Having identified and discussed the main subsystems in a communication system and certain characteristics of transmission media, let us now look at the techniques at our disposal for systems analysis and design.

Time and Frequency-Domain Analyses

From circuits courses or prior courses in linear systems analysis, you are well aware that the electrical engineer lives in the two worlds, so to speak, of time and frequency. Also, you should recall that dual time-frequency analysis techniques are especially valuable for linear systems for which the principle of superposition holds. Although many of the subsystems and operations encountered in communication systems are for the most part linear, many are not. Nevertheless, frequency-domain analysis is an extremely valuable tool to the communications engineer, more so perhaps than to other systems analysts. Since the communications engineer is concerned primarily with signal bandwidths and signal locations in the frequency domain, rather than with transient analysis, the essentially steady-state approach of the Fourier series and transforms is used. Accordingly, we will discuss the Fourier series, the Fourier integral, and their role in systems analysis in Chapter 2. We postpone a more complete discussion of these tools for systems analysis until then.

Modulation and Communication Theories

Modulation theory employs time and frequency-domain analyses to analyze and design systems for modulation and demodulation of information-bearing signals. To be specific, consider the message signal $m(t)$, which is to be trans-

---

* See Joseph Katz, "Semiconductor Optoelectric Devices for Free-Space Optical Communications," *IEEE Communications Society Magazine*, Vol. 21, pp. 20–27, Sept. 1983.

mitted through a channel using the method of double-sideband modula-tion. The modulated carrier for double-sideband modulation is of the form $x_c(t) = A_c m(t) \cos \omega_c t$, where ω_c is the carrier frequency in radians per second and A_c is the carrier amplitude. Not only must a modulator be built that can multiply two signals, but amplifiers are required to provide the proper power level of the transmitted signal. The exact design of such amplifiers is not of concern in a systems approach. However, the frequency content of the mod-ulated carrier, for example, is important to their design and therefore must be specified. The dual time-frequency analysis approach is especially helpful in providing such information.

At the other end of the channel, there must be a receiver configuration capable of extracting a replica of $m(t)$ from the modulated signal, and one can again apply time and frequency-domain techniques to good effect.

The analysis of the effect of interfering signals on system performance and the subsequent modifications in design to improve performance in the face of such interfering signals are part of *communication theory*, which, in turn, makes use of modulation theory.

This discussion, although mentioning interfering signals, has not explicitly emphasized the uncertainty aspect of the information-transfer problem. Indeed, much can be done without applying probabilistic methods. However, as pointed out previously, the application of probabilistic methods, coupled with optimization procedures, has been one of the key ingredients of the modern communications era and led to the development during the 1950s and 1960s of new techniques and systems totally different in concept from those which existed before World War II.

We will now survey several approaches to statistical optimization of com-munication systems.

1.4 PROBABILISTIC APPROACHES TO SYSTEM OPTIMIZATION

The works of Wiener and Shannon, previously cited, were the beginning of modern statistical communication theory. Both these investigators applied probabilistic methods to the problem of extracting information-bearing sig-nals from noisy backgrounds, but they worked from different standpoints. In this section we briefly examine these two approaches to optimum system design.

Statistical Signal Detection and Estimation Theory

Wiener considered the problem of optimally filtering signals from noise, where "optimum" is used in the sense of minimizing the average squared

error between the desired and the actual output. The resulting filter structure is referred to as the *Wiener optimum filter*. This type of approach is most
appropriate for analog communication systems in which the demodulated
output of the receiver is to be a faithful replica of the message input to the
transmitter.

Wiener's approach is reasonable for analog communications. However, in
the early 1940s, North [1943] provided a more fruitful approach to the digital communications problem, in which the receiver must distinguish
between a number of discrete signals in background noise. Actually, North
was concerned with radar, which requires only the detection of the presence
or absence of a pulse. Since fidelity of the detected signal at the receiver is
of no consequence in such signal-detection problems, North sought the filter
that would maximize the peak-signal-to-rms-noise ratio at its output. The
resulting optimum filter is called the *matched filter,* for reasons that will
become apparent in Chapter 7, where we consider digital data transmission.
Later adaptations of the Wiener and matched-filter ideas to time-varying
backgrounds resulted in *adaptive filters.* We will consider a subclass of such
filters in Chapter 7 when *equalization* of digital data signals is discussed.

The signal-extraction approaches of Wiener and North, formalized in the
language of statistics in the early 1950s by several researchers [see Middleton 1960, p. 832, for several references], were the beginnings of what is
today called *statistical signal detection* and *estimation theory.* Woodward and
Davies [1952], in considering the design of receivers utilizing *all* the information available at the channel output, determined that this so-called ideal
receiver computes the probabilities of the received waveform given the possible transmitted messages. These computed probabilities are known as *a
posteriori* probabilities. The ideal receiver then makes the decision that the
transmitted message was the one corresponding to the largest *a posteriori*
probability. Although perhaps somewhat vague at this point, this *maximum a
posteriori* (MAP) *principle,* as it is called, is one of the cornerstones of detection
and estimation theory. We will examine these ideas in more detail in Chapters 7 through 9.

Information Theory and Coding

The basic problem that Shannon considered is, "Given a message source,
how shall the messages produced be represented so as to maximize the
information conveyed through a given channel?" Although Shannon for-
mulated his theory for both discrete and analog sources, we will think here
in terms of discrete systems. Clearly, a basic consideration in this theory is
a measure of information. Once a suitable measure has been defined (and
we will do so in Chapter 10), the next step is to define the information-

carrying capacity, or simply capacity, of a channel as the maximum rate at which information can be conveyed through it. The obvious question that now arises is, "Given a channel, how closely can we approach the capacity of the channel, and what is the quality of the received message?" A most surprising, and the singularly most important, result of Shannon's theory is that by suitably restructuring the transmitted signal, we can transmit information through a channel *at any rate less than the channel capacity with arbitrarily small error,* despite the presence of noise, provided we have an arbitrarily long time available for transmission. This is the gist of Shannon's *second theorem.* Limiting our discussion at this point to binary discrete sources, a proof of Shannon's second theorem proceeds by selecting codewords at random from the set of 2^n possible binary sequences n digits long at the channel input. The probability of error in receiving a given n-digit sequence, when averaged over all possible code selections, becomes arbitrarily small as n becomes arbitrarily large. Thus many suitable codes exist, *but we are not told how to find these codes.* Indeed, this has been the dilemma of information theory since its inception and is an area of active research. In recent years, great strides have been made in finding good coding and decoding techniques that are implementable with a reasonable amount of hardware and require only a reasonable amount of time to decode. Several basic coding techniques will be discussed in Chapter 10.

1.5 PREVIEW OF THIS BOOK

From the previous discussion, the importance of probability and noise characterization in analysis of communication systems should be apparent. Accordingly, after presenting basic signal, system, and noiseless modulation theory in Chapters 2 and 3, we shall discuss probability and noise theory in Chapters 4 and 5. Following this, we apply these tools to the noise analysis of analog communications schemes in Chapter 6. In Chapters 7 and 8, we use probabilistic techniques to find optimum receivers when we consider digital data transmission. Various types of digital modulation schemes are analyzed in terms of error probability. In Chapter 9, we approach optimum signal detection and estimation techniques on a generalized basis and use signal-space techniques to provide insight as to why systems that have been analyzed previously perform as they do. As already mentioned, information theory and coding are the subjects of Chapter 10. This provides us with a means of comparing actual communication systems with the ideal. Such comparisons are then considered in Chapter 10 to provide a basis for selection of systems.

   In closing, we must note that large areas of communications technology—

such as optical, computer, and military communications—are not touched on in this book. However, one can apply the principles developed in this text in those areas as well.

FURTHER READING

The references for this chapter were chosen to indicate the historical development of modern communications theory and by and large are not easy reading. They are found in the Historical References section of the Bibliography. You also may consult the introductory chapters of the books listed in the Further Reading sections of Chapters 2 and 3. These books appear in the main portion of the Bibliography.

2 SIGNAL AND LINEAR SYSTEM ANALYSIS

The study of information transmission systems is inherently concerned with the transmission of signals through systems. Recall that in Chapter 1 a *signal* was defined as the time history of some quantity, usually a voltage or current. A *system* is a combination of devices and networks (subsystems) chosen to perform a desired function. Because of the sophistication of modern communication systems, a great deal of analysis and experimentation with trial subsystems occurs before actual building of the desired system. Thus the communications engineer's tools are mathematical models for signals and systems.

In this chapter, we review techniques useful for modeling and analysis of signals and systems used in communications engineering.* Of primary concern will be the dual time-frequency viewpoint for signal representation, and models for linear, time-invariant, two-port systems. It is important always to keep in mind that a model is not the signal or the system, but a mathematical idealization of certain characteristics of it that are most important to the problem at hand.

With this brief introduction, we now consider signal classifications and various methods for modeling signals and systems. These include frequency-domain representations for signals via the complex exponential Fourier series and the Fourier transform followed by linear system models and techniques for analyzing the effects of such systems on signals.

2.1 SIGNAL MODELS

Deterministic and Random Signals

In this book we are concerned with two broad classes of signals, referred to as deterministic and random. *Deterministic signals* can be modeled as com-

---

* More complete treatments of these subjects can be found in texts on linear system theory. (See, for example, Ziemer, Tranter, and Fannin (1993).)

pletely specified functions of time. For example, the signal

$$x(t) = A \cos \omega_0 t, \qquad -\infty < t < \infty \tag{2.1}$$

where A and ω_0 are constants, is a familiar example of a deterministic signal. Another example of a deterministic signal is the unit rectangular pulse, denoted as $\Pi(t)$ and defined as

$$\Pi(t) = \begin{cases} 1, & |t| < \frac{1}{2} \\ 0, & \text{otherwise} \end{cases} \tag{2.2}$$

Random signals are signals that take on random values at any given time instant and must be modeled probabilistically. They will be considered in Chapter 5. Figure 2.1 illustrates the various types of signals just discussed.

FIGURE 2.1   Examples of various types of signals. (a) Deterministic (sinusoidal) signal. (b) Unit rectangular pulse signal. (c) Random signal.

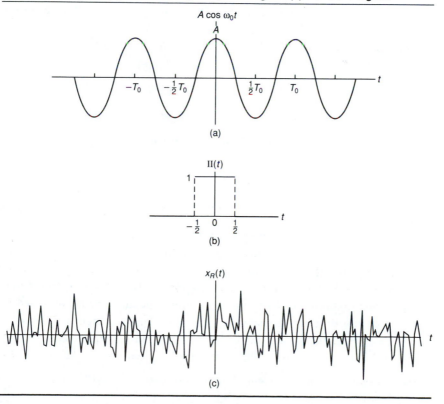

Periodic and Aperiodic Signals

The signal defined by (2.1) is an example of a *periodic signal*. A signal $x(t)$ is periodic if and only if

$$x(t + T_0) = x(t), \qquad -\infty < t < \infty \tag{2.3}$$

where the constant T_0 is the period. The smallest such number satisfying (2.3) is referred to as the *fundamental period* (the modifier "fundamental" is often excluded). Any signal not satisfying (2.3) is called *aperiodic*.

Phasor Signals and Spectra

A useful periodic signal in system analysis is the signal

$$\tilde{x}(t) = Ae^{j(\omega_0 t + \theta)}, \qquad -\infty < t < \infty \tag{2.4}$$

which is characterized by three parameters: amplitude A, phase θ in radians, and frequency ω_0 in radians per second or $f_0 = \omega_0/2\pi$ hertz. We will refer to $\tilde{x}(t)$ as a *rotating phasor* to distinguish it from the *phasor* $Ae^{j\theta}$, for which $e^{j\omega_0 t}$ is implicit. Using Euler's theorem,* we may readily show that $\tilde{x}(t) = \tilde{x}(t + T_0)$, where $T_0 = 2\pi/\omega_0$. Thus $\tilde{x}(t)$ is a periodic signal with period $2\pi/\omega_0$.

The rotating phasor $Ae^{j(\omega_0 t + \theta)}$ can be related to a real, sinusoidal signal $A \cos(\omega_0 t + \theta)$ in two ways. The first is by taking its real part,

$$x(t) = A \cos(\omega_0 t + \theta) = \text{Re } \tilde{x}(t)$$
$$= \text{Re } Ae^{j(\omega_0 t + \theta)} \tag{2.5}$$

and the second is by taking one-half of the sum of $\tilde{x}(t)$ and its complex conjugate,

$$A \cos(\omega_0 t + \theta) = \tfrac{1}{2}\tilde{x}(t) + \tfrac{1}{2}\tilde{x}*(t)$$
$$= \tfrac{1}{2}Ae^{j(\omega_0 t + \theta)} + \tfrac{1}{2}Ae^{-j(\omega_0 t + \theta)} \tag{2.6}$$

Figure 2.2 illustrates these two procedures graphically.

Equations (2.5) and (2.6), which give alternative representations of the sinusoidal signal $x(t) = A \cos(\omega_0 t + \theta)$ in terms of the rotating phasor $\tilde{x}(t) = A \exp[j(\omega_0 t + \theta)]$, are time-domain representations for $x(t)$. Two equivalent representations of $x(t)$ in the frequency domain may be obtained by noting that the rotating phasor signal is completely specified if the parameters A and θ are given for a particular f_0. Thus plots of the magnitude and angle of $Ae^{j\theta}$ versus frequency give sufficient information to characterize $\tilde{x}(t)$

* Euler's theorem, given in Appendix C, is $e^{\pm ju} = \cos u \pm j \sin u$. Also recall that $e^{j2\pi} = 1$.

FIGURE 2.2 Two ways of relating a phasor signal to a sinusoidal signal. (a) Projection of a rotating phasor onto the real axis. (b) Addition of complex conjugate rotating phasors.

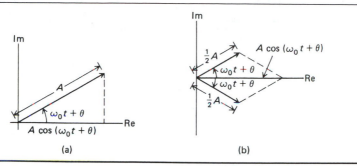

(a)   (b)

completely. Because $\tilde{x}(t)$ exists only at the single frequency f_0 for this case of a single sinusoidal signal, the resulting plots consist of single lines. Such plots, consisting of discrete lines and referred to as *amplitude and phase line spectra* for $\tilde{x}(t)$, are shown in Figure 2.3(a). These are *frequency-domain* representations not only of $\tilde{x}(t)$ but of $x(t)$ as well, by virtue of (2.5). The plots of Figure 2.3(a) are referred to as the *single-sided amplitude and phase spectra* of $x(t)$ because they exist only for positive frequency. For a signal consisting of a sum of sinusoids of differing frequencies, the single-sided spectra consist of a multiplicity of lines, one for each sinusoidal component of the sum.

By plotting the amplitude and phase of the complex conjugate phasors of (2.6) versus frequency, one obtains another frequency-domain representation for $x(t)$, referred to as the *double-sided amplitude and phase spectra*. This representation is shown in Figure 2.3(b). Two important observations may be made from Figure 2.3(b). First, the lines at the *negative* frequency $f = -f_0$ exist precisely because it is necessary to add complex conjugate (or oppositely rotating) phasor signals to obtain the real signal $A \cos (\omega_0 t + \theta)$. Second, we note that the amplitude spectrum has *even* symmetry about $f = 0$ and that the phase spectrum has *odd* symmetry. This symmetry is again a consequence of $x(t)$ being a real signal. As in the single-sided case, the two-sided spectra for a sum of sinusoids consist of a multiplicity of lines, with one pair for each sinusoidal component.

Figures 2.3(a) and 2.3(b) are therefore equivalent spectral representations for the signal $A \cos (\omega_0 t + \theta)$, consisting of lines at the frequency $f = f_0$ (and its negative). For this simple case, the use of spectral plots seems to be an unnecessary complication, but we will find shortly how the Fourier series and Fourier transform lead to spectral representations for more complex signals.

FIGURE 2.3   Amplitude and phase spectra for the signal $A \cos (\omega_0 t + \theta)$.
(a) Single-sided. (b) Double-sided.

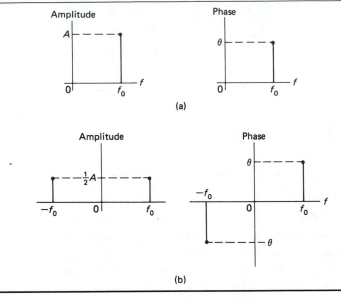

(a)

(b)

▼ **EXAMPLE 2.1**   (a) To sketch the single-sided and double-sided spectra of

$$x(t) = 2 \sin (10\pi t - \tfrac{1}{6}\pi) \qquad (2.7)$$

we note that $x(t)$ can be written as

$$x(t) = 2 \cos (10\pi t - \tfrac{1}{6}\pi - \tfrac{1}{2}\pi) = 2 \cos (10\pi t - \tfrac{2}{3}\pi)$$
$$= \text{Re } 2e^{j(10\pi t - 2\pi/3)} = e^{j(10\pi t - 2\pi/3)} + e^{-j(10\pi t - 2\pi/3)} \qquad (2.8)$$

Thus the single-sided and double-sided spectra are as shown in Figure 2.3, with $A = 2$, $\theta = -\tfrac{2}{3}\pi$ rad, and $f_0 = 5$ Hz.

(b) If more than one sinusoidal component is present in a signal, its spectra consist of multiple lines. For example, the signal

$$y(t) = 2 \sin (10\pi t - \tfrac{1}{6}\pi) + \cos (20\pi t) \qquad (2.9)$$

can be rewritten as

$$y(t) = 2 \cos (10\pi t - \tfrac{2}{3}\pi) + \cos (20\pi t) = \text{Re } [2e^{j(10\pi t - 2\pi/3)} + e^{j20\pi t}]$$
$$= e^{j(10\pi t - 2\pi/3)} + e^{-j(10\pi t - 2\pi/3)} + \tfrac{1}{2}e^{j20\pi t} + \tfrac{1}{2}e^{-j20\pi t}$$

$$(2.10)$$

Its single-sided amplitude spectrum consists of a line of amplitude 2 at $f = 5$ Hz and a line of amplitude 1 at $f = 10$ Hz. Its single-sided phase spectrum consists of a single line of amplitude $-2\pi/3$ at $f = 5$ Hz. To get the double-sided amplitude spectrum, one simply *halves* the amplitude of the lines in the single-sided amplitude spectrum and takes the mirror image of this result about $f = 0$. The double-sided phase spectrum is obtained by taking the mirror image of the single-sided phase spectrum about $f = 0$ and inverting the left-hand (negative frequency) portion.

Singularity Functions

An important subclass of aperiodic signals is the singularity functions. In this book we will be concerned with only two: the unit impulse function $\delta(t)$ and the unit step function $u(t)$. The unit impulse function is *defined* in terms of the integral

$$\int_{-\infty}^{\infty} x(t)\,\delta(t)\,dt = x(0) \tag{2.11}$$

where $x(t)$ is any test function that is continuous at $t = 0$. A change of variables and redefinition of $x(t)$ results in the *sifting property*

$$\int_{-\infty}^{\infty} x(t)\,\delta(t - t_0)\,dt = x(t_0) \tag{2.12}$$

where $x(t)$ is continuous at $t = t_0$. We will make considerable use of the sifting property in systems analysis. By considering the special cases $x(t) = 1$ and $x(t) = 0$ for $t < t_1$ and $t > t_2$ the two properties

$$\int_{t_1}^{t_2} \delta(t - t_0)\,dt = 1, \qquad t_1 < t_0 < t_2 \tag{2.13a}$$

and

$$\delta(t - t_0) = 0, \qquad t \neq t_0 \tag{2.13b}$$

are obtained that provide an alternative definition of the unit impulse. Equation (2.13b) allows the integrand in Equation (2.12) to be replaced by $x(t_0)$ $\delta(t - t_0)$, and the sifting property then follows from (2.13a).

Other properties of the delta function that can be proved from the definition (2.11) are the following:

1. $\delta(at) = \dfrac{1}{|a|}\,\delta(t)$

2. $\delta(-t) = \delta(t)$

3. $\int_{t_1}^{t_2} x(t)\, \delta(t - t_0)\, dt = \begin{cases} x(t_0), & t_1 < t_0 < t_2 \\ 0, & \text{otherwise} \end{cases}$

4. $x(t)\, \delta(t - t_0) = x(t_0)\, \delta(t - t_0)$

5. $\int_{t_1}^{t_2} x(t)\delta^{(n)}(t - t_0)\, dt = (-1)^n x^{(n)}(t_0), \qquad t_1 < t_0 < t_2.$ [In this equation, the superscript n denotes the nth derivative; $x(t)$ and its first n derivatives are assumed continuous at $t = t_0$.]

6. If $f(t) = g(t)$, where $f(t) = a_0\, \delta(t) + a_1\, \delta^{(1)}(t) + \cdots + a_n\, \delta^{(n)}(t)$ and $g(t) = b_0\, \delta(t) + b_1\, \delta^{(1)}(t) + \cdots + b_n\, \delta^{(n)}(t)$, this implies that $a_0 = b_0$, $a_1 = b_1, \ldots, a_n = b_n$.

It is reassuring to note that (2.13a) and (2.13b) correspond to the intuitive notion of a unit impulse as the limit of a suitably chosen conventional function having unity area in an infinitesimally small width. An example is the signal

$$\delta_\epsilon(t) = \frac{1}{2\epsilon}\, \Pi\left(\frac{t}{2\epsilon}\right) = \begin{cases} \dfrac{1}{2\epsilon}, & |t| < \epsilon \\ 0, & \text{otherwise} \end{cases} \tag{2.14}$$

which is shown in Figure 2.4(a). It seems apparent that any signal having unity area and zero width in the limit as some parameter approaches zero is a suitable representation for $\delta(t)$, for example, the signal

$$\delta_{1\epsilon}(t) = \epsilon\left(\frac{1}{\pi t}\, \sin\frac{\pi t}{\epsilon}\right)^2 \tag{2.15}$$

which is sketched in Figure 2.4(b).

Other singularity functions may be defined as integrals or derivatives of unit impulses. We will need only the unit step $u(t)$, defined to be the integral of the unit impulse. Thus

$$u(t) \triangleq \int_{-\infty}^{t} \delta(\lambda)\, d\lambda = \begin{cases} 0, & t < 0 \\ 1, & t > 0 \\ \text{undefined}, & t = 0 \end{cases} \tag{2.16}$$

or

$$\delta(t) = \frac{du(t)}{dt} \tag{2.17}$$

You are no doubt familiar with the usefulness of the unit step for "turning on" signals of doubly infinite duration and for representing signals of the

FIGURE 2.4 Two representations for the unit impulse function in the limit as $\epsilon \to 0$. (a) $(1/2\epsilon)\Pi(t/2\epsilon)$. (b) $\epsilon[(1/\pi t)\sin(\pi t/\epsilon)]^2$.

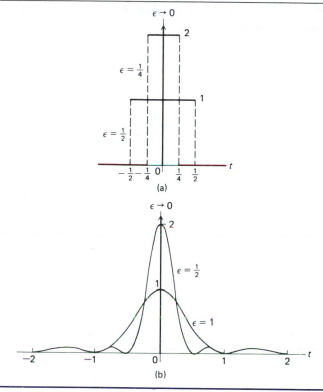

(a)

(b)

staircase type. For example, the unit rectangular pulse function defined by (2.2) can be written in terms of unit steps as

$$\Pi(t) = u(t + \tfrac{1}{2}) - u(t - \tfrac{1}{2}) \tag{2.18}$$

We are now ready to consider power and energy signal classifications.

2.2 SIGNAL CLASSIFICATIONS

Because the particular representation used for a signal depends on the type of signal involved, it is useful to pause at this point and introduce signal classifications. In this chapter we will be considering two signal classes, those with finite energy and those with finite power.* As a specific example, sup-

---

* Signals that are neither energy nor power signals are easily found. For example, $x(t) = t^{-1/4}$, $t \geq t_0 > 0$ and zero otherwise.

pose $e(t)$ is the voltage across a resistance R producing a current $i(t)$. The instantaneous power per ohm is $p(t) = e(t)i(t)/R = i^2(t)$. Integrating over the interval $|t| \leq T$, the total energy and the average power on a per-ohm basis are obtained as the limits

$$E = \lim_{T \to \infty} \int_{-T}^{T} i^2(t) \, dt \qquad (2.19a)$$

and

$$P = \lim_{T \to \infty} \frac{1}{2T} \int_{-T}^{T} i^2(t) \, dt \qquad (2.19b)$$

respectively.

For an arbitrary signal $x(t)$, which may, in general, be complex, we define total (normalized) energy as

$$E \triangleq \lim_{T \to \infty} \int_{-T}^{T} |x(t)|^2 \, dt = \int_{-\infty}^{\infty} |x(t)|^2 \, dt \qquad (2.20a)$$

and (normalized) power as

$$P \triangleq \lim_{T \to \infty} \frac{1}{2T} \int_{-T}^{T} |x(t)|^2 \, dt \qquad (2.20b)$$

Based on the definitions (2.20a) and (2.20b), we can define two distinct classes of signals:

1. We say $x(t)$ is an *energy signal* if and only if $0 < E < \infty$, so that $P = 0$.
2. We classify $x(t)$ as a *power signal* if and only if $0 < P < \infty$, thus implying that $E = \infty$.

▼| **EXAMPLE 2.2**   As an example of determining the classification of a signal, consider

$$x_1(t) = Ae^{-\alpha t} u(t), \qquad \alpha > 0 \qquad (2.21)$$

where A and α are constants. Using (2.20a), we may readily verify that $x_1(t)$ is an energy signal, since $E = A^2/2\alpha$. Letting $\alpha \to 0$, we obtain the signal $x_2(t) = Au(t)$, which has infinite energy. Applying (2.20b), we find that $P = $
▲| $\frac{1}{2}A^2$, thus verifying that $x_2(t)$ is a power signal.

▼ EXAMPLE 2.3   Consider the rotating phasor signal given by Equation (2.4). We may verify that $\tilde{x}(t)$ is a power signal, since

$$P = \lim_{T \to \infty} \frac{1}{2T} \int_{-T}^{T} |\tilde{x}(t)|^2 \, dt = \lim_{T \to \infty} \frac{1}{2T} \int_{-T}^{T} A^2 \, dt = A^2 \qquad (2.22)$$

▲ is finite.

We note that there is no need to carry out the limiting operation to find P for a periodic signal, since an average carried out over a single period gives the same result as (2.20b); that is, for a periodic signal $x_p(t)$,

$$P = \frac{1}{T_0} \int_{t_0}^{t_0 + T_0} |x_p(t)|^2 \, dt \qquad (2.23)$$

where T_0 is the period. The proof of (2.23) is left to the problems.

▼ EXAMPLE 2.4   The sinusoidal signal

$$x_p(t) = A \cos(\omega_0 t + \theta) \qquad (2.24)$$

has average power

$$
\begin{aligned}
P &= \frac{1}{T_0} \int_{t_0}^{t_0 + T_0} A^2 \cos^2(\omega_0 t + \theta) \, dt \\
&= \frac{\omega_0}{2\pi} \int_{t_0}^{t_0 + (2\pi/\omega_0)} \frac{A^2}{2} \, dt + \frac{\omega_0}{2\pi} \int_{t_0}^{t_0 + (2\pi/\omega_0)} \frac{A^2}{2} \cos 2(\omega_0 t + \theta) \, dt \\
&= \frac{A^2}{2} \qquad (2.25)
\end{aligned}
$$

where the identity $\cos^2 u = \frac{1}{2} + \frac{1}{2} \cos 2u$ has been used and the second integral is zero because the integration is over two complete periods of the ▲ integrand.

†2.3 GENERALIZED FOURIER SERIES*

Our discussion of the phasor signal given by (2.4) illustrated the dual time-frequency nature of such signals. Fourier series and transform representations for signals are the key to generalizing this dual nature, since they amount to expressing signals as superpositions of complex exponential functions $e^{j\omega t}$.

---

* Sections preceded by a dagger can be omitted without loss of continuity.

In anticipation of signal space concepts, to be introduced and applied to communication systems analysis in Chapters 8 and 9, the discussion in this section is concerned with the representation of signals as series of orthogonal functions or, as referred to here, generalized Fourier series. Such generalized Fourier series representations allow signals to be represented as points in a generalized vector space, referred to as *signal space,* thereby allowing information transmission to be viewed in a geometrical context. In the following section, the generalized Fourier series will be specialized to the complex exponential form of the Fourier series.

To begin our consideration of generalized Fourier series, we recall from vector analysis that any vector **A** in a three-dimensional space can be expressed in terms of any three vectors **a**, **b**, and **c** that do not all lie in the same plane and are not collinear:

$$\mathbf{A} = A_1\mathbf{a} + A_2\mathbf{b} + A_3\mathbf{c} \tag{2.26}$$

where A_1, A_2, and A_3 are appropriately chosen constants. The vectors **a**, **b**, and **c** are said to be *linearly independent,* for no one of them can be expressed as a linear combination of the other two. For example, it is impossible to write $\mathbf{a} = \alpha\mathbf{b} + \beta\mathbf{c}$, no matter what choice is made for the constants α and β.

Such a set of linearly independent vectors is said to form a *basis set* for three-dimensional vector space. Such vectors *span* a three-dimensional vector space in the sense that *any* vector **A** can be expressed as a linear combination of them.

We may, in an analogous fashion, consider the problem of representing a time function, or signal, $x(t)$ on a T-second interval $(t_0, t_0 + T)$ as a similar expansion. Thus we consider a set of time functions $\phi_1(t)$, $\phi_2(t)$, . . . , $\phi_N(t)$, which are specified independently of $x(t)$, and seek a series expansion of the form

$$x_a(t) = \sum_{n=1}^{N} X_n\phi_n(t), \qquad t_0 \leq t \leq t_0 + T \tag{2.27}$$

in which the N coefficients X_n are independent of time and the subscript a indicates that (2.27) is considered an approximation.

We assume that the $\phi_n(t)$'s in (2.27) are *linearly independent;* that is, no one of them can be expressed as a sum of the other $N - 1$. A set of linearly independent $\phi_n(t)$'s will be called a *basis function set.*

We now wish to examine the error in the approximation of $x(t)$ by $x_a(t)$. As in the case of ordinary vectors, the expansion (2.27) is easiest to use if the $\phi_n(t)$'s are orthogonal on the interval $(t_0, t_0 + T)$. That is,

$$\int_{t_0}^{t_0+T} \phi_m(t)\phi_n^*(t)\, dt = c_n\delta_{mn} \triangleq \begin{cases} c_n, & n = m \\ 0, & n \neq m \end{cases} \quad \text{(all } m \text{ and } n) \quad (2.28)$$

where, if $c_n = 1$ for all n, the $\phi_n(t)$'s are said to be *normalized*. A normalized orthogonal set of functions is called an *orthonormal basis set*. The asterisk in (2.28) denotes complex conjugate, since we wish to allow the possibility of complex-valued $\phi_n(t)$'s. The symbol δ_{mn}, called the *Kronecker delta function*, is defined as unity if $m = n$ and zero otherwise.

The error in the approximation of $x(t)$ by the series of (2.27) will be measured in the integral-squared sense:

$$\text{Error} = \epsilon_N = \int_T |x(t) - x_a(t)|^2\, dt \quad (2.29)$$

where $\int_T (\;) \, dt$ denotes integration over t from t_0 to $t_0 + T$. The *integral-squared error* (ISE) is an applicable measure of error only when $x(t)$ is an energy signal or a power signal. If $x(t)$ is an energy signal of infinite duration, the limit as $T \to \infty$ is taken.

We now find the set of coefficients X_n that minimizes the ISE. Substituting (2.27) in (2.29), expressing the magnitude squared of the integrand as the integrand times its complex conjugate and expanding, we obtain

$$\epsilon_N = \int_T |x(t)|^2\, dt - \sum_{n=1}^{N} \left[X_n^* \int_T x(t)\phi_n^*(t)\, dt + X_n \int_T x^*(t)\phi_n(t)\, dt \right]$$

$$+ \sum_{n=1}^{N} c_n |X_n|^2 \quad (2.30)$$

in which the orthogonality of the $\phi_n(t)$'s has been used after interchanging the orders of summation and integration. The details are left to the problems. To find the X_n's that minimize ϵ_N, we add and subtract the quantity

$$\sum_{n=1}^{N} \frac{1}{c_n} \left| \int_T x(t)\phi_n^*(t)\, dt \right|^2$$

which yields, after rearrangement of terms, the following result for ϵ_N:

$$\epsilon_N = \int_T |x(t)|^2\, dt - \sum_{n=1}^{N} \frac{1}{c_n} \left| \int_T x(t)\phi_n^*(t)\, dt \right|^2$$

$$+ \sum_{n=1}^{N} c_n \left| X_n - \frac{1}{c_n} \int_T x(t)\phi_n^*(t)\, dt \right|^2 \quad (2.31)$$

The first two terms on the right-hand side of (2.31) are independent of the coefficients X_n. Since the last sum on the right-hand side is nonnegative, we will minimize ϵ_N if we choose each X_n such that the corresponding term in the sum is zero. Thus, since $c_n > 0$, the choice of

$$X_n = \frac{1}{c_n} \int_T x(t)\phi_n^*(t) \, dt \tag{2.32}$$

for X_n minimizes the ISE. The resulting minimum-error coefficients will be referred to as the *Fourier coefficients*.

The minimum value for ϵ_N, from (2.31), is obviously

$$(\epsilon_N)_{min} = \int_T |x(t)|^2 \, dt - \sum_{n=1}^{N} \frac{1}{c_n} \left| \int_T^T x(t)\phi_n^*(t) \, dt \right|^2$$

$$= \int_T |x(t)|^2 \, dt - \sum_{n=1}^{N} c_n |X_n|^2 \tag{2.33}$$

If we can find an infinite set of orthonormal functions such that

$$\lim_{N \to \infty} (\epsilon_N)_{min} = 0 \tag{2.34}$$

for any signal that is integrable square,

$$\int_T |x(t)|^2 \, dt < \infty \tag{2.35}$$

we say that the $\phi_n(t)$'s are *complete*. In the sense that the ISE is zero, we may then write

$$x(t) = \sum_{n=1}^{\infty} X_n \phi_n(t) \qquad (\text{ISE} = 0) \tag{2.36}$$

although there may be a number of isolated points of discontinuity where actual equality does not hold. For almost all points in the interval $(t_0, t_0 + T)$, Equation (2.34) requires that $x(t)$ be equal to $x_a(t)$ as $N \to \infty$.

Assuming a complete orthogonal set of functions, we obtain from (2.33) the relation

$$\int_T |x(t)|^2 \, dt = \sum_{n=1}^{\infty} c_n |X_n|^2 \tag{2.37}$$

This equation is known as *Parseval's theorem*.

▼ | **EXAMPLE 2.5**   Consider the set of two orthonormal functions shown in Figure 2.5(a). The signal

$$x(t) = \begin{cases} \sin \pi t, & 0 \le t \le 2 \\ 0, & \text{otherwise} \end{cases} \tag{2.38}$$

is to be approximated by a two-term generalized Fourier series of the form given by (2.27). The Fourier coefficients, from (2.32), are given by

$$X_1 = \int_0^2 \phi_1(t) \sin \pi t \, dt = \int_0^1 \sin \pi t \, dt = \frac{2}{\pi} \tag{2.39a}$$

$$X_2 = \int_0^2 \phi_2(t) \sin \pi t \, dt = \int_1^2 \sin \pi t \, dt = -\frac{2}{\pi} \tag{2.39b}$$

Thus the generalized two-term Fourier series approximation for this signal is

$$x_a(t) = \frac{2}{\pi} \phi_1(t) - \frac{2}{\pi} \phi_2(t) = \frac{2}{\pi} \left[\Pi \left(t - \frac{1}{2} \right) - \Pi \left(t - \frac{3}{2} \right) \right] \tag{2.40}$$

where $\Pi(t)$ is the unit rectangular pulse defined by (2.1). The signal $x(t)$ and the approximation $x_a(t)$ are compared in Figure 2.5(b). Figure 2.5(c) emphasizes the signal space interpretation of $x_a(t)$ by representing it as the point $(2/\pi, -2/\pi)$ in the two-dimensional space spanned by the orthonormal functions $\phi_1(t)$ and $\phi_2(t)$. Representation of an arbitrary $x(t)$ exactly ($\epsilon_N = 0$) would require an infinite set of properly chosen orthogonal functions (that is, a complete set).

▲ | The minimum ISE, from (2.33), is

$$(\epsilon_2)_{\min} = \int_0^2 \sin^2 \pi t \, dt - 2 \left(\frac{2}{\pi} \right)^2 = 1 - \frac{8}{\pi^2} \cong 0.189 \tag{2.41}$$

2.4   FOURIER SERIES

Complex Exponential Fourier Series

Given a signal $x(t)$ defined over the interval $(t_0, t_0 + T_0)$ with the definition

$$\omega_0 = 2\pi f_0 = \frac{2\pi}{T_0} \tag{2.42}$$

we define the *complex exponential Fourier series* as

$$x(t) = \sum_{n=-\infty}^{\infty} X_n e^{jn\omega_0 t}, \qquad t_0 \le t \le t_0 + T_0 \tag{2.43}$$

FIGURE 2.5 Approximation of a sinewave pulse with a generalized Fourier series. (a) Orthonormal functions. (b) Sinewave and approximation. (c) Signal space representation.

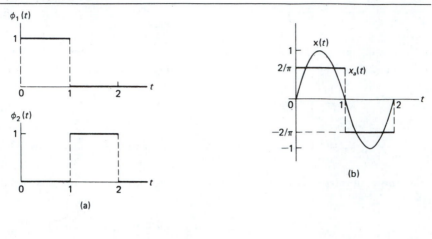

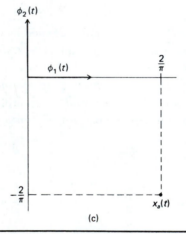

where

$$X_n = \frac{1}{T_0} \int_{t_0}^{t_0 + T_0} x(t) e^{-jn\omega_0 t} \, dt \qquad (2.44)$$

It can be shown to represent the signal $x(t)$ exactly in the interval $(t_0, t_0 + T_0)$, except at a point of jump discontinuity where it converges to the arithmetic mean of the left- and right-hand limits.* Outside the interval $(t_0,$

---

* Dirichlet's conditions state that sufficient conditions for convergence are that $x(t)$ be defined and bounded on the range $(t_0, t_0 + T_0)$ and have only a finite number of maxima and minima and a finite number of discontinuities on this range.

$t_0 + T_0$), of course, nothing is guaranteed. However, we note that the right-hand side of (2.43) is periodic with period T_0, since it is the sum of periodic rotating phasors with harmonic frequencies. Thus, if $x(t)$ is periodic with period T_0, the Fourier series of (2.43) is an accurate representation for $x(t)$ for *all* t (except at points of discontinuity). The integration of (2.44) can then be taken over any period.

   A useful observation about a complete orthonormal-series expansion of a signal is that the series is *unique*. For example, if we somehow find a Fourier expansion for a signal $x(t)$, we know that no other Fourier expansion for that $x(t)$ exists, since $\{e^{jn\omega_0 t}\}$ forms a complete set. The usefulness of this observation is illustrated with the following example.

▼

EXAMPLE 2.6   Consider the signal

$$x(t) = \cos \omega_0 t + \sin^2 2\omega_0 t \qquad (2.45)$$

where $\omega_0 = 2\pi/T_0$. Find the complex exponential Fourier series.

SOLUTION   We could compute the Fourier coefficients using (2.44), but by using appropriate trigonometric identities and Euler's theorem, we obtain

$$
\begin{aligned}
x(t) &= \cos \omega_0 t + \tfrac{1}{2} - \tfrac{1}{2} \cos 4\omega_0 t \\
&= \tfrac{1}{2}e^{j\omega_0 t} + \tfrac{1}{2}e^{-j\omega_0 t} + \tfrac{1}{2} - \tfrac{1}{4}e^{j4\omega_0 t} - \tfrac{1}{4}e^{-j4\omega_0 t} \qquad (2.46)
\end{aligned}
$$

Invoking uniqueness and equating the second line term by term with $\sum_{n=-\infty}^{\infty} X_n e^{jn\omega_0 t}$, we find that

$$
\begin{aligned}
X_0 &= \tfrac{1}{2} \\
X_1 &= \tfrac{1}{2} = X_{-1} \qquad (2.47) \\
X_4 &= -\tfrac{1}{4} = X_{-4}
\end{aligned}
$$

with all other X_n's equal to zero. Thus considerable labor is saved by noting
▲ that the Fourier series of a signal is unique.

The Fourier Coefficients as Time Averages and Their Symmetry Properties

The expression for X_n (2.44) can be interpreted as the time average of $x(t)e^{-jn\omega_0 t}$, where, for future convenience, we define the *time average* of a signal $v(t)$ as

$$\langle v(t) \rangle = \begin{cases} \lim_{T \to \infty} \dfrac{1}{2T} \displaystyle\int_{-T}^{T} v(t)\, dt & [v(t) \text{ not periodic}] \\[2ex] \dfrac{1}{T_0} \displaystyle\int_{T_0} v(t)\, dt & [v(t) \text{ periodic}] \end{cases}$$

$$(2.48a)$$

$$(2.48b)$$

where $\int_{T_0}$ indicates integration over any period. For $n = 0$, we see that $X_0 = \langle x(t) \rangle$ is the average, or dc, value of $x(t)$.

Using Euler's theorem, we can write (2.44) as

$$X_n = \langle x(t) \cos n\omega_0 t \rangle - j\langle x(t) \sin n\omega_0 t \rangle \qquad (2.49)$$

and assuming $x(t)$ is real, it follows that

$$X_n^* = X_{-n} \qquad (2.50)$$

by replacing n by $-n$ in (2.49). Writing X_n as

$$X_n = |X_n| e^{j/X_n} \qquad (2.51)$$

we obtain

$$|X_n| = |X_{-n}| \qquad \text{and} \qquad \underline{/X_n} = -\underline{/X_{-n}} \qquad (2.52)$$

Thus, for real signals, the magnitude of the Fourier coefficients is an even function of n, and the argument is odd.

Several symmetry properties can be derived for the Fourier coefficients, depending on the symmetry of $x(t)$. For example, suppose $x(t)$ is even; that is, $x(t) = x(-t)$. Then the second term in (2.49) is zero, since $x(t) \sin n\omega_0 t$ is an odd function. Thus X_n is purely real, and furthermore, X_n is an even function of n since $\cos n\omega_0 t$ is an even function of n. These consequences of $x(t)$ being even are illustrated by Example 2.6.

On the other hand, if $x(t) = -x(-t)$ [that is, $x(t)$ is odd], it readily follows that X_n is purely imaginary, since $\langle x(t) \cos n\omega_0 t \rangle = 0$ by virtue of $x(t) \cos n\omega_0 t$ being odd. In addition, X_n is an odd function of n, since $\sin n\omega_0 t$ is an odd function of n.

Another type of symmetry is *halfwave symmetry*, defined as

$$x(t \pm \tfrac{1}{2}T_0) = -x(t) \qquad (2.53)$$

where T_0 is the period of $x(t)$. For signals with halfwave symmetry,

$$X_n = 0, \qquad n = 0, \pm 2, \pm 4, \ldots \qquad (2.54)$$

which states that the Fourier series for such a signal consists only of odd-indexed terms. The proof of this is left to the problems.

Trigonometric Form of the Fourier Series

Using (2.52) and assuming $x(t)$ real, we can regroup the complex exponential Fourier series by pairs of terms of the form

$$X_n e^{jn\omega_0 t} + X_{-n} e^{-jn\omega_0 t} = |X_n| e^{j(n\omega_0 t + \underline{/X_n})} + |X_n| e^{-j(n\omega_0 t + \underline{/X_n})}$$
$$= 2|X_n| \cos(n\omega_0 t + \underline{/X_n}) \tag{2.55}$$

Hence (2.43) can be written in the equivalent trigonometric form:

$$x(t) = X_0 + \sum_{n=1}^{\infty} 2|X_n| \cos(n\omega_0 t + \underline{/X_n}) \tag{2.56}$$

Expanding the cosine in (2.56), we obtain still another equivalent series of the form

$$x(t) = X_0 + \sum_{n=1}^{\infty} A_n \cos n\omega_0 t + \sum_{n=1}^{\infty} B_n \sin n\omega_0 t \tag{2.57}$$

where

$$A_n = 2|X_n| \cos \underline{/X_n}$$
$$= 2\langle x(t) \cos n\omega_0 t \rangle$$
$$= \frac{2}{T_0} \int_{t_0}^{t_0 + T_0} x(t) \cos(n\omega_0 t)\, dt \tag{2.58}$$

and

$$B_n = -2|X_n| \sin \underline{/X_n}$$
$$= 2\langle x(t) \sin n\omega_0 t \rangle$$
$$= \frac{2}{T_0} \int_{t_0}^{t_0 + T_0} x(t) \sin(n\omega_0 t)\, dt \tag{2.59}$$

In either the trigonometric or the exponential form of the Fourier series, X_0 represents the average or dc component of $x(t)$. The term for $n = 1$ is called the *fundamental*, the term for $n = 2$ is called the *second harmonic*, and so on.

Parseval's Theorem

Using (2.23) for average power of a periodic signal, substituting (2.43) for $x(t)$, and interchanging the order of integration and summation, we find Parseval's theorem to be

$$P = \frac{1}{T_0} \int_{T_0} |x(t)|^2 \, dt = \sum_{n=-\infty}^{\infty} |X_n|^2 \tag{2.60a}$$

$$= X_0^2 + 2 \sum_{n=1}^{\infty} |X_n|^2 \tag{2.60b}$$

which is a specialization of (2.37). In words, (2.60a) simply states that the average power of a periodic signal $x(t)$ is the sum of the powers in the phasor components of its Fourier series.

Examples of Fourier Series—The Sinc Function

Table 2.1 gives Fourier series for several commonly occurring periodic waveforms. The left-hand column specifies the signal over one period. The definition of periodicity,

$$x(t) = x(t + T_0)$$

specifies it for all t. The derivation of the Fourier coefficients given in the right-hand column of Table 2.1 is left to the problems. Note that the full-rectified sinewave actually has the period $\frac{1}{2}T_0$.

For the periodic pulse train, it is convenient to express the coefficients in terms of the *sinc function*, defined as

$$\text{sinc } z = \frac{\sin \pi z}{\pi z} \tag{2.61}$$

The sinc function will arise often in our discussion. It is shown in Figure 2.6 along with $\text{sinc}^2 z$.

▼ **EXAMPLE 2.7**   Specialize the results of Table 2.1 to the complex exponential and trigonometric Fourier series of a squarewave with even symmetry and amplitudes zero and A.

SOLUTION   The solution proceeds by letting $t_0 = 0$ and $\tau = \frac{1}{2}T_0$ in item 1 of Table 2.1. Thus

$$X_n = \frac{1}{2}A \text{ sinc } (\frac{1}{2}n) \tag{2.62}$$

TABLE 2.1   Fourier Series for Several Periodic Signals

Signal: $-\frac{1}{2}T_0 \leq t \leq \frac{1}{2}T_0$ (One Period)	Coefficients for Exponential Fourier Series

1. Asymmetrical pulse train:

$$x(t) = A\Pi\left(\frac{t - t_0}{\tau}\right), \tau < T_0$$

$$x(t) = x(t + T_0), \quad \text{all } t$$

$$X_n = \frac{A\tau}{T_0} \text{sinc } (nf_0\tau)e^{-j2\pi nf_0t_0}$$

$$n = 0, \pm 1, \pm 2, \ldots$$

2. Half-rectified sinewave:

$$x(t) = \begin{cases} A \sin \omega_0 t, & 0 \leq t \leq \frac{1}{2}T_0 \\ 0, & -\frac{1}{2}T_0 \leq t \leq 0 \end{cases}$$

$$x(t) = x(t + T_0), \quad \text{all } t$$

$$X_n = \begin{cases} \dfrac{A}{\pi(1 - n^2)}, & n = 0, \pm 2, \pm 4, \ldots \\ 0, & n = \pm 3, \pm 5, \ldots \\ -\frac{1}{4}jA, & n = 1 \\ \frac{1}{4}jA, & n = -1 \end{cases}$$

3. Full-rectified sinewave:

$$x(t) = A|\sin \omega_0 t|$$

$$X_n = \begin{cases} \dfrac{2A}{\pi(1 - n^2)}, & n = 0, \pm 2, \pm 4, \ldots \\ 0, & n = \pm 1, \pm 3, \ldots \end{cases}$$

4. Triangular wave:

$$x(t) = \begin{cases} \dfrac{4A}{T_0}t + A, & -\frac{1}{2}T_0 \leq t < 0 \\ -\dfrac{4A}{T_0}t + A, & 0 \leq t \leq \frac{1}{2}T_0 \end{cases}$$

$$x(t) = x(t + T_0), \quad \text{all } t$$

$$X_n = \begin{cases} \dfrac{4A}{\pi^2 n^2}, & n \text{ odd} \\ 0, & n \text{ even} \end{cases}$$

FIGURE 2.6   Graphs of sinc z and sinc^2 z

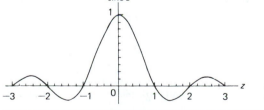

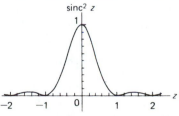

But

$$\text{sinc}\left(\tfrac{1}{2}n\right) = \frac{\sin\left(\tfrac{1}{2}n\pi\right)}{\tfrac{1}{2}n\pi} \tag{2.63}$$

$$= \begin{cases} 1, & n = 0 \\ 0, & n = \text{even} \\ |2/n\pi|, & n = \pm 1, \pm 5, \pm 9, \ldots \\ -|2/n\pi|, & n = \pm 3, \pm 7, \ldots \end{cases}$$

Thus

$$x(t) = \cdots \frac{A}{5\pi}e^{-j5\omega_0 t} - \frac{A}{3\pi}e^{-j3\omega_0 t} + \frac{A}{\pi}e^{-j\omega_0 t}$$

$$+ \frac{A}{2} + \frac{A}{\pi}e^{j\omega_0 t} - \frac{A}{3\pi}e^{j3\omega_0 t} + \frac{A}{5\pi}e^{j5\omega_0 t} - \cdots$$

$$= \frac{A}{2} + \frac{2A}{\pi}\left(\cos\omega_0 t - \frac{1}{3}\cos 3\omega_0 t + \frac{1}{5}\cos 5\omega_0 t - \cdots\right) \tag{2.64}$$

The first equation is the complex exponential form of the Fourier series, and the second equation is the trigonometric form. The dc component of this squarewave is $X_0 = \tfrac{1}{2}A$. Setting this term to zero in the preceding Fourier series, we have the Fourier series of a squarewave of amplitudes $\pm\tfrac{1}{2}A$. Such a squarewave has halfwave symmetry, and this is precisely the reason that no even harmonics are present in its Fourier series.

Line Spectra

The complex exponential Fourier series (2.43) of a signal is simply a summation of phasors. In Section 2.1 we showed how a phasor could be characterized in the frequency domain by two plots: one showing its amplitude versus frequency and one showing its phase. Similarly, a periodic signal can be characterized in the frequency domain by making two plots: one showing amplitudes of the separate phasor components versus frequency and the other showing their phases versus frequency. The resulting plots are called the *two-sided amplitude* and phase spectra*, respectively, of the signal. From (2.52) it follows that, for a real signal, the amplitude spectrum is even and the phase spectrum is odd, which is simply a result of the addition of complex conjugate phasors to get a real sinusoidal signal.

Figure 2.7(a) shows the double-sided spectrum for a half-rectified sine-

---

* *Magnitude spectrum* would be a more accurate term, although *amplitude spectrum* is the customary term.

FIGURE 2.7 Line spectra for half-rectified sinewave. (a) Double-sided. (b) Single-sided.

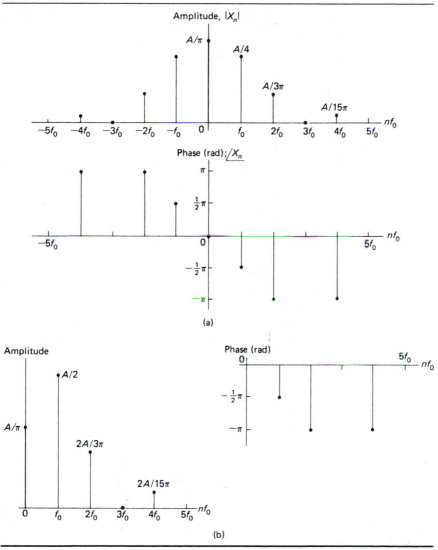

wave as plotted from the results given in Table 2.1. For $n = 2, 4, \ldots$, X_n is represented as follows:

$$X_n = -\left|\frac{A}{\pi(1-n^2)}\right| = \frac{A}{\pi(n^2-1)} e^{-j\pi} \qquad (2.65a)$$

For $n = -2, -4, \ldots$, it is represented as

$$X_n = -\left|\frac{A}{\pi(1 - n^2)}\right| = \frac{A}{\pi(n^2 - 1)} e^{j\pi} \tag{2.65b}$$

to ensure that the phase is odd, as it must be. (Note that $e^{\pm j\pi} = -1$.) Thus

$$|X_n| = \begin{cases} \frac{1}{4}A, & n = \pm 1 \\ \left|\dfrac{A}{\pi(1 - n^2)}\right|, & \text{all even } n \end{cases} \tag{2.66a}$$

$$\underline{/X_n} = \begin{cases} -\pi, & n = 2, 4, \ldots \\ -\frac{1}{2}\pi, & n = 1 \\ 0, & n = 0 \\ \frac{1}{2}\pi, & n = -1 \\ \pi, & n = -2, -4, \ldots \end{cases} \tag{2.66b}$$

The *single-sided line spectra* are obtained by plotting the amplitudes and phase angles of the terms in the trigonometric Fourier series (2.56) versus nf_0. Because the series (2.56) has only nonnegative frequency terms, the single-sided spectra exist only for $nf_0 \geq 0$. From (2.56) it is readily apparent that the single-sided phase spectrum of a periodic signal is identical to its double-sided phase spectrum for $nf_0 \geq 0$ and zero for $nf_0 > 0$. The single-sided amplitude spectrum is obtained from the double-sided amplitude spectrum by doubling the amplitude of all lines for $nf_0 > 0$. The line at $nf_0 = 0$ stays the same. The single-sided spectra for the half-rectified sinewave are shown in Figure 2.7(b).

As a second example, consider the pulse train

$$x(t) = \sum_{n=-\infty}^{\infty} A\Pi\left(\frac{t - nT_0 - \frac{1}{2}\tau}{\tau}\right) \tag{2.67}$$

From Table 2.1, with $t_0 = \frac{1}{2}\tau$ substituted in item 1, the Fourier coefficients are

$$X_n = \frac{A\tau}{T_0} \text{sinc}(nf_0\tau) e^{-j\pi nf_0\tau} \tag{2.68}$$

The Fourier coefficients can be put in the form $|X_n| \exp(j\underline{/X_n})$, where

$$|X_n| = \frac{A\tau}{T_0} |\text{sinc}(nf_0\tau)| \tag{2.69a}$$

and

$$
\underline{/X_n} = \begin{cases} -\pi n f_0 \tau & \text{if} & \text{sinc } (n f_0 \tau) > 0 \\ -\pi n f_0 \tau + \pi & \text{if} \quad n f_0 > 0 \quad \text{and} \quad \text{sinc } (n f_0 \tau) < 0 \quad \text{(2.69b)} \\ -\pi n f_0 \tau - \pi & \text{if} \quad n f_0 < 0 \quad \text{and} \quad \text{sinc } (n f_0 \tau) < 0 \end{cases}
$$

The $\pm\pi$ on the right-hand side of (2.69b) accounts for $|\text{sinc } n f_0 \tau| = -\text{sinc } n f_0 \tau$ whenever sinc $n f_0 \tau < 0$. Since the phase spectrum must have odd symmetry if $x(t)$ is real, π is subtracted if $n f_0 < 0$ and added if $n f_0 > 0$. The reverse could have been done—the choice is arbitrary. With these considerations, the double-sided amplitude and phase spectra can now be plotted. They are shown in Figure 2.8 for several choices of τ and T_0. Note that appropriate multiples of 2π are subtracted from the lines in the phase spectrum.

Comparing Figures 2.8(a) and 2.8(b), we note that the zeros of the envelope of the amplitude spectrum, which occur at multiples of $1/\tau$ Hz, move out along the frequency axis as the pulse width decreases. That is, *the time duration of a signal and its spectral width are inversely proportional,* a property that will be shown to be true in general later. Second, comparing Figures 2.8(a) and 2.8(c), we note that the separation between lines in the spectra is $1/T_0$. Thus the density of the spectral lines with frequency increases as the period of $x(t)$ increases.

2.5   THE FOURIER TRANSFORM

To generalize the Fourier series representation (2.43) to a representation valid for aperiodic signals, we consider the two basic relationships (2.43) and (2.44). Suppose that $x(t)$ is nonperiodic but is an energy signal, so that it is integrable square in the interval $(-\infty, \infty)$.* In the interval $|t| < \frac{1}{2}T_0$, we can represent $x(t)$ as the Fourier series

$$
x(t) = \sum_{n=-\infty}^{\infty} \left[\frac{1}{T_0} \int_{-T_0/2}^{T_0/2} x(\lambda) e^{-j2\pi n f_0 \lambda} \, d\lambda \right] e^{j2\pi n f_0 t}, \qquad |t| < \frac{T_0}{2} \quad \text{(2.70)}
$$

---

* This means that $x(t)$ should be an energy signal. Dirichlet's conditions give sufficient conditions for a signal to have a Fourier transform. They are that $x(t)$ be (1) single-valued with a finite number of maxima and minima and a finite number of discontinuities in any finite time interval, and (2) absolutely integrable, that is, $\int_{-\infty}^{\infty} |x(t)| \, dt < \infty$. These conditions include all energy signals.

FIGURE 2.8 Spectra for a periodic pulse train signal. (a) $\tau = \frac{1}{4}T_0$. (b) $\tau = \frac{1}{8}T_0$; T_0 same as in (a). (c) $\tau = \frac{1}{8}T_0$; τ same as in (a).

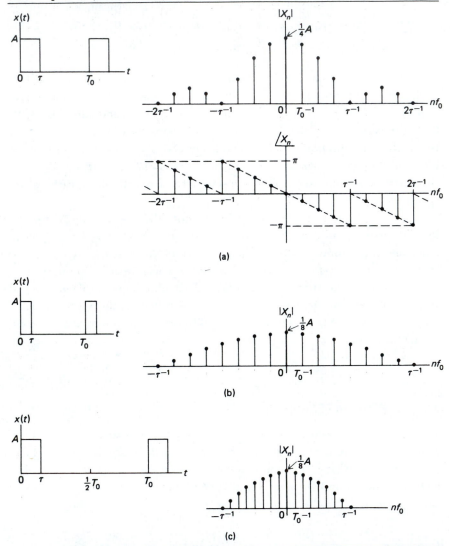

where $f_0 = 1/T_0$. To represent $x(t)$ for all time, we simply let $T_0 \rightarrow \infty$ such that $nf_0 = n/T_0$ becomes the continuous variable f, $1/T_0$ becomes the differential df, and the summation becomes an integral. Thus

$$x(t) = \int_{-\infty}^{\infty} \left[\int_{-\infty}^{\infty} x(\lambda) e^{-j2\pi f \lambda} \, d\lambda \right] e^{j2\pi ft} \, df \qquad (2.71)$$

Defining the inside integral as

$$X(f) = \int_{-\infty}^{\infty} x(\lambda)e^{-j2\pi f\lambda}\,d\lambda \tag{2.72}$$

we can write (2.71) as

$$x(t) = \int_{-\infty}^{\infty} X(f)e^{j2\pi ft}\,df \tag{2.73}$$

The existence of these integrals is assured, since $x(t)$ is an energy signal. We note that

$$X(f) = \lim_{T_0 \to \infty} T_0 X_n \tag{2.74}$$

which avoids the problem that $|X_n| \to 0$ as $T_0 \to \infty$.

The frequency-domain description of $x(t)$ provided by (2.72) is referred to as the *Fourier transform* of $x(t)$, written symbolically as $X(f) = \mathcal{F}[x(t)]$. Conversion back to the time domain is achieved via the *inverse Fourier transform* (2.73), written symbolically as $x(t) = \mathcal{F}^{-1}[X(f)]$.

Expressing (2.72) and (2.73) in terms of $f = \omega/2\pi$ results in easily remembered symmetrical expressions. Integrating (2.73) with respect to the variable ω requires a factor of $(2\pi)^{-1}$.

Amplitude and Phase Spectra

Writing $X(f)$ in terms of amplitude and phase as

$$X(f) = |X(f)|e^{j\theta(f)}, \qquad \theta(f) = \underline{/X(f)} \tag{2.75}$$

we can show, for real $x(t)$, that

$$|X(f)| = |X(-f)| \qquad \text{and} \qquad \theta(f) = -\theta(-f) \tag{2.76}$$

just as for Fourier series. This is done by using Euler's theorem to write

$$R = \text{Re } X(f) = \int_{-\infty}^{\infty} x(t)\cos 2\pi ft\,dt \tag{2.77a}$$

and

$$I = \text{Im } X(f) = -\int_{-\infty}^{\infty} x(t)\sin 2\pi ft\,dt \tag{2.77b}$$

Thus the real part of $X(f)$ is even and the imaginary part is odd if $x(t)$ is a real signal. Since $|X(f)|^2 = R^2 + I^2$ and $\tan \theta(f) = +I/R$, the symmetry properties (2.76) follow. A plot of $|X(f)|$ versus f is referred to as the *amplitude spectrum** of $x(t)$, and a plot of $\underline{/X(f)} = \theta(f)$ versus f is known as the *phase spectrum*.

Symmetry Properties

If $x(t) = x(-t)$, that is, if $x(t)$ is even, then $x(t) \sin 2\pi ft$ is odd and Im $X(f)$ = 0. Furthermore, Re $X(f)$ is an even function of f. Thus the Fourier transform of a real, even function is real and even.

On the other hand, if $x(t)$ is odd, $x(t) \cos 2\pi ft$ is odd and Re $X(f)$ = 0. Thus the Fourier transform of a real, odd function is imaginary. In addition, Im $X(f)$ is an odd function of frequency because $\sin 2\pi ft$ is an odd function.

▼ EXAMPLE 2.8   Consider the pulse

$$x(t) = A\Pi \left(\frac{t - t_0}{\tau} \right) \tag{2.78}$$

Its Fourier transform is

$$X(f) = \int_{-\infty}^{\infty} A\Pi \left(\frac{t - t_0}{\tau} \right) e^{-j2\pi ft} \, dt$$

$$= A \int_{t_0 - \tau/2}^{t_0 + \tau/2} e^{-j2\pi ft} \, dt = A\tau \text{ sinc } f\tau \, e^{-j2\pi ft_0} \tag{2.79}$$

Its amplitude spectrum is

$$|X(f)| = A\tau \, |\text{sinc } f\tau| \tag{2.80a}$$

and its phase spectrum is

$$\theta(f) = \begin{cases} -2\pi t_0 f & \text{if } \text{sinc } f\tau > 0 \\ -2\pi t_0 f \pm \pi & \text{if } \text{sinc } f\tau < 0 \end{cases} \tag{2.80b}$$

The term $\pm \pi$ is used to account for sinc $f\tau$ being negative, and if $+\pi$ is used for $f > 0$, $-\pi$ is used for $f < 0$, or vice versa, to ensure that $\theta(f)$ is odd. When $|\theta(f)|$ exceeds 2π, an appropriate multiple of 2π may be subtracted from $\theta(f)$. Figure 2.9 shows the amplitude and phase spectra for this signal. The similarity to Figure 2.8 is to be noted, especially the *inverse relationship between spectral width and pulse duration.*

▲

---

* *Amplitude density spectrum* would be more correct, since its dimensions are (amplitude units)(time) = (amplitude units)/(frequency), but we will use the term *amplitude spectrum* for simplicity.

FIGURE 2.9   Amplitude and phase spectra for a pulse signal. (a) Amplitude spectrum. (b) Phase spectrum ($t_0 = \frac{1}{2}\tau$ is assumed).

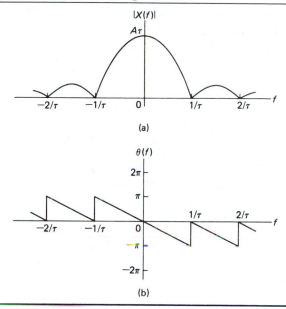

(a)

(b)

Energy Spectral Density

The energy of a signal, defined by (2.20a), can be expressed in the frequency domain as follows:

$$E \triangleq \int_{-\infty}^{\infty} |x(t)|^2 \, dt$$

$$= \int_{-\infty}^{\infty} x^*(t) \left[\int_{-\infty}^{\infty} X(f)e^{j2\pi ft} \, df \right] dt \qquad (2.81)$$

where $x(t)$ has been written in terms of its Fourier transform. Reversing the order of integration, we obtain

$$E = \int_{-\infty}^{\infty} X(f) \left[\int_{-\infty}^{\infty} x^*(t)e^{j2\pi ft} \, dt \right] df$$

$$= \int_{-\infty}^{\infty} X(f) \left[\int_{-\infty}^{\infty} x(t)e^{-j2\pi ft} \, dt \right]^* df = \int_{-\infty}^{\infty} X(f)X^*(f) \, df$$

or

$$E = \int_{-\infty}^{\infty} |x(t)|^2 \, dt = \int_{-\infty}^{\infty} |X(f)|^2 \, df \qquad (2.82)$$

This is referred to as *Rayleigh's energy theorem* or Parseval's theorem for Fourier transforms.

Examining $|X(f)|^2$ and recalling the definition of $X(f)$ given by (2.72), we note that the former has the units of (volts-seconds)^2 or, since we are considering power on a per-ohm basis, (watts-seconds)/hertz = joules/hertz. Thus we see that $|X(f)|^2$ has the units of energy density, and we define the energy spectral density of a signal as

$$G(f) \triangleq |X(f)|^2 \tag{2.83}$$

By integrating $G(f)$ over all frequency, we obtain the total energy.

▼ **EXAMPLE 2.9**   Rayleigh's energy theorem is convenient for finding the energy in a signal whose square is not easily integrated in the time domain, or vice versa. For example, the signal

$$x(t) = 40 \text{ sinc } (20t) \tag{2.84}$$

has energy density

$$G(f) = |X(f)|^2 = \left[2\Pi \left(\frac{f}{20} \right) \right]^2 = 4\Pi \left(\frac{f}{20} \right) \tag{2.85}$$

where $\Pi(f/20)$ need not be squared because it has unity amplitude. Using Rayleigh's energy theorem, we find that the energy in $x(t)$ is

$$E = \int_{-\infty}^{\infty} G(f) \, df = \int_{-10}^{10} 4 \, df = 80 \text{ J} \tag{2.86}$$

This checks with the result that is obtained by integrating $x^2(t)$ over all t using the definite integral $\int_{-\infty}^{\infty} \text{sinc}^2 u \, du = 1$.

The energy contained in the frequency interval $(0, W)$ can be found from the integral

$$E_W = \int_{-W}^{W} G(f) \, df = 2 \int_{0}^{W} \left[2\Pi \left(\frac{f}{20} \right) \right]^2 df \tag{2.87}$$

$$= \begin{cases} 8W, & W \leq 10 \\ 80, & W > 10 \end{cases}$$

▲

Convolution

We digress somewhat from our consideration of the Fourier transform to define the convolution operation and illustrate it by example.

The convolution of two signals, $x_1(t)$ and $x_2(t)$, is a new function of time, $x(t)$, written symbolically in terms of x_1 and x_2 as

$$x(t) = x_1(t) * x_2(t) = \int_{-\infty}^{\infty} x_1(\lambda)x_2(t - \lambda) \, d\lambda \qquad (2.88)$$

Note that t is a parameter as far as the integration is concerned. The integrand is formed from x_1 and x_2 by three operations: (1) time reversal to obtain $x_2(-\lambda)$, (2) time shifting to obtain $x_2(t - \lambda)$, and (3) multiplication of $x_1(\lambda)$ and $x_2(t - \lambda)$ to form the integrand. An example will illustrate the implementation of these operations to form $x_1 * x_2$. Note that the dependence on time is often suppressed.

▼| EXAMPLE 2.10   Find the convolution of the two signals

$$x_1(t) = e^{-\alpha t}u(t) \quad \text{and} \quad x_2(t) = e^{-\beta t}u(t), \qquad \alpha > \beta > 0 \qquad (2.89)$$

SOLUTION   The steps involved in the convolution are illustrated in Figure 2.10 for $\alpha = 4$ and $\beta = 2$. Mathematically, we can form the integrand by direct substitution:

$$x(t) = x_1 * x_2 = \int_{-\infty}^{\infty} e^{-\alpha\lambda}u(\lambda)e^{-\beta(t-\lambda)}u(t - \lambda) \, d\lambda \qquad (2.90)$$

FIGURE 2.10   The operations involved in the convolution of two exponentially decaying signals

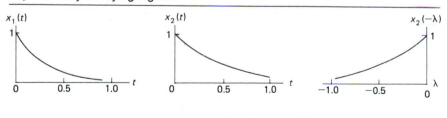

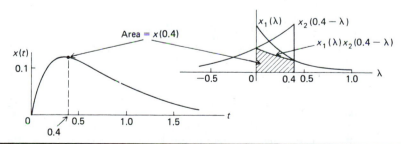

But

$$u(\lambda)u(t - \lambda) = \begin{cases} 0, & \lambda < 0 \\ 1, & 0 < \lambda < t \\ 0, & \lambda > t \end{cases} \qquad (2.91)$$

Thus

$$x(t) = \begin{cases} 0, \ t < 0 \\ \displaystyle\int_0^t e^{-\beta t} e^{-(\alpha - \beta)\lambda} \, d\lambda = \frac{1}{\alpha - \beta} (e^{-\beta t} - e^{-\alpha t}), & t \geq 0 \quad (2.92) \end{cases}$$

This result for $x(t)$ is also shown in Figure 2.10.

Transform Theorems: Proofs and Applications*

Several useful theorems involving Fourier transforms can be proved. These are useful for deriving Fourier transform pairs as well as deducing general frequency domain relationships. The notation $x(t) \leftrightarrow X(f)$ will be used to denote a Fourier transform pair.

Each theorem will be stated along with a proof in most cases. Several examples giving applications will be given after the statements of all the theorems. In the statement of the theorems, $x(t)$, $x_1(t)$, and $x_2(t)$ denote signals with $X(f)$, $X_1(f)$, and $X_2(f)$ denoting their respective Fourier transforms. Constants are denoted by a, a_1, a_2, t_0, and f_0.

▷ SUPERPOSITION THEOREM

$$a_1 x_1(t) + a_2 x_2(t) \leftrightarrow a_1 X_1(f) + a_2 X_2(f) \qquad (2.93)$$

PROOF   By the defining integral for the Fourier transform,

$$\mathcal{F}\{a_1 x_1(t) + a_2 x_2(t)\} = \int_{-\infty}^{\infty} [a_1 x_1(t) + a_2 x_2(t)] e^{-j2\pi ft} \, dt$$

$$= a_1 \int_{-\infty}^{\infty} x_1(t) e^{-j2\pi ft} \, dt + a_2 \int_{-\infty}^{\infty} x_2(t) e^{-j2\pi ft} \, dt$$

$$= a_1 X_1(f) + a_2 X_2(f)$$

$$(2.94)$$

▷ TIME-DELAY THEOREM

$$x(t - t_0) \leftrightarrow X(f) e^{-j2\pi ft_0} \qquad (2.95)$$

* See Tables C.1 and C.2 in Appendix C for a listing of Fourier transform pairs and theorems.

PROOF   Using the defining integral for the Fourier transform, we have

$$\mathcal{F}\{x(t - t_0)\} = \int_{-\infty}^{\infty} x(t - t_0)e^{-j2\pi ft}\, dt$$

$$= \int_{-\infty}^{\infty} x(\lambda)e^{-j2\pi f(\lambda + t_0)}\, dt$$

$$= e^{-j2\pi ft_0} \int_{-\infty}^{\infty} x(\lambda)e^{-j2\pi f\lambda}\, d\lambda$$

$$= X(f)e^{-j2\pi ft_0} \tag{2.96}$$

where the substitution $\lambda = t - t_0$ was used in the first integral.

▷ SCALE-CHANGE THEOREM

$$x(at) \leftrightarrow \frac{1}{|a|} X\left(\frac{f}{a}\right) \tag{2.97}$$

PROOF   First, assume that $a > 0$. Then

$$F\{x(at)\} = \int_{-\infty}^{\infty} x(at)e^{-j2\pi ft}\, dt = \int_{-\infty}^{\infty} x(\lambda)e^{-j2\pi f\lambda/a}\, \frac{d\lambda}{a} = \frac{1}{a} X\left(\frac{f}{a}\right) \tag{2.98}$$

where the substitution $\lambda = at$ has been used. Next considering $a < 0$, we write

$$\mathcal{F}\{x(at)\} = \int_{-\infty}^{\infty} x(-|a|t)e^{-j2\pi ft}\, dt = \int_{-\infty}^{\infty} x(\lambda)e^{+j2\pi f\lambda/|a|}\, \frac{d\lambda}{|a|}$$

$$= \frac{1}{|a|} X\left(-\frac{f}{|a|}\right) = \frac{1}{|a|} X\left(\frac{f}{a}\right) \tag{2.99}$$

where use has been made of the relation $-|a| = a$.

▷ DUALITY THEOREM

$$X(t) \leftrightarrow x(-f) \tag{2.100}$$

That is, if the Fourier transform of $x(t)$ is $X(f)$, then the Fourier transform of $X(f)$ with f replaced by t is the original time-domain signal with t replaced by $-f$.

PROOF   The proof of this theorem follows by virtue of the fact that the only difference between the Fourier transform integral and the inverse Fourier transform integral is a minus sign in the exponent of the integrand.

▷ FREQUENCY TRANSLATION THEOREM

$$x(t)e^{j2\pi f_0 t} \leftrightarrow X(f - f_0) \tag{2.101}$$

PROOF   To prove the frequency translation theorem, note that

$$\int_{-\infty}^{\infty} x(t)e^{j2\pi f_0 t}e^{-j2\pi ft} \, dt = \int_{-\infty}^{\infty} x(t)e^{-j2\pi(f-f_0)t} \, dt = X(f - f_0) \tag{2.102}$$

▷ MODULATION THEOREM

$$x(t)\cos(2\pi f_0 t) \leftrightarrow \tfrac{1}{2}X(f - f_0) + \tfrac{1}{2}X(f + f_0) \tag{2.103}$$

PROOF   The proof of this theorem follows by writing $\cos(2\pi f_0 t)$ in exponential form as $(e^{j2\pi f_0 t} + e^{-j2\pi f_0 t})/2$ and applying the superposition and frequency translation theorems.

▷ DIFFERENTIATION THEOREM

$$\frac{d^n x(t)}{dt^n} \leftrightarrow (j2\pi f)^n X(f) \tag{2.104}$$

PROOF   We prove the theorem for $n = 1$ by using integration by parts on the defining Fourier transform integral as follows:

$$\mathscr{F}\left\{\frac{dx}{dt}\right\} = \int_{-\infty}^{\infty} \frac{dx(t)}{dt} e^{-j2\pi ft} \, dt$$

$$= x(t)e^{-j2\pi ft}\Big|_{-\infty}^{\infty} + j2\pi f \int_{-\infty}^{\infty} x(t)e^{-j2\pi ft} \, dt$$

$$= j2\pi f \, X(f) \tag{2.105}$$

where $u = e^{-j2\pi ft}$ and $dv = (dx/dt)dt$ have been used in the integration-by-parts formula, and the first term of the middle equation vanishes at each end point by virtue of $x(t)$ being an energy signal. The proof for values of $n > 1$ follows by induction.

▷ INTEGRATION THEOREM

$$\int_{-\infty}^{t} x(\lambda) \, d\lambda \leftrightarrow (j2\pi f)^{-1}X(f) + \frac{1}{2}X(0)\delta(f) \tag{2.106}$$

PROOF   If $X(0) = 0$ the proof of the integration theorem can be carried out by using integration by parts as in the case of the differentiation theorem. We obtain

$$\mathscr{F}\left\{ \int_{-\infty}^{t} x(\lambda)\, d\lambda \right\}$$

$$= \left[\int_{-\infty}^{t} x(\lambda)\, d\lambda \right] \left(-\frac{1}{j2\pi f}e^{-j2\pi ft} \right)\bigg|_{-\infty}^{\infty} + \frac{1}{j2\pi f}\int_{-\infty}^{\infty} x(t)e^{-j2\pi ft}\, dt$$

$$(2.107)$$

The first term vanishes if $X(0) = \int_{-\infty}^{\infty} x(t)\, dt = 0$, and the second term is just $X(f)/j2\pi f$. For $X(0) \neq 0$, a limiting argument must be used to account for the Fourier transform of the nonzero average value of $x(t)$.

▷ **CONVOLUTION THEOREM**

$$\int_{-\infty}^{\infty} x_1(\lambda)x_2(t - \lambda)\, d\lambda = \int_{-\infty}^{\infty} x_1(t - \lambda)x_2(\lambda)\, d\lambda \leftrightarrow X_1(f)X_2(f) \quad (2.108)$$

PROOF   To prove the convolution theorem of Fourier transforms, we represent $x_2(t - \lambda)$ in terms of the inverse Fourier transform integral as

$$x_2(t - \lambda) = \int_{-\infty}^{\infty} X_2(f)e^{j2\pi f(t-\lambda)}\, df \qquad (2.109)$$

Denoting the convolution operation as $x_1(t) * x_2(t)$, we have

$$x_1(t) * x_2(t) = \int_{-\infty}^{\infty} x_1(\lambda) \left[\int_{-\infty}^{\infty} X_2(f)e^{j2\pi f(t-\lambda)}\, df \right] d\lambda$$

$$= \int_{-\infty}^{\infty} X_2(f) \left[\int_{-\infty}^{\infty} x_1(\lambda)e^{-j2\pi f\lambda}\, d\lambda \right] e^{j2\pi ft}\, df \quad (2.110)$$

where the last step results from reversing the orders of integration. The bracketed term inside the integral is $X_1(f)$, the Fourier transform of $x_1(t)$. Thus

$$x_1 * x_2 = \int_{-\infty}^{\infty} X_1(f)X_2(f)e^{j2\pi ft}\, df \qquad (2.111)$$

which is the inverse Fourier transform of $X_1(f)X_2(f)$. Taking the Fourier transform of this result yields the desired transform pair.

▷ MULTIPLICATION THEOREM

$$x_1(t)x_2(t) \leftrightarrow X_1(f) * X_2(f) = \int_{-\infty}^{\infty} X_1(\lambda)X_2(f - \lambda) \, d\lambda \qquad (2.112)$$

PROOF   The proof of the multiplication theorem proceeds in a manner analogous to the proof of the convolution theorem.

▼ EXAMPLE 2.11   Use the duality theorem to show that

$$A \text{ sinc } 2Wt \leftrightarrow \frac{A}{2W} \Pi \left(\frac{f}{2W} \right) \qquad (2.113)$$

SOLUTION   From Example 2.8, we know that

$$x(t) = A\Pi \left(\frac{t}{\tau} \right) \leftrightarrow A\tau \text{ sinc } f\tau = X(f)$$

Considering $X(t)$, and using the duality theorem, we obtain

$$X(t) = A\tau \text{ sinc } \tau t \leftrightarrow A\Pi \left(\frac{-f}{\tau} \right) = x(-f) \qquad (2.114)$$

where τ is a parameter with dimension $(\text{s})^{-1}$, which may be somewhat confusing at first sight! By letting $\tau = 2W$, the given relationship follows.

▼ EXAMPLE 2.12   Obtain the following Fourier transform pairs:

1. $A\delta(t) \leftrightarrow A$
2. $A\delta(t - t_0) \leftrightarrow Ae^{-j2\pi f t_0}$
3. $A \leftrightarrow A\delta(f)$
4. $Ae^{j2\pi f_0 t} \leftrightarrow A\delta(f - f_0)$

SOLUTION   Even though these signals are not energy signals, we can formally derive the Fourier transform of each by obtaining the Fourier transform of a "proper" energy signal that approaches the given signal in the limit as some parameter approaches zero or infinity. For example, formally,

$$\mathcal{F}[A\delta(t)] = \mathcal{F}\left[\lim_{\tau \to 0} \left(\frac{A}{\tau} \right) \Pi \left(\frac{t}{\tau} \right) \right] = \lim_{\tau \to 0} A \text{ sinc } (f\tau) = A \qquad (2.115)$$

We can use a formal procedure such as this to define Fourier transforms for the other three signals as well. It is easier, however, to use the sifting property of the delta function and the appropriate Fourier transform theorems.

The same results are obtained. For example, we obtain the first transform pair directly by writing down the Fourier transform integral with $x(t) = \delta(t)$ and invoking the sifting property:

$$\mathcal{F}[A\delta(t)] = A \int_{-\infty}^{\infty} \delta(t)e^{-j\omega t}\, dt = A \qquad (2.116)$$

Transform pair 2 follows by application of the time-delay theorem to pair 1.

Transform pair 3 can be obtained by using the inverse-transform relationship or the first transform pair and the duality theorem. Using the latter, we obtain

$$X(t) = A \leftrightarrow A\delta(-f) = A\delta(f) = x(-f) \qquad (2.117)$$

where the impulse function is assumed to be even.

Transform pair 4 follows by applying the frequency-translation theorem to pair 3. The Fourier transform pairs of Example 2.12 will be used often in the discussion of modulation.

▲

▼ **EXAMPLE 2.13**   Use the differentiation theorem to obtain the Fourier transform of the triangular signal, defined as

$$\Lambda\left(\frac{t}{\tau}\right) \triangleq \begin{cases} 1 - |t|/\tau, & |t| < \tau \\ 0, & \text{otherwise} \end{cases} \qquad (2.118)$$

SOLUTION   Differentiating $\Lambda(t/\tau)$ twice, we obtain, as shown in Figure 2.11,

$$\frac{d^2\Lambda(t/\tau)}{dt^2} = \frac{1}{\tau}\delta(t + \tau) - \frac{2}{\tau}\delta(t) + \frac{1}{\tau}\delta(t - \tau) \qquad (2.119)$$

Using the differentiation, superposition, and time-shift theorems and the result of Example 2.12, we obtain

$$\mathcal{F}\left[\frac{d^2\Lambda(t/\tau)}{dt^2}\right] = (j2\pi f)^2\mathcal{F}\left[\Lambda\left(\frac{t}{\tau}\right)\right]$$

$$= \frac{1}{\tau}(e^{j2\pi f\tau} - 2 + e^{-j2\pi f\tau}) \qquad (2.120a)$$

or

$$\mathcal{F}\left[\Lambda\left(\frac{t}{\tau}\right)\right] = \frac{2\cos 2\pi f\tau - 2}{\tau(j2\pi f)^2} = \tau\frac{\sin^2 \pi f\tau}{(\pi f\tau)^2} \qquad (2.120b)$$

FIGURE 2.11 Triangular signal and its first two derivatives. (a) Triangular signal. (b) First derivative of triangular signal. (c) Second derivative of triangular signal.

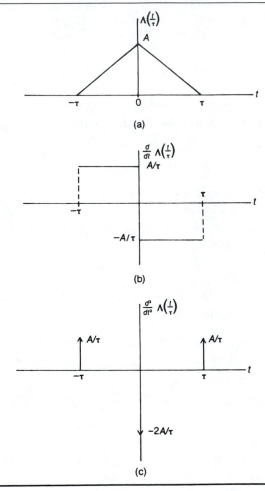

where we have used $\frac{1}{2}(1 - \cos 2\pi ft) = \sin^2 \pi ft$. Summarizing, we have shown that

$$\Lambda\left(\frac{t}{\tau}\right) \leftrightarrow \tau \, \text{sinc}^2 f\tau \tag{2.121}$$

where $(\sin \pi f\tau)/(\pi f\tau)$ has been replaced by $\text{sinc } f\tau$.

▼ **EXAMPLE 2.14**   As another example of obtaining Fourier transforms of signals involving impulses, let us consider the signal

$$y_s(t) = \sum_{m=-\infty}^{\infty} \delta(t - mT_s) \tag{2.122}$$

It is a periodic waveform referred to as the *ideal sampling waveform* and consists of an infinite sequence of impulses spaced by T_s seconds.

SOLUTION   To obtain the Fourier transform of $y_s(t)$, we note that it is periodic and, in a formal sense, therefore can be represented by a Fourier series. Thus

$$y_s(t) = \sum_{m=-\infty}^{\infty} \delta(t - mT_s) = \sum_{n=-\infty}^{\infty} Y_n e^{jn\omega_s t}, \qquad \omega_s = \frac{2\pi}{T_s} \tag{2.123}$$

where

$$Y_n = \frac{1}{T_s} \int_{T_s} \delta(t) e^{-jn\omega_s t} \, dt = f_s, \qquad f_s = \frac{\omega_s}{2\pi} \tag{2.124}$$

by the sifting property of the impulse function. Therefore,

$$y_s(t) = f_s \sum_{n=-\infty}^{\infty} e^{jn\omega_s t} \tag{2.125}$$

Fourier-transforming term by term, we obtain

$$Y_s(f) = f_s \sum_{n=-\infty}^{\infty} \mathcal{F}[1 \cdot e^{jn\omega_s t}] = f_s \sum_{n=-\infty}^{\infty} \delta(f - nf_s) \tag{2.126}$$

where we have used the results of Example 2.12. Summarizing, we have shown that

$$\sum_{m=-\infty}^{\infty} \delta(t - mT_s) \leftrightarrow f_s \sum_{n=-\infty}^{\infty} \delta(f - nf_s) \tag{2.127}$$

▲

   The transform pair (2.127) is useful in spectral representations of periodic signals by the Fourier transform, which will be considered shortly.

▼ **EXAMPLE 2.15**   The convolution theorem can be used to obtain the Fourier transform of the triangle $\Lambda(t/\tau)$ defined by (2.118).

SOLUTION   We proceed by first showing that the convolution of two pulses is a triangle. The steps in computing

$$y(t) = \int_{-\infty}^{\infty} \Pi\left(\frac{t-\lambda}{\tau}\right) \Pi\left(\frac{\lambda}{\tau}\right) d\lambda \tag{2.128}$$

are carried out in Table 2.2. Summarizing the results, we have

$$\tau\Lambda\left(\frac{t}{\tau}\right) = \Pi\left(\frac{t}{\tau}\right) * \Pi\left(\frac{t}{\tau}\right) = \begin{cases} 0, & t < -\tau \\ \tau - |t|, & |t| \le \tau \\ 0, & t > \tau \end{cases} \tag{2.129}$$

Using the transform pair

$$\Pi\left(\frac{t}{\tau}\right) \leftrightarrow \tau \operatorname{sinc} ft \tag{2.130}$$

and the convolution theorem, we obtain the transform pair

$$\Lambda\left(\frac{t}{\tau}\right) \leftrightarrow \tau \operatorname{sinc}^2 f\tau \tag{2.131}$$

▲ as in Example 2.13.

Table 2.2 Computation of $\Pi(t/\tau) * \Pi(t/\tau)$

Range on t	Integrand	Limits	Area
$-\infty < t < -\tau$			0
$-\tau < t < 0$		$-\frac{1}{2}\tau$ to $t + \frac{1}{2}\tau$	$\tau + t$
$0 < t < \tau$		$t - \frac{1}{2}\tau$ to $\frac{1}{2}\tau$	$\tau - t$
$\tau < t < \infty$			0

A useful result is the convolution of an impulse $\delta(t - t_0)$ with a signal $x(t)$. Carrying out the operation, we obtain

$$\delta(t - t_0) * x(t) = \int_{-\infty}^{\infty} \delta(\lambda - t_0)x(t - \lambda) \, d\lambda$$

$$= x(t - t_0) \tag{2.132}$$

by the sifting property of the delta function. That is, convolution of $x(t)$ with an impulse occurring at time t_0 simply shifts $x(t)$ to t_0.

▼ **EXAMPLE 2.16**   Consider the Fourier transform of the cosinusoidal pulse

$$x(t) = A\Pi\left(\frac{t}{\tau}\right) \cos \omega_0 t, \qquad \omega_0 = 2\pi f_0 \tag{2.133}$$

Using the transform pair

$$e^{\pm j2\pi f_0 t} \leftrightarrow \delta(f \mp f_0) \tag{2.134}$$

obtained earlier and Euler's theorem, we find that

$$\cos 2\pi f_0 t \leftrightarrow \tfrac{1}{2}\delta(f - f_0) + \tfrac{1}{2}\delta(f + f_0) \tag{2.135}$$

We have shown also that

$$A\Pi\left(\frac{t}{\tau}\right) \leftrightarrow A\tau \operatorname{sinc} f\tau$$

Therefore, using the multiplication theorem, we obtain

$$X(f) = \tfrac{1}{2}A\tau \, (\operatorname{sinc} f\tau) * [\delta(f - f_0) + \delta(f + f_0)]$$

$$= \tfrac{1}{2}A\tau \, [\operatorname{sinc} (f - f_0)\tau + \operatorname{sinc} (f + f_0)\tau] \tag{2.136}$$

where $\delta(f - f_0) * Z(f) = Z(f - f_0)$ has been used. Figure 2.12(c) shows
▲ $X(f)$. The same result can be obtained via the modulation theorem.

Fourier Transforms of Periodic Signals

The Fourier transform of a periodic signal, in a strict mathematical sense, does not exist, since periodic signals are not energy signals. However, using the transform pairs derived in Example 2.12 for a constant and a phasor signal, we could, in a formal sense, write down the Fourier transform of a periodic signal by Fourier-transforming its complex Fourier series term by term.

A somewhat more useful form for the Fourier transform of a periodic signal is obtained by applying the convolution theorem and the transform

FIGURE 2.12   (a)–(c) Application of the multiplication theorem. (c)–(e) Application of the convolution theorem. Note: × denotes multiplication; ∗ denotes convolution, ↔ denotes transform pairs.

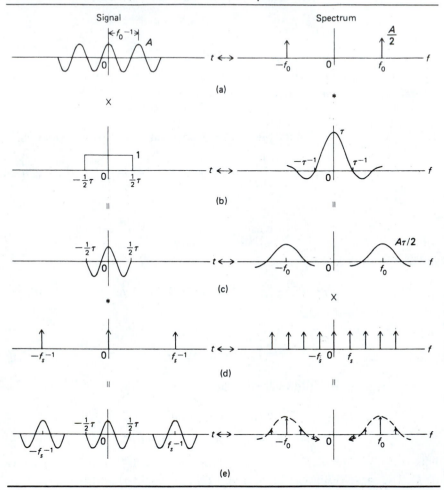

pair (2.127) for the ideal sampling wave. To obtain it, consider the result of convolving the ideal sampling waveform with a pulse-type signal $p(t)$ to obtain a new signal $x(t)$, where $x(t)$ is a periodic power signal. This is apparent when one carries out the convolution with the aid of (2.132):

$$x(t) = \left[\sum_{m=-\infty}^{\infty} \delta(t - mT_s) \right] * p(t) = \sum_{m=-\infty}^{\infty} p(t - mT_s) \quad (2.137)$$

Applying the convolution theorem and the Fourier transform pair of (2.127), we find that the Fourier transform of $x(t)$ is

$$X(f) = \mathcal{F}\left\{ \sum_{m=-\infty}^{\infty} \delta(t - mT_s) \right\} P(f)$$

$$= f_s P(f) \sum_{n=-\infty}^{\infty} \delta(f - nf_s) = \sum_{n=-\infty}^{\infty} f_s P(nf_s)\, \delta(f - nf_s) \quad (2.138)$$

where $P(f) = \mathcal{F}[p(t)]$. Summarizing, we have obtained the Fourier transform pair

$$\sum_{m=-\infty}^{\infty} p(t - mT_s) \leftrightarrow \sum_{n=-\infty}^{\infty} f_s P(nf_s)\, \delta(f - nf_s) \quad (2.139)$$

The usefulness of (2.139) is illustrated with an example.

▼ **EXAMPLE 2.17**   The Fourier transform of a single cosinusoidal pulse was found in Example 2.16 and is shown in Figure 2.12(c). The Fourier transform of a periodic cosinusoidal pulse train, which could represent the output of a radar transmitter, for example, is obtained by writing it as

$$y(t) = \left[\sum_{m=-\infty}^{\infty} \delta(t - mT_s) \right] * \Pi\left(\frac{t}{\tau}\right) \cos \omega_0 t, \qquad \omega_0 \gg 2\pi/\tau$$

$$= \sum_{m=-\infty}^{\infty} \Pi\left(\frac{t - mT_s}{\tau}\right) \cos \omega_0(t - mT_s), \qquad f_s \leq \tau^{-1} \quad (2.140)$$

This signal is illustrated in Figure 2.12(e). According to (2.139), the Fourier transform of $y(t)$ is

$$Y(f) = \sum_{n=-\infty}^{\infty} \frac{Af_s \tau}{2} [\text{sinc } (nf_s - f_0)\tau + \text{sinc } (nf_s + f_0)\tau]\, \delta(f - nf_s) \quad (2.141)$$

▲ The spectrum is illustrated on the right-hand side of Figure 2.12(e).

Poisson Sum Formula

We can develop the *Poisson sum formula* by taking the inverse Fourier transform of the right-hand side of (2.139). When we use the transform pair $\exp(-j2\pi nf_s t) \leftrightarrow \delta(f - nf_s)$ (see Example 2.12), it follows that

$$\mathcal{F}^{-1}\left\{ \sum_{n=-\infty}^{\infty} f_s P(nf_s)\, \delta(f - nf_s) \right\} = f_s \sum_{n=-\infty}^{\infty} P(nf_s) e^{j2\pi nf_s t} \quad (2.142)$$

Equating this to the left-hand side of (2.139), we obtain the Poisson sum formula:

$$\sum_{m=-\infty}^{\infty} p(t - mT_s) = f_s \sum_{n=-\infty}^{\infty} P(nf_s)e^{j2\pi nf_s t} \tag{2.143}$$

The Poisson sum formula is useful when one goes from the Fourier transform to sampled approximations of it. For example, Equation (2.143) says that the sample values $P(nf_s)$ of $P(f) = \mathcal{F}\{p(t)\}$ are the Fourier series coefficients of the periodic function $T_s \sum_{n=-\infty}^{\infty} p(t - mT_s)$.

†2.6 POWER SPECTRAL DENSITY AND CORRELATION

Recalling the definition of energy spectral density, Equation (2.83), we see that it is of use only for energy signals for which the integral of $G(f)$ over all frequencies gives total energy, a finite quantity. For power signals, it is meaningful to speak in terms of *power spectral density*. Analogous to $G(f)$, we define the power spectral density $S(f)$ of a signal $x(t)$ as a real, even, non-negative function of frequency, which gives total average power per ohm when integrated; that is,

$$P = \int_{-\infty}^{\infty} S(f) \, df = \langle x^2(t) \rangle \tag{2.144}$$

Since $S(f)$ is a function that gives the variation of density of power with frequency, we conclude that it must consist of a series of impulses for the periodic power signals that we have so far considered. Later, in Chapter 5, we will consider random signals, which have power spectral densities that are continuous functions of frequency.

▼| EXAMPLE 2.18   Considering the sinusoidal signal

$$x(t) = A \cos(\omega_0 t + \theta) \tag{2.145}$$

we note that its average power per ohm, $\frac{1}{2}A^2$, is concentrated at the single frequency $f_0 = \omega_0/2\pi$. However, since the power spectral density must be an even function of frequency, we split this power equally between $+f_0$ and $-f_0$. Thus the power spectral density of $x(t)$ is, from intuition, given by

$$S(f) = \tfrac{1}{4}A^2 \, \delta(f - f_0) + \tfrac{1}{4}A^2 \, \delta(f + f_0) \tag{2.146}$$

Checking this by using (2.144), we see that integration over all frequencies
▲| results in the average power per ohm, $\frac{1}{2}A^2$.

The Time-Average Autocorrelation Function

To introduce the time-average autocorrelation function, we return to the energy spectral density of an energy signal. Without any apparent reason, suppose we take the inverse Fourier transform of $G(f)$, letting the independent variable be τ:

$$\phi(\tau) \triangleq \mathcal{F}^{-1}[G(f)] = \mathcal{F}^{-1}[X(f)X^*(f)]$$
$$= \mathcal{F}^{-1}[X(f)] * \mathcal{F}^{-1}[X^*(f)] \tag{2.147}$$

The last step follows by application of the convolution theorem. Applying the time-reversal theorem (item 3b in Table C.2 in Appendix C) to write $\mathcal{F}^{-1}[X^*(f)] = x(-\tau)$ and then the convolution theorem, we obtain

$$\phi(\tau) = x(\tau) * x(-\tau) = \int_{-\infty}^{\infty} x(\lambda)x(\lambda + \tau)\, d\lambda$$

$$= \lim_{T \to \infty} \int_{-T}^{T} x(\lambda)x(\lambda + \tau)\, d\lambda \qquad \text{(energy signal)} \tag{2.148}$$

Equation (2.148) will be referred to as the *time-average autocorrelation function for energy signals*. We see that it gives a measure of the similarity, or coherence, between a signal and a delayed version of the signal. Note that $\phi(0) = E$, the signal energy. The similarity of the correlation operation to convolution also should be noted. The major point of (2.147) is that the autocorrelation function and energy spectral density are Fourier transform pairs. We forgo further discussion of the time-average autocorrelation function for energy signals in favor of analogous results for power signals.

The time-average autocorrelation function $R(\tau)$ of a power signal $x(t)$ is defined as the time average

$$R(\tau) = \langle x(t)x(t + \tau) \rangle$$

$$= \lim_{T \to \infty} \frac{1}{2T} \int_{-T}^{T} x(t)x(t + \tau)\, dt \qquad \text{(power signal)} \tag{2.149}$$

If $x(t)$ is periodic with period T_0, the integrand of (2.149) is periodic, and the time average can be taken over a single period:

$$R(\tau) = \frac{1}{T_0} \int_{T_0} x(t)x(t + \tau)\, dt \qquad [x(t) \text{ periodic}] \tag{2.150}$$

Just like $\phi(\tau)$, $R(\tau)$ gives a measure of the similarity between the signal at time t and at time $t + \tau$; it is a function of the delay variable τ, since time is the variable of integration. In addition to being a measure of the similarity between a signal and its time displacement, we note that

$$R(0) = \langle x^2(t) \rangle = \int_{-\infty}^{\infty} S(f) \, df \qquad (2.151)$$

Thus we suspect that the time-average autocorrelation function and power spectral density of a power signal are closely related, just as they are for energy signals. This relationship is stated formally by the *Wiener-Khinchine theorem*, which says that the time-average autocorrelation function of a signal and its power spectral density are Fourier transform pairs:

$$S(f) = \mathcal{F}[R(\tau)] = \int_{-\infty}^{\infty} R(\tau)e^{-j2\pi f \tau} \, d\tau \qquad (2.152a)$$

and

$$R(\tau) = \mathcal{F}^{-1}[S(f)] = \int_{-\infty}^{\infty} S(f)e^{j2\pi f \tau} \, df \qquad (2.152b)$$

A formal proof of the Wiener-Khinchine theorem will be given in Chapter 5. We simply take (2.152a) as the definition of power spectral density at this point. We note that (2.151) follows immediately from (2.152b) by setting $\tau = 0$.

Properties of R(τ)

The time-average autocorrelation function has several useful properties, which are listed below:

1. $R(0) = \langle x^2(t) \rangle \geq |R(\tau)|$, for all τ; that is, a relative maximum of $R(\tau)$ exists at $\tau = 0$.
2. $R(-\tau) = \langle x(t)x(t - \tau) \rangle = R(\tau)$; that is, $R(\tau)$ is even.
3. $\lim_{|\tau| \to \infty} R(\tau) = \langle x(t) \rangle^2$ if $x(t)$ does not contain a periodic component.
4. If $x(t)$ is periodic in t with period T_0, then $R(\tau)$ is periodic in τ with period T_0.
5. The time-average autocorrelation function of any signal has a Fourier transform that is nonnegative.

Property 5 results by virtue of the fact that normalized power is a nonnegative quantity. These properties will be proved in Chapter 5.

▼ EXAMPLE 2.19   Returning to the sinusoidal signal of Example 2.18, we see that by the Wiener-Khinchine theorem its autocorrelation function is

$$R(\tau) = \mathscr{F}^{-1} \left[\frac{A^2}{4} \delta(f + f_0) + \frac{A^2}{4} \delta(f - f_0) \right]$$

$$= \frac{A^2}{2} \cos \omega_0 \tau, \quad \omega_0 = 2\pi f_0 \tag{2.153}$$

Checking this result by computing the time average using (2.149), we find that

$$R(\tau) = \frac{1}{T_0} \int_{T_0} A^2 \cos(\omega_0 t + \theta) \cos(\omega_0 t + \omega_0 \tau + \theta) \, dt$$

$$= \frac{1}{T_0} \int_{T_0} \frac{A^2}{2} \cos \omega_0 \tau \, dt + \frac{1}{T_0} \int_{T_0} \frac{A^2}{2} \cos(2\omega_0 t + \omega_0 \tau + 2\theta) \, dt$$

$$= \frac{A^2}{2} \cos \omega_0 t \tag{2.154}$$

where the second integral of the second equation is zero by virtue of the periodicity of the integrand. Note that all the properties listed above for $R(\tau)$ are satisfied by $R(\tau) = \frac{1}{2}A^2 \cos \omega_0 \tau$ except Property 3, which does not
▲ apply.

The autocorrelation function and power spectral density are important tools for systems analysis involving random signals.

2.7   SIGNALS AND LINEAR SYSTEMS

In this section we are concerned with the characterization of systems and their effect on signals. In system modeling, the actual elements, such as resistors, capacitors, inductors, springs, and masses, that compose a particular system are usually not of concern. Rather, we view a system in terms of the operation it performs on an input to produce an output. Symbolically, this is accomplished, for a single-input, single-output system, by writing

$$y(t) = \mathcal{H}[x(t)] \tag{2.155}$$

where $\mathcal{H}$ is the operator that produces the output $y(t)$ from the input $x(t)$, as illustrated in Figure 2.13. We now consider certain classes of systems, the first of which is linear time-invariant systems.

FIGURE 2.13   Operator representation of a linear system

Definition of a Linear Time-Invariant System

If a system is *linear*, superposition holds. That is, if $x_1(t)$ results in the output $y_1(t)$ and $x_2(t)$ results in the output $y_2(t)$, then the output due to $\alpha_1 x_1(t) + \alpha_2 x_2(t)$, where α_1 and α_2 are constants, is given by

$$y(t) = \mathcal{H}[\alpha_1 x_1(t) + \alpha_2 x_2(t)] = \alpha_1 \mathcal{H}[x_1(t)] + \alpha_2 \mathcal{H}[x_2(t)]$$
$$= \alpha_1 y_1(t) + \alpha_2 y_2(t) \tag{2.156}$$

If the system is *time-invariant*, or *fixed*, the delayed input $x(t - t_0)$ gives the delayed output $y(t - t_0)$; that is,

$$y(t - t_0) = \mathcal{H}[x(t - t_0)] \tag{2.157}$$

With these properties explicitly stated, we are now ready to obtain a more concrete description of a linear time-invariant (LTI) system.

Impulse Response and the Superposition Integral

The *impulse response* $h(t)$ of an LTI system is defined to be the response of the system to an impulse applied at $t = 0$, that is

$$h(t) \triangleq \mathcal{H}[\delta(t)] \tag{2.158}$$

By the time-invariant property of the system, the response to an impulse applied at any time t_0 is $h(t - t_0)$, and the response to the linear combination of impulses $\alpha_1 \delta(t - t_1) + \alpha_2 \delta(t - t_2)$ is $\alpha_1 h(t - t_1) + \alpha_2 h(t - t_2)$ by the superposition property and time-invariance. Through induction, we may therefore show that the response to the input

$$x(t) = \sum_{n=1}^{N} \alpha_n \delta(t - t_n) \tag{2.159}$$

is

$$y(t) = \sum_{n=1}^{N} \alpha_n h(t - t_n) \tag{2.160}$$

We will use (2.160) to obtain the *superposition integral,* which expresses the response of an LTI system to an arbitrary input (with suitable restrictions) in terms of the impulse response of the system. Considering the arbitrary input signal $x(t)$ of Figure 2.14(a), we can represent it as

$$x(t) = \int_{-\infty}^{\infty} x(\lambda)\delta(t - \lambda)\, d\lambda \qquad\qquad (2.161)$$

by the sifting property of the unit impulse. Approximating the integral of (2.161) as a sum, we obtain

$$x(t) \cong \sum_{n=N_1}^{N_2} x(n\,\Delta t)\,\delta(t - n\,\Delta t)\,\Delta t, \qquad \Delta t \ll 1 \qquad (2.162)$$

where $t_1 = N_1\,\Delta t$ is the starting time of the signal and $t_2 = N_2\,\Delta t$ is the ending time. The output, using (2.160) with $\alpha_n = x(n\,\Delta t)\,\Delta t$ and $t_n = n\,\Delta t$, is

$$\tilde{y}(t) = \sum_{n=N_1}^{N_2} x(n\,\Delta t)h(t - n\,\Delta t)\,\Delta t \qquad\qquad (2.163)$$

where the tilde denotes the output resulting from the approximation to the input given by (2.162). In the limit as Δt approaches zero and $n\,\Delta t$

FIGURE 2.14   A signal and an approximate representation. (a) Signal. (b) Approximation with a sequence of impulses.

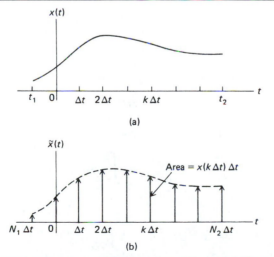

approaches the continuous variable λ, the sum becomes an integral, and we obtain

$$y(t) = \int_{-\infty}^{\infty} x(\lambda)h(t - \lambda)\, d\lambda \qquad (2.164a)$$

where the limits have been changed to $\pm\infty$ to allow arbitrary starting and ending times for $x(t)$. Making the substitution $\sigma = t - \lambda$, we obtain the equivalent result

$$y(t) = \int_{-\infty}^{\infty} x(t - \sigma)h(\sigma)\, d\sigma \qquad (2.164b)$$

Because these equations were obtained by superposition of a number of elementary responses due to each individual impulse, they are referred to as *superposition integrals*. A simplification results if the system under consideration is *causal,* that is, is a system that does not respond before an input is applied. For a causal system, $h(t - \lambda) = 0$ for $t < \lambda$ and the upper limit on (2.164a) can be set equal to t. Furthermore, if $x(t) = 0$ for $t < 0$, the lower limit becomes zero.

Stability

A fixed, linear system is bounded-input, bounded-output (BIBO) stable if every bounded input results in a bounded output. It can be shown* that a system is BIBO stable if and only if

$$\int_{-\infty}^{\infty} |h(t)|\, dt < \infty \qquad (2.165)$$

Transfer Function

Applying the convolution theorem of Fourier transforms, item 8 of Table C.2 in Appendix C, to either (2.164a) or (2.164b), we obtain

$$Y(f) = H(f)X(f) \qquad (2.166)$$

where $X(f) = \mathscr{F}\{x(t)\}$, $Y(f) = \mathscr{F}\{y(t)\}$, and

$$H(f) = \mathscr{F}\{h(t)\} = \int_{-\infty}^{\infty} h(t)e^{-j2\pi ft}\, dt \qquad (2.167a)$$

or

$$h(t) = \mathscr{F}^{-1}\{H(f)\} = \int_{-\infty}^{\infty} H(f)e^{j2\pi ft}\, df \qquad (2.167b)$$

---

* See Ziemer, Tranter, and Fannin (1993), Chapter 2.

$H(f)$ is referred to as the *transfer function* of the system. We see that either $h(t)$ or $H(f)$ is an equally good characterization of the system. By an inverse Fourier transform on (2.166), the output becomes

$$y(t) = \int_{-\infty}^{\infty} X(f)H(f)e^{j2\pi ft}\, df \tag{2.168}$$

Causality

A system is causal if it does not anticipate the input. In terms of the impulse response, it follows that for a time-invariant causal system,

$$h(t) = 0, \qquad t < 0 \tag{2.169}$$

When causality is viewed from the standpoint of the transfer function of the system, a celebrated theorem by Wiener and Paley* states that if

$$\int_{-\infty}^{\infty} |h(t)|^2\, dt = \int_{-\infty}^{\infty} |H(f)|^2\, df < \infty \tag{2.170a}$$

with $h(t) \equiv 0$ for $t < 0$, it is then necessary that

$$\int_{-\infty}^{\infty} \frac{|\ln|H(f)||}{1+f^2}\, df < \infty \tag{2.170b}$$

Conversely, if $|H(f)|$ is square-integrable, and if the integral in (2.170b) is unbounded, then we cannot make $h(t) \equiv 0$, $t < 0$, no matter what we choose for $\underline{/H(f)}$. Consequences of (2.170) are that no filter can have $|H(f)| \equiv 0$ over a finite band of frequencies (i.e., a filter cannot perfectly reject any band of frequencies). In fact, the Paley-Wiener criterion restricts the rate at which $|H(f)|$ for a linear causal time-invariant system can vanish. For example,

$$|H(f)| = e^{-k_1|f|} \tag{2.171a}$$

and

$$|H(f)| = e^{-k_2 f^2} \tag{2.171b}$$

are not allowable amplitude responses for causal filters because (2.170b) does not give a finite result in either case.

The sufficiency statement of the Payley-Wiener criterion is: Given any square-integrable function $|H(f)|$ for which (2.170) is satisfied, there exists an $\underline{/H(f)}$ such that $H(f) = |H(f)|e^{j\underline{/H(f)}}$ is the Fourier transform of $h(t)$ for a causal filter.

---

* See William Siebert, *Circuits, Signals, and Systems*, McGraw-Hill, New York, 1986, p. 476.

Symmetry Properties of H(f)

The transfer function $H(f)$ is, in general, a complex quantity. We therefore write it in terms of magnitude and argument as

$$H(f) = |H(f)|e^{j\underline{/H(f)}} \qquad (2.172)$$

where $|H(f)|$ is called the *amplitude-response function* and $\underline{/H(f)}$ is called the *phase-shift function* of the network. Also, $H(f)$ is the Fourier transform of a real-time function $h(t)$. Therefore, it follows that

$$|H(f)| = |H(-f)| \qquad (2.173a)$$

and

$$\underline{/H(f)} = -\underline{/H(-f)} \qquad (2.173b)$$

▼ EXAMPLE 2.20   Consider the lowpass RC filter shown in Figure 2.15. We may find its transfer function by a number of methods. First, we may write down the governing differential equation (integral-differential equations, in general) as

$$RC\frac{dy}{dt} + y(t) = x(t) \qquad (2.174)$$

and Fourier-transform it, obtaining

$$(j2\pi f RC + 1)Y(f) = X(f)$$

or

$$H(f) = \frac{Y(f)}{X(f)} = \frac{1}{1 + j(f/f_3)}$$

$$= \frac{1}{\sqrt{1 + (f/f_3)^2}}\, e^{-j\,\tan^{-1}(f/f_3)} \qquad (2.175)$$

where $f_3 = 1/2\pi RC$ is the 3-dB frequency, or half-power frequency. Second, we can use Laplace transform theory with s replaced by $j2\pi f$. Third, we can

FIGURE 2.15   An RC network

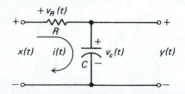

FIGURE 2.16   Amplitude and phase responses of the lowpass RC filter.
(a) Amplitude response. (b) Phase shift.

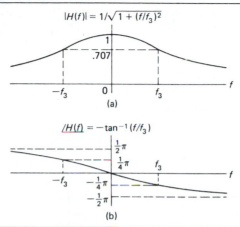

use ac sinusoidal steady-state analysis. The amplitude response and phase
response of this system are illustrated in Figures 2.16(a) and (b).

Using the transform pair

$$\alpha e^{-\alpha t} u(t) \leftrightarrow \frac{\alpha}{\alpha + j2\pi f} \tag{2.176}$$

we find the impulse response of the filter to be

$$h(t) = \frac{1}{RC} e^{-t/RC} u(t) \tag{2.177}$$

Finally, we consider the response of the filter to the pulse

$$x(t) = A\Pi \left(\frac{t - \frac{1}{2}T}{T} \right) \tag{2.178}$$

Using appropriate Fourier transform pairs, we can readily find $Y(f)$, but its
inverse Fourier transformation requires some effort. Thus it appears that the
superposition integral is the best approach in this case. Choosing the form

$$y(t) = \int_{-\infty}^{\infty} h(t - \sigma)x(\sigma) \, d\sigma \tag{2.179}$$

we find, by direct substitution in $h(t)$, that

$$h(t - \sigma) = \frac{1}{RC} e^{-(t-\sigma)/RC} u(t - \sigma) = \begin{cases} \dfrac{1}{RC} e^{-(t-\sigma)/RC}, & \sigma < t \\[2mm] 0, & \sigma > t \end{cases} \tag{2.180}$$

Since $x(\sigma)$ is zero for $\sigma < 0$ and $\sigma > T$, we find that

$$
y(t) = \begin{cases} 0, & t < 0 \\ \int_0^t \dfrac{A}{RC} e^{-(t-\sigma)/RC}\, d\sigma, & 0 < t < T \\ \int_0^T \dfrac{A}{RC} e^{-(t-\sigma)/RC}\, d\sigma, & t > T \end{cases} \tag{2.181}
$$

Carrying out the integrations, we obtain

$$
y(t) = \begin{cases} 0, & t < 0 \\ A(1 - e^{-t/RC}), & 0 < t < T \\ A(e^{-(t-T)/RC} - e^{-t/RC}), & t > T \end{cases} \tag{2.182}
$$

This result is plotted in Figure 2.17 for several values of T/RC. Also shown are $|X(f)|$ and $|H(f)|$. Note that $T/RC = 2\pi f_3/T^{-1}$ is proportional to the ratio of the 3-dB frequency of the filter to the spectral width (T^{-1}) of the pulse. When this ratio is large, the spectrum of the input pulse is essentially passed undistorted by the system, and the output looks like the input. On the other hand, for $2\pi f_3/T^{-1} \ll 1$, the system distorts the input signal spectrum, and $y(t)$ looks nothing like the input. These ideas will be put on a firmer basis when signal distortion is discussed.

Input-Output Relationships for Spectra

Consider a fixed linear two-port system with transfer function $H(f)$, input $x(t)$, and output $y(t)$. If $x(t)$ and, therefore, $y(t)$ are energy signals, their energy spectral densities are $G_x(f) = |X(f)|^2$ and $G_y(f) = |Y(f)|^2$, respectively. Since $Y(f) = H(f)X(f)$, it follows that

$$
G_y(f) = |H(f)|^2 G_x(f) \tag{2.183}
$$

A similar relationship holds for power signals and spectra:

$$
S_y(f) = |H(f)|^2 S_x(f) \tag{2.184}
$$

This will be proved in Chapter 5.

Response to Periodic Inputs

Consider the steady-state response of a fixed linear system to the complex exponential input signal $Ae^{j2\pi f_0 t}$. Using the superposition integral, we obtain

$$
y_{ss}(t) = \int_{-\infty}^{\infty} h(\lambda) Ae^{j2\pi f_0(t - \lambda)}\, d\lambda
$$

FIGURE 2.17   (a) Waveforms and (b)–(d) spectra for a lowpass RC filter with pulse input. (a) Input and output signals. (b) $T/RC = 0.5$. (c) $T/RC = 2$. (d) $T/RC = 10$.

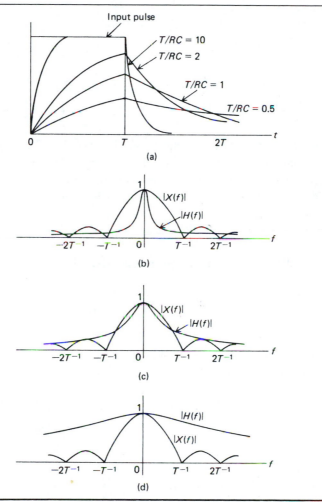

$$= Ae^{j2\pi f_0 t} \int_{-\infty}^{\infty} h(\lambda) e^{-j2\pi f_0 \lambda} \, d\lambda$$

$$= H(f_0) A e^{j2\pi f_0 t} \tag{2.185}$$

That is, the output is a complex exponential signal of the same frequency but with amplitude scaled by $|H(f_0)|$ and phase-shifted by $\underline{/H(f_0)}$ relative to the amplitude and phase of the input. Using superposition, we conclude that the steady-state output due to an arbitrary periodic input is represented by the complex exponential Fourier series

$$y(t) = \sum_{n=-\infty}^{\infty} X_n H(nf_0) e^{jn2\pi f_0 t} \tag{2.186}$$

or

$$y(t) = \sum_{n=-\infty}^{\infty} |X_n||H(nf_0)| \exp \{j[n2\pi f_0 t + \underline{/X_n} + \underline{/H(nf_0)}]\} \tag{2.187}$$

Thus, for a periodic input, the magnitude of each spectral component of the input is attenuated (or amplified) by the amplitude-response function *at the frequency of the particular spectral component,* and the phase of each spectral component is shifted by the value of the phase-shift function of the system *at the frequency of the spectral component.*

▼ **EXAMPLE 2.21**   Consider the response of a filter having the transfer function

$$H(f) = 2\Pi \left(\frac{f}{42}\right) e^{-j\pi f/10} \tag{2.188}$$

to a unit-amplitude triangular signal with period 0.1 s. From Table 2.1 and Equation (2.43), the exponential Fourier series of the input signal is

$$
\begin{aligned}
x(t) &= \cdots \frac{4}{25\pi^2} e^{-j100\pi t} + \frac{4}{9\pi^2} e^{-j60\pi t} + \frac{4}{\pi^2} e^{-j20\pi t} + \frac{4}{\pi^2} e^{j20\pi t} \\
&\quad + \frac{4}{9\pi^2} e^{j60\pi t} + \frac{4}{25\pi^2} e^{j100\pi t} + \cdots \\
&= \frac{8}{\pi^2} \left[\cos(20\pi t) + \frac{1}{9} \cos(60\pi t) + \frac{1}{25} \cos(100\pi t) + \cdots \right] \tag{2.189}
\end{aligned}
$$

The filter eliminates all harmonics above 21 Hz and passes all those below 21 Hz, imposing an amplitude scale factor of 2 and a phase shift of $-\pi f/10$ rad. The only harmonic of the triangular wave to be passed by the filter is the fundamental, which has a frequency of 10 Hz, giving a phase shift of $-\pi(10)/10 = -\pi$ rad. The output is therefore

$$y(t) = \frac{16}{\pi^2} \cos 20\pi \left(t - \frac{1}{20}\right) \tag{2.190}$$

▲ where the phase shift is seen to be equivalent to a delay of $\frac{1}{20}$ s.

Distortionless Transmission

Equation (2.187) shows that both the amplitudes and phases of the spectral components of a periodic signal will, in general, be altered as the signal is sent through a two-port system. This modification may be desirable in signal *processing* applications, but it amounts to distortion in signal *transmission* applications. While it may appear at first that ideal signal transmission results only if there is *no* attenuation and phase of the spectral components of the input, this requirement is too stringent. A system will be classified as distortionless if it introduces the same attenuation and time delay to all spectral components of the input, for then the output *looks like* the input. In particular, if the output of a system is given in terms of the input as

$$y(t) = H_0 x(t - t_0) \tag{2.191}$$

where H_0 and t_0 are constants, the output is a scaled, delayed replica of the input ($t_0 > 0$ for causality). Employing the time-delay theorem to Fourier-transform (2.191) and using the relation $H(f) = Y(f)/X(f)$, we obtain

$$H(f) = H_0 e^{-j2\pi f t_0} \tag{2.192}$$

as the transfer function of a distortionless system; that is, the amplitude response of a distortionless system is constant and the phase shift is linear with frequency. Of course, these restrictions are necessary only within the frequency ranges where the input has significant spectral content. Figure 2.18 and Example 2.22 will illustrate these comments.

In general, we can isolate three major types of distortion. First, if the system is linear but the amplitude response is not constant with frequency, the system is said to introduce *amplitude distortion*. Second, if the system is linear but the phase shift is not a linear function of frequency, the system introduces *phase,* or *delay, distortion*. Third, if the system is not linear, we have *nonlinear distortion*. Of course, these three types of distortion may occur in combination with one another.

Group and Phase Delay

One can often identify phase distortion in a linear system by considering the derivative of phase with respect to frequency. A distortionless system exhibits a phase response in which phase is directly proportional to frequency. Thus the derivative of phase with respect to frequency of a distortionless system is a constant. The negative of this constant is called the *group delay* of the system. In other words, the *group delay* is defined by the equation

$$T_g(f) = -\frac{1}{2\pi}\frac{d\theta(f)}{df} \tag{2.193}$$

FIGURE 2.18 Amplitude and phase response and group and phase delays of the filter for Example 2.22. (a) Amplitude response. (b) Phase response. (c) Group delay. (d) Phase delay.

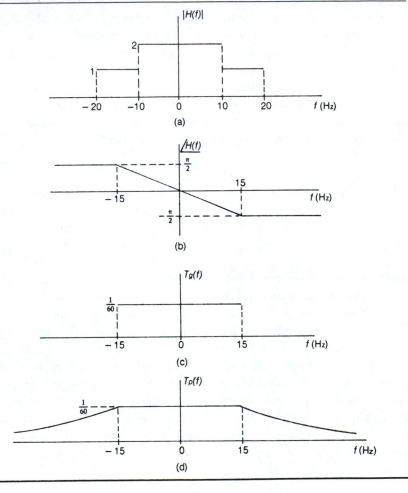

in which $\theta(f)$ is the phase response of the system. For a distortionless system, the phase is given by (2.192) as

$$\theta(f) = -2\pi f t_0 \qquad (2.194)$$

This yields a group delay of

$$T_g(f) = -\frac{1}{2\pi}\frac{d}{df}(-2\pi f t_0)$$

or

$$T_g(f) = t_0 \qquad (2.195)$$

This confirms the preceding observation that the group delay of a distortionless system is a constant.

Group delay is the delay that a group of two or more frequency components undergo in passing through a linear system. If a linear system has a single-frequency component as the input, the system is always distortionless, since the output can be written as an amplitude-scaled and phase-shifted (time-delayed) version of the input. As an example, assume that the input to a linear system is given by

$$x(t) = A \cos 2\pi f_1 t \qquad (2.196)$$

It follows from (2.187) that the output can be written as

$$y(t) = A|H(f_1)| \cos [2\pi f_1 t + \theta(f_1)] \qquad (2.197)$$

where $\theta(f_1)$ is the phase response of the system evaluated at $f = f_1$. Equation (2.197) can be written as

$$y(t) = A|H(f_1)| \cos 2\pi f_1 \left[t + \frac{\theta(f_1)}{2\pi f_1} \right] \qquad (2.198)$$

The delay of the single component is defined as the phase delay:

$$T_p(f) = - \frac{\theta(f)}{2\pi f} \qquad (2.199)$$

Thus (2.198) can be written as

$$y(t) = A|H(f_1)| \cos 2\pi f_1 [t - T_p(f_1)] \qquad (2.200)$$

Use of (2.194) shows that for a distortionless system, the phase delay is given by

$$T_p(f) = - \frac{1}{2\pi f} (-2\pi f t_0) = t_0 \qquad (2.201)$$

Thus we see that distortionless systems have equal group and phase delays. The following example should clarify the preceding definitions.

▼| **EXAMPLE 2.22**   Consider a system with amplitude response and phase shift as shown in Figure 2.18 and the following four inputs:

1. $x_1(t) = \cos 10\pi t + \cos 12\pi t$
2. $x_2(t) = \cos 10\pi t + \cos 26\pi t$

3. $x_3(t) = \cos 26\pi t + \cos 34\pi t$
4. $x_4(t) = \cos 32\pi t + \cos 34\pi t$

Although this system is somewhat unrealistic from a practical standpoint, we can use it to illustrate various combinations of amplitude and phase distortion. Using (2.187) and superposition, we obtain the following corresponding outputs:

1. $y_1(t) = 2 \cos (10\pi t - \frac{1}{6}\pi) + 2 \cos (12\pi t - \frac{1}{5}\pi)$
$= 2 \cos 10\pi(t - \frac{1}{60}) + 2 \cos 12\pi(t - \frac{1}{60})$
2. $y_2(t) = 2 \cos (10\pi t - \frac{1}{6}\pi) + \cos (26\pi t - \frac{13}{30}\pi)$
$= 2 \cos 10\pi(t - \frac{1}{60}) + \cos 26\pi(t - \frac{1}{60})$
3. $y_3(t) = \cos (26\pi t - \frac{13}{30}\pi) + \cos (34\pi t - \frac{1}{2}\pi)$
$= \cos 26\pi(t - \frac{1}{60}) + \cos 34\pi(t - \frac{1}{68})$
4. $y_4(t) = \cos (32\pi t - \frac{1}{2}\pi) + \cos (34\pi t - \frac{1}{2}\pi)$
$= \cos 32\pi(t - \frac{1}{64}) + \cos 34\pi(t - \frac{1}{68})$

Checking these results with (2.191), we see that only the input $x_1(t)$ is passed without distortion by the system. For $x_2(t)$, amplitude distortion results, and for $x_3(t)$ and $x_4(t)$, phase (delay) distortion is introduced.

The group delay and phase delay are also illustrated in Figure 2.18. It can be seen that for $|f| \leq 15$ Hz, the group and phase delays are both equal to $\frac{1}{60}$ s. For $|f| > 15$ Hz, the group delay is zero, and the phase delay is

$$T_p(f) = \frac{1}{4f} \tag{2.202}$$

Nonlinear Distortion

To illustrate the idea of nonlinear distortion, let us consider a nonlinear system with the input-output characteristic

$$y(t) = a_1 x(t) + a_2 x^2(t) \tag{2.203}$$

and with the input

$$x(t) = A_1 \cos \omega_1 t + A_2 \cos \omega_2 t \tag{2.204}$$

The output is therefore

$$y(t) = a_1(A_1 \cos \omega_1 t + A_2 \cos \omega_2 t)$$
$$+ a_2(A_1 \cos \omega_1 t + A_2 \cos \omega_2 t)^2 \tag{2.205}$$

Using trigonometric identities, we can write the output as

$$y(t) = a_1(A_1 \cos \omega_1 t + A_2 \cos \omega_2 t)$$
$$+ \tfrac{1}{2}a_2(A_1{}^2 + A_2{}^2) + \tfrac{1}{2}a_2(A_1{}^2 \cos 2\omega_1 t + A_2{}^2 \cos 2\omega_2 t)$$
$$+ a_2 A_1 A_2[\cos (\omega_1 + \omega_2)t + \cos (\omega_1 - \omega_2)t] \tag{2.206}$$

As can be seen in Figure 2.19, the system has produced frequencies in the output other than the frequencies of the input. In addition to the first term in (2.206), which may be considered the desired output, there are distortion terms at harmonics of the input frequencies as well as distortion terms involving sums and differences of the harmonics of the input frequencies. The former are referred to as *harmonic distortion terms,* and the latter are referred to as *intermodulation distortion terms.* Note that a second-order non-linearity could be used as a device to *double* the frequency of an input sinusoid. Third-order nonlinearities can be used as *triplers,* and so forth.

A general input signal can be handled by applying the multiplication theorem given in Table C.2 in Appendix C. Thus, for the nonlinear system with the transfer characteristic given by (2.203), the output spectrum is

$$Y(f) = a_1 X(f) + a_2 X(f) * X(f) \tag{2.207}$$

The second term, considered as distortion, is seen to give interference at all frequencies occupied by the desired output (the first term), and it is impos-

FIGURE 2.19   Input and output spectra for nonlinear system with discrete frequency input. (a) Input spectrum. (b) Output spectrum.

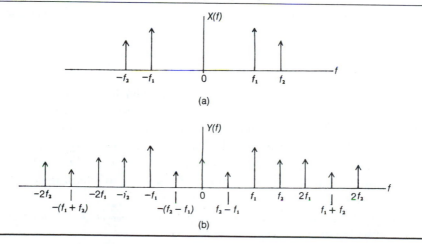

FIGURE 2.20 Input and output spectra for nonlinear system with continuous frequency input. (a) Input spectrum. (b) Output spectrum.

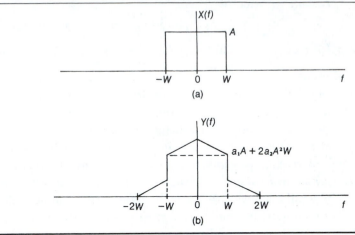

sible to isolate harmonic and intermodulation distortion components as before. For example, if

$$X(f) = A\Pi\left(\frac{f}{2W}\right) \tag{2.208}$$

Then the distortion term is

$$a_2 X(f) * X(f) = 2a_2 W A^2 \Lambda\left(\frac{f}{2W}\right) \tag{2.209}$$

The input and output spectra are shown in Figure 2.20. Note that the spectral width of the distortion term is *double* that of the input.

Ideal Filters

It is often convenient to work with filters having idealized transfer functions with rectangular amplitude-response functions that are constant within the passband and zero elsewhere. We will consider three general types of ideal filters: lowpass, highpass, and bandpass. Within the passband, a linear phase response characteristic is assumed. Thus, if B is the single-sided bandwidth *(width of the stopband* for the highpass filter)* of the filter in ques-

---

* The *stopband* of a filter will be defined here as the frequency range(s) for which $|H(f)|$ is below 3 dB of its maximum value.

tion, the transfer functions of these three ideal filters can be written as follows:

1. $H_{LP}(f) = H_0 \Pi(f/2B)e^{-j2\pi f t_0}$ (lowpass)
2. $H_{HP}(f) = H_0[1 - \Pi(f/2B)]e^{-j2\pi f t_0}$ (highpass)
3. $H_{BP}(f) = [H_l(f - f_0) + H_l(f + f_0)]e^{-j2\pi f t_0}$, where $H_l(f) = H_0 \Pi(f/B)$ (bandpass)

The amplitude- and phase-response functions for these filters are shown in Figure 2.21.

The corresponding impulse responses are obtained by inverse Fourier transformation. For example, the impulse response of an ideal lowpass filter is, from Example 2.11 and the time-delay theorem, given by

$$h_{LP}(t) = 2BH_0 \text{ sinc } [2B(t - t_0)] \qquad (2.210)$$

FIGURE 2.21 Amplitude- and phase-response functions for ideal filters.

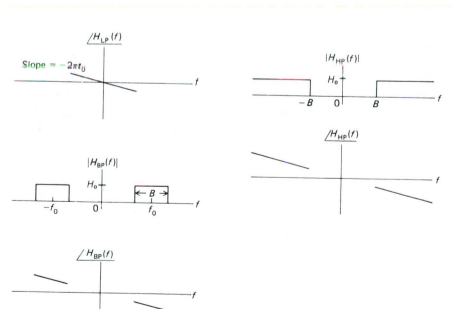

Since $h_{LP}(t)$ is not zero for $t < 0$, we see that an ideal lowpass filter is noncausal. Nevertheless, ideal filters are useful concepts because they simplify calculations and can give satisfactory results.

Turning to the ideal bandpass filter, we may use the modulation theorem to write its impulse response as

$$h_{BP}(t) = 2h_l(t - t_0) \cos 2\pi f_0(t - t_0) \qquad (2.211)$$

where

$$h_l(t) = \mathcal{F}^{-1}[H_l(f)] = H_0 B \text{ sinc } Bt \qquad (2.212)$$

Thus the impulse response of an ideal bandpass filter is the oscillatory signal

$$h_{BP}(t) = 2H_0 B \text{ sinc } B(t - t_0) \cos 2\pi f_0(t - t_0) \qquad (2.213)$$

Figure 2.22 illustrates $h_{LP}(t)$ and $h_{BP}(t)$. If $f_0 \gg B$, it is convenient to view $h_{BP}(t)$ as the slowly varying envelope $2H_0$ sinc Bt *modulating* the high-frequency oscillatory signal $\cos 2\pi f_0 t$ and shifted to the right by t_0 seconds.

*Approximation of Ideal Lowpass Filters by Realizable Filters**

Although ideal filters are noncausal and therefore unrealizable devices, there are several practical filter types that may be designed to approximate ideal filter characteristics as closely as desired. In this section we consider three such approximations for the lowpass case. Bandpass and highpass approximations may be obtained through suitable frequency transformation. The three filter types to be considered are (1) Butterworth, (2) Chebyshev, and (3) Bessel.

The Butterworth filter is a filter design chosen to maintain a constant amplitude response in the passband at the cost of less stopband attenuation. An nth-order Butterworth filter is characterized by a transfer function, in terms of the complex frequency s, of the form

$$H_{BU}(s) = \frac{\omega_3^n}{(s - s_1)(s - s_2) \cdots (s - s_n)} \qquad (2.214)$$

where the poles $s_1, s_2, \ldots, s_n$ are symmetrical with respect to the real axis and equally spaced about a semicircle of radius ω_3 in the left half s-plane and $f_3 = \omega_3/2\pi$ is the 3-dB cutoff frequency.† Typical pole locations are

---

* See Williams and Taylor (1988), Chapter 2, for a detailed discussion of classical filter designs.

† From basic circuit theory courses you will recall that the poles and zeros of a rational function of s, $H(s) = N(s)/D(s)$, are those values of complex frequency $s \triangleq \sigma + j\omega$ for which $D(s) = 0$ and $N(s) = 0$, respectively. See, for example, Cunningham and Stuller (1991).

FIGURE 2.22 Impulse responses for ideal lowpass and bandpass filters. (a) $h_{LP}(t)$. (b) $h_{BP}(t)$.

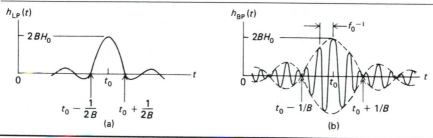

(a)

(b)

shown in Figure 2.23(a). The amplitude response for an nth-order Butterworth filter is of the form

$$|H_{BU}(f)| = \frac{1}{\sqrt{1 + (f/f_3)^{2n}}} \tag{2.215}$$

Note that as n approaches infinity, $|H_{BU}(f)|$ approaches an ideal lowpass filter characteristic. However, the filter delay also approaches infinity.

The Chebyshev lowpass filter has an amplitude response chosen to maintain a maximum allowable attenuation in the passband while maximizing the attenuation in the stopband. A typical pole-zero diagram is shown in Figure 2.23(b). The amplitude response of a Chebyshev filter is of the form

$$|H_C(f)| = \frac{1}{\sqrt{1 + \epsilon^2 C_n^2(f)}} \tag{2.216}$$

The parameter ϵ is specified by the maximum allowable attenuation in the passband, and $C_n(\ )$, known as a *Chebyshev polynomial,* is given by the recursion relation

$$C_n(f) = 2 \left(\frac{f}{f_c} \right) C_{n-1}(f) - C_{n-2}(f), \qquad n = 2, 3, \ldots \tag{2.217}$$

where

$$C_1(f) = \frac{f}{f_c} \qquad \text{and} \qquad C_0(f) = 1 \tag{2.218}$$

Regardless of the value of n, it turns out that $C_n(f_c) = 1$, so that $|H_C(f_c)| = (1 + \epsilon^2)^{-1/2}$. (Note that f_c is not necessarily the 3-dB frequency here.)

The Bessel lowpass filter is a design that attempts to maintain a linear phase response in the passband at the expense of the amplitude response.

FIGURE 2.23 Pole locations and amplitude responses for fourth-order Butterworth and Chebyshev filters. (a) Butterworth filter. (b) Chebyshev filter.

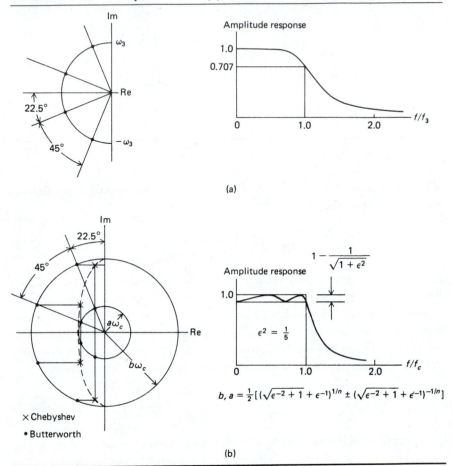

The cutoff frequency of a Bessel filter is defined by

$$f_c = (2\pi t_0)^{-1} = \frac{\omega_c}{2\pi} \tag{2.219}$$

where t_0 is the nominal delay of the filter. The transfer function of an nth-order Bessel filter is given by

$$H_{\mathrm{BE}}(f) = \frac{K_n}{B_n(f)} \tag{2.220}$$

where K_n is a constant chosen to yield $H(0) = 1$, and $B_n(f)$ is a Bessel polynomial of order n defined by

$$B_n(f) = (2n - 1)B_{n-1}(f) - \left(\frac{f}{f_c}\right)^2 B_{n-2}(f) \qquad (2.221)$$

where

$$B_0(f) = 1 \quad \text{and} \quad B_1(f) = 1 + j\left(\frac{f}{f_c}\right) \qquad (2.222)$$

Figure 2.24 illustrates the amplitude and group delay characteristics of third-order Butterworth, Bessel, and Chebyshev filters. All four filters are normalized to have 3-dB amplitude attenuation at a frequency of f_c Hz. The amplitude responses show that the Chebyshev filters have more attenuation than the Butterworth and Bessel filters do for frequencies exceeding the 3-dB frequency. Increasing the passband ($f < f_c$) ripple increases the stopband ($f > f_c$) attenuation.

The group delay characteristics shown in Figure 2.24(b) illustrate, as expected, that the Bessel filter has the most constant group delay. Comparison of the Butterworth and the 0.1-dB ripple Chebyshev group delays shows that although the group delay of the Chebyshev filter has a higher peak, it has a more constant group delay for frequencies less than about $0.3f_c$.

Relation of Pulse Resolution and Risetime to Bandwidth

In our consideration of signal distortion, we assumed bandlimited signal spectra. We found that the input signal to a filter is merely delayed and attenuated if the filter has constant amplitude response and linear phase response throughout the passband of the signal. But suppose the input signal is not bandlimited. What rule of thumb can we use to estimate the required bandwidth? This is a particularly important problem in pulse transmission, where the detection and resolution of pulses at a filter output are of interest.

A satisfactory definition for pulse duration and bandwidth, and the relationship between them, is obtained by consulting Figure 2.25. In Figure 2.25(a), a pulse with a single maximum, taken at $t = 0$ for convenience, is shown with a rectangular approximation of height $x(0)$ and duration T. It is required that the approximating pulse and $|x(t)|$ have equal areas. Thus

$$Tx(0) = \int_{-\infty}^{\infty} |x(t)|\ dt \geq \int_{-\infty}^{\infty} x(t)\ dt = X(0) \qquad (2.223a)$$

FIGURE 2.24 Comparison of third-order Butterworth, Chebyshev (0.1-dB ripple), and Bessel filters. (a) Amplitude response. (b) Group delay. All filters are designed to have a 1-Hz, 3-dB bandwidth.

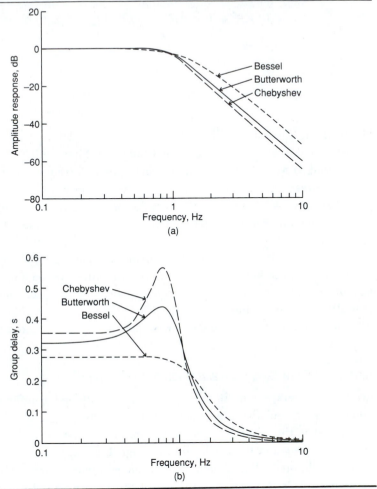

(a)

(b)

where we have used the relationship

$$X(0) = \mathcal{F}[x(t)]\Big|_{f=0} = \int_{-\infty}^{\infty} x(t)e^{-j2\pi t \cdot 0}\, dt \qquad (2.223b)$$

Turning to Figure 2.25(b), we obtain a similar inequality for the rectangular approximation to the pulse spectrum. Specifically, we may write

$$2WX(0) = \int_{-\infty}^{\infty} |X(f)|\, df \geq \int_{-\infty}^{\infty} X(f)\, df = x(0) \qquad (2.224)$$

FIGURE 2.25   Arbitrary pulse signal and spectrum. (a) Pulse and rectangular approximation. (b) Amplitude spectrum and rectangular approximation.

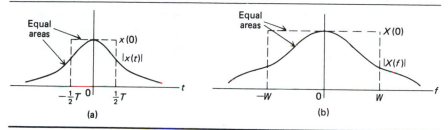

where we have used the relationship

$$x(0) = \mathcal{F}^{-1}[X(f)]\Big|_{t=0} = \int_{-\infty}^{\infty} X(f)e^{j2\pi f \cdot 0}\, df \qquad (2.225)$$

Thus we have the pair of inequalities

$$\frac{x(0)}{X(0)} \geq \frac{1}{T} \qquad \text{and} \qquad 2W \geq \frac{x(0)}{X(0)} \qquad\qquad (2.226)$$

which, when combined, result in the relationship of pulse duration and bandwidth

$$2W \geq \frac{1}{T} \qquad\qquad (2.227a)$$

or

$$W \geq \frac{1}{2T}\ \text{Hz} \qquad\qquad (2.227b)$$

Other definitions of pulse duration and bandwidth could have been used, but a relationship similar to (2.227) would have resulted.

This inverse relationship between pulse duration and bandwidth has been illustrated by all the examples involving pulse spectra that we have considered so far (for example, Examples 2.8, 2.11, 2.13). If pulses with bandpass spectra are considered, the relationship is

$$W \geq \frac{1}{T} \qquad\qquad (2.228)$$

This is illustrated by Example 2.16.

A result similar to (2.227) also holds between the *risetime* T_R and bandwidth of a pulse. A suitable definition of *risetime* is the time required for a pulse's leading edge to go from 10% to 90% of its final value. For the band-pass case, (2.228) holds with T replaced by T_R, where T_R is the risetime of the *envelope* of the pulse.

Risetime can be used as a measure of a system's distortion. To see how this is accomplished, we will express the step response of a filter in terms of its impulse response. From the superposition integral of (2.164b), with $x(t - \sigma) = u(t - \sigma)$, the step response of a filter with impulse response $h(t)$ is

$$y_s(t) = \int_{-\infty}^{\infty} h(\sigma)u(t - \sigma) \, d\sigma$$

$$= \int_{-\infty}^{t} h(\sigma) \, d\sigma \tag{2.229}$$

This follows because $u(t - \sigma) = 0$ for $\sigma > t$. Therefore, the step response of a linear system is the integral of its impulse response. This is not too surprising, since the unit step is the integral of a unit impulse.*

Examples 2.23 and 2.24 demonstrate how the risetime of a system's output due to a step input is a measure of the fidelity of the system.

▼ **EXAMPLE 2.23** The impulse response of a lowpass RC filter is given by

$$h(t) = \frac{1}{RC} e^{-t/RC} u(t) \tag{2.230}$$

for which the step response is found to be

$$y_s(t) = (1 - e^{-2\pi f_3 t})u(t) \tag{2.231}$$

where the 3-dB bandwidth of the filter, defined following (2.175), has been used. The step response is plotted in Figure 2.26(a), where it is seen that the 10% to 90% risetime is approximately

$$T_R \cong \frac{0.35}{f_3} = 2.2RC \tag{2.232}$$

which demonstrates the inverse relationship between bandwidth and rise-
▲ time.

---

* This result is a special case of a more general result for an LTI system: if the response of a system to a given input is known and that input is modified through a linear operation, such as integration, then the output to the modified input is obtained by performing the same linear operation on the output to the original input.

FIGURE 2.26   Step response of (a) a lowpass RC filter and (b) an ideal lowpass filter, illustrating 10% to 90% risetime of each.

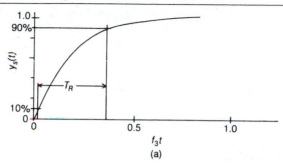

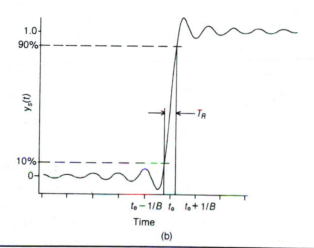

▼ **EXAMPLE 2.24**   Using (2.210) with $H_0 = 1$, the step response of an ideal lowpass filter is

$$y_s(t) = \int_{-\infty}^{t} 2B \, \text{sinc} \, [2B(\sigma - t_0)] \, d\sigma$$

$$= \int_{-\infty}^{t} 2B \, \frac{\sin \, [2\pi B(\sigma - t_0)]}{2\pi B(\sigma - t_0)} \, d\sigma \qquad (2.233)$$

By changing variables in the integrand to $u = 2\pi B(\sigma - t_0)$, the step response becomes

$$y_s(t) = \frac{1}{\pi} \int_{-\infty}^{2\pi B(t - t_0)} \frac{\sin u}{u} \, du = \frac{1}{2} + \frac{1}{\pi} \, \text{Si} \, [2\pi B(t - t_0)] \qquad (2.234)$$

where $\mathrm{Si}\,(x) = \int_0^x (\sin u/u)\,du = -\mathrm{Si}\,(-x)$ is the sine-integral function.*
A plot of $y_s(t)$ for an ideal lowpass filter, such as is shown in Figure 2.26(b),
reveals that the 10% to 90% risetime is approximately

$$T_R \cong \frac{0.44}{B} \tag{2.235}$$

Again, the inverse relationship between bandwidth and risetime is
demonstrated. ▲

2.8 SAMPLING THEORY

In many applications it is useful to represent a signal in terms of sample
values taken at appropriately spaced intervals. Such sample-data systems find
application in feedback control, digital computers, and pulse-modulation
communication systems.

In this section we consider the representation of a signal $x(t)$ by a so-called
ideal instantaneous sampled waveform of the form

$$x_\delta(t) = \sum_{n=-\infty}^{\infty} x(nT_s)\,\delta(t - nT_s) \tag{2.236}$$

where T_s is the sampling interval. Two questions to be answered in connection with such sampling are, "What are the restrictions on $x(t)$ and T_s to allow
perfect recovery of $x(t)$ from $x_\delta(t)$?" and "How is $x(t)$ recovered from $x_\delta(t)$?"
Both questions are answered by the *uniform sampling theorem for lowpass signals,*
which may be stated as follows:

▷THEOREM

If a signal $x(t)$ contains no frequency components for frequencies above $f =
W$ hertz, then it is completely described by instantaneous sample values uni-
formly spaced in time with period $T_s < 1/2W$. The signal can be exactly
reconstructed from the sampled waveform given by (2.236) by passing it
through an ideal lowpass filter with bandwidth B, where $W < B < f_s - W$
with $f_s = T_s^{-1}$. The frequency $2W$ is referred to as the *Nyquist frequency.*

To prove the sampling theorem, we find the spectrum of (2.236). Since
$\delta(x)$ is zero everywhere except at $x = 0$, (2.236) can be written as

$$x_\delta(t) = x(t) \sum_{n=-\infty}^{\infty} \delta(t - nT_s) \tag{2.237}$$

* See M. Abramowitz and I. Stegun, *Handbook of Mathematical Functions,* Dover Publications,
New York, 1972, pp. 238ff (Copy of the 10th National Bureau of Standards Printing).

Applying the multiplication theorem of Fourier transforms, (2.112), the Fourier transform of (2.237) is

$$X_\delta(f) = X(f) * \left[f_s \sum_{n=-\infty}^{\infty} \delta(f - nf_s) \right] \tag{2.238}$$

where the transform pair (2.127) has been used. Interchanging the orders of summation and convolution, and noting that

$$X(f) * \delta(f - nf_s) = \int_{-\infty}^{\infty} X(u)\,\delta(f - u - nf_s)\,du = X(f - nf_s) \tag{2.239}$$

by the sifting property of the delta function, we obtain

$$X_\delta(f) = f_s \sum_{n=-\infty}^{\infty} X(f - nf_s) \tag{2.240}$$

Thus, assuming that the spectrum of $x(t)$ is bandlimited to W Hz and that $f_s > 2W$ as stated in the sampling theorem, we may readily sketch $X_\delta(f)$. Figure 2.27 shows a typical choice for $X(f)$ and the corresponding $X_\delta(f)$. We note that sampling simply results in a periodic repetition of $X(f)$ in the frequency domain with a spacing f_s. If $f_s < 2W$, the separate terms in (2.240) overlap, and there is no apparent way to recover $x(t)$ from $x_\delta(t)$ without distortion. On the other hand, if $f_s > 2W$, the term in (2.240) for $n = 0$ is easily separated from the rest by ideal lowpass filtering. Assuming an ideal lowpass filter with the transfer function

$$H(f) = H_0 \Pi\left(\frac{f}{2B}\right) e^{-j2\pi f t_0}, \qquad W \le B \le f_s - W \tag{2.241}$$

the output spectrum, with $x_\delta(t)$ at the input, is

$$Y(f) = f_s H_0 X(f) e^{-j2\pi f t_0} \tag{2.242}$$

and by the time-shift theorem, the output waveform is

$$y(t) = f_s H_0 x(t - t_0) \tag{2.243}$$

FIGURE 2.27 Signal spectra for lowpass sampling. (a) Assumed spectrum for $x(t)$. (b) Spectrum of sampled signal.

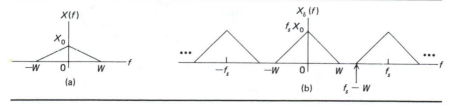

Thus, if the conditions of the sampling theorem are satisfied, we see that distortionless recovery of $x(t)$ from $x_\delta(t)$ is possible. Conversely, if the conditions of the sampling theorem are not satisfied, either because $x(t)$ is not bandlimited or because $f_s < 2W$, we see that distortion at the output of the reconstruction filter is inevitable. Such distortion, referred to as *aliasing*, is illustrated in Figure 2.28(a). It can be combated by filtering the signal before sampling or by increasing the sampling rate. A second type of error, illustrated in Figure 2.28(b), occurs in the reconstruction process and is due to the nonideal frequency response characteristics of practical filters. This type of error can be minimized by choosing reconstruction filters with sharper rolloff characteristics or by increasing the sampling rate. Note that the error due to aliasing and the error due to imperfect reconstruction filters are both *proportional to signal level*. Thus increasing the signal level does not improve the signal-to-error ratio.

An alternative expression for the reconstructed output from the ideal lowpass filter can be obtained by noting that when (2.236) is passed through a filter with impulse response $h(t)$, the output is

$$y(t) = \sum_{n=-\infty}^{\infty} x(nT_s)h(t - nT_s) \tag{2.244}$$

But $h(t)$ corresponding to (2.241) is given by (2.210). Thus

$$y(t) = 2BH_0 \sum_{n=-\infty}^{\infty} x(nT_s) \, \text{sinc} \, 2B(t - t_0 - nT_s) \tag{2.245}$$

FIGURE 2.28 Spectra illustrating two types of errors encountered in reconstruction of sampled signals. (a) Illustration of aliasing error in the reconstruction of sampled signals. (b) Illustration of error due to nonideal reconstruction filter.

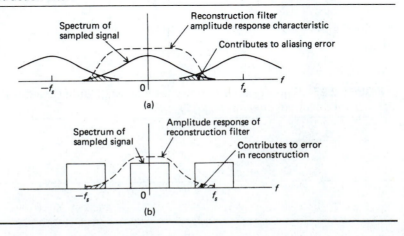

and we see that just as a periodic signal can be completely represented by its Fourier coefficients, a *bandlimited signal can be completely represented by its sample values.*

By setting $B = \frac{1}{2}f_s$, $H_0 = T_s$, and $t_0 = 0$ for simplicity, (2.245) becomes

$$y(t) = \sum_n x(nT_s) \; \text{sinc} \; (f_s t - n) \tag{2.246}$$

This expansion is equivalent to a generalized Fourier series of the form given by (2.36), for we may show that

$$\int_{-\infty}^{\infty} \text{sinc} \; (f_s t - n) \; \text{sinc} \; (f_s t - m) \; dt = K_n \delta_{nm} \tag{2.247}$$

Turning next to bandpass spectra, for which the upper limit on frequency f_u is much larger than the single-sided bandwidth W, one may naturally inquire as to the feasibility of sampling at rates less than $f_s > 2f_u$. The *uniform sampling theorem for bandpass spectra* gives the conditions for which this is possible, as follows:

▷ THEOREM

If a signal has a spectrum of bandwidth W Hz and upper frequency limit f_u, then a rate f_s at which the signal can be sampled is $2f_u/m$, where m is the largest integer not exceeding f_u/W. All higher sampling rates are not necessarily usable unless they exceed $2f_u$.

▼ EXAMPLE 2.25   Consider the bandpass signal $x(t)$ with the spectrum shown in Figure 2.29. According to the bandpass sampling theorem, it is possible to reconstruct $x(t)$ from sample values taken at a rate of

$$f_s = \frac{2f_u}{m} = \frac{2(3)}{2} = 3 \text{ samples per second} \tag{2.248}$$

To show that this is possible, we sketch the spectrum of the sampled signal. According to (2.240), which holds in general,

$$X_\delta(f) = 3 \sum_{-\infty}^{\infty} X(f - 3n) \tag{2.249}$$

The resulting spectrum is shown in Figure 2.29(b), and we see that it is theoretically possible to recover $x(t)$ from $x_\delta(t)$ by bandpass filtering.

Another way of sampling a bandpass signal of bandwidth W is to resolve it into two lowpass quadrature signals of bandwidth $\frac{1}{2}W$. Both of these may

FIGURE 2.29 Signal spectra for bandpass sampling. (a) Assumed
bandpass signal spectrum. (b) Spectrum of sampled signal.

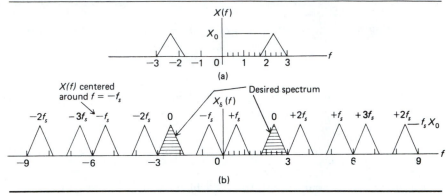

then be sampled at a minimum rate of $2(\frac{1}{2}W) = W$ samples per second, thus
resulting in an overall minimum sampling rate of $2W$ samples per second.

2.9 THE HILBERT TRANSFORM*

Definition

Consider a filter that simply phase-shifts all frequency components of its
input by $-\frac{1}{2}\pi$ radians; that is, its transfer function is

$$H(f) = -j \operatorname{sgn} f \qquad (2.250)$$

where the sgn function (read "signum f") is defined as

$$\operatorname{sgn} f = \begin{cases} 1, & f > 0 \\ 0, & f = 0 \\ -1, & f < 0 \end{cases} \qquad (2.251)$$

We note that $|H(f)| = 1$ and $\underline{/H(f)}$ is odd, as it must be. If $X(f)$ is the
input spectrum to the filter, the output spectrum is $-j \operatorname{sgn}(f)X(f)$, and the
corresponding time function is

$$\hat{x}(t) = \mathcal{F}^{-1}[-j \operatorname{sgn}(f)X(f)]$$
$$= h(t) * x(t) \qquad (2.252)$$

---

* It may be advantageous to postpone this section until consideration of single-sideband
systems in Chapter 3.

where $h(t) = -j\mathscr{F}^{-1}[\text{sgn } f]$ is the impulse response of the filter. To obtain $\mathscr{F}^{-1}[\text{sgn } f]$ without resorting to contour integration, we consider the inverse transform of the function

$$G(f; \alpha) = \begin{cases} e^{-\alpha f}, & f > 0 \\ -e^{\alpha f}, & f < 0 \end{cases} \qquad (2.253)$$

We note that $\lim_{\alpha \to 0} G(f; \alpha) = \text{sgn } f$. Thus our procedure will be to inverse Fourier-transform $G(f; \alpha)$ and take the limit of the result as α approaches zero. Performing the inverse transformation, we obtain

$$g(t; \alpha) = \mathscr{F}^{-1}[G(f; \alpha)]$$

$$= \int_0^\infty e^{-\alpha f} e^{j2\pi ft} \, df - \int_{-\infty}^0 e^{\alpha f} e^{j2\pi ft} \, df = \frac{j4\pi t}{\alpha^2 + (2\pi t)^2} \qquad (2.254)$$

Taking the limit as α approaches zero, we get the transform pair

$$\frac{j}{\pi t} \leftrightarrow \text{sgn } f \qquad (2.255)$$

Using this result in (2.252), we obtain the output of the filter:

$$\hat{x}(t) = \int_{-\infty}^\infty \frac{x(t')}{\pi(t - t')} \, dt' = \int_{-\infty}^\infty \frac{x(t - t')}{\pi t'} \, dt' \qquad (2.256)$$

The function $\hat{x}(t)$ is defined as the *Hilbert transform* of $x(t)$.

Since the Hilbert transform corresponds to a phase shift of $-\frac{1}{2}\pi$, we note that the Hilbert transform of $\hat{x}(t)$ corresponds to the transfer function $(-j \text{ sgn } f)^2 = -1$, or a phase shift of π radians. Thus

$$\hat{\hat{x}}(t) = -x(t) \qquad (2.257)$$

▼ **EXAMPLE 2.26**   For the following input to a Hilbert transform filter:

$$x(t) = \cos 2\pi f_0 t \qquad (2.258)$$

we obtain the output spectrum

$$\hat{X}(f) = \tfrac{1}{2}\delta(f - f_0)e^{-j\pi/2} + \tfrac{1}{2}\delta(f + f_0)e^{j\pi/2} \qquad (2.259)$$

Taking the inverse Fourier transform, we find the output signal to be

$$\hat{x}(t) = \cos(2\pi f_0 t - \tfrac{1}{2}\pi) \quad \text{or} \quad \widehat{\cos 2\pi f_0 t} = \sin 2\pi f_0 t \qquad (2.260)$$

Of course, the Hilbert transform could have been found directly in this case by adding $-\frac{1}{2}\pi$ to the argument of the cosine. Doing this for the signal sin $\omega_0 t$, we find that

$$\widehat{\sin 2\pi f_0 t} = -\cos 2\pi f_0 t \qquad (2.261)$$

We may use the two results obtained in Example 2.26 to show that

$$\widehat{e^{j2\pi f_0 t}} = -j \, \text{sgn} \, (2\pi f_0) e^{j2\pi f_0 t} \qquad (2.262)$$

This is done by considering the two cases $f_0 > 0$ and $f_0 < 0$ and using Euler's theorem in conjunction with the results of (2.260) and (2.261).

Properties

The Hilbert transform has several useful properties that will be used later. Three of these properties will be proved here:

1. The energy (or power) in a signal $x(t)$ and its Hilbert transform $\hat{x}(t)$ are equal.

   To show this, we consider the energy spectral densities at the input and output of a Hilbert transform filter. Since $H(f) = -j \, \text{sgn} \, f$, these densities are related by

   $$|\hat{X}(f)|^2 \triangleq |\mathcal{F}[\hat{x}(t)]|^2 = |-j \, \text{sgn} \, f|^2 \, |X(f)|^2 = |X(f)|^2 \qquad (2.263)$$

   where $\hat{X}(f) = \mathcal{F}[\hat{x}(t)]$. Thus, since the energy spectral densities at input and output are equal, so are the total energies. A similar proof holds for power signals.

2. A signal and its Hilbert transform are orthogonal; that is,

   $$\int_{-\infty}^{\infty} x(t)\hat{x}(t) \, dt = 0 \qquad \text{(energy signals)} \qquad (2.264a)$$

   or

   $$\lim_{T \to \infty} \frac{1}{2T} \int_{-T}^{T} x(t)\hat{x}(t) \, dt = 0 \qquad \text{(power signals)} \qquad (2.264b)$$

   Considering (2.264a), we note that the left-hand side can be written as

   $$\int_{-\infty}^{\infty} x(t)\hat{x}(t) \, dt = \int_{-\infty}^{\infty} X(f)\hat{X}^*(f) \, df \qquad (2.265)$$

   by Parseval's theorem generalized, where $\hat{X}(f) = \mathcal{F}[\hat{x}(t)]$. It therefore follows that

   $$\int_{-\infty}^{\infty} x(t)\hat{x}(t) \, dt = \int_{-\infty}^{\infty} (+j \, \text{sgn} \, f)|X(f)|^2 \, df \qquad (2.266)$$

But the integrand of the right-hand side of (2.266) is odd, being the product of the even function $|X(f)|^2$ and the odd function $+j \operatorname{sgn} f$. Therefore, the integral is zero, and (2.264a) is proved. A similar proof holds for (2.264b).

3. If $c(t)$ and $m(t)$ are signals with nonoverlapping spectra, where $m(t)$ is lowpass and $c(t)$ is highpass, then

$$\widehat{m(t)c(t)} = m(t)\hat{c}(t) \tag{2.267}$$

To prove this relationship, we use the Fourier integral to represent $m(t)$ and $c(t)$ in terms of their spectra, $M(f)$ and $C(f)$, respectively. Thus

$$m(t)c(t) = \int_{-\infty}^{\infty} \int_{-\infty}^{\infty} M(f)C(f') \exp\left[j2\pi(f + f')t\right] df\, df' \tag{2.268}$$

where we assume $M(f) = 0$ for $|f| > W$ and $C(f') = 0$ for $|f'| < W$. The Hilbert transform of (2.268) is

$$\widehat{m(t)c(t)} = \int_{-\infty}^{\infty} \int_{-\infty}^{\infty} M(f)C(f') \widehat{\exp\left[j2\pi(f + f')t\right]}\, df\, df'$$

$$= \int_{-\infty}^{\infty} \int_{-\infty}^{\infty} M(f)C(f')[-j \operatorname{sgn} (f + f')] \exp\left[j2\pi(f + f')t\right]\, df\, df' \tag{2.269}$$

where (2.262) has been used. However, the product $M(f)C(f')$ is nonvanishing only for $|f| < W$ and $|f'| > W$, and we may replace $\operatorname{sgn} (f + f')$ by $\operatorname{sgn} f'$ in this case. Thus

$$\widehat{m(t)c(t)}$$

$$= \int_{-\infty}^{\infty} M(f) \exp (j2\pi ft)\, df \int_{-\infty}^{\infty} C(f')[-j \operatorname{sgn} f' \exp (j2\pi f't)]\, df' \tag{2.270}$$

However, the first integral on the right-hand side is just $m(t)$, and the second integral is $\hat{c}(t)$, since

$$c(t) = \int_{-\infty}^{\infty} C(f') \exp (j2\pi f't)\, df'$$

and

$$\hat{c}(t) = \int_{-\infty}^{\infty} \widehat{C(f') \exp (j2\pi f't)}\, df'$$

$$= \int_{-\infty}^{\infty} C(f')[-j \operatorname{sgn} f' \exp (j2\pi f't)]\, df' \tag{2.271}$$

Hence (2.270) is equivalent to (2.267), which was the relationship to be proved.

▼ EXAMPLE 2.27   Given that $m(t)$ is a lowpass signal with $M(f) = 0$ for $|f| > W$. We may directly apply (2.267) in conjunction with (2.266) and (2.261) to show that

$$\widehat{m(t) \cos \omega_0 t} = m(t) \sin \omega_0 t \qquad (2.272)$$

and

$$\widehat{m(t) \sin \omega_0 t} = -m(t) \cos \omega_0 t \qquad (2.273)$$

▲ if $f_0 = \omega_0/2\pi > W$.

Analytic Signals

The Hilbert transform is used to define the analytic signal $x_p(t)$ in terms of the real signal $x(t)$:

$$x_p(t) = x(t) + j\hat{x}(t) \qquad (2.274)$$

We now consider several properties of the analytic signal.

We used the term *envelope* in connection with the ideal bandpass filter. The *envelope* of a signal is defined mathematically as the magnitude of the analytic signal $x_p(t)$. The concept of an envelope will acquire more importance when we discuss modulation in Chapter 3.

▼ EXAMPLE 2.28   In a previous section we showed that the impulse response of an ideal bandpass filter with bandwidth B, delay t_0, and center frequency f_0 is given by

$$h_{BP}(t) = 2H_0B \text{ sinc } B(t - t_0) \cos \omega_0(t - t_0) \qquad (2.275)$$

Since $B < f_0$, we can use the result of Example 2.27 to determine the Hilbert transform of $h_{BP}(t)$:

$$\hat{h}_{BP}(t) = 2H_0B \text{ sinc } B(t - t_0) \sin \omega_0(t - t_0) \qquad (2.276)$$

Thus the envelope is

$$|x_p(t)| = \sqrt{[2H_0B \text{ sinc } B(t - t_0)]^2[\cos^2 \omega_0(t - t_0) + \sin^2 \omega_0(t - t_0)]} \qquad (2.277)$$

or

$$|x_p(t)| = 2H_0B |\text{sinc } B(t - t_0)| \qquad (2.278)$$

as shown in Figure 2.22(b).

The envelope is obviously easy to identify if the signal is composed of a lowpass signal multiplied by a high-frequency sinusoid. Note, however, that the envelope is mathematically defined for any signal.

The spectrum of the analytic signal is also of interest. We will use it to advantage in Chapter 3 when we investigate single-sideband modulation. Since the analytic signal can be written as

$$x_p(t) = x(t) + j\hat{x}(t) \tag{2.279}$$

it follows that the Fourier transform of $x_p(t)$ is

$$X_p(f) = X(f) + j\{-j \text{ sgn } (f)X(f)\} \tag{2.280}$$

where the term in braces is the Fourier transform of $\hat{x}(t)$. Thus

$$X_p(f) = X(f)[1 + \text{sgn } f] \tag{2.281}$$

or

$$X_p(f) = \begin{cases} 2X(f), & f > 0 \\ 0, & f < 0 \end{cases} \tag{2.282}$$

The subscript p is used to denote that the spectrum is nonzero only for positive frequencies.

Similarly, we can show that the signal

$$x_n(t) = x(t) - j\hat{x}(t) \tag{2.283}$$

is nonzero only for negative frequencies. Replacing $\hat{x}(t)$ by $-\hat{x}(t)$ in the preceding discussion results in

$$X_n(f) = X(f)[1 - \text{sgn } f] \tag{2.284}$$

or

$$X_n(f) = \begin{cases} 0, & f > 0 \\ 2X(f), & f < 0 \end{cases} \tag{2.285}$$

These spectra are shown in Figure 2.30.

Two observations must be made at this point. First, if $X(f)$ is nonzero at $f = 0$, then $X_p(f)$ and $X_n(f)$ will be discontinuous at $f = 0$. Also, we should not be confused that $|X_n(f)|$ and $|X_p(f)|$ are not even, since the corresponding time-domain signals are not real.

Complex Envelope Representation of Bandpass Signals

If $X(f)$ in (2.280) corresponds to a signal with a bandpass spectrum, as shown in Fig. 2.31(a), it then follows by (2.282) that $X_p(f)$ is just twice the positive-

FIGURE 2.30 Spectra of analytic signals. (a) Spectrum of $x(t)$. (b) Spectrum of $x(t) + j\hat{x}(t)$. (c) Spectrum of $x(t) - j\hat{x}(t)$.

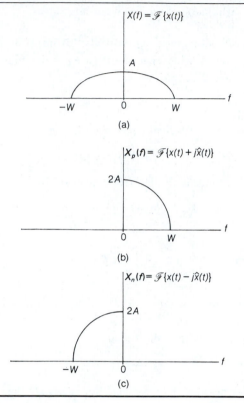

(a)

(b)

(c)

frequency portion of $X(f) = \mathcal{F}\{x(t)\}$, as shown in Fig. 2.31(b). By the frequency-translation theorem, it follows that $x_p(t)$ can be written as

$$x_p(t) = \tilde{x}(t)e^{j2\pi f_0 t} \tag{2.286}$$

where $\tilde{x}(t)$ is a complex-valued lowpass signal (hereafter referred to as the *complex envelope*). The amplitude spectrum of $\tilde{x}(t)$ is shown in Fig. 2.31 (c).

To find $\tilde{x}(t)$, we may proceed along two paths. First, using (2.274), we find the analytic signal and then solve (2.286) for $\tilde{x}(t)$. Second, we can find $\tilde{x}(t)$ by using a frequency domain approach to obtain $X(f)$, which is then scaled by a factor of 2 to give $X_p(f)$ and translated to baseband. The inverse Fourier transform of this spectrum is then $\tilde{x}(t)$.

Since $x_p(t) = x(t) + j\hat{x}(t)$, where $x(t)$ and $\hat{x}(t)$ are the real and imaginary

FIGURE 2.31 Spectra pertaining to the formation of a complex envelope of a signal $x(t)$. (a) A bandpass signal spectrum. (b) Twice the positive-frequency portion of $X(f)$, corresponding to $\mathcal{F}[x(t) + j\hat{x}(t)]$. (c) Spectrum of $\tilde{x}(t)$.

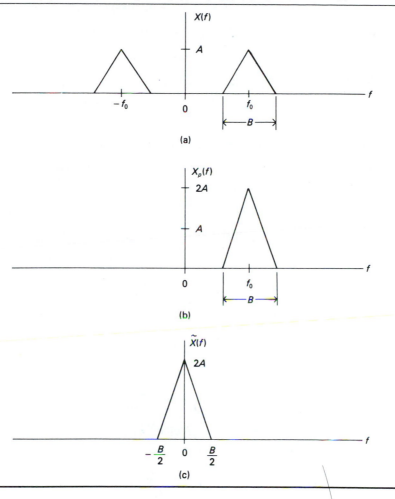

parts, respectively, of $x_p(t)$, it follows from (2.286) that

$$x_p(t) = \tilde{x}(t)e^{j2\pi f_0 t} \triangleq x(t) + j\hat{x}(t) \tag{2.287}$$

or

$$x(t) = \text{Re}[\tilde{x}(t)e^{j2\pi f_0 t}] \tag{2.288a}$$

and

$$\hat{x}(t) = \text{Im}[\tilde{x}(t)e^{j2\pi f_0 t}] \tag{2.288b}$$

Thus, from (2.288a), the real signal $x(t)$ can be expressed in terms of its complex envelope as

$$x(t) = x_R(t) \cos (2\pi f_0 t) - x_I(t) \sin (2\pi f_0 t) \tag{2.289}$$

where

$$\tilde{x}(t) \triangleq x_R(t) + jx_I(t) \tag{2.290}$$

The signals $x_R(t)$ and $x_I(t)$ are known as the *inphase* and *quadrature components* of $x(t)$.

Complex Envelope Representation of Bandpass Systems

Consider a bandpass system with impulse response $h(t)$ which is represented in terms of a complex envelope $\tilde{h}(t)$ as

$$h(t) = \text{Re}[\tilde{h}(t)e^{j2\pi f_0 t}] \tag{2.291}$$

where $\tilde{h}(t) = h_R(t) + jh_I(t)$. If the input is bandpass also with representation (2.288a), the output is

$$y(t) = x(t) * h(t) = \int_{-\infty}^{\infty} h(\lambda)x(t - \lambda)\, d\lambda \tag{2.292}$$

Since we can represent $h(t)$ and $x(t)$ as

$$h(t) = \tfrac{1}{2}\tilde{h}(t)e^{j2\pi f_0 t} + \text{c.c.} \tag{2.293a}$$

and

$$x(t) = \tfrac{1}{2}\tilde{x}(t)e^{j2\pi f_0 t} + \text{c.c.} \tag{2.293b}$$

where c.c. stands for the complex conjugate of the immediately preceding term, the output can be expressed as

$$
\begin{aligned}
y(t) &= \int_{-\infty}^{\infty} [\tfrac{1}{2}\tilde{h}(t)e^{j2\pi f_0 \lambda} + \text{c.c.}][\tfrac{1}{2}\tilde{x}(t - \lambda)e^{j2\pi f_0(t-\lambda)} + \text{c.c.}]\, d\lambda \\
&= \tfrac{1}{4} \int_{-\infty}^{\infty} \tilde{h}(\lambda)\tilde{x}(t - \lambda)\, d\lambda\; e^{j2\pi f_0 t} + \text{c.c} \\
&\quad + \tfrac{1}{4} \int_{-\infty}^{\infty} \tilde{h}(\lambda)\tilde{x}^*(t - \lambda)e^{j4\pi f_0 \lambda}\, d\lambda\; e^{-j2\pi f_0 t} + \text{c.c.}
\end{aligned} \tag{2.294}
$$

The second pair of terms is approximately zero by virtue of the factor $e^{j4\lambda f_0 \lambda}$ in the integrand ($\tilde{h}$ and $\tilde{x}$ are slowly varying with respect to this complex exponential, and therefore, the integrand cancels to zero half-cycle by half-cycle). Thus

$$y(t) = \tfrac{1}{2}\text{Re}\{[\tilde{h}(t) * \tilde{x}(t)]e^{j2\pi f_0 t}\} \qquad (2.295)$$

where

$$\tilde{y}(t) = \tilde{h}(t) * \tilde{x}(t) = \mathcal{F}^{-1}[\tilde{H}(f)\tilde{X}(f)] \qquad (2.296)$$

in which $\tilde{H}(f)$ and $\tilde{X}(f)$ are the respective Fourier transforms of $\tilde{h}(t)$ and $\tilde{x}(t)$.

▼ **EXAMPLE 2.29**   As an example, consider the input

$$x(t) = \Pi(t/\tau) \cos (2\pi f_0 t) \qquad (2.297)$$

to a filter with impulse response

$$h(t) = \alpha e^{-\alpha t} u(t) \cos (2\pi f_0 t) \qquad (2.298)$$

Using the complex envelope analysis just developed with $\tilde{x}(t) = \Pi(t/\tau)$ and $\tilde{h}(t) = \alpha e^{-\alpha t} u(t)$, we have as the complex envelope of the filter output

$$\tilde{y}(t) = \Pi(t/\tau) * \alpha e^{-\alpha t} u(t)$$
$$= [1 - e^{-\alpha(t+\tau/2)}]u(t + \tau/2) - [1 - e^{-(t-\tau/2)}]u(t - \tau/2) \quad (2.299)$$

Multiplying this by $\tfrac{1}{2}e^{j2\pi f_0 t}$ and taking the real part results in the output of the filter in accordance with (2.295).

To check this result, we convolve (2.297) and (2.298) directly. The super-position integral becomes

$$y(t) = x(t) * h(t)$$
$$= \int_{-\infty}^{\infty} \Pi(\lambda/\tau) \cos (2\pi f_0 \lambda) \alpha e^{-\alpha(t-\lambda)} u(t - \lambda) \cos [2\pi f_0 (t - \lambda)] \, d\lambda$$
$$\qquad (2.300)$$

But

$$\cos (2\pi f_0 \lambda) \cos [2\pi f_0 (t - \lambda)]$$
$$= \tfrac{1}{2} \cos (2\pi f_0 t) + \tfrac{1}{2} \cos [2\pi f_0 (t - 2\lambda)] \quad (2.301)$$

so that the superposition integral becomes

$$y(t) = \tfrac{1}{2} \int_{-\infty}^{\infty} \Pi(\lambda/\tau) \alpha e^{-\alpha(t-\lambda)} u(t - \lambda) \, d\lambda \, \cos (2\pi f_0 t)$$

$$+ \tfrac{1}{2} \int_{-\infty}^{\infty} \Pi(\lambda/\tau) \alpha e^{-\alpha(t-\lambda)} u(t - \lambda) \cos [2\pi f_0 (t - 2\lambda)] \, d\lambda \qquad (2.302)$$

If $f_0^{-1} \ll \tau$ and $f_0^{-1} \ll \alpha^{-1}$, the second integral is approximately zero, so that we have only the first integral, which is $\Pi(t/\tau)$ and $\alpha e^{-\alpha t} u(t)$ convolved and the result multiplied by $\frac{1}{2} \cos (2\pi f_0 t)$, which is the same as before.

2.10   THE DISCRETE FOURIER TRANSFORM AND FAST FOURIER TRANSFORM

In order to compute the Fourier spectrum of a signal by means of a digital computer, the time-domain signal must be represented by sample values, and the spectrum must be computed at a discrete number of frequencies. It can be shown that the following sum gives an approximation to the Fourier spectrum of a signal at frequencies $k/(NT_s)$, $k = 0, 1, \ldots, N - 1$:

$$X_k = \sum_{n=0}^{N-1} x_n e^{-j2\pi nk/N}, \qquad k = 0, 1, \ldots, N - 1 \qquad (2.303)$$

where $x_0, x_1, x_2, \ldots, x_{N-1}$ are N sample values of the signal taken at T_s-second intervals for which the Fourier spectrum is desired. The sum (2.303) is called the *discrete Fourier transform (DFT)* of the sequence $\{x_n\}$. According to the sampling theorem, if the samples are spaced by T_s seconds, the spectrum repeats every $f_s = T_s^{-1}$ Hz. Since there are N frequency samples in this interval, it follows that the frequency resolution of (2.303) is $f_s/N = 1/(NT_s) \triangleq 1/T$. To obtain the sample sequence $\{x_n\}$ from the DFT sequence $\{X_k\}$, the sum

$$x_n = \frac{1}{N} \sum_{k=0}^{N-1} X_k e^{j2\pi nk/N}, \qquad n = 0, 1, 2, \ldots, N - 1 \qquad (2.304)$$

is used. That (2.303) and (2.304) form a transform pair can be shown by substituting (2.303) into (2.304) and using the sum formula for a geometric series:

$$S_N \equiv \sum_{k=0}^{N-1} x^k = \frac{1 - x^N}{1 - x}, \qquad x \neq 1 \qquad (2.305a)$$

and

$$S_N = N, \qquad x = 1 \qquad (2.305b)$$

As indicated above, the DFT and inverse DFT are approximations to the true Fourier spectrum of a signal $x(t)$, at the discrete set of frequencies $\{0, 1/T, 2/T, \ldots, (N - 1)/T\}$. The error can be small if the DFT and its inverse are applied properly to a signal. To indicate the approximations involved, we must visualize the spectrum of a sampled signal that is truncated to a finite number of sample values and whose spectrum is then sampled at a discrete number N of points. To see the approximations involved, we use the following Fourier transform theorems:

1. The Fourier transform of an ideal sampling waveform (Example 2.14):

$$y_s(t) = \sum_{m=-\infty}^{\infty} \delta(t - mT_s) \leftrightarrow f_s^{-1} \sum_{n=-\infty}^{\infty} \delta(f - nf_s), \qquad f_s = T_s^{-1}$$

2. The Fourier transform of a rectangular window function:

$$\Pi(t/T) \leftrightarrow T \text{ sinc } (fT)$$

3. The convolution theorem of Fourier transforms:

$$x_1(t) * x_2(t) \leftrightarrow X_1(f)X_2(f)$$

4. The multiplication theorem of Fourier transforms:

$$x_1(t)x_2(t) \leftrightarrow X_1(f) * X_2(f)$$

The approximations involved are illustrated by the following example.

▼ **EXAMPLE 2.30**   An exponential signal is to be sampled, the samples truncated to a finite number, and the result represented by a finite number of samples of the Fourier spectrum of the sampled truncated signal. The continuous-time signal and its Fourier transform are

$$x(t) = e^{-|t|/\tau} \leftrightarrow X(f) = \frac{2\tau}{1 + 2(\pi f \tau)^2} \tag{2.306}$$

These signals are shown in Fig. 2.32(a). However, we are representing the signal by sample values spaced by T_s seconds, which entails multiplying the original signal by the ideal sampling waveform $y_s(t)$. The resulting spectrum of this signal is the convolution of $X(f)$ with $y_s(t) = 2\tau f_s \sum_{n=-\infty}^{\infty} \delta(f - nf_s)$. The result of this convolution, shown in Fig. 2.32(b), is

$$X_s(f) = f_s \sum_{n=-\infty}^{\infty} \frac{2\tau}{1 + [2\pi\tau(f - f_s)]^2} \tag{2.307}$$

In calculating the DFT, only a T-second chunk of $x(t)$ can be used (N samples spaced by $T_s = T/N$). This means that the sampled time domain signal is effectively multiplied by a window function $\Pi(t/T)$. In the frequency domain, this corresponds to convolution with the Fourier transform of the

FIGURE 2.32 Signals and spectra illustrating the computation of the DFT. (a) Signal to be sampled and its spectrum ($\tau = 1$ s). (b) Sampled signal and its spectrum ($f_s = 1$ Hz). (c) Windowed, sampled signal and its spectrum ($T = 4^+$ s). (d) Sampled signal spectrum and corresponding periodic repetition of the sampled, windowed signal.

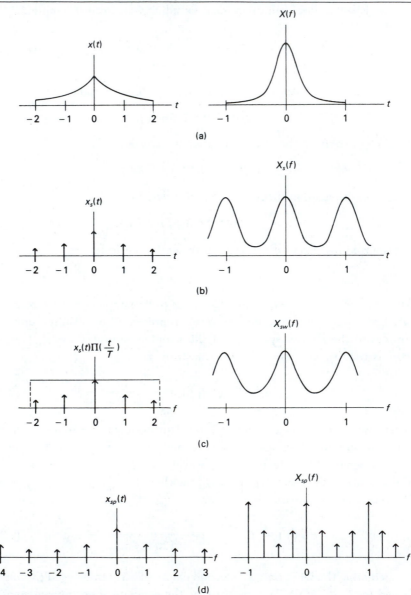

rectangular window function, which is $T \, \text{sinc} \, (fT)$. The resulting windowed, sampled signal and its spectrum are sketched in Fig. 2.32(c). Finally, the spectrum is available only at N discrete frequencies separated by the reciprocal of the window duration $1/T$. This corresponds to convolution in the time domain with a sequence of delta functions, and this result is shown in Fig. 2.32(d). It can be seen that unless one is careful, there is indeed a considerable likelihood that the DFT spectrum will look nothing like the spectrum of the original continuous-time signal. Means for minimizing these errors are discussed in several references on the subject.*

A little thought will indicate that to compute the complete DFT spectrum of a signal, approximately N^2 complex multiplications are required in addition to a number of complex additions. It is possible to find algorithms that allow the computation of the DFT spectrum of a signal using only approximately $N \log_2 N$ complex multiplications. Such algorithms are referred to as *fast Fourier transform (FFT) algorithms*.

One such algorithm is based on a procedure called *decimation in time* (DIT). A flow graph for an eight-point DIT algorithm is shown in Figure 2.33, where the lower case x's denote input samples and the uppercase X's denote output samples. The arrows denote the flow of computation, with $W_N^p = e^{-j2\pi p/N}$ alongside the arrowhead indicating multiplication by this factor. Fortunately, FFT algorithms are included in most mathematics packages (see the footnote after the Computer Exercises for reference to some of these), so it is not necessary to design one's own FFT subroutine in doing computations requiring the FFT.

SUMMARY

1. Two general classes of signals are deterministic signals and random signals. A deterministic signal can be written as a completely known function of time whereas the amplitudes of random signals must be described probabilistically.

2. A periodic signal of period T_0 is one for which $x(t) = x(t + T_0)$.

3. A single-sided spectrum for a rotating phasor $\tilde{x}(t) = Ae^{j(\omega_0 t + \theta)}$ shows A (amplitude) and θ (phase) versus f (frequency). The real, sinusoidal signal corresponding to this phasor is obtained by taking the real part of $\tilde{x}(t)$.

---

* Ziemer, Tranter, and Fannin (1993), Chapter 10.

FIGURE 2.33 A flow graph for computation of an eight-point FFT using the DIT technique

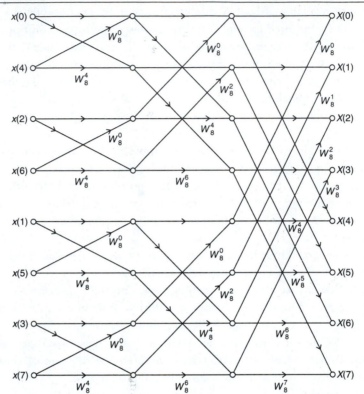

A double-sided spectrum results if we think of forming $x(t) = \frac{1}{2}\tilde{x}(t) + \frac{1}{2}\tilde{x}^*(t)$. Graphs of amplitude and phase (two plots) of this rotating phasor sum versus f are known as the two-sided amplitude and phase spectra, respectively. Such spectral plots are referred to as frequency-domain representations of the signal $A \cos(\omega_0 t + \theta)$.

4. The unit impulse function, $\delta(t)$, can be thought of as a zero-width, infinite-height pulse with unity area. Many functions with these properties, in the limit as some parameter approaches zero, are suitable representations for a unit impulse. The sifting property, $\int_{-\infty}^{\infty} x(\lambda)\,\delta(\lambda - t_0)\,d\lambda = x(t_0)$, is a generalization of the defining relation for a unit impulse. The unit step function, $u(t)$, is the integral of a unit impulse.

5. A signal $x(t)$ for which $E = \int_{-\infty}^{\infty} |x(t)|^2 \, dt$ is finite is called an *energy signal*. If $x(t)$ is such that

$$P = \lim_{T \to \infty} \frac{1}{2T} \int_{-T}^{T} |x(t)|^2 \, dt$$

is finite, the signal is known as a *power signal*.

6. A set of orthogonal functions, $\phi_1(t), \phi_2(t), \ldots, \phi_N(t)$, can be used as a series approximation of the form

$$x_a(t) = \sum_{n=1}^{N} X_n \phi_n(t)$$

to a signal $x(t)$, which is integral-square in the interval $(t_0, t_0 + T)$. The integral-squared error between $x_a(t)$ and $x(t)$ is minimized if the coefficients are chosen as

$$X_n = \frac{1}{c_n} \int_{t_0}^{t_0 + T} x(t) \phi_n^*(t) \, dt$$

where

$$\int_{t_0}^{t_0 + T} \phi_n(t) \phi_m^*(t) \, dt = c_n \delta_{nm}, \quad c_n = \text{real constant}$$

For a complete set of $\phi_n(t)$'s, the integral-squared error approaches zero as N approaches infinity, and Parseval's theorem then holds:

$$\int_{T} |x(t)|^2 \, dt = \sum_{n=1}^{\infty} c_n |X_n|^2$$

7. If $\phi_n(t) = e^{jn\omega_0 t}$, where $\omega_0 = 2\pi/T_0$ and T_0 is the expansion interval, is used in an orthogonal function series, the result is the complex exponential Fourier series. If $x(t)$ is periodic with period T_0, the exponential Fourier series represents $x(t)$ exactly for all t, except at points of discontinuity.

8. For exponential Fourier series of real signals, the Fourier coefficients obey $X_n = X_{-n}^*$, which implies that $|X_n| = |X_{-n}|$ and $\underline{/X_n} = -\underline{/X_{-n}}$. Plots of $|X_n|$ and $\underline{/X_n}$ versus nf_0 are referred to as the discrete, double-sided amplitude and phase spectra, respectively, of $x(t)$. If $x(t)$ is real, the amplitude spectrum is even and the phase spectrum is odd as functions of nf_0.

9. Parseval's theorem for periodic signals is

$$\frac{1}{T_0} \int_{T_0} |x(t)|^2 \, dt = \sum_{n=-\infty}^{\infty} |X_n|^2$$

10. The Fourier transform of a signal $x(t)$ is

$$X(f) = \int_{-\infty}^{\infty} x(t) e^{-j2\pi ft} \, dt$$

and the inverse Fourier transform is

$$x(t) = \int_{-\infty}^{\infty} X(f) e^{j2\pi ft} \, df$$

For real signals, $|X(f)| = |X(-f)|$ and $\underline{/X(f)} = -\underline{/X(-f)}$.

11. Plots of $|X(f)|$ and $\underline{/X(f)}$ versus f are referred to as the *double-sided amplitude* and *phase spectra,* respectively, of $x(t)$. Thus, as functions of frequency, the amplitude spectrum of a real signal is even, and the phase spectrum is odd.

12. The energy of a signal is

$$\int_{-\infty}^{\infty} |x(t)|^2 \, dt = \int_{-\infty}^{\infty} |X(f)|^2 \, df$$

This is known as *Rayleigh's energy theorem.* The energy spectral density of a signal is $G(f) = |X(f)|^2$. It is the density of energy with frequency of the signal.

13. The convolution of two signals, $x_1(t)$ and $x_2(t)$, is

$$x(t) = x_1 * x_2 = \int_{-\infty}^{\infty} x_1(\lambda) x_2(t - \lambda) \, d\lambda = \int_{-\infty}^{\infty} x_1(t - \lambda) x_2(\lambda) \, d\lambda$$

The convolution theorem of Fourier transforms states that $X(f) = X_1(f)X_2(f)$, where $X(f)$, $X_1(f)$, and $X_2(f)$ are the Fourier transforms of $x(t)$, $x_1(t)$, and $x_2(t)$, respectively.

14. The Fourier transform of a periodic signal can be obtained formally by Fourier-transforming its exponential Fourier series term by term, using $Ae^{j2\pi f_0 t} \leftrightarrow A\delta(f - f_0)$, even though, mathematically speaking, Fourier transforms of power signals do not exist. (The latter may be found as the limit of the Fourier transform of a properly Fourier-transformable function as a parameter approaches zero or infinity.)

15. The power spectrum $S(f)$ of a power signal $x(t)$ is a real, even, nonnegative function that integrates to give total average power: $\langle x^2(t) \rangle = \int_{-\infty}^{\infty} S(f)\, df$. The time-average autocorrelation function of a power signal is defined as $R(\tau) = \langle x(t)x(t + \tau) \rangle$. The Wiener-Khinchine theorem states that $S(f)$ and $R(\tau)$ are Fourier transform pairs.

16. A linear system, denoted operationally as $\mathcal{H}(\ )$, is one for which superposition holds; that is, if $y_1 = \mathcal{H}(x_1)$ and $y_2 = \mathcal{H}(x_2)$, then $\mathcal{H}(\alpha_1 x_1 + \alpha_2 x_2) = \alpha_1 y_1 + \alpha_2 y_2$, where x_1 and x_2 are inputs and y_1 and y_2 are outputs. A system is fixed, or time-invariant, if, given $y(t) = \mathcal{H}[x(t)]$, the input $x(t - t_0)$ results in the output $y(t - t_0)$.

17. The impulse response $h(t)$ of a linear, fixed system is its response to an impulse applied at $t = 0$: $h(t) = \mathcal{H}[\delta(t)]$. The output of a fixed, linear system to an input $x(t)$ is given by $y(t) = h(t) * x(t)$.

18. A *causal system* is one which does not anticipate its input. For such a system, $h(t) = 0$ for $t < 0$. A *stable system* is one for which every bounded input results in a bounded output. A system is stable if and only if $\int_{-\infty}^{\infty} |h(t)|\, dt < \infty$.

19. The transfer function $H(f)$ of a linear fixed system is the Fourier transform of $h(t)$. The Fourier transform of the system output $y(t)$ due to an input $x(t)$ is $Y(f) = H(f)X(f)$, where $X(f)$ is the Fourier transform of the input. $|H(f)| = |H(-f)|$ is called the *amplitude response* of the system, and $\underline{/H(f)} = -\underline{/H(-f)}$ is called the *phase response*.

20. For a fixed linear system with a periodic input, the Fourier coefficients of the output are given by $Y_n = H(nf_0)X_n$, where X_n represents the Fourier coefficients of the input.

21. Input and output spectral densities for a fixed linear system are related by

$$G_y(f) = |H(f)|^2 G_x(f) \qquad \text{(energy signals)}$$
$$S_y(f) = |H(f)|^2 S_x(f) \qquad \text{(power signals)}$$

22. A system is distortionless if its output looks like its input except for a time delay and amplitude scaling: $y(t) = H_0 x(t - t_0)$. The transfer function of a distortionless system is $H(f) = H_0 e^{-j2\pi f t_0}$. Such a system's amplitude response is $|H(f)| = H_0$ and its phase response is $\underline{/H(f)} = -2\pi t_0 f$ over the band of frequencies occupied by the input. Three types of distortion that a system may introduce are amplitude, phase (or delay), and nonlinear, depending on whether $|H(f)| \neq$ constant,

$\underline{/H(f)} \neq -$ constant $\times f$, or the system is nonlinear, respectively. Two other important properties of a linear system are the group and phase delays. These are defined by

$$T_g(f) = -\frac{1}{2\pi}\frac{d}{df}\theta(f) \qquad \text{and} \qquad T_p(f) = -\frac{\theta(f)}{2\pi f}$$

respectively, in which $\theta(f)$ is the phase response of the network. Distortionless systems have equal group and phase delays.

23. Ideal filters are convenient in communication system analysis, even though they are unrealizable. Three types of ideal filters are lowpass, bandpass, and highpass. Throughout their passband, ideal filters have constant amplitude response and linear phase response. Outside their passband, ideal filters perfectly reject all spectral components of the input.

24. Approximations to ideal filters are Butterworth, Chebyshev, and Bessel filters. The first two are attempts at approximating the amplitude response of an ideal filter, and the latter is an attempt to approximate the linear phase response of an ideal filter.

25. An inequality relating the duration T of a pulse and its single-sided bandwidth W is $W \geq 1/2T$. Pulse risetime T_R and signal bandwidth are related approximately by $W = 1/2T_R$. These relationships hold for the lowpass case. For bandpass filters and signals, the required bandwidth is doubled, and the risetime is that of the envelope of the signal.

26. The sampling theorem for lowpass signals of bandwidth W states that a signal can be perfectly recovered by lowpass filtering from sample values taken at a rate of $f_s > 2W$ samples per second. The spectrum of an impulse-sampled signal is

$$X_\delta(f) = f_s \sum_{n=-\infty}^{\infty} X(f - nf_s)$$

where $X(f)$ is the spectrum of the original signal. For bandpass signals, lower sampling rates than specified by the lowpass sampling theorem may be possible.

27. The Hilbert transform $\hat{x}(t)$ of a signal $x(t)$ corresponds to a $-90°$ phase shift of all the signal's positive-frequency components. Mathematically,

$$\hat{x}(t) = \int_{-\infty}^{\infty} \frac{x(t')}{\pi(t - t')} \, dt'$$

In the frequency domain, $\hat{X}(f) = -j \operatorname{sgn}(f)X(f)$, where $\operatorname{sgn}(f)$ is the signum function, $X(f) = \mathcal{F}[x(t)]$, and $\hat{X}(f) = \mathcal{F}[\hat{x}(t)]$. The Hilbert transform of $\cos \omega_0 t$ is $\sin \omega_0 t$, and the Hilbert transform of $\sin \omega_0 t$ is $-\cos \omega_0 t$. The power (or energy) in a signal and its Hilbert transform are equal. A signal and its Hilbert transform are orthogonal in the range $(-\infty, \infty)$. If $m(t)$ is a lowpass signal and $c(t)$ is a highpass signal with nonoverlapping spectra,

$$\widehat{m(t)c(t)} = m(t)\hat{c}(t)$$

The Hilbert transform can be used to define the analytic signal

$$z(t) = x(t) \pm j\hat{x}(t)$$

The magnitude of the analytic signal, $|z(t)|$, is the envelope of the real signal $x(t)$. The Fourier transform of an analytic signal, $Z(f)$, is identically zero for $f < 0$ or $f > 0$, respectively, depending on if the $+$ sign or $-$ sign is chosen for the imaginary part of $z(t)$.

28. The complex envelope $\tilde{x}(t)$ of a bandpass signal is defined by

$$x(t) + j\hat{x}(t) = \tilde{x}(t)e^{j2\pi f_0 t}$$

where f_0 is the center frequency of the signal. Similarly, the complex envelope $\tilde{h}(t)$ of the impulse response of a bandpass system is defined by

$$h(t) + j\hat{h}(t) = \tilde{h}(t)e^{j2\pi f_0 t}$$

Its output is conveniently obtained in terms of the complex envelope of the output which can be found from either of the operations

$$\tilde{y}(t) = \tilde{h}(t) * \tilde{x}(t)$$

or

$$\tilde{y}(t) = \mathcal{F}^{-1}[\tilde{H}(f)\tilde{X}(f)]$$

where $\tilde{H}(f)$ and $\tilde{X}(f)$ are the Fourier transforms of $\tilde{h}(t)$ and $\tilde{x}(t)$, respectively. The actual (real) output is then given by

$$y(t) = \operatorname{Re}[\tilde{y}(t)e^{j2\pi f_0 t}]$$

29. The discrete Fourier transform (DFT) is defined as

$$X_k = \sum_{n=0}^{N-1} x_n e^{j2\pi nk/N} = \operatorname{DFT}[\{x_n\}], \qquad k = 0, 1, \ldots, N-1$$

and the inverse DFT can be found from

$$x_n = \frac{1}{N} \text{DFT}[\{X_k^*\}]^*, \qquad n = 0, 1, \ldots, N - 1$$

The DFT can be used to digitally compute spectra of sampled signals and to approximate operations carried out by the normal Fourier transform, for example, filtering.

30. FFT algorithms can be obtained in various ways. The flow graph for a *decimation in time* (DIT) algorithm is shown in Figure 2.33.

FURTHER READING

Fourier techniques from the standpoint of the electrical engineer are dealt with in many texts. Lathi (1991) and Stremler (1982) treat them in the context of communication systems. Bracewell (1986) and Papoulis (1977) are concerned exclusively with Fourier theory and applications.

Introductory reading on Fourier series, transfer functions, and impulse response is found in most introductory circuit-analysis texts. See, for example, Hayt and Kemmerly (1993), Cunningham and Stuller (1991), or Smith and Dorf (1992). Most of the books cited above provide good treatments of systems analysis. Ziemer, Tranter, and Fannin (1993) and Frederick and Carlson (1988) are devoted to continuous and discrete signal and system theory. Schwartz and Friedland (1965) provide an excellent exposition of system characterization at the senior-graduate level, especially of the relationship of transfer function and impulse response to state-variable formulations. Sampling theory and the Hilbert transform are treated by Carlson (1985) and Haykin (1983), among others, at about the same level as presented here. Treatments of the discrete Fourier and fast Fourier transforms are available in Ziemer, Tranter, and Fannin (1993) and Brigham (1988).

PROBLEMS

Section 2.1

2.1 Sketch the single-sided and double-sided amplitude and phase spectra of

$$x(t) = 10 \cos (4\pi t + \pi/6) + 3 \sin (10\pi t + 2\pi/3)$$

2.2 A signal has the double-sided amplitude and phase spectra shown in Figure 2.34. Write a time-domain expression for the signal.

FIGURE 2.34

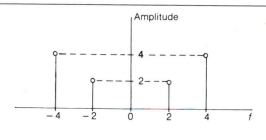

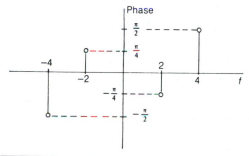

2.3 The sum of two or more sinusoids may or may not be periodic depending on the relationship of their separate frequencies. For the sum of two sinusoids, let the frequencies of the individual terms be f_1 and f_2, respectively. For the sum to be periodic, f_1 and f_2 must be commensurable; i.e., there must be a number f_0 contained in each an integral number of times. Thus, if f_0 is the largest such number,

$$f_1 = n_1 f_0 \qquad \text{and} \qquad f_2 = n_2 f_0$$

where n_1 and n_2 are integers; f_0 is the fundamental frequency. Which of the signals given below are periodic? Find the periods of those that are periodic.

(a) $x_1(t) = 2 \cos (2t) + 3 \sin (5\pi t)$
(b) $x_2(t) = \cos (5t) + 5 \cos (15t)$
(c) $x_3(t) = 3 \sin (10\pi t) + \sin (12\pi t)$
(d) $x_4(t) = 4 \cos (2\pi t) + 3 \cos (4\pi t) + 5 \sin (26\pi t)$

2.4 Sketch the single-sided and double-sided spectra of

(a) $x_a(t) = 6 \cos (6\pi t - \pi/3)$

(b) $x_b(t) = 4 \sin (10\pi t) + 2 \cos (12\pi t)$

2.5   (a) Show that the function $\delta_\epsilon(t)$ sketched in Figure 2.4(b) has unity
         area.

      (b) Show that

$$\delta_\epsilon(t) = \epsilon^{-1} e^{-t/\epsilon} u(t)$$

has unity area. Sketch this function for $\epsilon = 1, \frac{1}{2}$, and $\frac{1}{4}$. Comment on
its suitability as an approximation for the unit impulse function.

2.6   Show that a suitable approximation for the unit impulse function as
      $\epsilon \rightarrow 0$ is given by

$$\delta_\epsilon(t) = \begin{cases} \epsilon^{-1}(1 - |t|/\epsilon), & |t| \leq \epsilon \\ 0, & \text{otherwise} \end{cases}$$

2.7   Use the properties of the Riemann integral to prove the properties
      of the unit impulse function given after (2.13b).

2.8   Use the properties of the unit impulse function given after (2.13b) to
      evaluate the following integrals.

      (a) $\int_{-\infty}^{\infty} [t^2 + \cos(\pi t)] \delta(t - 8) \, dt$
      (b) $\int_{-5^-}^{15^+} (t^2 + 1) [\sum_{n=-\infty}^{\infty} \delta(t - 5n)] \, dt$
      (c) $\int_{-1}^{10} (t^2 + t) \delta(t + 3) \, dt$
      (d) $\int_{-\infty}^{\infty} [e^{-\pi t} + \sin(10\pi t)] \delta(2t + 5) \, dt$
      (e) $\int_{-\infty}^{\infty} [\cos(2\pi t) + e^{-10t}] \delta^{(2)}(t) \, dt$

2.9   Which of the following signals are periodic and which are aperiodic?
      Find the periods of those which are periodic. Sketch all signals.

      (a) $x_1(t) = \cos(4\pi t) + \sin(7\pi t)$
      (b) $x_2(t) = e^{-10t} u(t)$
      (c) $x_3(t) = \sum_{n=-\infty}^{\infty} \delta(t - 6n)$
      (d) $x_4(t) = \sin(3t) + \cos(2\pi t)$
      (e) $x_5(t) = \sum_{n=-\infty}^{\infty} \Pi(t - 2n)$
      (f) $x_6(t) = 1 + \cos(6\pi t)$

2.10 Write the signal $x(t) = 2\cos(5\pi t) + 4\sin(11\pi t)$ as

      (a) The real part of a sum of rotating phasors.
      (b) A sum of rotating phasors plus their complex conjugates.
      (c) From your results in parts (a) and (b), sketch the single-sided and
          double-sided amplitude and phase spectra of $x(t)$.

Section 2.2

2.11 Find the normalized power for each signal below that is a power sig-
      nal and the normalized energy for each signal that is an energy signal.

If a signal is neither a power signal nor an energy signal, so designate it. Sketch each signal (α is a positive constant).

(a) $x_1(t) = 4 \sin (6\pi t + \pi/6)$ (b) $x_2(t) = e^{-\alpha t} u(t)$

(c) $x_3(t) = e^{\alpha t} u(-t)$ (d) $x_4(t) = (\alpha^2 + t^2)^{-1/4}$

(e) $x_5(t) = e^{-\alpha |t|}$ (f) $x_6(t) = \cos (5\pi t) u(t)$

2.12   Classify each of the following signals as an energy signal or a power signal by calculating E, the energy, or P, the power (A, θ, ω, and τ are positive constants).

(a) $A | \sin (\omega t + \theta)|$ (b) $A\tau / \sqrt{\tau + jt}$, $j = \sqrt{-1}$

(c) $At^2 e^{-t/\tau} u(t)$ (d) $\Pi(t/\tau) + \Pi(t/2\tau)$

2.13   Sketch each of the following periodic waveforms and compute their average powers.

(a) $x_1(t) = \sum_{n=-\infty}^{\infty} \Pi[(t - 8n)/3]$

(b) $x_2(t) = \sum_{n=-\infty}^{\infty} \Lambda[(t - 4n)/2]$

(c) $x_3(t) = \sum_{n=-\infty}^{\infty} \Lambda[(t - 2n)/2] u(t - 2n)$

(d) $x_4(t) = | \sin (5\pi t)|$

2.14   For each of the following signals, determine both the normalized energy and power. (*Note:* 0 and ∞ are possible answers.)

(a) $x_1(t) = 8e^{(2 + j4\pi)t}$ (b) $x_2(t) = \Pi[(t - 3)/2] + \Pi(t - 3)$

(c) $x_3(t) = 8e^{j4\pi t} u(t)$ (d) $x_4(t) = 2 + 3 \cos (8\pi t)$

2.15   Show that the signal

$$x(t) = t^{-1/4} u(t - 2)$$

is neither an energy nor a power signal.

2.16   (a) Prove Equation (2.23).

     (b) Use Equation (2.23) to find the power in each of the signals given in Problem 2.13.

2.17   Show that the following are energy signals. Sketch each signal.

(a) $x_1(t) = \Pi(t/10) \cos (3\pi t)$

(b) $x_2(t) = e^{-|t|/2} \cos (10\pi t)$

(c) $x_3(t) = 2u(t) - 2u(t - 5)$

(d) $x_4(t) = \int_{-\infty}^{t} u(\lambda) \, d\lambda - 2 \int_{-\infty}^{t-10} u(\lambda) \, d\lambda + \int_{-\infty}^{t-20} u(\lambda) \, d\lambda$

Section 2.3

2.18   (a) Fill in the steps for obtaining (2.30) from (2.29).

     (b) Obtain (2.31) from (2.30).

2.19 (a) Given the set of orthogonal functions

$$\phi_n(t) = \Pi \left\{ \frac{4[t - (2n - 1)T/8]}{T} \right\}, \quad n = 1, 2, 3, 4$$

sketch and dimension accurately these functions.

(b) Approximate the ramp signal

$$x(t) = (t/T)\Pi \left[\frac{(t - T/2)}{T} \right]$$

by a generalized Fourier series using this set.

(c) Do the same for the set

$$\phi_n(t) = \Pi \left\{ \frac{2[t - (2n - 1)T/4]}{T} \right\}, \quad n = 1, 2$$

(d) Compute the integral-squared error for both part (b) and part (c). What do you conclude about the dependence of ϵ_N on N?

Section 2.4

2.20 Using the uniqueness property of the Fourier series, find exponential Fourier series for the following signals:

(a) $x_1(t) = \cos^2(\omega_0 t)$ (b) $x_2(t) = \cos(\omega_0 t) + \sin(\omega_0 t)$
(c) $x_3(t) = \sin^2(\omega_0 t) \cos(2\omega_0 t)$ (d) $x_4(t) = \cos^3(\omega_0 t)$

(*Hint:* Use appropriate trigonometric identities and Euler's theorem.)

2.21 Expand the signal $x(t) = 4|t|$ in a complex exponential Fourier series over the interval $|t| \leq 4$. Sketch the signal to which the Fourier series converges for all t.

2.22 Fill in all the steps to show that

(a) $|X_n| = |X_{-n}|$ and $\underline{/X_n} = -\underline{/X_{-n}}$ for $x(t)$ real.
(b) X_n is a real, even function of n for $x(t)$ real and even.
(c) X_n is imaginary and an odd function of n for $x(t)$ real and odd.
(d) Halfwave symmetry implies that $X_n = 0$, n even.

2.23 Obtain the complex exponential Fourier series coefficients for the (a) pulse train, (b) half-rectified sinewave, (c) full-rectified sinewave, and (d) triangular waveform given in Table 2.1.

2.24 Find the ratio of the power contained in a pulse train for $|nf_0| \leq \tau^{-1}$ to the total power for each of the following cases:

(a) $\tau/T_0 = \frac{1}{2}$ (b) $\tau/T_0 = \frac{1}{5}$

2.25   (a) If $x(t)$ has the Fourier series

$$x(t) = \sum_{n=-\infty}^{\infty} X_n e^{j2\pi n f_0 t}$$

and $y(t) = x(t - t_0)$, show that

$$Y_n = X_n e^{-j2\pi n f_0 t_0}$$

where the Y_n's are the Fourier coefficients for $y(t)$.

(b) Verify the theorem proved in part (a) by examining the Fourier coefficients for $x(t) = \cos(\omega_0 t)$ and $y(t) = \sin(\omega_0 t)$. *Hint:* Use the uniqueness property of the Fourier series.

2.26   Use the Fourier series expansion of a periodic triangular signal to find the sum of the following series:

   (a) $1 + \frac{1}{9} + \frac{1}{25} + \frac{1}{49} + \cdots$

   (b) $1 - \frac{1}{3} + \frac{1}{5} - \frac{1}{7} + \cdots$

2.27   Using the results given in Table 2.1 for the Fourier coefficients of a pulse train, plot the double-sided amplitude and phase spectra for the waveforms shown in Figure 2.35. *Hint:* Note that $x_b(t) = -x_a(t) + A$. How is a sign change and dc level shift manifested in the spectrum of the waveform?

2.28   (a) Plot the single-sided and double-sided amplitude and phase spectra of the square wave shown in Figure 2.36(a). How do they differ from those plotted in Problem 2.27 for Figure 2.35(b)?

   (b) Obtain an expression relating the complex exponential Fourier series coefficients of the triangular waveform shown in Figure 2.36(b) and those of $x_a(t)$ shown in Figure 2.36(a). *Hint:* Note that $x_a(t) = K[dx_b(t)/dt]$, where K is an appropriate scale change.

   (c) Plot the double-sided amplitude and phase spectra for $x_b(t)$.

FIGURE 2.35

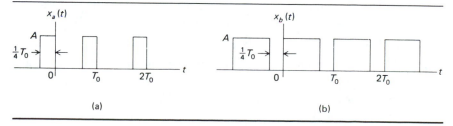

(a)   (b)

FIGURE 2.36

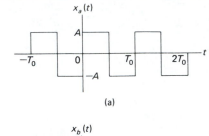

(a)

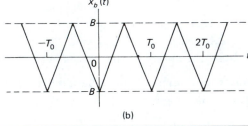

(b)

Section 2.5

2.29 Sketch each signal given below, and find its Fourier transform. Plot the amplitude and phase spectra of each signal (A and α are positive constants).

(a) $x_1(t) = Ae^{-\alpha t}u(t)$

(b) $x_2(t) = Ae^{\alpha t}u(-t)$

(c) $x_3(t) = x_1(t) - x_2(t)$

(d) Use the result of part (c) to find the Fourier transform of the signum function defined as

$$\text{sgn}(t) = \begin{cases} 1, & t > 0 \\ -1, & t < 0 \end{cases}$$

2.30 (a) Use the result of Problem 2.29(d) and the relation $u(t) = \frac{1}{2}[\text{sgn}(t) + 1]$ to find the Fourier transform of the unit step.

(b) Use the integration theorem and the Fourier transform of the unit impulse function to find the Fourier transform of the unit step. Compare the result with part (a).

2.31 Using only the Fourier transform of the unit impulse function and appropriate transform theorems, find the Fourier transforms of the signals shown in Figure 2.37.

FIGURE 2.37

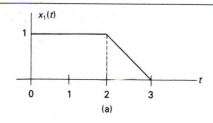

(a)

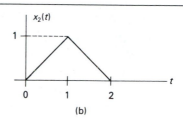

(b)

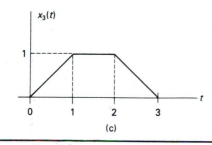

(c)

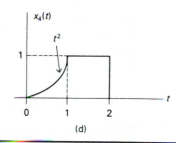

(d)

2.32   Prove the multiplication theorem of Fourier transforms, showing all the steps.

2.33   Find the Fourier transforms of the signals given below. Discuss how any symmetry properties a given signal may have affect its Fourier transform in terms of being real or purely imaginary.

(a) $x_1(t) = \delta(t + 10) + 2\delta(t) + \delta(t - 10)$
(b) $x_2(t) = \delta(t + 5) - \delta(t - 5)$
(c) $x_3(t) = \sum_{n=0}^{4} (n + 1)\delta(t - 2n)$

2.34   Find and plot the energy spectral densities of the following signals. Dimension your plots fully.

(a) $x_1(t) = 2e^{-3|t|}$ (b) $x_2(t) = 20 \, \text{sinc} \, (30t)$
(c) $x_3(t) = 4\Pi(5t)$ (d) $x_4(t) = 4\Pi(5t) \cos (40\pi t)$

2.35   Evaluate the following integrals using Rayleigh's energy theorem (Parseval's theorem for Fourier transforms).

(a) $I_1 = \int_{-\infty}^{\infty} \dfrac{df}{\alpha^2 + (2\pi f)^2}$ (b) $I_2 = \int_{-\infty}^{\infty} \text{sinc}^2 \, (\tau f) \, df$

(c) $I_3 = \int_{-\infty}^{\infty} \dfrac{df}{[\alpha^2 + (2\pi f)^2]^2}$ (d) $I_4 = \int_{-\infty}^{\infty} \text{sinc}^4 \, (\tau f) \, df$

2.36 Obtain and sketch the convolutions of the following signals.

(a) $y_1(t) = e^{-\alpha t}u(t)*\Pi(t - \tau)$, α and τ positive constants
(b) $y_2(t) = [\Pi(t/2) + \Pi(t)]*\Pi(t)$
(c) $y_3(t) = e^{-\alpha|t|}*\Pi(t)$, $\alpha > 0$
(d) $y_4(t) = x(t)*u(t)$, where $x(t)$ is any energy signal [you will have to assume a particular form for $x(t)$ to sketch this one, but obtain the general result before doing so].

2.37 Obtain the Fourier transforms of the signals $y_1(t)$, $y_2(t)$, and $y_3(t)$ in Problem 2.36 using the convolution theorem of Fourier transforms.

2.38 Given the following signals, suppose that all energy spectral components *outside* the bandwidth $|f| \leq W$ are removed by an ideal filter, while all energy spectral components within this bandwidth are kept. Find the ratio of energy to total energy kept in each case.

(a) $x_1(t) = e^{-\alpha t}u(t)$
(b) $x_2(t) = \Pi(t/\tau)$ (requires numerical integration)

2.39 (a) Find the Fourier transform of the cosine pulse

$$x(t) = A\Pi\left(\frac{2t}{T_0}\right) \cos \omega_0 t, \qquad \text{where } \omega_0 = \frac{2\pi}{T_0}$$

Express your answer in terms of the sinc function and sketch $|X(f)|$.

(b) Obtain the Fourier transform of the raised cosine pulse

$$y(t) = \tfrac{1}{2}A\Pi\left(\frac{2t}{T_0}\right)(1 + \cos 2\omega_0 t)$$

Sketch the amplitude spectrum, and compare with that of $x(t)$.

(c) Use Equation (2.139) with the result of part (a) to find the Fourier transform of the half-rectified cosine wave.

2.40 Show that the Fourier transform of $x(t) = e^{-\pi t^2}$ is $X(f) = e^{-\pi f^2}$. This is an example of a signal whose Fourier transform is of the same functional form as the signal itself.
[*Hint:* Complete the square in the exponent of the Fourier transform integral and use $\int_{-\infty}^{\infty} e^{-\alpha u^2}\, du = \sqrt{\pi/\alpha}$.]

Section 2.6

2.41 Show that the time-average correlation operation can be written in terms of convolution as

$$R(\tau) = \lim_{T \to \infty} \frac{1}{2T} [x(t) * x(-t)] \Big|_{t=\tau}$$

2.42 Obtain the time-average correlation function for a squarewave. Does it make any difference whether the squarewave has even or odd symmetry?

Section 2.7

2.43 A system is governed by the differential equation

(a) Find $H(f)$.

$$\frac{dy}{dt} + ay = b \frac{dx}{dt} + cx$$

(b) Find and plot $|H(f)|$ and $\underline{/H(f)}$ for $c = 0$.
(c) Find and plot $|H(f)|$ and $\underline{/H(f)}$ for $b = 0$.

2.44 For each of the following transfer functions, determine the unit-impulse response of the system.

(a) $H_1(f) = \dfrac{j2\pi f}{10 + j2\pi f}$ (b) $H_2(f) = \dfrac{3e^{-j6\pi f}}{4 + j20\pi f}$

2.45 A filter has transfer function $H(f) = \Pi(f/2B)$ and input $x(t) = 2W \operatorname{sinc}(2Wt)$.

(a) Find the output $y(t)$ for $W < B$.
(b) Find the output $y(t)$ for $W > B$.
(c) In which case does the output suffer distortion? What influenced your answer?

2.46 A second-order active bandpass filter (BPF), known as a *Sallen-Key cir-cuit*, is shown in Figure 2.38.

(a) Show that the transfer function of this filter is given by

$$H(j\omega) = \frac{(K\omega_0 / \sqrt{2})(j\omega)}{-\omega^2 + (\omega_0/Q)(j\omega) + \omega_0^2}, \qquad \omega = 2\pi f$$

FIGURE 2.38

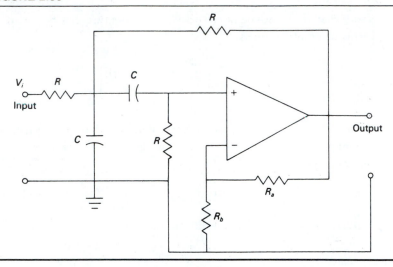

where

$$\omega_0 = \sqrt{2}(RC)^{-1}$$

$$Q = \frac{\sqrt{2}}{4 - K}$$

$$K = 1 + \frac{R_a}{R_b}$$

(b) Plot $|H(f)|$.

(c) Show that the 3-dB bandwidth of the filter can be expressed as $B = f_0/Q$, where $f_0 = \omega_0/2\pi$.

(d) Design a BPF using this circuit with center frequency $f_0 = 1000$ Hz and 3-dB bandwidth of 300 Hz. Find values of R_a, R_b, R, and C that will give these desired specifications.

2.47 For the two circuits shown in Figure 2.39, determine $H(f)$ and $h(t)$. Sketch accurately the amplitude and phase responses. Plot the amplitude response in decibels. Use a logarithmic frequency axis.

2.48 Using the Paley-Wiener criterion, show that

$$|H(f)| = e^{-\beta f^2}$$

is not a suitable amplitude response for a causal LTI.

2.49 Determine whether or not the filters shown in Figure 2.40 are BIBO stable.

FIGURE 2.39

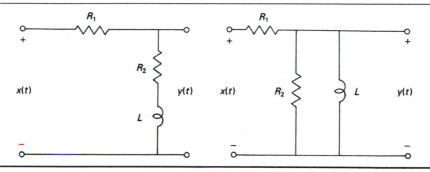

2.50 Given a filter with transfer function

$$H(f) = \frac{20}{9 + (2\pi f)^2}$$

and input $x(t) = e^{-2t}u(t)$, obtain and plot accurately the energy spectral densities of the input and output.

2.51 A filter with transfer function

$$H(f) = 3\Pi(f/26)$$

has as an input a half-rectified cosine waveform of fundamental frequency 10 Hz. Determine the output of the filter.

2.52 Another definition of bandwidth for a signal is the 90% energy containment bandwidth. For a signal with energy spectral density $G(f) = |X(f)|^2$, it is given by B_{90} in the relation

$$0.9E_{Total} = \int_{-B_{90}}^{B_{90}} G(f)\,df = 2\int_0^{B_{90}} G(f)\,df$$

FIGURE 2.40

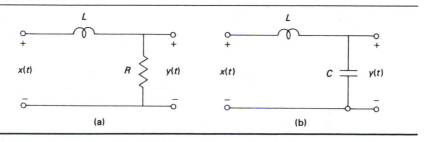

(a)   (b)

Obtain B_{90} for the following signals if it is defined. If it is not defined for a particular signal, state why it is not.

(a) $x_1(t) = \sqrt{2\alpha}\, e^{-\alpha t} u(t)$,   where α is a positive constant
(b) $x_2(t) = 2W \operatorname{sinc}(2Wt)$
(c) $x_3(t) = \Pi(t/\tau)$ (requires numerical integration)

2.53   An ideal quadrature phase shifter has

$$H(f) = \begin{cases} e^{-j\pi/2}, & f > 0 \\ e^{+j\pi/2}, & f < 0 \end{cases}$$

If the input is a squarewave, sketch an approximation for the output waveform using the first four terms of its trigonometric Fourier series as obtained in Example 2.7.

2.54   A filter has amplitude response and phase shift shown in Figure 2.41. Find the output for each of the inputs given below. For which cases is the transmission distortionless? Tell what type of distortion is imposed for the others.

(a) $\cos 50\pi t + 5 \cos 120\pi t$ (b) $\cos 120\,\pi t + 0.5 \cos 160\pi t$
(c) $\cos 120\pi t + 3 \cos 140\pi t$ (d) $2 \cos 20\pi t + 4 \cos 40\pi t$

2.55   A system has the unit impulse response

$$h(t) = 4e^{-5t}u(t)$$

Determine and accurately plot, on the same set of axes, the group delay and the phase delay.

2.56   A system has the transfer function

$$H(f) = \frac{3 + j2\pi f}{10 + j2\pi f}$$

Determine and accurately plot the group delay and the phase delay.

2.57   The nonlinear system defined by

$$y(t) = x(t) + 0.3x^2(t)$$

FIGURE 2.41

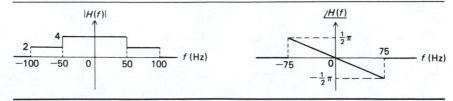

has an input signal with the bandpass spectrum

$$X(f) = 4\Pi\left(\frac{f-20}{10}\right) + 4\Pi\left(\frac{f+20}{10}\right)$$

Sketch the spectrum of the output, labeling all important frequencies and amplitudes.

2.58 (a) Consider a nonlinear device with the transfer characteristic $y(t) = x(t) + 0.1x^3(t)$. The frequency of the input signal $x(t) = \cos 2000\pi t$ is to be tripled by passing the signal through the non-linearity and then through a second-order bandpass filter with a transfer function approximated by

$$H(f) = \frac{1}{1 + j2Q(f - 3000)} + \frac{1}{1 + j2Q(f + 3000)}$$

Neglecting negative frequency contributions, compute, in terms of the parameter Q, the *total harmonic distortion* (THD) at the tripler output, defined as

$$THD = \frac{\text{total power in all output distortion terms}}{\text{power in desired output component}} \times 100\%$$

Note that the desired output component in this case is the third harmonic of the input frequency.

(b) Find the minimum value of Q that will result in THD $\leq 0.01\%$.

2.59 A nonlinear device has $y(t) = a_0 + a_1 x(t) + a_2 x^2(t) + a_3 x^3(t)$. If $x(t) = \cos \omega_1 t + \cos \omega_2 t$, list all the frequency components present in $y(t)$. Discuss the use of this device as a frequency multiplier.

2.60 (a) For a Butterworth filter, show that

$$\lim_{n \to \infty} |H_{BU}(f)| = \begin{cases} 1, & |f| < f_3 \\ 0, & |f| > f_3 \end{cases}$$

(b) For a Chebyshev filter, show that

$$\epsilon = \sqrt{10^{R_{dB}/10} - 1}$$

where R_{dB} is the ripple in dB.

2.61 Find the impulse response of an ideal highpass filter with the transfer function

$$H_{HP}(f) = H_0 \left[1 - \Pi\left(\frac{f}{2W}\right)\right] e^{-j2\pi f t_0}$$

2.62   Sketch the amplitude response and the group delay for a second-order Chebyshev filter for $\epsilon = 0.5$, 0.707, and 1.0.

2.63   Verify the pulsewidth-bandwidth relationship of Equation (2.227) for the following signals. Sketch each signal and its spectrum.

(a) $x(t) = Ae^{-t^2/2\tau^2}$ (Gaussian pulse) (see Problem 2.40)
(b) $x(t) = Ae^{-\alpha|t|}$, $\alpha > 0$ (double-sided exponential)

Section 2.8

2.64   A sinusoidal signal of frequency 1 Hz is to be sampled periodically.

(a) Find the maximum allowable time interval between samples.
(b) Samples are taken at $\frac{1}{3}$-s intervals. Show graphically, to your satisfaction, that no other sinewave or signal with bandwidth less than 1.5 Hz can be represented by these samples.
(c) The samples are spaced $\frac{2}{3}$ s apart. Show graphically that these may represent another sinewave of frequency less than 1.5 Hz.

2.65   A flat-top sampler can be represented as the block diagram of Figure 2.42.

(a) Assuming $\tau \ll T_s$, sketch the output for a typical $x(t)$.
(b) Find the spectrum of the output, $Y(f)$, in terms of the spectrum of the input, $X(f)$. What must be the relationship between τ and T_s to minimize distortion in the recovered waveform?

2.66   Figure 2.43 illustrates so-called *zero-order-hold reconstruction*.

(a) Sketch $y(t)$ for a typical $x(t)$. Under what conditions is $y(t)$ a good approximation to $x(t)$?
(b) Find the spectrum of $y(t)$ in terms of the spectrum of $x(t)$. Discuss the approximation of $y(t)$ to $x(t)$ in terms of frequency-domain arguments.

2.67   The signal $x(t) = \cos 20\pi t$ is sampled at 15 samples per second. Show that the signal $\cos 10\pi t$ impersonates it as far as sample values are concerned.

FIGURE 2.42

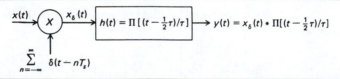

FIGURE 2.43

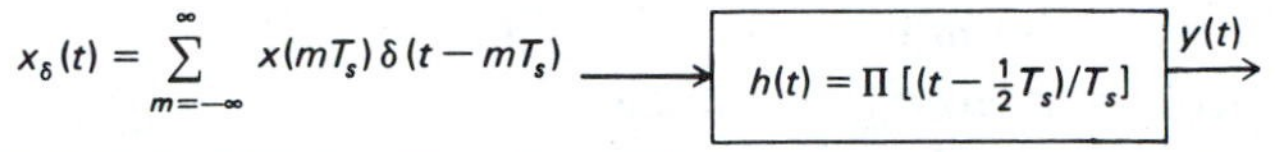

$x_\delta(t) = \sum_{m=-\infty}^{\infty} x(mT_s)\,\delta(t - mT_s)$
$h(t) = \Pi\,[(t - \frac{1}{2}T_s)/T_s]$
$y(t)$

2.68 Determine the range of permissible cutoff frequencies for the ideal lowpass filter used to reconstruct the signal

$$x(t) = 10 \cos(600\pi t) \cos^2(1600\pi t)$$

which is sampled at 4000 samples per second. Sketch $X(f)$ and $X_\delta(f)$. Find the minimum allowable sampling frequency.

2.69 Given the bandpass signal spectrum shown in Figure 2.44, sketch spectra for the following sampling rates f_s and indicate which ones are suitable.

(a) $2B$ (b) $2.5B$ (c) $3B$
(d) $4B$ (e) $5B$ (f) $6B$

Section 2.9

2.70 Using the integral expression for the Hilbert transform, Equation (2.256), show that

$$\widehat{\cos \omega_0 t} = \sin \omega_0 t$$

2.71 Using appropriate Fourier transform theorems and pairs, express the spectrum $Y(f)$ of

$$y(t) = x(t) \cos \omega_0 t + \hat{x}(t) \sin \omega_0 t$$

in terms of the spectrum $X(f)$ of $x(t)$, where $X(f)$ is lowpass with bandwidth

$$B < f_0 = \frac{\omega_0}{2\pi}$$

Sketch $Y(f)$ for a typical $X(f)$.

FIGURE 2.44

X(f)
A
−3B
−2B
−B
0
B
2B
3B
f (Hz)

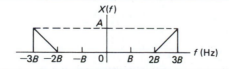

2.72 Show that $x(t)$ and $\hat{x}(t)$ are orthogonal for the following signals:

    (a) $x_a(t) = \cos \omega_0 t$

    (b) $x_b(t) = 2 \cos \omega_0 t + \sin \omega_0 t \cos^2 2\omega_0 t$

    (c) $x_c(t) = A\Pi(t/\tau)$

2.73 (a) Using the results of Problem 2.71, sketch $Y(f)$ for $x(t) = \cos 10\pi t$ and $f_0 = \omega_0/2\pi = 20$ Hz.

    (b) Obtain the spectrum for the same signal by using trigonometric identities and Equations (2.260) and (2.261).

2.74 Assume that the Fourier transform of $x(t)$ is real and has the shape shown in Figure 2.45. Determine and plot the spectrum of each of the following signals:

    (a) $x_1(t) = \frac{3}{4}x(t) + \frac{1}{4j}\hat{x}(t)$

    (b) $x_2(t) = [\frac{3}{4}x(t) + \frac{3}{4j}\hat{x}(t)]e^{j2\pi f_0 t}, \qquad f_0 \gg W$

    (c) $x_3(t) = [\frac{3}{4}x(t) + \frac{1}{4j}\hat{x}(t)]e^{j2\pi Wt}$

    (d) $x_4(t) = [\frac{3}{4}x(t) - \frac{1}{4j}\hat{x}(t)]e^{j\pi Wt}$

2.75 Consider the signal

$$x(t) = 2W \operatorname{sinc} (2Wt) \cos (2\pi f_0 t)$$

    (a) Obtain and sketch the spectrum of $x_p(t) = x(t) + j\hat{x}(t)$.

    (b) Obtain and sketch the spectrum of the complex envelope, $\tilde{x}(t)$, where the complex envelope is defined by (2.286).

    (c) Find the complex envelope $\tilde{x}(t)$.

2.76 Consider the input

$$x(t) = \Pi(t/\tau) \cos [2\pi(f_0 + \Delta f)t], \qquad \Delta f \ll f_0$$

to a filter with impulse response

$$h(t) = \alpha e^{-\alpha t} \cos (2\pi f_0 t)u(t)$$

Find the output using complex envelope techniques.

FIGURE 2.45

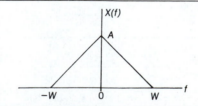

Section 2.10

2.77 Prove that (2.303) and (2.304) are indeed inverses of each other by substituting (2.303) into (2.304), reversing the orders of summation, and using the sum formula for a geometric series to show that the final result is $x_n = x_n$.

COMPUTER EXERCISES*

1. (a) Write a computer program to obtain the generalized Fourier series for an energy signal using the orthonormal basis set

$$\Phi_n(t) = \Pi(t - 0.5 - n), \quad n = 0, 1, 2, \cdot \cdot \cdot, T - 1, T \text{ integer}$$

where the signal extent is $(0, T)$ with T assumed to be integer-valued. Your program should compute the generalized Fourier coefficients, the integral-squared error, and make a plot of the signal being approximated and the approximating waveform. Test your program with the signal $e^{-2t}u(t), 0 \le t \le 5$.

   (b) Repeat part (a) with the orthonormal basis set

$$\Phi_n(t) = \sqrt{2}\Pi\left(\frac{t - 0.5 - n}{0.5}\right), \quad n = 0, 1, 2, \cdot \cdot \cdot, 2T - 1, T \text{ integer}$$

What is the integral-squared error now?

   (c) Can you deduce whether or not the basis set resulting from repeatedly halving the pulse width and doubling the amplitude is complete?

2. Write a computer program to evaluate the coefficients of the complex exponential Fourier series of a signal. Include a plot of the amplitude and phase spectrum of the signal for which the Fourier series coefficients are evaluated. Check by evaluating the Fourier series coefficients of a squarewave. Plot the squarewave approximation by summing the series through the seventh harmonic.

3. (a) Write a computer program to evaluate the coefficients of the complex exponential Fourier series of a signal by using the fast Fourier

---

* When doing these computer exercises, we suggest that the student make use of a mathematics package such as Mathcad, MATLAB, or Mathematica. Although the resulting programs may run slower than those written in C, Fortran, or some other high-level language, considerable time will be saved in being able to use the plotting capability of these mathematics packages.

transform (FFT). Note that you do not have to write a program to evaluate the FFT if the suggestion in the footnote is followed. Check it by evaluating the Fourier series coefficients of a squarewave and comparing your results with Computer Exercise 2.

(b) How would you use the same approach to evaluate the Fourier transform of a pulse-type signal. How do the two outputs differ? Compute an approximation to the Fourier transform of a square pulse signal 1 unit wide and compare with the theoretical result.

4. (a) Write a computer program to find the bandwidth of a lowpass energy signal that contains a certain specified percentage of its total energy, for example, 95%. In other words, write a program to find W in the equation

$$E_W = \frac{\displaystyle\int_0^W G_x(f)df}{\displaystyle\int_0^\infty G_x(f)df} \times 100\%$$

with E_W set equal to a specified value, where $G_x(f)$ is the energy spectral density of the signal.

(b) Write a computer program to find the time duration of a lowpass energy signal that contains a certain specified percentage of its total energy, for example, 95%. In other words, write a program to find T in the equation

$$E_T = \frac{\displaystyle\int_0^T |x(t)|^2\, dt}{\displaystyle\int_0^\infty |x(t)|^2\, dt} \times 100\%$$

with E_T set equal to a specified value, where it is assumed that the signal is zero for $t < 0$.

(c) Use the program you developed in parts (a) and (b) to check the pulse-duration bandwidth relationship (2.227).

5. Write a computer program to evaluate and plot the amplitude and phase response functions for a given transfer function. Use it to plot the amplitude and phase response functions for the Sallen-Key circuit of Problem 2.46 with the parameters chosen to realize a second-order Butterworth filter.

6. Using the FFT, write a program to find the output of a bandpass system for an arbitrary bandpass input using the complex envelope approach summarized by (2.295) and (2.296). Your program should include plots of the envelopes of the input and output as well as their spectra.

3 BASIC MODULATION TECHNIQUES

Before an information-bearing signal is transmitted through a communication channel, some type of modulation process is typically utilized to produce a signal that can easily be accommodated by the channel. In this chapter we will discuss various types of modulation techniques. The modulation process commonly translates an information-bearing signal, usually referred to as the *message signal,* to a new spectral location. For example, if the signal is to be transmitted through the atmosphere or free space, frequency translation is necessary to raise the signal spectrum to a frequency that can be radiated efficiently with antennas of reasonable size. If more than one signal utilizes a channel, modulation allows translation of the different signals to different spectral locations, thus allowing the receiver to select the desired signal. Multiplexing allows two or more message signals to be transmitted by a single transmitter and received by a single receiver simultaneously.

The logical choice of a modulation technique for a specific application is influenced by the characteristics of the message signal, the characteristics of the channel, the performance desired from the overall communication system, the use to be made of the transmitted data, and the economic factors that are always important in practical applications. This text will place emphasis on the first three of these factors. Through experience, the communications engineer develops sufficient insight to place the last two factors in proper perspective for a particular application.

The two basic types of analog modulation are continuous-wave modulation and pulse modulation. In *continuous-wave modulation,* a parameter of a high-frequency carrier is varied proportionally to the message signal such that a one-to-one correspondence exists between the parameter and the message signal. The carrier is usually assumed to be sinusoidal, but as will be illustrated, this is not a necessary restriction. For a sinusoidal carrier, a general modulated carrier can be represented mathematically as

$$x_c(t) = A(t) \cos [\omega_c t + \phi(t)]$$ (3.1)

where ω_c is known as the *carrier frequency*. Since a sinusoid is completely specified by its amplitude and argument, it follows that once the carrier frequency is specified, only two parameters are candidates to be varied in the modulation process: the instantaneous amplitude $A(t)$ and the instantaneous phase deviation $\phi(t)$. When the instantaneous amplitude $A(t)$ is linearly related to the modulating signal, the result is *linear modulation*. Letting $\phi(t)$ or the time derivative of $\phi(t)$ be linearly related to the modulating signal yields phase or frequency modulation, respectively. Collectively, phase and frequency modulation are referred to as *angle modulation*, since the phase angle of the modulated carrier conveys the information.

In *analog pulse modulation*, the message waveform is sampled at discrete time intervals and the amplitude, width, or position of a pulse is varied in one-to-one correspondence with the values of the samples. Since the samples are taken at discrete times, the periods between the samples are available for other uses, such as insertion of samples from other message signals. This is referred to as *time-division multiplexing*. If the value of each sample is quantized and encoded, *pulse-code modulation results*. We also briefly consider *delta modulation*. Pulse-code modulation and delta modulation are digital rather than analog modulation techniques, but are considered in this chapter for completeness.

3.1   LINEAR MODULATION

A general linearly modulated carrier is represented by setting the instantaneous phase deviation $\phi(t)$ in (3.1) equal to zero. Thus a linearly modulated carrier is represented by

$$x_c(t) \;=\; A(t) \cos \omega_c t \tag{3.2}$$

in which the carrier amplitude $A(t)$ varies in one-to-one correspondence with the message signal. We next discuss several different types of linear modulation as well as techniques that can be used for demodulation.

Double-Sideband Modulation

Double-sideband (DSB) modulation results when $A(t)$ is proportional to the message signal $m(t)$. Thus the output of a DSB modulator can be represented as

$$x_c(t) \;=\; A_c m(t) \cos \omega_c t \tag{3.3}$$

which illustrates that DSB modulation is simply the multiplication of a carrier, $A_c \cos \omega_c t$, by the message signal. It follows from the modulation theorem that the spectrum of a DSB signal is given by

$$X_c(f) = \tfrac{1}{2}A_cM(f + f_c) + \tfrac{1}{2}A_cM(f - f_c), \qquad f_c = \frac{\omega_c}{2\pi} \qquad (3.4)$$

The process of DSB modulation is described in Figure 3.1. Figure 3.1(a) illustrates a DSB system and shows that a DSB signal is demodulated by multiplying the received signal, denoted by $x_r(t)$, by the demodulation carrier $2\cos\omega_c t$ and lowpass filtering. For the idealized system that we are considering here, the received signal, $x_r(t)$, is identical to the transmitted signal, $x_c(t)$. The output of the multiplier is

$$d(t) = 2A_c[m(t)\cos\omega_c t]\cos\omega_c t \qquad (3.5)$$

or

$$d(t) = A_cm(t) + A_cm(t)\cos 2\omega_c t \qquad (3.6)$$

where we have used the trigonometric identity $2\cos^2 x = 1 + \cos 2x$.

The time-domain signals are shown in Figure 3.1(b) for an assumed $m(t)$. The message signal $m(t)$ forms the envelope of $x_c(t)$. The waveform for $d(t)$ can be best understood by realizing that since $\cos^2\omega_c t$ is nonnegative for all t, $d(t)$ is positive if $m(t)$ is positive and $d(t)$ is negative if $m(t)$ is negative. Also note that $m(t)$ (appropriately scaled) forms the envelope of $d(t)$ and that the frequency of the sinusoid under the envelope is $2f_c$ rather than f_c.

The spectra of the signals $m(t)$, $x_c(t)$, and $d(t)$ are shown in Figure 3.1(c) for an assumed $M(f)$ having a bandwidth W. The spectra $M(f + f_c)$ and $M(f - f_c)$ are simply the message spectrum translated to $f = \pm f_c$. The portion of $M(f - f_c)$ above the carrier frequency is called the *upper sideband* (USB), and the portion below the carrier frequency is called the *lower sideband* (LSB). Since the carrier frequency f_c is typically much greater than the bandwidth of the message signal W, the spectra of the two terms in $d(t)$ do not overlap. Thus $d(t)$ can be lowpass-filtered and amplitude-scaled by A_c to yield the demodulated output $y_D(t)$. In practice, any amplitude scaling factor can be used since, as we saw in Chapter 2, multiplication by a constant does not

FIGURE 3.1 Double-sideband (DSB) modulation. (a) System. (b) Waveforms. (c) Spectra.

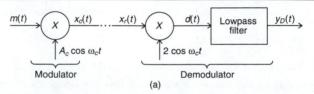

(a)

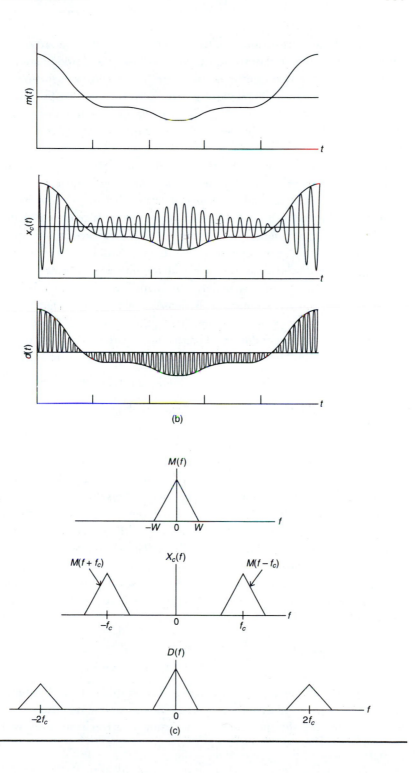

(b)

(c)

induce amplitude distortion. Thus, for convenience, A_c is usually set equal to unity at the demodulator output. For this case, the demodulated output $y_D(t)$ will equal the message signal $m(t)$. The lowpass filter that removes the term at $2f_c$ must have a bandwidth greater than or equal to the bandwidth of the message signal W. We will see in Chapter 6 that when noise is present, this lowpass filter, known as the *postdetection* filter, should have the smallest possible bandwidth.

The basic difficulty with the use of DSB modulation is the need for a demodulation carrier at the receiver that has the same frequency as the original modulation carrier and is phase-coherent with it. Demodulation that requires a coherent reference is known as *synchronous* or *coherent demodulation*. The derivation of a coherent demodulation carrier requires careful attention, for if the demodulation carrier is out of synchronism by even a small amount, serious distortion of the demodulated message waveform can result. This effect will be thoroughly analyzed in Chapter 6, but a simplified analysis can be carried out by assuming a demodulation carrier in Figure 3.1(a) of the form $2 \cos [\omega_c t + \theta(t)]$, where $\theta(t)$ is a time-varying phase error. By making use of the trigonometric identity $\cos x \cos y = \frac{1}{2} \cos (x + y) + \frac{1}{2} \cos (x - y)$, it can be shown that this choice of demodulation carrier yields

$$d(t) = A_c m(t) \cos \theta(t) + A_c m(t) \cos [2\omega_c t + \theta(t)] \qquad (3.7)$$

which, after lowpass filtering and amplitude scaling, becomes

$$y_D(t) = m(t) \cos \theta(t) \qquad (3.8)$$

assuming, once again, that the spectra of the two terms of $d(t)$ do not overlap. If the phase error $\theta(t)$ is a constant, the effect of the phase error is an attenuation of the demodulated message signal. This does not represent distortion, since the effect of the phase error can be removed by amplitude scaling unless $\theta(t)$ is exactly 90°. However, if $\theta(t)$ is time-varying in an unknown and unpredictable manner, the effect of the phase error can be serious distortion of the demodulated output.

There are several techniques that can be utilized to generate a coherent demodulation carrier. For one technique, the first step is to square the received DSB signal, which yields

$$\begin{aligned} x_r^2(t) &= A_c^2 m^2(t) \cos^2 \omega_c t \\ &= \tfrac{1}{2} A_c^2 m^2(t) + \tfrac{1}{2} A_c^2 m^2(t) \cos 2\omega_c t \end{aligned} \qquad (3.9)$$

If $m(t)$ is a power signal, $m^2(t)$ has a nonzero dc value. Thus, by the modulation theorem, $x_r^2(t)$ has a discrete frequency component at $2\omega_c$, which can be extracted from the spectrum of $x_r^2(t)$ using a narrowband bandpass filter.

The frequency of this component can be halved to yield the desired demodulation carrier. Later we will discuss a convenient technique for implementing the required frequency divider.

A very convenient technique for demodulating DSB signals is to use a Costas phase-lock loop. This system will be presented in Section 3.4.

The analysis of DSB illustrates that the spectrum of a DSB signal does not contain a discrete spectral component at the carrier frequency unless $m(t)$ has a dc component. For this reason, DSB systems with no carrier frequency component present are often referred to as *suppressed carrier systems*. However, if a carrier component is transmitted along with the DSB signal, demodulation can be simplified. The received carrier component can be extracted using a narrowband bandpass filter and can be used as the demodulation carrier. Alternatively, if the carrier amplitude is sufficiently large, the need for generating a demodulation carrier can be completely avoided. This naturally leads to the subject of amplitude modulation.

Amplitude Modulation

Amplitude modulation (AM) results when a dc bias A is added to $m(t)$ prior to the modulation process. The result of the dc bias is that a carrier component is present in the transmitted signal. For AM, the transmitted signal is defined as

$$x_c(t) = [A + m(t)]A_c' \cos \omega_c t \tag{3.10}$$

or

$$x_c(t) = A_c[1 + am_n(t)] \cos \omega_c t \tag{3.11}$$

In (3.11), $A_c = AA_c'$, $m_n(t)$ is the message signal normalized such that the minimum value of $m_n(t)$ is -1, and the parameter a is known as the modulation index.* We shall assume that $m(t)$ has zero dc value so that the carrier component in the transmitted signal arises entirely from the bias, A. The time-domain representation of AM is illustrated in Figures 3.2(a) and 3.2(b), and the block diagram of the modulator for producing AM is shown in Figure 3.2(c).

The fact that coherent demodulation can be used for AM is easily shown and is left to the problems (see Problem 3.2). The advantage of AM over DSB, however, is that a very simple technique, known as *envelope detection*, can be used if sufficient carrier power is transmitted. An envelope detector

---

* The parameter a as used here is sometimes called the *negative modulation factor*. Also, the quantity $a \times (100\%)$ is often referred to as the *percent modulation*.

FIGURE 3.2 AM modulation. (a) Message signal. (b) Modulator output for $a < 1$. (c) Modulator.

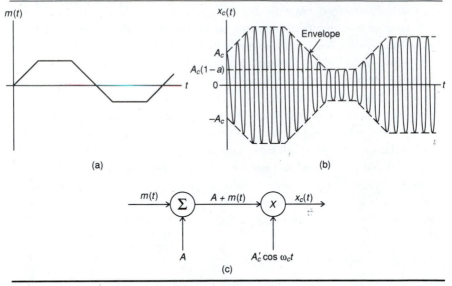

is implemented as shown in Figure 3.3(a). It can be seen from Figure 3.3(b) that as the carrier frequency is increased, the envelope, defined as $A_c[1 + am_n(t)]$, becomes easier to observe. It also follows from observation of Figure 3.3(b) that if $A_c[1 + am_n(t)]$ goes negative, distortion will result. The modulation index is defined such that if $a = 1$, the minimum value of $A_c[1 + am_n(t)]$ is zero. We therefore define

$$m_n(t) = \frac{m(t)}{|\min m(t)|} \tag{3.12}$$

and

$$a = \frac{|\min m(t)|}{A} \tag{3.13}$$

Clearly, any value of index less than 1 will result in $A_c[1 + am_n(t)] > 0$ for all t. It should be pointed out that most practical information-bearing message signals, such as speech or music signals, have nearly symmetrical maximum and minimum values.

In order for the envelope-detection process to operate properly, the RC time constant of the detector, shown in Figure 3.3(a), must be chosen carefully. The appropriate value for the time constant is related to the carrier frequency and to the bandwidth of $m(t)$. In practice, satisfactory operation requires a carrier frequency of at least 10 times the bandwidth of $m(t)$, W. Also, the cutoff frequency of the RC circuit must lie between f_c and W and must be well separated from both. This is illustrated in Figure 3.3(c).

All information in the modulator output is contained in the sidebands. Thus the carrier component of (3.11), $A_c \cos \omega_c t$, is wasted power as far as information transfer is concerned. This fact can be of considerable importance in an environment where power is limited and can completely preclude the use of AM as a modulation technique.

The total (normalized) power contained in the AM modulator output is

$$\langle x_c^2(t) \rangle = \langle [A + m(t)]^2 (A_c')^2 \cos^2 \omega_c t \rangle \tag{3.14}$$

where $\langle\ \rangle$ denotes the time average value. If $m(t)$ is slowly varying with respect to $\cos \omega_c t$, then

$$\langle x_c^2(t) \rangle = [A^2 + 2A\langle m(t)\rangle + \langle m^2(t)\rangle]\tfrac{1}{2}(A_c')^2 \tag{3.15a}$$

which reduces to

$$\langle x_c^2(t) \rangle = [A^2 + \langle m^2(t)\rangle]\tfrac{1}{2}(A_c')^2 \tag{3.15b}$$

FIGURE 3.3   Envelope detection. (a) Circuit. (b) Waveforms. (c) Effect of RC time constant.

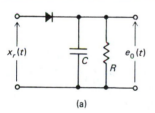

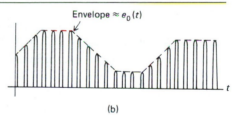

(a)   (b)

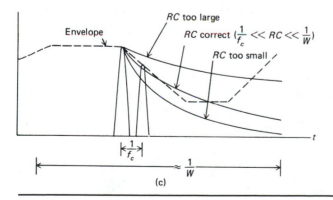

(c)

if the time average value of $m(t)$ is assumed zero. The *efficiency*, E, is defined as the percentage of total power that conveys information. Thus

$$E = \frac{\langle m^2(t)\rangle}{A^2 + \langle m^2(t)\rangle} \ (100\%) \tag{3.16}$$

Since $m(t) = aAm_n(t)$, efficiency can be written in terms of the modulation index as

$$E = \frac{a^2\langle m_n^2(t)\rangle}{1 + a^2\langle m_n^2(t)\rangle} \ (100\%) \tag{3.17}$$

If the message signal has symmetrical maximum and minimum values, such that $|\min m(t)|$ and $|\max m(t)|$ are equal, then $\langle m_n^2(t)\rangle \leq 1$. It follows that for $a \leq 1$, the maximum efficiency is 50% and is achieved for squarewave-type message signals. If $m(t)$ is a sinewave, $\langle m_n^2(t)\rangle = \frac{1}{2}$ and the efficiency is 33.3% for $a = 1$. Efficiency obviously declines rapidly as the index is reduced below unity. If the message signal does not have symmetrical maximum and minimum values, then higher values of efficiency can be achieved (see Problem 3.6).

The main advantage of AM is that since a coherent reference is not needed for demodulation, the demodulator becomes simple and inexpensive. In many applications, such as commercial radio, this fact alone is sufficient to justify its use.

The AM modulator output $x_c(t)$ is shown in Figure 3.4 for three values of the modulation index: $a = 0.5$, $a = 1.0$, and $a = 1.5$. The message signal $m(t)$ is assumed to be a unity amplitude sinusoid with a frequency of 1 Hz. A unity amplitude carrier is also assumed. The envelope detector output, $e_o(t)$, as identified in Figure 3.3, is also shown for each value of the modulation index. Note that for $a = 0.5$ the envelope is always positive. For $a = 1.0$ the envelope reaches zero. Thus, envelope detection can be used for both of these cases. For $a = 1.5$ the envelope goes negative and $e_o(t)$, which is the absolute value of the envelope, is a badly distorted replica of the message signal.

▼ **EXAMPLE 3.1**   Let us determine the efficiency and the output spectrum for an AM modulator operating with a modulation index of 0.5. The carrier power is 50 W, and the message signal is

$$m(t) = 4 \cos\left[2\pi f_m t - \frac{\pi}{9}\right] + 2 \sin[4\pi f_m t] \tag{3.18}$$

FIGURE 3.4 Modulated carrier and envelope detector outputs for various values of the modulation index. (a) *a* = 0.5. (b) *a* = 1.0. (c) *a* = 1.5.

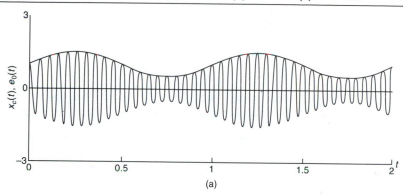

(a)

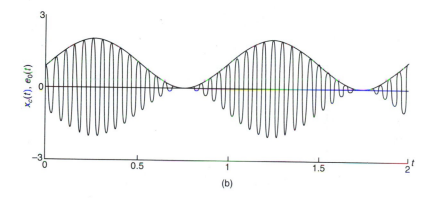

(b)

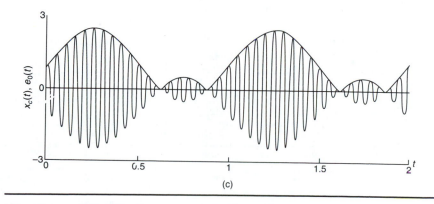

(c)

SOLUTION The first step is to determine the minimum value of $m(t)$. The message signal is shown in Figure 3.5(a). The minimum value of $m(t)$ is -4.364 and falls at $f_m t = 0.435$, as shown. The normalized message signal is given by

$$m_n(t) = \frac{1}{4.364}\left\{4\cos\left[2\pi f_m t - \frac{\pi}{9}\right] + 2\sin\left[4\pi f_m t\right]\right\} \qquad (3.19)$$

FIGURE 3.5 Waveform and spectra for Example 3.1 (a) Message signal. (b) Amplitude spectrum of modulator output. (c) Phase spectrum of modulator output.

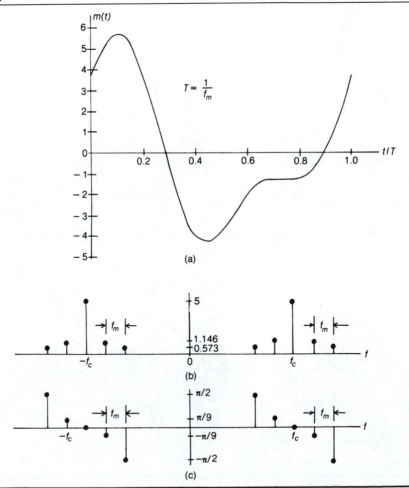

(a)

(b)

(c)

or

$$m_n(t) = 0.9166 \cos\left[2\pi f_m t - \frac{\pi}{9}\right] + 0.4583 \sin\left[4\pi f_m t\right] \qquad (3.20)$$

The mean-square value of $m_n(t)$ is

$$\langle m_n{}^2(t)\rangle = \tfrac{1}{2}(0.9166)^2 + \tfrac{1}{2}(0.4583)^2 = 0.5251 \qquad (3.21)$$

Thus the efficiency is

$$E = \frac{(0.25)(0.5251)}{1 + (0.25)(0.5251)} \qquad (3.22)$$

which yields

$$E = 11.60\% \qquad (3.23)$$

Since the carrier power is 50 W, we have

$$\tfrac{1}{2}(A_c)^2 = 50 \qquad (3.24)$$

from which

$$A_c = 10 \qquad (3.25)$$

Also, since $\sin x = \cos(x - \pi/2)$, we can write $x_c(t)$ as

$$x_c(t) = 10\left\{1 + 0.5\left[0.9166 \cos\left(2\pi f_m t - \frac{\pi}{9}\right)\right.\right.$$
$$\left.\left. + 0.4583 \cos\left(4\pi f_m t - \frac{\pi}{2}\right)\right]\right\} \cos(2\pi f_c t) \qquad (3.26)$$

In order to plot the spectrum of $x_c(t)$, we write the preceding equation as

$$x_c(t) = 10\cos(2\pi f_c t)$$
$$+ 2.292\left\{\cos\left[2\pi(f_c + f_m)t - \frac{\pi}{9}\right]\right.$$
$$\left. + \cos\left[2\pi(f_c - f_m)t + \frac{\pi}{9}\right]\right\}$$
$$+ 1.146\left\{\cos\left[2\pi(f_c + 2f_m)t - \frac{\pi}{2}\right]\right.$$
$$\left. + \cos\left[2\pi(f_c - 2f_m)t + \frac{\pi}{2}\right]\right\} \qquad (3.27)$$

Parts (b) and (c) of Figure 3.5 show the amplitude and phase spectra of $x_c(t)$. Note that the amplitude spectrum has even symmetry about the carrier frequency and that the phase spectrum has odd symmetry about the carrier frequency for $f > 0$ or $f < 0$. Of course, the overall amplitude spectrum is even about $f = 0$, and the overall phase spectrum is odd about $f = 0$.

Single-Sideband Modulation

In our development of DSB, we saw that the upper sideband (USB) and lower sideband (LSB) have even amplitude and odd phase symmetry about the carrier frequency. Thus transmission of both sidebands is not necessary, since either sideband contains sufficient information to reconstruct the message signal $m(t)$. Elimination of one of the sidebands prior to transmission results in single sideband (SSB), which reduces the bandwidth of the modulator output from $2W$ to W, where W is the bandwidth of $m(t)$. However, this bandwidth savings is accompanied by a considerable increase in complexity.

On the following pages, two different methods are used to derive the time-domain expression for the signal at the output of an SSB modulator. Although the two methods are equivalent, they do present different view-

FIGURE 3.6   Generation of SSB by sideband filtering. (a) SSB modulator. (b) Spectra (single-sided).

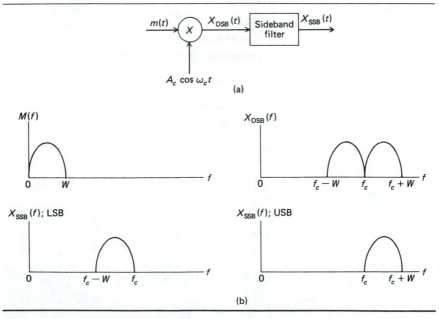

points. In the first method, the transfer function of the filter used to generate an SSB signal from a DSB signal is derived using the Hilbert transform. The second method derives the SSB signal directly from $m(t)$ using the results illustrated in Figure 2.30 and the frequency-translation theorem.

The generation of an SSB signal by sideband filtering is illustrated in Figure 3.6. First, a DSB signal, $x_{\mathrm{DSB}}(t)$, is formed. Sideband filtering of the DSB signal then yields an upper-sideband or a lower-sideband SSB signal, depending on the filter passband selected.

The filtering process that yields lower-sideband SSB is illustrated in detail in Figure 3.7. A lower-sideband SSB signal can be generated by passing a DSB signal through an ideal filter that passes the lower sideband and rejects the upper sideband. It follows from Figure 3.7(b) that the transfer function of this filter is

$$H_L(f) = \tfrac{1}{2}[\mathrm{sgn}\ (f + f_c) - \mathrm{sgn}\ (f - f_c)] \tag{3.28}$$

Since the Fourier transform of a DSB signal is

$$X_{\mathrm{DSB}}(f) = \tfrac{1}{2}A_c M(f + f_c) + \tfrac{1}{2}A_c M(f - f_c) \tag{3.29}$$

the transform of the lower-sideband SSB signal is

$$X_c(f) = \tfrac{1}{4}A_c[M(f + f_c)\ \mathrm{sgn}\ (f + f_c) + M(f - f_c)\ \mathrm{sgn}\ (f + f_c)]$$
$$- \tfrac{1}{4}A_c[M(f + f_c)\ \mathrm{sgn}\ (f - f_c) + M(f - f_c)\ \mathrm{sgn}\ (f - f_c)] \tag{3.30}$$

which is

$$X_c(f) = \tfrac{1}{4}A_c[M(f + f_c) + M(f - f_c)]$$
$$+ \tfrac{1}{4}A_c[M(f + f_c)\ \mathrm{sgn}\ (f + f_c) - M(f - f_c)\ \mathrm{sgn}\ (f - f_c)] \tag{3.31}$$

From our study of DSB, we know that

$$\tfrac{1}{2}A_c m(t)\ \cos\ \omega_c t \leftrightarrow \tfrac{1}{4}A_c[M(f + f_c) + M(f - f_c)] \tag{3.32}$$

and from our study of Hilbert transforms in Chapter 2, we recall that

$$\hat{m}(t) \leftrightarrow -j\ \mathrm{sgn}\ (f)M(f)$$

By the frequency-translation theorem, we have

$$m(t)e^{\pm j2\pi f_c t} \leftrightarrow M(f \mp f_c) \tag{3.33}$$

Replacing $m(t)$ by $\hat{m}(t)$ in the foregoing yields

$$\hat{m}(t)e^{\pm j2\pi f_c t} \leftrightarrow -jM(f \mp f_c)\ \mathrm{sgn}\ (f \mp f_c) \tag{3.34}$$

FIGURE 3.7 Generation of lower-sideband SSB. (a) Sideband filtering process. (b) Generation of lower-sideband filter.

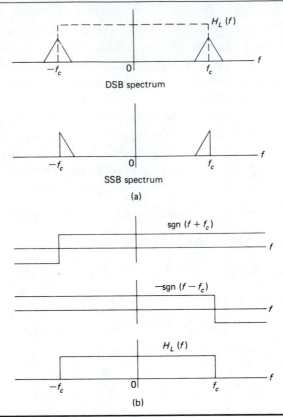

Thus

$$\mathscr{F}^{-1}\{\tfrac{1}{4}A_c[M(f+f_c)\text{ sgn }(f+f_c) - M(f-f_c)\text{ sgn }(f-f_c)]\}$$

$$= -A_c\frac{1}{4j}\hat{m}(t)e^{-j2\pi f_c t} + A_c\frac{1}{4j}\hat{m}(t)e^{+j2\pi f_c t}$$

$$= \tfrac{1}{2}A_c\hat{m}(t)\sin \omega_c t \qquad\qquad (3.35)$$

Combining (3.32) and (3.35), we get the general form of a lower-sideband SSB signal:

$$x_c(t) = \tfrac{1}{2}A_c m(t)\cos \omega_c t + \tfrac{1}{2}A_c\hat{m}(t)\sin \omega_c t \qquad (3.36)$$

A similar development can be carried out for upper-sideband SSB. The result is

$$x_c(t) = \tfrac{1}{2}A_c m(t) \cos \omega_c t - \tfrac{1}{2}A_c \hat{m}(t) \sin \omega_c t \qquad (3.37)$$

which shows that LSB and USB modulators have the same defining equations except for the sign of the term representing the Hilbert transform of the modulation. Observation of the spectrum of an SSB signal illustrates that SSB systems do not have dc response.

The generation of SSB by the method of sideband filtering the output of DSB modulators requires the use of filters that are very nearly ideal if low-frequency information is contained in $m(t)$. These filters are typically imple-mented using crystal filter technology.

Another method for generating an SSB signal, known as *phase-shift modu-lation,* is illustrated in Figure 3.8. This system is a term-by-term realization of (3.36) or (3.37). Like the ideal filters required for sideband filtering, the ideal wideband phase shifter, which performs the Hilbert transforming oper-ation, is impossible to implement exactly. However, since the frequency at which the discontinuity occurs is $f = 0$ instead of $f = f_c$, ideal phase shifters can be closely approximated.

An alternative derivation of $x_c(t)$ for an SSB signal is based on the discus-sion surrounding Figure 2.30. As shown in Figure 3.9, the positive-frequency portion of $M(f)$ is given by

$$M_p(f) = \tfrac{1}{2}\mathscr{F}\{m(t) + j\hat{m}(t)\} \qquad (3.38)$$

and the negative-frequency portion of $M(f)$ is given by

$$M_n(f) = \tfrac{1}{2}\mathscr{F}\{m(t) - j\hat{m}(t)\} \qquad (3.39)$$

FIGURE 3.8   Phase-shift modulator

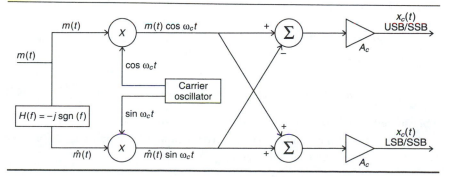

FIGURE 3.9 Alternative derivation of SSB signals. (a) $M(f)$, $M_p(f)$, and $M_n(f)$. (b) Upper-sideband SSB signal. (c) Lower-sideband SSB signal.

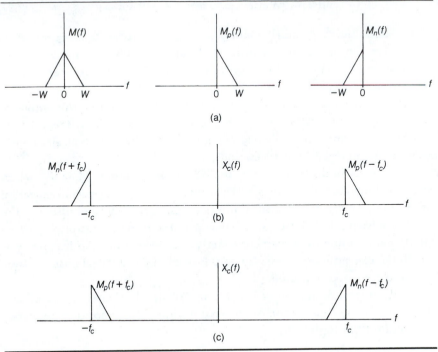

By definition, an upper-sideband SSB signal is given in the frequency domain by

$$X_c(f) = \tfrac{1}{2}A_c M_p(f - f_c) + \tfrac{1}{2}A_c M_n(f + f_c) \tag{3.40}$$

Inverse Fourier-transforming yields

$$x_c(t) = \tfrac{1}{4}A_c[m(t) + j\hat{m}(t)]e^{j2\pi f_c t} + \tfrac{1}{4}A_c[m(t) - j\hat{m}(t)]e^{-j2\pi f_c t} \tag{3.41}$$

which is

$$x_c(t) = \tfrac{1}{4}A_c m(t)[e^{j2\pi f_c t} + e^{-j2\pi f_c t}] + j\tfrac{1}{4}A_c \hat{m}(t)[e^{j2\pi f_c t} - e^{-j2\pi f_c t}] \tag{3.42}$$

The preceding expression is clearly equivalent to (3.37).

The lower-sideband SSB signal is derived in a similar manner. By definition, for a lower-sideband SSB signal,

$$X_c(f) = \tfrac{1}{2}A_c M_p(f + f_c) + \tfrac{1}{2}A_c M_n(f - f_c) \tag{3.43}$$

This becomes, after inverse Fourier-transforming,

$$x_c(t) = \tfrac{1}{4}A_c[m(t) + j\hat{m}(t)]e^{-j2\pi f_c t} + \tfrac{1}{4}A_c[m(t) - j\hat{m}(t)]e^{j2\pi f_c t} \quad (3.44)$$

which can be written as

$$x_c(t) = \tfrac{1}{4}A_c m(t)[e^{j2\pi f_c t} + e^{-j2\pi f_c t}] - j\tfrac{1}{4}A_c \hat{m}(t)[e^{j2\pi f_c t} - e^{-j2\pi f_c t}] \quad (3.45)$$

This expression is clearly equivalent to (3.36). Figures 3.9(b) and 3.9(c) show the four signals used in this development: $M_p(f + f_c)$, $M_p(f - f_c)$, $M_n(f + f_c)$, and $M_n(f - f_c)$.

There are several methods that can be employed to demodulate SSB. The simplest technique is to multiply $x_c(t)$ by a demodulation carrier and lowpass filter the result, as illustrated in Figure 3.1(a). We assume a demodulation carrier having a phase error $\theta(t)$ that yields

$$d(t) = [\tfrac{1}{2}A_c m(t) \cos \omega_c t \pm \tfrac{1}{2}A_c \hat{m}(t) \sin \omega_c t]\{4 \cos [\omega_c t + \theta(t)]\} \quad (3.46)$$

where the factor of 4 is chosen for mathematical convenience. The preceding expression can be written as

$$
\begin{aligned}
d(t) = {} & A_c m(t) \cos \theta(t) + A_c m(t) \cos [2\omega_c t + \theta(t)] \\
& \mp A_c \hat{m}(t) \sin \theta(t) \pm A_c \hat{m}(t) \sin [2\omega_c t + \theta(t)] \quad (3.47)
\end{aligned}
$$

Lowpass filtering and amplitude scaling yield

$$y_D(t) = m(t) \cos \theta(t) \mp \hat{m}(t) \sin \theta(t) \quad (3.48)$$

for the demodulated output. Observation of (3.48) illustrates that for $\theta(t)$ equal to zero, the demodulated output is the desired message signal. However, if $\theta(t)$ is nonzero, the output consists of the sum of two terms. The first term is a time-varying attenuation of the message signal and is the output present in a DSB system operating in a similar manner. The second term is a crosstalk term and can represent serious distortion if $\theta(t)$ is not small.

Another useful technique for demodulating an SSB signal is carrier reinsertion, which is illustrated in Figure 3.10. The output of a local oscillator is added to the received signal $x_r(t)$. This yields

$$e(t) = [\tfrac{1}{2}A_c m(t) + K] \cos \omega_c t \pm \tfrac{1}{2}A_c \hat{m}(t) \sin \omega_c t \quad (3.49)$$

which is the input to the envelope detector.

The output of the envelope detector must next be computed. This is slightly more difficult for signals of the form of (3.49) than for signals of the form of (3.10) because both cosine and sine terms are present. In order to derive the desired result, consider the signal

$$x(t) = a(t) \cos \omega_c t - b(t) \sin \omega_c t \quad (3.50)$$

FIGURE 3.10   Demodulation using carrier reinsertion

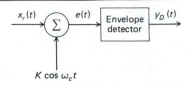

which can be represented as illustrated in Figure 3.11. Figure 3.11 shows
the amplitude of the direct component $a(t)$, the amplitude of the quadrature
component $b(t)$, and the resultant $R(t)$. It follows from Figure 3.11 that

$$a(t) = R(t) \cos \theta(t) \quad \text{and} \quad b(t) = R(t) \sin \theta(t)$$

This yields

$$x(t) = R(t)[\cos \theta(t) \cos \omega_c t - \sin \theta(t) \sin \omega_c t] \tag{3.51}$$

which is

$$x(t) = R(t) \cos [\omega_c t + \theta(t)] \tag{3.52}$$

where

$$\theta(t) = \tan^{-1} \frac{b(t)}{a(t)} \tag{3.53}$$

The instantaneous amplitude $R(t)$, which is the envelope of the signal, is
given by

$$R(t) = \sqrt{a^2(t) + b^2(t)} \tag{3.54}$$

and will be the output of an envelope detector with $x(t)$ on the input if $a(t)$
and $b(t)$ are slowly varying with respect to $\cos \omega_c t$.

A comparison of (3.49) and (3.54) illustrates that the envelope of an SSB

FIGURE 3.11   Signal representation

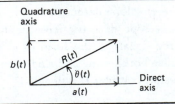

signal, after carrier reinsertion, is given by

$$y_D(t) = \sqrt{[\tfrac{1}{2}A_c m(t) + K]^2 + [\tfrac{1}{2}A_c \hat{m}(t)]^2} \qquad (3.55)$$

which is the demodulated output $y_D(t)$ in Figure 3.10. If K is chosen large enough such that

$$[\tfrac{1}{2}A_c m(t) + K]^2 \gg [\tfrac{1}{2}A_c \hat{m}(t)]^2$$

the output of the envelope detector becomes

$$y_D(t) \cong \tfrac{1}{2}A_c m(t) + K \qquad (3.56)$$

from which the message signal can easily be extracted. The development shows that carrier reinsertion requires that the locally generated carrier must be phase-coherent with the original modulation carrier. This is easily accomplished in speech-transmission systems. The frequency and phase of the demodulation carrier can be manually adjusted until intelligibility is obtained.

▼| **EXAMPLE 3.2**   As we saw in the preceding analysis, the concept of single sideband is probably best understood by using frequency-domain analysis. However, the SSB time-domain waveforms are also interesting and are the subject of this example. Assume that the message signal is given by

$$m(t) = \cos(\omega_1 t) - 0.4 \cos(2\omega_1 t) + 0.9 \cos(3\omega_1 t) \qquad (3.57)$$

The Hilbert transform of $m(t)$ is

$$\hat{m}(t) = \sin(\omega_1 t) - 0.4 \sin(2\omega_1 t) + 0.9 \sin(3\omega_1 t) \qquad (3.58)$$

These two waveforms are shown in Figures 3.12(a) and (b). As we have seen, the SSB signal is given by

$$x_c(t) = \frac{A_c}{2} [m(t) \cos \omega_c t \pm \hat{m}(t) \sin \omega_c t] \qquad (3.59)$$

with the choice of sign depending upon the sideband to be used for transmission. Using (3.39), (3.40), and (3.41), we can place $x_c(t)$ in the standard form of (3.1). This gives

$$x_c(t) = R(t) \cos[\omega_c t + \theta(t)] \qquad (3.60)$$

where the envelope $R(t)$ is

$$R(t) = \frac{A_c}{2} \sqrt{m^2(t) + \hat{m}^2(t)} \qquad (3.61)$$

FIGURE 3.12 Time-domain signals for SSB system. (a) Message signal. (b) Hilbert transform of message signal. (c) Envelope of SSB signal. (d) Upper-sideband SSB signal with message signal. (e) Lower-sideband SSB signal with message signal.

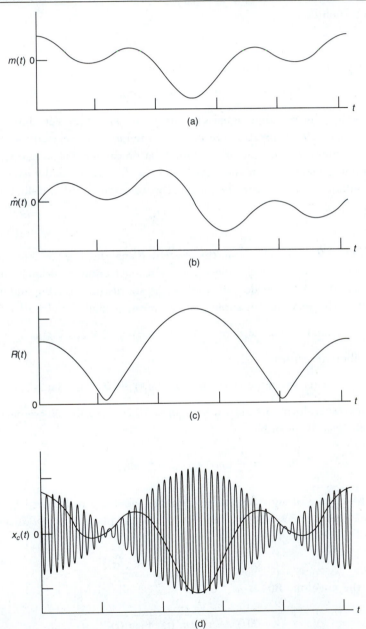

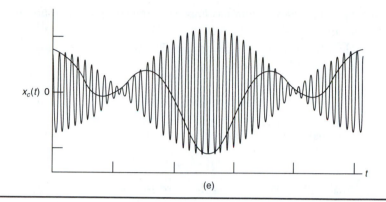

(e)

and $\theta(t)$, which is the phase deviation of $x_c(t)$, is given by

$$\theta(t) = \pm \tan^{-1}\left[\frac{\hat{m}(t)}{m(t)}\right] \tag{3.62}$$

The instantaneous frequency of $x_c(t)$ is therefore

$$\frac{d}{dt}\{\omega_c t + \theta(t)\} = \omega_c \pm \frac{d}{dt}\left\{\tan^{-1}\left[\frac{\hat{m}(t)}{m(t)}\right]\right\} \tag{3.63}$$

From (3.61) we see that the envelope of the SSB signal is independent of the choice of the sideband. The instantaneous frequency, however, is a rather complicated function of the message signal and also depends upon the choice of sideband. We therefore see that the message signal $m(t)$ affects both the envelope and phase of the modulated carrier $x_c(t)$. In DSB and AM the message signal affected only the envelope of $x_c(t)$.

The envelope of the SSB signal is shown in Figure 3.12(c). The upper-sideband SSB signal is illustrated in Figure 3.12(d) and the lower-sideband SSB signal is shown in Figure 3.12(c). It is easily seen that both the upper-sideband and lower-sideband SSB signals have the envelope shown in Figure 3.12(c). The message signal $m(t)$ is also shown in parts (d) and (e) in Figure 3.12.

Vestigial-Sideband Modulation

Vestigial-sideband (VSB) modulation overcomes two of the difficulties present in SSB modulation. By allowing a vestige of the unwanted sideband to appear at the output of an SSB modulator, we may simplify the design of the sideband filter, since the need for sharp cutoff at the carrier frequency is eliminated. In addition, a VSB system has improved low-frequency response and can even have dc response.

An example will illustrate the technique. For simplicity, let the message signal be the sum of two sinusoids:

$$m(t) = A \cos \omega_1 t + B \cos \omega_2 t \qquad (3.64)$$

This message signal is then multiplied by a carrier, $\cos \omega_c t$, to form the DSB signal

$$
\begin{aligned}
e_{DSB}(t) = &\tfrac{1}{2}A \cos (\omega_c + \omega_1)t + \tfrac{1}{2}A \cos (\omega_c - \omega_1)t \\
&+ \tfrac{1}{2}B \cos (\omega_c + \omega_2)t + \tfrac{1}{2}B \cos (\omega_c - \omega_2)t \qquad (3.65)
\end{aligned}
$$

Figure 3.13(a) shows the single-sided spectrum of this signal. A vestigial sideband filter is then used to generate the VSB signal. Figure 3.13(b) shows the assumed form of the VSB filter. The skirt of the VSB filter must be sym-

FIGURE 3.13 Generation of vestigial sideband. (a) DSB spectrum (single-sided). (b) VSB filter characteristic near f_c. (c) VSB spectrum.

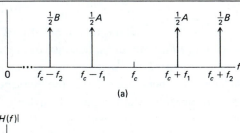

(a)

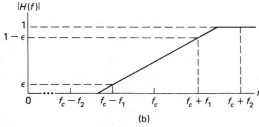

(b)

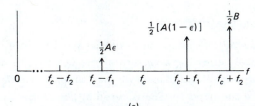

(c)

metrical about the carrier as shown. Figure 3.13(c) shows the single-sided spectrum of the filter output. This is the spectrum of the VSB signal.

The spectrum shown in Figure 3.13(c) corresponds to the VSB signal

$$x_c(t) = \tfrac{1}{2}A\epsilon \cos(\omega_c - \omega_1)t$$
$$+ \tfrac{1}{2}A(1 - \epsilon) \cos(\omega_c + \omega_1)t + \tfrac{1}{2}B \cos(\omega_c + \omega_2)t \quad (3.66)$$

This signal can be demodulated by multiplying by $4\cos \omega_c t$ and lowpass filtering. The result is

$$e(t) = A\epsilon \cos \omega_1 t + A(1 - \epsilon) \cos \omega_1 t + B \cos \omega_2 t \quad (3.67)$$

or

$$e(t) = A \cos \omega_1 t + B \cos \omega_2 t \quad (3.68)$$

which is the assumed message signal.

The slight increase in bandwidth required for VSB over that required for SSB is often more than offset by the resulting electronic simplifications. As a matter of fact, if a carrier component is added to a VSB signal, envelope detection can be used. The development of this technique is similar to the development of envelope detection of SSB with carrier reinsertion and is relegated to the problems. The process, however, is demonstrated in the following example.

▼ **EXAMPLE 3.3** In this example we consider the time domain waveforms corresponding to VSB modulation and consider demodulation using envelope detection or carrier reinsertation.

We assume the same message signal as was assumed in the previous example. In other words

$$m(t) = \cos \omega_1 t - 0.4 \cos 2\omega_1 t + 0.9 \cos 3\omega_1 t \quad (3.69)$$

The message signal $m(t)$ is shown in Figure 3.14(a).

The VSB signal can be expressed as

$$x_c(t) = A_c[\epsilon_1 \cos(\omega_c + \omega_1)t + (1 - \epsilon_1) \cos(\omega_c - \omega_1)t$$
$$- 0.4\epsilon_2 \cos(\omega_c + 2\omega_1)t - 0.4(1 - \epsilon_2) \cos(\omega_c - 2\omega_1)t$$
$$+ 0.9\epsilon_3 \cos(\omega_c + 3\omega_1)t + 0.9(1 - \epsilon_3) \cos(\omega_c - 3\omega_1)] \quad (3.70)$$

The modulated carrier, along with the message signal, is shown in Figure 3.14(b) for $\epsilon_1 = 0.64$, $\epsilon_2 = 0.78$, and $\epsilon_3 = 0.92$.

The result of carrier reinsertion and envelope detection is shown in Figure 3.14(c). The message signal, biased by the amplitude of the carrier component, is clearly shown and will be the output of an envelope detector.

FIGURE 3.14 Time-domain signals for VSB system. (a) Message signal. (b) VSB signal and message signal. (c) Sum of VSB signal and carrier signal.

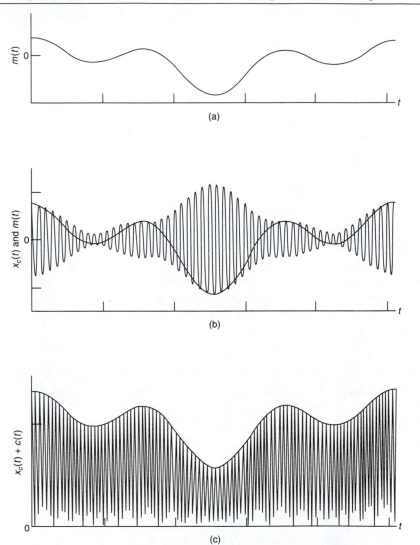

(a)

(b)

(c)

   VSB is currently used in the United States for transmission of the video signal in commercial television broadcasting. However, exact shaping of the vestigial sideband is not carried out at the transmitter, but at the receiver, where signal levels are low. The filter in the transmitter simply bandlimits the video signal, as shown in Figure 3.15(a). The video carrier frequency is

denoted f_v, and, as can be seen, the bandwidth of the video signal is approximately 5.25 MHz. The spectrum of the audio signal is centered about the
audio carrier, which is 4.5 MHz above the video carrier. The modulation
method used for audio transmission is FM. When we study FM in the following sections, you will understand the shape of the audio spectrum. Since the
spectrum centered on the audio carrier is a line spectrum in Figure 3.15(a),
a periodic audio signal is implied. This was done for clarity, and in practice
the audio signal will have a continuous spectrum. Figure 3.15(b) shows the
amplitude response of the receiver VSB filter.

 Also shown in Figure 3.15(a), at a frequency 3.58 MHz above the video
carrier, is the color carrier. Quadrature multiplexing, which we shall study
in Section 3.6, is used with the color subcarrier so that two different signals
are transmitted with the color carrier. These two signals, commonly referred
to as the I-channel and Q-channel signals, carry luminance and chrominance
(color) information necessary to reconstruct the image at the receiver.

 One problem faced by the designers of a system for the transmission of a
color TV signal was that the transmitted signal was required to be compatible

FIGURE 3.15   Transmitted spectrum and VSB filtering for television. (a)
Spectrum of transmitted signal. (b) VSB filter in receiver.

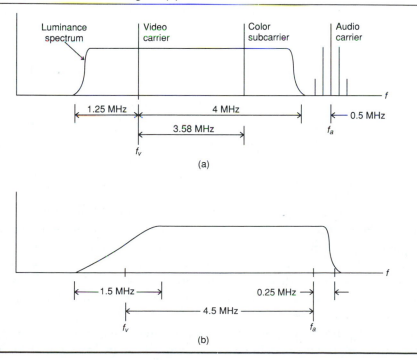

with existing black-and-white television receivers. Such a consideration is a significant constraint in the design process. A similar problem was faced by the designers of stereophonic radio receivers. We shall study this system in Section 3.6 and, since the audio system is simpler than a color TV system, we can see how the compatibility problem was solved.

Switching Modulators

In all our previous developments, the carrier has been assumed to be sinusoidal. However, if the modulator is followed by a bandpass filter, any periodic waveform can serve as a carrier. For example, if the carrier is periodic with fundamental frequency, f_c, it can be represented by the Fourier series

$$c(t) = \sum_{n=-\infty}^{\infty} C_n e^{jn2\pi f_c t} \tag{3.71}$$

Multiplying this carrier by the message signal $m(t)$ yields

$$m(t)c(t) = \sum_{n=-\infty}^{\infty} C_n m(t) e^{jn2\pi f_c t} \tag{3.72}$$

From the frequency-translation theorem, the Fourier transform of $m(t)c(t)$ is

$$\mathscr{F}[m(t)c(t)] = \sum_{n=-\infty}^{\infty} C_n M(f - nf_c) \tag{3.73}$$

which shows that the effect of multiplying $m(t)$ by a periodic waveform is to translate the spectrum of $m(t)$ to a carrier frequency and all harmonics of the carrier frequency. A DSB signal can be obtained from (3.73) using a bandpass filter with center frequency f_c and bandwidth $2W$, where W is the highest frequency present in the message signal.

Considerable simplification of the product device results if a squarewave carrier, which varies between zero and unity, is used. When the carrier is unity, the multiplier output is $m(t)$, and when the carrier is zero, the multiplier output is zero. Thus the multiplier can be replaced by a switch that opens and closes at the carrier frequency. Switching modulators are much easier to implement than are analog product devices, especially at higher frequencies. A switching modulator is illustrated in Figure 3.16(c), together with appropriate spectra.

Switching modulators are often implemented using diode bridges. Two examples are the shunt and series modulators illustrated in Figure 3.17. The carrier waveform, having frequency f_c, alternately short-circuits and open-

FIGURE 3.16   Modulation using periodic carriers. (a) Modulation with a periodic carrier. (b) Spectrum of multiplier output. (c) Switching modulator.

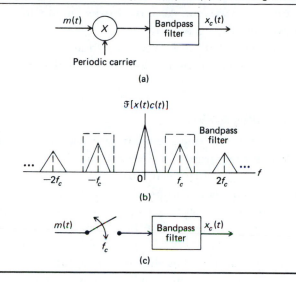

(a)

(b)

(c)

FIGURE 3.17   (a) Shunt modulator. (b) Series modulator.

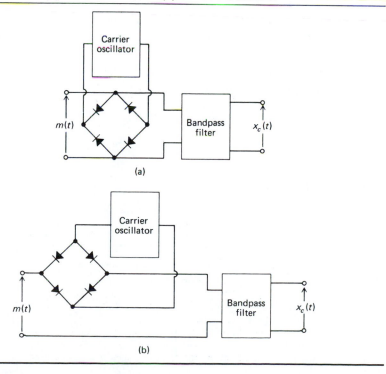

(a)

(b)

circuits the diodes so that they operate as a switch. If DSB is to be generated, care must be exercised to ensure that the diodes are matched, thereby balancing the carrier component. Also, the carrier amplitude must be large, so that diode switching occurs properly.

The switching modulator is an example of the use of nonlinear elements to implement linear modulation. Additional examples are presented in the problems.

Frequency Translation and Mixing

We have seen that multiplying a lowpass signal by a high-frequency periodic signal translates the spectrum of the lowpass signal to all frequencies present in the periodic signal. Quite often it is desirable to translate a bandpass signal to a new center frequency. This process also can be accomplished by multiplication of the bandpass signal by a periodic signal and is called *mixing* or *converting*. A block diagram of a mixer is given in Figure 3.18. As an example, the bandpass signal $m(t) \cos \omega_1 t$ can be translated to a new carrier frequency ω_2 by multiplying it by a local oscillator signal of the form $2 \cos (\omega_1 \pm \omega_2)t$. By using appropriate trigonometric identities, we can easily show that the result of the multiplication is

$$e(t) = m(t) \cos \omega_2 t + m(t) \cos (2\omega_1 \pm \omega_2)t \qquad (3.74)$$

The undesired term is removed by filtering. The filter should have a bandwidth at least $2W$ for the assumed DSB modulation, where W is the bandwidth of $m(t)$.

A common problem with mixers is that inputs of the form $k(t) \cos (\omega_1 \pm 2\omega_2)t$ are also translated to ω_2, since

$$2k(t) \cos [(\omega_1 \pm 2\omega_2)t] \cos [(\omega_1 \pm \omega_2)t]$$
$$= k(t) \cos \omega_2 t + k(t) \cos (2\omega_1 \pm 3\omega_2)t \qquad (3.75)$$

FIGURE 3.18 Mixer

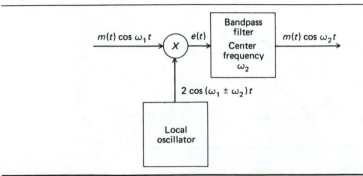

FIGURE 3.19 Superheterodyne receiver

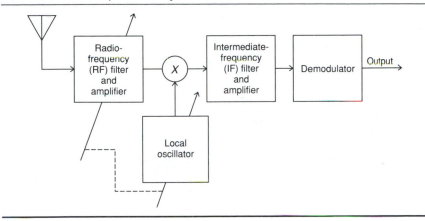

In (3.75), all three signs must be plus or all three must be minus. The input frequency $\omega_1 \pm 2\omega_2$, which results in an output at ω_2, is referred to as an *image frequency* of the desired frequency ω_1.

To realize that image frequencies are a problem, consider the superheterodyne receiver shown in Figure 3.19. The carrier frequency of the signal to be demodulated is ω_c, and the intermediate-frequency (IF) filter is a bandpass filter with center frequency ω_{IF}. The superheterodyne receiver has good sensitivity (the ability to detect weak signals) and selectivity (the ability to separate closely spaced signals). This results because the IF filter, which provides most of the predetection filtering, need not be tunable. Thus it can be a rather complex filter. Tuning of the receiver is accomplished by varying the frequency of the local oscillator. The superheterodyne receiver of Figure 3.19 is the mixer of Figure 3.18 with $\omega_c = \omega_1$ and $\omega_{IF} = \omega_2$. The mixer translates the input frequency ω_c to the IF frequency ω_{IF}. As shown previously, the image frequency $\omega_c \pm 2\omega_{IF}$, where the sign depends on the choice of local oscillator frequency, also will appear at the IF output. This means that if we are attempting to receive a signal having carrier frequency ω_c, we can also receive a signal at $\omega_c + 2\omega_{IF}$ if the local oscillator frequency is $\omega_c + \omega_{IF}$ or a signal at $\omega_c - 2\omega_{IF}$ if the local oscillator frequency is $\omega_c - \omega_{IF}$. There is only one image frequency, and it is always separated from the desired frequency by $2\omega_{IF}$. Figure 3.20 shows the desired signal and image signal for a local oscillator having the frequency

$$\omega_{LO} = \omega_c + \omega_{IF} \tag{3.76}$$

FIGURE 3.20   Illustration of image frequency (high-side tuning)

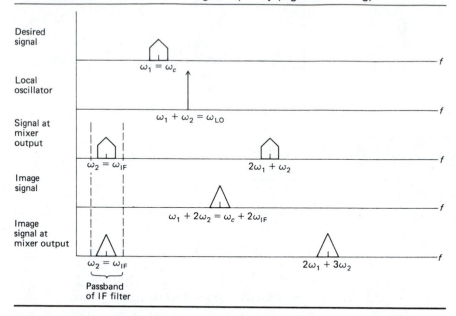

The image frequency can be eliminated by the radio-frequency (RF) filter. A standard IF frequency for AM radio is 455 kHz. Thus the image frequency is separated from the desired signal by almost 1 MHz. This shows that the RF filter need not be narrowband. Furthermore, since the AM broadcast band occupies the frequency range 540 kHz to 1.6 MHz, it is apparent that a tunable RF filter is not required, provided that stations at the high end of the band are not located geographically near stations at the low end of the band. Some inexpensive receivers take advantage of this fact. Additionally, if the RF filter is made tunable, it need be tunable only over a narrow range of frequencies.

One decision to be made when designing a superheterodyne receiver is whether the frequency of the local oscillator is to be below the frequency of the input carrier *(low-side tuning)* or above the frequency of the input carrier *(high-side tuning)*. A simple example based on the standard AM broadcast band illustrates one major consideration. The standard AM broadcast band extends from 540 kHz to 1600 kHz. For this example, let us choose a common intermediate frequency, 455 kHz. As shown in Table 3.1, for low-side tuning, the frequency of the local oscillator must be variable from 85 to 1600 kHz, which represents a frequency range in excess of 13 to 1. If high-side tuning is used, the frequency of the local oscillator must be variable from 995 to 2055 kHz, which represents a frequency range slightly in excess of 2 to 1. Oscillators whose frequency must vary over a large ratio

TABLE 3.1   Comparison of Low-Side and High-Side Tuning for AM Broadcast Band with f_{IF} = 455 kHz

	Lower Frequency	Upper Frequency	Tuning Range of Local Oscillator
Standard AM broadcast band	540 kHz	1600 kHz	—
Frequencies of local oscillator for low-side tuning	540 kHz — 455 kHz = 85 kHz	1600 kHz — 455 kHz = 1145 kHz	13.47 to 1
Frequencies of local oscillator for high-side tuning	540 kHz + 455 kHz = 995 kHz	1600 kHz + 455 kHz = 2055 kHz	2.07 to 1

are much more difficult to implement than are those whose frequency varies over a small ratio.

The relationship between the desired signal to be demodulated and the image signal is summarized in Figure 3.21 for low-side and high-side tuning. The desired signal to be demodulated has a carrier frequency of ω_c and the image signal has a carrier frequency of ω_i.

FIGURE 3.21   Relationship between ω_c and ω_i for (a) low-side tuning and (b) high-side tuning

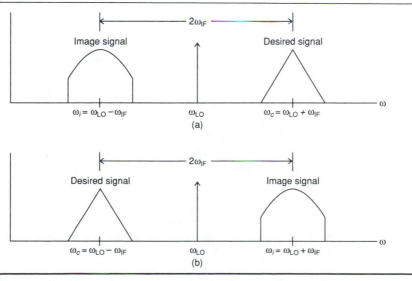

3.2 ANGLE MODULATION

To generate angle modulation, the amplitude of the modulated carrier is held constant and either the phase or the time derivative of the phase of the carrier is varied linearly with the message signal $m(t)$. Thus the general angle-modulated signal is given by

$$x_c(t) = A_c \cos [\omega_c t + \phi(t)] \tag{3.77}$$

The instantaneous phase of $x_c(t)$ is defined as

$$\theta_i(t) = \omega_c t + \phi(t) \tag{3.78}$$

and the instantaneous frequency is defined as

$$\omega_i(t) = \frac{d\theta_i}{dt} = \omega_c + \frac{d\phi}{dt} \tag{3.79}$$

The functions $\phi(t)$ and $d\phi/dt$ are known as the *phase deviation* and *frequency deviation*, respectively.

The two basic types of angle modulation are phase modulation (PM) and frequency modulation (FM). *Phase modulation* implies that the phase deviation of the carrier is proportional to the message signal. Thus, for phase modulation,

$$\phi(t) = k_p m(t) \tag{3.80}$$

where k_p is the *deviation constant* in radians per unit of $m(t)$. Similarly, *frequency modulation* implies that the frequency deviation of the carrier is proportional to the modulating signal. This yields

$$\frac{d\phi}{dt} = k_f m(t) \tag{3.81}$$

The phase deviation of a frequency-modulated carrier is given by

$$\phi(t) = k_f \int_{t_0}^{t} m(\alpha) \, d\alpha + \phi_0 \tag{3.82}$$

in which ϕ_0 is the phase deviation at $t = t_0$. It follows from (3.81) that k_f is the frequency-deviation constant, expressed in radians per second per unit of $m(t)$. Since it is often more convenient to measure frequency deviation in hertz, we define

$$k_f = 2\pi f_d \tag{3.83}$$

where f_d is known as the *frequency-deviation constant* of the modulator and is expressed in hertz per unit of $m(t)$.

With these definitions, the phase modulator output is

$$x_c(t) = A_c \cos [\omega_c t + k_p m(t)] \qquad pm \qquad (3.84)$$

and the frequency modulator output is

$$x_c(t) = A_c \cos \left[\omega_c t + 2\pi f_d \int^t m(\alpha) \, d\alpha \right] \qquad fm \qquad (3.85)$$

The lower limit of the integral is typically not specified, since to do so would require the inclusion of an initial condition as shown in (3.82).

Figures 3.22 and 3.23 illustrate the outputs of PM and FM modulators for a step function and a sinusoidal message signal, respectively. With a unit-step message signal, the instantaneous frequency of the PM modulator output is f_c for both $t < t_0$ and $t > t_0$. The phase of the unmodulated carrier is advanced by $k_p = \pi/2$ radians for $t > t_0$, giving rise to a signal that is discontinuous at $t = t_0$. The frequency of the output of the FM modulator is f_c for $t < t_0$ and $f_c + f_d$ for $t > t_0$. The modulator output phase is, however, continuous at $t = t_0$.

With a sinusoidal message signal, the phase deviation of the PM modulator output is proportional to $m(t)$. The frequency deviation is proportional to the derivative of the phase deviation. Thus the instantaneous frequency of the output of the PM modulator is maximum when the *slope* of $m(t)$ is maximum and minimum when the *slope* of $m(t)$ is minimum. The frequency deviation of the FM modulator output is proportional to $m(t)$. Thus the instantaneous frequency of the FM modulator ouput is maximum when $m(t)$ is maximum and minimum when $m(t)$ is minimum. It should be noted that if $m(t)$ were not shown along with the modulator outputs, it would not be possible to distinguish the PM and FM modulator outputs. In the following sections we will devote considerable attention to the case in which $m(t)$ is sinusoidal.

Narrowband Angle Modulation

An angle-modulated carrier can be represented in exponential form by writing (3.77) as

$$x_c(t) = \text{Re} \, (A_c e^{j\omega_c t} e^{j\phi(t)}) \qquad (3.86)$$

where Re $(\cdot)$ implies that the real part of the argument is to be taken. Expanding $e^{j\phi(t)}$ in a power series yields

$$x_c(t) = \text{Re} \left\{ A_c e^{j\omega_c t} \left[1 + j\phi(t) - \frac{\phi^2(t)}{2!} - \cdots \right] \right\} \qquad (3.87)$$

If the maximum value of $|\phi(t)|$ is much less than unity, the modulated carrier can be approximated as

$$x_c(t) \cong \mathrm{Re}\,[A_c e^{j\omega_c t} + A_c \phi(t) j e^{j\omega_c t}]$$

Taking the real part yields

$$x_c(t) \cong A_c \cos \omega_c t - A_c \phi(t) \sin \omega_c t \qquad (3.88)$$

The form of (3.88) is reminiscent of AM. The modulator output contains a carrier component and a term in which a function of $m(t)$ multiplies a 90°

FIGURE 3.22 Comparison of PM and FM modulator outputs for a unit-step input. (a) Message signal. (b) Unmodulated carrier. (c) Phase modulator ouput ($k_p = \frac{1}{2}\pi$). (d) Frequency modulator output.

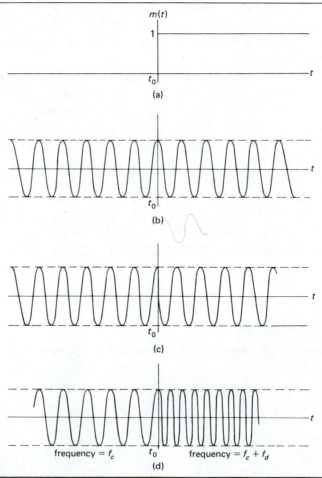

FIGURE 3.23   Angle modulation with sinusoidal message signal.
(a) Message signal $m(t)$. (b) Unmodulated carrier $A_c \cos 2\pi f_c t$. (c) Output of
phase modulator with $m(t)$. (d) Output of frequency modulator with $m(t)$.

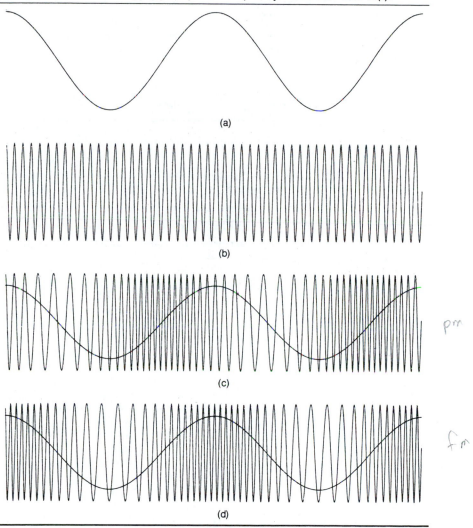

(a)

(b)

(c) pm

(d) fm

phase-shifted carrier. This multiplication generates a pair of sidebands. Thus,
if $\phi(t)$ has a bandwidth W, the bandwidth of a narrowband angle modulator
output is $2W$. It is important to note, however, that the carrier and the *resul-*
tant of the sidebands for narrowband angle modulation with sinusoidal modulation are in phase quadrature, whereas for AM they are not. This will be
illustrated in Example 3.4.

FIGURE 3.24   Generation of narrowband angle modulation

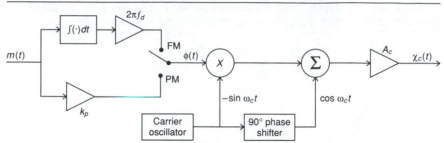

The generation of narrowband angle modulation is easily accomplished using the method shown in Figure 3.24. The switch allows for the generation of either narrowband FM or narrowband PM. We will show later that narrowband angle modulation is useful for the generation of angle-modulated signals that are not necessarily narrowband through a process called narrowband to wideband conversion.

▼| EXAMPLE 3.4   Consider an FM system operating with

$$m(t) = A \cos \omega_m t \tag{3.89}$$

From (3.82), with t_0 equal to zero,

$$\phi(t) = k_f \int_0^t A \cos \omega_m \alpha \, d\alpha = \frac{A k_f}{\omega_m} \sin \omega_m t = \frac{A f_d}{f_m} \sin \omega_m t \tag{3.90}$$

so that

$$x_c(t) = A_c \cos \left(\omega_c t + \frac{A f_d}{f_m} \sin \omega_m t \right) \tag{3.91}$$

If $f_d/f_m \ll 1$, the modulator output can be approximated as

$$x_c(t) = A_c \left(\cos \omega_c t - \frac{A f_d}{f_m} \sin \omega_c t \sin \omega_m t \right) \tag{3.92}$$

which is

$$x_c(t) = A_c \cos \omega_c t + \frac{A_c}{2} \frac{A f_d}{f_m} [\cos (\omega_c + \omega_m)t - \cos (\omega_c - \omega_m)t] \tag{3.93}$$

Thus $x_c(t)$ can be written as

$$x_c(t) = A_c \, \text{Re} \left\{ e^{j2\pi f_c t} \left[1 + \frac{A f_d}{2 f_m} (e^{j2\pi f_m t} - e^{-j2\pi f_m t}) \right] \right\} \qquad (3.94)$$

It is interesting to compare this result with the equivalent result for an AM signal. Since sinusoidal modulation is assumed, the AM signal can be written as

$$x_c(t) = A_c(1 + a \cos 2\pi f_m t) \cos 2\pi f_c t \qquad (3.95)$$

where a is the modulation index. Combining the two cosine terms yields

$$x_c(t) = A_c \cos 2\pi f_c t + \frac{A_c}{2} a[\cos 2\pi (f_c + f_m)t + \cos 2\pi (f_c - f_m)t] \qquad (3.96)$$

This can be written in exponential form as

$$x_c(t) = A_c \, \text{Re} \left\{ e^{j2\pi f_c t} \left[1 + \frac{a}{2} a(e^{j2\pi f_m t} + e^{-j2\pi f_m t}) \right] \right\} \qquad (3.97)$$

Comparing (3.94) and (3.97) illustrates the similarity between the two signals. The first, and most important, difference is the sign of the term at frequency $f_c - f_m$, which represents the lower sideband. The other difference is that the index a in the AM signal is replaced by f_d/f_m in the narrowband FM signal. We will see in the following section that f_d/f_m determines the index for an FM signal. Thus these two parameters are in a sense equivalent.

Additional insight is gained by sketching the phasor diagrams and the amplitude and phase spectra for both signals. These are given in Figure 3.25. The phasor diagrams are drawn using the carrier phase as a reference. The difference between AM and narrowband angle modulation with a sinusoidal message signal lies in the fact that the phasor resulting from the lower-sideband (LSB) and upper-sideband (USB) phasors adds to the carrier for AM but is in phase quadrature with the carrier for angle modulation. This difference results from the minus sign in the LSB component and is also clearly seen in the phase spectra of the two signals. The amplitude spectra are equivalent.

Spectrum of an Angle-Modulated Signal

The derivation of the spectrum of an angle-modulated signal is typically a very difficult task. However, if the message signal is sinusoidal, the instantaneous phase deviation of the modulated carrier is sinusoidal for both FM and PM, and the spectrum can be obtained with ease. Thus this is the case

FIGURE 3.25 Comparison of AM and narrowband angle modulation. (a) Phasor diagrams. (b) Single-sided amplitude spectra. (c) Single-sided phase spectra.

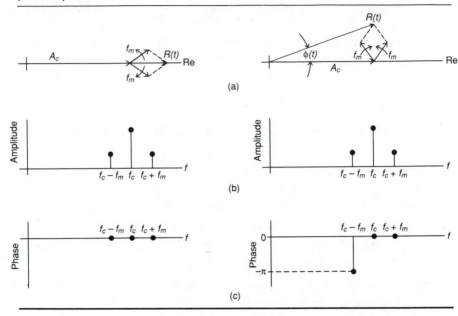

we will consider. Even though we are restricting our attention to a very special case, the results provide much insight into the frequency-domain behavior of angle modulation.

In order to compute the spectrum of an angle-modulated signal with a sinusoidal message signal, we assume that

$$\phi(t) = \beta \sin \omega_m t \qquad (3.98)$$

The parameter β is known as the *modulation index* and is the maximum value of phase deviation for both FM and PM. The signal

$$x_c(t) = A_c \cos (\omega_c t + \beta \sin \omega_m t) \qquad (3.99)$$

can be expressed as

$$x_c(t) = A_c \, \text{Re} \, (e^{j\omega_c t} e^{j\beta \sin \omega_m t}) \qquad (3.100)$$

The function $e^{j\beta \sin \omega_m t}$ is periodic with frequency ω_m and can therefore be expanded in a Fourier series. The Fourier coefficients are given by

$$\frac{\omega_m}{2\pi} \int_{-\pi/\omega_m}^{\pi/\omega_m} e^{j\beta \sin \omega_m t} e^{-jn\omega_m t} \, dt = \frac{1}{2\pi} \int_{-\pi}^{\pi} e^{-j(nx - \beta \sin x)} \, dx \qquad (3.101)$$

This integral cannot be evaluated in closed form. However, it has been well tabulated. The integral is a function of n and β and is known as the *Bessel function* of the first kind of order n and argument β. It is denoted $J_n(\beta)$ and is tabulated for several values of n and β in Table 3.2. The significance of the underlining of various values in the table will be explained later.

Thus, with the aid of Bessel functions, the Fourier series for $e^{j\beta\sin\omega_m t}$ can be written as

$$e^{j\beta\sin\omega_m t} = \sum_{n=-\infty}^{\infty} J_n(\beta)e^{jn\omega_m t} \tag{3.102}$$

which allows the modulated carrier to be written as

$$x_c(t) = A_c \, \text{Re} \left[e^{j\omega_c t} \sum_{n=-\infty}^{\infty} J_n(\beta)e^{jn\omega_m t} \right] \tag{3.103}$$

Taking the real part yields

$$x_c(t) = A_c \sum_{n=-\infty}^{\infty} J_n(\beta) \cos (\omega_c + n\omega_m)t \tag{3.104}$$

from which the spectrum of $x_c(t)$ can be determined by inspection. The spectrum has components at the carrier frequency and has an infinite number of sidebands separated from the carrier frequency by integer multiples of the modulation frequency ω_m. The amplitude of each spectral component can be determined from a table of values of the Bessel function. Such tables typically give $J_n(\beta)$ only for positive values of n. However, from the definition of $J_n(\beta)$, it can be determined that

$$J_{-n}(\beta) = J_n(\beta) \tag{3.105}$$

for n even and

$$J_{-n}(\beta) = -J_n(\beta) \tag{3.106}$$

for n odd. The single-sided spectra of (3.104) are shown in Figure 3.26.

A useful relationship between values of $J_n(\beta)$ for various values of n is the recursion formula

$$J_{n-1}(\beta) + J_{n+1}(\beta) = \frac{2n}{\beta} J_n(\beta) \tag{3.107}$$

Thus we can compute $J_{n+1}(\beta)$ from knowledge of $J_n(\beta)$ and $J_{n-1}(\beta)$. This enables us to compute a table of values of the Bessel function, similar to Table 3.2, up to any value of n from $J_0(\beta)$ and $J_1(\beta)$.

TABLE 3.2 Table of Bessel Functions

n	β = 0.05	β = 0.1	β = 0.2	β = 0.3	β = 0.5	β = 0.7	β = 1.0	β = 2.0	β = 3.0	β = 5.0	β = 7.0	β = 8.0	β = 10.0
0	0.999	0.998	0.990	0.978	0.938	0.881	0.765	0.224	−0.260	−0.178	0.300	0.172	−0.246
1	0.025	0.050	0.100	0.148	0.242	0.329	0.440	0.577	0.339	−0.328	−0.005	0.235	0.043
2		0.001	0.005	0.011	0.031	0.059	0.115	0.353	0.486	0.047	−0.301	−0.113	0.255
3				0.001	0.003	0.007	0.020	0.129	0.309	0.365	−0.168	−0.291	0.058
4						0.001	0.002	0.034	0.132	0.391	0.158	−0.105	−0.220
5								0.007	0.043	0.261	0.348	0.186	−0.234
6								0.001	0.011	0.131	0.339	0.338	−0.014
7									0.003	0.053	0.234	0.321	0.217
8										0.018	0.128	0.223	0.318
9										0.006	0.059	0.126	0.292
10										0.001	0.024	0.061	0.207
11											0.008	0.026	0.123
12											0.003	0.010	0.063
13											0.001	0.003	0.029
14												0.001	0.012
15													0.005
16													0.002
17													0.001

FIGURE 3.26 Spectra of an angle-modulated signal. (a) Single-sided amplitude spectrum. (b) Single-sided phase spectrum.

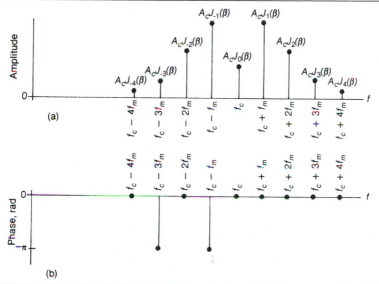

Figure 3.27 illustrates the behavior of the Fourier Bessel coefficients $J_n(\beta)$, for $n = 0, 1, 2, 4$, and 6 with $0 \leq \beta \leq 9$. Several interesting observations can be made. First, for $\beta \ll 1$, it is clear that $J_0(\beta)$ predominates, giving rise to narrowband angle modulation. It also can be seen that $J_n(\beta)$ oscillates for increasing β, but that the amplitude of oscillation decreases with increasing β. Also of interest is the fact that the maximum value of $J_n(\beta)$ decreases with increasing n.

As Figure 3.27 shows, $J_n(\beta)$ is equal to zero at several values of β. Denoting these values of β by β_{nk}, where $k = 0, 1, 2$, we have the results in Table 3.3. As an example, $J_0(\beta)$ is zero for β equal to 2.4048, 5.5201, and 8.6537. Of course, there are an infinite number of points at which $J_n(\beta)$ is zero for any n, but consistent with Figure 3.27, only the values in the range $0 \leq \beta \leq 9$ are shown in Table 3.3. It follows that since $J_0(\beta)$ is zero at β equal to 2.4048, 5.5201, and 8.6537, the spectrum of the modulator output will not contain a component at the carrier frequency for these values of the modulation index. These points are referred to as *carrier nulls*. In a similar manner, the components at $f = f_c \pm f_m$ are zero if $J_1(\beta)$ is zero. The values of the modulation index giving rise to this condition are 0, 3.8317, and 7.0156. It should be obvious why only $J_0(\beta)$ is nonzero at $\beta = 0$. If the modulation index is zero, then either $m(t)$ is zero or the deviation constant f_d is zero. In

FIGURE 3.27 $J_n(\beta)$ as a function of β

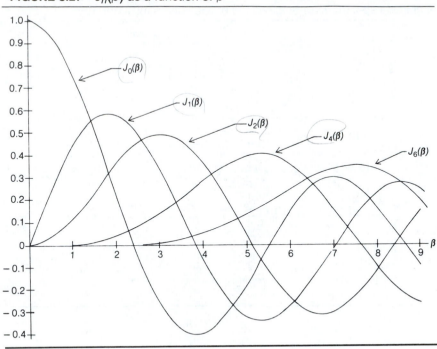

either case, the modulator output is the unmodulated carrier, which has frequency components only at the carrier frequency.

In computing the spectrum of the modulator output, our starting point was the assumption that

$$\phi(t) = \beta \sin \omega_m t \tag{3.108}$$

We did not specify the modulator type. The assumed $\phi(t)$ could represent the phase deviation of a PM modulator with $m(t) = A \sin \omega_m t$ and an index $\beta = k_p A$.

TABLE 3.3   Values of β for which $J_n(\beta) = 0$ for $0 \le \beta \le 9$

n		β_{n0}	β_{n1}	β_{n2}
0	$J_0(\beta) = 0$	2.4048	5.5201	8.6537
1	$J_1(\beta) = 0$	0.0000	3.8317	7.0156
2	$J_2(\beta) = 0$	0.0000	5.1356	8.4172
4	$J_4(\beta) = 0$	0.0000	7.5883	—
6	$J_6(\beta) = 0$	0.0000	—	—

Alternatively, an FM modulator with $m(t) = A \cos \omega_m t$ yields the assumed $\phi(t)$ with

$$\beta = \frac{2\pi f_d A}{\omega_m} = \frac{f_d A}{f_m} \qquad (3.109)$$

Thus the modulation index for FM is a function of the modulation frequency. This is not the case for PM. The behavior of the spectrum of an FM signal is illustrated in Figure 3.28 as f_m is decreased while holding $A f_d$ constant. For large values of f_m, the signal is narrowband FM, since only two sidebands are significant. For small values of f_m, many sidebands have significant value.

FIGURE 3.28   Amplitude spectrum of an FM signal as β is increased by decreasing f_m

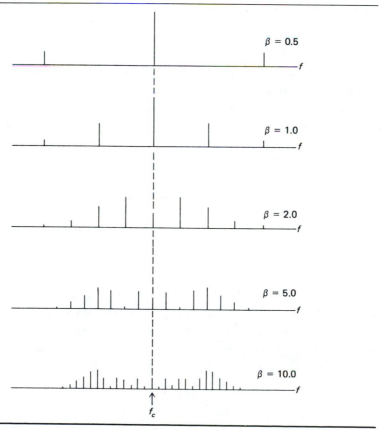

Power in an Angle-Modulated Signal

The power in an angle-modulated signal is easily computed from (3.77). Squaring (3.77) and taking the time-average value yields

$$\langle x_c^2(t) \rangle = A_c^2 \langle \cos^2 [\omega_c t + \phi(t)] \rangle \tag{3.110}$$

which can be written as

$$\langle x_c^2(t) \rangle = \tfrac{1}{2}A_c^2 + \tfrac{1}{2}A_c^2 \langle \cos 2[\omega_c t + \phi(t)] \rangle \tag{3.111}$$

If the carrier frequency is large so that $x_c(t)$ has negligible frequency content in the region of dc, the second term in (3.111) is negligible and

$$\langle x_c^2(t) \rangle = \tfrac{1}{2}A_c^2 \tag{3.112}$$

Thus the power contained in the output of an angle modulator is independent of the message signal. This is one important difference between angle modulation and linear modulation.

Bandwidth of Angle-Modulated Signals

Strictly speaking, the bandwidth of an angle-modulated signal is infinite, since angle modulation of a carrier results in the generation of an infinite number of sidebands. However, it can be seen from the series expansion of $J_n(\beta)$ (Appendix C, Table C.4) that for large n

$$J_n(\beta) \approx \frac{\beta^n}{2^n n!} \tag{3.113}$$

Thus, for fixed β,

$$\lim_{n \to \infty} J_n(\beta) = 0 \tag{3.114}$$

This behavior can also be seen from the values of $J_n(\beta)$ given in Table 3.2.

Since the values of $J_n(\beta)$ become negligible for sufficiently large n, the bandwidth of an angle-modulated signal can be defined by considering only those terms that contain significant power. The power ratio, P_r, is defined as the ratio of the power contained in the carrier ($n = 0$) component and the k components each side of the carrier to the total power in $x_c(t)$. Thus

$$P_r = \frac{\tfrac{1}{2}A_c^2 \sum_{n=-k}^{k} J_n^2(\beta)}{\tfrac{1}{2}A_c^2} \tag{3.115a}$$

or simply

$$P_r = J_0^2(\beta) + 2 \sum_{n=1}^{k} J_n^2(\beta) \qquad (3.115b)$$

Bandwidth for a particular application can be determined by defining an acceptable power ratio, solving for the required value of k using a table of Bessel functions, and then recognizing that the resulting bandwidth is

$$B = 2kf_m \qquad (3.116)$$

The acceptable value of the power ratio is dictated by the particular application of the system. Two power ratios are depicted in Table 3.2: $P_r \geq 0.7$ and $P_r \geq 0.98$. The value of n corresponding to k for $P_r \geq 0.7$ is indicated by a single underscore, and the value of n corresponding to k for $P_r \geq 0.98$ is indicated by a double underscore. For $P_r \geq 0.98$, it is noted that n is equal to the integer part of $1 + \beta$, so that

$$B \cong 2(\beta + 1)f_m \qquad (3.117)$$

The preceding expression assumes sinusoidal modulation, since the modulation index β is defined only for sinusoidal modulation. For arbitrary $m(t)$, a generally accepted expression for bandwidth results if the deviation ratio D is defined as

$$D = \frac{\text{peak frequency deviation}}{\text{bandwidth of } m(t)} \qquad (3.118)$$

which is

$$D = \frac{f_d}{W} [\max |m(t)|] \qquad (3.119)$$

The deviation ratio plays the same role for nonsinusoidal modulation as the modulation index plays for sinusoidal systems. Replacing β by D and replacing f_m by W in (3.117), we obtain

$$B = 2(D + 1)W \qquad (3.120)$$

This expression for bandwidth is generally referred to as *Carson's rule*. If $D \ll 1$, the bandwidth is approximately $2W$, and the signal is known as a *narrowband angle-modulated signal*. Conversely, if $D \gg 1$, the bandwidth is approximately $2DW = 2f_d[\max |m(t)|]$, which is twice the peak frequency deviation. Such a signal is known as a *wideband angle-modulated signal*.

▼ **EXAMPLE 3.5**   In this example we consider an FM modulator with output

$$x_c(t) = 100 \cos [2\pi(1000)t + \phi(t)] \qquad (3.121)$$

The modulator operates with $f_d = 8$ and has the input message signal

$$m(t) = 5 \cos 2\pi(8)t \qquad (3.122)$$

The modulator is followed by a bandpass filter with a center frequency of 1000 Hz and a bandwidth of 56 Hz, as shown in Figure 3.29(a). Our problem is to determine the power at the filter output.

FIGURE 3.29   System and spectra for Example 3.5. (a) FM system. (b) Single-sided spectrum of modulator output. (c) Amplitude response of bandpass filter.

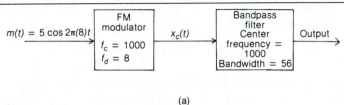

(a)

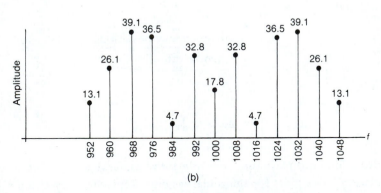

(b)

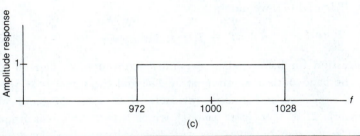

(c)

SOLUTION   The peak deviation is $5f_d$ or 40 Hz, and $f_m = 8$ Hz. Thus the modulation index is

$$\beta = \tfrac{40}{8} = 5 \tag{3.123}$$

This yields the single-sided amplitude spectrum shown in Figure 3.29(b).

Figure 3.31(c) shows the passband of the bandpass filter. The filter passes the component at the carrier frequency and three components on each side of the carrier. Thus the power ratio is

$$P_r = J_0^2(5) + 2[J_1^2(5) + J_2^2(5) + J_3^2(5)] \tag{3.124}$$

which is

$$P_r = (0.178)^2 + 2[(0.328)^2 + (0.047)^2 + (0.365)^2] \tag{3.125}$$

This yields

$$P_r = 0.518 \tag{3.126}$$

The power at the output of the modulator is

$$\overline{x_c^2} = \tfrac{1}{2}A_c^2 = \tfrac{1}{2}(100)^2 = 5000 \text{ W} \tag{3.127}$$

The power at the filter output is the power of the modulator output multiplied by the power ratio. Thus the power at the filter output is

$$P_r\overline{x_c^2} = 2589 \text{ W} \tag{3.128}$$

▼ EXAMPLE 3.6   In the development of the spectrum of an angle-modulated signal, it was assumed that the message signal was a single sinusoid. We now consider a somewhat more general problem in which the message signal is the sum of two sinusoids. Let the message signal be

$$m(t) = A \cos \omega_1 t + B \cos \omega_2 t \tag{3.129}$$

For FM modulation the phase deviation is therefore given by

$$\phi(t) = \beta_1 \sin \omega_1 t + \beta_2 \sin \omega_2 t \tag{3.130}$$

where $\beta_1 = Af_d/f_1$ and $\beta_2 = Bf_d/f_2$. The modulator output for this case becomes

$$x_c(t) = A_c \cos (\omega_c t + \beta_1 \sin \omega_1 t + \beta_2 \sin \omega_2 t) \tag{3.131}$$

which can be expressed

$$x_c(t) = A_c \text{Re}\{e^{j\omega_c t} e^{j\beta_1 \sin \omega_1 t} e^{j\beta_2 \sin \omega_2 t}\} \tag{3.132}$$

Using (3.102) we can write

$$e^{j\beta_1 \sin \omega_1 t} = \sum_{n=-\infty}^{\infty} J_n(\beta_1) e^{jn\omega_1 t} \tag{3.133a}$$

and

$$e^{j\beta_2 \sin \omega_2 t} = \sum_{m=-\infty}^{\infty} J_m(\beta_2) e^{jm\omega_2 t} \tag{3.133b}$$

The modulator output can therefore be written

$$x_c(t) = A_c \text{Re} \left\{ e^{j\omega_c t} \sum_{n=-\infty}^{\infty} J_n(\beta_1) e^{jn\omega_1 t} \sum_{m=-\infty}^{\infty} J_m(\beta_2) e^{jm\omega_2 t} \right\} \tag{3.134}$$

which, upon taking the real part, can be expressed

$$x_c(t) = A_c \sum_{n=-\infty}^{\infty} \sum_{m=-\infty}^{\infty} J_n(\beta_1) J_m(\beta_2) \cos(\omega_c + n\omega_1 + m\omega_2)t \tag{3.135}$$

Examination of the signal $x_c(t)$ shows that it not only contains frequency components at $\omega_c + n\omega_1$ and $\omega_c + m\omega_2$, but contains frequency components at $\omega_c + n\omega_1 + m\omega_2$ for all combinations of n and m. Therefore, the spectrum of the modulator output due to a message signal consisting of the sum of two sinusoids contains additional components over the spectrum formed by the superposition of the two spectra resulting from the individual message components. This example therefore illustrates the nonlinear nature of angle modulation. The spectrum resulting from a message signal consisting of the sum of two sinusoids is shown in Figure 3.30 for the case in which $\beta_1 = \beta_2$ and $f_2 = 12f_1$.

FIGURE 3.30 Aplitude spectrum for (3.135) with $\beta_1 = \beta_2 = 2$ and $f_2 = 12f_1$.

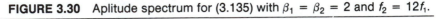

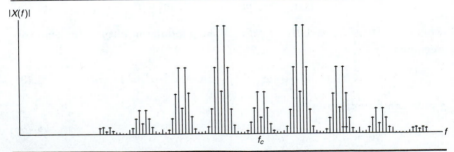

FIGURE 3.31   Frequency modulation utilizing narrowband-to-wideband conversion

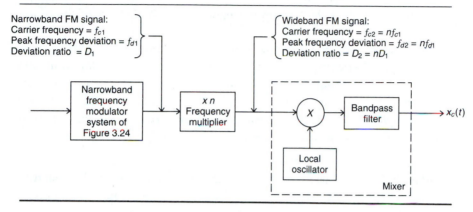

Narrowband-to-Wideband Conversion

One technique for generating wideband FM is illustrated in Figure 3.31. The carrier frequency of the narrowband frequency modulator is f_{c1}, and the peak frequency deviation is f_{d1}. The frequency multiplier multiplies the argument of the input sinusoid by n. In other words, if the input of a frequency multiplier is

$$x(t) = A_c \cos [\omega_0 t + \phi(t)] \tag{3.136}$$

the output of the frequency multiplier is

$$y(t) = A_c \cos [n\omega_0 t + n\phi(t)] \tag{3.137}$$

Assuming that the output of the local oscillator is

$$e_{LO}(t) = 2 \cos \omega_{LO} t \tag{3.138}$$

results in

$$e(t) = A_c \cos [(n\omega_0 + \omega_{LO})t + n\phi(t)]$$
$$+ A_c \cos [(n\omega_0 - \omega_{LO})t + n\phi(t)] \tag{3.139}$$

for the multiplier output. This signal is then filtered, using a bandpass filter having center frequency ω_c, given by

$$\omega_c = n\omega_0 + \omega_{LO} \qquad \text{or} \qquad \omega_c = n\omega_0 - \omega_{LO}$$

This yields the output

$$x_c(t) = A_c \cos [\omega_c t + n\phi(t)] \tag{3.140}$$

The bandwidth of the bandpass filter is chosen in order to pass the desired term in (3.139). One can use Carson's rule to determine the bandwidth of the bandpass filter if the transmitted signal is to contain 98% of the power in $x_c(t)$.

The central idea in narrowband-to-wideband conversion is that the frequency multiplier changes both the carrier frequency and the deviation ratio by a factor of n, whereas the mixer changes the effective carrier frequency but does not affect the deviation ratio.

This technique of implementing wideband frequency modulation is known as *indirect frequency modulation*.

▼ EXAMPLE 3.7   A narrowband-to-wideband converter is implemented as shown in Figure 3.31. The output of the narrowband frequency modulator is given by (3.136) with ω_0 given by 2π (100,000). The peak frequency deviation of $\phi(t)$ is 50 Hz and the bandwidth of $\phi(t)$ is 500 Hz. The wideband output $x_c(t)$ is to have a carrier frequency of 85 MHz and a deviation ratio of 5. Determine the frequency multiplier factor, n. Also determine two possible local oscillator frequencies. Finally, determine the center frequency and the bandwidth of the bandpass filter.

SOLUTION   The deviation ratio at the output of the narrowband modulator is

$$D_1 = \frac{f_{d1}}{W} = \frac{50}{500} = 0.1 \tag{3.141}$$

The frequency multiplier factor is

$$n = \frac{D_2}{D_1} = \frac{5}{0.1} = 50 \tag{3.142}$$

Thus, the carrier frequency at the output of the narrowband modulator is

$$nf_{c1} = 50(100,000) = 5 \text{ MHz} \tag{3.143}$$

The two permissible frequencies for the local oscillator are

$$85 + 5 = 90 \text{ MHz} \tag{3.144a}$$

and

$$85 - 5 = 80 \text{ MHz} \tag{3.144b}$$

The center frequency of the bandpass filter must be equal to the desired carrier frequency of the wideband output. Thus the center frequency of bandpass filter is 85 MHz. The bandwidth of the bandpass filter is established using Carson's rule. From (3.120) we have

$$B = 2(D + 1) W = 2(5 + 1)(500) \qquad (3.145a)$$

Thus

$$B = 6{,}000 \text{ Hz} \qquad (3.145b)$$

Demodulation of Angle-Modulated Signals

The demodulation of an FM signal requires a circuit that yields an output proportional to the frequency deviation of the input. Such circuits are known as *discriminators*. If the input to an ideal discriminator is the angle-modulated signal

$$x_r(t) = A_c \cos [\omega_c t + \phi(t)] \qquad (3.146)$$

the output of the ideal discriminator is

$$y_D(t) = \frac{1}{2\pi} K_D \frac{d\phi}{dt} \qquad (3.147)$$

For FM, $\phi(t)$ is given by

$$\phi(t) = 2\pi f_d \int^t m(\alpha) \, d\alpha \qquad (3.148)$$

so that (3.147) becomes

$$y_D(t) = K_D f_d m(t) \qquad (3.149)$$

The constant K_D is known as the *discriminator constant* and has units of volts per hertz. Since an ideal discriminator yields an output signal proportional to the frequency deviation from a carrier, it has a linear frequency-to-voltage transfer function, which passes through zero at $f = f_c$. This is illustrated in Figure 3.32.

FIGURE 3.32   Ideal discriminator characteristic

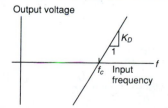

The system characterized by Figure 3.32 also can be used to demodulate PM signals. Since $\phi(t)$ is proportional to $m(t)$ for PM, $y_D(t)$ given by (3.147) is proportional to the time derivative of $m(t)$ for PM inputs. Integration of the discriminator output yields a signal proportional to $m(t)$. Thus a demodulator for PM can be implemented as an FM discriminator followed by an integrator. We define the output of a PM discriminator as

$$y_D(t) = K_D k_p m(t) \tag{3.150}$$

It will be clear from the text whether $y_D(t)$ and K_D refer to an FM or a PM system.

An approximation to the characteristic illustrated in Figure 3.32 can be obtained by the use of a differentiator followed by an envelope detector, as shown in Figure 3.33. If the input to the differentiator is

$$x_r(t) = A_c \cos [\omega_c t + \phi(t)] \tag{3.151}$$

the output of the differentiator is

$$e(t) = -A_c \left(\omega_c + \frac{d\phi}{dt} \right) \sin [\omega_c t + \phi(t)] \tag{3.152}$$

This is exactly the same form as an AM signal, except for the phase deviation $\phi(t)$. Thus, after differentiation, envelope detection can be used to recover the message signal. The envelope of $e(t)$ is

$$y(t) = A_c \left(\omega_c + \frac{d\phi}{dt} \right) \tag{3.153}$$

and is always positive if

$$\omega_c > - \frac{d\phi}{dt}, \text{ all } t$$

which is certainly the typical case. With this assumption, the output of the envelope detector is

$$y_D(t) = A_c \frac{d\phi}{dt} = 2\pi A_c f_d m(t) \tag{3.154}$$

FIGURE 3.33   FM discriminator

FIGURE 3.34   FM discriminator with bandpass limiter

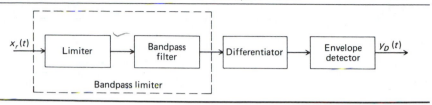

assuming that the dc term, $A_c\omega_c$, is removed. Comparing (3.154) and (3.149) shows that the discriminator constant for this discriminator is

$$K_D = 2\pi A_c \tag{3.155}$$

We will see later that interference and channel noise perturb the amplitude A_c of $x_r(t)$. In order to ensure that the amplitude at the input to the differentiator is constant, a *limiter* is placed before the differentiator. The output of the limiter is a signal of squarewave type, which is $K \operatorname{sgn}[x_r(t)]$. A bandpass filter having center frequency ω_c is then placed after the limiter to convert the signal back to the sinusoidal form required by the differentiator to yield the response defined by (3.152). The cascade combination of a limiter and a bandpass filter is known as a *bandpass limiter*. The complete discriminator is illustrated in Figure 3.34.

The process of differentiation can often be realized using a time-delay implementation, as shown in Figure 3.35. The signal $e(t)$, which is the input to the envelope detector, is given by

$$e(t) = x_r(t) - x_r(t - \tau) \tag{3.156}$$

which can be written

$$\frac{e(t)}{\tau} = \frac{x_r(t) - x_r(t - \tau)}{\tau} \tag{3.157}$$

Since, by definition,

$$\lim_{\tau \to 0} \frac{e(t)}{\tau} = \lim_{\tau \to 0} \frac{x_r(t) - x_r(t - \tau)}{\tau} = \frac{dx_r(t)}{dt} \tag{3.158}$$

it follows that for small τ,

$$e(t) \cong \tau \frac{dx_r(t)}{dt} \tag{3.159}$$

FIGURE 3.35   Discriminator implementation using delay and envelope detection

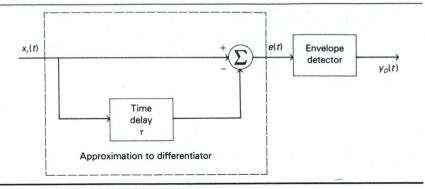

This is, except for the constant factor τ, identical to the envelope detector input shown in Figure 3.33 and defined by (3.152). The resulting discriminator constant K_D is $2\pi A_c\tau$.

There are many other techniques that can be used to implement a discriminator. In Section 3.4 we will examine the phase-lock loop, which is an especially attractive implementation.

▼| **EXAMPLE 3.8**   Consider the simple RC network shown in Figure 3.36(a). The transfer function is

$$H(f) = \frac{R}{R + \dfrac{1}{j2\pi fC}} = \frac{j2\pi fRC}{1 + j2\pi fRC} \tag{3.160}$$

The amplitude response is shown in Figure 3.36(b). If all frequencies present in the input are low, so that

$$f \ll \frac{1}{2\pi RC}$$

the transfer function can be approximated by

$$H(f) = j2\pi fRC \tag{3.161}$$

Thus, for small f, the RC network has the linear amplitude-frequency characteristic required of an ideal discriminator.

Equation (3.161) illustrates that for small f, the RC filter acts as a differentiator with gain RC. Thus, the RC network can be used in place of the differentiator in Figure 3.34 to yield a discriminator with

$$K_D = 2\pi A_c RC \qquad (3.162)$$

Example 3.8 illustrates the essential components of a frequency discriminator, a circuit that has an amplitude response linear with frequency and an envelope detector. However, a highpass filter does not in general yield a practical implementation. This can be seen from the expression for K_D. Clearly the 3-dB frequency of the filter, $1/2\pi RC$, must exceed the carrier frequency f_c. In commercial FM broadcasting, the carrier frequency at the discriminator input, i.e., the IF frequency, is on the order of 10 MHz. As a result, the discriminator constant K_D is very small indeed.

A solution to the problem of a very small K_D is to use a bandpass filter, as illustrated in Figure 3.37. But, as shown in Figure 3.37(a), the region of linear operation is often unacceptably small. In addition, use of a bandpass filter results in a dc bias on the discriminator output. This dc bias could of

FIGURE 3.36   Implementation of a simple discriminator. (a) RC network. (b) Transfer function. (c) Simple discriminator.

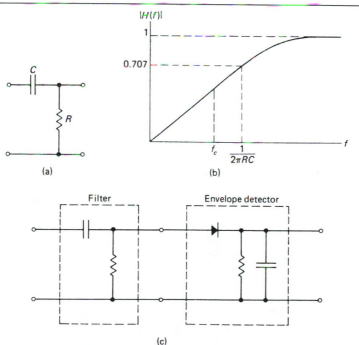

(a)

(b)

(c)

FIGURE 3.37 Derivation of balanced discriminator. (a) Bandpass filter. (b) Stagger-tuned bandpass filters. (c) Amplitude response $H(f)$ of balanced discriminator. (d) Balanced discriminator.

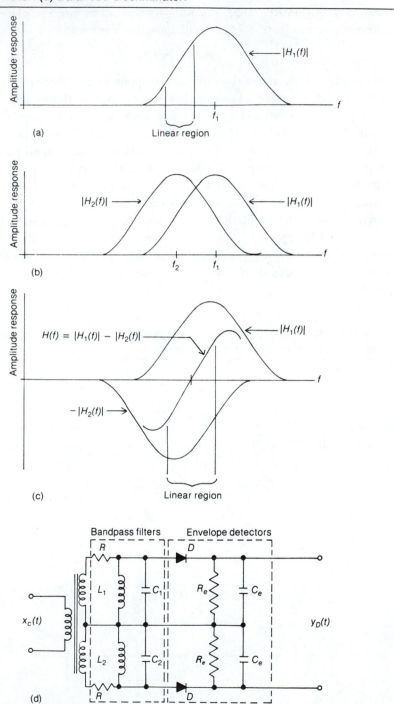

course be removed by a blocking capacitor, but the blocking capacitor would negate an inherent advantage of FM—namely, that FM has dc response. One can solve these problems by using two filters with staggered center frequencies f_1 and f_2, as shown in Figure 3.37(b). The magnitudes of the envelope detector outputs following the two filters are proportional to $|H_1(f)|$ and $|H_2(f)|$. Subtracting these two outputs yields the overall characteristic

$$H(f) = |H_1(f)| - |H_2(f)| \qquad (3.163)$$

as shown in Figure 3.37(c). The combination is linear over a wider frequency range than would be the case for either filter used alone, and it is clearly possible to make $H(f_c) = 0$.

There are several techniques that can be used to combine the outputs of two envelope detectors. A differential amplifier can be used, for example. Another alternative, using a strictly passive circuit, is shown in Figure 3.37(d). A center-tapped transformer supplies the input signal $x_c(t)$ to the inputs of the two bandpass filters. The center frequencies of the two bandpass filters are given by

$$f_i = \frac{1}{2\pi\sqrt{L_i C_i}} \qquad (3.164)$$

for $i = 1, 2$. The envelope detectors are formed by the diodes and the resistor-capacitor combinations $R_e C_e$. The output of the upper envelope detector is proportional to $|H_1(f)|$, and the output of the lower envelope detector is proportional to $|H_2(f)|$. The output of the upper envelope detector is the positive portion of its input envelope, and the output of the lower envelope detector is the negative portion of its input envelope. Thus $y_D(t)$ is proportional to $|H_1(f)| - |H_2(f)|$. This system is known as a *balanced discriminator* because the response to the undeviated carrier is balanced so that the net response is zero.

3.3   INTERFERENCE

We now consider the effect of interference in communication systems. In real-world systems interference occurs from various sources, such as radio-frequency emissions from transmitters having carrier frequencies close to that of the carrier being demodulated. We also study interference because the analysis of systems in the presence of interference provides us with important insights into the behavior of systems operating in the presence of noise, which is the topic of Chapter 6. In this section we consider both linear modulation and angle modulation. It is important to understand the very different manner in which these two systems behave in the presence of interference.

Interference in Linear Modulation

As a simple case of linear modulation in the presence of interference, we consider the received signal having the spectrum (single-sided) shown in Figure 3.38. The received signal consists of three components: a carrier component, a pair of sidebands representing a sinusoidal message signal and an undesired interfering tone of frequency $f_c + f_i$. The input to the demodulator is therefore

$$x_c(t) = A_c \cos \omega_c t + A_i \cos (\omega_c + \omega_i)t + A_m \cos \omega_m t \cos \omega_c t \quad (3.165)$$

Multiplying $x_c(t)$ by $2 \cos \omega_c t$ and lowpass-filtering (coherent demodulation) yields

$$y_D(t) = A_m \cos \omega_m t + A_i \cos \omega_i t \quad (3.166)$$

where we have assumed that the interference component is passed by the filter and that the dc term resulting from the carrier is blocked. From this simple example we see that the signal and interference are additive at the receiver output if the interference is additive at the receiver input. This result was obtained because the coherent demodulator operates as a linear demodulator.

The effect of interference with envelope detection is quite different because of the nonlinear nature of the envelope detector. The analysis with envelope detection is much more difficult than the coherent demodulation case. Some insight can be gained by writing $x_c(t)$ in a form that leads to the phasor diagram. In order to develop the phasor diagram we write (3.165) in the form

$$x_c(t) = \text{Re}[A_c e^{j\omega_c t} + A_i e^{j\omega_c t} e^{j\omega_i t} + \tfrac{1}{2}A_m e^{j\omega_c t} e^{j\omega_m t} + \tfrac{1}{2}A_m e^{j\omega_c t} e^{-j\omega_m t}] \quad (3.167)$$

FIGURE 3.38 Assumed received-signal spectrum

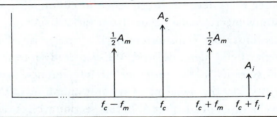

FIGURE 3.39 Phasor diagrams illustrating interference. (a) Phasor diagram without interference. (b) Phasor diagram with interference.

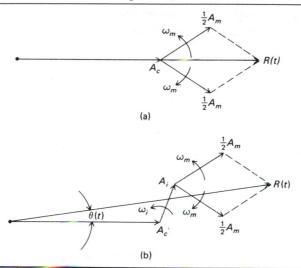

which also can be written as

$$x_r(t) = \text{Re} \left\{ e^{j\omega_c t}[A_c + A_i e^{j\omega_i t} + \tfrac{1}{2} A_m e^{j\omega_m t} + \tfrac{1}{2} A_m e^{-j\omega_m t}] \right\} \qquad (3.168)$$

The phasor diagram is constructed with respect to the carrier by taking the carrier frequency as equal to zero. The phasor diagrams are illustrated in Figure 3.39, both with and without interference. The output of an ideal envelope detector is $R(t)$ in both cases. The phasor diagrams illustrate that interference induces both an amplitude distortion and a phase deviation.

The effect of interference with envelope detection is determined by writing (3.165) as

$$x_r(t) = A_c \cos \omega_c t + A_m \cos \omega_m t \cos \omega_c t$$
$$+ A_i[\cos \omega_c t \cos \omega_i t - \sin \omega_c t \sin \omega_i t] \qquad (3.169)$$

which is

$$x_r(t) = [A_c + A_m \cos \omega_m t + A_i \cos \omega_i t] \cos \omega_c t - A_i \sin \omega_i t \sin \omega_c t \qquad (3.170)$$

If $A_c \gg A_i$, which is the usual case of interest, the second term in (3.170) is negligible compared to the first term. For this case, the output of the envelope detector is

$$y_D(t) \cong A_m \cos \omega_m t + A_i \cos \omega_i t \qquad (3.171)$$

assuming that the dc term is blocked. Thus, for the small interference case, envelope detection and coherent demodulation are essentially equivalent.

If $A_c \ll A_i$, the assumption cannot be made that the second term of (3.170) is negligible, and the output is significantly different. To show this, (3.165) is rewritten as

$$x_r(t) = A_c \cos(\omega_c + \omega_i - \omega_i)t + A_i \cos(\omega_c + \omega_i)t$$
$$+ A_m \cos\omega_m t \cos(\omega_c + \omega_i - \omega_i)t \qquad (3.172)$$

which, when we use appropriate trigonometric identities, becomes

$$x_r(t) = A_c[\cos(\omega_c + \omega_i)t \cos\omega_i t + \sin(\omega_c + \omega_i)t \sin\omega_i t]$$
$$+ A_i \cos(\omega_c + \omega_i)t + A_m \cos\omega_m t[\cos(\omega_c + \omega_i)t \cos\omega_i t$$
$$+ \sin(\omega_c + \omega_i)t \sin\omega_i t] \qquad (3.173)$$

Equation (3.173) can also be written as

$$x_r(t) = [A_i + A_c \cos\omega_i t + A_m \cos\omega_m t \cos\omega_i t] \cos(\omega_c + \omega_i)t$$
$$+ [A_c \sin\omega_i t + A_m \cos\omega_m t \sin\omega_i t] \sin(\omega_c + \omega_i)t \qquad (3.174)$$

If $A_i \gg A_c$, the second term in (3.174) is negligible with respect to the first term. It follows that the envelope detector output is approximated by

$$y_D(t) \cong A_c \cos\omega_i t + A_m \cos\omega_m t \cos\omega_i t \qquad (3.175)$$

At this point, several observations are in order. In envelope detectors, the largest high-frequency component is treated as the carrier. If $A_c \gg A_i$, the effective demodulation carrier has a frequency ω_c, whereas if $A_i \gg A_c$, the *effective* carrier frequency becomes the interference frequency $\omega_c + \omega_i$.

The spectra of the envelope detector output are illustrated in Figure 3.40 for $A_c \gg A_i$ and for $A_c \ll A_i$. For $A_c \gg A_i$ the interfering tone simply appears as a sinusoidal component, having frequency f_i, at the output of the envelope detector. This illustrates that for $A_c \gg A_i$, the envelope detector performs as a linear demodulator. The situation is much different for $A_c \ll A_i$, as can be seen from (3.175) and Figure 3.40(b). For this case we see that the sinusoidal message signal, having frequency f_m, modulates the interference tone. The output of the envelope detector has a spectrum that reminds us of the spectrum of an AM signal with carrier frequency f_i and sideband components at $f_i + f_m$ and $f_i - f_m$. The message signal is effectively lost. This degradation of the desired signal is called the *threshold effect* and is a consequence of the nonlinear nature of the envelope detector. We shall study the threshold effort in detail in Chapter 6 when we investigate the effect of noise in analog systems.

FIGURE 3.40 Envelope detector output spectra. (a) $A_c \gg A_i$, (b) $A_c \ll A_i$.

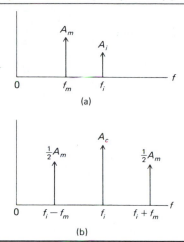

(a)

(b)

Interference in Angle Modulation

We now consider the effect of interference in angle modulation. We will see that the effect of interference in angle modulation is quite different from what was observed in linear modulation. Furthermore, we will see that the effect of interference in an FM system can be reduced by placing a lowpass filter at the discriminator output. We will consider this problem in considerable detail since the results will provide significant insight into the behavior of FM discriminators operating in the presence of noise.

Assume that the input to a PM or FM ideal discriminator is an unmodulated carrier plus an interfering tone at frequency $\omega_c + \omega_i$. Thus the input to the discriminator is assumed to have the form

$$x_r(t) = A_c \cos \omega_c t + A_i \cos (\omega_c + \omega_i)t \tag{3.176}$$

which can be written as

$$x_r(t) = A_c \cos \omega_c t + A_i \cos \omega_i t \cos \omega_c t - A_i \sin \omega_i t \sin \omega_c t \tag{3.177}$$

Using Equations (3.50) through (3.54), the preceding expression can be written as

$$x_r(t) = R(t) \cos [\omega_c t + \psi(t)] \tag{3.178}$$

where

$$R(t) = \sqrt{(A_c + A_i \cos \omega_i t)^2 + (A_i \sin \omega_i t)^2} \tag{3.179}$$

and

$$\psi(t) = \tan^{-1} \frac{A_i \sin \omega_i t}{A_c + A_i \cos \omega_i t} \tag{3.180}$$

If $A_c \gg A_i$, Equations (3.179) and (3.180) can be approximated

$$R(t) = A_c + A_i \cos \omega_i t \tag{3.181}$$

and

$$\psi(t) = \frac{A_i}{A_c} \sin \omega_i t \tag{3.182}$$

Thus (3.178) is

$$x_r(t) = A_c \left(1 + \frac{A_i}{A_c} \cos \omega_i t \right) \cos \left(\omega_c t + \frac{A_i}{A_c} \sin \omega_i t \right) \tag{3.183}$$

Since the instantaneous phase deviation $\psi(t)$ is given by

$$\psi(t) = \frac{A_i}{A_c} \sin \omega_i t \tag{3.184}$$

the ideal discriminator output for PM is

$$y_D(t) = K_D \frac{A_i}{A_c} \sin \omega_i t \tag{3.185}$$

and for FM is

$$y_D(t) = \frac{1}{2\pi} K_D \frac{d}{dt} \frac{A_i}{A_c} \sin \omega_i t \tag{3.186}$$

$$= K_D \frac{A_i}{A_c} f_i \cos \omega_i t$$

For both cases, the discriminator output is a sinusoid of frequency f_i. The amplitude of the discriminator output, however, is proportional to the frequency f_i for the FM case. It can be seen that for small f_i, the interfering tone has less effect on the FM system than on the PM system and that the opposite is true for large values of f_i. Values of $f_i > W$, the bandwidth of $m(t)$, are of little interest, since they can be removed by a lowpass filter following the discriminator.

For larger values of A_i the assumption that $A_i \ll A_c$ cannot be made and (3.186) no longer can describe the discriminator output. If the condition

$A_i \ll A_c$ does not hold, the discriminator is not operating above threshold and the analysis becomes much more difficult. Some insight into this case can be obtained from the phasor diagram, which is obtained by writing (3.176) in the form

$$x_r(t) = \text{Re}\{[A_c + A_i \, e^{j\omega_i t}] \, e^{j\omega_c t}\} \tag{3.187}$$

The term in square brackets defines the phasor. The phasor diagram is shown in Figure 3.41(a). The carrier phase is taken as the reference and the interference phase is

$$\theta(t) = \omega_i t \tag{3.188}$$

Approximations to the phase of the resultant $\psi(t)$ can be determined using the phasor diagram.

From Figure 3.41(b) we see that the magnitude of the discriminator output will be small when $\theta(t)$ is near zero. This results because for $\theta(t)$ near zero, a given change in $\theta(t)$ will result in a much smaller change in $\psi(t)$. Using the relationship between arc length, s, angle, θ, and radius, r, which is $s = \theta r$, we obtain

$$s = \theta(t)A_i \approx (A_c + A_i) \, \psi(t), \qquad \theta(t) \approx 0 \tag{3.189}$$

Solving for $\psi(t)$ yields

$$\psi(t) \approx \frac{A_i}{A_c + A_i} \, \omega_i t \tag{3.190}$$

FIGURE 3.41 Phasor diagram for carrier plus single-tone interference. (a) Phasor diagram for general $\theta(t)$. (b) Phasor diagram for $\theta(t) \approx 0$. (c) Phasor diagram for $\theta(t) \approx \pi$ and $A_i \lesssim A_c$. (d) Phasor diagram for $\theta(t) \approx \pi$ and $A_i \gtrsim A_c$.

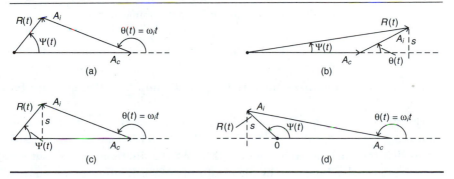

Since the discriminator output is defined by

$$y_D(t) = \frac{K_D}{2\pi} \frac{d\psi}{dt} \tag{3.191}$$

we have

$$y_D(t) = K_D \frac{A_i}{A_c + A_i} f_i, \qquad \theta(t) \approx 0 \tag{3.192}$$

This is a positive quantity for $f_i > 0$ and a negative quantity for $f_i < 0$.

If A_i is slightly less than A_c, denoted $A_i \lesssim A_c$, and $\theta(t)$ is near π, a small positive change in $\theta(t)$ will result in a large negative change in $\psi(t)$. The result will be a negative spike appearing at the discriminator output. From Figure 3.41(c) we can write

$$s = A_i(\pi - \theta(t)) \approx (A_c - A_i)\psi(t), \qquad \theta(t) \approx \pi \tag{3.193}$$

which can be expressed

$$\psi(t) \approx \frac{A_i(\pi - \omega_i t)}{A_c - A_i} \tag{3.194}$$

Using (3.191) we see that the discriminator output is

$$y_D(t) \approx - K_D \frac{A_i}{A_c - A_i} f_i, \qquad \theta(t) \approx \pi \tag{3.195}$$

This is a negative quantity for $f_i > 0$ and a positive quantity for $f_i < 0$.

If A_i is slightly greater than A_c, denoted $A_i \gtrsim A_c$, and $\theta(t)$ is near π, a small positive change in $\theta(t)$ will result in a large positive change in $\psi(t)$. The result will be a positive spike appearing at the discriminator output. From Figure 3.41(d) we can write

$$s = A_i[\pi - \theta(t)] \approx (A_i - A_c)[\pi - \psi(t)], \quad \theta(t) \approx \pi \tag{3.196}$$

Solving for $\psi(t)$ and differentiating gives the discriminator output

$$y_D(t) \approx K_D \frac{A_i}{A_i - A_c} f_i \tag{3.197}$$

Note that this is a positive quantity for $f_i > 0$ and a negative quantity for $f_i < 0$.

The phase deviation and discriminator output waveforms are shown in Figure 3.42 for $A_i = 0.1A_c$, $A_i = 0.9A_c$, and $A_i = 1.1A_c$. Figure 3.42(a) illustrates that for small A_i the phase deviation and the discriminator output are nearly sinusoidal as predicted by the results of the small interference analysis given in (3.184) and (3.185). For $A_i = 0.9A_c$, we see that we have a negative

FIGURE 3.42 Phase deviation and discriminator outputs due to interference. (a) Phase deviation and discriminator output for $A_i = 0.2A_c$. (b) Phase deviation and discriminator output for $A = 0.9A_c$. (c) Phase deviation and discriminator output for $A = 1.1A_c$.

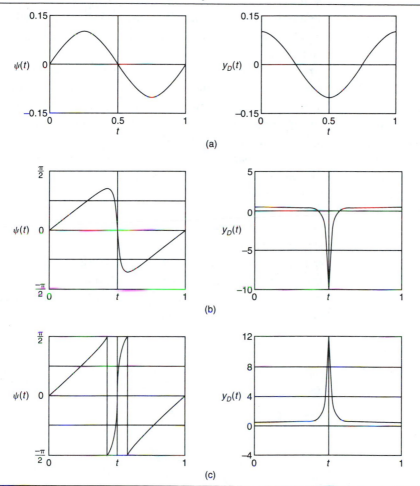

spike at the discriminator output as predicted by (3.195). For $A_c = 1.1A_c$, we have a positive spike at the discriminator output as predicted by (3.197).

Note that for $A_i > A_c$ the origin of the phasor diagram is encircled as $\theta(t)$ goes from 0 to 2π. In other words $\psi(t)$ goes from 0 to 2π as $\theta(t)$ goes from 0 to 2π. The origin is not encircled if $A_i < A_c$. Thus the integral

$$\int_T \left(\frac{d\psi}{dt}\right) dt = \begin{cases} 2\pi, & A_i > A_c \\ 0, & A_i < A_c \end{cases} \qquad (3.198)$$

where T is the time required for $\theta(t)$ to go from $\theta(t) = 0$ to $\theta(t) = 2\pi$. In other words $T = 2\pi/\omega_i$. Thus the area under the discriminator output curve is 0 for parts (a) and (b) of Figure 3.42 and $2\pi K_D$ for the discriminator output curve in Figure 3.42(c).

The origin encirclement phenomenon will be revisited in Chapters 5 and 6. An understanding of the results presented here will prove beneficial when we examine noise effects in FM demodulation.

For operation above threshold $A_i \ll A_c$, the severe effect of interference on FM for large f_i can be reduced by placing a filter, called a *deemphasis filter*, at the FM discriminator output. This filter is typically a simple RC lowpass filter with a 3-dB frequency considerably less than the modulation bandwidth W. The deemphasis filter effectively reduces the interference for large f_i, as shown in Figure 3.43. For large frequencies, the magnitude of the transfer function of a first-order filter is approximately $1/f$. Since the amplitude of the interference increases linearly with f_i for FM, the output is constant for large f_i, as shown in Figure 3.43.

Since $f_3 < W$, the deemphasis filter distorts the message signal in addition to combating interference. The distortion can be avoided by passing the message through a *preemphasis filter* that has a transfer function equal to the reciprocal of the transfer function of the deemphasis filter. Since the transfer function of the cascade combination of the preemphasis and deemphasis filters is unity, there is no detrimental effect on the modulation. This yields the system shown in Figure 3.44.

FIGURE 3.43   Amplitude of discriminator output due to interference

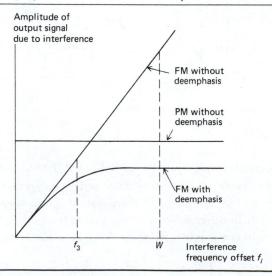

FIGURE 3.44   FM system with preemphasis and deemphasis

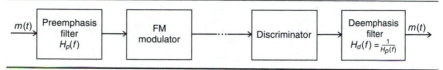

The improvement offered by the use of preemphasis and deemphasis is not gained without a price. The preemphasis filter amplifies the high-frequency components, and this results in increased deviation. The end result is an increase in bandwidth. However, many signals of practical interest have little power in the higher-frequency components, so the increased deviation resulting from the use of preemphasis is negligible. We shall see in Chapter 6, when noise is studied, that the use of preemphasis and deemphasis often provides sufficient improvement to be definitely worth the price.

3.4   FEEDBACK DEMODULATORS

We have already studied the technique of FM to AM conversion for demodulating an angle-modulated signal. We shall see in Chapter 6 that improved performance in the presence of noise can be gained by utilizing a feedback demodulator. In this section we examine the basic operation of several types of feedback demodulators. Such systems are widely used in today's communication systems not only because of their superior performance but also because of their ease of implementation using inexpensive integrated circuits. The phase-lock loop is the basic building block.

Phase-Lock Loops for FM Demodulation

A block diagram of a phase-lock loop (PLL) is shown in Figure 3.45. Such a system generally contains four basic elements:

1. A phase detector
2. A loop filter
3. A loop amplifier
4. A voltage-controlled oscillator (VCO)

In order to understand the operation of a PLL, assume that the input signal is given by

$$x_r(t) = A_c \cos \left[\omega_c t + \phi(t) \right] \tag{3.199}$$

FIGURE 3.45   Phase-lock loop

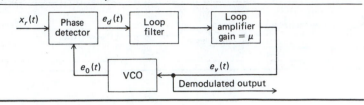

and that the VCO output signal is given by

$$e_0(t) = A_v \sin [\omega_c t + \theta(t)] \qquad (3.200)$$

There are many different types of phase detectors, all having different operating properties. For our application, we assume that the phase detector is a multiplier followed by a lowpass filter to remove the second harmonic of the carrier. We also assume that an inverter is present to remove the minus sign resulting from the multiplication. With these assumptions, the output of the phase detector becomes

$$e_d(t) = \tfrac{1}{2} A_c A_v K_d \sin [\phi(t) - \theta(t)] \qquad (3.201)$$

where K_d is the constant associated with the multiplier in the phase detector.

The output of the phase detector is filtered, amplified, and applied to the VCO. A VCO is essentially a frequency modulator, the frequency deviation of the output, $d\theta/dt$, being proportional to the input signal. In other words,

$$\frac{d\theta}{dt} = K_v e_v(t) \quad \text{rad/s} \qquad (3.202)$$

which yields

$$\theta(t) = K_v \int^t e_v(\alpha) \, d\alpha \qquad (3.203)$$

The parameter K_v is known as the VCO constant and is measured in radians per second per unit of input.

Since the output of the phase detector is determined by the phase deviation of the PLL input and VCO output, we can model a PLL without regard to the carrier frequency ω_c. Such a model is shown in Figure 3.46. This model is known as the *nonlinear model* because of the sinusoidal nonlinearity in the phase detector. When the PLL is operating *in lock*, the VCO phase $\theta(t)$ is a good estimate of the input-phase deviation $\phi(t)$. For this mode of operation, the phase error $\phi(t) - \theta(t)$ is small, and

$$\sin [\phi(t) - \theta(t)] \cong \phi(t) - \theta(t) \qquad (3.204)$$

FIGURE 3.46   Nonlinear PLL model

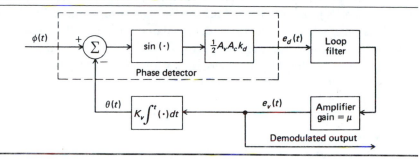

Thus, if the phase error is small, the sinusoidal nonlinearity can be neglected and the PLL becomes a linear feedback control system, which is easily analyzed. The linear model that results is illustrated in Figure 3.47. While both the nonlinear and linear models involve $\theta(t)$ and $\phi(t)$ rather than $x_r(t)$ and $e_0(t)$, knowledge of $\theta(t)$ and $\phi(t)$ fully determines $x_r(t)$ and $e_0(t)$, as can be seen from (3.199) and (3.200).

If $\theta(t) \cong \phi(t)$, it follows that

$$\frac{d\theta(t)}{dt} \cong \frac{d\phi(t)}{dt} \tag{3.205}$$

and the VCO frequency deviation is a good estimate of the input frequency deviation. For an FM system, the input frequency deviation is proportional

FIGURE 3.47   Linear PLL model

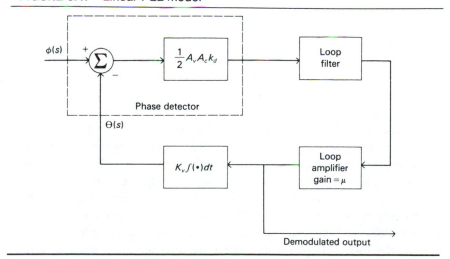

to the message signal $m(t)$. Since the VCO frequency deviation is proportional to the VCO input $e_v(t)$, it follows that $e_v(t)$ is proportional to $m(t)$ if (3.205) is satisfied. Thus $e_v(t)$ is the demodulated output for FM systems.

   We will now show that the phase error signal tends to drive the PLL into lock. In order to simplify the analysis, we assume that the loop filter is replaced by a short circuit. For this case,

$$e_v(t) = \tfrac{1}{2}\mu A_c A_v K_d \sin [\phi(t) - \theta(t)] \tag{3.206}$$

which yields

$$\theta(t) = K_t \int^t \sin [\phi(\alpha) - \theta(\alpha)] \, d\alpha \tag{3.207}$$

where K_t is the total effective loop gain, which is

$$K_t = \tfrac{1}{2}\mu A_c A_v K_d K_v \tag{3.208}$$

The expression for $\theta(t)$ can be differentiated to yield

$$\frac{d\theta}{dt} = K_t \sin [\phi(t) - \theta(t)] \tag{3.209}$$

Assume that the input to the FM modulator is a unit step so that the frequency deviation $d\phi/dt$ is a unit step of magnitude $\Delta\omega$. Let the phase error $\phi(t) - \theta(t)$ be denoted $\psi(t)$. This yields

$$\frac{d\theta}{dt} = \frac{d\phi}{dt} - \frac{d\psi}{dt} = \Delta\omega - \frac{d\psi}{dt} = K_t \sin \psi(t), \qquad t \geq 0 \tag{3.210}$$

or

$$\frac{d\psi}{dt} + K_t \sin \psi(t) = \Delta\omega \tag{3.211}$$

This equation is sketched in Figure 3.48. It relates the frequency error and the phase error.

   A plot of the derivative of a function versus the function is known as a *phase-plane plot* and tells us much about the operation of a nonlinear system. The PLL must operate with a phase error $\psi(t)$ and a frequency error $d\psi/dt$ that are consistent with (3.211). To demonstrate that the PLL achieves lock, assume that it is operating with zero phase and frequency errors prior to the application of the frequency step. When the step in frequency is applied, the frequency error becomes $\Delta\omega$. This establishes the initial operating point, point B in Figure 3.46, assuming $\Delta\omega > 0$. In order to determine the trajectory of the operating point, we need only recognize that since dt, a time

FIGURE 3.48   Phase-plane plot

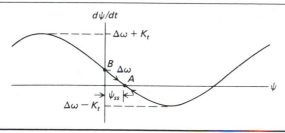

increment, is always a positive quantity, $d\psi$ must be positive if $d\psi/dt$ is positive. Thus, in the upper half plane, ψ increases. In other words, the operating point moves from left to right in the upper half plane. In the same manner, the operating point moves from right to left in the lower half plane, the region for which $d\psi/dt$ is less than zero. Thus the operating point must move from point B to point A. When the operating point attempts to move from point A by a small amount, it is forced back to point A. Thus point A is a stable operating point and is the steady-state operating point of the system. The steady-state phase error is ψ_{ss}, and the steady-state frequency error is zero.

The preceding analysis illustrates that the loop locks only if there is an intersection of the operating curve with the $d\psi/dt = 0$ axis. Thus, if the loop is to lock, $\Delta\omega$ must be less than K_t. For this reason, K_t is shown as the *lock range* for the first-order PLL.

The phase-plane plot for a first-order PLL with a frequency-step input is illustrated in Figure 3.49. The loop gain is $2\pi(50)$, and four values for the frequency step are shown: $\Delta f = 12, 24, 48,$ and 55 Hz. The steady-state phase errors are indicated by A, B, and C for frequency-step values of 12, 24, and 48 Hz, respectively. For $\Delta f = 55$, the loop does not lock, as shown.

From the previous paragraphs we saw that the steady-state phase errors resulting from changes in the input frequency can be made as small as desired by increasing the total loop gain K_t. For $K_t \gg \Delta\omega$, the phase errors will be small and the linear model can be used. The relationship between $\theta(t)$ and $\phi(t)$ is, from (3.207) with the linear approximation made,

$$\theta(t) = K_t \int^t [\phi(\alpha) - \theta(\alpha)] \, d\alpha \qquad (3.212)$$

which yields the differential equation

$$\frac{d\theta(t)}{dt} + K_t\theta(t) = K_t\phi(t) \qquad (3.213)$$

FIGURE 3.49 Phase-plane plot of first-order PLL for several initial frequency errors

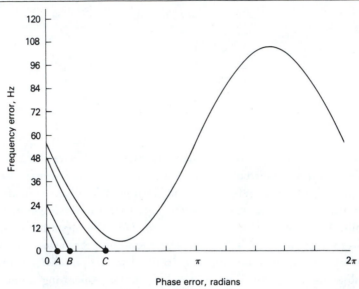

The loop transfer function, in terms of the Laplace transform variable s, is

$$H(s) = \frac{\Theta(s)}{\Phi(s)} = \frac{K_t}{s + K_t} \tag{3.214}$$

The unit impulse response is therefore

$$h(t) = K_t e^{-K_t t} u(t) \tag{3.215}$$

The limit of $h(t)$ as the loop gain K_t tends to infinity satisfies all properties of the delta function. Therefore,

$$\lim_{K_t \to \infty} K_t e^{-K_t t} u(t) = \delta(t) \tag{3.216}$$

which proves that for large loop gain $\theta(t) = \phi(t)$.

The PLL also can be used as a demodulator for phase-modulated signals by integrating the VCO input. Since the VCO input signal is proportional to the frequency deviation of the PLL input, the integral of this signal is proportional to the phase deviation of the PLL input.

We also have seen that operation of the PLL as a discriminator requires a large loop gain. This implies a large bandwidth. In general, very large values of loop gain cannot be used in practical applications without difficulty. However, the use of appropriate loop filters allows good performance to be achieved with reasonable values of loop gain and bandwidth. These filters make the analysis more complicated than our simple example, as we shall soon see.

There is one final observation we should make. The expression for loop gain shows that it is a function of the amplitude of the input signal. This means that a PLL must be designed for a given signal level. If that signal level changes, a new design may be necessary. In most practical applications, the dependence on loop gain is removed by placing a limiter on the loop input.

▼ **EXAMPLE 3.9**   The input to an FM modulator is $m(t) = Au(t)$. The resulting modulated carrier

$$x_c(t) = A_c \cos\left[\omega_c t + k_f A \int^t u(\alpha) \, d\alpha\right] \tag{3.217}$$

is to be demodulated using a first-order PLL. Determine the demodulated output.

SOLUTION   This problem will be solved using linear analysis and the Laplace transform. The Laplace transform of the loop transfer function, Equation (3.214), is

$$\frac{\Theta(s)}{\Phi(s)} = \frac{K_t}{K_t + s} \tag{3.218}$$

The phase deviation $\phi(t)$ is

$$\phi(t) = A k_f \int^t u(\alpha) \, d\alpha \tag{3.219}$$

which yields

$$\Phi(s) = \frac{A k_f}{s^2} \tag{3.220}$$

This gives

$$\Theta(s) = \frac{A K_t k_f}{s^2(s + K_t)} \tag{3.221}$$

The Laplace transform of the defining equation of the VCO, Equation (3.203), yields

$$E_v(s) = \frac{s}{K_v}\Theta(s)$$

(3.222)

so that

$$E_v(s) = \frac{Ak_f}{K_v}\frac{K_t}{s(s + K_t)}$$

(3.223)

Partial fraction expansion yields

$$E_v(s) = \frac{Ak_f}{K_v}\left(\frac{1}{s} - \frac{1}{s + K_t}\right)$$

(3.224)

Thus the demodulated output is given by

$$e_v(t) = \frac{Ak_f}{K_v}(1 - e^{-K_t t})u(t)$$

(3.225)

which is sketched in Figure 3.50 for k_f equal to K_v.

As the loop gain K_t is increased, the demodulated output clearly becomes a closer approximation to $m(t)$. It should be remembered that we have assumed that linear analysis is valid. This requires that the phase error be small. The validity of this assumption is easily checked by solving for the phase error, an exercise that is left to the problems.

The first-order PLL has a number of drawbacks that limit its use for most applications. Among these drawbacks are the limited lock range and the nonzero steady-state phase error to a step-frequency input. Both these problems can be solved by using a second-order PLL, which is obtained by using a loop filter of the form

$$F(s) = \frac{s + a}{s}$$

(3.226)

FIGURE 3.50 Demodulated output for $m(t) = Au(t)$

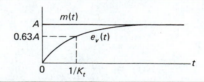

FIGURE 3.51   Linear PLL model

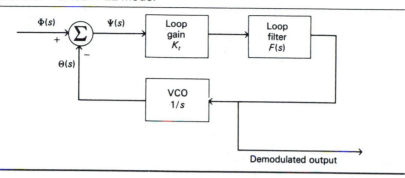

This choice of loop filter results in what is generally referred to as a *perfect second-order PLL*.

If the phase error is small, the linear model applies and the transfer function of the second-order PLL is easily derived. The linear model is redrawn in Figure 3.51 in terms of the Laplace transform variable s. From Figure 3.51, it follows that

$$\Theta(s) = \frac{K_t F(s)}{s} [\Phi(s) - \Theta(s)] \tag{3.227}$$

so that the PLL transfer function is

$$H(s) = \frac{\Theta(s)}{\Phi(s)} = \frac{K_t F(s)}{s + K_t F(s)} \tag{3.288}$$

With $F(s)$ given by (3.226), the transfer function becomes

$$\frac{\Theta(s)}{\Phi(s)} = \frac{K_t(s + a)}{s^2 + K_t s + K_t a} \tag{3.229}$$

We also can write the relationship between the phase error $\Psi(s)$ and the input phase $\Phi(s)$ from Figure 3.51. The result is

$$\frac{\Psi(s)}{\Phi(s)} = \frac{s}{s + K_t F(s)} \tag{3.230}$$

which for the loop filter defined by (3.226) yields

$$\frac{\Psi(s)}{\Phi(s)} = \frac{s^2}{s^2 + K_t a s + K_t a} \tag{3.231}$$

We now place this in the standard form for a second-order system. The result is

$$\frac{\Psi(s)}{\Phi(s)} = \frac{s^2}{s^2 + 2\zeta\omega_n s + \omega_n^2} \tag{3.232}$$

in which ζ is the damping factor and ω_n is the natural frequency. It follows that the natural frequency is

$$\omega_n = \sqrt{K_t a} \tag{3.233}$$

and that the damping factor is

$$\zeta = \frac{1}{2}\sqrt{\frac{K_t}{a}} \tag{3.234}$$

A typical value of the damping factor is 0.707.

Using the linear model, we can easily show that the steady-state phase error due to a step in the input frequency is zero. Since phase is the integral of frequency and integration is equivalent to division by s, the input phase due to a step in frequency of magnitude $\Delta\omega$ is

$$\Phi(s) = \frac{\Delta\omega}{s^2} \tag{3.235}$$

From (3.232) we see that the Laplace transform of the phase error $\psi(t)$ is

$$\psi(s) = \frac{\Delta\omega}{s^2 + 2\zeta\omega_n s + \omega_n^2} \tag{3.236}$$

Inverse transforming yields, for $\zeta < 1$,

$$\psi(t) = \frac{\Delta\omega}{\omega_n\sqrt{1 - \zeta^2}} e^{-\zeta\omega_n t} \sin(\omega_n\sqrt{1 - \zeta^2}t) \tag{3.237}$$

and we see that $\psi(t) \to 0$ as $t \to \infty$. Thus the steady-state phase error is zero. We now turn our attention to the lock range by examining the phase-plane plot of the second-order PLL.

A mathematical development of the phase-plane plot of a second-order PLL is well beyond the level of our treatment here. However, the phase-plane plot was obtained, using computer simulation, for a second-order PLL having a damping factor ζ of 0.707 and a natural frequency f_n of 10 Hz. For these parameters, the loop gain K_t is 88.9 and the filter parameter a is 44.4. The input to the PLL is assumed to be a step change in frequency at

FIGURE 3.52   Phase-plane plot for second-order PLL

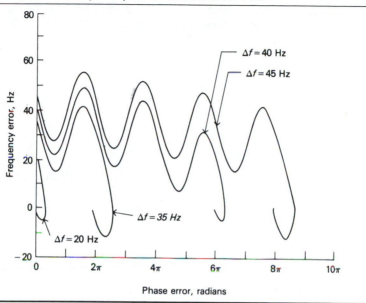

Phase error, radians

time $t = t_0$. Four values were used for the step change in frequency $\Delta \omega = 2\pi(\Delta f)$. These were $\Delta f = 20$, 35, 40, and 45 Hz.

The results are illustrated in Figure 3.52. Note that for $\Delta f = 20$ Hz, the operating point returns to a steady-state value for which the frequency and phase error are both zero. For $\Delta f = 35$ Hz, the phase-plane is somewhat more complicated. The steady-state frequency error is zero, but the steady-state phase error is 2π rad. We say that the PLL has slipped one cycle. The cycle-slipping phenomenon accounts for the nonzero steady-state phase error. The responses for $\Delta f = 40$ and 45 Hz illustrate that three and four cycles are slipped, respectively. The instantaneous VCO frequency is shown in Figure 3.53 for these four cases. The cycle-slipping behavior is clearly shown.

The second-order PLL does indeed have an infinite lock range, and cycle slipping occurs until the phase error is within π rad of the steady-state value.

Phase-lock loops also allow for simple implementation of frequency multipliers and dividers. There are two basic schemes. In the first scheme, harmonics of the input are generated and the VCO tracks one of these harmonics. This scheme is most useful for implementing frequency multipliers. The second scheme is to generate harmonics of the VCO output and to phase-lock one of these frequency components to the input. This scheme

FIGURE 3.53 VCO frequency for four values of input frequency step.
(a) VCO frequency for Δf = 20 Hz. (b) VCO frequency for Δf = 35 Hz.
(c) VCO frequency for Δf = 40 Hz. (d) VCO frequency for Δf = 45 Hz.

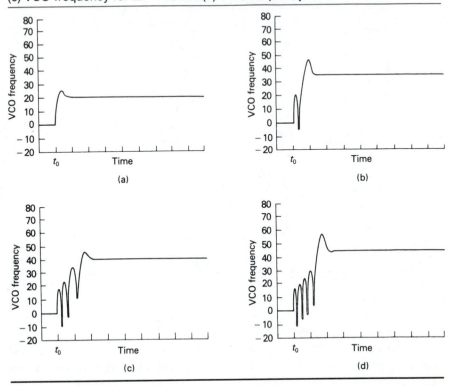

can be used to implement either frequency multipliers or frequency dividers.

Figure 3.54 illustrates the first technique. The limiter is a nonlinear device and therefore generates harmonics of the input frequency. If the input is sinusoidal, the output of the limiter is a squarewave; therefore, odd harmonics are present. In the example illustrated, the VCO quiescent frequency [VCO output frequency f_c with $e_v(t)$ equal to zero] is set equal to $5f_0$. The result is that the VCO phase-locks to the fifth harmonic of the input. Thus the system shown multiplies the input frequency by 5.

Figure 3.55 illustrates frequency division by a factor of 2. The VCO quiescent frequency is $\frac{1}{2}f_0$ Hz, but the VCO output waveform is a narrow pulse that has the spectrum shown. The component at frequency f_0 phase-locks to the input. A bandpass filter can be used to select the component desired from the VCO output spectrum. For the example shown, the center fre-

FIGURE 3.54   Phase-lock loop used as a frequency multiplier

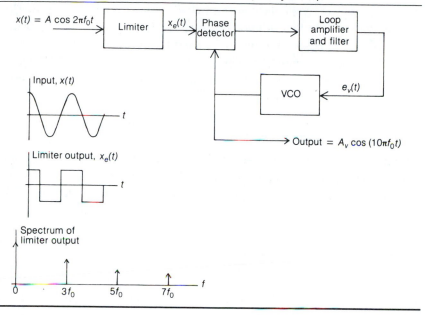

FIGURE 3.55   Phase-lock loop used as a frequency divider

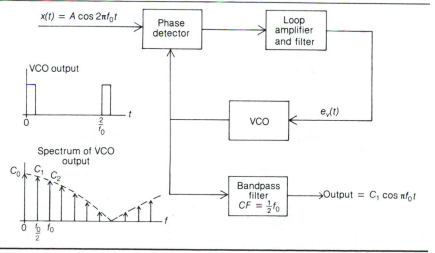

quency of the bandpass filter should be $\frac{1}{2}f_0$. The bandwidth of the bandpass filter must be less than the spacing between the components in the VCO output spectrum; in this case, this spacing is $\frac{1}{2}f_0$. It is worth noting that the system shown in Figure 3.55 could be used to multiply the input frequency by 5 by setting the center frequency of the bandpass filter to $5f_0$. Thus this system could also serve as a $\times 5$ frequency multiplier, like the first example. Many variations are possible.

Frequency-Compressive Feedback

Another system that can be used to demodulate FM signals is the frequency-compressive feedback demodulator, which is illustrated in Figure 3.56. The system is similar to a PLL except for the bandpass filter and discriminator in the loop and the offset in the VCO center frequency. As in the case of the PLL, assume that the input is

$$x_r(t) = A_c \cos [\omega_c t + \phi(t)] \tag{3.238}$$

Also assume that the center frequency of the VCO is $\omega_c - \omega_0$. The frequency ω_0 is the VCO offset frequency and the center frequency of the discriminator. The VCO output is given by

$$e_0(t) = A_v \sin [(\omega_c - \omega_0)t + \theta(t)] \tag{3.239}$$

The quantity $\theta(t)$ is the phase deviation of the VCO and is related to the VCO input $e_v(t)$:

$$\theta(t) = K_v \int^t e_v(\alpha) \, d\alpha \tag{3.240}$$

where K_v is the VCO constant. It should be noted that except for the VCO offset, the expressions for $x_r(t)$, $e_0(t)$, and $\theta(t)$ are identical to the equivalent expressions for the PLL. The multiplier output $e_d(t)$ is given by

$$\begin{aligned} e_d(t) &= \tfrac{1}{2}A_c A_v \sin [(2\omega_c - \omega_0)t + \phi(t) + \theta(t)] \\ &\quad - \tfrac{1}{2}A_c A_v \sin [\omega_0 t + \phi(t) - \theta(t)] \end{aligned} \tag{3.241}$$

FIGURE 3.56   Frequency-compressive feedback demodulator

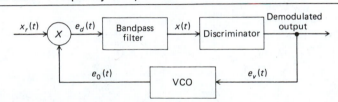

The phase deviation of the discriminator input is

$$\phi(t) - K_v \int^t e_v(\alpha) \, d\alpha \tag{3.242}$$

Thus the discriminator output can be written as

$$e_v(t) = \frac{1}{2\pi} K_D \frac{d}{dt} \left[\phi(t) - K_v \int^t e_v(\alpha) \, d\alpha \right] \tag{3.243}$$

which is

$$e_v(t) = \frac{(1/2\pi)K_D}{1 + (1/2\pi)K_v K_D} \frac{d\phi}{dt} = \frac{K_D f_d}{1 + (1/2\pi)K_v K_D} m(t) \tag{3.244}$$

Thus the system is an FM demodulator.

The advantage of this technique can be understood by using (3.244) to write an expression for $x(t)$. This yields

$$x(t) = -\tfrac{1}{2} A_c A_v \sin \left[\omega_0 t + \frac{1}{1 + (1/2\pi)K_D K_v} \phi(t) \right] \tag{3.245}$$

This shows that for large values of the product $K_D K_v$, the phase deviation can be made small, thereby reducing significantly the bandwidth of the signal at the discriminator input. It is even possible to compress the bandwidth of a wideband FM signal to that of a narrowband FM signal at the input to the discriminator. It is this bandwidth compression that gives the scheme the name *frequency-compressive feedback*. The advantages of this bandwidth compression will be examined in Chapter 6.

Costas Phase-Lock Loops

We have seen that systems utilizing feedback can be used to demodulate angle-modulated carriers. A feedback system also can be used to generate the coherent demodulation carrier necessary for the demodulation of DSB signals. One system that accomplishes this is the Costas phase-lock loop illustrated in Figure 3.57. The input to the loop is the assumed DSB signal

$$x_r(t) = m(t) \cos \omega_c t \tag{3.246}$$

The signals at the various points within the loop are easily derived from the assumed input and VCO output and are included in Figure 3.57. The lowpass filter preceding the VCO is assumed sufficiently narrow so that the output is $K \sin 2\theta$, essentially the dc value of the input. This signal drives the VCO such that θ is reduced. For sufficiently small θ, the output of the top lowpass filter is the demodulated output.

FIGURE 3.57   Costas phase-lock loop

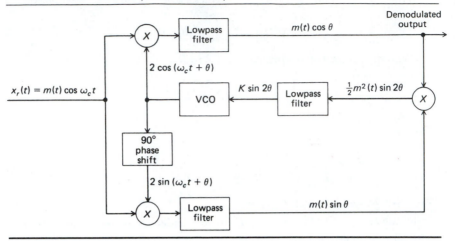

We will see in Chapter 7 that the Costas phase-lock loop is useful in the implementation of digital data systems.

3.5 PULSE MODULATION

In Section 2.8 we saw that continuous bandlimited signals can be represented by a sequence of discrete samples and that the continuous signal can be reconstructed with negligible error if the sampling rate is sufficiently high. Consideration of sampled signals immediately leads us to the topic of pulse modulation.

Analog Pulse Modulation

Analog pulse modulation results when some characteristic of a pulse is made to vary in one-to-one correspondence with the message signal. We shall see that since a pulse is characterized by three quantities—amplitude, width, and position—there are three types of analog pulse modulation. These are pulse-amplitude modulation, pulse-width modulation, and pulse-position modulation.

Pulse-Amplitude Modulation (PAM)   As illustrated in Figure 3.58, a PAM waveform consists of a sequence of flat-topped pulses. The amplitude of each pulse corresponds to the value of the message signal $m(t)$ at the leading edge of the pulse. Thus this type of modulation is essentially a sampling

FIGURE 3.58   Pulse-amplitude modulation

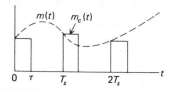

operation with the sample values represented by the leading edge of each pulse.

The difference between pulse-amplitude modulation and sampling as developed in Chapter 2 is slight. We saw in Chapter 2 that a sampled signal can be represented as

$$m_\delta(t) = \sum_{n=-\infty}^{\infty} m(nT_s)\, \delta(t - nT_s) \tag{3.247}$$

From Figure 3.58, the PAM signal can be written as

$$m_c(t) = \sum_{n=-\infty}^{\infty} m(nT_s)\Pi\left[\frac{t - (nT_s + \frac{1}{2}\tau)}{\tau}\right] \tag{3.248}$$

where $\Pi[\cdot]$ is the time-domain unit pulse function defined in Chapter 2. The sampled signal $m_\delta(t)$ can be transformed into the PAM waveform defined by (3.248) by passing $m_\delta(t)$ through a network, as shown in Figure 3.59(a). The network simply holds the value of the sample for τ seconds. The $n = 0$ term in (3.247) and (3.248) illustrates that if the input to the holding network is $A\delta(t)$, the output must be $A\Pi[(t - \frac{1}{2}\tau)/\tau]$. Thus the impulse response of the holding network is

$$h(t) = \Pi\left[\frac{t - \frac{1}{2}\tau}{\tau}\right] \tag{3.249}$$

as shown in Figure 3.59(b). The transfer function $H(f)$ of the holding network is

$$H(f) = \tau \, \text{sinc} \, f\tau \, e^{-j\pi f \tau} \tag{3.250}$$

as derived in Example 2.8 with $t_0 = \frac{1}{2}\tau$. The amplitude response and phase response are illustrated in Figures 3.59(c) and (d), respectively.

It should be remembered from Section 2.8 that $m(t)$ can be recovered from $m_\delta(t)$ by lowpass filtering. Thus a PAM signal can be demodulated using a two-step process. First $m_c(t)$ is passed through an equalization filter that

FIGURE 3.59 Generation of PAM. (a) Holding network. (b) Impulse response of holding network. (c) Amplitude response of holding network. (d) Phase response of holding network.

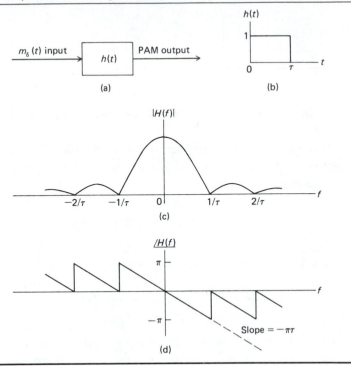

has the transfer function $H_e(f) = 1/H(f)$. If $\tau \ll T_s$, equalization is usually unnecessary. The message signal $m(t)$ can be recovered from the filter output $m_\delta(t)$ by lowpass filtering.

Pulse-Width Modulation (PWM)   A PWM waveform consists of a sequence of pulses, each of which having a width proportional to the values of a message signal at the sampling instants. The generation of a PWM waveform is illustrated in Figure 3.60. Since the width of a pulse cannot be negative, a dc bias must be added to $m(t)$ prior to modulation. A PAM waveform $x_1(t)$ is then generated from the biased signal $m(t) + K$. To $x_1(t)$ is added a sequence of synchronized triangular pulses as illustrated. The signal, $x_1(t) + p(t)$, is level-sliced by a circuit that gives an output A when $x_1(t) + p(t)$ is above the slicing level. This signal is the desired PWM waveform.

A PWM waveform can be demodulated by lowpass filtering. The proof of

FIGURE 3.60   Generation of PWM

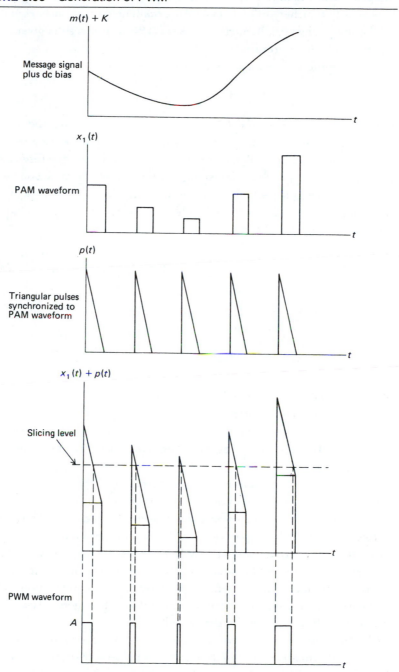

this statement requires derivation of the spectrum of the PWM signal, which is generally a rather difficult task. An excellent derivation is contained in the book by Schwartz, Bennett, and Stein (1966). The results given here are taken from that reference.

If we assume the message signal

$$m(t) = A \cos 2\pi f_m t \tag{3.251}$$

there are two fundamental frequencies present. These are the signal frequency f_m and the sampling frequency $f_s = 1/T_s$. If these two frequencies are incommensurable, which is certainly the general case, the resulting PWM waveform is nonperiodic, even though the message signal is assumed periodic and sampling is periodic. Despite this difficulty, Schwartz, Bennett, and Stein illustrate a technique that leads to a trigonometric series expression for the PWM signal. The result is

$$x(t) = \tfrac{1}{2}h - \tfrac{1}{2}h\beta \cos 2\pi f_m t + \frac{1}{\pi} h \sum_{m=1}^{\infty} [(-1)^m - J_0(m\pi\beta)] \frac{1}{m} \sin 2\pi mf_s t$$

$$- \frac{1}{\pi} h \sum_{m=1}^{\infty} \sum_{n=1}^{\infty} \frac{1}{m} J_0(m\pi\beta) \left\{ \sin \left[2\pi(mf_s + nf_m)t + \frac{n\pi}{2} \right] \right.$$

$$+ \sin \left[2\pi(mf_s - nf_m)t + \frac{n\pi}{2} \right] \right\} \tag{3.252}$$

In this expression, h denotes the amplitude of the PWM pulses and β denotes the modulation index. The modulation index is specified by defining $\beta = 1$ as the value of β resulting in the maximum possible pulse width without causing interference with adjacent pulses. Note that the component at f_m is proportional to β.

The expression for $x(t)$ illustrates the character of the spectrum of a PWM signal with sinusoidal modulation. A dc component having value $\tfrac{1}{2}h$ is present. A component is also present at the message frequency f_m. The amplitude of this component is proportional to the modulation index β. Lowpass filtering allows the component at f_m, which represents the message $m(t)$, to be recovered from the PWM signal. Also present in the spectrum of $x(t)$ are spectral components at frequencies $mf_s + nf_m$ for all combinations of m and n. These spectral components are intermodulation products between the signal and sampling frequencies and their harmonics.

The spectrum is illustrated in Figure 3.61 for a modulation index $\beta = 0.1$. The sampling frequency f_s is $8f_m$. It can be seen that the spectrum consists of Fourier-Bessel spectra centered about the sampling frequency and harmonics of the sampling frequency. The Fourier-Bessel spectrum centered

FIGURE 3.61 Spectra of PWM signal with $\beta = 0.1$

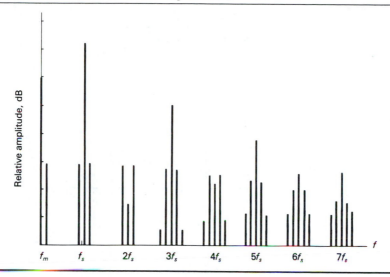

about $f = f_s$ consists of a single component for each sideband corresponding to narrowband modulation ($\beta = 0.1$). The Fourier-Bessel spectrum centered about $f = 7f_s$ consists of a pair of components in each sideband and therefore does not correspond to narrowband modulation. It can be seen from Figure 3.61 and from (3.252) that the effective index is a function of m, the harmonic of f_s. The process is very similar to that encountered in the narrowband to wideband conversion process.

Pulse-Position Modulation (PPM)   A PPM signal consists of pulses in which the pulse displacement from a specified time reference is proportional to the sample values of the information-bearing signal. Just as for PWM, a dc bias must be added to the message signal prior to modulation so that the input to the PPM modulator is nonnegative for all values of time. As a matter of fact PPM is easily generated from PWM, as illustrated in Figure 3.62(a). The PWM signal is placed on the input of a monostable multivibrator, which is triggered on by negative-going transitions of the input signal. The on-time of the monostable multivibrator is fixed. The resulting PPM waveform for an assumed PWM waveform is illustrated in Figure 3.62(b).

The PPM signal can be represented by the expression

$$x(t) = \sum_{n=-\infty}^{\infty} g(t - t_n) \tag{3.253}$$

FIGURE 3.62 Generation of PPM from PWM. (a) Generation.
(b) Waveforms.

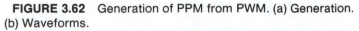

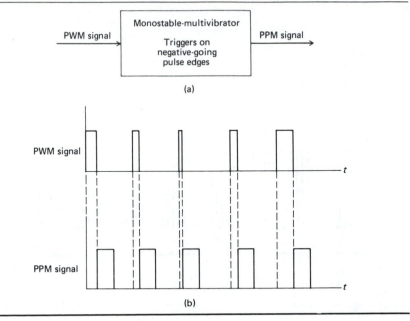

where $g(t)$ represents the shape of the individual pulses and the occurrence times t_n are related to the values of the message signal $m(t)$ at the sampling instants nT_s, as discussed in the preceding paragraph. Determining the spectrum of a PPM signal is a process very similar to determine that of a PWM signal, which we just discussed. This problem is also considered by Schwartz, Bennett, and Stein (1966). If we assume the message signal

$$m(t) = A \cos 2\pi f_m t \tag{3.254}$$

we can write the PPM signal as

$$x(t) = f_s - \beta f_m \sin 2\pi f_m t + \sum_{m=1}^{\infty} \sum_{n=-\infty}^{\infty} \frac{1}{2m} (mf_s + nf_m) J_n(m\pi\beta)$$

$$\cdot \cos \left[2\pi(mf_s + nf_m)t + \frac{n\pi}{2} \right] \tag{3.255}$$

assuming that $g(t)$ can be represented by an impulse function. If one uses a pulse shape $g(t)$ that cannot be approximated by an impulse function, one obtains the resulting expression for $x(t)$ by weighting each term in the preceding expression by the Fourier transform of $g(t)$ evaluated at the frequency of the respective term. For example, if $g(t)$ is an even pulse with real transform $G(f)$, the resulting expression for $x(t)$ is

$$x(t) = f_s G(0) - \beta f_m G(f_m) \sin 2\pi f_m t$$
$$+ \sum_{m=1}^{\infty} \sum_{n=-\infty}^{\infty} \frac{1}{2m} (mf_s + nf_m) J_n(m\pi\beta) G(mf_s + nf_m)$$
$$\cdot \cos \left[2\pi(mf_s + nf_m)t + \frac{n\pi}{2} \right] \tag{3.256}$$

As in the case of PWM, β is the modulation index. A value of 1 for this index results in the largest allowable value of t_n without interference with adjacent pulse positions.

It can be seen from the expression for $x(t)$ that the frequency component at f_m is proportional to the derivative of the message signal $m(t)$. Thus, if the spectral component at f_m is extracted from the spectrum of $x(t)$ by lowpass filtering (the dc term can be removed by a blocking capacitor), the output of the filter is the derivative of $m(t)$ scaled by a constant. We can obtain the message signal $m(t)$ by integrating the filter output. In cases in which the sampling frequency is sufficiently high, the lowpass filter can be omitted, since an integrator also performs lowpass filtering.

The integration can be eliminated by converting the PPM signal to a PWM signal and then demodulating the PWM signal by lowpass filtering. PPM-to-PWM conversion is essentially the reverse of the operations shown in Figure 3.62. A train of pulses is generated at frequency f_s. The leading edges of these pulses are synchronized to the zero-displacement position of the PPM pulses. The equation for $x(t)$ illustrates that a frequency component is present at the sampling frequency if $J_0(\pi\beta) \neq 0$. This component can be used to generate the pulses at frequency f_s. These pulses are then turned off by the leading edge of the PPM pulses. It also can be seen from the equation for $x(t)$ that the frequencies present in a PPM signal are exactly the same as the frequencies present in a PWM signal.

Digital Pulse Modulation

In analog pulse modulation systems, the amplitude, width, or position of a pulse can vary over a continuous range in accordance with the message amplitude at the sampling instant. In systems utilizing digital pulse modulation, the transmitted samples take on only discrete values. We now examine two types of digital pulse modulation: delta modulation and pulse-code modulation.

Delta Modulation (DM)   *Delta modulation (DM)* is a modulation technique in which the message signal is encoded into a sequence of binary symbols. These binary symbols are represented by the polarity of impulse functions at the modulator output. The electronic circuits to implement both the modulator and the demodulator are extremely simple. It is this electronic simplicity that makes DM an attractive technique.

A block diagram of a delta modulator is illustrated in Figure 3.63(a). The input to the pulse modulator portion of the circuit is

$$d(t) = m(t) - m_s(t) \tag{3.257}$$

where $m(t)$ is the message signal and $m_s(t)$ is a reference waveform. The signal $d(t)$ is hard-limited and multiplied by the pulse-generator output. This yields

$$x_c(t) = \Delta(t) \sum_{n=-\infty}^{\infty} \delta(t - nT_s) \tag{3.258}$$

where $\Delta(t)$ is a hard-limited version of $d(t)$. The preceding expression can be written as

$$x_c(t) = \sum_{n=-\infty}^{\infty} \Delta(nT_s)\delta(t - nT_s) \tag{3.259}$$

Thus the output of the delta modulator is a series of impulses, each having positive or negative polarity depending on the sign of $d(t)$ at the sampling instants. In practical applications, the output of the pulse generator is not, of course, a sequence of impulse functions but rather a sequence of pulses that are narrow with respect to their periods. Impulse functions are assumed here because of the resulting mathematical simplicity.

The reference signal $m_s(t)$ is generated by integrating $x_c(t)$. This yields

$$m_s(t) = \sum_{n=-\infty}^{\infty} \Delta(nT_s) \int^{t} \delta(\alpha - nT_s)\, d\alpha \tag{3.260}$$

which is a stairstep approximation of $m(t)$. The reference signal $m_s(t)$ is shown in Figure 3.63(b) for an assumed $m(t)$. The transmitted waveform $x_c(t)$ is illustrated in Figure 3.63(c).

Demodulation of DM is accomplished by integrating $x_c(t)$ to form the stairstep approximation $m_s(t)$. This signal can then be lowpass-filtered to suppress the discrete jumps in $m_s(t)$. Since a lowpass filter approximates an integrator, it is often possible to eliminate the integrator portion of the

FIGURE 3.63   Delta modulation. (a) Delta modulator. (b) Modulation waveform and stairstep approximation. (c) Modulator output.

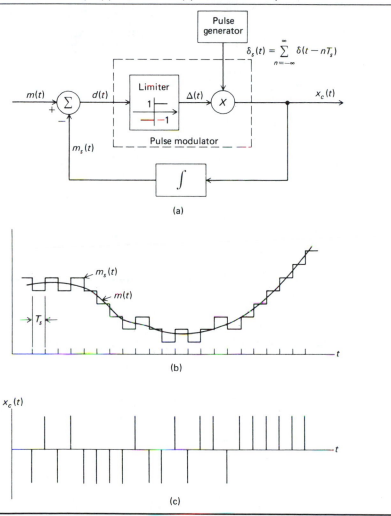

(a)

(b)

(c)

demodulator and to demodulate DM simply by lowpass filtering, as was done for PAM and PWM.

A difficulty with DM is the problem of slope overload. *Slope overload* occurs when the message signal $m(t)$ has a slope greater than can be followed by the stairstep approximation $m_s(t)$. This effect is illustrated in Figure 3.64(a), which shows a step change in $m(t)$ at time t_0. Assuming that each pulse in $x_c(t)$ has weight δ_0, the maximum slope that can be followed by $m_s(t)$ is δ_0/T_s, as shown. Figure 3.64(b) shows the resulting error signal due to a step

FIGURE 3.64   Illustration of slope overload. (a) Illustration of $m(t)$ and $m_s(t)$ with step change in $m(t)$. (b) Error between $m(t)$ and $m_s(t)$.

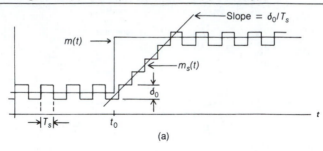

(a)

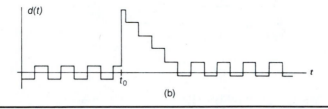

(b)

change in $m(t)$ at t_0. It can be seen that significant error exists for some time following the step change in $m(t)$. The duration of the error due to slope overload depends on the amplitude of the step, the impulse weights δ_0, and the sampling period T_s.

A simple analysis can be carried out assuming that the message signal $m(t)$ is a sinusoid:

$$m(t) = A \sin 2\pi f_1 t \qquad (3.261)$$

The maximum slope that $m_s(t)$ can follow is

$$S_m = \frac{\delta_0}{T_s} \qquad (3.262)$$

The derivative of $m(t)$ is

$$\frac{d}{dt} m(t) = 2\pi A f_1 \cos 2\pi f_1 t \qquad (3.263)$$

It follows that $m_s(t)$ can follow $m(t)$ without slope overload if

$$\frac{\delta_0}{T_s} \geq 2\pi A f_1 \qquad (3.264)$$

This example implies that a bandwidth constraint exists on $m(t)$ if slope overload is to be avoided.

One technique for overcoming the problem of slope overload is to modify the modulator, as shown in Figure 3.65. The result is known as *adaptive delta modulation*. The system is explained by recognizing that the weights δ_0, and consequently the step size of $m_s(t)$, can be very small if $m(t)$ is nearly constant. A rapidly changing $m(t)$ requires a larger value of δ_0 if slope overload is to be avoided. A lowpass filter is used as shown, with $x_c(t)$ as its input. If $m(t)$ is constant or nearly constant, the pulses constituting $x_c(t)$ will alternate in sign. Thus the dc value, determined over the time constant of the lowpass filter, is nearly zero. This small value controls the gain of the variable-gain amplifier such that it is very small under this condition. Thus δ_0 is made small at the integrator input. The square-law or magnitude device is used to ensure that the control voltage and amplifier gain $g(t)$ are always positive. If $m(t)$ is increasing or decreasing rapidly, the pulses $x_c(t)$ will have the same polarity over this period. Thus the magnitude of the output of the lowpass filter will be relatively large. The result is an increase in the gain of the variable-gain amplifier and consequently an increase in δ_0. This in turn reduces the time span of significant slope overload.

FIGURE 3.65 Adaptive delta modulator

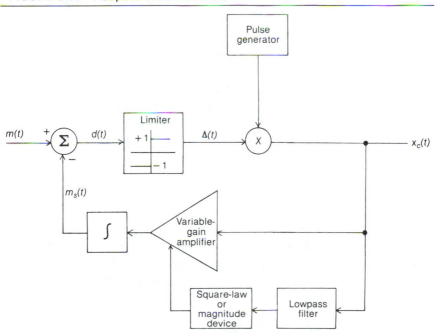

FIGURE 3.66 Adaptive DM receiver

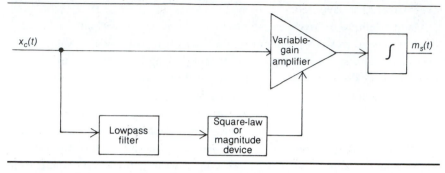

The use of an adaptive delta modulator requires that the receiver be adaptive also, so that the step size at the receiver changes to match the changes in δ_0 at the modulator. This is illustrated in Figure 3.66.

Pulse-Code Modulation (PCM)   The generation of PCM is a three-step process, as illustrated in Figure 3.67(a). The message signal $m(t)$ is first sampled. The sample values are then quantized. In PCM, the quantization level of each sample is the transmitted quantity instead of the sample value. Typically, the quantization level is encoded into a binary sequence, as shown in Figure 3.67(b). The modulator output is a pulse representation of the binary sequence, which is shown in Figure 3.67(c). A binary "one" is represented as a pulse, and a binary "zero" is represented as the absence of a pulse. This absence of a pulse is indicated by a dashed line in Figure 3.67(c). The PCM waveform of Figure 3.67(c) shows that a PCM system requires synchronization so that the starting points of the digital words can be determined at the demodulator.

To consider the bandwidth requirements of a PCM system, suppose that q quantization levels are used, satisfying

$$q = 2^n \tag{3.265}$$

where n is an integer. For this case, $n = \log_2 q$ binary pulses must be transmitted for each sample of the message signal. If this signal has bandwidth W and the sampling rate is $2W$, then $2nW$ binary pulses must be transmitted per second. Thus the maximum width of each binary pulse is

$$(\Delta\tau)_{max} = \frac{1}{2nW} \tag{3.266}$$

FIGURE 3.67   Generation of PCM. (a) PCM modulator. (b) Quantization and encoding. (c) Transmitted output.

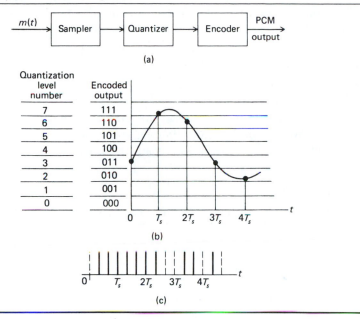

We saw in Section 2.7 that the bandwidth required for transmission of a pulse is inversely proportional to the pulse width, so that

$$B = knW \tag{3.267}$$

where B is the required bandwidth of the PCM system and k is a constant of proportionality. This yields

$$B \cong 2Wk \log_2 q \tag{3.268}$$

This represents a lower bound on bandwidth, since we have assumed both a minimum sampling rate and a minimum value of bandwidth for transmitting a pulse. Equation (3.268) shows that the PCM signal bandwidth is proportional to W and the logarithm of q.

If the major source of error in the system is quantization error, it follows that a small error requirement dictates large transmission bandwidth. Thus, in a PCM system, error can be exchanged for bandwidth. We shall see that this behavior is typical of many nonlinear systems operating in noisy environments. However, before noise effects can be analyzed, we must take a detour and develop the theory of probability and random processes. Knowledge of this area enables one to accurately model realistic and practical communication systems operating in everyday, nonidealized environments.

The pulses representing the sample values of a PCM waveform can be transmitted on an RF carrier by the use of amplitude, phase, or frequency modulation. This is illustrated in Figure 3.68 for the case in which the data bits are represented by an NRZ (non–return to zero) waveform. Six bits are shown, corresponding to the data sequence 101001. For digital amplitude modulation, known as *amplitude shift keying (ASK)*, the carrier amplitude is determined by the data bit for that interval. For digital phase modulation, known as *phase-shift keying (PSK)*, the phase of the carrier is established by the data bit. The phase changes can be clearly seen in Figure 3.68. For digital

FIGURE 3.68   An example of digital modulation schemes

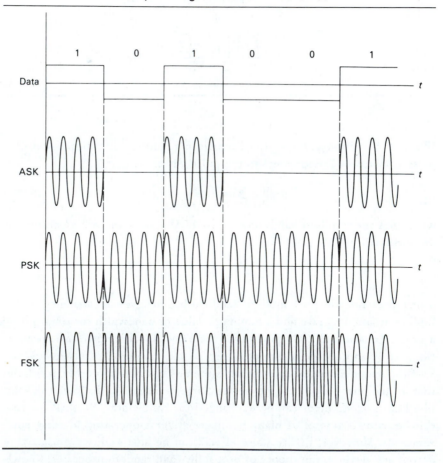

frequency modulation, known as *frequency-shift keying (FSK)*, the carrier frequency is established by the data bit. These techniques will be studied in detail in Chapter 7.

3.6   MULTIPLEXING

In many applications, a large number of data sources are located at a common point, and it is desirable to transmit these signals simultaneously using a single communication channel. This is accomplished using multiplexing. We will now examine several different types of multiplexing, each having advantages and disadvantages.

Frequency-Division Multiplexing (FDM)

Frequency-division multiplexing (FDM) is a technique whereby several message signals are translated, using modulation, to different spectral locations and added to form a baseband signal. The carriers used to form the baseband are usually referred to as *subcarriers*. Then, if desired, the baseband signal can be transmitted over a single channel using a single modulation process. Several different types of modulation can be used to form the baseband, as illustrated in Figure 3.69. In this example, there are N information signals contained in the baseband. Observation of the baseband spectrum in Figure 3.69(c) suggests that baseband modulator 1 is a DSB modulator with subcarrier frequency f_1. Modulator 2 is an upper-sideband SSB modulator, and modulator N is an angle modulator.

An FDM demodulator is shown in Figure 3.69(b). The RF demodulator output is ideally the baseband signal. The individual channels in the baseband are extracted using bandpass filters. The bandpass filter outputs are demodulated in the conventional manner.

Observation of the baseband spectrum illustrates that the baseband bandwidth is equal to the sum of the bandwidths of the modulated signals plus the sum of the *guardbands*, the empty spectral bands between the channels necessary for filtering. This bandwidth is lower-bounded by the sum of the bandwidths of the message signals. This bandwidth,

$$B = \sum_{i=1}^{N} W_i \tag{3.269}$$

where W_i is the bandwidth of $m_i(t)$, is achieved when all baseband modulators are SSB and all guardbands have zero width.

FIGURE 3.69 Frequency-division multiplexing. (a) FDM modulator. (b) FDM demodulator. (c) Baseband spectrum.

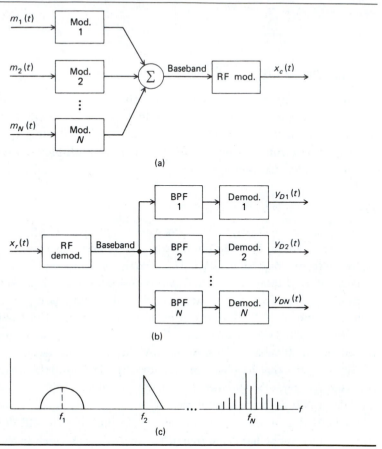

(a)

(b)

(c)

An Example of Frequency-Division Multiplexing: Stereophonic FM Broadcasting

As an example of frequency-division multiplexing we now consider stereophonic FM broadcasting. A necessary condition established in the development of stereophonic FM is that stereo FM be compatible with monophonic FM receivers. In other words, the output from a monophonic FM receiver must be the composite (left-channel plus right-channel) stereo signal.

The scheme adopted for stereophonic broadcasting is shown in Figure 3.70(a). As can be seen, the first step in the generation of a stereo FM signal is to first form the left-plus-right channel signal, $l(t) + r(t)$, and the left-minus-right channel signal, $l(t) - r(t)$. The left-minus-right channel signal is

FIGURE 3.70   Stereophonic FM transmitter and receiver. (a) Stereophonic FM transmitter. (b) Single-sided spectrum of FM baseband signal. (c) Stereophonic FM receiver.

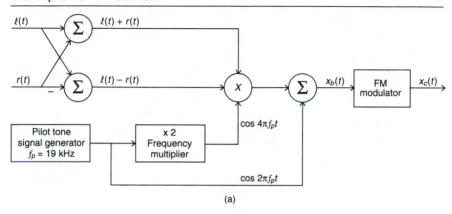

(a)

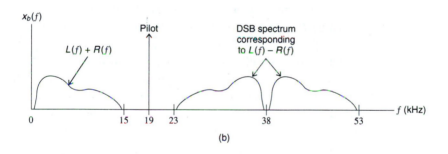

(b)

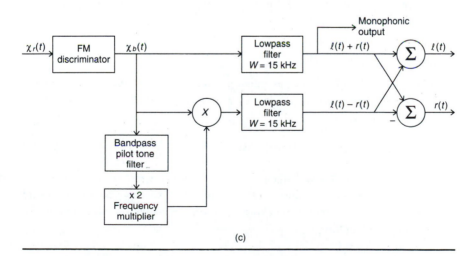

(c)

then translated to 38 kHz using DSB modulation with a carrier derived from
a 19-kHz oscillator. A frequency doubler is used to generate a 38-kHz carrier from a 19-kHz oscillator. We previously saw that a phase-lock loop could
be used to implement a frequency doubler.

The baseband signal is found by adding the left-plus-right channel signal,
the left-minus-right channel signal, and the 19-kHz pilot tone. The spectrum
of the baseband signal is shown in Figure 3.70(b) for assumed left-channel and
right-channel signals. The baseband signal is the input to the FM modulator.
It is important to note that if a monophonic FM transmitter, having a message
bandwidth of 15 kHz, and a stereophonic FM transmitter, having a message
bandwidth of 53 kHz, both have the same constraint on the peak deviation,
the deviation ratio, D, of the stereophonic FM transmitter is reduced by a
factor of $53/15 = 3.53$. The impact of this reduction in the deviation ratio
will be seen when we consider noise effects in Chapter 6.

The block diagram of a stereophonic FM receiver is shown in Figure
3.70(c). The output of the FM discriminator is the baseband signal $x_b(t)$
which, under ideal conditions, is identical to the baseband signal at the input
to the FM modulator. As can be seen from the spectrum of the baseband
signal, the left-plus-right channel signal can be generated by filtering the
baseband signal with a lowpass filter having a bandwidth of 15 kHz. Note
that this signal constitutes the monophonic output. The left-minus-right
channel signal is obtained by coherently demodulating the DSB signal using
a 38-kHz demodulation carrier. This coherent demodulation carrier is
obtained by recovering the 19-kHz pilot using a bandpass filter and then
using a frequency doubler as was done in the modulator. The left-plus-right
channel signal and the left-minus-right channel signal are added and subtracted, as shown in Figure 3.70(c) to generate the left channel signal and
the right channel signal.

Quadrature Multiplexing (QM)

Another type of multiplexing is *quadrature multiplexing (QM)*, in which quadrature carriers are used for frequency translation. For the system shown in
Figure 3.71, the signal

$$x_c(t) = A_c[m_1(t) \cos \omega_c t + m_2(t) \sin \omega_c t] \qquad (3.270)$$

is a quadrature-multiplexed signal. By sketching the spectra of $m_1(t) \cos \omega_c t$
and $m_2(t) \sin \omega_c t$, we see these spectra overlap if the spectra of $m_1(t)$ and
$m_2(t)$ overlap. Thus, strictly speaking, quadrature multiplexing is not a frequency-division technique.

FIGURE 3.71 Quadrature multiplexing

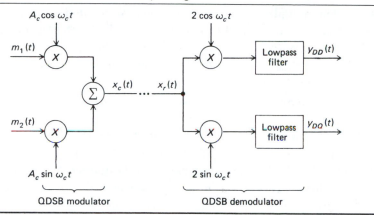

A QM signal is demodulated by using quadrature demodulation carriers. To show this, multiply $x_r(t)$ by $2 \cos(\omega_c t + \theta)$. This yields

$$2x_r(t) \cos(\omega_c t + \theta) = A_c[m_1(t) \cos\theta - m_2(t) \sin\theta$$
$$+ m_1(t) \cos(2\omega_c t + \theta) + m_2(t) \sin(2\omega_c t + \theta)] \tag{3.271}$$

The last two terms in the preceding expression can be removed by using a lowpass filter. The output of the lowpass filter is

$$y_{DD}(t) = A_c[m_1(t) \cos\theta - m_2(t) \sin\theta] \tag{3.272}$$

which yields $m_1(t)$, the desired output for $\theta = 0$. The quadrature channel is demodulated using a demodulation carrier of the form $2 \sin \omega_c t$.

The preceding result illustrates the effect of a demodulation phase error on QM. The result of this phase error is both an attenuation, which can be time-varying, of the desired signal and crosstalk from the quadrature channel. It should be noted that QM can be used to represent both DSB and SSB with appropriate definitions of $m_1(t)$ and $m_2(t)$.

Frequency-division multiplexing can be used with quadrature multiplexing by translating pairs of signals, using quadrature carriers, to each subcarrier frequency. Each channel has bandwidth $2W$ and accommodates two message signals, each having bandwidth W. Thus, assuming zero-width guardbands, a baseband of bandwidth NW can accommodate N message signals, each of bandwidth W, and requires $\frac{1}{2}N$ separate subcarrier frequencies.

Time-Division Multiplexing (TDM)

Time-division multiplexing is best understood by considering Figure 3.72(a).
The data sources are assumed to have been sampled at the Nyquist rate or
higher. The commutator then interlaces the samples to form the baseband
signal shown in Figure 3.72(b). At the channel output, the baseband signal
is demultiplexed by using a second commutator as illustrated. Proper operation of this system depends on proper synchronization between the two
commutators.

  If all message signals have equal bandwidth, then the samples are transmitted sequentially, as shown in Figure 3.72(b). If the sampled data signals
have unequal bandwidths, more samples must be transmitted per unit time
from the wideband channels. This is easily accomplished if the bandwidths
are harmonically related. For example, assume that a TDM system has four
channels of data. Also assume that the bandwidth of the first and second
data sources, $s_1(t)$ and $s_2(t)$, is W Hz, the bandwidth of $s_3(t)$ is $2W$ Hz, and
the bandwidth of $s_4(t)$ is $4W$ Hz. It is easy to show that a permissible

FIGURE 3.72   Time-division multiplexing. (a) Time-division multiplexing
system. (b) Baseband signal.

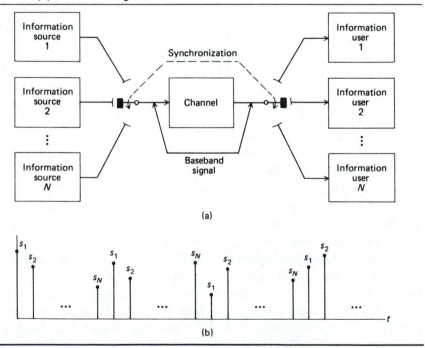

(a)

(b)

sequence of baseband samples is a periodic sequence, one period of which is . . . $s_1 s_4 s_3 s_4 s_2 s_4 s_3 s_4$ . . . .

The minimum bandwidth of a TDM baseband is easy to determine using the sampling theorem. Assuming Nyquist rate sampling, the baseband contains $2W_i T$ samples from the ith channel in each T-second interval, where W_i is the bandwidth of the ith channel. Thus the total number of baseband samples in a T-second interval is

$$n_s = \sum_{i=1}^{N} 2W_i T \tag{3.273}$$

Assuming that the baseband is a lowpass signal of bandwidth B, the required sampling rate is $2B$. In a T-second interval, we then have $2BT$ total samples. Thus

$$n_s = 2BT = \sum_{i=1}^{N} 2W_i T \tag{3.274}$$

or

$$B = \sum_{i=1}^{N} W_i \tag{3.275}$$

which is the same as the minimum required bandwidth obtained for FDM.

An Example: The Digital Telephone System

As an example of a digital TDM system, we consider a multiplexing scheme common to many telephone systems. The sampling format is illustrated in Figure 3.73(a). A voice signal is sampled at 8000 samples per second, and each sample is quantized into seven binary digits. An additional binary digit, known as a *signaling bit,* is added to the basic seven bits that represent the sample value. The signaling bit is used in establishing calls and for synchronization. Thus eight bits are transmitted for each sample value, yielding a bit rate of 64,000 bits per second (64 kbps). Twenty-four of these 64 kbps voice channels are grouped together to yield a T1 carrier. The T1 frame consists of $24(8) + 1 = 193$ bits. The extra bit is used for frame synchronization. The frame duration is the reciprocal of the fundamental sampling frequency, or 0.125 ms. Since the frame rate is 8000 frames per second, with 193 bits per frame, the T1 data rate is 1.544 Mbps.

FIGURE 3.73 Digital multiplexing scheme for digital telephone. (a) T1 frame. (b) Digital multiplexing.

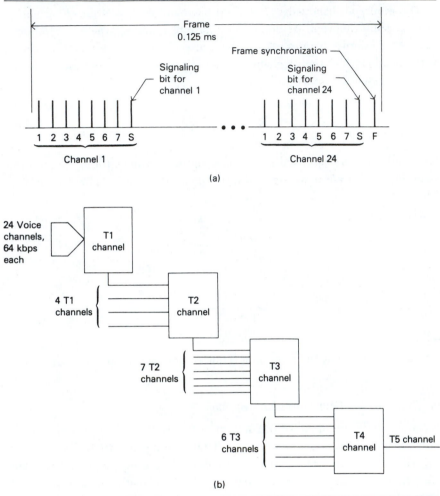

(a)

(b)

As shown in Figure 3.73(b), four T1 carriers can be multiplexed to yield a T2 carrier, which consists of 96 voice channels. Seven T2 carriers yield a T3 carrier, and 6 T3 carriers yield a T4 carrier. The bit rate of a T4 channel, consisting of 4032 voice channels with signaling bits and framing bits, is 274.176 Mbps. A T1 link is typically used for short transmission distances in areas of heavy usage. T4 and T5 channels are used for long transmission distances.

Comparison of Multiplexing Schemes

We have seen that for all three types of multiplexing studied, the baseband bandwidth is lower-bounded by the total information bandwidth. However, there are advantages and disadvantages to each multiplexing technique.

The basic advantage of FDM is simplicity of implementation, and if the channel is linear, disadvantages are difficult to identify. However, many channels have small, but nonnegligible nonlinearities. As we saw in Chapter 2, nonlinearities lead to intermodulation distortion. In FDM systems, the result of intermodulation distortion is crosstalk between channels in the baseband. This problem is avoided in TDM systems.

However, TDM also has inherent disadvantages. Samplers are required, and if continuous data are required by the data user, the continuous waveforms must be reconstructed from the samples. One of the biggest difficulties with TDM is maintaining synchronism between the multiplexing and demultiplexing commutators.

The basic advantage of QM is that QM allows simple DSB modulation to be used while at the same time making efficient use of baseband bandwidth. It also allows dc response, which SSB does not. The basic problem with QM is crosstalk between the quadrature channels, which results if perfectly coherent demodulation carriers are not available. Of course, if QM and FDM are used together, intermodulation becomes a problem.

Other advantages and disadvantages of FDM, QM, and TDM will become apparent when we study performance in the presence of noise in Chapter 6.

3.7   COMPARISON OF MODULATION SYSTEMS

In this chapter, a large number of different techniques for information transmission have been introduced. It is important that these techniques be compared so that logical choices can be made between the many available systems when various application needs arise. Unfortunately, at this point in our study, such a comparison cannot be accomplished with any rigor. The reason for this is that we have only studied systems in a highly idealized environment. Specifically, the assumption has been made that the received signal at the demodulator input $x_r(t)$ is *exactly* the transmitted signal $x_c(t)$.

In a practical environment, the transmitted signal is subjected to many undesirable perturbations prior to demodulation. The most important perturbation is usually *noise*, which is inadvertently added to the signal at several points in the system. Noise is always present in varying degrees in practical systems, and its effect must be considered in any meaningful comparison of systems.

The noise performance of modulation systems is often specified by comparing the signal-to-noise ratios (SNR) at the input and output of the demodulator. This is illustrated in Figure 3.74. The *predetection filter* is typically the combination of the radio-frequency (RF) and intermediate-frequency (IF) filters. As we shall see in Chapter 6, the parameter of interest at the output of the predetection filter is

$$(\text{SNR})_T = \frac{P_T}{\langle n_i^2(t) \rangle} \qquad (3.277)$$

where P_T is the power of the received signal and $\langle n_i^2(t) \rangle$ is the noise power measured in the bandwidth of the message signal.

The signal at the output of the demodulator is

$$y_D(t) = m(t) + n(t) \qquad (3.278)$$

where $n(t)$ is the output noise component. The signal-to-noise ratio at this point is

$$(\text{SNR})_D = \frac{\langle m^2(t) \rangle}{\langle n^2(t) \rangle} \qquad (3.279)$$

For thermal noise sources, $\langle n_i^2(t) \rangle$ can be obtained from knowledge of the system noise figure. This system figure of merit is determined using techniques developed in Appendix A.

The noise performance of several important systems is illustrated in Figure 3.75. The curves shown give the performance of nonlinear systems only above threshold. The approximate point where threshold occurs is indicated by a heavy dot. Below threshold, the performance degrades rapidly.

Figure 3.75 is not intended to be complete. It is placed here only to illustrate the importance of considering noise effects and to illustrate the superior performance of nonlinear systems over certain ranges of signal-to-

FIGURE 3.74   Receiver block diagram

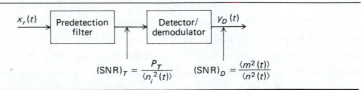

FIGURE 3.75 Noise performance curves

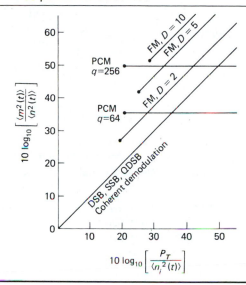

noise ratios. The noise performance of these systems is the subject of Chapter 6. At that time, the results shown in Figure 3.75 will be derived in detail.

For completeness, we include Table 3.4, which compares continuous-wave analog systems. The list of applications is by no means complete but does indicate typical uses.

SUMMARY

1. Modulation is the process by which a parameter of a carrier is varied in one-to-one correspondence with an information-bearing signal usually referred to as the message. Several uses of modulation are to achieve efficient transmission, to allocate channels, and for multiplexing.

2. If the carrier is continuous, the modulation is continuous-wave modulation. If the carrier is a sequence of pulses, the modulation is pulse modulation.

3. There are two basic types of continuous-wave modulation: linear modulation and angle modulation.

4. Assume that a general modulated carrier is given by

$$x_c(t) = A(t) \cos [\omega_c t + \phi(t)]$$

TABLE 3.4   A Comparison of Analog Systems

	Bandwidth	DC Response	Efficiency	Complexity	Typical Applications
DSB	$2W$	Yes	100%	*Moderate.* Coherent demodulation is required	Low bandwidth communication systems
AM	$2W$	No	<50%	*Minor.* Simple modulators and envelope detection.	Broadcast radio
SSB	W	No	100%	*Major.* Phase shift modulators and coherent demodulators are required.	Voice communication systems
VSB	$W+$	Yes	100%	*Major.* Symmetric filters and coherent demodulation are required.	Wideband systems
VSB + Carrier	$W+$	No	<50%	*Moderate.* Symmetric filter required, but envelope detection can be used.	TV video
FM	$2(D+1)W$	Yes	Not applicable	*Moderate.* Simple phase-lock loop demodulators can be used.	High-fidelity broadcast radio
PM	$2(D+1)W$	Yes (with calibration)	Not applicable	*Moderate.* Essentially the same as FM.	Data transmission; often used in the generation of FM

If $A(t)$ is proportional to the message signal, the result is linear modulation. If $\phi(t)$ is proportional to the message signal, the result is phase modulation (PM). If the time derivative of $\phi(t)$ is proportional to the message signal, the result is frequency modulation (FM). Both PM and FM are examples of angle modulation. Angle modulation is a nonlinear process.

5. The simplest example of linear modulation is double sideband (DSB). DSB is implemented as a simple product device, and coherent demodulation must be used.

6. If a carrier component is added to a DSB signal, the result is amplitude modulation (AM). This is a useful modulation technique because it allows simple envelope detection to be used.

7. The efficiency of a modulation process is defined as the percentage of total power that conveys information. For AM, this is given by

$$E = \frac{a^2 \langle m_n^2(t) \rangle}{1 + a^2 \langle m_n^2(t) \rangle} \ (100\%)$$

where the parameter a is known as the *modulation index* and $m_n(t)$ is $m(t)$ normalized so that the peak value is unity. If envelope demodulation is used, the index must be less than unity.

8. A single-sideband (SSB) signal is generated by transmitting only one of the sidebands in a DSB signal. SSB signals are generated either by sideband filtering a DSB signal or by using a phase-shift modulator. SSB signals can be written as

$$x_c(t) = \tfrac{1}{2} A_c m(t) \cos \omega_c t \pm \tfrac{1}{2} A_c \hat{m}(t) \sin \omega_c t$$

in which the plus sign is used for lower-sideband SSB and the minus sign is used for upper-sideband SSB. These signals can be demodulated either through the use of coherent demodulation or through the use of carrier reinsertion.

9. Vestigial sideband (VSB) results when a vestige of one sideband appears on an otherwise SSB signal. VSB is easier to generate than SSB. Demodulation can be coherent, or carrier reinsertion can be used.

10. Switching modulators, followed by filters, can be used as implementations of product devices.

11. Frequency translation is accomplished by multiplying a signal by a carrier and filtering. These systems are known as *mixers*. The concept of mixing is used in superheterodyne receivers. Mixing results in *image frequencies*, which can be troublesome.

12. The general expression for an angle-modulated signal is

$$x_c(t) = A_c \cos [\omega_c t + \phi(t)]$$

For a PM signal, $\phi(t)$ is given by

$$\phi(t) = k_p m(t)$$

and for an FM signal, it is

$$\phi(t) = 2\pi f_d \int^t m(\alpha) \, d\alpha$$

where k_p and f_d are the phase and frequency deviation constants, respectively.

13. Angle modulation results in an infinite number of sidebands. If only a single pair of sidebands is significant, the result is narrowband angle modulation. Narrowband angle modulation, with sinusoidal message, has approximately the same spectrum as a DSB signal except for a 180° phase shift of the lower sideband.

14. An angle-modulated carrier with a sinusoidal message signal can be expressed as

$$x_c(t) = A_c \sum_{n=-\infty}^{\infty} J_n(\beta) \cos (\omega_c + n\omega_m)t$$

The term $J_n(\beta)$ is the Bessel function of the first kind of order n and argument β. The parameter β is known as the *modulation index*. If $m(t) = A \sin \omega_m t$, then $\beta = k_p A$ for PM, and $\beta = f_d A/f_m$ for FM.

15. The power contained in an angle-modulated carrier is $\langle x_c^2(t) \rangle = \frac{1}{2}A_c^2$ if the carrier frequency is large compared to the bandwidth of the modulated carrier.

16. The bandwidth of an angle-modulated signal is, strictly speaking, infinite. However, a measure of the bandwidth can be obtained by defining the power ratio

$$P_r = J_0^2(\beta) + 2 \sum_{n=1}^{k} J_n^2(\beta)$$

which is the ratio of the total power $\frac{1}{2}A_c^2$ to the power in the bandwidth $B = 2kf_m$. A power ratio of 0.98 yields $B = 2(\beta + 1)f_m$.

17. The deviation ratio of an angle-modulated signal is

$$D = \frac{\text{peak frequency deviation}}{\text{bandwidth of } m(t)}$$

18. Carson's rule for estimating the bandwidth of an angle-modulated carrier with an arbitrary message signal is $B = 2(D + 1)W$.

19. Narrowband-to-wideband conversion is a technique whereby a wideband FM signal is generated from a narrowband FM signal. The system makes use of a frequency multiplier, which, unlike a mixer, multiplies the deviation as well as the carrier frequency.

20. Demodulation of an angle-modulated signal is accomplished through the use of a discriminator. This device yields an output signal propor-

tional to the frequency deviation of the input signal. Placing an integrator at the discriminator output allows PM signals to be demodulated.

21. An FM discriminator can be implemented as a differentiator followed by an envelope detector. Bandpass limiters are used at the differentiator input to eliminate amplitude variations.

22. *Interference,* the presence of undesired signal components, can be a problem in demodulation. Interference at the input of a demodulator results in undesired components at the demodulator output. If the interference is large and if the demodulator is nonlinear, thresholding can occur. The result of this is a complete loss of the signal component.

23. Interference is also a problem in angle modulation. In FM systems, the effect of interference is a function of both the amplitude and frequency of the interfering tone. In PM systems, the effect of interference is a function only of the amplitude of the interfering tone. For FM interference can be reduced by the use of preemphasis and deemphasis.

24. A phase-lock loop (PLL) is a simple and practical system for the demodulation of angle-modulated signals. It is a feedback control system and is analyzed as such. Phase-lock loops also provide simple implementations of frequency multipliers and frequency dividers.

25. Placing a bandpass filter and a discriminator within a phase-lock loop yields a frequency-compressive feedback system. This system is also useful for demodulation of angle-modulated signals.

26. The Costas phase-lock loop, which is a variation of the phase-lock loop, is a system for the demodulation of DSB signals.

27. Analog pulse modulation results when the message signal is sampled and a pulse train carrier is used. A parameter of each pulse is varied in one-to-one correspondence with the value of each sample.

28. Pulse-amplitude modulation (PAM) results when the *amplitude* of each carrier pulse is proportional to the value of the message signal at each sampling instant. PAM is essentially a sample-and-hold operation. Demodulation of PAM is accomplished by lowpass filtering.

29. Pulse-width modulation (PWM) results when the *width* of each carrier pulse is proportional to the value of the message signal at each sampling instant. Demodulation of PWM is also accomplished by lowpass filtering.

30. Pulse-position modulation (PPM) results when the *position* of each carrier pulse, as measured by the displacement of each pulse from a fixed ref-

erence, is proportional to the value of the message signal at each sampling instant.

31. Digital pulse modulation results when the sample values of the message signal are quantized and encoded prior to transmission.

32. Delta modulation (DM) is an easily implemented form of digital pulse modulation. In DM, the message signal is encoded into a sequence of binary symbols. The binary symbols are represented by the polarity of impulse functions at the modulator output. Demodulation is ideally accomplished by integration, but lowpass filtering is often a simple and satisfactory substitute.

33. Pulse-code modulation (PCM) results when the message signal is sampled and quantized, and each quantized sample value is encoded as a sequence of binary symbols. PCM differs from DM in that in PCM each quantized sample value is transmitted but in DM the transmitted quantity is the polarity of the change in the message signal from one sample to the next.

34. Multiplexing is a scheme allowing two or more message signals to be communicated simultaneously using a single system.

35. Frequency-division multiplexing (FDM) results when simultaneous transmission is accomplished by translating message spectra, using modulation, to *nonoverlapping* locations in a baseband spectrum. The baseband signal is then transmitted using any carrier modulation method.

36. Quadrature multiplexing (QM) results when two message signals are translated, using linear modulation with quadrature carriers, to the *same* spectral locations. Demodulation is accomplished coherently using quadrature demodulation carriers. A phase error in a demodulation carrier results in serious distortion of the demodulated signal. This distortion has two components: a time-varying attenuation of the desired output signal and crosstalk from the quadrature channel.

37. Time-division multiplexing (TDM) results when samples from two or more data sources are interlaced, using commutation, to form a baseband signal. Demultiplexing is accomplished by using a second commutator, which must be synchronous with the multiplexing commutator.

38. Any meaningful comparison of systems must take into consideration a large number of factors. Examples are cost or complexity of both transmitter and receiver, the bandwidth requirements, compatibility with existing systems, and performance, especially in the presence of noise.

FURTHER READING

One can find basic treatments of modulation theory at about the same technical level of this text in a wide variety of books. Examples are Carlson (1986), Stremler (1982), Haykin (1989), Lathi (1983), Roden (1985), Shanmugam (1979), and Couch (1990). Taub and Schilling (1986) have an excellent treatment of phase-lock loops. The performance of phase-lock loops in the absence of noise is discussed by Viterbi (1966, Chapters 2 and 3) and Gardner (1979). Also, as we mentioned in this chapter, Schwartz, Bennett, and Stein (1966) have an excellent section of PWM and PPM.

PROBLEMS

Section 3.1

3.1 Assume that a DSB signal

$$x_c(t) = A_c m(t) \cos (\omega_c t + \phi_0)$$

is demodulated using the demodulation carrier $2 \cos [\omega_c t + \theta(t)]$. Determine, in general, the demodulated output $y_D(t)$. Let $\theta(t) = \theta_0$, where θ_0 is a constant, and determine the mean-square error between $m(t)$ and the demodulated output as a function of ϕ_0 and θ_0. Now let $\theta(t) = \omega_0 t$ and compute the mean-square error between $m(t)$ and the demodulated output.

3.2 Show that an AM signal can be demodulated using coherent demodulation by assuming a demodulation carrier of the form

$$2 \cos [\omega_c t + \theta(t)]$$

where $\theta(t)$ is the demodulation phase error. Why is coherent demodulation of AM not used in practice?

3.3 Design an envelope detector that uses a full-wave rectifier rather than the half-wave rectifier shown in Figure 3.3. Sketch the resulting waveforms, as was done in Figure 3.3(b), for the full-wave rectifier. What are the advantages of the full-wave rectifier?

3.4 Three message signals are periodic with period T, as shown in Figure 3.76. Each of the three message signals is applied to an AM modulator. For each message signal, determine the modulation efficiency for $a = 0.7$ and $a = 1$.

3.5 The positive portion of the envelope of the output of an AM modulator is shown in Figure 3.77. The message signal is a waveform having

FIGURE 3.76

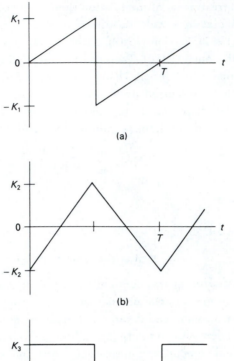

(a)

(b)

(c)

FIGURE 3.77

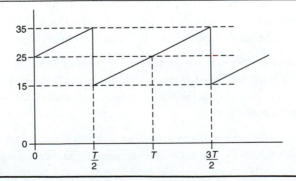

FIGURE 3.78

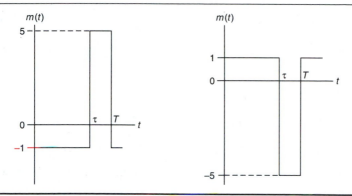

zero dc value. Determine the modulation index, the carrier power, the efficiency, and the power in the sidebands.

3.6 In this problem we examine the efficiency of AM modulation for the case in which the message signal does not have symmetrical maximum and minimum values. Two message signals are shown in Figure 3.78. Each is periodic with period T, and τ is chosen such that the dc value of $m(t)$ is zero. Calculate the efficiency for each $m(t)$ for $a = 1$.

3.7 An AM modulator operates with the message signal

$$m(t) = 7 \cos 20\pi t - 9 \cos 60\pi t$$

The unmodulated carrier is given by $100 \cos 200\pi t$, and the system operates with an index of $\frac{1}{2}$.

(a) Write the equation for $m_n(t)$, the normalized signal with a minimum value of -1.
(b) Determine $\langle m_n^2(t) \rangle$, the power in $m_n(t)$.
(c) Determine the efficiency of the modulator.
(d) Sketch the double-sided spectrum of $x_c(t)$, the modulator output, giving the weights and frequencies of all components.

3.8 Rework Problem 3.7 for the message signal

$$m(t) = 7 \cos 20\pi t + 9 \cos 60\pi t$$

3.9 An AM modulator has output

$$x_c(t) = 40 \cos 2\pi(200)t + 6 \cos 2\pi(180)t + 6 \cos 2\pi(220)t$$

Determine the modulation index and the efficiency.

3.10   An AM modulator has output

$$x_c(t) = A \cos 2\pi(200)t + B \cos 2\pi(180)t + B \cos 2\pi(220)t$$

The carrier power is 100 W, and the efficiency is 40%. Determine A, B, and the modulation index.

3.11   An AM modulator has output

$$x_c(t) = 20 \cos 2\pi(150)t + 6 \cos 2\pi(160)t + 6 \cos 2\pi(140)t$$

Determine the modulation index and the efficiency.

3.12   An AM modulator is operating with an index of 0.8. The modulation input is

$$m(t) = 2 \cos (2\pi f_m t) + \cos (4\pi f_m t) + 2 \cos (10\pi f_m t)$$

(a) Sketch the spectrum of the modulator output showing the weights of all impulse functions.
(b) What is the efficiency of the modulation process?

3.13   Consider the system shown in Figure 3.79. Assume that the average value of $m(t)$ is zero and that the maximum value of $|m(t)|$ is M. Also assume that the square-law device is defined by $y(t) = 4x(t) + 2x^2(t)$.

(a) Write the equation for $y(t)$.
(b) Describe the filter that yields an AM signal for $g(t)$. Give the necessary filter type and the frequencies of interest.
(c) What value of M yields a modulation index of 0.1?
(d) What is an advantage of this method of modulation?

3.14   Redraw Figure 3.7 to illustrate the generation of upper-sideband SSB. Give the equation defining the upper-sideband filter. Complete the analysis by deriving the expression for the output of an upper-sideband SSB modulator.

3.15   Show that both AM and DSB can be demodulated using the synchronous switching system shown in Figure 3.80.

FIGURE 3.79

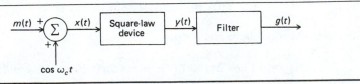

FIGURE 3.80

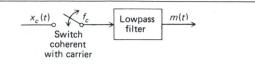

3.16 An upper-sideband SSB modulator has the message signal $m(t) = A_m \cos \omega_m t$. The unmodulated carrier is given by $c(t) = A_c \sin \omega_c t$. Sketch the modulator output $x_c(t)$. Repeat for a lower-sideband SSB modulator.

3.17 The VSB filter illustrated in Figure 3.13 is characterized by

$$H(f_c - f_1) = \epsilon e^{j\phi}$$
$$H(f_c + f_1) = (1 - \epsilon)e^{j\theta_1}$$
$$H(f_c + f_2) = 1 \cdot e^{j\theta_2}$$

The message signal is given by Equation (3.64) and is to be demodulated by multiplying by $4 \cos \omega_c t$ and lowpass filtering. It is required that the demodulated output be distortionless. Derive the expressions for θ_1 and θ_2 as functions of ϕ. Generalize your result by giving the ideal phase response of a VSB filter.

3.18 Prove that carrier reinsertion with envelope detection can be used for demodulation of VSB.

3.19 Sketch Figure 3.20 for the case where $\omega_{LO} = \omega_c - \omega_{IF}$.

3.20 A mixer is used in a short-wave superheterodyne receiver. The receiver is designed to receive transmitted signals between 5 and 25 MHz. High-side tuning is to be used. Determine the tuning range of the local oscillator for IF frequencies varying between 400 kHz and 2 MHz. Plot the ratio defined by the tuning range over this range of IF frequencies as in Table 3.1.

3.21 A superheterodyne receiver uses an IF frequency of 455 kHz. The receiver is tuned to a transmitter having a carrier frequency of 1120 kHz. Give two permissible frequencies of the local oscillator and the image frequency for each.

3.22 Repeat the preceding problem assuming that the IF frequency is 2500 kHz.

Section 3.2

3.23 Let the input to a phase modulator be $m(t) = u(t - t_0)$, as shown in Figure 3.22(a). Assume that the unmodulated carrier is $A_c \cos \omega_c t$ and that $\omega_c t_0 = 2\pi n$, where n is an integer. Sketch accurately the phase modulator output for $k_p = \pi$, $\frac{1}{4}\pi$, and $\frac{1}{8}\pi$, as was done in Figure 3.22(c) for $k_p = \frac{1}{2}\pi$.

3.24 The carrier power of an FM signal is 40 W, and the carrier frequency is $f_c = 40$ Hz. A sinusoidal message signal is used with index $\beta = 10$. The sinusoidal message signal has a frequency of 5 Hz. Determine the average value of $x_c(t)$. We usually assume $\langle x_c(t) \rangle = 0$. By drawing appropriate spectra, explain this apparent contradiction.

3.25 Given that $J_0(3) = -0.2601$ and that $J_1(3) = 0.3391$, determine $J_4(3)$.

3.26 Determine and sketch the spectrum (amplitude and phase) of an angle-modulated signal assuming that the instantaneous phase deviation is $\phi(t) = \beta \sin 2\pi f_m t$. Also assume $\beta = 8$, $f_m = 20$ Hz, and $f_c = 1000$ Hz.

3.27 Repeat the preceding problem assuming that the instantaneous phase deviation is $\phi(t) = \beta \cos 2\pi f_m t$.

3.28 An FM modulator has input $m(t) = 4 \cos 10\pi t$. The peak frequency deviation is 25 Hz. The modulator is followed by an ideal bandpass filter with a center frequency given by the carrier frequency and a bandwidth of 54 Hz. Determine the power at the filter output assuming that the power of the modulator output is 100 W. Sketch the amplitude and phase spectrum of the filter output.

3.29 A modulated signal is given by

$$x_c(t) = 5 \cos [2\pi(80)t] + 10 \cos [2\pi(100)t] + 3 \cos [2\pi(120)t]$$

Assuming a carrier frequency of 100 Hz, write this signal in the form of (3.1). Give equations for the envelope, $R(t)$, and the phase deviation, $\phi(t)$.

3.30 Repeat the preceding problem assuming that

$$x_c(t) = 5 \cos [2\pi(80)t] + 10 \cos [2\pi(100)t] + 3 \sin [2\pi(120)t]$$

3.31 A transmitter uses a carrier frequency of 1000 hz. Determine both the phase and frequency deviation for each of the following transmitter outputs:

(a) $x_c(t) = \cos [2\pi(1000)t + 10t^2]$

(b) $x_c(t) = \cos[2\pi(500)t^2]$
(c) $x_c(t) = \cos[2\pi(1100)t]$
(d) $x_c(t) = \cos[2\pi(1000)t + 10\sqrt{t}]$

3.32 An FM modulator has output

$$x_c(t) = 100 \cos\left[\omega_c t + 2\pi f_d \int^t m(\alpha)\, d\alpha\right]$$

where $f_d = 30$ Hz/V. Assume that $m(t)$ is the rectangular pulse $m(t) = 8\Pi[\frac{1}{4}(t - 2)]$.

(a) Sketch the phase deviation in radians.
(b) Sketch the frequency deviation in hertz.
(c) Determine the peak frequency deviation in hertz.
(d) Determine the peak phase deviation in radians.
(e) Determine the power at the modulator output.

3.33 Repeat the preceding problem assuming that $m(t)$ is the triangular pulse $4\Lambda[\frac{1}{3}(t - 6)]$.

3.34 An FM modulator with $f_d = 25$ Hz/V has the message signal shown in Figure 3.81. Plot the frequency deviation in hertz and the phase deviation in radians.

3.35 An FM modulator with $f_d = 15$ Hz/V has the message signal shown in Figure 3.82. Plot the frequency deviation in hertz and the phase deviation in radians.

3.36 An FM modulator with $f_d = 10$ Hz/V has the message signal shown in Figure 3.83. Plot the frequency deviation in hertz and the phase deviation in radians.

FIGURE 3.81

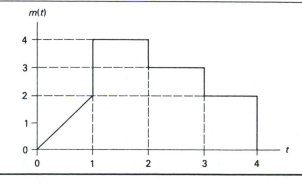

FIGURE 3.82

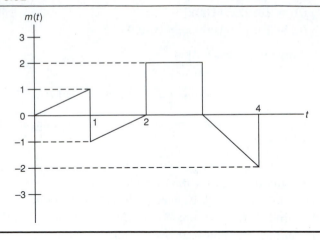

3.37   An FM modulator has $f_c = 1000$ Hz and $f_d = 12.5$. The modulator
        has input $m(t) = 4 \cos 2\pi(10)t$.

   (a) What is the modulation index?
   (b) Sketch, approximately to scale, the magnitude spectrum of the
       modulator output. Show all frequencies of interest.
   (c) Is this narrowband FM? Why?
   (d) If the same $m(t)$ is used for a phase modulator, what must k_p be
       to yield the index given in (a)?

FIGURE 3.83

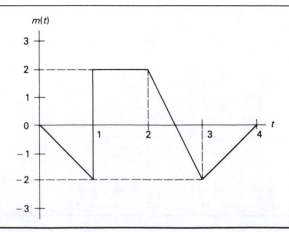

3.38   An audio signal has a bandwidth of 25 kHz. The maximum value of
       $|m(t)|$ is 5 V. This signal frequency modulates a carrier. Estimate the
       bandwidth of the modulator output, assuming that the deviation con-
       stant of the modulator is (a) 20 Hz/V, (b) 200 Hz/V, (c) 2 kHz/V, and
       (d) 20 kHz/V.

3.39   By making use of Equations (3.104) and (3.112), show that

$$\sum_{n=-\infty}^{\infty} J_n^2(\beta) = 1$$

3.40   Prove that $J_n(\beta)$ can be expressed as

$$J_n(\beta) = \frac{1}{\pi} \int_0^{\pi} \cos(\beta \sin x - nx)\, dx$$

3.41   Using the result of Problem 3.40, show that

$$J_{-n}(\beta) = (-1)^n J_n(\beta)$$

3.42   An FM modulator is followed by an ideal bandpass filter having a
       center frequency of 500 Hz and a bandwidth of 70 Hz. The gain of
       the filter is 1 in the passband. The unmodulated carrier is given by
       10 cos $(1000\pi t)$ and the message signal is $m(t) = 10 \cos(20\pi t)$. The
       transmitter constant f_d is 8 Hz/V.

       (a) Determine the peak frequency deviation in hertz.
       (b) Determine the peak phase deviation in radians.
       (c) Determine the modulation index.
       (d) Determine the power at the filter input and the filter output.
       (e) Draw the single-sided spectrum of the signal at the filter input
           and the filter output. Label the amplitude and frequency of each
           spectral component.

3.43   An FM modulator operates with a peak deviation of 700 Hz. The
       message signal is given by

$$m(t) = 5 \cos 200\pi t$$

       The power at the modulator ouput is 20 kW. The modulator is fol-
       lowed by a bandpass filter having a center frequency equal to the
       carrier frequency and a bandwidth of 1.3 kHz. Given that $J_0(7) = 0.3001$ and $J_1(7) = -0.0047$, compute the power at the filter out-
       put.

3.44   Reconstruct Figure 3.28 for the case where the five values of the
       modulation index are achieved by increasing the modulator deviation
       constant while holding f_m constant.

3.45   A sinusoidal message signal has a frequency of 150 Hz. This signal is
       the input to an FM modulator with an index of 10. Determine the
       bandwidth of the modulator output if a power ratio, P_r, of 0.8 is
       needed. Repeat for a power ratio of 0.9.

3.46   A narrowband FM signal has a carrier frequency of 110 kHz and a
       deviation ratio of 0.05. The modulation bandwidth is 10 kHz. This
       signal is used to generate a wideband FM signal with a deviation ratio
       of 20 and a carrier frequency of 100 MHz. The scheme utilized to
       accomplish this is illustrated in Figure 3.31. Give the required value
       of frequency multiplication, n. Also, fully define the mixer by
       giving *two* permissible frequencies for the local oscillator, and
       define the required bandpass filter (center frequency and band-
       width).

3.47   Consider the FM discriminator shown in Figure 3.84. The envelope
       detector can be considered ideal with an infinite input impedance.
       Plot the magnitude of the transfer function $E(f)/X_r(f)$. From this
       plot, determine a suitable carrier frequency and the discriminator
       constant, K_D, and estimate the allowable peak frequency deviation of
       the input signal.

3.48   By adjusting the values of R, L, and C in Problem 3.47, design a dis-
       criminator for a carrier frequency of 100 MHz, assuming that the
       peak frequency deviation is 6 MHz. What is the discriminator con-
       stant, K_D, for your design?

3.49   Solving system problems in which the input frequency is varying is
       often a difficult task. Approximate solutions can often be obtained by
       using sinusoidal steady-state techniques with the system-transfer
       function evaluated at the instantaneous input frequency. This is
       known as the *quasi-steady-state approximation*. In other words, if the
       input to a system is

$$x(t) = A \cos [\omega_c t + \phi(t)]$$

FIGURE 3.84

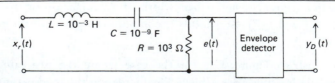

the system output is approximated by using

$$H\left(f_c + \frac{1}{2\pi}\frac{d\phi}{dt}\right)$$

for the transfer function. Rework Example 3.8 using this technique.

Section 3.3

3.50 Assume that an FM discriminator operates in the presence of sinusoidal interference. Show that the discriminator output is a nonzero constant (dc) for each of the following cases: $A_i = A_c$, $A_i = -A_c$, and $A_i \gg A_c$. Determine the discriminator output for each of these cases.

Section 3.4

3.51 Sketch the input phase deviation $\phi(t)$, the phase deviation of the VCO output $\theta(t)$, and the phase error for Example 3.9. Determine the relationship between the various parameters (A, k_f, K_t, K_v) that must exist for the maximum phase error to be less than 0.2 rad.

3.52 Rework Example 3.9 for $m(t) = A \cos \omega_m t$.

3.53 Using $x_r(t) = m(t) \cos \omega_c t$ and $e_0(t) = 2 \cos (\omega_c t + \theta)$ for the assumed Costas PLL input and VCO output, respectively, verify that all signals shown at the various points in Figure 3.57 are correct. Assuming that the VCO frequency deviation is defined by $d\theta/dt = -K_v e_v(t)$, where $e_v(t)$ is the VCO input and K_v is a positive constant, derive the phase plane. Using the phase plane, verify that the loop locks.

3.54 Using a single phase-lock loop, design a system that has an output frequency equal to $\frac{7}{3} f_0$, where f_0 is the input frequency. Describe fully, by sketching, the output of the VCO for your design. Draw the spectrum at the VCO output and at any other point in the system necessary to explain the operation of your design. Describe any filters used in your design by defining the center frequency and the appropriate bandwidth of each.

3.55 A first-order PLL is operating with zero frequency and phase error when a step in frequency of magnitude $\Delta\omega$ is applied. The loop gain K_t is $2\pi(100)$. Determine the steady-state phase error, in degrees, for $\Delta\omega = 2\pi(30)$, $2\pi(50)$, $2\pi(80)$, and $-2\pi(80)$ rad/s. What happens if $\Delta\omega = 2\pi(120)$ rad/s?

3.56   The imperfect second-order PLL is defined as a PLL with the loop filter

$$F(s) = \frac{s + a}{s + \epsilon}$$

Use the linear model of the PLL and derive the transfer function for $\Theta(s)/\Phi(s)$. Derive expressions for ω_n and ζ in terms of K_t, a, and ϵ.

3.57   Using the results of the previous problem, determine K_t, a, and ϵ for a second-order PLL having a natural frequency of 100 Hz and a damping factor of 0.707. Assume that $\epsilon = 0.1 K_t$.

3.58   A frequency-compressive feedback system is used to demodulate an FM signal having a deviation ratio of 5. The VCO constant is 25 Hz/V. Determine the discriminator constant K_D so that the deviation ratio at the discriminator input is 0.4.

Section 3.5

3.59   Sketch a circuit using a capacitor that can convert PWM to PAM. The charge on the capacitor should be proportional to the area under the PWM pulse. Show typical signals at all points in your block diagram. How is the PAM signal demodulated?

3.60   A continuous data signal is quantized and transmitted using a PCM system. If each data sample at the receiving end of the system must be known to within $\pm 0.5\%$ of the peak-to-peak full-scale value, how many binary symbols must each transmitted digital word contain? Assume that the message signal is speech and has a bandwidth of 4 kHz. Estimate the bandwidth of the resulting PCM signal (choose k).

3.61   Extend Problem 3.60 by plotting the bandwidth of the PCM signal as a function of accuracy. Let the accuracy vary from 50% of the peak-to-peak full-scale value of the input signal to 0.1% of this value. Let the bandwidth of the message signal be W.

3.62   A delta modulator has the message signal

$$m(t) = 9 \sin 2\pi(10)t + 5 \sin 2\pi(20)t$$

Determine the minimum sampling frequency required to prevent slope overload, assuming that the impulse weights δ_0 are 0.05π.

3.67   Repeat Problem 3.62, assuming that

$$m(t) = 9 \cos 2\pi(10)t + 5 \cos 2\pi(20)t$$

Section 3.6

3.64 Five messages bandlimited to W, W, $2W$, $4W$, and $4W$ Hz, respectively, are to be time-division multiplexed. Devise a commutator configuration such that each signal is periodically sampled at its own minimum rate and the samples are properly interlaced. What is the *minimum* transmission bandwidth required for this TDM signal?

3.65 In an FDM communication system, the transmitted baseband signal is

$$x(t) = m_1(t) \cos \omega_1 t + m_2(t) \cos \omega_2 t$$

This system has a second-order nonlinearity between transmitter input and receiver output. Thus the received baseband signal $y(t)$ can be expressed as

$$y(t) = a_1 x(t) + a_2 x^2(t)$$

Assuming that the two message signals, $m_1(t)$ and $m_2(t)$, have the spectra

$$M_1(f) = M_2(f) = \Pi\left(\frac{f}{W}\right)$$

sketch the spectrum of $y(t)$. Discuss the difficulties encountered in demodulating the received baseband signal. In many FDM systems, the subcarrier frequencies ω_1 and ω_2 are harmonically related. Describe any additional problems this presents.

COMPUTER EXERCISES

The following exercises require that waveforms and the spectra of waveforms be plotted. Although these exercises can be developed using any high-level language, a special-purpose computer package, having significant computational and graphics capabilities, such as Mathcad or MATLAB, is recommended.

3.1 Develop a computer program to demonstrate DSB and AM modulation. Both the waveforms and the spectra corresponding to the waveforms at each point in the system are to be generated and plotted within the program. You may use coherent demodulation for AM. An interesting message signal, with a known spectrum, can easily be generated using either three-term or a four-term Fourier series.

3.2 The purpose of the exercise is to demonstrate the properties of SSB modulation. Develop a computer program to generate both upper-sideband and lower-sideband SSB signals and display both the time-domain signals and the amplitude spectra of these signals. Plot the envelope of the SSB signals and show that both the upper-sideband and the lower-sideband SSB signals have the same envelope. It would also be interesting to look at the phase deviation and frequency deviation of the modulated carrier. Show that carrier reinsertion can be used to demodulate a SSB signal. Illustrate the effect of using a demodulation carrier with insufficient amplitude when using the carrier reinsertion technique.

3.3   In this computer exercise we investigate the properties of VSB modulation. Develop a computer program to generate and plot a VSB signal and the corresponding amplitude spectrum. Using the program, show that VSB can be demodulated using carrier reinsertion.

3.4   The purpose of this computer exercise is to study angle modulation using a sinusoidal message signal. Develop a program that allows you to generate an angle-modulated signal using a single sinusoid as the message signal. Using this message signal, plot the time-domain signal and both the amplitude and phase spectra of the signal for several values of the modulation index. Using several values of β show that the spectra shown in Figure 3.28 are correct.

3.5   Repeat the preceding problem using a message signal consisting of the sum of two sinusoids. It is not necessary to determine and plot the phase spectrum of $x_c(t)$. Using your program convince yourself that the result of Exercise 3.6 is correct.

3.6   Develop a computer program to generate the amplitude spectrum at the output of an FM modulator assuming a square-wave message signal. Plot the output for various values of peak detection. Comment on your observations.

PROBABILITY AND RANDOM VARIABLES 4

The objective of this chapter is to provide the necessary background in probability theory to make possible the mathematical description or modeling of random signals. Our ultimate goal is to develop the mathematical tools necessary to deal with random signals, or *random processes*, in systems analysis, although this will not be accomplished until Chapter 5.

Because of their unpredictable nature, we cannot assign exact values to random signals or noise waveforms at a given instant in time but must describe them in terms of averages or the probability of a particular outcome. Thus we begin this chapter with an elementary consideration of probability theory. Next we introduce the idea of random variables and their statistical description. If we think of random variables as depending on a parameter, such as time, we are really speaking of random signals, the topic of Chapter 5.

4.1 WHAT IS PROBABILITY?

Each of us has an intuitive notion about probability from prior experience. Let us therefore begin by considering some of these intuitive approaches to probability.

Equally Likely Outcomes

Suppose you match coin tosses with someone. This is an example of a random, or chance, experiment for which there are two possible outcomes, or happenings. Either the coins will match when flipped or they will not match. Since there is no prior reason to believe that one of these outcomes is favored over the other, you would most likely guess, before flipping, that your chances of winning or losing are equally likely.

Stated more generally, if there are N possible *equally likely* and *mutually exclusive* outcomes (that is, the occurrence of one outcome precludes the occurrence of any of the others) to a random, or chance, experiment and if N_A of these outcomes correspond to an event A of interest, then the probability of event A, or $P(A)$, is

$$P(A) = \frac{N_A}{N} \tag{4.1}$$

However, there are practical difficulties with this definition of probability, which is often referred to as the *classical* (or *equally likely*) *definition of probability*. One must be able to break the chance experiment up into two or more equally likely outcomes and this is not always possible. The most obvious experiments fitting these conditions are card games, dice, and coin tossing.

Philosophically, there is difficulty with this definition in that use of the words "equally likely" really amounts to saying something about being equally probable, which means we are using probability to define probability.

To summarize, there are difficulties with the classical, or equally likely, definition of probability, but it is useful in engineering problems when it is reasonable to list N equally likely, *mutually exclusive** outcomes. The following example will illustrate its usefulness in a situation where it applies.

▼ EXAMPLE 4.1   Given a deck of 52 playing cards, (a) What is the probability of drawing the ace of spades? (b) What is the probability of drawing a spade?

SOLUTION

(a) Using the principle of equal likelihood, we have one favorable outcome in 52 possible outcomes. Therefore, $P(\text{ace of spades}) = \frac{1}{52}$.

(b) Again using the principle of equal likelihood, we have 13 favorable out-
▲   comes in 52, and $P(\text{spade}) = \frac{13}{52} = \frac{1}{4}$.

Relative Frequency

Suppose we wish to assess the probability of an unborn childs being a boy. Using the classical definition, we predict a probability of $\frac{1}{2}$, since there are two possible mutually exclusive outcomes, which from outward appearances appear equally probable. However, yearly birth statistics for the United States consistently indicate that the ratio of males to total births is about 0.51. In the years 1987 to 1989, for example, the ratio of male births to

---

* *Mutually exclusive* means that if a given outcome occurs, it precludes the occurrence of any other possible outcomes.

female births in the United States was 1050 to 1000, respectively, or 51.22% of all births were male. This is an example of the relative-frequency approach to probability.

In the *relative-frequency approach,* we consider a random experiment, enumerate all possible outcomes, repeatedly perform the experiment (often conceptually), and take the ratio of the number of outcomes, N_A, favorable to an event of interest, A, to the total number of trials, N, as an approximation of the probability of A, $P(A)$. We define the limit of N_A/N, called the *relative frequency* of A, as $N \to \infty$, as $P(A)$:

$$P(A) \triangleq \lim_{N \to \infty} \frac{N_A}{N} \tag{4.2}$$

This definition of probability can be used to estimate $P(A)$. However, since the infinite number of experiments implied by (4.2) cannot be performed, only an approximation to $P(A)$ is obtained. Thus the relative-frequency notion of probability is useful for estimating a probability but is not satisfactory as a mathematical basis for probability.

The following example fixes these ideas and will be referred to later in this chapter.

▼ **EXAMPLE 4.2**   Consider the simultaneous tossing of two fair coins. Thus, on any given trial, we have the possible outcomes HH, HT, TH, and TT, where, for example, HT denotes a head on the first coin and a tail on the second coin. (We imagine that numbers are painted on the coins so we can tell them apart.) What is the probability of two heads on any given trial?

SOLUTION   By distinguishing between the coins, the correct answer, using equal likelihood, is $\frac{1}{4}$. Similarly, it follows that $P(HT) = P(TH) = P(TT) = \frac{1}{4}$.

▲

Sample Spaces and the Axioms of Probability

Because of the difficulties mentioned for the preceding two definitions of probability, mathematicians prefer to approach probability on an axiomatic basis. We now briefly describe this axiomatic approach, which is general enough to encompass all the previously given definitions.

We can view a chance experiment geometrically by representing its possible outcomes as elements of a space referred to as a *sample space* $\mathcal{S}$. An *event* is defined as a collection of outcomes. An impossible collection of outcomes is referred to as the *null event*. Figure 4.1(a) shows a representation of

FIGURE 4.1   Sample spaces. (a) Pictorial representation of an arbitrary sample space. Points show outcomes; circles show events. (b) Sample space representation for the tossing of two coins.

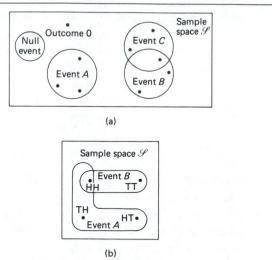

(a)

(b)

a sample space. Three events of interest, *A*, *B*, and *C*, which do not encompass the entire sample space, are shown.

A specific example of a chance experiment might consist of measuring the dc voltage at the output terminals of a power supply. The sample space for this experiment would be the collection of all possible numerical values for this voltage. On the other hand, if the experiment is the tossing of two coins, as in Example 4.2, the sample space would consist of the four outcomes HH, HT, TH, and TT enumerated earlier. A sample-space representation for this experiment is shown in Figure 4.1(b). Two events of interest, *A* and *B*, are shown. Event *A* denotes at least one head, and event *B* consists of the coins matching. Note that *A* and *B* encompass all possible outcomes for this particular example.

In the axiomatic approach, a *measure,* called *probability,* is somehow assigned to the events of a sample space* such that this measure possesses the properties of probability. The properties or axioms of this probability measure are chosen to yield a satisfactory theory such that results from applying the theory will be consistent with experimentally observed phenomena. A set of satisfactory axioms is the following:

1. $P(A) \geq 0$ for all events *A* in the sample space $\mathscr{S}$.

---

* For example, by the relative-frequency or the equally likely approaches.

2. The probability of all posssible events occurring is unity, $P(\mathcal{S}) = 1$.
3. If the occurrence of A precludes the occurrence of B, and vice versa (that is, A and B are *mutually exclusive*), then $P(A \text{ or } B) = P(A) + P(B)$.*

It is emphasized that this approach to probability does not give us the number $P(A)$; it must be obtained by some other means.

Before we proceed further, it is convenient to introduce some useful notation. The event "A or B or both" will be denoted as $(A + B)$. The event "both A and B" will be denoted either as (AB) or as (A, B). We will sometimes be interested in the event "not A," which will be denoted $\overline{A}$. An event such as $(A + B)$, which is composed of two or more events, will be referred to as a *compound event*. Other notations often used for the events just defined are $A \cup B$ ("A union B"), $A \cap B$ ("A intersection B"), and A^c ("A complement"), respectively.

Some Useful Probability Relationships

We now employ concepts of relative frequency to show the reasonableness of some useful relationships among probabilities of other compound events. We could proceed directly from the axioms just stated, but the relative-frequency approach taken here is more intuitively satisfying.

Consider a chance experiment having, among other possible outcomes, two events of immediate interest, A and B, which may or may not happen together. For example, the experiment could be the simultaneous tossing of two fair coins, as in Example 4.2, with event A being the occurrence of *at least* one head (that is, HH, HT, or TH) and event B denoting that the outcomes on each coin match (that is, either HH or TT). Although, in this particular example, the compound event "A or B or both" encompasses the entire sample space, there may, in general, be other possible events.

To continue, suppose we repeat our experiment N times and record the information given in Table 4.1. According to the relative-frequency approach, as N becomes large without bound,

$$P(A) \cong \frac{N_{\overline{A}B} + N_{AB}}{N} \tag{4.3a}$$

$$P(B) \cong \frac{N_{\overline{A}B} + N_{AB}}{N} \tag{4.3b}$$

---

* To the third axiom is added the condition of countable additivity if the number of outcomes is countably infinite. We will not include it here in the interest of brevity.

TABLE 4.1

Case 1	$N_{A\bar{B}}$ =	number of times A occurred alone.
Case 2	$N_{\bar{A}B}$ =	number of times B occurred alone.
Case 3	N_{AB} =	number of times A and B occurred together.
Case 4	$N_{\bar{A}\bar{B}}$ =	number of times neither A nor B occurred.
Thus	N =	$N_{\bar{A}B} + N_{A\bar{B}} + N_{AB} + N_{\bar{A}\bar{B}}$.

and

$$P(AB) \cong \frac{N_{AB}}{N} \tag{4.3c}$$

Consider now the probability of A or B or both, $P(A + B)$. Clearly, the number of occurrences of this compound event is $N_{A\bar{B}} + N_{\bar{A}B} + N_{AB}$. Thus, in a large number of trials,

$$P(A + B) \cong \frac{N_{A\bar{B}} + N_{\bar{A}B} + N_{AB}}{N} \tag{4.4a}$$

$$= \frac{N_{A\bar{B}} + N_{AB}}{N} + \frac{N_{\bar{A}B} + N_{AB}}{N} - \frac{N_{AB}}{N}$$

Using (4.3) and taking the limit as $N \to \infty$, this becomes

$$P(A + B) = P(A) + P(B) - P(AB) \tag{4.4b}$$

Comparing this with axiom 3, we see that if A and B are not mutually exclusive, the probability $P(AB)$ must be subtracted from the right-hand side. Equation (4.4b) could have been derived directly from the axioms by defining appropriate compound events (see Problem 4.2).

Next consider the probability that A occurs, given that B occurred, $P(A \mid B)$, and the probability that B occurs, given that A occurred, $P(B \mid A)$. These probabilities are referred to as *conditional probabilities*. Considering $P(A \mid B)$ first, we see from Table 4.1 that given that event B occurred, we must consider only cases 2 and 3 for a total of $N_{\bar{A}B} + N_{AB}$ outcomes. Of these, only N_{AB} outcomes are favorable to event A occurring, so

$$P(A \mid B) = \frac{N_{AB}}{N_{\bar{A}B} + N_{AB}} = \frac{N_{AB}/N}{(N_{\bar{A}B} + N_{AB})/N} \tag{4.5}$$

In terms of previously defined probabilities, therefore,

$$P(A \mid B) = \frac{P(AB)}{P(B)} \tag{4.6a}$$

Similar reasoning results in the relationship

$$P(B|A) = \frac{P(AB)}{P(A)} \qquad (4.6b)$$

In an axiomatic approach, conditional probabilities could be *defined* in this fashion, and we will take the preceding relations as definitions. However, the development given here shows their reasonableness.

Putting Equations (4.6a) and (4.6b) together, we obtain

$$P(A|B)P(B) = P(B|A)P(A) \qquad (4.7a)$$

or

$$P(B|A) = \frac{P(B)P(A|B)}{P(A)} \qquad (4.7b)$$

This is a special case of *Bayes' rule*.

Finally, suppose that the occurrence or nonoccurrence of B in no way influences the occurrence or nonoccurrence of A. If this is true, A and B are said to be *statistically independent*. Thus, if we are given B, this tells us nothing about A, and therefore, $P(A|B) = P(A)$. Similarly, $P(B|A) = P(B)$. From Equation (4.6a) or (4.6b) it follows that, for such events,

$$P(AB) = P(A)P(B) \qquad (4.8)$$

Equation (4.8) will be taken as the definition of statistically independent events.

▼ **EXAMPLE 4.3**   Referring to Example 4.2, suppose A denotes at least one head and B denotes a match. The sample space is shown in Figure 4.1(b). To find $P(A)$ and $P(B)$, we may proceed in several different ways.

SOLUTION   First, if we use equal likelihood, there are three outcomes favorable to A (that is, HH, HT, and TH) among four possible outcomes, yielding $P(A) = \frac{3}{4}$. For B, there are two favorable outcomes in four possibilities, giving $P(B) = \frac{1}{2}$.

As a second approach, we note that the outcomes on separate coins are statistically independent with $P(H) = P(T) = \frac{1}{2}$. (Why?) Also, event A consists of any of the mutually exclusive outcomes HH, TH, and HT, giving

$$P(A) = (\tfrac{1}{2} \cdot \tfrac{1}{2}) + (\tfrac{1}{2} \cdot \tfrac{1}{2}) + (\tfrac{1}{2} \cdot \tfrac{1}{2}) = \tfrac{3}{4} \qquad (4.9)$$

by Equation (4.8) and axiom 3. Similarly, since B consists of the mutually exclusive outcomes HH and TT,

$$P(B) = (\tfrac{1}{2} \cdot \tfrac{1}{2}) + (\tfrac{1}{2} \cdot \tfrac{1}{2}) = \tfrac{1}{2} \tag{4.10}$$

again through the use of (4.8) and axiom 3. Also, $P(AB) = P(\text{at least one head and a match}) = P(\text{HH}) = \tfrac{1}{4}$.

Next, consider the probability of at least one head given that we had a match, $P(A \mid B)$. Using Bayes' rule, we obtain

$$P(A \mid B) = \frac{P(AB)}{P(B)} = \frac{\tfrac{1}{4}}{\tfrac{1}{2}} = \frac{1}{2} \tag{4.11}$$

which is reasonable, since given B, the only outcomes under consideration are HH and TT, only one of which is favorable to event A. Next, finding $P(B \mid A)$, the probability of a match given at least one head, we obtain

$$P(B \mid A) = \frac{P(AB)}{P(A)} = \frac{\tfrac{1}{4}}{\tfrac{3}{4}} = \frac{1}{3} \tag{4.12}$$

Checking this result using the principle of equal likelihood, we have one favorable event among three candidate events (HH, TH, and HT), which yields a probability of $\tfrac{1}{3}$. We note that

$$P(AB) \neq P(A)P(B) \tag{4.13}$$

Thus events A and B are not statistically independent, although the events H and T on either coin are independent.

Finally, consider the joint probability $P(A + B)$. Using (4.4b), we obtain

$$P(A + B) = \tfrac{3}{4} + \tfrac{1}{2} - \tfrac{1}{4} = 1 \tag{4.14}$$

Remembering that $P(A + B)$ is the probability of at least one head, *or a match, or both*, we see that this includes all possible outcomes. Thus our result is correct.

▲

▼ **EXAMPLE 4.4**   This example illustrates the reasoning to be applied when trying to determine if two events are independent. A single card is drawn at random from a deck of cards. Which of the following pairs of events are independent? (a) The card is a club, and the card is black. (b) The card is a king, and the card is black.

SOLUTION   We use the relationship $P(AB) = P(A \mid B)P(B)$ (always valid) and check it against the relation $P(AB) = P(A)P(B)$ (valid only for independent events). For part (a), we let A be the event that the card is a club and B be the event that it is black. Since there are 26 black cards in an ordinary deck of cards, 13 of which are clubs, the conditional probability $P(A \mid B)$ is $\frac{13}{26}$ (given we are considering only black cards, we have 13 favorable outcomes for the card being a club). The probability that the card is black is $\frac{26}{52}$, because half the cards in the 52-card deck are black. The probability of a club (event A), on the other hand, is $\frac{13}{52}$ (13 cards in a 52-card deck are clubs). In this case,

$$P(A \mid B)P(B) = \tfrac{13}{26}\tfrac{26}{52} \neq P(A)P(B) = \tfrac{13}{52}\tfrac{26}{52} \tag{4.15}$$

and the events are not independent.

For part (b), we let A be the event that a king is drawn, and event B be that it is black. In this case, the probability of a king given that the card is black is $\frac{2}{26}$ (two cards of the 26 black cards are kings). The probability of a king is simply $\frac{4}{52}$ (4 kings in the 52-card deck). Hence,

$$P(A \mid B)P(B) = \tfrac{2}{26}\tfrac{26}{52} = P(A)P(B) = \tfrac{4}{52}\tfrac{26}{52} \tag{4.16}$$

which shows that the events king and black are statistically independent.

EXAMPLE 4.5   As an example more closely related to communications, consider the transmission of binary digits through a channel as might occur, for example, in computer systems. As is customary, we denote the two possible symbols as 0 and 1. Let the probability of receiving a zero, given a zero was sent, $P(0r \mid 0s)$, and the probability of receiving a 1, given a 1 was sent, $P(1r \mid 1s)$, be

$$P(0r \mid 0s) = P(1r \mid 1s) = 0.9 \tag{4.17}$$

Thus the probabilities $P(1r \mid 0s)$ and $P(0r \mid 1s)$ must be

$$P(1r \mid 0s) = 1 - P(0r \mid 0s) = 0.1 \tag{4.18a}$$

and

$$P(0r \mid 1s) = 1 - P(1r \mid 1s) = 0.1 \tag{4.18b}$$

respectively. These probabilities characterize the channel and would be obtained through experimental measurement or analysis. Techniques for calculating them for particular situations will be discussed in Chapter 7.

In addition to these probabilities, suppose that we have determined through measurement that the probability of sending a zero is

$$P(0s) = 0.8 \tag{4.19a}$$

and therefore the probability of sending a 1 is

$$P(1s) = 1 - P(0s) = 0.2 \tag{4.19b}$$

Note that once $P(0r|0s)$, $P(1r|1s)$, and $P(0s)$ are specified, the remaining probabilities are calculated using axioms 2 and 3.

The next question we ask is, "If a 1 was received, what is the probability, $P(1s|1r)$, that a 1 was sent?" Applying Bayes' rule, we find that

$$P(1s|1r) = \frac{P(1r|1s)P(1s)}{P(1r)} \qquad (4.20)$$

To find $P(1r)$, we note that

$$P(1r, 1s) = P(1r|1s)P(1s) = 0.18 \qquad (4.21a)$$

and

$$P(1r, 0s) = P(1r|0s)P(0s) = 0.08 \qquad (4.21b)$$

Thus

$$P(1r) = P(1r, 1s) + P(1r, 0s) = 0.18 + 0.08 = 0.26 \qquad (4.22)$$

and

$$P(1s|1r) = \frac{(0.9)(0.2)}{0.26} = 0.69 \qquad (4.23)$$

Similarly, one can calculate $P(0s|1r) = 0.31$, $P(0s|0r) = 0.97$, and $P(1s|0r)$ ▲ $= 0.03$. For practice, you should go through the necessary calculations.

Venn Diagrams and Tree Diagrams

It is sometimes convenient to visualize the relationships between various events for a chance experiment in terms of a *Venn diagram*. In such diagrams, the sample space is indicated as a rectangle, with the various events indicated by circles or ellipses. Such a diagram looks exactly as shown in Figure 4.1(a), where it is seen that events B and C are not mutually exclusive, as indicated by the overlap between them, whereas event A is mutually exclusive of events B and C.

Another handy device for determining probabilities of compound events is a *tree diagram,* particularly if the compound event can be visualized as happening in time sequence. This device is illustrated by the following example.

▼ EXAMPLE 4.6   Suppose five cards are drawn without replacement from a standard 52-card deck. What is the probability that three of a kind results?

SOLUTION   The tree diagram for this chance experiment is shown in Figure 4.2. On the first draw, we focus on a particular card, denoted as X. The

FIGURE 4.2   A card-drawing problem illustrating the construction of a tree diagram

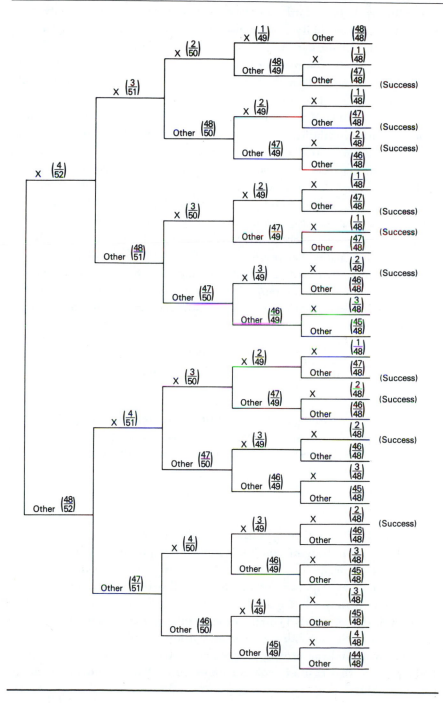

second draw results in four possible events of interest: a card is drawn that matches the first card with probability $\frac{3}{51}$, or a match is not obtained with probability $\frac{48}{51}$. If some card other than X was drawn on the first draw, then X results with probability $\frac{4}{51}$ on the second draw (lower half of Figure 4.2). At this point, 50 cards are left in the deck. If we follow the upper branch, which corresponds to a match of the first card, two events of interest are again possible: another match that will be referred to as a *triple* with probability of $\frac{2}{50}$ *on that draw*, or a card that does not match the first two with probability $\frac{48}{50}$. If a card other than X was obtained on the second draw, then X occurs with probability $\frac{4}{50}$ if X was obtained on the first draw, and probability $\frac{46}{50}$ if it was not. The remaining branches are filled in similarly. Each path through the tree will either result in success or failure, and the probability of drawing the cards along a particular path will be the product of the separate probabilities along each path. Since a particular sequence of draws resulting in success is mutually exclusive of the sequence of draws resulting in any other success, we simply add up all the products of probabilities along all paths that result in success. In addition to these sequences involving card X, there are 12 others involving other face values which result in three of a kind. Thus we multiply the result obtained from Figure 4.2 by 13. The probability of drawing three cards of the same value, in any order, is then given by

$$P(3 \text{ of a kind}) = 13\frac{(10)(4)(3)(2)(48)(47)}{(52)(51)(50)(49)(48)}$$

$$= 0.02257 \tag{4.24}$$

(Note that drawing four of a kind is termed a failure, even though in a game of poker one would be very happy to have four of a kind. In this particular example, we are interested in the probability of obtaining three of a kind only.)

▲

▼ **EXAMPLE 4.7**   Another type of problem very closely related to those amenable to tree-diagram solution is a reliability problem. Reliability problems can result from considering the overall failure of a system composed of several components each of which may fail with a certain probability p. An example is shown in Figure 4.3, where a battery is connected to a load through the series-parallel combination of relay switches each of which may fail to close with probability p (or close with probability $q = 1 - p$). The problem is to find the probability that current flows in the load. From the diagram, it is clear that a circuit is completed if S1 or S2 and S3 are closed.

FIGURE 4.3   Circuit illustrating the calculation of reliability

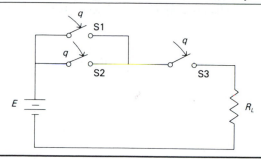

Therefore,

$$P(\text{success}) = P(\text{S1 or S2 and S3 closed})$$
$$= P(\text{S1 or S2 or both closed})P(\text{S3 closed})$$
$$= [1 - P(\text{both switches open})]P(\text{S3 closed})$$
$$= (1 - p^2)q \qquad (4.25)$$

where it is assumed that the separate switch actions are statistically independent.

Some More General Relationships

We now obtain some useful formulas for a case somewhat more general than that given above. We consider an experiment, each mutually exclusive outcome of which is composed of a compound event (A_i, B_j). The totality of all these compound events, $i = 1, 2, \ldots, M$, $j = 1, 2, \ldots, N$, composes the entire sample space (that is, the events are exhaustive).

For example, the experiment might consist of rolling a pair of dice with $(A_i, B_j) = $ (number of spots showing on die 1, number of spots showing on die 2).

Suppose the probability of the joint event (A_i, B_j) is $P(A_i, B_j)$. We can think of each compound event as a simple event, and if we sum the probabilities of all these mutually exclusive, exhaustive events, we will obtain a probability of 1, since we have included all possible outcomes. That is

$$\sum_{i=1}^{M} \sum_{j=1}^{N} P(A_i, B_j) = 1 \qquad (4.26)$$

Now let us consider a particular event B_j. Associated with this particular event, we have M possible mutually exclusive, but not exhaustive outcomes

(A_1, B_j), (A_2, B_j), . . . , (A_M, B_j). If we sum over the corresponding probabilities, we will obtain the probability of B_j irrespective of the outcome on A. Thus

$$P(B_j) = \sum_{i=1}^{M} P(A_i, B_j) \tag{4.27}$$

Similar reasoning leads to the result

$$P(A_i) = \sum_{j=1}^{N} P(A_i, B_j) \tag{4.28}$$

$P(A_i)$ and $P(B_j)$ are referred to as *marginal probabilities*.

Suppose we wish to find the conditional probability of B_m given A_n, $P(B_m \mid A_n)$. In terms of the joint probabilities $P(A_i, B_j)$, we can write this conditional probability as

$$P(B_m \mid A_n) = \frac{P(A_n, B_m)}{\sum_{j=1}^{N} P(A_n, B_j)} \tag{4.29}$$

which is a more general form of Bayes' rule than that given by (4.7).

▼ EXAMPLE 4.8   A certain experiment has the joint and marginal probabilities shown in Table 4.2. Find the missing probabilities.

SOLUTION   Using $P(B_1) = P(A_1, B_1) + P(A_2, B_1)$, we obtain $P(B_1) = 0.1 + 0.1 = 0.2$. Also, since $P(B_1) + P(B_2) + P(B_3) = 1$, we have $P(B_3) = 1 - 0.2 - 0.5 = 0.3$. Finally, using $P(A_1, B_3) + P(A_2, B_3) = P(B_3)$, we get $P(A_1, B_3) = 0.3 - 0.1 = 0.2$, and therefore, $P(A_1) = 0.1 + 0.4 + 0.2$
▲ $= 0.7$.

TABLE 4.2 $P(A_i, B_j)$

A_i ⟍ B_j	B_1	B_2	B_3	$P(A_i)$
A_1	0.1	0.4		
A_2	0.1	0.1	0.1	0.3
$P(B_j)$		0.5		1

TABLE 4.3 Possible Random Variables

Outcome: S_i	R.V. No. 1: $X_1(S_i)$	R.V. No. 2: $X_2(S_i)$
S_1 = heads	$X_1(S_1) = 1$	$X_2(S_1) = \pi$
S_2 = tails	$X_1(S_2) = -1$	$X_2(S_2) = \sqrt{2}$

4.2 RANDOM VARIABLES, DISTRIBUTION FUNCTIONS, AND DENSITY FUNCTIONS

Random Variables

In the applications of probability it is often more convenient to work in terms of numerical outcomes (for example, the number of errors in a digital data message) rather than nonnumerical outcomes (for example, failure of a component). Because of this, we introduce the idea of a *random variable,* which is defined as a rule that assigns a numerical value to each possible outcome of a chance experiment. (The term *random variable* is a misnomer; a random variable is really a function, since it is a rule that assigns the members of one set to those of another.)

As an example, consider the tossing of a coin. Possible assignments of random variables are given in Table 4.3. These are examples of *discrete random variables* and are illustrated in Figure 4.4(a).

As an example of a *continuous random variable,* consider the spinning of a pointer, such as is typically found in children's games. A possible assignment of a random variable would be the angle Θ_1, in radians, that the pointer makes with the vertical when it stops. Defined in this fashion, Θ_1 has values that continuously increase with rotation of the pointer. A second possible random variable, Θ_2, would be Θ_1 minus integer multiples of 2π radians, such that $0 \leq \Theta_2 < 2\pi$, which is commonly denoted as Θ_1 modulo 2π. These random variables are illustrated in Figure 4.4(b).

At this point, we introduce a convention that will be adhered to throughout this book (although not in all books). Capital letters (X, Θ, and so on) denote random variables, and the corresponding lower-case letters (x, θ, and so on) *denote the values that the random variables take on.*

Probability (Cumulative) Distribution Functions

We need some way of probabilistically describing random variables that works equally well for discrete and continuous random variables. One way of accomplishing this is by means of the *cumulative distribution function (cdf).*

FIGURE 4.4   Pictorial representation of sample spaces and random variables. (a) Coin-tossing experiment. (b) Pointer-spinning experiment.

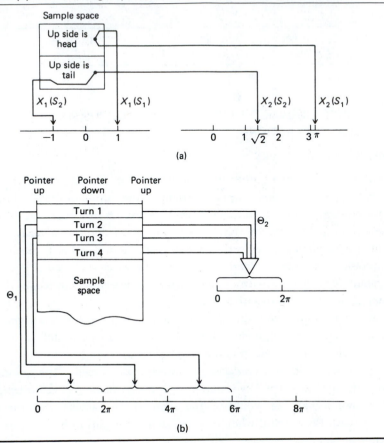

Consider a chance experiment with which we have associated a random variable X. We define the cdf $F_X(x)$ as

$$F_X(x) = \text{probability that } X \leq x = P(X \leq x) \qquad (4.30)$$

We note that $F_X(x)$ is a *function* of x, not of the random variable X. But $F_X(x)$ also depends on the assignment of the random variable X, which accounts for the subscript.

The cdf has the following properties:

(a) $0 \leq F_X(x) \leq 1$, with $F_X(-\infty) = 0$ and $F_X(\infty) = 1$.
(b) $F_X(x)$ is continuous from the right; that is, $\lim_{x \to x_0+} F_X(x) = F_X(x_0)$.

(c) $F_X(x)$ is a nondecreasing function of x; that is, $F_X(x_1) \leq F_X(x_2)$ if $x_1 < x_2$.

The reasonableness of the preceding properties is shown by the following considerations.

Since $F_X(x)$ is a probability, it must, by our previous axioms, lie between 0 and 1, inclusive. Since $X = -\infty$ excludes all possible outcomes of our experiment, $F_X(-\infty) = 0$, and since $X = \infty$ includes all possible outcomes, $F_X(\infty) = 1$.

For $x_1 < x_2$, the events $X \leq x_1$ and $x_1 < X \leq x_2$ are mutually exclusive; furthermore, $X \leq x_2$ implies $X \leq x_1$ or $x_1 < X \leq x_2$. By axiom 3, therefore,

$$P(X \leq x_2) = P(X \leq x_1) + P(x_1 < X \leq x_2)$$

or

$$P(x_1 < X \leq x_2) = F_X(x_2) - F_X(x_1) \tag{4.31}$$

Since probabilities are nonnegative, the left-hand side of (4.31) is nonnegative. Thus we see that property (c) holds.

The reasonableness of the right-continuity property is shown as follows. Suppose the random variable X takes on the value x_0 with probability P_0. Consider $P(X \leq x)$. If $x < x_0$, the event $X = x_0$ is not included, no matter how close x is to x_0. When $x = x_0$, we include the event $X = x_0$, which occurs with probability P_0. Since the events $X \leq x < x_0$ and $X = x_0$ are mutually exclusive, $P(X \leq x)$ must jump by an amount P_0 when $x = x_0$, as shown in Figure 4.5. Thus $F_X(x) = P(X \leq x)$ is continuous from the right. This is illustrated in Figure 4.5 by the dot on the curve to the right of the jump. What is more useful for our purposes, however, is that the *magnitude of any jump of $F_X(x)$, say at x_0, is equal to the probability that $X = x_0$.*

FIGURE 4.5   Illustration of the jump property of $F_X(x)$

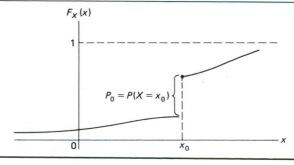

Probability Density Function

From (4.31) we see that the cdf of a random variable is a complete and useful description for the computation of probabilities. However, we will also be interested in averages when we consider noise in communication systems. For the purpose of computing statistical averages, the *probability density function* (*pdf*) $f_X(x)$ of a random variable X is more convenient. The pdf of X is defined in terms of the cdf of X by

$$f_X(x) = \frac{dF_X(x)}{dx} \tag{4.32}$$

Since the cdf of a discrete random variable is discontinuous, its pdf, mathematically speaking, does not exist. By representing the derivative of a jump-discontinuous function at a point of discontinuity by a delta function of area equal to the magnitude of the jump, we can define pdf's for discrete random variables. In some books, this problem is avoided by defining a *probability mass function* for a discrete random variable, which consists simply of lines equal in magnitude to the probabilities that the random variable will take on its possible values.

Recalling that $F_X(-\infty) = 0$, we see from (4.32) that

$$F_X(x) = \int_{-\infty}^{x} f_X(x') \, dx' \tag{4.33}$$

That is, the *area* under the pdf from $-\infty$ to x is the probability that the observed value will be less than or equal to x.

From (4.32), (4.33), and the properties of $F_X(x)$, we see that the pdf has the following properties:

$$f_X(x) = \frac{dF_X(x)}{dx} \geq 0 \tag{4.34}$$

$$\int_{-\infty}^{\infty} f_X(x) \, dx = 1 \tag{4.35}$$

$$P(x_1 < X \leq x_2) = F_X(x_2) - F_X(x_1) = \int_{x_1}^{x_2} f_X(x) \, dx \tag{4.36}$$

To obtain another enlightening and very useful interpretation of $f_X(x)$, we consider (4.36) with $x_1 = x - dx$ and $x_2 = x$. The integral then becomes $f_X(x) \, dx$, so

$$f_X(x) \, dx = P(x - dx < X \leq x) \tag{4.37}$$

TABLE 4.4

Outcome	X	$P(X = x_j)$
TT	$x_1 = 0$	$\frac{1}{4}$
TH ⎫	$x_2 = 1$	$\frac{1}{2}$
HT ⎭		
HH	$x_3 = 2$	$\frac{1}{4}$

That is, the ordinate at any point x on the pdf curve multiplied by dx gives the probability of the random variable X lying in an infinitesimal range around the point x assuming that $f_X(x)$ is continuous at x.

The following two examples illustrate cdf's and pdf's for discrete and continuous cases, respectively.

▼ **EXAMPLE 4.9**   Suppose two fair coins are tossed and X denotes the number of heads that turn up. The possible outcomes, the corresponding values of X, and the respective probabilities are summarized in Table 4.4. The cdf and pdf for this experiment and random variable definition are shown in Figure 4.6. The properties of the cdf and pdf for discrete random variables are demonstrated by this figure, as a careful examination will reveal. It is emphasized that the cdf and pdf change if the definition of the random variable or the probability assigned is changed. As an exercise, you should plot the cdf and pdf for the case where X denotes the number of heads
▲ divided by 2.

FIGURE 4.6   The cdf and pdf for a coin-tossing experiment

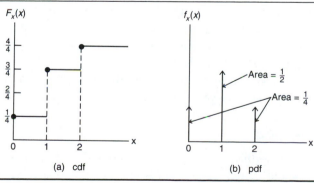

(a) cdf   (b) pdf

FIGURE 4.7   The pdf (a) and cdf (b) for a pointer-spinning experiment.

(a)   (b)

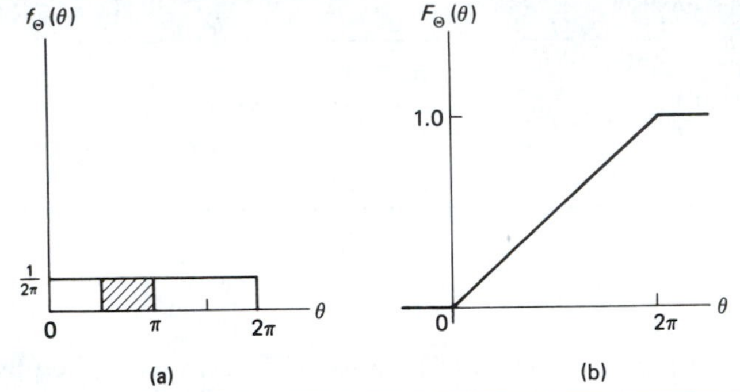

▼ **EXAMPLE 4.10**   Consider the pointer-spinning experiment described earlier. We assume that any one stopping point is not favored over any other and that our random variable Θ is defined as the angle that the pointer makes with the vertical, modulo 2π. Thus Θ is limited to the range $(0, 2\pi)$, and for any two angles θ_1 and θ_2 in $(0, 2\pi)$, we have

$$P(\theta_1 - \Delta\theta < \Theta \leq \theta_1) = P(\theta_2 - \Delta\theta < \Theta \leq \theta_2) \qquad (4.38)$$

by the assumption that the pointer is equally likely to stop at any angle in $(0, 2\pi)$. In terms of the pdf $f_\Theta(\theta)$, this can be written, using (4.37), as

$$f_\Theta(\theta_1) = f_\Theta(\theta_2), \qquad 0 \leq \theta_1, \theta_2 < 2\pi \qquad (4.39)$$

Thus, in the interval $(0, 2\pi)$, $f_\Theta(\theta)$ is a constant, and outside $(0, 2\pi)$, $f_\Theta(\theta)$ is zero by the modulo 2π condition (this means that angles less than 0 or greater than 2π are impossible). By (4.35), it follows that $f_\Theta(\theta) = 1/(2\pi)$ in $(0, 2\pi)$; $f_\Theta(\theta)$ is shown graphically in Figure 4.7(a). The cdf $F_\Theta(\theta)$ is easily obtained by performing a graphical integration of $f_\Theta(\theta)$ and is shown in Figure 4.7(b).

To illustrate the use of these graphs, suppose we wish to find the probability of the pointer landing anyplace in the interval $(\frac{1}{2}\pi, \pi)$. The desired probability is given either as the area under the pdf curve from $\frac{1}{2}\pi$ to π, shaded in Figure 4.7(a), or as the value of the ordinate at $\theta = \pi$ minus the value of the ordinate at $\theta = \frac{1}{2}\pi$ on the cdf curve. The probability of the
▲ pointer's landing *exactly* at $\frac{1}{2}\pi$, however, is 0.

Joint cdf's and pdf's

Some chance experiments must be characterized by two or more random variables. The cdf or pdf description is readily extended to such cases. For simplicity, we will consider only the case of two random variables.

To give a specific example, consider the chance experiment in which darts are repeatedly thrown at a target, as shown schematically in Figure 4.8. The point at which the dart lands on the target must be described in terms of two numbers. In this example, we denote the impact point by the two random variables X and Y, whose values are the xy coordinates of the point where the dart sticks, with the origin being fixed at the bull's eye.

The *joint cdf* of X and Y is defined as

$$F_{XY}(x, y) = P(X \le x, \quad Y \le y) \tag{4.40}$$

where the comma is interpreted as "and." The *joint pdf* of X and Y is defined as

$$f_{XY}(x, y) = \frac{\partial^2 F_{XY}(x, y)}{\partial x \, \partial y} \tag{4.41}$$

Just as we did in the case of single random variables, we can show that

$$P(x_1 < X \le x_2, y_1 < Y \le y_2) = \int_{y_1}^{y_2} \int_{x_1}^{x_2} f_{XY}(x, y) \, dx \, dy \tag{4.42}$$

which is the two-dimensional equivalent of (4.36). Letting $x_1 = y_1 = -\infty$ and $x_2 = y_2 = \infty$, we include the entire sample space. Thus

$$F_{XY}(\infty, \infty) = \int_{-\infty}^{\infty} \int_{-\infty}^{\infty} f_{XY}(x, y) \, dx \, dy = 1 \tag{4.43}$$

Letting $x_1 = x - dx$, $x_2 = x$, $y_1 = y - dy$, and $y_2 = y$, we obtain the following enlightening special case of (4.42):

$$f_{XY}(x, y) \, dx \, dy = P(x - dx < X \le x, \quad y - dy < Y \le y) \tag{4.44}$$

Thus the probability of finding X in an infinitesimal interval around x while simultaneously finding Y in an infinitesimal interval around y is $f_{XY}(x, y) \, dxdy$.

Given a joint cdf or pdf, we can obtain the cdf or pdf of one of the random

FIGURE 4.8   The dart-throwing experiment

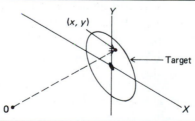

variables using the following considerations. The cdf for X irrespective of the value Y takes on is simply

$$F_X(x) = P(X \leq x, -\infty < Y < \infty)$$
$$= F_{XY}(x, \infty) \tag{4.45}$$

By similar reasoning, the cdf for Y alone is

$$F_Y(y) = F_{XY}(\infty, y) \tag{4.46}$$

$F_X(x)$ and $F_Y(y)$ are referred to as *marginal cdf's*. Using (4.40) and (4.42), we can express (4.45) and (4.46) as

$$F_X(x) = \int_{-\infty}^{\infty} \int_{-\infty}^{x} f_{XY}(x', y') \, dx' \, dy' \tag{4.47}$$

and

$$F_Y(y) = \int_{-\infty}^{y} \int_{-\infty}^{\infty} f_{XY}(x', y') \, dx' \, dy' \tag{4.48}$$

respectively. Since

$$f_X(x) = \frac{dF_X(x)}{dx} \qquad \text{and} \qquad f_Y(y) = \frac{dF_Y(y)}{dy} \tag{4.49}$$

we obtain

$$f_X(x) = \int_{-\infty}^{\infty} f_{XY}(x, y') \, dy' \tag{4.50}$$

and

$$f_Y(y) = \int_{-\infty}^{\infty} f_{XY}(x', y) \, dx' \tag{4.51}$$

from (4.47) and (4.48), respectively. Thus, to obtain the marginal pdf's $f_X(x)$ and $f_Y(y)$ from the joint pdf $f_{XY}(x, y)$, we simply integrate out the undesired variable (or variables for more than two random variables). Hence the joint cdf or pdf contains all the information possible about the joint random variables X and Y. Similar results hold for more than two random variables.

Two random variables are *statistically independent* (or simply independent) if the values each takes on do not influence the values of the other. Thus, for any x and y, it must be true that

$$P(X \leq x, Y \leq y) = P(X \leq x)P(Y \leq y) \tag{4.52a}$$

or, in terms of cdf's,

$$F_{XY}(x, y) = F_X(x)F_Y(y) \qquad (4.52b)$$

That is, the joint cdf of independent random variables factors into the product of the separate marginal cdf's. Differentiating both sides of (4.52b) with respect to first x and then y, and using the definition of the pdf, we obtain

$$f_{XY}(x, y) = f_X(x)f_Y(y) \qquad (4.53)$$

which shows that the joint pdf of independent random variables also factors.

If two random variables are not independent, we can write their joint pdf in terms of conditional pdf's $f_{X|Y}(x|y)$ and $f_{Y|X}(y|x)$ as

$$f_{XY}(x, y) = f_X(x)f_{Y|X}(y|x)$$
$$= f_Y(y)f_{X|Y}(x|y) \qquad (4.54)$$

These relations *define* the conditional pdf's of two random variables. An intuitively satisfying interpretation of $f_{X|Y}(x|y)$ is

$$f_{X|Y}(x|y)\, dx = P[x - dx < X \le x \text{ given } Y = y] \qquad (4.55)$$

There is a similar interpretation for $f_{Y|X}(y|x)$. Equation (4.55) is reasonable in that if X and Y are dependent, a given value of Y should influence the probability distribution for X. On the other hand, if X and Y are independent, information about one of the random variables tells us nothing about the other. Thus, for independent random variables,

$$f_{X|Y}(x|y) = f_X(x) \qquad \text{and} \qquad f_{Y|X}(y|x) = f_Y(y) \qquad (4.56)$$

which could serve as an alternative definition of statistical independence. The following example illustrates the preceding ideas.

▼| **EXAMPLE 4.11**   Two random variables X and Y have the joint pdf

$$f_{XY}(x, y) = \begin{cases} Ae^{-(2x+y)}, & x, y \ge 0 \\ 0, & \text{otherwise} \end{cases} \qquad (4.57)$$

where A is a constant. We evaluate A from

$$\int_{-\infty}^{\infty} \int_{-\infty}^{\infty} f_{XY}(x, y)\, dx\, dy = 1 \qquad (4.58)$$

Since

$$\int_{0}^{\infty} \int_{0}^{\infty} e^{-(2x+y)}\, dx\, dy = \tfrac{1}{2} \qquad (4.59)$$

$A = 2$. We find the marginal pdf's from (4.50) and (4.51) as follows:

$$f_X(x) = \int_{-\infty}^{\infty} f_{XY}(x, y) \, dy = \begin{cases} \int_0^{\infty} 2e^{-(2x+y)} \, dy, & x \geq 0 \\ 0, & x < 0 \end{cases}$$

$$= \begin{cases} 2e^{-2x}, & x \geq 0 \\ 0, & x < 0 \end{cases}$$

(4.60)

$$f_Y(y) = \begin{cases} e^{-y}, & y \geq 0 \\ 0, & y < 0 \end{cases}$$

(4.61)

These joint and marginal pdf's are shown in Figure 4.9. From these results, we note that X and Y are statistically independent since $f_{XY}(x, y) = f_X(x)f_Y(y)$.

FIGURE 4.9   Joint and marginal pdf's for two random variables. (a) Joint pdf. (b) Marginal pdf for X. (c) Marginal pdf for Y.

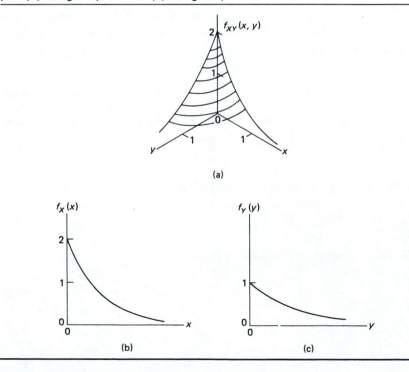

We find the joint cdf by integrating the joint pdf on both variables, using (4.42) and (4.40), which gives

$$F_{XY}(x, y) = \int_{-\infty}^{y} \int_{-\infty}^{x} f_{XY}(x', y') \, dx' \, dy'$$

$$= \begin{cases} (1 - e^{-2x})(1 - e^{-y}), & x, y \geq 0 \\ 0, & \text{otherwise} \end{cases} \tag{4.62}$$

Dummy variables are used in the integration to avoid confusion. Note that $F_{XY}(-\infty, -\infty) = 0$ and $F_{XY}(\infty, \infty) = 1$, as they should, since in the first case we have the probability of an impossible event and in the latter we include all possible outcomes. We also can use the result for $F_{XY}(x, y)$ to obtain

$$F_X(x) = F_{XY}(x, \infty) = \begin{cases} (1 - e^{-2x}), & x \geq 0 \\ 0, & \text{otherwise} \end{cases} \tag{4.63a}$$

and

$$F_Y(y) = F_{XY}(\infty, y) = \begin{cases} (1 - e^{-y}), & y \geq 0 \\ 0, & \text{otherwise} \end{cases} \tag{4.63b}$$

Also note that the joint cdf factors into the product of the marginal cdf's, as it should be for statistically independent random variables.

The conditional pdf's are

$$f_{X|Y}(x|y) = \frac{f_{XY}(x, y)}{f_Y(y)} = \begin{cases} 2e^{-2x}, & x \geq 0 \\ 0, & x < 0 \end{cases} \tag{4.64a}$$

and

$$f_{Y|X}(y|x) = \frac{f_{XY}(x, y)}{f_X(x)} = \begin{cases} e^{-y}, & y \geq 0 \\ 0, & y < 0 \end{cases} \tag{4.64b}$$

They are equal to the respective marginal pdf's, as they should be for independent random variables.

▼ **EXAMPLE 4.12**   To illustrate how to check a probability density function for statistical independence of the corresponding random variables, we consider the joint pdf

$$f_{XY}(x, y) = \begin{cases} \beta xy, & 0 \leq x \leq y, 0 \leq y \leq 4 \\ 0, & \text{otherwise} \end{cases} \tag{4.65}$$

For independence, the joint pdf should be the product of the marginal pdf's.

FIGURE 4.10 Probability density function for Example 4.12

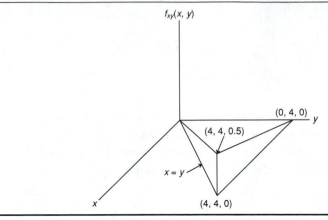

SOLUTION   This example is somewhat tricky because of the limits, so a diagram of the pdf is given in Figure 4.10. Although not absolutely necessary to answer the question of independence, we proceed by first finding the constant β which is chosen to normalize the pdf to unity. Integration over all x and y gives $\beta = \frac{1}{32}$. We next proceed to find the marginal pdf's. Integrating over x first and checking the figure to obtain the proper limits of integration, we obtain

$$
\begin{aligned}
f_Y(y) &= \int_0^y \frac{xy}{32}\,dx, \qquad && 0 \le y \le 4 \\
&= \begin{cases} y^3/64, & 0 \le y \le 4 \\ 0, & \text{otherwise} \end{cases} && \text{(4.66)}
\end{aligned}
$$

The pdf on X is similarly obtained as

$$
\begin{aligned}
f_X(x) &= \int_x^4 \frac{xy}{32}\,dy, \qquad && 0 \le x \le 4 \\
&= \begin{cases} (x/4)[1 - (x/4)^2], & 0 \le x \le 4 \\ 0, & \text{otherwise} \end{cases} && \text{(4.67)}
\end{aligned}
$$

It is clear that the product of the marginal pdf's is not equal to the joint pdf, so the random variables X and Y are not statistically independent.

Transformation of Random Variables

Situations are often encountered in which we know the pdf (or cdf) of a random variable X and desire the pdf of a second random variable Y defined as a function of X, for example,

$$Y = g(X) \tag{4.68}$$

We initially consider the case where $g(X)$ is a monotonic function of its argument (for example, it is either nondecreasing or nonincreasing as the independent variable ranges from $-\infty$ to ∞), a restriction that may be relaxed if necessary.

A typical function is shown in Figure 4.11. The probability that X lies in the range $(x - dx, x)$ is the same as the probability that Y lies in the range $(y - dy, y)$, where $y = g(x)$. Using (4.37), we obtain

$$f_X(x) \, dx = f_Y(y) \, dy \tag{4.69a}$$

if $g(X)$ is monotonically increasing, and

$$f_X(x) \, dx = -f_Y(y) \, dy \tag{4.69b}$$

if $g(X)$ is monotonically decreasing, since an *increase* in x results in a *decrease* in y. Both cases are taken into account by writing

$$f_Y(y) = f_X(x) \left| \frac{dx}{dy} \right|_{x = g^{-1}(y)} \tag{4.70}$$

where $x = g^{-1}(y)$ denotes the inversion of (4.68) for x in terms of y.

FIGURE 4.11   A typical monotonic transformation of a random variable

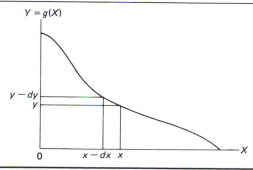

▼ EXAMPLE 4.13   To illustrate the use of (4.70), let us consider the pdf of
Example 4.10, namely

$$f_\Theta(\theta) = \begin{cases} \dfrac{1}{2\pi}, & 0 \le \theta \le 2\pi \\ 0, & \text{otherwise} \end{cases} \tag{4.71}$$

and the transformation

$$Y = -\left(\frac{1}{\pi}\right)\Theta + 1 \tag{4.72}$$

Since $dy/d\theta = -1/\pi$, the pdf of Y is

$$f_Y(y) = f_\Theta(\theta = -\pi y + \pi)|-\pi| = \begin{cases} \frac{1}{2}, & -1 \le y \le 1 \\ 0, & \text{otherwise} \end{cases} \tag{4.73}$$

▲

Consider next the case of $g(x)$ nonmonotonic. A typical case is shown
in Figure 4.12. For the case shown, the infinitesimal interval $(y - dy, y)$
corresponds to three infinitesimal intervals on the x axis: $(x_1 - dx_1, x_1)$,
$(x_2 - dx_2, x_2)$, and $(x_3 - dx_3, x_3)$. The probability that X lies in any one
of these intervals is equal to the probability that Y lies in the interval
$(y - dy, y)$. This can be generalized to the case of N disjoint intervals where
it follows that

$$P(y - dy, y) = \sum_{i=1}^{N} P(x_i - dx_i, x_i) \tag{4.74}$$

where we have generalized to N intervals on the X axis corresponding to the
interval $(y - dy, y)$ on the Y axis. Since

$$P(y - dy, y) = f_Y(y)\,dy \tag{4.75}$$

FIGURE 4.12 A nonmonotonic transformation of a random variable

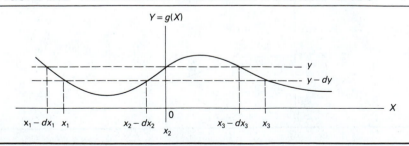

and

$$P(x_i - dx_i, x_i) = f_X(x_i)\, dx_i \qquad (4.76)$$

we have

$$f_Y(y) = \sum_{n=1}^{N} f_X(x_i) \left| \frac{dx_i}{dy} \right|_{x_i\,=\,g_i^{-1}(y)} \qquad (4.77)$$

where the absolute value signs are used because a probability must be positive and $x_i = g_i^{-1}(y)$ is the ith solution to $g(y) = x$.

▼ **EXAMPLE 4.14**   Consider the transformation

$$y = x^2 \qquad (4.78)$$

If $f_X(x) = 0.5 \exp(-|x|)$, find $f_Y(y)$.

SOLUTION   There are two solutions to $x^2 = y$; these are

$$x_1 = \sqrt{y} \quad \text{and} \quad x_2 = -\sqrt{y}, \qquad y > 0 \qquad (4.79)$$

Their derivatives are

$$\frac{dx_1}{dy} = 1/(2\sqrt{y}) \quad \text{and} \quad \frac{dx_2}{dy} = -1/(2\sqrt{y}), \qquad y > 0 \quad (4.80)$$

Using these results in (4.77), we obtain $f_Y(y)$ to be

$$f_Y(y) = \frac{e^{-\sqrt{y}}}{2\sqrt{y}}, \qquad y > 0 \qquad (4.81)$$

▲ Since Y cannot be negative, $f_Y(y) = 0,\ y < 0$.

For two or more random variables, we consider only one-to-one transformations and the probability of the joint occurrence of random variables lying within infinitesimal areas (or volumes for more than two random variables). Thus, suppose two new random variables U and V are defined in terms of two old random variables X and Y by the relations

$$U = g_1(X, Y) \quad \text{and} \quad V = g_2(X, Y) \qquad (4.82)$$

The new pdf $f_{UV}(u, v)$ is obtained from the old pdf $f_{XY}(x, y)$ by using (4.44) to write

$$P(u - du < U \le u, \quad v - dv < V \le v)$$
$$= P(x - dx < X \le x, \quad y - dy < Y \le y)$$

or

$$f_{UV}(u, v) \, dA_{uv} = f_{XY}(x, y) \, dA_{xy} \tag{4.83}$$

where dA_{uv} is the infinitesimal area in the uv plane corresponding to the infinitesimal area dA_{xy} in the xy plane through the transformation (4.82).

The ratio of elementary area dA_{xy} to dA_{uv} is given by the Jacobian

$$\frac{\partial(x, y)}{\partial(u, v)} = \begin{vmatrix} \dfrac{\partial x}{\partial u} & \dfrac{\partial x}{\partial v} \\ \dfrac{\partial y}{\partial u} & \dfrac{\partial y}{\partial v} \end{vmatrix} \tag{4.84}$$

so that

$$f_{UV}(u, v) = f_{XY}(x, y) \left| \frac{\partial(x, y)}{\partial(u, v)} \right|_{\substack{x=g_1^{-1}(u, v) \\ y=g_2^{-1}(u, v)}} \tag{4.85}$$

where the inverse functions $g_1^{-1}(u, v)$ and $g_2^{-1}(u, v)$ exist because the transformation (4.82) is assumed to be one-to-one. An example will help clarify this discussion.

▼ EXAMPLE 4.15   Consider the dart-throwing game discussed in connection with joint cdf's and pdf's. We assume that the joint pdf in terms of rectangular coordinates for the impact point is

$$f_{XY}(x, y) = \frac{\exp[-(x^2 + y^2)/2\sigma^2]}{2\pi\sigma^2}, \qquad -\infty < x, y < \infty \tag{4.86}$$

where σ^2 is a constant. This is a special case of the *joint Gaussian pdf*, which we will discuss in more detail shortly.

Instead of rectangular coordinates, we wish to use polar coordinates R and Θ, defined by

$$R = \sqrt{X^2 + Y^2} \tag{4.87a}$$

and

$$\Theta = \tan^{-1}\left(\frac{Y}{X}\right) \tag{4.87b}$$

so that

$$X = R \cos \Theta = g_1^{-1}(R, \Theta) \tag{4.88a}$$

and

$$Y = R \sin \Theta = g_2^{-1}(R, \Theta) \tag{4.88b}$$

where

$$0 \leq \Theta < 2\pi, \qquad 0 \leq R < \infty$$

Under this transformation, the infinitesimal area $dx\,dy$ in the xy plane transforms to the area $r\,dr\,d\theta$ in the $r\theta$ plane, as determined by the Jacobian, which is

$$\frac{\partial(x, y)}{\partial(r, \theta)} = \begin{vmatrix} \cos\theta & -r\sin\theta \\ \sin\theta & r\cos\theta \end{vmatrix} = r \tag{4.89}$$

Thus the joint pdf of R and θ is

$$f_{R\Theta}(r, \theta) = \frac{re^{-r^2/2\sigma^2}}{2\pi\sigma^2}, \qquad \begin{matrix} 0 \leq \theta < 2\pi \\ 0 \leq r < \infty \end{matrix} \tag{4.90}$$

which follows from (4.85), which for this case takes the form

$$f_{R\Theta}(r, \theta) = rf_{XY}(x, y) \Big|_{\substack{x=r\,\cos\,\theta \\ y=r\,\sin\,\theta}} \tag{4.91}$$

If we integrate $f_{R\Theta}(r, \theta)$ over θ to get the pdf for R alone, we obtain

$$f_R(r) = \frac{re^{-r^2/2\sigma^2}}{\sigma^2}, \qquad 0 \leq r < \infty \tag{4.92}$$

which is referred to as the *Rayleigh pdf*. The probability that the dart lands in a ring of radius r from the bull's eye and of thickness dr is given by $f_R(r)\,dr$. From the sketch of the Rayleigh pdf given in Figure 4.13, we see that the most probable distance for the dart to land from the bull's eye is $R = \sigma$. By integrating (4.90) over r, it can be shown that the pdf of θ is uniform in (0, 2π).

FIGURE 4.13   The Rayleigh pdf

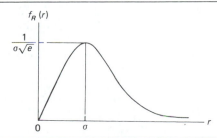

4.3 STATISTICAL AVERAGES

The probability functions (cdf and pdf) we have just discussed provide us with all the information possible about a random variable or a set of random variables. Often we do not need such a complete description, and in many cases, we are not able to obtain the cdf or pdf. A partial description of a random variable is given in terms of various statistical averages, or mean values.

Average of a Discrete Random Variable

To introduce the idea of a statistical average, we consider a discrete random variable X which takes on the possible values $x_1, x_2, \ldots, x_M$ with the respective probabilities $P_1, P_2, \ldots, P_M$. The statistical average, or *expectation*, of X is defined as

$$\overline{X} = E\{X\} = \sum_{j=1}^{M} x_j P_j \tag{4.93}$$

To show the reasonableness of this definition, we look at it in terms of relative frequency. If the underlying chance experiment is repeated a large number of times N and $X = x_1$ is observed n_1 times and $X = x_2$ is observed n_2 times, etc., the arithmetical average of the observed values is

$$\frac{n_1 x_1 + n_2 x_2 + \cdots + n_M x_M}{N} = \sum_{j=1}^{M} x_j \frac{n_j}{N} \tag{4.94}$$

But, by the relative-frequency interpretation of probability, n_j/N approaches P_j, the probability of the event $X = x_j$, as N becomes large. Thus, in the limit as N approaches infinity, (4.94) becomes (4.93).

▼ EXAMPLE 4.16   Suppose that a class of 100 students is given a test and The following scores result:

Score:	90	85	80	75	70	65	60
No. of students:	1	4	20	50	20	4	1

The average score, as we have known how to compute since our grammar school days, is

$$\frac{(90 \times 1) + (85 \times 4) + \cdots + (60 \times 1)}{100} = 75 \tag{4.95}$$

Now consider the same problem using (4.93). Defining the random variable X as taking on values numerically equal to the test scores, we have, using the relative-frequency approximation, the following probabilities:

$$P(X = 90) \cong 0.01$$
$$P(X = 85) \cong 0.04$$
$$P(X = 80) \cong 0.2$$
$$P(X = 75) \cong 0.5 \qquad (4.96)$$
$$P(X = 70) \cong 0.2$$
$$P(X = 65) \cong 0.04$$
$$P(X = 60) \cong 0.01$$

Using (4.93), we obtain, for the expectation of X, the same result as before:

$$E\{X\} = (0.01 \times 90) + (0.04 \times 85) + \cdots + (0.01 \times 60) = 75$$
$$(4.97)$$

Average of a Continuous Random Variable

Now we look at the case where X is a continuous random variable with the pdf $f_X(x)$. In order to extend the definition given by (4.93) to this case, we consider the range of values that X may take on, say x_0 to x_M, to be broken up into a large number of small subintervals of length Δx, as shown in Figure 4.14.

The probability that X lies between $x_i - \Delta x$ and x_i is, from (4.37), given by

$$P(x_i - \Delta x < X \leq x_i) \cong f_X(x_i) \, \Delta x, \qquad i = 1, 2, \ldots, M \qquad (4.98)$$

for Δx small. Thus we have approximated X by a discrete random variable that takes on the values $x_0, x_1, \ldots, x_M$ with probabilities $f_X(x_0)\Delta x, \ldots,$

FIGURE 4.14   A discrete approximation for a continuous random variable X

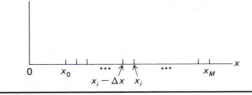

$f_X(x_M) \, \Delta x$, respectively. Using (4.93), the expectation of this random variable is

$$E\{X\} \cong \sum_{i=0}^{M} x_i f_X(x_i) \, \Delta x \qquad (4.99a)$$

As Δx approaches zero, this becomes a better and better approximation for $E\{X\}$. In the limit, as Δx approaches dx, the sum becomes an integral, giving

$$E\{X\} = \int_{-\infty}^{\infty} x f_X(x) \, dx \qquad (4.99b)$$

for the expectation of X.

Average of a Function of a Random Variable

We are interested not only in $E\{X\}$, which is referred to as the *mean* or *first moment* of X, but also in statistical averages of functions of X. Letting $Y = g(X)$, the statistical average or expectation of the new random variable Y could be obtained as

$$E\{Y\} = \int_{-\infty}^{\infty} y f_Y(y) \, dy \qquad (4.100)$$

where $f_Y(y)$ is the pdf of Y, which can be found from $f_X(x)$ by application of (4.70). However, it is often more convenient simply to find the expectation of the function $g(X)$ as given by

$$\overline{g(X)} \triangleq E\{g(X)\} = \int_{-\infty}^{\infty} g(x) f_X(x) \, dx \qquad (4.101)$$

which is identical to $E\{Y\}$ as given by (4.100). Two examples follow to illustrate the use of (4.100) and (4.101).

▼ EXAMPLE 4.17   Suppose the random variable Θ has the pdf

$$f_\Theta(\theta) = \begin{cases} \dfrac{1}{2\pi}, & |\theta| < \pi \\ 0, & \text{otherwise} \end{cases} \qquad (4.102)$$

Then $E\{\Theta^n\}$ is referred to as the nth moment of Θ and is given by

$$E\{\Theta^n\} = \int_{-\infty}^{\infty} \theta^n f_\Theta(\theta) \, d\theta = \int_{-\pi}^{\pi} \theta^n \frac{d\theta}{2\pi} \qquad (4.103)$$

Since the integrand is odd if n is odd, $E\{\Theta^n\} = 0$ for n odd. For n even,

$$E\{\Theta^n\} = \frac{1}{\pi} \int_0^\pi \theta^n \, d\theta = \frac{\pi^n}{n+1} \qquad (4.104)$$

The first moment or mean of Θ, $E\{\Theta\}$, is a measure of the location of $f_\Theta(\theta)$ (that is, the "center of mass"). Since $f_\Theta(\theta)$ is symmetrically located about $\theta = 0$, it is not surprising that $\Theta = 0$.

EXAMPLE 4.18   Later we shall consider certain random waveforms that can be modeled as sinusoids with random phase angles having uniform pdf in $[-\pi, \pi]$. In this example, we consider a random variable X that is defined in terms of the uniform random variable Θ considered in Example 4.17 by

$$X = \cos \Theta \qquad (4.105)$$

The density function of X, $f_X(x)$, is found as follows. First $-1 \le \cos \theta \le 1$, so $f_X(x) = 0$ for $|x| > 1$. Second, the transformation is not one-to-one, there being two values of Θ for each value of X, since $\cos \theta = \cos (-\theta)$. However, we can still apply (4.70) by noting that positive and negative angles have equal probabilities and writing

$$f_X(x) = 2 f_\Theta(\theta) \left| \frac{d\theta}{dx} \right|, \qquad |x| < 1 \qquad (4.106)$$

Now $\theta = \cos^{-1} x$ and $|d\theta/dx| = (1 - x^2)^{-1/2}$, which yields

$$f_X(x) = \begin{cases} \dfrac{1}{\pi \sqrt{1 - x^2}}, & |x| \le 1 \\ 0, & |x| > 1 \end{cases} \qquad (4.107)$$

This pdf is illustrated in Figure 4.15.

The mean and second moment of X can be calculated using either (4.100) or (4.101). Using (4.100), we obtain

$$\overline{X} = \int_{-1}^1 \frac{x}{\pi \sqrt{1 - x^2}} \, dx = 0 \qquad (4.108)$$

because the integrand is odd, and

$$\overline{X^2} = \int_{-1}^1 \frac{x^2 \, dx}{\pi \sqrt{1 - x^2}} = \frac{1}{2} \qquad (4.109)$$

FIGURE 4.15   Probability density function of a sinusoid with uniform random phase

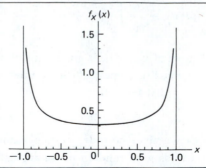

by a table of integrals. Using (4.101), we find that

$$\bar{X} = \int_{-\pi}^{\pi} \cos\theta \, \frac{d\theta}{2\pi} = 0 \tag{4.110}$$

and

$$\overline{X^2} = \int_{-\pi}^{\pi} \cos^2\theta \, \frac{d\theta}{2\pi} = \int_{-\pi}^{\pi} \frac{1}{2}(1 + \cos 2\theta) \, \frac{d\theta}{2\pi} = \frac{1}{2} \tag{4.111}$$

as obtained before.

Average of a Function of More Than One Random Variable

The expectation of a function $g(X, Y)$ of two random variables X and Y is defined in a manner analogous to the case of a single random variable. If $f_{XY}(x, y)$ is the joint pdf of X and Y, the expectation of $g(X, Y)$ is

$$E\{g(X, Y)\} = \int_{-\infty}^{\infty} \int_{-\infty}^{\infty} g(x, y) f_{XY}(x, y) \, dx \, dy \tag{4.112}$$

The generalization to more than two random variables should be obvious.

Equation (4.112) and its generalization to more than two random variables include the single-random-variable case, for suppose $g(X, Y)$ is replaced by a function of X alone, say $h(X)$. Then using (4.50), we obtain the following from (4.112):

$$E\{h(X)\} = \int_{-\infty}^{\infty} \int_{-\infty}^{\infty} h(x) f_{XY}(x, y) \, dx \, dy$$

$$= \int_{-\infty}^{\infty} h(x) f_X(x) \, dx \tag{4.113}$$

▼ **EXAMPLE 4.19**   Consider the joint pdf of Example 4.11 and the expectation of $g(X, Y) = XY$. From (4.112), this expectation is

$$E\{XY\} = \int_{-\infty}^{\infty} \int_{-\infty}^{\infty} xy f_{XY}(x, y) \, dx \, dy = \int_{0}^{\infty} \int_{0}^{\infty} 2xy e^{-(2x+y)} \, dx \, dy$$

$$= 2 \int_{0}^{\infty} xe^{-2x} \, dx \int_{0}^{\infty} ye^{-y} \, dy = \tfrac{1}{2} \qquad (4.114)$$

We recall from Example 4.11 that X and Y are statistically independent. From the last line of the preceding equation for $E\{XY\}$, we see that

$$E\{XY\} = E\{X\}E\{Y\} \qquad (4.115)$$

a result that holds in general for *statistically independent* random variables. In fact, for independent random variables, it readily follows that

$$E\{h(X)g(Y)\} = E\{h(X)\}E\{g(Y)\} \qquad (4.116)$$

where $h(X)$ and $g(Y)$ are two functions of X and Y, respectively. In the special case where $h(X) = X^m$ and $g(Y) = Y^n$, the average $E\{h(X)g(Y)\} = E\{X^m Y^n\}$. The expectations $E\{X^m Y^n\}$ are referred to as the *joint moments* of order $m + n$ of X and Y. According to (4.116), the *joint moments of statistically independent random variables factor.*
▲

▼ **EXAMPLE 4.20**   When finding the expectation of a function of more than one random variable, you may find it easier to use the concept of conditional expectation. Consider, for example, a function $g(X, Y)$ of two random variables X and Y, with the joint pdf $f_{XY}(x, y)$. The expectation of $g(X, Y)$ is

$$E\{g(X, Y)\} = \int_{-\infty}^{\infty} \int_{-\infty}^{\infty} g(x, y) f_{XY}(x, y) \, dx \, dy$$

$$= \int_{-\infty}^{\infty} \left[\int_{-\infty}^{\infty} g(x, y) f_{X|Y}(x \mid y) \, dx \right] f_Y(y) \, dy = E\{E[g(X, Y) \mid Y]\}$$

$$(4.117)$$

where $f_{X|Y}(x \mid y)$ is the conditional pdf of X, given Y, and $E[g(X, Y) \mid Y]$ denotes the conditional expectation of $g(X, Y)$, given Y.

As a specific application, let us consider the firing of projectiles at a target. Projectiles are fired until the target is hit for the first time, after which firing ceases. Assume that the probability of a projectile's hitting the target is p and that the firings are independent of one another. Find the average number of projectiles fired at the target.

SOLUTION   To solve this problem, let N be a random variable denoting the number of projectiles fired at the target. Let the random variable H be 1 if the first projectile hits the target and 0 if it does not. Using the concept of conditional expectation, we find the average value of N is given by

$$E\{N\} = E\{E[N|H]\} = pE[N|H = 1] + (1 - p)E[N|H = 0]$$
$$= p \times 1 + (1 - p)(1 + E[N]) \tag{4.118}$$

where $E[N \mid = 0] = 1 + E[N]$ because $N \geq 1$ if a miss occurs on the first firing. By solving the last expression for $E[N]$, we obtain

$$E[N] = \frac{1}{p} \tag{4.119}$$

If $E[N]$ is evaluated directly, it is necessary to sum the series:

$$E[N] = 1 \times p + 2 \times (1 - p)p + 3 \times (1 - p)^2 p + \cdots \tag{4.120}$$

which is not too difficult in this instance. However, the conditional-expectation method clearly makes it easier to keep track of the bookkeeping.

Variance of a Random Variable

The statistical average

$$\sigma_x^2 \triangleq E[(X - \bar{X})^2] = E\{[X - E(X)]^2\} \tag{4.121}$$

is called the *variance* of the random variable X; σ_x is called the *standard deviation* of X and is a measure of the concentration of the pdf of X, or $f_X(x)$, about the mean. The notation var $\{X\}$ for σ_x^2 is sometimes used. A useful relation for obtaining σ_x^2 is

$$\sigma_x^2 = E\{X^2\} - E^2\{X\} \tag{4.122}$$

which, in words, says that the variance of X is simply its second moment minus its mean, squared. To prove (4.122), we let $E\{X\} = m_x$. Then

$$\sigma_x^2 = \int_{-\infty}^{\infty} (x - m_x)^2 f_X(x) \, dx = \int_{-\infty}^{\infty} (x^2 - 2xm_x + m_x^2) f_X(x) \, dx$$
$$= E\{X^2\} - 2m_x^2 + m_x^2 = E\{X^2\} - E^2\{X\} \tag{4.123}$$

which follows because $\int_{-\infty}^{\infty} x f_X(x) \, dx = m_x$.

▼ | EXAMPLE 4.21   Let X have the uniform pdf

$$f_X(x) = \begin{cases} \dfrac{1}{b-a}, & a \le x \le b \\ 0, & \text{otherwise} \end{cases} \qquad (4.124)$$

Then

$$E\{X\} = \int_a^b x \frac{dx}{b-a} = \tfrac{1}{2}(a+b) \qquad (4.125)$$

and

$$E\{X^2\} = \int_a^b x^2 \frac{dx}{b-a} = \tfrac{1}{3}(b^2 + ab + a^2) \qquad (4.126)$$

Thus

$$\sigma_x^2 = \tfrac{1}{3}(b^2 + ab + a^2) - \tfrac{1}{4}(a^2 + 2ab + b^2) = \tfrac{1}{12}(a-b)^2 \qquad (4.127)$$

Consider the following special cases:

1. $a = 1$ and $b = 2$, for which $\sigma_x^2 = \frac{1}{12}$
2. $a = 0$ and $b = 1$, for which $\sigma_x^2 = \frac{1}{12}$
3. $a = 0$ and $b = 2$, for which $\sigma_x^2 = \frac{1}{3}$

For cases 1 and 2, the pdf of X has the same width but is centered about different means; the variance is the same for both cases. In case 3, the pdf is wider than it is for cases 1 and 2, which is manifested by the larger variance.

▲ |

Average of a Linear Combination of N Random Variables

It is easily shown that the expected value, or average, of an arbitrary linear combination of random variables is the same as the linear combination of their respective means. That is,

$$E\left\{\sum_{i=1}^N a_i X_i\right\} = \sum_{i=1}^N a_i E\{X_i\} \qquad (4.128)$$

where $X_1, X_2, \ldots, X_N$ are random variables and $a_1, a_2, \ldots, a_N$ are arbitrary constants. Equation (4.128) will be demonstrated for the special case $N = 2$; generalization to the case $N > 2$ is not difficult, but results in unwieldy notation.

Let $f_{X_1X_2}(x_1, x_2)$ be the joint pdf of X_1 and X_2. Then, using the definition of the expectation of a function of two random variables in (4.112), it follows that

$$E\{a_1X_1 + a_2X_2\} \triangleq \int_{-\infty}^{\infty} \int_{-\infty}^{\infty} (a_1x_1 + a_2x_2)f_{X_1X_2}(x_1, x_2) \, dx_1 \, dx_2$$

$$= a_1 \int_{-\infty}^{\infty} \int_{-\infty}^{\infty} x_1 f_{X_1X_2}(x_1, x_2) \, dx_1 \, dx_2$$

$$+ a_2 \int_{-\infty}^{\infty} \int_{-\infty}^{\infty} x_2 f_{X_1X_2}(x_1, x_2) \, dx_1 \, dx_2 \qquad (4.129)$$

Considering the first double integral and using (4.50) and (4.99), we find that

$$\int_{-\infty}^{\infty} \int_{-\infty}^{\infty} x_1 f_{X_1X_2}(x_1, x_2) \, dx_1 \, dx_2 = \int_{-\infty}^{\infty} x_1 \left\{ \int_{-\infty}^{\infty} f_{X_1X_2}(x_1, x_2) \, dx_2 \right\} dx_1$$

$$= \int_{-\infty}^{\infty} x_1 f_{X_1}(x_1) \, dx_1$$

$$\triangleq E\{X_1\} \qquad (4.130)$$

Similarly, it can be shown that the second double integral reduces to $E\{X_2\}$. Thus (4.128) has been proved for the case $N = 2$.

We emphasize that (4.128) holds regardless of whether or not the X_i terms are independent. Also, it should be noted that a similar result holds for a linear combination of functions of N random variables.

Variance of a Linear Combination of Independent Random Variables

If $X_1, X_2, \ldots, X_N$ are *statistically independent* random variables, then

$$\text{var} \left\{ \sum_{i=1}^{N} a_i X_i \right\} = \sum_{i=1}^{N} a_i^2 \, \text{var} \{X_i\} \qquad (4.131)$$

where $a_1, a_2, \ldots, a_N$ are arbitrary constants and $\text{var} \{X_i\} \triangleq E\{(X_i - \bar{X}_i)^2\}$. This relation will be demonstrated for the case $N = 2$. Let $Z = a_1X_1 + a_2X_2$ and let $f_{X_i}(x_i)$ be the marginal pdf of X_i. Then the joint pdf of X_1 and X_2 is $f_{X_1}(x_1)f_{X_2}(x_2)$ by the assumption of statistical independence. Also, $\bar{Z} =$

$a_1 \bar{X}_1 + a_2 \bar{X}_2$ by (4.128). Also, var $\{Z\} = E\{(Z - \bar{Z})^2\}$. But, since $Z = a_1 X_1 + a_2 X_2$, we may write var $\{Z\}$ as

$$
\begin{aligned}
\text{var } \{Z\} &= E\{[(a_1 X_1 + a_2 X_2) - (a_1 \bar{X}_1 + a_2 \bar{X}_2)]^2\} \\
&= E\{[a_1(X_1 - \bar{X}_1) + a_2(X_2 - \bar{X}_2)]^2\} \\
&= a_1{}^2 E\{(X_1 - \bar{X}_1)^2\} + 2a_1 a_2 E\{(X_1 - \bar{X}_1)(X_2 - \bar{X}_2)\} \\
&\quad + a_2{}^2 E\{(X_2 - \bar{X}_2)^2\}
\end{aligned}
\tag{4.132}
$$

The first and last terms in the preceding equation are $a_1{}^2$ var $\{X_1\}$ and $a_2{}^2$ var $\{X_2\}$, respectively. The middle term is zero, since

$$
\begin{aligned}
E\{(X_1 - \bar{X}_1)(X_2 - \bar{X}_2)\} \\
= \int_{-\infty}^{\infty} \int_{-\infty}^{\infty} (x_1 - \bar{X}_1)(x_2 - \bar{X}_2) f_{X_1}(x_1) f_{X_2}(x_2) \, dx_1 \, dx_2 \\
= \int_{-\infty}^{\infty} (x_1 - \bar{X}_1) f_{X_1}(x_1) \, dx_1 \int_{-\infty}^{\infty} (x_2 - \bar{X}_2) f_{X_2}(x_2) \, dx_2 \\
= (\bar{X}_1 - \bar{X}_1)(\bar{X}_2 - \bar{X}_2) = 0
\end{aligned}
\tag{4.133}
$$

We reiterate that the assumption of *statistical independence* was used to show that the middle term above is zero.

Another Special Average—The Characteristic Function

If we let $g(X) = e^{jvX}$ in (4.101), we obtain an average known as the *characteristic function* of X, or $M_X(jv)$, defined as

$$
M_X(jv) \triangleq E\{e^{jvX}\} = \int_{-\infty}^{\infty} f_X(x) e^{+jvx} \, dx
\tag{4.134}
$$

It is seen that $M_X(jv)$ would be the *Fourier transform* of $f_X(x)$, as we have defined the Fourier transform in Chapter 2, provided a minus sign were used in the exponent instead of a plus sign. Thus, if we remember to replace $j\omega$ by $-jv$ in Fourier transform tables, we can use them to obtain characteristic functions from pdf's.

A pdf is obtained from the corresponding characteristic function by the inverse transform relationship

$$
f_X(x) = \frac{1}{2\pi} \int_{-\infty}^{\infty} M_X(jv) e^{-jvx} \, dv
\tag{4.135}
$$

This illustrates one possible use of the characteristic function. It is sometimes easier to obtain the characteristic function than the pdf, and the latter is then obtained by inverse Fourier transformation.

Another use for the characteristic function is seen when we differentiate (4.134) with respect to v:

$$\frac{\partial M_X(jv)}{\partial v} = j \int_{-\infty}^{\infty} x f_X(x) e^{jvx} \, dx \tag{4.136}$$

Setting $v = 0$ after differentiation and dividing by j, we obtain

$$E\{X\} = (-j) \left. \frac{\partial M_X(jv)}{\partial v} \right|_{v=0} \tag{4.137}$$

For the nth moment, the relation

$$E\{X^n\} = (-j)^n \left. \frac{\partial^n M_X(jv)}{\partial v^n} \right|_{v=0} \tag{4.138}$$

can be proved by repeated differentiation.

▼| EXAMPLE 4.22   When we use a table of Fourier transforms, the one-sided exponential pdf

$$f_X(x) = \exp(-x) u(x) \tag{4.139}$$

has the corresponding characteristic function

$$M_X(jv) = \frac{1}{1 - jv} \tag{4.140}$$

When we use repeated differentiation or when we expand the characteristic
▲| function in a power series in jv, it follows from (4.138) that $E\{X^n\} = n!$

The pdf of the Sum of Two Independent Random Variables

The following problem is often encountered: given two *statistically independent* random variables X and Y with known pdf's $f_X(x)$ and $f_Y(y)$, respectively, what is the pdf of their sum $Z = X + Y$? To illustrate the use of the characteristic function, we will use it to find the pdf of Z, or $f_Z(z)$, even though we could find this pdf directly.

By definition of the characteristic function of Z, we can write

$$M_Z(jv) = E\{e^{jvZ}\} = E\{e^{jv(X+Y)}\}$$
$$= \int_{-\infty}^{\infty} \int_{-\infty}^{\infty} e^{jv(x+y)} f_X(x) f_Y(y) \, dx \, dy \tag{4.141}$$

since the joint pdf of X and Y is $f_X(x)f_Y(y)$ by the assumption of statistical independence of X and Y. We can write (4.141) as the product of two integrals, since $e^{jv(x+y)} = e^{jvx}e^{jvy}$. This results in

$$M_Z(jv) = \int_{-\infty}^{\infty} f_X(x)e^{jvx}\,dx \int_{-\infty}^{\infty} f_Y(y)e^{jvy}\,dy$$

$$= E\{e^{jvX}\}E\{e^{jvY}\} \tag{4.142}$$

But, from the definition of the characteristic function, Equation (4.134), we see that

$$M_Z(jv) = M_X(jv)M_Y(jv) \tag{4.143}$$

where $M_X(jv)$ and $M_Y(jv)$ are the characteristic functions of X and Y, respectively. Remembering that the characteristic function is the Fourier transform of the corresponding pdf and that a product in the frequency domain corresponds to convolution in the time domain (even with the minus sign in the exponent!), it follows that

$$f_Z(z) = f_X(x) * f_Y(y) = \int_{-\infty}^{\infty} f_X(z-u)f_Y(u)\,du \tag{4.144}$$

The following example illustrates the use of (4.144).

▼ EXAMPLE 4.23   We consider the sum of four identically distributed, independent random variables:

$$Z = X_1 + X_2 + X_3 + X_4 \tag{4.145}$$

where the pdf of X_i is

$$f_{X_i}(x_i) = \Pi(x_i) = \begin{cases} 1, & |x_i| \leq \frac{1}{2} \\ 0, & \text{otherwise}, \ i = 1, 2, 3, 4 \end{cases} \tag{4.146}$$

where $\Pi(x_i)$ is the unit rectangular pulse function defined by Equation (2.2). We find $f_Z(z)$ by applying (4.144) twice. Thus we have

$$Z_1 = X_1 + X_2 \quad \text{and} \quad Z_2 = X_3 + X_4 \tag{4.147}$$

The pdf's of Z_1 and Z_2 are identical, both being the convolution of a uniform density with itself. From Table 2.2, we can immediately write down the result:

$$f_{Z_i}(z_i) = \Lambda(z_i) = \begin{cases} [1 - |z_i|], & |z_i| \leq 1 \\ 0, & \text{otherwise} \end{cases} \tag{4.148}$$

where $f_{Z_i}(z_i)$ is the pdf of Z_i, $i = 1, 2$. To find $f_Z(z)$, we simply convolve $f_{Z_i}(z_i)$ with itself. Thus

$$f_Z(z) = \int_{-\infty}^{\infty} f_{Z_i}(z - u) f_{Z_i}(u) \, du \qquad (4.149)$$

The factors in the integrand are sketched in Figure 4.16(a). Clearly, $f_Z(z) = 0$ for $z < -2$ or $z > 2$. Since $f_{Z_i}(z_i)$ is even, $f_Z(z)$ is also even. Thus we need not consider $f_Z(z)$ for $z < 0$. From Figure 4.16(a) it follows that for $1 \leq z \leq 2$,

$$f_Z(z) = \int_{z-1}^{1} (1 - u)(1 + u - z) \, du = \tfrac{1}{6}(2 - z)^3 \qquad (4.150)$$

FIGURE 4.16 The pdf for the sum of four independent uniformly distributed random variables. (a) Convolution of two triangular pdf's. (b) Comparison of actual and Gaussian pdf's.

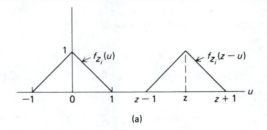

(a)

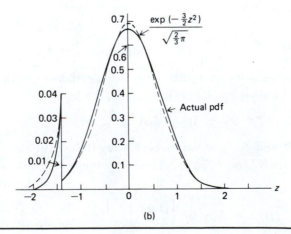

(b)

and for $0 \leq z \leq 1$, we obtain

$$
f_Z(z) = \int_{z-1}^0 (1 + u)(1 + u - z)\, du + \int_0^z (1 - u)(1 + u - z)\, du
$$

$$
+ \int_z^1 (1 - u)(1 - u + z)\, du
$$

$$
= (1 - z) - \tfrac{1}{3}(1 - z)^3 + \tfrac{1}{6}z^3 \tag{4.151}
$$

A graph of $f_Z(z)$ is shown in Figure 4.16(b) along with the graph of the function

$$
\frac{\exp\left(-\tfrac{3}{2}z^2\right)}{\sqrt{\tfrac{2}{3}\pi}} \tag{4.152}
$$

which represents a marginal Gaussian pdf of mean 0 and variance $\tfrac{1}{3}$, the same variance as $Z = X_1 + X_2 + X_3 + X_4$ [the results of Example (4.21) and Equation (4.131) can be used to obtain the variance of Z]. We will describe the Gaussian pdf more fully later.

The reason for the striking similarity of the two pdf's shown in Figure 4.16(b) will become apparent when we discuss the central-limit theorem.

Covariance and the Correlation Coefficient

Two useful joint averages of a pair of random variables X and Y are their *covariance* μ_{XY}, defined as

$$
\mu_{XY} = E\{(X - \overline{X})(Y - \overline{Y})\} \tag{4.153}
$$

and their *correlation coefficient* ρ_{XY}, which is written in terms of the covariance as

$$
\rho_{XY} = \frac{\mu_{XY}}{\sigma_X \sigma_Y} \tag{4.154}
$$

Both are measures of the interdependence of X and Y. The correlation coefficient is more convenient because it is normalized such that $-1 \leq \rho_{XY} \leq 1$. If $\rho_{XY} = 0$, X and Y are said to be *uncorrelated*.

It is easily shown that $\rho_{XY} = 0$ for statistically independent random variables. If X and Y are independent, their joint pdf $f_{XY}(x, y)$ is the product of the respective marginal pdf's; that is, $f_{XY}(x, y) = f_X(x)f_Y(y)$. Thus

$$
\begin{aligned}
\mu_{XY} &= \int_{-\infty}^{\infty} \int_{-\infty}^{\infty} (x - \bar{X})(y - \bar{Y}) f_X(x) f_Y(y) \, dx \, dy \\
&= \int_{-\infty}^{\infty} (x - \bar{X}) f_X(x) \, dx \int_{-\infty}^{\infty} (y - \bar{Y}) f_Y(y) \, dy \\
&= (\bar{X} - \bar{X})(\bar{Y} - \bar{Y}) = 0 \qquad (4.155)
\end{aligned}
$$

Considering next the cases $X = \pm \alpha Y$, where α is a positive constant, we obtain

$$
\begin{aligned}
\mu_{XY} &= \int_{-\infty}^{\infty} \int_{-\infty}^{\infty} (\pm \alpha y \mp \alpha \bar{Y})(y - \bar{Y}) f_{XY}(x, y) \, dx \, dy \\
&= \pm \alpha \int_{-\infty}^{\infty} \int_{-\infty}^{\infty} (y - \bar{Y})^2 f_{XY}(x, y) \, dx \, dy = \pm \alpha \sigma_y^2 \quad (4.156)
\end{aligned}
$$

Using (4.131) with $N = 1$, we can write the variance of X as $\sigma_x^2 = \alpha^2 \sigma_Y^2$. Thus the correlation coefficient is

$$
\rho_{XY} = +1, \quad \text{for } X = +\alpha Y \qquad \text{and} \qquad \rho_{XY} = -1, \quad \text{for } X = -\alpha Y
$$

To summarize, the correlation coefficient of two independent random variables is zero. When two random variables are linearly related, their correlation is $+1$ or -1 depending on whether one is a positive or a negative constant times the other.

4.4   SOME USEFUL pdf's*

We have already considered several often used probability distributions in the examples. These have included the Rayleigh pdf (Example 4.15), the pdf of a sinewave of random phase (Example 4.18), and the uniform pdf (Example 4.21). Some others, which will be useful in our future considerations, are given below.

Binomial Distribution

One of the most common discrete distributions in the application of probability to systems analysis is the binomial distribution. We consider a chance experiment with two mutually exclusive, exhaustive outcomes A and $\bar{A}$, with

---

* A number of useful probability distributions are given in a table in the summary at the end of this chapter.

probabilities $P(A) = p$ and $P(\overline{A}) = q = 1 - p$, respectively. Assigning the discrete random variable K to be numerically equal to the number of times event A occurs in N trials of our chance experiment, we seek the probability that exactly $k \leq n$ occurrences of the event A occur in n repetitions of the experiment. (Thus our actual chance experiment is the replication of the basic experiment n times.) The resulting distribution is called the *binomial distribution*.

Specific examples in which the binomial distribution is the result are the following: In n tosses of a coin, what is the probability of $k \leq n$ heads? In the transmission of n messages through a channel, what is the probability of $k \leq n$ errors?

Note that in all cases we are interested in *exactly* k occurrences of the event, not, for example, at least k of them, although we may find the latter probability if we have the former.

Although the problem we are considering is very general, let us solve it by visualizing the coin-tossing experiment. We wish to obtain the probability of k heads in n tosses of the coin if the probability of a head on a single toss is p and the probability of a tail is $1 - p = q$. One of the possible sequences of k heads in n tosses is

$$\underbrace{H\,H\,\cdots\,H}_{k \text{ heads}}\;\underbrace{T\,T\,\cdots\,T}_{n\,-\,k \text{ tails}}$$

Since the trials are independent, the probability of this particular sequence is

$$\underbrace{p \cdot p \cdots p}_{k \text{ factors}} \cdot \underbrace{q \cdot q \cdots q}_{n\,-\,k \text{ factors}} = p^k q^{n-k} \tag{4.157}$$

But the preceding sequence of k heads in n trials is only one of

$$\binom{n}{k} \triangleq \frac{n!}{k!(n-k)!} \tag{4.158}$$

possible sequences, where $\binom{n}{k}$ is the binomial coefficient. To see this, we consider the number of ways k *identifiable* heads can be arranged in n slots. The first head can fall in any of the n slots, the second in any of $n - 1$ slots (the first head already occupies one slot), the third in any of $n - 2$ slots, and so on for a total of

$$n(n - 1)(n - 2) \cdots (n - k + 1) = \frac{n!}{(n - k)!} \tag{4.159}$$

possible arrangements in which we identify each head. However, we are not
concerned about which head occupies which slot. For each possible identifiable arrangement, there are $k!$ arrangements for which we can switch the
heads around and still have the same slots occupied. Thus the total number
of arrangments, if we do not identify the particular coin occupying each slot,
is

$$\frac{n(n-1) \cdots (n-k+1)}{k!} = \binom{n}{k} \tag{4.160}$$

Since the occurrence of any of these $\binom{n}{k}$ possible arrangements precludes
the occurrence of any other [that is, the $\binom{n}{k}$ outcomes of our experiment are
mutually exclusive], and since each occurs with probability $p^k q^{n-k}$, the probability of *exactly* k heads in n trials in *any* order is

$$P(K = k) \triangleq P_n(k) = \binom{n}{k} p^k q^{n-k}, \qquad k = 0, 1, \ldots, n \tag{4.161}$$

Equation (4.161), known as the *binomial probability distribution* (note that it is
not a pdf or a cdf), is plotted in Figure 4.17(a)–(d) for four different values
of p and n.

The mean of a binomially distributed random variable K, by Equation
(4.93), is given by

$$E\{K\} = \sum_{k=0}^{n} k \frac{n!}{k!(n-k)!} p^k q^{n-k} \tag{4.162}$$

Noting that the sum can be started at $k = 1$ since the first term is zero, we
can write

$$E\{K\} = \sum_{k=1}^{n} \frac{n!}{(k-1)!(n-k)!} p^k q^{n-k} \tag{4.163}$$

Letting $m = k - 1$ yields the sum

$$E\{K\} = \sum_{m=0}^{n-1} \frac{n!}{m!(n-m-1)!} p^{m+1} q^{n-m-1}$$

$$= np \sum_{m=0}^{n-1} \frac{(n-1)!}{m!(n-m-1)!} p^m q^{n-m-1} \tag{4.164}$$

Finally, letting $\ell = n - 1$ and recalling that, by the binomial theorem,

$$(x + y)^\ell = \sum_{m=0}^{\ell} \binom{\ell}{m} x^m y^{\ell-m} \tag{4.165}$$

FIGURE 4.17   The binomial distribution with comparison to Laplace and
Poisson approximations. (a) $n = 1$, $p = 0.5$. (b) $n = 2$, $p = 0.5$. (c) $n = 3$,
$p = 0.5$. (d) $n = 4$, $p = 0.5$. (e) $n = 5$, $p = 0.5$. Circles are Laplace
approximations. (f) $n = 5$, $p = \frac{1}{10}$. Circles are Poisson approximations.

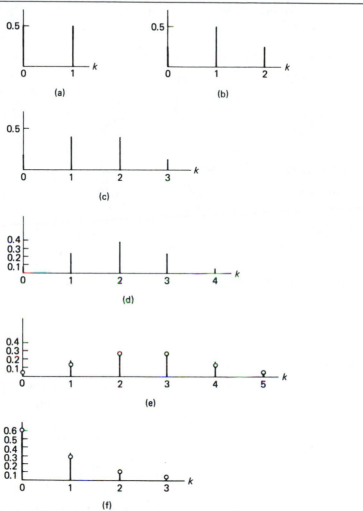

we obtain

$$\overline{K} = E\{K\} = np(p + q)^{\ell} = np \tag{4.166}$$

since $p + q = 1$. The result is reasonable; in a long sequence of n tosses of
a coin, we would expect about $np = \frac{1}{2}n$ heads.

We can go through a similar series of manipulations to show that $E\{K^2\}$ $= np(np + q)$. Using this result, it follows that the variance of a binomially distributed random variable is

$$\sigma_K{}^2 = E\{K^2\} - E^2\{K\} = npq = \overline{K}(1 - p) \tag{4.167}$$

▼ EXAMPLE 4.24   The probability of having two boys in a four-child family, assuming single births and the probability of a male $\cong 0.5$, from (4.161), is

$$P_4(2) = \binom{4}{2}\left(\frac{1}{2}\right)^4 = \frac{3}{8} \tag{4.168}$$

▲

Laplace Approximation to the Binomial Distribution

When n becomes large, computations using (4.161) become unmanageable. In the limit as $n \to \infty$, it can be shown that for $|k - np| \le \sqrt{npq}$

$$P_n(k) \cong \frac{1}{\sqrt{2\pi npq}} \exp\left[-\frac{(k - np)^2}{2npq}\right] \tag{4.169}$$

A comparison of the Laplace approximation with the actual distribution is given in Figure 4.17(e)

Poisson Distribution and Poisson Approximation to the Binomial Distribution

If we consider a chance experiment in which an event whose probability of occurrence in a very small time interval ΔT is $P = \alpha\Delta T$, where α is a constant of proportionality, and if successive occurrences are statistically independent, then the probability of k events in time T is

$$P_T(k) = \frac{(\alpha T)^k}{k!} e^{-\alpha T} \tag{4.170}$$

For example, the emission of electrons from a hot metal surface obeys this law, which is called the *Poisson distribution*.

The Poisson distribution can be used to approximate the binomial distribution when the number of trials n is large, the probability of each event p is small, and the product $np \cong npq$. The approximation is

$$P_n(k) \cong \frac{(\overline{K})^k}{k!} e^{-\overline{K}} \tag{4.171}$$

where, as calculated previously, $\overline{K} = E\{K\} = np$ and $\sigma_k^2 = E\{K\}q \cong E\{K\}$. This approximation is compared with the binomial distribution in Figure 4.17(f).

▼ **EXAMPLE 4.25**   The probability of error on a single transmission in a digital communication system is $P_e = 10^{-4}$. What is the probability of more than three errors in 1000 transmissions?

SOLUTION   We find the probability of three errors or less from (4.171):

$$P(K \leq 3) = \sum_{k=0}^{3} \frac{(\overline{K})^k}{k!} e^{-\overline{K}} \tag{4.172}$$

where $\overline{K} = (10^{-4})(1000) = 0.1$. Hence

$$P(K \leq 3) = e^{-0.1} \left[\frac{(0.1)^0}{0!} + \frac{(0.1)^1}{1!} + \frac{(0.1)^2}{2!} + \frac{(0.1)^3}{3!} \right] \cong 0.999996 \tag{4.173}$$

▲ Therefore, $P(K > 3) = 1 - P(K \leq 3) \cong 4 \times 10^{-6}$.

Geometric Distribution

Suppose we are interested in the probability of the first head in a series of coin tossings, or the first error in a long string of digital signal transmissions occurring on the kth trial. The distribution describing such experiments is called the *geometric distribution* and is

$$P(k) = pq^{k-1}, \qquad 1 \leq k < \infty \tag{4.174}$$

where p is the probability of the event of interest occurring (i.e., head, error, etc.) and q is the probability of it not occurring.

▼ **EXAMPLE 4.26**   The probability of the first error occurring at the 1000th transmission in a digital data transmission system where the probability of error is $p = 10^{-6}$ is

▲ $$P(1000) = 10^{-6}(1 - 10^{-6})^{999} = 9.99 \times 10^{-7} \cong 10^{-6}$$

▼ **EXAMPLE 4.27**   Find the probability of the first head in a sequence of 10 tossings of a fair coin occurring on the 1st, 2nd, . . . , 10th tossings.

SOLUTION   The desired probabilities are

$$P(k) = 0.5(1 - 0.5)^{k-1} = 0.5^k, \qquad 1 \leq k \leq 10$$

The first five are listed below. The student should compute the rest of the values.

k	1	2	3	4	5
P_k	0.5	0.25	0.125	0.063	0.031

▲

Gaussian Distribution

In our future considerations, the Gaussian pdf will be used repeatedly. There are at least two reasons for this. One is that the assumption of Gaussian statistics for random phenomena often makes an intractable problem tractable. The other, and more fundamental reason, is that because of a remarkable phenomenon summarized by a theorem called the *central-limit theorem,* many naturally occurring random quantities, such as noise or measurement errors, are Gaussianly distributed. The following is a statement of the central-limit theorem.

▷ CENTRAL-LIMIT THEOREM

Let X_1, X_2, ... be independent, identically distributed random variables, each with finite mean m and finite variance σ^2. Let Z_n be a sequence of unit-variance, zero-mean random variables, defined as

$$Z_n \triangleq \frac{\sum\limits_{i=1}^{n} X_i - nm}{\sigma \sqrt{n}} \tag{4.175}$$

Then

$$\lim_{n \to \infty} P(Z_n \leq z) = \int_{-\infty}^{z} \frac{e^{-t^2/2}}{\sqrt{2\pi}} \, dt \tag{4.176}$$

In other words, the cdf of the normalized sum (4.175) approaches a Gaussian cdf, no matter what the distribution of the component random variables. The only restriction is that they be independent and identically distributed and that their means and variances be finite. In some cases the independence and identically distributed assumptions can be relaxed. It is important, however, that no one of the component random variables or a finite combination of them dominate the sum.

We will not prove the central-limit theorem or use it in later work. We state it here simply to give partial justification for our almost exclusive assumption of Gaussian statistics from now on. For example, electrical noise is often the result of a superposition of voltages due to a large number of charge carriers. Turbulent boundary-layer pressure fluctuations on an aircraft skin are the superposition of minute pressures due to numerous eddies. Random errors in experimental measurements are due to many irregular fluctuating causes. In all these cases, the Gaussian approximation for the fluctuating quantity is useful and valid. Example 4.23 illustrates that surprisingly few terms in the sum are required to give a Gaussian-appearing pdf, even where the component pdf's are far from Gaussian.

The generalization of the joint Gaussian pdf first introduced in Example 4.15 is

$$
f_{XY}(x, y)
= \frac{\exp\left\{ -\dfrac{\left(\dfrac{x - m_x}{\sigma_x}\right)^2 - 2\rho\left(\dfrac{x - m_x}{\sigma_x}\right)\left(\dfrac{y - m_y}{\sigma_y}\right) + \left(\dfrac{y - m_y}{\sigma_y}\right)^2}{2(1 - \rho^2)} \right\}}{2\pi\sigma_x\sigma_y\sqrt{1 - \rho^2}}
$$

$$(4.177)$$

where, through straightforward but tedious integrations, it can be shown that

$$m_x = E\{X\} \qquad (4.178a)$$

$$m_y = E\{Y\} \qquad (4.178b)$$

$$\sigma_x^2 = \text{var}\,\{X\} \qquad (4.178c)$$

$$\sigma_y^2 = \text{var}\,\{Y\} \qquad (4.178d)$$

and

$$\rho = \frac{E\{(X - m_x)(Y - m_y)\}}{\sigma_x\sigma_y} \qquad (4.178e)$$

The joint pdf for $N > 2$ Gaussian random variables may be written in a compact fashion through the use of matrix notation. The general form is given in Appendix B.

The marginal pdf for X (or Y) can be obtained by integrating (4.177) over *y* (or *x*). Again, the integration is tedious and will be left to the problems. The marginal pdf for *X* is

$$n(m_x, \sigma_x) = \frac{\exp\left[-(x - m_x)^2/2\sigma_x^2\right]}{\sqrt{2\pi\sigma_x^2}} \tag{4.179}$$

where the notation $n(m_x, \sigma_x)$ has been introduced to denote a Gaussian pdf of mean m_x and standard deviation σ_x. This function is shown in Figure 4.18.

We will sometimes assume in the discussions to follow that $m_x = m_y = 0$ in (4.177) and (4.179), for if they are not zero, we can consider new random variables X' and Y' defined as $X' = X - m_x$ and $Y' = Y - m_y$ which do have zero means. Thus no generality is lost in assuming zero means.

For $\rho = 0$, that is, *X* and *Y* uncorrelated, the cross term in the exponent of (4.177) is zero, and $f_{XY}(x, y)$, with $m_x = m_y = 0$, can be written as

$$f_{XY}(x, y) = \frac{\exp\left(-x^2/2\sigma_x^2\right)}{\sqrt{2\pi\sigma_x^2}} \frac{\exp\left(-y^2/2\sigma_y^2\right)}{\sqrt{2\pi\sigma_y^2}} = f_X(x)f_Y(y) \tag{4.180}$$

Thus *uncorrelated Gaussian random variables are also statistically independent.* We emphasize that this does not hold for all pdf's, however.

It can be shown that the sum of any number of Gaussian random variables, independent or not, is Gaussian. The sum of two independent Gaussian random variables is easily shown to be Gaussian. Let $Z = X_1 + X_2$, where the pdf of X_i is $n(m_i, \sigma_i)$. Using a table of Fourier transforms or completing the square and integrating, we find that the characteristic function of X_i is

$$M_{X_i}(jv) = \int_{-\infty}^{\infty} (2\pi\sigma_i^2)^{-1/2} \exp\left[\frac{-(x_i - m_i)^2}{2\sigma_i^2}\right] \exp\left(jvx_i\right) dx_i$$

$$= \exp\left(jm_iv - \frac{\sigma_i^2 v^2}{2}\right) \tag{4.181}$$

FIGURE 4.18 The Gaussian pdf with mean m_x and variance σ_x^2

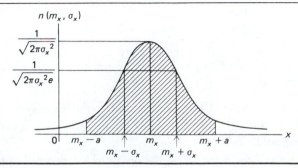

Thus the characteristic function of Z is

$$M_Z(jv) = M_{X_1}(jv)M_{X_2}(jv) = \exp\left[j(m_1 + m_2)v - \frac{(\sigma_1^2 + \sigma_2^2)v^2}{2}\right] \quad (4.182)$$

which is the characteristic function of a Gaussian random variable of mean $m_1 + m_2$ and variance $\sigma_1^2 + \sigma_2^2$.

Gaussian Q-Function

As Figure 4.18 shows, $n(m_x, \sigma_x)$ describes a continuous random variable that may take on any value in $(-\infty, \infty)$ but is most likely to be found near $X = m_x$. The even symmetry of $n(m_x, \sigma_x)$ about $x = m_x$ leads to the conclusion that $P(X \leq m_x) = P(X \geq m_x) = \frac{1}{2}$.

Suppose we wish to find the probability that X lies in the interval $[m_x - a, m_x + a]$. Using (4.36), we can write this probability as

$$P[m_x - a \leq X \leq m_x + a] = \int_{m_x-a}^{m_x+a} \frac{\exp\left[-(x - m_x)^2/2\sigma_x^2\right]}{\sqrt{2\pi\sigma_x^2}}\, dx \quad (4.183)$$

which is the shaded area in Figure 4.18. With the change of variables $y = (x - m_x)/\sigma_x$, this gives

$$P[m_x - a \leq X \leq m_x + a] = \int_{-a/\sigma_x}^{a/\sigma_x} \frac{e^{-y^2/2}}{\sqrt{2\pi}}\, dy$$

$$= 2\int_{0}^{a/\sigma_x} \frac{e^{-y^2/2}}{\sqrt{2\pi}}\, dy \quad (4.184)$$

where the last integral follows by virtue of the integrand being even. Unfortunately, this integral cannot be evaluated in closed form.

The Gaussian Q-function, or simply Q-function, is defined as

$$Q(u) = \int_{u}^{\infty} \frac{e^{-y^2/2}}{\sqrt{2\pi}}\, dy \quad (4.185)$$

It has been evaluated numerically, and rational and asymptotic approximations are available to evaluate it for moderate and large arguments, respectively.* Using this transcendental function definition, we may rewrite (4.184) as

$$P[m_x - a \leq X \leq m_x + a] = 2 \left[\frac{1}{2} - \int_{a/\sigma_x}^{\infty} \frac{e^{-y^2/2}}{\sqrt{2\pi}} \, dy \right]$$

$$= 1 - 2Q\left(\frac{a}{\sigma_x}\right) \tag{4.186}$$

A useful approximation for the Q-function for large arguments is

$$Q(u) \cong \frac{e^{-u^2/2}}{u\sqrt{2\pi}}, \, u \gg 1 \tag{4.187}$$

Numerical comparison of (4.185) and (4.187) shows that less than a 6% error results for $u \geq 3$ in using this approximation. This, and other results for the Q-function, are given in Appendix C.

A related integral is the error function, defined as

$$erf(u) = \frac{2}{\sqrt{\pi}} \int_0^u e^{-y^2} \, dy \tag{4.188}$$

It can be shown to be related to the Q-function by

$$erf(u) = 1 - 2Q(\sqrt{2}u) \tag{4.189}$$

Chebyshev's Inequality

The difficulties encountered above in evaluating (4.183) make an approximation to such probabilities desirable. *Chebyshev's inequality* gives us a lower bound, *regardless of the specific form of the pdf involved,* provided its second moment is finite. The probability of finding a random variable X within $\pm k$ standard deviations of its mean is at least $1 - 1/k^2$, according to Chebyshev's inequality. That is,

$$P[|X - m_x| \leq k\sigma_x] \geq 1 - \frac{1}{k^2}, \quad k > 0 \tag{4.190}$$

Considering $k = 3$, we obtain

$$P[|X - m_x| \leq 3\sigma_x] \geq \tfrac{8}{9} \cong 0.889 \tag{4.191}$$

---

* These are provided in M. Abramowitz and I. Stegun (eds.), *Handbook of Mathematical Functions with Formulas, Graphs, and Mathematical Tables.* National Bureau of Standards, Applied Mathematics Series No. 55, Issued June 1964 (pp. 931 ff). Also New York: Dover, 1972.

Assuming X is Gaussian and using the Q-function, this probability can be computed exactly to be 0.9973. In words, the probability that a random variable deviates from its mean by more than ± 3 standard deviations is not greater than 0.111, regardless of its pdf. (There is the restriction that its second moment must be finite.) Note that the bound for this example is not very tight.

SUMMARY

1. The objective of probability theory is to attach real numbers between zero and one called *probabilities* to the *outcomes* of chance experiments— that is, experiments in which the outcomes are not uniquely determined by the causes but depend on chance—and to interrelate probabilities of *events,* which are defined to be combinations of outcomes.

2. Two events are *mutually exclusive* if the occurrence of one of them precludes the occurrence of the other. A set of events is said to be *exhaustive* if one of them must occur in the performance of a chance experiment. The *null event* cannot happen, and the *certain event* must happen in the performance of a chance experiment.

3. The *equally likely definition* of the probability $P(A)$ of an event A states that if a chance experiment can result in a number N of mutually exclusive, equally likely outcomes, then $P(A)$ is the ratio of the number of outcomes favorable to A, or N_A, to the total number. It is a circular definition in that probability is used to define probability, but it is nevertheless useful in many situations such as drawing cards from well-shuffled decks.

   The *relative-frequency definition* of the probability of an event A assumes that the chance experiment is replicated a large number of times N and

   $$P(A) = \lim_{N \to \infty} \frac{N_A}{N}$$

   where N_A is the number of replications resulting in the occurrence of A. The *axiomatic approach* defines the probability $P(A)$ of an event A as a real number satisfying the following axioms:

   (a) $P(A) \geq 0$.
   (b) $P(\text{certain event}) = 1$.
   (c) If A and B are mutually exclusive events, $P(A \text{ or } B \text{ or both}) = P(A) + P(B)$.

   The axiomatic approach encompasses the first two definitions given.

4. Given two events A and B, the compound event "A or B or both" is denoted as $A + B$, the compound event "both A and B" is denoted as (AB) or as (A, B), and the event "not A" is denoted as $\bar{A}$. If A and B are not necessarily mutually exclusive, the axioms of probability may be used to show that $P(A + B) = P(A) + P(B) - P(AB)$. Letting $P(A \mid B)$ denote the probability of A occurring given that B occurred and $P(B \mid A)$ denote the probability of B given A, these probabilities are defined, respectively, as

$$P(A \mid B) = \frac{P(AB)}{P(B)} \quad \text{and} \quad P(B \mid A) = \frac{P(AB)}{P(A)}$$

A special case of *Bayes' rule* results by putting these two definitions together:

$$P(B \mid A) = \frac{P(A \mid B)P(B)}{P(A)}$$

Statistically independent events are events for which $P(AB) = P(A)P(B)$.

5. A random variable is a rule that assigns a real number to every outcome of a chance experiment. For example, in flipping a coin, assigning $X = +1$ to the occurrence of a head and $X = -1$ to the occurrence of a tail constitutes the assignment of a discrete-valued random variable.

6. The cumulative distribution function (cdf) $F_X(x)$ of a random variable X is defined as the probability that $X \leq x$ where x is a running variable. $F_X(x)$ lies between zero and 1 with $F_X(-\infty) = 0$ and $F_X(\infty) = 1$, is continuous from the right, and is a nondecreasing function of its argument. Discrete random variables have step-discontinuous cdf's, and continuous random variables have continuous cdf's.

7. The probability density function (pdf) $f_X(x)$ of a random variable X is defined to be the derivative of the cdf. Thus

$$F_X(x) = \int_{-\infty}^{x} f_X(x') \, dx'$$

The pdf is nonnegative and integrates over all x to unity. A useful interpretation of the pdf is that $f_X(x) \, dx$ is the probability of the random variable X lying in an infinitesimal range dx about x.

8. The joint cdf $F_{XY}(x, y)$ of two random variables X and Y is defined as the probability that $X \leq x$ and $Y \leq y$ where x and y are particular values of X and Y. Their joint pdf $f_{XY}(x, y)$ is the second partial derivative of the cdf first with respect to x and then with respect to y. The cdf of X (Y)

alone (that is, the marginal cdf) is found by setting y (x) to infinity in the argument of F_{XY}. The pdf of X (Y) alone (that is, the marginal pdf) is found by integrating f_{XY} over all y (x).

9. Two statistically independent random variables have joint cdf's and pdf's that factor into the respective marginal cdf's or pdf's.

10. The conditional pdf of X given Y is defined as

$$f_{X|Y}(x|y) = \frac{f_{XY}(x, y)}{f_Y(y)}$$

with a similar definition for $f_{Y|X}(y|x)$. The expression $f_{X|Y}(x|y)\, dx$ can be interpreted as the probability that $x - dx < X \le x$ given $Y = y$.

11. Given $Y = g(X)$ where $g(X)$ is a monotonic function,

$$f_Y(y) = f_X(x) \left|\frac{dx}{dy}\right|_{x = g^{-1}(y)}$$

where $g^{-1}(y)$ is the inverse of $y = g(x)$. Joint pdf's of functions of more than one random variable can be transformed similarly.

12. Important probability functions defined in Chapter 4 are the Rayleigh pdf [Equation (4.92)], the pdf of a random-phased sinusoid (Example 4.18), the uniform pdf (Example 4.21), the binomial probability distribution [Equation (4.161)], the Laplace and Poisson approximations to the binomial distribution [Equations (4.169) and (4.171)], and the Gaussian pdf [Equations (4.177) and (4.179)].

13. The statistical average, or expectation, of a function $g(X)$ of a random variable X with pdf $f_X(x)$ is defined as

$$E\{g(X)\} = \overline{g(X)} = \int_{-\infty}^{\infty} g(x) f_X(x)\, dx$$

The average of $g(X) = X^n$ is called the nth moment of X. The first moment is known as the *mean* of X. Averages of functions of more than one random variable are found by integrating the function times the joint pdf over all values of its arguments. The averages $\overline{g(X, Y)} = \overline{X^m Y^n}$ are called the *joint moments* of the order $m + n$. The variance of a random variable X is the average $\overline{(X - \overline{X})^2} = \overline{X^2} - \overline{X}^2$.

14. The average of $\overline{\sum a_i X_i}$ is $\sum a_i \overline{X_i}$; that is, summations and averaging can be interchanged. The variance of a sum of random variables is the sum of the respective variances *if the random variables are statistically independent.*

15. The characteristic function $M_X(jv)$ of a random variable X that has the pdf $f_X(x)$ is the expectation of exp (jvX) or, equivalently, the Fourier transform of $f_X(x)$ with a plus sign in the exponential of the Fourier transform integral. Thus the pdf is the inverse Fourier transform of the characteristic function.

    The nth moment of X can be found from $M_X(jv)$ by differentiating with respect to v for n times, multiplying by $(-j)^n$, and setting $v = 0$. The characteristic function of $Z = X + Y$, where X and Y are independent, is the product of the characteristic functions of X and Y. Thus, by the convolution theorem of Fourier transforms, the pdf of Z is the convolution of the pdf's of X and Y.

16. The covariance μ_{XY} of two random variables X and Y is the average

$$\mu_{XY} = E[(X - \bar{X})(Y - \bar{Y})]$$

    The correlation coefficient ρ_{XY} is

$$\frac{\mu_{XY}}{\sigma_X \sigma_Y}$$

    Both give a measure of the linear interdependence of X and Y, but ρ_{XY} is handier because it is bounded by ± 1.

17. The central-limit theorem states that, under suitable conditions, the sum of a large number N of random variables (not necessarily with the same pdf's) tends to a Gaussian pdf as N becomes large.

18. The Q-function can be used to compute probabilities of Gaussian random variables being in certain ranges. The Q-function is tabulated in Appendix C, and rational and asymptotic approximations are given for computing it.

19. Chebyshev's inequality gives the lower bound of the probability that a random variable is within k standard deviations of its mean as $1 - k^2$, *regardless of the pdf of the random variable* (its second moment must be finite).

20. Table 4.5 summarizes a number of useful probability distributions with their means and variances.

FURTHER READING

Several books are available that deal with probability theory from an engineering viewpoint. Among these are Beckman (1967), Breipohl (1970), Helstrom (1990), Leon-Garcia (1994), Peebles (1987), and Williams (1991).

TABLE 4.5 Probability Distributions of Some Random Variables and Their Means and Variances

	Probability Density or Distribution Function	Mean	Variance
Uniform: $$f_x(x) = \begin{cases} \dfrac{1}{b-a}, & a \le x \le b \\ 0, & x < a \text{ or } x > b \end{cases}$$		$\dfrac{1}{2}(b+a)$	$\dfrac{1}{12}(b-a)^2$
Gaussian: $$f_x(x) = \dfrac{1}{\sqrt{2\pi\sigma^2}} e^{(x-m)^2/2\sigma^2}$$		m	σ^2
Rayleigh: $$f_R(r) = \dfrac{r e^{-r^2/2\sigma^2}}{\sigma^2}, \quad 0 \le r < \infty$$		$\sqrt{\dfrac{\pi}{2}}\,\sigma$	$\dfrac{1}{2}(4-\pi)\sigma^2$
Laplacian: $$f_x(x) = \dfrac{\alpha}{2} e^{-\alpha|x|}, \quad \alpha > 0$$		0	$\dfrac{2}{\alpha^2}$
One-sided exponential: $$f_x(x) = \alpha e^{-\alpha x} u(x)$$		$\dfrac{1}{\alpha}$	$\dfrac{1}{\alpha^2}$
Hyperbolic: $$f_x(x) = \dfrac{(m-1)h^{m-1}}{2(|x|+h)^m}, \quad m > 3, h > 0$$		0	$\dfrac{2h^2}{(m-3)(m-2)}$
Nakagami-m: $$f_x(x) = \dfrac{2m^m}{\Gamma(m)} x^{2m-1} e^{-mx^2}, \quad x \ge 0$$		$\dfrac{1 \times 3 \times \ldots \times (2m-2)}{2^m \Gamma(m)}$	$\dfrac{\Gamma(m+1)}{\Gamma(m)\sqrt{m}}$
Lognormal:* $$f_x(x) = \dfrac{\exp\left[-\dfrac{(\ln x - m_y)^2}{2\sigma_y^2}\right]}{x\sqrt{2\pi\sigma_y^2}}$$		$e^{m_y + 2\sigma_y^2}$	$e^{2m_y + \sigma_y^2}(e^{\sigma_y^2} - 1)$
Binomial: $$P_n(k) = \binom{n}{k} p^k q^{n-k}, \quad k = 0, 1, \ldots, n \\ p + q = 1$$		np	npq
Poisson: $$P(k) = \dfrac{\lambda^k}{k!} e^{-\lambda}, \quad 0 \le k < \infty$$		λ	λ
Geometric: $$P(k) = pq^{k-1}, \quad 1 \le k < \infty$$		$\dfrac{1}{p}$	$\dfrac{q}{p^2}$

* The lognormal random variable results from the transformation $y = \ln x$ where Y is a Gaussian random variable with mean m_y and variance σ_y^2.

More mathematical in nature are the following: Barr and Zehna (1983), Papoulis (1990), Papoulis (1984), and Roberts (1992). A good overview with many examples is Ash (1992).

PROBLEMS

Section 4.1

4.1 A circle is divided into 25 equal parts. A pointer is spun until it stops on one of the parts, which are numbered from 1 through 25. Describe the sample space and, assuming equally likely outcomes, find (a) P(an even number), (b) P(the number 25), (c) P(the numbers 4, 5, or 9), and (d) P(a number greater than 10).

4.2 (a) Prove (4.4) using the axioms of probability. (b) Repeat the derivation for three events A, B, and C, and show that

$$P(A \text{ or } B \text{ or } C) = P(A) + P(B) + P(C)$$
$$- P(A, B) - P(B, C) - P(A, C) + P(A, B, C)$$

4.3 If five cards are drawn without replacement from an ordinary deck of cards, what is the probability that (a) three kings and two aces result, (b) four of a kind result, (c) all are of the same suit, (d) an ace, king, queen, jack, and ten of the same suit result, and (e) given that an ace, king, jack, and ten have been drawn, what is the probability that the next card drawn will be a queen (not all of the same suit)?

4.4 What equations must be satisified in order for three events A, B, and C to be independent?

4.5 (a) Show that if two events are mutually exclusive and statistically independent, the probability of one, or both, is zero.
     (b) Given three events A, B, and C, with A and B mutually exclusive and B and C independent, what can be said about the relationship between A and C?

4.6 Figure 4.19 is a graph that represents a communication network, where the nodes are receiver/repeater boxes and the edges (or links) represent communication channels which, if connected, convey the message perfectly. However, there is the probability p that a link will be broken and the probability $q = 1 - p$ that it will be whole. (*Hint:* Use a tree diagram like Fig. 4.2).

     (a) What is the probability that at least one working path is available between the nodes labeled A and B?

FIGURE 4.19

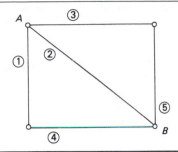

(b) Remove link 4. Now what is the probability that at least one working path is available between nodes A and B?

(c) Remove link 2. What is the probability that at least one working path is available between nodes A and B?

(d) Which is the more serious situation, the removal of link 4 or link 2? Why?

4.7 Given a binary communication channel where A = input and B = output, let $P(A) = 0.4$, $P(B|A) = 0.9$, and $P(\bar{B}|\bar{A}) = 0.6$. Find $P(A|B)$ and $P(A|\bar{B})$.

4.8 (a) Find the probabilities omitted from Figure 4.20.
   (b) Find the probabilities $P(A_3|B_3)$, $P(B_2|A_1)$, and $P(B_3|A_2)$.

Section 4.2

4.9 Two dice are tossed.
   (a) Let X_1 be a random variable that is numerically equal to the square of the total number of spots on the up faces of the dice. Construct a table that defines this random variable.

FIGURE 4.20 Table of joint and marginal probabilities $P(A_iB_j)$

	B_1	B_2	B_3	$P(A_i)$
A_1	0.05		0.45	0.55
A_2		0.15	0.10	
A_3	0.05	0.05		0.15
$P(B_j)$				1.0

(b) Let X_2 be a random variable that has the value of 1 if the sum of the number of spots up on both dice is less than 7 and the value zero if it is greater than or equal to 7. Repeat part (a) for this case.

4.10   Three fair coins are tossed simultaneously. Define a random variable as the total number of heads up on the coins. Plot the cumulative-distribution function and the probability-density function corresponding to this random variable.

4.11   A certain continuous random variable has the cumulative distribution function

$$F_X(x) = \begin{cases} 0, & x < 0 \\ Ax^3, & 0 \le x \le 10 \\ B, & x > 10 \end{cases}$$

(a) Find the proper values for A and B.
(b) Plot $F_X(x)$.
(c) Obtain and plot the pdf $f_X(x)$.
(d) Compute $P(X > 7)$.
(e) Compute $P(3 \le X < 7)$.

4.12   The following functions can be pdf's if the constants are chosen properly. Find the proper conditions on the constants so that they are. (A, B, C, D, α, β, γ, and τ are positive constants.)

(a) $f(x) = Ae^{-\alpha x} u(x)$,   where $u(x)$ is the unit step
(b) $f(x) = Be^{\beta x} u(-x)$
(c) $f(x) = Ce^{-\gamma |x|}$
(d) $f(x) = \dfrac{D}{\tau^2 + x^2}$

4.13   Test X and Y for independence if

(a) $f_{XY}(x, y) = Ae^{-|x| - 2|y|}$

(b) $f_{XY}(x, y) = C(1 - x - y)$, $0 \le x \le 1 - y$ and $0 \le y \le 1$
Prove your answers.

4.14   The joint pdf of two random variables is

$$f_{XY}(x, y) = \begin{cases} C(1 + xy), & 0 \le x \le 6, 0 \le y \le 5 \\ 0, & \text{otherwise} \end{cases}$$

Find the following:

(a) The constant C.

(b) $F_{XY}(0.1, 1.5)$.

(c) $f_{XY}(x, 3)$.

(d) $f_{X|Y}(x \mid 3)$.

4.15 Repeat the previous problem for

$$f_{XY}(x, y) = \begin{cases} C(1 + xy), & 0 \le x \le 6,\, 0 \le y \le x \\ 0, & \text{otherwise} \end{cases}$$

4.16 The joint pdf of the random variables X and Y is

$$f_{XY}(x, y) = Axye^{-(x+y)}, \qquad x \ge 0 \quad \text{and} \quad y \ge 0$$

(a) Find the constant A.

(b) Find the marginal pdf's of X and Y, $f_X(x)$ and $f_Y(y)$.

(c) Are X and Y statistically independent? Justify your answer.

4.17 (a) For what value of α is the function

$$f(x) = \alpha x^{-2} u(x - \alpha)$$

a probability density function? [$u(x)$ is the unit step function.]

(b) Find the corresponding cumulative distribution function.

(c) Compute $P(X \ge 10)$.

4.18 Given the Gaussian random variable with the pdf

$$f_X(x) = \frac{e^{-x^2/2\sigma^2}}{\sqrt{2\pi}\,\sigma}$$

let $Y = X^2$. Find the pdf of Y. (*Hint:* Note that $Y = X^2$ is symmetrical about $X = 0$ and that it is impossible for Y to be less than zero.)

4.19 If X is Gaussian with the pdf given in Problem 4.18 and the random variable Y is given in terms of X as

$$Y = \begin{cases} aX, & X \ge 0 \\ 0, & X < 0 \end{cases}$$

find the pdf of Y. (*Hint:* When $X < 0$, what is Y? How is this manifested in the pdf for Y?)

4.20 A nonlinear system has input X and output Y. The pdf for the input is Gaussian as given in Problem 4.18. Determine the pdf of the output, assuming that the nonlinear system has the following input/output relationship: (a) $Y = |X|$; (b) $Y = X - X^3/3$.

Section 4.3

4.21 Let $f_X(x) = A \exp(-b|x|)$ for all x.

(a) Find the relationship between A and b such that this function is a pdf.
(b) Calculate $E(X)$ for this random variable.
(c) Find $E(X^2)$ for this random variable.
(d) What is the variance for this random variable?

4.22 Show that for any random variable it is true that $E(X^2) > E^2(X)$ unless the random variable is zero almost always.

4.23 Verify the entries in Table 4.5 for the mean and variances of the probability distributions given there.

4.24 A random variable X has the pdf

$$f_X(x) = A e^{-bx}[u(x) - u(x - B)]$$

where $u(x)$ is the unit step function and A, B, and b are positive constants.

(a) Find the proper relationship between the constants A, b, and B.
(b) Determine and plot the cdf.
(c) Compute $E(X)$.
(d) Determine $E(X^2)$.
(e) What is the variance of X?

4.25 If

$$f_X(x) = (2\pi\sigma^2)^{-1/2} \exp\left(\frac{-x^2}{2\sigma^2}\right)$$

show that

(a) $E\{X^{2n}\} = 1 \cdot 3 \cdot 5 \cdots (2n - 1)\sigma^{2n}, \qquad \text{for } n = 1, 2, \ldots$
(b) $E\{X^{2n-1}\} = 0, \qquad \text{for } n = 1, 2, \ldots$

4.26 The random variable has pdf

$$f_x(x) = \tfrac{1}{2}\delta(x - 4) + \tfrac{1}{8}[u(x - 3) - u(x - 7)]$$

where $u(x)$ is the unit step. Determine the mean and the variance of the random variable thus defined.

4.27 Find the mean of $Z = X + R$, where X has the pdf given in Problem 4.21 and R has the Rayleigh pdf given in Equation (4.92).

4.28 Two random variables X and Y have means and variances given below:

$$m_x = 2 \quad \sigma_x^2 = 3 \quad m_y = 1 \quad \sigma_y^2 = 5$$

A new random variable Z is defined as

$$Z = 3X - 4Y$$

Determine the mean and variance of Z for each of the following cases of correlation between the random variables X and Y:

(a) $\rho_{XY} = 0$: (b) $\rho_{XY} = 0.25$; (c) $\rho_{XY} = 0.6$; (d) $\rho_{XY} = 1.0$.

4.29 Two Gaussian random variables X and Y, with zero mean and variance σ^2, between which there is a correlation coefficient ρ, have a joint probability density given by

$$f(x, y) = \frac{1}{2\pi\sigma^2 \sqrt{1 - \rho^2}} \exp\left[-\frac{x^2 - 2\rho xy + y^2}{2\sigma^2(1 - \rho^2)} \right]$$

Verify that the symbol ρ in the expression for $f(x, y)$ is the correlation coefficient. That is, evalaute $E\{XY\}/\sigma^2$.

4.30 Using the definition of a conditional pdf given by Equation (4.54) and the expressions for the marginal and joint Gaussian pdf's, show that for two jointly Gaussian random variables X and Y, the conditional density function of X given Y has the form of a Gaussian density with conditional mean and the conditional variance given by

$$E(X|Y) = m_x + \frac{\rho\sigma_x}{\sigma_y}(Y - m_y)$$

$$\text{var}(X|Y) = \sigma_x^2(1 - \rho^2)$$

respectively.

4.31 (a) Find the characteristic function corresponding to the pdf of Problem 4.21.
   (b) Find $E\{X\}$ and $E\{X^2\}$ from the characteristic function.
   (c) Do the results of (b) correspond with those of Problem 4.21?

4.32 The independent random variables X and Y have the probability densities

$$f(x) = e^{-x}u(x)$$
$$f(y) = \tfrac{1}{2}e^{-|y|}$$

Find and plot the probability density of the random variable $Z = X + Y$.

4.33   The random variable X has a probability density uniform in the range $-\frac{1}{2} \leq x \leq \frac{1}{2}$ and zero elsewhere. The independent variable Y has a density uniform in the range $1 \leq y \leq 3$ and zero elsewhere. Find the plot the density of $Z = X + Y$.

4.34   A random variable X is defined by

$$f_X(x) = 4e^{-8|x|}$$

The random variable Y is related to X by $Y = 2 + 3X$.

(a) Determine $E\{X\}$, $E\{X^2\}$, and σ_x^2.
(b) Determine $f_Y(y)$.
(c) Determine $E\{Y\}$, $E\{Y^2\}$, and σ_y^2.
(d) If you used $f_Y(y)$ in part (c), repeat that part using only $f_X(x)$.

4.35   A random variable X has the uniform probability density function

$$f_X(x) = \begin{cases} \dfrac{1}{b-a}, & a \leq x \leq b \\ \\ 0, & \text{otherwise} \end{cases}$$

(a) Determine the characteristic function $M_x(jv)$.
(b) Use the characteristic function to compute $E\{X\}$.
(c) Use the characteristic function to compute $E(X^2)$.
(d) What is the variance σ_x^2?

4.36   A random variable X has the probability density function

$$f_X(x) = \begin{cases} ae^{-ax}, & x \geq 0 \\ 0, & x < 0 \end{cases}$$

where a is an arbitrary positive constant.

(a) Determine the characteristic function $M_X(jv)$.
(b) Use the characteristic function to determine $E\{X\}$ and $E\{X^2\}$. Check your results by computing

$$\int_{-\infty}^{\infty} x^n f_X(x)\, dx$$

for $n = 1$ and 2.
(c) Compute σ_x^2.

Section 4.4

4.37 Compare the binomial, Laplace, and Poisson distributions for

(a) $n = 3$ and $p = \frac{1}{5}$,   (b) $n = 3$ and $p = \frac{1}{10}$,
(c) $n = 10$ and $p = \frac{1}{5}$,   (d) $n = 10$ and $p = \frac{1}{10}$

4.38 An honest coin is flipped 20 times. (a) Determine the probability of the occurrence of 5 or 6 heads. (b) Determine the probability of the first head occurring at toss number 15.

4.39 An honest coin is flipped 100 times. Determine the probability of the occurrence of between 50 and 60 heads inclusive.

4.40 Passwords in a computer installation take the form $X_1X_2X_3X_4$, where each character X_i is one of the 26 letters of the alphabet. Determine the maximum possible number of different passwords available for assignment for each of the two following conditions:

(a) A given letter of the alphabet can be used only once in a password.
(b) Letters can be repeated if desired, so that each X_i is completely arbitrary.
(c) If selection of letters for a given password is completely random, what is the probability that your competitor could access, on a single try, your computer in part (a)? part (b)?

4.41 (a) By applying the binomial distribution, find the probability that there will be fewer than 3 heads when 10 honest coins are tossed.
(b) Do the same computation using the Laplace approximation.
(c) Compare the results of parts (a) and (b) by computing the percent error of the Laplace approximation.

4.42 A digital data transmission system has an error probability of 10^{-6} per digit. Find the probability of 3 or more errors in 10^6 digits.

4.43 (a) Using the expression for the joint Gaussian pdf [Equation (4.177)], obtain the marginal pdf's $f_X(x)$ and $f_Y(y)$. Hint: Complete the square in the exponent in x and use the integral

$$\int_{-\infty}^{\infty} \exp\left(-u^2/2\right) du = \sqrt{2\pi}$$

after making an appropriate substitution to obtain $f_X(x)$. Follow a similar procedure for $f_Y(y)$.
(b) From the results of (a), find $E\{X\}$, $E\{Y\}$, var $\{X\}$, and var $\{Y\}$.

4.44   Consider the Cauchy density function

$$f_X(x) = \frac{K}{1 + x^2}, \qquad -\infty \le x \le \infty$$

(a) Find K.
(b) Find $E\{X\}$ and show that var $\{X\}$ is not finite.
(c) Show that the characteristic function of a Cauchy random variable is $M_x(jv) = \pi K e^{-|v|}$.
(d) Now consider $Z = X_1 + \cdots + X_N$ where the X_i's are Cauchy and independent. Thus their characteristic function is

$$M_Z(jv) = (\pi K)^N e^{-N|v|}$$

Show that $f_Z(z)$ is Cauchy, *Comment:* $f_Z(z)$ is not Gaussian as $N \to \infty$ because var $\{X_i\}$ is not finite and the conditions of the central-limit theorem are violated.

4.45   (Chi-squared pdf) Consider the random variable $Y = \sum_{i=1}^{N} X_i^2$ where the X_i's are independent Gaussian random variables with pdf's $n(0, \sigma)$.

(a) Show that the characteristic function of X_i^2 is

$$M_{X_i^2}(jv) = (1 - 2jv\sigma^2)^{-1/2}$$

(b) Show that the pdf of Y is

$$f_Y(y) = \begin{cases} \dfrac{y^{N/2-1} e^{-y/2\sigma^2}}{2^{N/2}\sigma^N \Gamma(N/2)}, & y \ge 0 \\ 0, & y < 0 \end{cases}$$

where $\Gamma(x)$ is the gamma function, which, for $x = n$ an integer, is $\Gamma(n) = (n - 1)!$. This pdf is known as the χ^2 (chi-squared) pdf with N degrees of freedom. *Hint:* Use the Fourier transform pair

$$\frac{y^{N/2-1} e^{-y/\alpha}}{\alpha^{N/2}\Gamma(N/2)} \leftrightarrow (1 - j\alpha v)^{-N/2}$$

(c) Show that for N large, the χ^2 pdf can be approximated as

$$f_Y(y) \cong \exp\left[-\frac{1}{2}\left(\frac{y - N\sigma^2}{\sqrt{4N\sigma^4}}\right)^2\right] \Big/ \sqrt{4N\pi\sigma^4}, \qquad N \gg 1$$

Hint: Use the central-limit theorem. Since the X_i's are independent,

$$\bar{Y} = \sum_{i=1}^{N} \overline{X_i^2} = N\sigma^2$$

and

$$\text{var}(Y) = \sum_{i=1}^{N} \text{var}(X_i^2) = N \text{ var}(X_i^2)$$

(d) Compare the approximation obtained in part (c) with $f_Y(y)$ for $N = 2, 4, 8$.

(e) Let $R^2 = Y$. Show that the pdf of R for $N = 2$ is Rayleigh.

4.46 Compare the Q-function and the approximation to it for large arguments given by (4.187) by plotting both expressions on a log-log graph. (*Note:* A mathematics package, such as Mathcad is handy for this problem.)

4.47 Determine the cdf for a Gaussian random variable of mean m and variance σ^2. Express in terms of the Q-function. Plot the resulting cdf for $m = 0$, and $\sigma = 0.5, 1$, and 2.

4.48 Let X be uniformly distributed over $|x| \leq 1$. Plot $P(|X| \leq k\sigma_x)$ versus k and the corresponding bound given by Chebyshev's inequality.

4.49 A random variable X has the probability density function

$$f_X(x) = \frac{e^{-(x-12)^2/30}}{\sqrt{30\pi}}$$

Express the following probabilities in terms of the Q function:

(a) $P(X \leq 11)$ (b) $P(10 < X \leq 12)$
(c) $P(11 < X \leq 13)$ (d) $P(9 < X \leq 12)$

4.50 Prove Chebyshev's inequality. *Hint:* Let $Y = (X - m_X)/\sigma_X$ and find a bound for $P(|Y| < k)$ in terms of k.

4.51 If the random variable X is Gaussian, with zero mean and variance σ^2, obtain numerical values for the following probabilities:

(a) $P(|X| > \sigma)$ (b) $P(|X| > 2\sigma)$
(c) $P(|X| > 3\sigma)$

4.52   Speech is sometimes idealized as having a Laplacian-amplitude pdf.
       That is, the amplitude is distributed according to

$$f_X(x) = \left(\frac{a}{2}\right) \exp\left(-a|x|\right)$$

Express the variance of X in terms of a and find the three probabilities
found in Problem 4.51. Compare your answers with those you
obtained for Problem 4.51.

COMPUTER EXERCISES

4.1.   (a) In preparation for this computer exercise, prove the following the-
       orem: If X is a continuous random variable with cdf $F_X(x)$, then the
       random variable

$$Y = F_X(X)$$

is a uniformly distributed random variable in the interval $(0,1)$.

(b) Use the theorem proved in part (1) to generate a continuous random
variable with desired pdf (or cdf) given that most programming lan-
guages (including the mathematics packages mentioned in the footnote
to the computer exercises in Chapter 2) have a uniform random number
generator.

(c) Design a random generator to generate a sequence of exponentially
distributed random variables having the pdf

$$f_X(x) = \alpha e^{-\alpha x} u(x)$$

where $u(x)$ is the unit step. Plot histograms of the random numbers
generated to check the validity of the random number generator you
designed.

4.2   (a) Let U and V be two statistically independent random numbers uni-
       formly distributed in $[0, 1]$. Show that the following transformation
       generates two statistically independent Gaussian random numbers with
       unit variance and zero mean:

$$X = R \cos\left(2\pi U\right)$$
$$Y = R \sin\left(2\pi U\right)$$

where

$$R = \sqrt{-2 \ln (V)}$$

(*Hint:* First show that R is Rayleigh.)

(b) Generate 1000 random variable pairs according to the above algorithm. Plot histograms for each set (i.e., X and Y), and compare with Gaussian pdf's after properly scaling the histograms.

4.3. Using the results of Problem 4.30 and the Gaussian random number generator designed in Computer Exercise 4.2, design a Gaussian random number generator that will provide a specified correlation between adjacent samples. Let

$$\rho(\tau) = e^{-\alpha|\tau|}$$

and plot sequences of Gaussian random numbers for various choices of α. Show how stronger correlation between adjacent samples affects the variation from sample to sample. (*Note:* To get memory over more than adjacent samples, a digital filter should be used with independent Gaussian samples at the input.)

4.4. Design a computer program to repeatedly convolve two functions. You might wish to use the FFT to do this efficiently. Use it to verify that the pdf of a sum of independent random variables (with finite first and second moments) approaches a Gaussian pdf, thus giving a plausibility proof of the central-limit theorem.

4.5. Check the validity of the central-limit theorem by repeatedly generating n independent uniformly distributed random variables in the interval $(-0.5, 0.5)$, forming the sum given by (4.175), and plotting the histogram. Do this for $n = 5$, 10, and 20. Can you say anything qualitatively and quantitatively about the approach of the sums to Gaussian random numbers? Repeat for exponentially distributed component random variables (do Computer Exercise 4.1 first). Can you think of any drawbacks to this approach to generating Gaussian random variables?

5 RANDOM SIGNALS AND NOISE

With the mathematical background developed in Chapter 4, we have finally arrived at the point where we can consider the statistical description of random waveforms. The importance of considering such waveforms, as pointed out in Chapter 1, lies in the fact that noise in communication systems is due to the random motion of charge carriers, and we observe macroscopic properties such as the current through a conductor that is due to the superposition of individual microscopic currents.

In the relative-frequency approach to probability, we imagined repeating the underlying chance experiment many times, the implication being that the replication process was carried out sequentially in time. In the study of random waveforms, however, the outcomes of our chance experiments are mapped into functions of time, or waveforms, rather than numbers, as in the case of random variables. The particular waveform is not predictable in advance of the experiment, just as the particular value of a random variable is not predictable before the chance experiment is performed. The problem we now address is the statistical description of chance experiments resulting in waveforms as outputs. To visualize how we may accomplish this, we again think in terms of relative frequency.

5.1 A RELATIVE-FREQUENCY DESCRIPTION OF RANDOM PROCESSES

For simplicity, we consider a binary digital waveform generator whose output randomly switches between $+1$ and -1 in T_0-second intervals, as shown in Figure 5.1. Let $X(t, \zeta_i)$ be the random waveform corresponding to the output of the ith generator. Suppose we use relative frequency to estimate $P(X = +1)$ by examining the outputs of all generators at a particular time. Since the outputs are functions of time, we must specify the time when

FIGURE 5.1   A statistically identical set of binary waveform generators with typical outputs

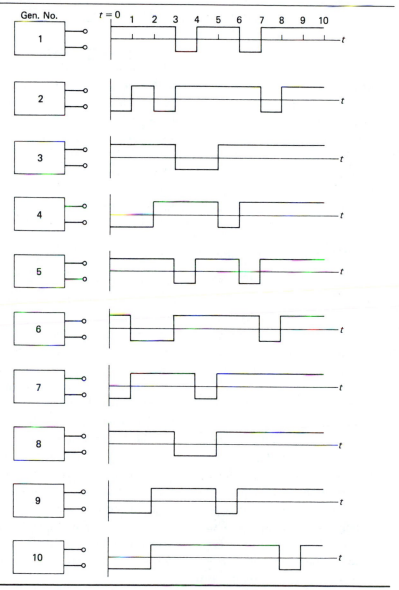

writing down the relative frequency. From an examination of the generator outputs we construct the following table:

Time Interval:	(0,1)	(1,2)	(2,3)	(3,4)	(4,5)	(5,6)	(6,7)	(7,8)	(8,9)	(9,10)
Relative Frequency:	$\frac{5}{10}$	$\frac{6}{10}$	$\frac{8}{10}$	$\frac{6}{10}$	$\frac{7}{10}$	$\frac{8}{10}$	$\frac{8}{10}$	$\frac{8}{10}$	$\frac{9}{10}$	$\frac{10}{10}$

From this table it is seen that the relative frequencies change with the time interval. Although this variation in relative frequency could be the result of *statistical irregularity,* we highly suspect that some phenomenon is making $X = +1$ more probable as time increases. To reduce the possibility that statistical irregularity is the culprit, we might repeat the experiment with 100 generators or 1000 generators. This is obviously a mental experiment in that it would be very difficult to obtain a set of identical generators and prepare them all in identical fashions. For some random processes, known as *ergodic,* it is possible to replace measurements on a set of generators at the same time with measurements across time on a single generator output.

5.2   SOME TERMINOLOGY OF RANDOM PROCESSES

Sample Functions and Ensembles

In the same fashion as is illustrated in Figure 5.1, we could imagine performing any chance experiment many times simultaneously. If, for example, the random quantity of interest is the voltage at the terminals of a noise generator, the random variable X_1 may be assigned to represent the possible values of this voltage at time t_1 and the random variable X_2 the values at time t_2. As in the case of the digital waveform generator, we can imagine many noise generators all constructed in an identical fashion, insofar as we can make them, and run under identical conditions. Figure 5.2(a) shows typical waveforms generated in such an experiment. Each waveform $X(t, \zeta_i)$, is referred to as a *sample function,* where ζ_i is a member of a sample space $\mathcal{S}$. The totality of all sample functions is called an *ensemble.* The underlying chance experiment that gives rise to the ensemble of sample functions is called a *random,* or *stochastic, process.* Thus, to every outcome ζ we assign, according to a certain rule, a time function $X(t, \zeta)$. For a specific ζ, say ζ_i, $X(t, \zeta_i)$ signifies a single time function. For a specific time t_j, $X(t_j, \zeta)$ denotes a random variable. For fixed $t = t_j$ and fixed $\zeta = \zeta_i$, $X(t_j, \zeta_i)$ is a *number.* In what follows, we often suppress the ζ.

To summarize, the difference between a random variable and a random process is that for a random variable an outcome in the sample space is

FIGURE 5.2   Typical sample functions of a random process and illustration of the relative-frequency interpretation of its joint pdf. (a) Ensemble of sample functions. (b) Superposition of the sample functions shown in (a).

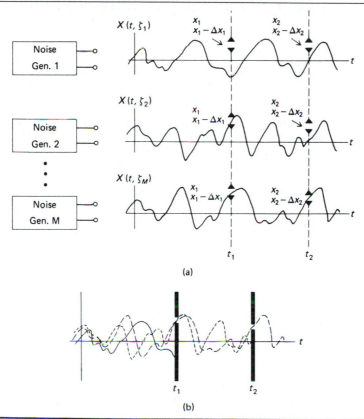

(a)

(b)

mapped into a number, whereas for a random process it is mapped into a function of time.

Description of Random Processes in Terms of Joint pdf's

A complete description of a random process $\{X(t, \zeta)\}$ is given by the N-fold joint pdf that probabilistically describes the possible values assumed by a typical sample function at times $t_N > t_{N-1} > \cdots > t_1$, where N is arbitrary. For $N = 1$, we can interpret this joint pdf $f_{X_1}(x_1, t_1)$ as

$$f_{X_1}(x_1, t_1)dx_1 = P(x_1 - dx_1 < X_1 \leq x_1 \text{ at time } t_1) \tag{5.1}$$

where $X_1 = X(t_1, \zeta)$. Similarly, for $N = 2$, we can interpret the joint pdf $f_{X_1 X_2}(x_1, t_1; x_2, t_2)$ as

$$f_{X_1 X_2}(x_1, t_1; x_2, t_2)dx_1 dx_2 = P(x_1 - dx_1 < X_1 \leq x_1 \text{ at time } t_1$$
$$\text{and } x_2 - dx_2 < X_2 \leq x_2 \text{ at time } t_2) \quad (5.2)$$

where $X_2 = X(t_2, \zeta)$.

To help visualize the interpretation of (5.2), Figure 5.2(b) shows the three sample functions of Figure 5.2(a) superimposed with barriers placed at $t = t_1$ and $t = t_2$. According to the relative-frequency interpretation, the joint probability given by (5.2) is the number of sample functions that pass through the slits in both barriers divided by the total number M of sample functions as M becomes large without bound.

Stationarity

We have indicated the possible dependence of $f_{X_1 X_2}$ on t_1 and t_2 by including them in its argument. If $\{X(t)\}$ were a Gaussian random process, for example, its values at time t_1 and t_2 would be described by Equation (4.177) where m_X, m_Y, σ_X^2, σ_Y^2, and ρ would, in general, depend on t_1 and t_2.* Note that we need a general N-fold pdf to completely describe the random process $\{X(t)\}$. In general, such a pdf depends on N time instants $t_1, t_2, \ldots, t_N$. In some cases, these joint pdf's depend only on the time differences $t_2 - t_1$, $t_3 - t_1, \ldots, t_N - t_1$; that is, the choice of time origin for the random process is immaterial. Such random processes are said to be *statistically stationary in the strict sense*, or simply *stationary*.

For stationary processes, means and variances are independent of time, and the correlation coefficient (or covariance) depends only on the time difference $t_2 - t_1$.† Figure 5.3 contrasts sample functions of stationary and nonstationary processes. It may happen that in some cases the mean and variance of a random process are time-independent and the covariance is a function only of the time difference, but the N-fold joint pdf depends on the time origin. Such random processes are called *wide-sense stationary processes* to distinguish them from strictly stationary processes (that is, processes whose N-fold pdf is independent of time origin). Strict-sense stationarity implies wide-sense stationarity, but the reverse is not necessarily true. An exception occurs for *Gaussian random processes for which wide-sense stationarity does imply strict-sense stationarity*, since the joint Gaussian pdf is completely

---

* For a stationary process, all joint moments are independent of time origin. We are interested primarily in the covariance, however.

† At N instants of time its values would be described by (B.1) of Appendix B.

FIGURE 5.3   Sample functions of nonstationary processes contrasted with a sample function of a stationary process. (a) Time-varying mean. (b) Time-varying variance. (c) Stationary.

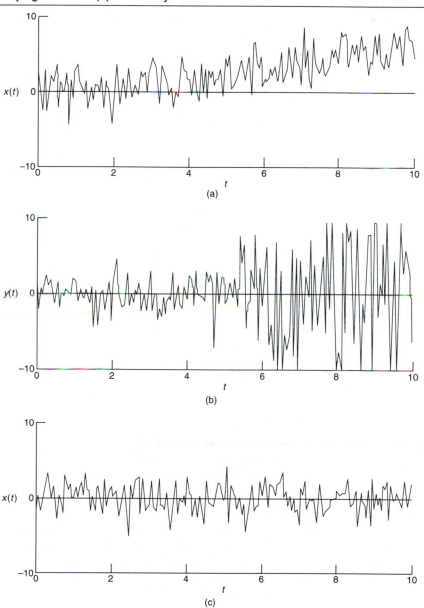

specified in terms of the means, variances, and covariances of $X(t_1)$, $X(t_2)$, $\ldots$, $X(t_N)$.

Partial Description of Random Processes: Ergodicity

As in the case of random variables, we may not always require a complete statistical description of a random process, or we may not be able to obtain the N-fold joint pdf even if desired. In such cases, we work with various moments, either by choice or by necessity. The most important averages are the mean,

$$m_X(t) = E\{X(t)\} = \overline{X(t)} \tag{5.3}$$

the variance,

$$\sigma_X^2(t) = E\{[X(t) - \overline{X(t)}]^2\} = \overline{X^2(t)} - [\overline{X(t)}]^2 \tag{5.4}$$

and the covariance,

$$\mu_X(t, t + \tau) = E\{[X(t) - \overline{X(t)}][X(t + \tau) - \overline{X(t + \tau)}]\} \tag{5.5}$$
$$= E\{X(t)X(t + \tau)\} - \overline{X(t)}\ \overline{X(t + \tau)}$$

In (5.5), we let $t = t_1$ and $t + \tau = t_2$. The first term on the right-hand side is the *autocorrelation function* computed as a *statistical*, or *ensemble*, *average* (that is, the average is across the sample functions at times t and $t + \tau$). In terms of the joint pdf of the random process, the autocorrelation function is

$$R_X(t_1, t_2) = \int_{-\infty}^{\infty} \int_{-\infty}^{\infty} x_1 x_2 f_{X_1 X_2}(x_1, t_1; x_2, t_2)\ dx_1\ dx_2 \tag{5.6}$$

where $X_1 = X(t_1)$ and $X_2 = X(t_2)$. If the process is stationary, $f_{X_1 X_2}$ does not depend on t but rather on the time difference, $\tau = t_2 - t_1$ and as a result, $R_X(t_1, t_2) = R_X(\tau)$ is a function only of τ. The following question arises: "If we compute the autocorrelation function using the definition of a time average given in Chapter 2, Equation (2.149), will the result be the same as the statistical average given by (5.6)?" For many processes, referred to as *ergodic*, the answer is affirmative. Ergodic processes are processes for which *time and ensemble averages are interchangeable*. Thus, if $X(t)$ is an ergodic process, *all time and corresponding ensemble averages are interchangeable*. In particular,

$$m_X = E\{X(t)\} = \langle X(t) \rangle$$

$$\sigma_X^2 = E\{[X(t) - \overline{X(t)}]^2\} = \langle [X(t) - \langle X(t) \rangle]^2 \rangle$$

and

$$R_X(\tau) = E\{X(t)X(t + \tau)\} = \langle X(t)X(t + \tau)\rangle$$

where

$$\langle v(t)\rangle \triangleq \lim_{T\to\infty} \frac{1}{2T} \int_{-T}^{T} v(t)\, dt$$

as defined in Chapter 2. We emphasize that for ergodic processes *all* time and ensemble averages are interchangeable, not just the mean, variance, and autocorrelation function.

▼ **EXAMPLE 5.1**   Consider the random process with sample functions*

$$n(t) = A \cos(\omega_0 t + \Theta)$$

where ω_0 is a constant and Θ is a random variable with the pdf

$$f_\Theta(\theta) = \begin{cases} \dfrac{1}{2\pi}, & |\theta| \le \pi \\[2mm] 0, & \text{otherwise} \end{cases} \tag{5.7}$$

Computed as statistical averages, the first and second moments are

$$\overline{n(t)} = \int_{-\infty}^{\infty} A\cos(\omega_0 t + \theta)\, f_\Theta(\theta)\, d\theta = \int_{-\pi}^{\pi} A\cos(\omega_0 t + \theta)\, \frac{d\theta}{2\pi} = 0 \tag{5.8a}$$

and

$$\overline{n^2(t)} = \int_{-\pi}^{\pi} A^2 \cos^2(\omega_0 t + \theta)\, \frac{d\theta}{2\pi} = \frac{A^2}{2} \tag{5.8b}$$

respectively. The variance is equal to the second moment, since the mean is zero.

Computed as time averages, the first and second moments are

$$\langle n(t)\rangle = \lim_{T\to\infty} \frac{1}{2T} \int_{-T}^{T} A\cos(\omega_0 t + \theta)\, dt = 0 \tag{5.9a}$$

---

* In this example we violate our earlier established convention that sample functions are denoted by capital letters. This is quite often done if confusion will not result.

and

$$\langle n^2(t) \rangle = \lim_{T \to \infty} \frac{1}{2T} \int_{-T}^{T} A^2 \cos^2(\omega_0 t + \theta) \, dt = \frac{A^2}{2} \tag{5.9b}$$

respectively. In general, the time average of some function of an ensemble member of a random process is a random variable. In this example, $\langle n(t) \rangle$ and $\langle n^2(t) \rangle$ are constants! We suspect that this random process is stationary and ergodic, even though the preceding results do not *prove* this. It turns out that this is indeed true.

To continue the example, consider the pdf

$$f_\Theta(\theta) = \begin{cases} \dfrac{2}{\pi}, & |\theta| \leq \tfrac{1}{4}\pi \\[2mm] 0, & \text{otherwise} \end{cases} \tag{5.10}$$

For this case, the expected value, or mean, of the random process computed at an arbitrary time t is

$$\overline{n(t)} = \int_{-\pi/4}^{\pi/4} A \cos(\omega_0 t + \theta) \frac{2}{\pi} \, d\theta$$

$$= \frac{2}{\pi} A \sin(\omega_0 t + \theta) \Big|_{-\pi/4}^{\pi/4} = \frac{2\sqrt{2}A}{\pi} \cos \omega_0 t \tag{5.11a}$$

The second moment, computed as a statistical average, is

$$\overline{n^2(t)} = \int_{-\pi/4}^{\pi/4} A^2 \cos^2(\omega_0 t + \theta) \frac{2}{\pi} \, d\theta$$

$$= \int_{-\pi/4}^{\pi/4} \frac{A^2}{\pi} [1 + \cos(2\omega_0 t + 2\theta)] \, d\theta = \frac{A^2}{2} + \frac{A^2}{\pi} \cos 2\omega_0 t \tag{5.11b}$$

Since stationarity of a random process implies that all moments are independent of time origin, the preceding results show that this process is not stationary. In order to comprehend the physical reason for this, you should sketch some typical sample functions. In addition, this process cannot be ergodic, since ergodicity requires stationarity. Indeed, the time-average first and second moments are still $\langle n(t) \rangle = 0$ and $\langle n^2(t) \rangle = \frac{1}{2}A^2$, respectively. Thus we have exhibited two time averages that are not equal to the corresponding statistical averages.

Meanings of Various Averages for Ergodic Processes

It is useful to pause at this point and summarize the meanings of various averages for an ergodic process:

1. The mean $\overline{X(t)} = \langle X(t) \rangle$ is the dc component.
2. $\overline{X(t)}^2 = \langle X(t) \rangle^2$ is the dc power.
3. $\overline{X^2(t)} = \langle X^2(t) \rangle$ is the total power.
4. $\sigma_X^2 = \overline{X^2(t)} - \overline{X(t)}^2 = \langle X^2(t) \rangle - \langle X(t) \rangle^2$ is the power in the ac (time-varying) component.
5. The total power $\overline{X^2(t)} = \sigma_X^2 + \overline{X(t)}^2$ is the ac power plus the dc power.

Thus, in the case of ergodic processes, we see that these moments are measurable quantities in the sense that they can be replaced by the corresponding time averages and a finite-time approximation to these time averages can be measured in the laboratory.

▼ **EXAMPLE 5.2**   To illustrate some of the definitions given above with regard to correlation functions, let us consider a random telegraph waveform $X(t)$, as illustrated in Figure 5.4. The sample functions of this random process have the following properties:

1. The values taken on at any time instant t_0 are either $X(t_0) = A$ or $X(t_0) = -A$ with equal probability.
2. The number k of switching instants in any time interval T obeys a Poisson distribution, as defined by Equation (4.170), with the attendant assumptions leading to this distribution. (That is, the probability of more than one switching instant occurring in an infinitesimal time interval dt is zero, with the probability of exactly one switching instant occurring in dt being $\alpha \, dt$, where α is a constant. Furthermore, successive switching occurrences are independent.)

FIGURE 5.4   Sample function of a random telegraph waveform

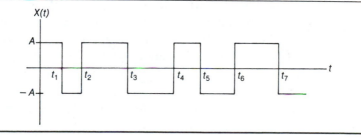

If τ is any positive time interval, the autocorrelation function of the random process defined by the preceding properties can be calculated as

$$R_X(\tau) = E[X(t)X(t + \tau)]$$

$$= A^2 P\{X(t) \text{ and } X(t + \tau) \text{ have the same sign}\}$$
$$+ (-A^2)P\{X(t) \text{ and } X(t + \tau) \text{ have different signs}\}$$

$$= A^2 P\{\text{even number of switching instants in } (t, t + \tau)\}$$
$$- A^2 P\{\text{odd number of switching instants in } (t, t + \tau)\}$$

$$= A^2 \sum_{\substack{k=0 \\ k \text{ even}}}^{\infty} \frac{(\alpha\tau)^k}{k!} \exp(-\alpha\tau) - A^2 \sum_{\substack{k=0 \\ k \text{ odd}}}^{\infty} \frac{(\alpha\tau)^k}{k!} \exp(-\alpha\tau)$$

$$= A^2 \exp(-\alpha\tau) \sum_{k=0}^{\infty} \frac{(-\alpha\tau)^k}{k!}$$

$$= A^2 \exp(-\alpha\tau) \exp(-\alpha\tau) = A^2 \exp(-2\alpha\tau) \tag{5.12}$$

The preceding development was carried out under the assumption that τ was positive. It could have been similarly carried out with τ negative, such that

$$R_X(\tau) = E\{X(t)X(t - |\tau|)\} = E\{X(t - |\tau|)X(t)\} = A^2 \exp(-2\alpha|\tau|) \tag{5.13}$$

This is a result that holds for all τ. That is, $R_X(\tau)$ is an even function of τ, which we will show in general shortly.

5.3 CORRELATION FUNCTIONS AND POWER SPECTRA

The autocorrelation function, computed as a statistical average, has been defined by (5.6). If a process is ergodic, the autocorrelation function computed as a time average, as in Equation (2.149) of Chapter 2, is equal to the statistical average of (5.6). In Chapter 2, we defined the power spectral density $S(f)$ as the Fourier transform of the autocorrelation function $R(\tau)$. The *Weiner-Khinchine theorem* is a formal statement of this result for stationary random processes, for which $R(t_1, t_2) = R(t_2 - t_1) = R(\tau)$. For such processes, previously defined as wide-sense stationary, the power spectral density and autocorrelation function are Fourier transform pairs:

$$S(f) \leftrightarrow R(\tau) \tag{5.14}$$

If the process is ergodic, $R(\tau)$ can be calculated as either a time or an ensemble average.

Since $R_X(0) = \overline{X^2(t)}$ is the average power contained in the process, we have from the inverse Fourier transform that

$$\text{Average power} = R_X(0) = \int_{-\infty}^{\infty} S_X(f)\, df \qquad (5.15)$$

which is reasonable, since the definition of $S_X(f)$ is that it is *power density* with respect to frequency.

Power Spectral Density

An intuitively satisfying, and in some cases computationally useful, expression for the power spectral density of a random process can be obtained by the following approach. Consider a particular sample function, $n(t, \zeta_i)$, of a random process. To obtain a function giving power density versus frequency using the Fourier transform, we consider a truncated version, $n_T(t, \zeta_i)$, defined as*

$$n_T(t, \zeta_i) = \begin{cases} n(t, \zeta_i), & |t| < \tfrac{1}{2}T \\ 0, & \text{otherwise} \end{cases} \qquad (5.16)$$

Since sample functions of stationary random processes are power signals, the Fourier transform of $n(t, \zeta_i)$ does not exist, which necessitates defining $n_T(t, \zeta_i)$. The Fourier transform of a truncated sample function is

$$N_T(f, \zeta_i) = \int_{-T/2}^{T/2} n(t, \zeta_i) e^{-j2\pi ft}\, dt \qquad (5.17)$$

and its energy spectral density, according to Equation (2.83), is $|N_T(f, \zeta_i)|^2$. The time-average power density over the interval $[-\tfrac{1}{2}T, \tfrac{1}{2}T]$ for this sample function is $|N_T(f, \zeta_i)|^2/T$. Since this time-average power density depends on the particular sample function chosen, we perform an ensemble average and take the limit as $T \to \infty$ to obtain the distribution of power density with frequency. This is defined as the power spectral density $S_n(f)$:

$$S_n(f) = \lim_{T \to \infty} \frac{\overline{|N_T(f, \zeta_i)|^2}}{T} \qquad (5.18)$$

The operations of taking the limit and taking the ensemble average in (5.18) cannot be interchanged.

---

* Again, we use a lower-case letter to denote a random process for the simple reason that we need to denote the Fourier transform of $n(t)$ by an upper-case letter.

▼ **EXAMPLE 5.3**   Let us find the power spectral density of the random process considered in Example 5.1 using (5.18). In this case,

$$n_T(t, \Theta) = A\Pi\left(\frac{t}{T}\right) \cos\left[\omega_0\left(t + \frac{\Theta}{\omega_0}\right)\right] \tag{5.19}$$

By the time-delay theorem of Fourier transforms and using the transform pair

$$\cos \omega_0 t \leftrightarrow \tfrac{1}{2}\delta(f - f_0) + \tfrac{1}{2}\delta(f + f_0) \tag{5.20}$$

we obtain

$$\mathscr{F}[\cos(\omega_0 t + \Theta)] = \tfrac{1}{2}\delta(f - f_0)e^{j\Theta} + \tfrac{1}{2}\delta(f + f_0)e^{-j\Theta} \tag{5.21}$$

We also recall from Chapter 2 (Example 2.8) that $\Pi(t/T) \leftrightarrow T \text{ sinc } Tf$, so, by the multiplication theorem of Fourier transforms,

$$N_T(f, \Theta) = (AT \text{ sinc } Tf) * [\tfrac{1}{2}\delta(f - f_0)e^{j\Theta} + \tfrac{1}{2}\delta(f + f_0)e^{-j\Theta}]$$
$$= \tfrac{1}{2}AT[e^{j\Theta} \text{ sinc } (f - f_0)T + e^{-j\Theta} \text{ sinc } (f + f_0)T] \tag{5.22}$$

Therefore, the energy spectral density of the sample function is

$$|N_T(f, \Theta)|^2 = (\tfrac{1}{2}AT)^2[\text{sinc}^2 \ T(f - f_0) + e^{j2\Theta} \text{ sinc } T(f - f_0) \text{ sinc } T(f + f_0)$$
$$+ e^{-j2\Theta} \text{ sinc } T(f - f_0) \text{ sinc } T(f + f_0) + \text{sinc}^2 \ T(f + f_0)] \tag{5.23}$$

In obtaining $\overline{[|N_T(f, \Theta)|^2]}$, we note that

$$\overline{\exp(\pm j2\Theta)} = \int_{-\pi}^{\pi} e^{\pm j2\Theta} \frac{d\theta}{2\pi} = \int_{-\pi}^{\pi} (\cos 2\theta \pm j \sin 2\theta) \frac{d\theta}{2\pi} = 0 \tag{5.24}$$

Thus we obtain

$$\overline{|N_T(f, \Theta)|^2} = (\tfrac{1}{2}AT)^2[\text{sinc}^2 \ T(f - f_0) + \text{sinc}^2 \ T(f + f_0)] \tag{5.25}$$

and the power spectral density is

$$S_n(f) = \lim_{T\to\infty} \tfrac{1}{4}A^2[T \text{ sinc}^2 \ T(f - f_0) + T \text{ sinc}^2 \ T(f + f_0)] \tag{5.26}$$

However, a representation of the delta function is $\lim_{T\to\infty} T \text{ sinc}^2 \ Tu = \delta(u)$. See Figure 2.4(b). Thus

$$S_n(f) = \tfrac{1}{4}A^2\delta(f - f_0) + \tfrac{1}{4}A^2\delta(f + f_0) \tag{5.27}$$

The average power is $\int_{-\infty}^{\infty} S_n(f) \, df = \tfrac{1}{4}A^2$, the same as that obtained in ▲ Example 5.1.

Wiener-Khinchine Theorem

To simplify the notation in the proof of the Wiener-Khinchine theorem, we rewrite (5.18) as

$$S_n(f) = \lim_{T \to \infty} \frac{E\{|\mathcal{F}[n_{2T}(t)]|^2\}}{2T} \tag{5.28}$$

where, for convenience, we have truncated over a $2T$-second interval and dropped ζ in the argument of $n_{2T}(t)$. Note that

$$|\mathcal{F}[n_{2T}(t)]|^2 = \left| \int_{-T}^{T} n(t)e^{-j\omega t}\, dt \right|^2, \qquad \omega = 2\pi f$$

$$= \int_{-T}^{T} \int_{-T}^{T} n(t)n(\sigma)e^{-j\omega(t-\sigma)}\, dt\, d\sigma \tag{5.29}$$

where the product of two integrals has been written as an iterated integral. Taking the ensemble average and interchanging the orders of averaging and integration, we obtain

$$E\{|\mathcal{F}[n_{2T}(t)]|^2\} = \int_{-T}^{T} \int_{-T}^{T} E\{n(t)n(\sigma)\}e^{-j\omega(t-\sigma)}\, dt\, d\sigma$$

$$= \int_{-T}^{T} \int_{-T}^{T} R_n(t - \sigma)e^{-j\omega(t-\sigma)}\, dt\, d\sigma \tag{5.30}$$

by the definition of the autocorrelation function. The change of variables $u = t - \sigma$ and $v = t$ is now made with the aid of Figure 5.5. In the uv plane we integrate over v first and then over u by breaking the integration

FIGURE 5.5   Regions of integration for Equation (5.30)

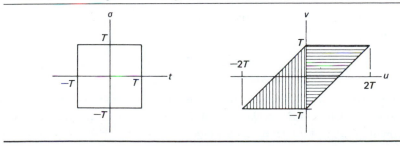

over u up into two integrals, one for u negative and one for u postiive. Thus

$$E\{|\mathscr{F}[n_{2T}(t)]|^2\}$$

$$= \int_{u=-2T}^0 R_n(u)e^{-j\omega u}\left(\int_{-T}^{u+T}dv\right)du + \int_{u=0}^{2T}R_n(u)e^{-j\omega u}\left(\int_{u-T}^T dv\right)du$$

$$= \int_{-2T}^0 (2T+u)R_n(u)e^{-j\omega u}\,du + \int_0^{2T}(2T-u)R_n(u)e^{-j\omega u}\,du$$

$$= 2T\int_{-2T}^{2T}\left(1 - \frac{|u|}{2T}\right)R_n(u)e^{-j\omega u}\,du \qquad (5.31)$$

The power spectral density is, by (5.28),

$$S_n(f) = \lim_{T\to\infty}\int_{-2T}^{2T}\left(1 - \frac{|u|}{2T}\right)R_n(u)e^{-j\omega u}\,du \qquad (5.32)$$

which is the limit as $T \to \infty$ results in (5.14).

▼ **EXAMPLE 5.4**   Since the power spectral density and the autocorrelation function are transform pairs, the autocorrelation function of the random process defined in Example 5.1 is, from the result of Example 5.3, given by

$$R_n(\tau) = \mathscr{F}^{-1}[\tfrac14 A^2\delta(f-f_0) + \tfrac14 A^2\delta(f+f_0)] = \tfrac12 A^2\cos\omega_0\tau \qquad (5.33)$$

Computing $R_n(\tau)$ as an ensemble average, we obtain

$$R_n(\tau) = E\{n(t)n(t+\tau)\} = \int_{-\pi}^\pi A^2\cos(\omega_0 t + \theta)\cos[\omega_0(t+\tau)+\theta]\frac{d\theta}{2\pi}$$

$$= \frac{A^2}{4\pi}\int_{-\pi}^\pi\{\cos\omega_0\tau + \cos[\omega_0(2t+\tau)+2\theta]\}\,d\theta = \tfrac12 A^2\cos\omega_0\tau$$

$$(5.34)$$

which is the same result as that obtained using the Wiener-Khinchine
▲ theorem.

Properties of the Autocorrelation Function

The properties of the autocorrelation function for a random process $X(t)$ were stated in Chapter 2, at the end of Section 2.6, and all time averages may now be replaced by statistical averages. These properties are now easily proved.

Property 1 states that $|R(\tau)| \leq R(0)$ for all τ. To show this, consider the nonnegative quantity

$$[X(t) \pm X(t + \tau)]^2 \geq 0 \qquad (5.35)$$

where $\{X(t)\}$ is a stationary random process. Squaring and averaging term by term, we obtain

$$\overline{X^2(t)} \pm \overline{2X(t)X(t + \tau)} + \overline{X^2(t + \tau)} \geq 0 \qquad (5.36a)$$

which reduces to

$$2R(0) \pm 2R(\tau) \geq 0 \qquad \text{or} \qquad -R(0) \leq R(\tau) \leq R(0) \qquad (5.36b)$$

because $\overline{X^2(t)} = \overline{X^2(t + \tau)} = R(0)$ by the stationarity of $\{X(t)\}$.

Property 2 states that $R(-\tau) = R(\tau)$. This is easily proved by noting that

$$R(\tau) \triangleq \overline{X(t)X(t + \tau)} = \overline{X(t' - \tau)X(t')} \triangleq R(-\tau) \qquad (5.37)$$

where the change of variables $t' = t + \tau$ has been made.

Property 3 states that $\lim_{|\tau| \to \infty} R(\tau) = \overline{X(t)}^2$ if $\{X(t)\}$ does not contain a periodic component. To show this, we note that

$$\begin{aligned} \lim_{|\tau| \to \infty} R(\tau) &\triangleq \lim_{|\tau| \to \infty} \overline{X(t)X(t + \tau)} \\ &\cong \overline{X(t)} \; \overline{X(t + \tau)}, \qquad \text{where } |\tau| \text{ is large} \\ &= \overline{X(t)}^2 \end{aligned} \qquad (5.38)$$

where the second step follows intuitively because the interdependence between $X(t)$ and $X(t + \tau)$ becomes smaller as $|\tau| \to \infty$ (if no periodic components are present), and the last step results from the stationarity of $\{X(t)\}$.

Property 4, which states that $R(\tau)$ is periodic if $\{X(t)\}$ is periodic, follows by noting from the time-average definition of the autocorrelation function given by Equation (2.149) that periodicity of the integrand implies periodicity of $R(\tau)$.

Finally, property 5, which says that $\mathcal{F}[R(\tau)]$ is nonnegative, is a direct consequence of the Wiener–Khinchine theorem and (5.18).

▼ **EXAMPLE 5.5**   Processes for which

$$S(f) = \begin{cases} \frac{1}{2}N_0, & |f| \leq B \\ 0, & \text{otherwise} \end{cases} \qquad (5.39)$$

are commonly referred to as *bandlimited white noise*, since, as $B \to \infty$, all frequencies are present, in which case the process is simply called *white*. N_0

is the single-sided power spectral density of the nonbandlimited process. For a bandlimited white-noise process,

$$R(\tau) = BN_0 \text{ sinc } 2B\tau \qquad (5.40)$$

As $B \to \infty$, $R(\tau) \to \frac{1}{2}N_0\delta(\tau)$. That is, no matter how close together we sample a white-noise process, the samples have zero correlation. If, in addition, the process is Gaussian, the samples are independent. A white-noise process has infinite power and is therefore a mathematical idealization, but it is nevertheless useful in systems analysis.

▼ EXAMPLE 5.6   As another example of calculating power spectra, let us consider a random process with sample functions that can be expressed as

$$X(t) = \sum_{k=-\infty}^{\infty} a_k p(t - kT - \Delta) \qquad (5.41)$$

where $\ldots a_{-1}, a_0, a_1, \ldots, a_k \ldots$ is a sequence of random variables with

$$E[a_k] = 0 \qquad \text{and} \qquad E[a_k a_{k+m}] = R_m \qquad (5.42)$$

The function $p(t)$ is a deterministic pulse-type waveform where T is the separation between pulses; Δ is a random variable that is independent of the value of a_k and uniformly distributed in the interval $(-T/2, T/2)$. The average value of this waveform is

$$E[X(t)] = \sum_{k=-\infty}^{\infty} E(a_k)E[p(t - kT - \Delta)] = 0 \qquad (5.43)$$

The autocorrelation function is

$$
\begin{aligned}
R_X(\tau) \\
= E[X(t)X(t + \tau)] \\
= E\left\{ \sum_{k=-\infty}^{\infty} \sum_{m=-\infty}^{\infty} a_k a_{k+m} p(t - kT - \Delta)\, p[t + \tau - (k + m)T - \Delta] \right\} \\
= \sum_{k=-\infty}^{\infty} \sum_{m=-\infty}^{\infty} E[a_k a_{k+m}]E\{p(t - kT - \Delta)\, p[t + \tau - (k + m)T - \Delta]\} \\
= \sum_{m=-\infty}^{\infty} R_m \sum_{k=-\infty}^{\infty} \int_{-T/2}^{T/2} p(t - kT - \Delta)\, p[t + \tau - (k + m)T - \Delta]\, \frac{d\Delta}{T} \\
= \sum_{m=-\infty}^{\infty} R_m \sum_{k=-\infty}^{\infty} \int_{t-(k+1/2)T}^{t-(k-1/2)T} p(u)(u + \tau - mT)\, \frac{du}{T}
\end{aligned}
$$

$$= \sum_{m=-\infty}^{\infty} R_m \left(\frac{1}{T}\right) \int_{-\infty}^{\infty} p(t + r - mT)p(t) \, dt$$

$$= \sum_{m=-\infty}^{\infty} R_m r(\tau - mT) \tag{5.44}$$

where

$$r(\tau) = \frac{1}{T} \int_{-\infty}^{\infty} p(t + \tau)p(t) \, dt \tag{5.45}$$

is the pulse-correlation function. We consider two cases as illustrations.

Case 1

Let $a_k = \pm A$, for all k, with $R_m = A^2$, $m = 0$, and $R_m = 0$, $m \neq 0$. Also, let $p(t) = \Pi(t/T)$, a unit pulse. This waveform is referred to as a *binary random waveform* and is typical of a baseband digital binary-modulated signal. It is useful in the computation of power spectra for some forms of modulation of binary digital data. The pulse-correlation function is

$$r(\tau) = \frac{1}{T} \int_{-\infty}^{\infty} \Pi\left(\frac{t + \tau}{T}\right) \Pi\left(\frac{t}{T}\right) dt$$

$$= \frac{1}{T} \int_{-T/2}^{T/2} \Pi\left(\frac{t + \tau}{T}\right) dt = \Lambda\left(\frac{\tau}{T}\right) \tag{5.46}$$

The autocorrelation function is

$$R_X(\tau) = A^2 \Lambda\left(\frac{\tau}{T}\right) \tag{5.47}$$

Using the Wiener-Khinchine theorem, we obtain the power spectral density:

$$S_X(f) = A^2 T \, \text{sinc}^2(fT) \tag{5.48}$$

Case 2

Memory is built into the sequence of a_k's by the relationship

$$a_k = g_0 A_k + g_1 A_{k-1} \tag{5.49}$$

where g_0 and g_1 are constants and the A_k's are random variables such that $A_k = \pm A$ and

$$E[A_k A_{k+m}] = \begin{cases} A^2, & m = 0 \\ 0, & m \neq 0 \end{cases} \tag{5.50}$$

Again it is assumed that $p(t)$ is a unit-pulse function, although other pulse shapes could be assumed. It can be shown that

$$E[a_k a_{k+m}] = \begin{cases} A^2(g_0{}^2 + g_1{}^2), & m = 0 \\ g_0 g_1 A^2, & m = \pm 1 \\ 0, & \text{otherwise} \end{cases} \tag{5.51}$$

As before, the pulse-correlation function is

$$r(\tau) = \Lambda\left(\frac{\tau}{T}\right) \tag{5.52}$$

The autocorrelation function is

$$R_X(\tau) = A^2\left\{[g_0{}^2 + g_1{}^2]\Lambda\left(\frac{\tau}{T}\right) + g_0 g_1\left[\Lambda\left(\frac{\tau + T}{T}\right) + \Lambda\left(\frac{\tau - T}{T}\right)\right]\right\} \tag{5.53}$$

The power spectral density of $x(t)$ is

$$S_X(f) = A^2 T \operatorname{sinc}^2(fT)\{g_0{}^2 + g_1{}^2 + g_0 g_1[\exp(j2\pi fT) + \exp(-j2\pi fT)]\} \tag{5.54}$$

Let both g_0 and g_1 be unity. The resulting power spectral density is

$$S_X(f) = 4A^2 T \operatorname{sinc}^2(fT) \cos^2(\pi fT) \tag{5.55}$$

Figure 5.6 compares the power spectra for the two cases considered here. Note the effect of memory. Other values for g_0 and g_1 can be assumed, and memory between more than adjacent pulses also can be assumed.

Cross-Correlation Function and Cross-Power Spectral Density

Suppose we wish to find the power in the sum of two noise voltages $X(t)$ and $Y(t)$. We might ask if we can simply add their separate powers. The answer is, in general, no. To see why, consider

$$n(t) = X(t) + Y(t) \tag{5.56}$$

where $X(t)$ and $Y(t)$ are two stationary random voltages that may be related (that is, that are not necessarily statistically independent). The power in the sum is

$$\begin{aligned} E\{n^2(t)\} &= E\{[X(t) + Y(t)]^2\} \\ &= E\{X^2(t)\} + 2E\{X(t)Y(t)\} + E\{Y^2(t)\} \\ &= P_X + 2P_{XY} + P_Y \end{aligned} \tag{5.57}$$

FIGURE 5.6   Power spectra of binary-valued waveforms. (a) Case in which there is no memory. (b) Case in which there is memory between adjacent pulses.

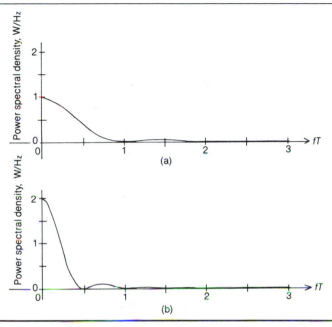

where P_X and P_Y are the powers of $X(t)$ and $Y(t)$, respectively, and P_{XY} is the cross power. More generally, we define the *cross-correlation function* as

$$R_{XY}(\tau) \; = \; E\{X(t)Y(t \, + \, \tau)\} \tag{5.58}$$

In terms of the cross-correlation function, $P_{XY} = R_{XY}(0)$. A *sufficient* condition for P_{XY} to be zero, so that we may simply add powers to obtain total power, is that

$$R_{XY}(0) \; = \; 0, \qquad \text{for all } \tau \tag{5.59}$$

Such processes are said to be *orthogonal*. If processes are statistically independent and at least one of them has zero mean, they are orthogonal. However, orthogonal processes are not necessarily statistically independent.

Cross-correlation functions can be defined for nonstationary processes also, in which case we have a function of two independent variables. We will not need to be this general in our considerations.

A useful symmetry property of the cross-correlation function for jointly stationary processes is

$$R_{XY}(\tau) = R_{YX}(-\tau) \tag{5.60}$$

which can be shown as follows. By definition,

$$R_{XY}(\tau) = E[X(t)Y(t + \tau)] \tag{5.61}$$

Defining $t' = t + \tau$, we obtain

$$R_{XY}(\tau) = E\{Y(t')X(t' - \tau)\} \triangleq R_{YX}(-\tau) \tag{5.62}$$

since the choice of time origin is immaterial for stationary processes.

The *cross-power spectral density* of two stationary random processes is defined as the Fourier transform of their cross-correlation function:

$$S_{XY}(f) = \mathcal{F}[R_{XY}(\tau)] \tag{5.63}$$

The cross-power spectral density provides, in the frequency domain, the same information about the random processes as does the cross-correlation function.

5.4 LINEAR SYSTEMS AND RANDOM PROCESSES*

Input-Output Relationships

In the consideration of the transmission of stationary random waveforms through fixed linear systems, a basic tool is the relationship of the output power spectral density to the input power spectral density, given as

$$S_y(f) = |H(f)|^2 S_x(f) \tag{5.64a}$$

The autocorrelation function of the output is the inverse Fourier transform of $S_y(f)$:

$$R_y(\tau) = \mathcal{F}^{-1}[S_y(f)] = \int_{-\infty}^{\infty} |H(f)|^2 S_x(f) e^{j2\pi f\tau} \, df \tag{5.64b}$$

$H(f)$ is the system's transfer function; $S_x(f)$ is the power spectral density of the input $x(t)$; $S_y(f)$ is the power spectral density of the output $y(t)$; and $R_y(\tau)$ is the autocorrelation function of the output. The analogous result for energy signals was proved in Chapter 2 [Equation (2.183)], and the result for power signals was simply stated.

---

* For the remainder of this chapter we use lower-case x and y to denote input and output random-process signals in keeping with Chapter 2 notation.

A proof of (5.64a) could be carried out by employing (5.18) in conjunction with Equation (2.166). We will take a somewhat longer route, however, and obtain several useful intermediate results. In addition, the proof provides practice in manipulating convolutions and expectations.

We begin by obtaining the cross-correlation function between input and output, $R_{xy}(\tau)$, defined as

$$R_{xy}(\tau) = E\{x(t)y(t + \tau)\} \tag{5.65}$$

Using the superposition integral, we have

$$y(t) = \int_{-\infty}^{\infty} h(u)x(t - u)\, du \tag{5.66}$$

where $h(t)$ is the system's impulse response. Equation (5.66) relates each sample function of the input and output processes, so we can write (5.65) as (transients are assumed to have died out)

$$R_{xy}(\tau) = E\left\{ x(t) \int_{-\infty}^{\infty} h(u)x(t + \tau - u)\, du \right\} \tag{5.67}$$

Since the integral does not depend on t, we can take $x(t)$ inside and interchange the operations of expectation and convolution. (Both are simply integrals over different variables.) Since $h(u)$ is not random, (5.67) becomes

$$R_{xy}(\tau) = \int_{-\infty}^{\infty} h(u)E\{x(t)x(t + \tau - u)\}\, du \tag{5.68}$$

By definition of the autocorrelation function of $x(t)$,

$$E\{x(t)x(t + \tau - u)\} = R_x(\tau - u) \tag{5.69}$$

Thus (5.68) can be written as

$$R_{xy}(\tau) = \int_{-\infty}^{\infty} h(u)R_x(\tau - u)\, du \triangleq h(\tau) * R_x(\tau) \tag{5.70}$$

That is, the cross-correlation function of input with output is *the autocorrelation function of the input convolved with the impulse response,* an easily remembered result. Since (5.70) is a convolution, the Fourier transform of $R_{xy}(\tau)$, the cross-power spectral density of $x(t)$ with $y(t)$, is

$$S_{xy}(f) = H(f)S_x(f) \tag{5.71}$$

From the time-reversal theorem, item 3b of Table C.2, the cross-power spectral density $S_{yx}(f)$ is

$$S_{yx}(f) = \mathcal{F}[R_{yx}(\tau)] = \mathcal{F}[R_{xy}(-\tau)] = S_{xy}^*(f) \qquad (5.72)$$

Employing (5.71) and using the relationships $H^*(f) = H(-f)$ and $S_x^*(f) = S_x(f)$ (where $S_x(f)$ is real), we obtain

$$S_{yx}(f) = H(-f)S_x(f) = H^*(f)S_x(f) \qquad (5.73)$$

where the order of the subscripts is important. Taking the inverse Fourier transform of (5.73) with the aid of the convolution theorem, item 8 of Table C.2, and again using the time-reversal theorem, we obtain

$$R_{yx}(\tau) = h(-\tau) * R_x(\tau) \qquad (5.74)$$

Let us pause to emphasize what we have obtained. By definition, $R_{xy}(\tau) \triangleq E\{x(t)y(t + \tau)\}$ can be written as

$$R_{xy}(\tau) \triangleq E\{x(t)\underbrace{[h(t) * x(t + \tau)]}_{y(t + \tau)}\} \qquad (5.75)$$

Combining this with (5.70), we have

$$E\{x(t)[h(t) * x(t + \tau)]\} = h(\tau) * R_x(\tau) \triangleq h(\tau) * E\{x(t)x(t + \tau)\} \qquad (5.76)$$

Similarly, (5.74) becomes

$$R_{yx}(\tau) \triangleq E\{\underbrace{[h(t) * x(t)]}_{y(t)}x(t + \tau)\} = h(-\tau) * R_x(\tau)$$

$$\triangleq h(-\tau) * E\{x(t)x(t + \tau)\} \qquad (5.77)$$

Thus, bringing the convolution operation outside the expectation gives a convolution of $h(\tau)$ with the autocorrelation function if $h(t) * x(t + \tau)$ is inside the expectation or a convolution of $h(-\tau)$ with the autocorrelation function if $h(t) * x(t)$ is inside the expectation.

These results are combined to obtain the autocorrelation function of the output of a linear system in terms of the input autocorrelation function as follows:

$$R_y(\tau) \triangleq E\{y(t)y(t + \tau)\} = E\{y(t)[h(t) * x(t + \tau)]\} \qquad (5.78)$$

which follows because $y(t + \tau) = h(t) * x(t + \tau)$. Using (5.76) with $x(t)$ replaced by $y(t)$, we obtain

$$R_y(\tau) = h(\tau) * E\{y(t)x(t + \tau)\}$$

$$= h(\tau) * R_{yx}(\tau)$$

$$= h(\tau) * \{h(-\tau) * R_x(\tau)\} \qquad (5.79)$$

where the last line follows by substituting from (5.74). Written in terms of integrals, (5.79) is

$$R_y(\tau) = \int_{-\infty}^{\infty} \int_{-\infty}^{\infty} h(u)h(v)R_x(\tau + v - u)\, dv\, du \qquad (5.80)$$

The Fourier transform of (5.79) is the output power spectral density and is easily obtained as follows:

$$S_y(f) \triangleq \mathcal{F}[R_y(\tau)] = \mathcal{F}[h(\tau) * R_{yx}(\tau)] = H(f)S_{yx}(f)$$
$$= |H(f)|^2 S_x(f) \qquad (5.81)$$

where (5.73) has been substituted to obtain the last line.

▼ **EXAMPLE 5.7**   The input to a filter with impulse response $h(t)$ and transfer function $H(f)$ is a white-noise process with power spectral density

$$S_x(f) = \tfrac{1}{2}N_0, \qquad -\infty < f < \infty \qquad (5.82)$$

The cross-power spectral density between input and output is

$$S_{xy}(f) = \tfrac{1}{2}N_0 H(f) \qquad (5.83)$$

and the cross-correlation function is

$$R_{xy}(\tau) = \tfrac{1}{2}N_0 h(\tau) \qquad (5.84)$$

Hence we could measure the impulse response of a filter by driving it with white noise and determining the cross-correlation function of input with output. Applications include system identification and channel
▲ measurement.

Filtered Gaussian Processes

Suppose the input to a linear system is a random process. What can we say about the output statistics? For general inputs and systems, this is usually a difficult question to answer. However, *if the input to a linear system is Gaussian, the output is also Gaussian.*

A nonrigorous demonstration of this is carried out as follows. The sum of two independent Gaussian random variables has already been shown to be Gaussian. By repeated application of this result, we can find that the sum of any number of independent Gaussian random variables is Gaussian.* For a fixed linear system, the output $y(t)$ in terms of the input $x(t)$ is given by

$$y(t) = \int_{-\infty}^{\infty} x(\tau)h(t - \tau) \, dt$$

$$= \lim_{\Delta\tau \to 0} \sum_{k=-\infty}^{\infty} x(k \, \Delta\tau)h(t - k \, \Delta\tau) \, \Delta\tau \qquad (5.85)$$

where $h(t)$ is the impulse response. By writing the integral as a sum, we have demonstrated that if $x(t)$ is a white Gaussian process, the output is also Gaussian (but not white) because, at any time t, the right-hand side of (5.85) is simply a linear combination of independent Gaussian random variables. (Recall Example 5.5, where the correlation function of white noise was shown to be an impulse. Also recall that uncorrelated Gaussian random variables are independent.)

If the input is not white, we can still show that the output is Gaussian by considering the cascade of two linear systems, as shown in Figure 5.7. The system in question is the one with the impulse response $h(t)$. To show that its output is Gaussian, we note that the cascade of $h_1(t)$ with $h(t)$ is a linear system with the impulse response

$$h_2(t) = h_1(t) * h(t) \qquad (5.86)$$

This system's input, $z(t)$, is Gaussian and white. Therefore, its output, $y(t)$, is also Gaussian by application of the theorem just proved. However, the output of the system with impulse response $h_1(t)$ is Gaussian by application of the same theorem, but not white. Hence the output of a linear system with nonwhite Gaussian input is Gaussian.

▼ EXAMPLE 5.8   The input to the lowpass RC filter shown in Figure 5.8 is white Gaussian noise with the power spectral density $S_{n_i}(f) = \frac{1}{2}N_0$, $-\infty < f < \infty$. The power spectral density of the output is

$$S_{n_0}(f) = S_{n_i}(f)|H(f)|^2 = \frac{\frac{1}{2}N_0}{1 + (f/f_3)^2} \qquad (5.87)$$

where $f_3 = (2\pi RC)^{-1}$ is the 3-dB cutoff frequency. Inverse Fourier transforming $S_{n_0}(f)$, we obtain $R_{n_0}(\tau)$, the output autocorrelation function:

$$R_{n_0}(\tau) = \frac{N_0}{4RC} e^{-|\tau|/RC} \qquad (5.88)$$

* This also follows from Appendix B, (B.11).

FIGURE 5.7   Cascade of two linear systems with Gaussian input

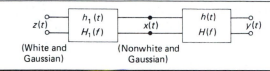

(White and   (Nonwhite and
Gaussian)   Gaussian)

The square of the mean of $n_0(t)$ is

$$\overline{n_0(t)}^2 = \lim_{|\tau| \to \infty} R_{n_0}(\tau) = 0 \tag{5.89}$$

and the mean-squared value, which is also equal to the variance since the mean is zero, is

$$\overline{n_0^2(t)} = \sigma_{n_0}^2 = R_{n_0}(0) = \frac{N_0}{4RC} \tag{5.90}$$

Alternatively, we can find the average power at the filter output by integrating the power spectral density of $n_0(t)$. The result is as above:

$$\overline{n_0^2(t)} = \int_{-\infty}^{\infty} \frac{\frac{1}{2}N_0}{1 + (f/f_3)^2}\, df = \frac{N_0}{2\pi RC} \int_0^{\infty} \frac{dx}{1 + x^2} = \frac{N_0}{4RC} \tag{5.91}$$

Since the input is Gaussian, the output is Gaussian as well. The first-order pdf is

$$f_{n_0}(y, t) = f_{n_0}(y) = \frac{e^{-2RCy^2/N_0}}{\sqrt{\pi N_0/2RC}} \tag{5.92}$$

by employing Equation (4.179). The second-order pdf at time t and $t + \tau$ is found by substitution into Equation (4.177). Letting X be a random variable that refers to the values the output takes on at time t and Y be a random

FIGURE 5.8   A lowpass RC filter with a white-noise input

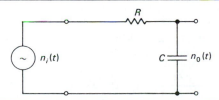

variable that refers to the values the output takes on at time $t + \tau$, we have, from the preceding results,

$$m_x = m_y = 0 \tag{5.93a}$$

$$\sigma_x^2 = \sigma_y^2 = \frac{N_0}{4RC} \tag{5.93b}$$

and the correlation coefficient is

$$\rho(\tau) = \frac{R_{n_0}(\tau)}{R_{n_0}(0)} = e^{-|\tau|/RC} \tag{5.93c}$$

Referring to Example 5.2, one can see that the random telegraph waveform has the same autocorrelation function as that of the output of the lowpass RC filter of Example 5.8. This demonstrates that processes with drastically different sample functions can have the same second-order averages.

Noise-Equivalent Bandwidth

If we pass white noise through a filter that has the transfer function $H(f)$, the average power at the output, by (5.64), is

$$P_{n_0} = \int_{-\infty}^{\infty} \tfrac{1}{2}N_0 \, |H(f)|^2 \, df = N_0 \int_0^{\infty} |H(f)|^2 \, df \tag{5.94}$$

where $\tfrac{1}{2}N_0$ is the two-sided power spectral density of the input. If the filter were ideal with bandwidth B_N and midband gain H_0, as shown in Figure 5.9, the noise power at the output would be

$$P_{n_0} = H_0^2(\tfrac{1}{2}N_0)(2B_N) = N_0 B_N H_0^2 \tag{5.95}$$

The question we now ask is the following: What is the bandwidth of an ideal, fictitious filter that has the same midband gain as $H(f)$ and that passes the

FIGURE 5.9 Comparison between $|H(f)|^2$ and an idealized approximation

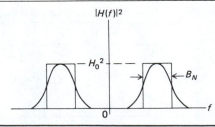

same noise power? If the midband gain of $H(f)$ is H_0, the answer is obtained by equating the preceding two results. Thus

$$B_N = \frac{1}{H_0^2} \int_0^\infty |H(f)|^2 \, df \qquad (5.96)$$

is the single-side bandwidth of the fictitious filter. B_N is called the *noise-equivalent bandwidth* of $H(f)$.

It is sometimes useful to determine the noise-equivalent bandwidth of a system using time-domain integration. By Rayleigh's energy theorem [see (2.82)], we have

$$\int_{-\infty}^\infty |H(f)|^2 \, df = \int_{-\infty}^\infty |h(t)|^2 \, dt \qquad (5.97)$$

Thus, (5.96) can be written as

$$B_N = \frac{1}{2H_0^2} \int_0^\infty |h(t)|^2 \, dt = \frac{\displaystyle\int_{-\infty}^\infty |h(t)|^2 \, dt}{2 \left[\displaystyle\int_{-\infty}^\infty h(t) \, dt \right]^2} \qquad (5.98)$$

where it is noted that

$$H_0 = H(f)\big|_{f=0} = \int_{-\infty}^\infty h(t) e^{-j2\pi ft} \, dt \big|_{f=0} = \int_{-\infty}^\infty h(t) \, dt \qquad (5.99)$$

For some systems, (5.98) is easier to evaluate than (5.96).

▼ **EXAMPLE 5.9** The noise-equivalent bandwidth of an nth-order Butterworth filter for which

$$|H_n(f)|^2 = \frac{1}{1 + (f/f_3)^{2n}} \qquad (5.100)$$

is

$$B_N(n) = \int_0^\infty \frac{1}{1 + (f/f_3)^{2n}} \, df = f_3 \int_0^\infty \frac{1}{1 + x^{2n}} \, dx$$

$$= \frac{\pi f_3 / 2n}{\sin(\pi/2n)}, \qquad n = 1, 2, \ldots \qquad (5.101)$$

where f_3 is the 3-dB frequency of the filter. For $n = 1$, (5.101) gives the result for a lowpass RC filter. As n approaches infinity, $H_n(f)$ approaches the transfer function of an ideal lowpass filter of single-sided bandwidth f_3. The noise-equivalent bandwidth is

$$\lim_{n \to \infty} B_N(n) = f_3 \tag{5.102}$$

as it should be by its definition. As the cutoff of a filter becomes sharper, its noise-equivalent bandwidth approaches its 3-dB bandwidth.

EXAMPLE 5.10   To illustrate the application of (5.98), consider the computation of the noise-equivalent bandwidth of a first-order Butterworth filter in the time domain. Its impulse response is

$$h(t) = \mathscr{F}^{-1} \left[\frac{1}{1 + j\dfrac{f}{f_3}} \right] = 2\pi f_3 e^{-2\pi f_3 t} u(t) \tag{5.103}$$

According to (5.98), the noise-equivalent bandwidth of this filter is

$$B_N = \frac{\displaystyle\int_0^\infty (2\pi f_3)^2 e^{-4\pi f_3 t}\, dt}{2\left[\displaystyle\int_0^\infty 2\pi f_3 e^{-2\pi f_3 t}\, dt \right]^2} = \frac{\pi f_3}{2}[-e^{4\pi f_3 t}]_0^\infty = \frac{\pi f_3}{2} \tag{5.104}$$

which checks with (5.101) if $n = 1$ is substituted.

5.5   NARROWBAND NOISE

Quadrature-Component and Envelope-Phase Representation

In most communication systems operating at a carrier frequency f_c, the bandwidth of the channel, B, is small compared with f_c. In such situations, it is convenient to represent the noise in terms of quadrature components as

$$n(t) = n_c(t) \cos(\omega_0 t + \theta) - n_s(t) \sin(\omega_0 t + \theta) \tag{5.105}$$

where $\omega_0 = 2\pi f_0$ and θ is an arbitrary phase angle. In terms of envelope and phase components, $n(t)$ can be written as

$$n(t) = R(t) \cos[\omega_0 t + \phi(t) + \theta] \tag{5.106}$$

where

$$R(t) = \sqrt{n_c^2 + n_s^2} \tag{5.107a}$$

FIGURE 5.10   A typical narrowband noise waveform

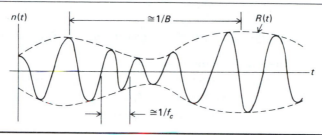

and

$$\phi(t) = \tan^{-1}\left[\frac{n_s(t)}{n_c(t)}\right] \tag{5.107b}$$

Actually, any random process can be represented in either of these forms, but if a process is narrowband, $R(t)$ and $\phi(t)$ can be interpreted as the slowly varying envelope and phase, respectively, as sketched in Figure 5.10.

Figure 5.11 shows the block diagram of a system for producing $n_c(t)$ and $n_s(t)$ where θ is, as yet, an arbitrary phase angle. Note that the composite operations used in producing $n_c(t)$ and $n_s(t)$ constitute linear systems (super-position holds from input to output). Thus, if $n(t)$ is a Gaussian process, so are $n_c(t)$ and $n_s(t)$. (The system of Figure 5.11 is to be interpreted as relating input and output processes sample function by sample function.)

We will prove several properties of $n_c(t)$ and $n_s(t)$. Most important, of

FIGURE 5.11   The operations involved in producing $n_c(t)$ and $n_s(t)$

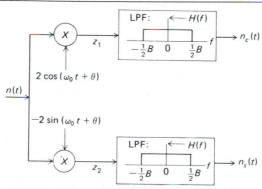

course, is whether equality really holds in (5.105) and in what sense. We will show that

$$E\{[n(t) - [n_c(t) \cos(\omega_0 t + \theta) - n_s(t) \sin(\omega_0 t + \theta)]]^2\} = 0 \quad (5.108)$$

That is, the mean-squared error between a sample function of the actual noise process and the right-hand side of (5.105) is zero (averaged over the ensemble of sample functions). More useful when using the representation in (5.105), however, are the following properties:

Mean and Variances

$$\overline{n(t)} = \overline{n_c(t)} = \overline{n_s(t)} = 0 \quad (5.109a)$$

$$\overline{n^2(t)} = \overline{n_c^2(t)} = \overline{n_s^2(t)} \triangleq N \quad (5.109b)$$

Power Spectral Densities

$$S_{n_c}(f) = S_{n_s}(f) = L_p[S_n(f - f_0) + S_n(f + f_0)] \quad (5.110a)$$

$$S_{n_c n_s}(f) = jL_p[S_n(f - f_0) - S_n(f + f_0)] \quad (5.110b)$$

where $L_p[\ \ ]$ denotes the lowpass part of the quantity in brackets; $S_n(f)$, $S_{n_c}(f)$, and $S_{n_s}(f)$ are the power spectral densities of $n(t)$, $n_c(t)$, and $n_s(t)$, respectively; and $S_{n_c n_s}(f)$ is the cross-power spectral density of $n_c(t)$ and $n_s(t)$.

From (5.110b), we see that

$$R_{n_c n_s}(\tau) \equiv 0, \quad \text{for all } \tau, \text{ if } L_p[S_n(f - f_0) - S_n(f + f_0)] = 0 \quad (5.110c)$$

This is an especially useful property in that it tells us that $n_c(t)$ and $n_s(t)$ are uncorrelated if the power spectral density of $n(t)$ is symmetrical about $f = f_0$ where $f > 0$. If, in addition, $n(t)$ is Gaussian, $n_c(t)$ and $n_s(t)$ will be *independent* Gaussian processes because they are uncorrelated, and the joint pdf of $n_c(t)$ and $n_s(t + \tau)$, for *any* delay τ, will simply be of the form

$$f(n_c, t; n_s, t + \tau) = \frac{e^{-(n_c^2 + n_s^2)/2N}}{2\pi N} \quad (5.111)$$

If $S_n(f)$ is not symmetrical about $f = f_0$, where $f > 0$, then (5.111) holds only for $\tau = 0$ or those values of τ for which $R_{n_c n_s}(\tau) = 0$.

Using the results of Example 4.15, the envelope and phase functions of (5.106) have the joint pdf

$$f(r, \phi) = \frac{r}{2\pi N} e^{-r^2/2N}, \quad \text{for } r > 0 \text{ and } |\phi| \leq \pi \quad (5.112)$$

which holds for the same conditions as for (5.111). The properties given by (5.110) will be proved first, followed by (5.108). Note that the results expressed by (5.109) follow from (5.110a).

The Power Spectral Density Function of $n_c(t)$ and $n_s(t)$

To prove (5.110a), we first find the power spectral density of $z_1(t)$, as defined in Figure 5.11, by computing its autocorrelation function and Fourier transforming the result. To simplify the derivation, it is assumed that θ is a uniformly distributed random variable in $(0, 2\pi)$ and is statistically independent of $n(t)$.

The autocorrelation function of $z_1(t) = 2n(t) \cos(\omega_0 t + \theta)$ is

$$
\begin{aligned}
R_{z_1}(\tau) &= E\{4n(t)n(t + \tau) \cos(\omega_0 t + \theta) \cos[\omega_0(t + \tau) + \theta]\} \\
&= 2E\{n(t)n(t + \tau)\} \cos \omega_0 \tau \\
&\quad + 2E\{n(t)n(t + \tau) \cos(2\omega_0 t + \omega_0 \tau + 2\theta)\} \qquad (5.113) \\
&= 2R_n(\tau) \cos \omega_0 \tau
\end{aligned}
$$

where $R_n(\tau)$ is the autocorrelation function of $n(t)$. In obtaining (5.113), we used appropriate trigonometric identities in addition to the independence of $n(t)$ and θ. Thus, by the multiplication theorem of Fourier transforms, the power spectral density of $z_1(t)$ is

$$
\begin{aligned}
S_{z_1}(f) &= S_n(f) * [\delta(f - f_0) + \delta(f + f_0)] \\
&= S_n(f - f_0) + S_n(f + f_0) \qquad (5.114)
\end{aligned}
$$

of which only the lowpass part is passed by $H(f)$. Thus the result for $S_{n_c}(f)$ expressed by (5.110a) follows. A similar proof can be carried out for $S_{n_s}(f)$. Equation (5.109b) follows by integrating (5.110a) over all f.

Next, let us consider (5.110b). To prove it, we need an expression for $R_{z_1 z_2}(\tau)$, the cross-correlation function of $z_1(t)$ and $z_2(t)$. (See Figure 5.11.) By definition, and from Figure 5.11,

$$
\begin{aligned}
R_{z_1 z_2}(\tau) &= E\{z_1(t)z_2(t + \tau)\} \\
&= -E\{4n(t)n(t + \tau) \cos(\omega_0 t + \theta) \sin[\omega_0(t + \tau) + \theta]\} \\
&= -2R_n(\tau) \sin \omega_0 \tau \qquad (5.115)
\end{aligned}
$$

where we again used appropriate trigonometric identities and the independence of $n(t)$ and θ. Letting $h(t)$ be the impulse response of the lowpass filters in Figure 5.11 and employing (5.76) and (5.77), the cross-correlation function of $n_c(t)$ and $n_s(t)$ can be written as

$$
\begin{aligned}
R_{n_c n_s}(\tau) &\triangleq E\{n_c(t)n_s(t + \tau)\} = E\{[h(t) * z_1(t)]n_s(t + \tau)\} \\
&= h(-\tau) * E\{z_1(t)n_s(t + \tau)\} = h(-\tau) * E\{z_1(t)[h(t) * z_2(t + \tau)]\} \\
&= h(-\tau) * h(\tau) * E\{z_1(t)z_2(t + \tau)\} = h(-\tau) * [h(\tau) * R_{z_1 z_2}(\tau)]
\end{aligned}
$$

$$
(5.116)
$$

The Fourier transform of $R_{n_c n_s}(\tau)$ is the cross-power spectral density, $S_{n_c n_s}(f)$, which, from the convolution theorem, is given by

$$S_{n_c n_s}(f) = H(f)\mathcal{F}[h(-\tau) * R_{z_1 z_2}(\tau)] = H(f)H^*(f)S_{z_1 z_2}(f)$$

$$= |H(f)|^2 S_{z_1 z_2}(f) \tag{5.117}$$

From (5.115) and the frequency-translation theorem, it follows that

$$S_{z_1 z_2}(f) = \mathcal{F}[jR_n(\tau)(e^{j\omega_0 \tau} - e^{-j\omega_0 \tau})]$$

$$= j[S_n(f - f_0) - S_n(f + f_0)] \tag{5.118}$$

Thus, from (5.117),

$$S_{n_c n_s}(f) = j|H(f)|^2[S_n(f - f_0) - S_n(f + f_0)]$$

$$= jL_p[S_n(f - f_0) - S_n(f + f_0)] \tag{5.119}$$

which proves (5.110b). Note that since the cross-power spectral density $S_{n_c n_s}(f)$ is imaginary, the cross-correlation function $R_{n_c n_s}(\tau)$ is odd. Thus $R_{n_c n_s}(0)$ is zero if the cross-correlation function is continuous at $\tau = 0$, which is the case for bandlimited signals.

▼ **EXAMPLE 5.11**   Let us consider a bandpass random process with the power spectral density shown in Figure 5.12(a). Choosing the center frequency of $f_0 = 7$ Hz results in $n_c(t)$ and $n_s(t)$ being uncorrelated. Figure 5.12(b) shows $S_{z_1}(f)$ [or $S_{z_2}(f)$] for $f_0 = 7$ Hz with $S_{n_c}(f)$ [or $S_{n_s}(f)$], that is, the lowpass part of $S_{z_1}(f)$, shaded. The integral of $S_n(f)$ is $2(6)(2) = 24$ W, which is the same result obtained from integrating the shaded portion of Figure 5.12(b).

Now suppose f_0 is chosen as 5 Hz. Then $S_{z_1}(f)$ and $S_{z_2}(f)$ are as shown in Figure 5.12(c), with $S_{n_c}(f)$ shown shaded. From Equation (5.110b), it follows that $-jS_{n_c n_s}(f)$ is the shaded portion of Figure 5.12(d). Because of the asymmetry that results from the choice of f_0, $n_c(t)$ and $n_s(t)$ are not uncorrelated. As a matter of interest, we can calculate $R_{n_c n_s}(\tau)$ easily by using the transform pair

$$2AW \text{ sinc } 2W\tau \leftrightarrow A\Pi\left(\frac{f}{2W}\right) \tag{5.120}$$

and the frequency-translation theorem. From Figure 5.12(d), it follows that

$$S_{n_c n_s}(f) = 2j\{-\Pi[\tfrac{1}{4}(f - 3)] + \Pi[\tfrac{1}{4}(f + 3)]\} \tag{5.121}$$

FIGURE 5.12   Spectra for Example 5.11. (a) Bandpass spectrum.
(b) Lowpass spectra for $f_0 = 7$ Hz. (c) Lowpass spectra for $f_0 = 5$ Hz.
(d) Cross spectra for $f_0 = 5$ Hz.

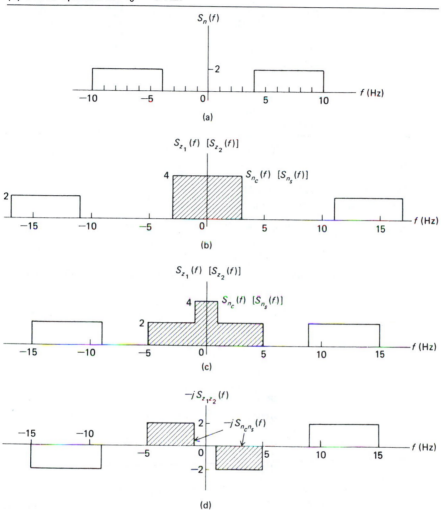

which results in the cross-correlation function

$$R_{n_c n_s}(\tau) = 2j(-4 \text{ sinc } 4\tau e^{j6\pi\tau} + 4 \text{ sinc } 4\tau e^{-j6\pi\tau}) = 16 \text{ sinc } 4\tau \sin 6\pi\tau$$

$$(5.122)$$

This cross-correlation function is shown in Figure 5.13. Although $n_c(t)$ and
$n_s(t)$ are not uncorrelated, we see that τ may be chosen such that $R_{n_c n_s}(\tau) = 0$ for particular values of τ.

FIGURE 5.13 Cross-correlation function of $n_c(t)$ and $n_s(t)$ for Example 5.11

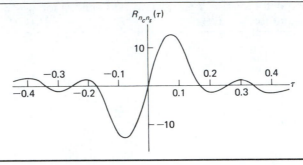

Proof That Equality Holds in Equation (5.105) in the
Sense of Zero Mean-Square Error

We now show that (5.105) holds. To simplify notation, let

$$\hat{n}(t) = n_c(t) \cos (\omega_0 t + \theta) - n_s(t) \sin (\omega_0 t + \theta) \qquad (5.123)$$

where $\hat{n}(t)$ is *not* to be confused with the Hilbert transform. Thus we must show that

$$E\{[n(t) - \hat{n}(t)]^2\} = 0 \qquad (5.124)$$

Expanding and taking the expectation term by term, we obtain

$$E\{(n - \hat{n})^2\} = \overline{n^2} - 2\overline{n\hat{n}} + \overline{\hat{n}^2} \qquad (5.125)$$

where the arguments have been dropped to simplify notation.

Let us consider the last term in (5.125) first. By the definition of $\hat{n}(t)$,

$$
\begin{aligned}
\overline{\hat{n}^2} &= \overline{E\{[n_c(t) \cos \omega_0 t + \theta) - n_s(t) \sin (\omega_0 t + \theta)]^2\}} \\
&= \overline{n_c^2} \, \overline{\cos^2 (\omega_0 t + \theta)} + \overline{n_s^2} \, \overline{\sin^2 (\omega_0 t + \theta)} \\
&\quad - \overline{2 n_c n_s} \, \overline{\cos (\omega_0 t + \theta) \sin (\omega_0 t + \theta)} \\
&= \tfrac{1}{2}\overline{n_c^2} + \tfrac{1}{2}\overline{n_s^2} \qquad\qquad\qquad\qquad (5.126) \\
&= \overline{n^2}
\end{aligned}
$$

where we have employed (5.109b) along with the averages

$$\overline{\cos^2 (\omega_0 t + \theta)} = \tfrac{1}{2} + \tfrac{1}{2} \overline{\cos 2(\omega_0 t + \theta)} = \tfrac{1}{2}$$

$$\overline{\sin^2(\omega_0 t + \theta)} = \tfrac{1}{2} - \tfrac{1}{2} \overline{\cos 2(\omega_0 t + \theta)} = \tfrac{1}{2}$$

and

$$\overline{\cos (\omega_0 t + \theta) \sin (\omega_0 t + \theta)} = \tfrac{1}{2} \overline{\sin 2(\omega_0 t + \theta)} = 0$$

Next, we consider $\overline{n\hat{n}}$. By definition of $\hat{n}(t)$, it can be written as

$$\overline{n\hat{n}} = E\{n(t)[n_c(t) \cos (\omega_0 t + \theta) - n_s(t) \sin (\omega_0 t + \theta)]\} \qquad (5.127)$$

But, from Figure 5.11,

$$n_c(t) = h(t') * [2n(t') \cos (\omega_0 t' + \theta)] \qquad (5.128a)$$

and

$$n_s(t) = h(t') * [2n(t') \sin (\omega_0 t' + \theta)] \qquad (5.128b)$$

where $h(t')$ is the impulse response of the lowpass filter in Figure 5.11. The argument t' has been used in (5.128) to remind us that the variable of integration in the convolution is different from the variable t in (5.127). Substituting (5.128) into (5.127), we obtain

$$
\begin{aligned}
\overline{n\hat{n}} &= E\{n(t)[h(t') * [2n(t') \cos (\omega_0 t' + \theta)] \cos (\omega_0 t + \theta) \\
&\quad + h(t') * [2n(t') \sin (\omega_0 t' + \theta)] \sin (\omega_0 t + \theta)]\} \\
&= E\{2n(t)h(t') * n(t')[\cos (\omega_0 t' + \theta) \cos (\omega_0 t + \theta) \\
&\quad + \sin (\omega_0 t' + \theta) \sin (\omega_0 t + \theta)]\} \\
&= E\{2n(t)h(t') * [n(t') \cos \omega_0(t - t')]\} \\
&= 2h(t') * [E\{n(t)n(t')\} \cos \omega_0(t - t')] \\
&= 2h(t') * [R_n(t - t') \cos \omega_0(t - t')] \\
&\triangleq 2 \int_{-\infty}^{\infty} h(t - t')R_n(t - t') \cos \omega_0(t - t')\, dt' \qquad (5.129)
\end{aligned}
$$

Letting $u = t - t'$, we obtain

$$\overline{n\hat{n}} = 2 \int_{-\infty}^{\infty} h(u) \cos (\omega_0 u)R_n(u)\, du \qquad (5.130)$$

Now, a general case of Parseval's theorem is

$$\int_{-\infty}^{\infty} x(t)y(t)\, dt = \int_{-\infty}^{\infty} X(f)Y^*(f)\, df \qquad (5.131)$$

where $x(t) \leftrightarrow X(f)$ and $y(t) \leftrightarrow Y(f)$. In (5.130) we note that

$$h(u) \cos \omega_0 u \leftrightarrow \tfrac{1}{2}H(f - f_0) + \tfrac{1}{2}H(f + f_0) \qquad (5.132)$$

and

$$R_n(u) \leftrightarrow S_n(f) \qquad (5.133)$$

Thus, by (5.131), we may write (5.130) as

$$\overline{n\hat{n}} = \int_{-\infty}^{\infty} [H(f - f_0) + H(f + f_0)]S_n(f) \, df \tag{5.134}$$

which follows because $S_n(f)$ is real. However, $S_n(f)$ is nonzero only where $H(f - f_0) + H(f + f_0) = 1$ because it was assumed narrowband. Thus (5.130) reduces to

$$\overline{n\hat{n}} = \int_{-\infty}^{\infty} S_n(f) \, df = \overline{n^2(t)} \tag{5.135}$$

Substituting (5.135) and (5.126) into (5.125), we obtain

$$E\{(n - \hat{n})^2\} = \overline{n^2} - 2\overline{n^2} + \overline{n^2} \equiv 0 \tag{5.136}$$

which shows that the mean-square error between $n(t)$ and $\hat{n}(t)$ is zero.

5.6 DISTRIBUTIONS ENCOUNTERED IN FM DEMODULATION OF SIGNALS IN GAUSSIAN NOISE

We now derive two probability distributions that arise in the consideration of demodulation of signals in Gaussian noise. These derivations make use of the quadrature-component and envelope-phase representations for noise described in the preceding section. They are the probability of a zero crossing of a bandlimited Gaussian process and the average rate of origin encirclement of a constant-amplitude sinusoid plus narrowband Gaussian noise.

The Zero-Crossing Problem

Consider a sample function of a lowpass, zero-mean Gaussian process $n(t)$, as illustrated in Figure 5.14. Let its effective noise bandwidth be W, its power spectral density be $S_n(f)$, and its autocorrelation function be $R_n(\tau)$.

Consider the probability of a zero crossing in a small time interval Δ seconds in duration. For Δ sufficiently small, so that more than one zero crossing is unlikely, the probability $P_{\Delta-}$ of a minus-to-plus zero crossing in a time interval $\Delta \ll 1/(2W)$ is the probability that $n_0 < 0$ and $n_0 + \dot{n}_0\Delta > 0$. That is,

$$\begin{aligned} P_{\Delta-} &= P(n_0 < 0 \text{ and } n_0 + \dot{n}_0\Delta > 0) \\ &= P(n_0 < 0 \text{ and } n_0 > -\dot{n}_0\Delta, \text{ all } \dot{n}_0 \geq 0) \\ &= P(-\dot{n}_0\Delta < n_0 < 0, \text{ all } \dot{n}_0 \geq 0) \end{aligned} \tag{5.137}$$

FIGURE 5.14   Sample function of a lowpass Gaussian process of bandwidth *W*

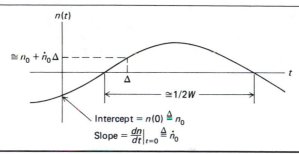

This can be written in terms of the joint pdf of n_0 and $\dot{n}_0$, $f_{n_0\dot{n}_0}(y, z)$, as

$$P_{\Delta-} = \int_0^\infty \left[\int_{-z\Delta}^0 f_{n_0\dot{n}_0}(y, z)\, dy \right] dz \qquad (5.138)$$

where y and z are running variables for n_0 and $\dot{n}_0$, respectively. Now $\dot{n}_0$ is a Gaussian random variable, since it involves a linear operation on $n(t)$, which is Gaussian by assumption. In Problem 5.39, it is shown that

$$E\{n_0\dot{n}_0\} = \left. \frac{dR_n(\tau)}{d\tau} \right|_{\tau=0} \qquad (5.139)$$

Thus, if the derivative of $R_n(\tau)$ exists at $\tau = 0$, it is zero because $R_n(\tau)$ must be even. It follows that

$$E\{n_0\dot{n}_0\} = 0 \qquad (5.140)$$

Therefore, n_0 and $\dot{n}_0$, which are samples of $n_0(t)$ and $dn_0(t)/dt$, respectively, are statistically independent, since uncorrelated Gaussian processes are independent. Thus, letting var $\{n_0\} = \overline{n_0^2}$ and var $\{\dot{n}_0\} = \overline{\dot{n}_0^2}$, the joint pdf of n_0 and $\dot{n}_0$ is

$$f_{n_0\dot{n}_0}(y, z) = \frac{\exp\left(-y^2/2\overline{n_0^2}\right)}{\sqrt{2\pi \overline{n_0^2}}} \frac{\exp\left(-z^2/2\overline{\dot{n}_0^2}\right)}{\sqrt{2\pi \overline{\dot{n}_0^2}}} \qquad (5.141)$$

which, when substituted into (5.138), yields

$$P_{\Delta-} = \int_0^\infty \frac{\exp\left(-z^2/2\overline{\dot{n}_0^2}\right)}{\sqrt{2\pi \overline{\dot{n}_0^2}}} \left[\int_{-z\Delta}^0 \frac{\exp\left(-y^2/2\overline{n_0^2}\right)}{\sqrt{2\pi \overline{n_0^2}}}\, dy \right] dz \qquad (5.142)$$

For Δ small, the inner integral of (5.142) can be approximated as $z\Delta/\sqrt{2\pi\overline{n_0^2}}$, which allows (5.142) to be simplified to

$$P_{\Delta-} \cong \frac{\Delta}{\sqrt{2\pi\overline{n_0^2}}} \int_0^\infty z \frac{\exp(-z^2/2\overline{n_0^2})}{\sqrt{2\pi\overline{n_0^2}}} \, dz \qquad (5.143)$$

Letting $\zeta = z^2/2\overline{\dot{n}_0^2}$ yields

$$P_{\Delta-} \cong \frac{\Delta}{2\pi\sqrt{\overline{n_0^2}\,\overline{\dot{n}_0^2}}} \int_0^\infty \overline{\dot{n}_0^2} e^{-\zeta} \, d\zeta$$

$$= \frac{\Delta}{2\pi} \sqrt{\frac{\overline{\dot{n}_0^2}}{\overline{n_0^2}}} \qquad (5.144)$$

for the probability of a minus-to-plus zero crossing in Δ seconds. By symmetry, the probability of a plus-to-minus zero crossing is the same. Thus, the probability of zero crossing in Δ seconds is

$$P_\Delta \cong \frac{\Delta}{\pi} \sqrt{\frac{\overline{\dot{n}_0^2}}{\overline{n_0^2}}} \qquad (5.145)$$

For example, suppose that $n(t)$ is an ideal lowpass process with the power spectral density

$$S_n(f) = \begin{cases} \frac{1}{2}N_0, & |f| \le W \\ 0, & \text{otherwise} \end{cases} \qquad (5.146)$$

Thus

$$R_n(\tau) = N_0 W \operatorname{sinc} 2W\tau \qquad (5.147)$$

which possesses a derivative at zero. Therefore, n_0 and $\dot{n}_0$ are independent. It follows that

$$\overline{n_0^2} = \operatorname{var}\{n_0\} = \int_{-\infty}^\infty S_n(f) \, df = R_n(0) = N_0 W \qquad (5.148)$$

and, since the transfer function of a differentiator is $H_d(f) = j2\pi f$, that

$$\overline{\dot{n}_0^2} = \operatorname{var}\{\dot{n}_0\} = \int_{-\infty}^\infty |H_d(f)|^2 S_n(f) \, df = \int_{-W}^W (2\pi f)^2 \tfrac{1}{2} N_0 \, df$$

$$= \tfrac{1}{3}(2\pi W)^2 (N_0 W) \qquad (5.149)$$

Substitution of these results into (5.145) gives

$$2P_{\Delta-} = 2P_{\Delta+} = P_{\Delta} = \frac{\Delta}{\pi} \frac{2\pi W}{\sqrt{3}} = \frac{2W\Delta}{\sqrt{3}} \qquad (5.150)$$

for the probability of a zero crossing in a small time interval Δ seconds in duration for a random process with an ideal rectangular lowpass spectrum.

Average Rate of Origin Encirclement for Sinusoid Plus Narrowband Gaussian Noise

Consider next the sum of a sinusoid plus narrowband Gaussian noise:

$$\begin{aligned} z(t) &= A \cos \omega_0 t + n(t) \\ &= A \cos \omega_0 t + n_c(t) \cos \omega_0 t - n_s(t) \sin \omega_0 t \end{aligned} \qquad (5.151)$$

where $n_c(t)$ and $n_s(t)$ are lowpass processes with statistical properties as described in Section 5.5. We may write $z(t)$ in terms of envelope $R(t)$ and phase $\theta(t)$ as

$$z(t) = R(t) \cos [\omega_0 t + \theta(t)] \qquad (5.152)$$

where

$$R(t) = \sqrt{[A + n_c(t)]^2 + n_s^2(t)} \qquad (5.153)$$

and

$$\theta(t) = \tan^{-1} \left[\frac{n_s(t)}{A + n_c(t)} \right] \qquad (5.154)$$

A phasor representation for this process is shown in Figure 5.15(a). In Figure 5.15(b), a possible trajectory for the tip of $R(t)$ that does not encircle the origin is shown along with $\theta(t)$ and $d\theta(t)/dt$. In Figure 5.15(c), a trajectory that encircles the origin is shown along with $\theta(t)$ and $d\theta(t)/dt$ for this case. For this latter case, the area under $d\theta/dt$ must be 2π radians. Recalling the definition of an ideal FM discriminator in Chapter 3, we see that the sketches for $d\theta/dt$ shown in Figure 5.15 represent the output of a discriminator in response to input of an unmodulated signal plus noise. For a high signal-to-noise ratio, the phasor will randomly fluctuate near the horizontal axis. Occasionally, however, it will encircle the origin as shown in Figure 5.15(c). Intuitively, these encirclements become more probable as the signal-to-noise ratio decreases. Because of its nonzero area, the impulsive type of output illustrated in Figure 5.15(c), caused by an encirclement of the origin, has a much more serious effect on the noise level of the discriminator output

FIGURE 5.15 Phasor diagrams showing possible trajectories for a sinusoid plus Gaussian noise. (a) Phasor representation of a sinusoid plus narrowband noise. (b) Trajectory that does not encircle origin. (c) Trajectory that does encircle origin.

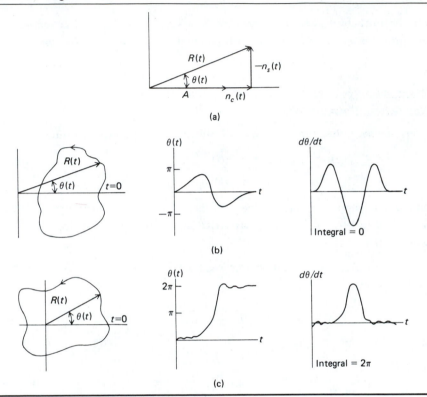

than does the noise excursion illustrated in Figure 5.15(b), which has zero area. We now derive an expression for the average number of noise spikes per second of the type illustrated in Figure 5.15(c). Only positive spikes, caused by counterclockwise origin encirclements, will be considered, since the average rate for negative spikes, which result from clockwise origin encirclements, is the same by symmetry.

Assume that if $R(t)$ crosses the horizontal axis when it is in the second quadrant, the origin encirclement will be completed. With this assumption, and considering a small interval Δ seconds in duration, the probability of a counterclockwise encirclement $P_{cc\Delta}$ in the interval $(0, \Delta)$ is

$$P_{cc\Delta} = P[A + n_c(t) < 0 \text{ and } n_s(t) \text{ makes } + \text{ to } - \text{ zero crossing in } (0, \Delta)]$$

$$= P[n_c(t) < -A]P_{\Delta-} \tag{5.155}$$

where $P_{\Delta -}$ is the probability of a minus-to-plus zero crossing in $(0, \Delta)$ as given by (5.150) with $n(t)$ replaced by $n_s(t)$, and the statistical independence of $n_c(t)$ and $n_s(t)$ has been used. Now by (5.109b), $\overline{n_c^2(t)} = \overline{n_s^2(t)} = \overline{n^2(t)}$. If $n(t)$ is an ideal bandpass process with single-sided bandwidth B and power spectral density N_0, then $\overline{n^2(t)} = N_0 B$, and

$$
P[n_c(t) < -A] = \int_{-\infty}^{-A} \frac{e^{-n_c^2/2N_0B}}{\sqrt{2\pi N_0 B}} \, dn_c
$$

$$
= Q(\sqrt{A^2/N_0 B}) \tag{5.156}
$$

where $Q(\cdot)$ is the Q-function defined by (4.185). From (5.150) with $W = \frac{1}{2}B$, which is the bandwidth of $n_s(t)$, we have

$$
P_{\Delta -} = \frac{\Delta B}{2\sqrt{3}} \tag{5.157}
$$

Substituting (5.156) and (5.157) into (5.155), we obtain

$$
P_{cc\Delta} = \frac{\Delta B}{2\sqrt{3}} Q\left(\sqrt{\frac{A^2}{N_0 B}}\right) \tag{5.158}
$$

The probability of a clockwise encirclement $P_{c\Delta}$ is the same by symmetry. Thus the expected number of encirclements per second, clockwise and counterclockwise, is

$$
\nu = \frac{1}{\Delta}(P_{c\Delta} + P_{cc\Delta})
$$

$$
= \frac{B}{\sqrt{3}} Q\left(\sqrt{\frac{A^2}{N_0 B}}\right) \tag{5.159}
$$

We note that the average number of encirclements per second increases in direct proportion to the bandwidth and decreases essentially exponentially with an increasing signal-to-noise ratio $A^2/2N_0 B$.

We can see this in Figure 5.16, which illustrates ν/B as a function of signal-to-noise ratio. Figure 5.16 also shows the asymptote $\nu/B = 1/2\sqrt{3} = 0.2887$.

The results derived above say nothing about the statistics of the number of impulses, N, in a time interval, T. In Problem 5.36, however, it is shown that the power spectral density of periodic impulse noise processes is given by

$$
S_I(f) = \nu \overline{a^2} \tag{5.160}
$$

FIGURE 5.16 Rate of origin encirclements as a function of signal-to-noise ratio

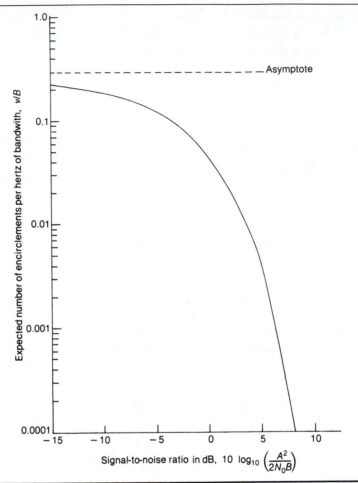

where ν is the average number of impulses per second ($\nu = f_s$ for a periodic impulse train) and $\overline{a^2}$ is the mean-squared value of the impulse weights a_k. A similar result can be shown for impulses which have exponentially distributed intervals between them (i.e., Poisson impulse noise). Approximating the impulse portion of $d\theta/dt$ as a Poisson impulse noise process with sample functions of the form

$$x(t) \triangleq \frac{d\theta(t)}{dt}\bigg|_{\text{impulse}} = \sum_{k=-\infty}^{\infty} \pm 2\pi\delta(t - t_k) \tag{5.161}$$

where t_k is a Poisson point process with average rate ν given by (5.159), we may approximate the power spectral density of this impulse noise process as white with a spectral level given by

$$S_x(f) = \nu(2\pi)^2$$

$$= \frac{4\pi^2 B}{\sqrt{3}} Q\left(\sqrt{\frac{A^2}{N_0 B}}\right), \qquad -\infty < f < \infty \qquad (5.162)$$

If the sinusoidal signal component in (5.151) is FM modulated, the average number of impulses per second is increased over that obtained for no modulation. Intuitively, the reason may be explained as follows. Consider a carrier that is FM modulated by a unit step. Thus

$$z(t) = A \cos 2\pi[f_c + f_d u(t)]t + n(t) \qquad (5.163)$$

where $f_d \le \frac{1}{2}B$ is the frequency-deviation constant in hertz per volt. Because of this frequency step, the carrier phasor shown in Figure 5.15(a) rotates counterclockwise at f_d Hz for $t > 0$. Since the noise is bandlimited to B Hz with center frequency f_c Hz, its average frequency is *less* than the instantaneous frequency of the modulated carrier when $t > 0$. Hence there will be a greater probability for a 2π clockwise rotation of $R(t)$ relative to the carrier phasor if it is frequency offset by f_d Hz (that is, modulated) than if it is not. In other words, the average rate for negative spikes will increase for $t > 0$ and that for positive spikes will decrease. Conversely, for a negative frequency step, the average rate for positive spikes will increase and that for negative spikes will decrease. It can be shown that the result is a net increase $\delta\nu$ in the spike rate over the case for no modulation, with the average increase approximated by (see Problems 5.34 and 5.35)

$$\overline{\delta\nu} = \overline{|\delta f|} \exp\left(\frac{-A^2}{2N_0 B}\right) \qquad (5.164)$$

where $\overline{|\delta f|}$ is the average of the magnitude of the frequency deviation. For the case just considered, $\overline{|\delta f|} = f_d$. The total average spike rate is then $\nu + \overline{\delta\nu}$. The power spectral density of the spike noise for modulated signals is obtained by substituting $\nu + \overline{\delta\nu}$ for ν in (5.162).

SUMMARY

1. A random process is completely described by the N-fold joint pdf of its amplitudes at the arbitrary times $t_1, t_2, \ldots, t_N$. If this pdf is invariant under a shift of the time origin, the process is said to be *statistically stationary in the strict sense.*

2. The autocorrelation function of a random process, computed as a statistical average, is defined as

$$R(t_1, t_2) = \int_{-\infty}^{\infty} \int_{-\infty}^{\infty} x_1 x_2 \, f_{X_1 X_2}(x_1, t_1; x_2, t_2) \, dx_1 \, dx_2$$

where $f_{X_1 X_2}$ is the joint amplitude pdf of the process at times t_1 and t_2. *If the process is stationary,*

$$R(t_1, t_2) = R(t_2 - t_1) = R(\tau)$$

where $\tau \triangleq t_2 - t_1$.

3. A process whose statistical average mean and variance are time-independent and whose autocorrelation function is a function only of $t_2 - t_1 = \tau$ is termed *wide-sense stationary. Strict-sense stationary processes are also wide-sense stationary. The converse is true only for special cases, one of which being Gaussian processes.*

4. A process for which statistical averages and time averages are equal is called *ergodic. Ergodicity implies stationarity, but the reverse is not necessarily true.*

5. The Wiener-Khinchine theorem states that the autocorrelation function and the power spectral density of a stationary random process are a Fourier transform pair.

   An expression for the power spectral density of a random process that is often useful is

$$S_n(f) = \lim_{T \to \infty} \frac{1}{T} E\{|\mathcal{F}[n_T(t)]|^2\}$$

where $n_T(t)$ is a sample function truncated to T seconds, centered about $t = 0$.

6. The autocorrelation function of a random process is a real, even function of the delay variable τ with an absolute maximum at $\tau = 0$. It is periodic for periodic random processes, and its Fourier transform is nonnegative for all frequencies. As $\tau \to \pm\infty$, the autocorrelation function approaches the square of the mean of the random process unless the random process is periodic. $R(0)$ gives the total average power in a process.

7. White noise has a constant power spectral density $\frac{1}{2}N_0$ for all f. Its autocorrelation function is $\frac{1}{2}N_0\delta(\tau)$. For this reason, it is sometimes called delta-correlated noise. It has infinite power and is therefore a mathematical idealization. However, it is, nevertheless, a useful approximation in many cases.

8. The cross-correlation function of two stationary random processes $X(t)$ and $Y(t)$ is defined as

$$R_{XY}(\tau) = E\{X(t)Y(t + \tau)\}$$

Their cross-power spectral density is

$$S_{XY}(f) = \mathcal{F}[R_{XY}(\tau)]$$

They are said to be *orthogonal* if $R_{XY}(\tau) = 0$ for all τ.

9. Consider a linear system with the impulse response $h(t)$ and the transfer function $H(f)$ with random input $x(t)$ and output $y(t)$. Then

$$S_Y(f) = |H(f)|^2 S_X(f)$$

$$R_Y(\tau) = \mathcal{F}^{-1}[S_Y(f)] = \int_{-\infty}^{\infty} |H(f)|^2 S_X(f) e^{j2\pi f \tau} \, df$$

$$R_{XY}(\tau) = h(\tau) * R_X(\tau)$$

$$S_{XY}(f) = H(f) S_X(f)$$

$$R_{YX}(\tau) = h(-\tau) * R_X(\tau)$$

$$S_{YX}(f) = H^*(f) S_X(f)$$

where $S(f)$ denotes the spectral density and $R(\tau)$ denotes the correlation function.

10. The output of a linear system with Gaussian input is Gaussian.

11. The noise-equivalent bandwidth of a linear system with a transfer function $H(f)$ is defined as

$$B_N = \frac{1}{H_0^2} \int_0^{\infty} |H(f)|^2 \, df$$

where $H_0 = $ maximum of $|H(f)|$. If the input is white noise with the single-sided power spectral density N_0, the output power is

$$P_0 = H_0^2 N_0 B_N$$

An equivalent expression for the noise-equivalent bandwidth written in terms of the impulse response of the filter is

$$B_N = \frac{\displaystyle\int_{-\infty}^{\infty} |h(t)|^2 \, dt}{2\left[\displaystyle\int_{-\infty}^{\infty} h(t) \, dt\right]^2}$$

12. The quadrature-component representation of a bandlimited random process $n(t)$ is

$$n(t) = n_c(t) \cos(\omega_0 t + \theta) - n_s(t) \sin(\omega_0 t + \theta)$$

where θ is an arbitrary phase angle. The envelope-phase representation is

$$n(t) = R(t) \cos(\omega_0 t + \phi(t) + \theta)$$

where $R^2(t) = n_c^2(t) + n_s^2(t)$ and $\tan[\phi(t)] = n_s(t)/n_c(t)$. If the process is narrowband, n_c, n_s, R, and ϕ vary slowly with respect to $\cos \omega_0 t$ and $\sin \omega_0 t$. If the power spectral density of $n(t)$ is $S_n(f)$, the power spectral densities of $n_c(t)$ and $n_s(t)$ are

$$S_{n_c}(f) = S_{n_s}(f) = L_p[S_n(f - f_0) + S_n(f + f_0)]$$

where $L_p[\ \ ]$ denotes the low-frequency part of the quantity in the brackets. If $L_p[S_n(f + f_0) - S_n(f - f_0)] = 0$, then $n_c(t)$ and $n_s(t)$ are orthogonal. The average powers of $n_c(t)$, $n_s(t)$, and $n(t)$ are equal. The processes $n_c(t)$ and $n_s(t)$ are given by

$$n_c(t) = L_p[2n(t) \cos(\omega_0 t + \theta)]$$

and

$$n_s(t) = -L_p[2n(t) \sin(\omega_0 t + \theta)]$$

Since these operations are linear, $n_c(t)$ and $n_s(t)$ will be Gaussian if $n(t)$ is Gaussian. Thus, $n_c(t)$ and $n_s(t)$ are independent if $n(t)$ is zero-mean Gaussian with a power spectral density that is symmetrical about $f = f_0$ for $f > 0$.

FURTHER READING

The references given in Chapter 4 also provide further reading on the subject matter of this chapter.

PROBLEMS

Section 5.1

5.1 A fair die is thrown. Depending on the number of spots on the up face, the following random processes are generated. Sketch several examples of sample functions for each case.

$$
\text{(a) } X(t, \zeta) = \begin{cases} A, & \text{1 or 2 spots up} \\ 0, & \text{3 or 4 spots up} \\ -A, & \text{5 or 6 spots up} \end{cases}
$$

$$
\text{(b) } X(t, \zeta) = \begin{cases} 5A, & \text{1 spot up} \\ 3A & \text{2 spots up} \\ A, & \text{3 spots up} \\ -A, & \text{4 spots up} \\ -3A, & \text{5 spots up} \\ -5A, & \text{6 spots up} \end{cases}
$$

$$
\text{(c) } X(t, \zeta) = \begin{cases} 4A, & \text{1 spot up} \\ 2A, & \text{2 spots up} \\ At, & \text{3 spots up} \\ -At, & \text{4 spots up} \\ -2A, & \text{5 spots up} \\ -4A, & \text{6 spots up} \end{cases}
$$

Section 5.2

5.2   Referring to Problem 5.1, what are the following probabilities for each case?

(a) $F_X(X \le 2A, t = 4)$
(b) $F_X(X \le 0, t = 4)$
(c) $F_X(X \le 2A, t = 2)$

5.3   A random process is composed of sample functions that are square-waves, each with constant amplitude A, period T_0, and random delay τ as sketched in Figure 5.17. The pdf of τ is

$$
f(\tau) = \begin{cases} 1/T_0, & |\tau| \le T_0/2 \\ 0, & \text{otherwise} \end{cases}
$$

(a) Sketch several typical sample functions.
(b) Write the first-order pdf for this random process at some arbitrary time t_0. (*Hint:* Because of the random delay τ, the pdf is independent of t_0. Also, it might be easier to deduce the cdf and differentiate it to get the pdf.)

FIGURE 5.17

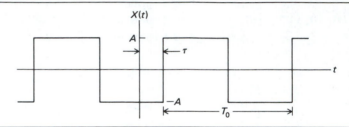

5.4   Let the sample functions of a random process be given by

$$X(t) = A \cos \omega_0 t$$

where ω_0 is fixed and A has the pdf

$$f_A(a) = \frac{e^{-a^2/2\sigma_a^2}}{\sqrt{2\pi}\sigma_a}$$

This random process is passed through an ideal integrator to give a random process $Y(t)$.

(a) Find an expression for the sample functions of the output process $Y(t)$.

(b) Write down an expression for the pdf of $Y(t)$ at time t_0. (*Hint:* Note that $\sin \omega_0 t_0$ is just a constant.)

(c) Is $Y(t)$ stationary? Is it ergodic?

5.5   (a) Find the time-average mean and the autocorrelation function for the random process of Problem 5.3.

(b) Find the ensemble-average mean and the autocorrelation function.

(c) Is this process wide-sense stationary? Why?

5.6   Consider the random process of Example 5.1 with the pdf of θ given by

$$p(\theta) = \begin{cases} 4/\pi, & \pi/4 \le \theta \le \pi/2 \\ 0, & \text{otherwise} \end{cases}$$

(a) Find the statistical-average and time-average mean and variance.

(b) Find the statistical-average and time-average autocorrelation functions.

(c) Is this process ergodic?

5.7 Let a random process be given as

$$Z(t) = X(t) \cos (\omega_0 t + \theta)$$

where $X(t)$ is a stationary random process with $E\{X(t)\} = 0$ and $E\{X^2(t)\} = \sigma_X^2$.

(a) If $\theta = 0$, find $E\{Z(t)\}$ and $E\{Z^2\}$. Is $Z(t)$ stationary?
(b) If θ is a random variable indpendent of $X(t)$ and uniformly distributed in the interval $(-\pi, \pi)$, show that

$$E\{Z(t)\} = 0 \quad \text{and} \quad E\{Z^2(t)\} = \sigma_X^2/2$$

(c) Is $Z(t)$ stationary?

5.8 The voltage of the output of a noise generator whose statistics are known to be closely Gaussian and stationary is measured with a dc voltmeter and a true root-mean-square (rms) voltmeter that is ac coupled. The dc meter reads 5 V, and the true rms meter reads 10 V. Write down an expression for the first-order pdf of the voltage at any time $t = t_0$. Sketch and dimension the pdf.

Section 5.3

5.9 Which of the following functions are suitable autocorrelation functions? Tell why or why not. (ω_0, τ_0, τ_1, A, B, and C are positive constants.)

(a) $A \cos \omega_0 \tau$
(b) $A\Lambda(\tau/\tau_0)$, where $\Lambda(x)$ is the unit-area triangular function defined in Chapter 2
(c) $A\Pi(\tau/\tau_0)$, where $\Pi(x)$ is the unit-area pulse function defined in Chapter 2
(d) $B\Pi(\tau/\tau_1) + C\Pi(\tau/2\tau_1)$
(e) $A \sin \omega_0 \tau$

5.10 A bandlimited white-noise process has a double-sided power spectral density of 2×10^{-10} W/Hz in the frequency range $|f| \leq 1$ MHz. Find its autocorrelation function and sketch and fully dimension it.

5.11 Consider a random binary pulse waveform as analyzed in Example 5.6, but with half-cosine pulses given by $p(t) = \cos (2\pi t/2T) \Pi(t/T)$. Obtain and sketch the autocorrelation function for the two cases considered in Example 5.6, namely,

(a) $a_k = \pm A$ for all k, where A is a constant, with $R_m = A^2$, $m = 0$, and $R_m = 0$ otherwise.

(b) $a_k = A_k + A_{k-1}$ with $A_k = \pm A$ and $E\{A_k A_{k+m}\} = A^2$, $m = 0$, and zero otherwise.

(c) Find and sketch the power spectral density for each preceding case.

5.12   Two random processes are given by

$$X(t) = n(t) + A \cos (\omega_0 t + \theta)$$

and

$$Y(t) = n(t) + A \sin (\omega_0 t + \theta)$$

where A and ω_0 are constants and θ is a random variable uniformly distributed in the interval $(-\pi, \pi)$. The first term, $n(t)$, represents a stationary random noise process with autocorrelation function $R_n(\tau) = B\Lambda(\tau/\tau_0)$, where B and τ_0 are nonnegative constants. Find and sketch the cross-correlation function of these two random processes.

5.13   Given two independent, wide-sense stationary random processes $X(t)$ and $Y(t)$ with autocorrelation functions $R_X(\tau)$ and $R_Y(\tau)$, respectively.

(a) Show that the autocorrelation function $R_Z(\tau)$ of their product $Z(t) = X(t)Y(t)$ is given by

$$R_Z(\tau) = R_X(\tau)R_Y(\tau)$$

(b) Express the power spectral density of $Z(t)$ in terms of the power spectral densities of $X(t)$ and $Y(t)$, denoted as $S_X(f)$ and $S_Y(f)$, respectively.

(c) Let $X(t)$ be a bandlimited stationary noise process with power spectral density $S_X(f) = 10\Pi(f/200)$, and let $Y(t)$ be the process defined by sample functions of the form

$$Y(t) = 5 \cos (50\pi t + \theta)$$

where θ is a uniformly distributed random variable in the interval $(0, 2\pi)$. Using the results derived in parts (a) and (b), obtain the autocorrelation function and power spectral density of $Z(t) = X(t)Y(t)$.

5.14   A random signal has the autocorrelation function

$$R(\tau) = 5 + 2\Lambda(\tau/10)$$

where $\Lambda(x)$ is the unit-area triangular function defined in Chapter 2. Determine the following:

(a) The ac power.

(b) The dc power.

(c) The total power.

(d) The power spectral density. Sketch it and label carefully.

5.15 A random process is defined as $Y(t) = X(t) + X(t - T)$, where $X(t)$ is a wide-sense stationary random process with autocorrelation function $R_X(\tau)$ and power spectral density $S_X(f)$.

(a) Show that $R_Y(\tau) = 2R_X(\tau) + R_X(\tau + T) + R_X(\tau - T)$.

(b) Show that $S_Y(f) = 4S_X(f) \cos^2 (\pi f T)$.

(c) If $X(t)$ has autocorrelation function $R_X(\tau) = 5\Lambda(\tau)$, where $\Lambda(\tau)$ is the unit-area triangular function, and $T = 0.5$, find and sketch the power spectral density of $Y(t)$ as defined in the problem statement.

5.16 The power spectral density of a wide-sense stationary random process is given by

$$S_X(f) = 10\delta(\tau) + 25 \text{ sinc}^2 (5f)$$

(a) Sketch and fully dimension this power spectral density function.

(b) Find the power in the dc component of the random process.

(c) Find the total power.

(d) Given that the area under the main lobe of the sinc-squared function is approximately 0.9 of the total area, which is unity if it has unity amplitude, find the fraction of the total power contained in this process for frequencies between 0 and 0.2 Hz.

5.17 Given the following autocorrelation functions $R_X(\tau)$ and power spectral densities $S_X(f)$ (α, K, K_1, K_2, and b are all positive constants):

$$R_X(\tau) = K \exp (-\alpha|\tau|) \cos 2\pi b\tau \tag{1}$$

$$R_X(\tau) = K_1 [\exp (-\alpha|\tau|) + 1] + K_2 \cos 2\pi b\tau \tag{2}$$

$$S_X(f) = K \exp (-\alpha f^2) + K_1\delta(f - 10) \tag{3}$$
$$+ K_1\delta(f + 10) + K_2\delta(\tau)$$

(a) Sketch each function and fully dimension.

(b) Determine the value of the dc power, if any, for each one.

(c) Determine the total power for each.

(d) Determine the frequency of the periodic component, if any, for each.

Section 5.4

5.18 A stationary random process $n(t)$ has a power spectral density of 10^{-6} W/Hz. It is passed through an ideal lowpass filter with transfer function $H(f) = \Pi(f/1 \text{ MHz})$, where $\Pi(x)$ is the unit-area pulse function defined in Chapter 2.

(a) What is the power spectral density of the output? Sketch.
(b) Obtain the autocorrelation function of the output. Sketch and label carefully.
(c) What is the power of the output process? Find it two different ways.

5.19 An ideal finite-time integrator is characterized by the input-output relationship

$$Y(t) = \frac{1}{T} \int_{t-T}^{t} X(\alpha) \, d\alpha$$

(a) What is its impulse response?
(b) What is its transfer function?
(c) The input is white noise with two-sided power spectral density $N_0/2$. Find the power spectral density of the output of the filter.
(d) Show that the autocorrelation function of the output is

$$R_0(\tau) = \frac{N_0}{2T} \Lambda(\tau/T)$$

where $\Lambda(x)$ is the unit-area triangular function defined in Chapter 2.

(e) What is the equivalent noise bandwidth of the integrator?
(f) Show that the result for the output noise power obtained using the equivalent noise bandwidth found in part (e) coincides with the result found from the autocorrelation function of the output found in part (d).

5.20 White noise with two-sided power spectral density $N_0/2$ drives a second-order Butterworth filter with transfer function magnitude

$$|H_{2bu}(f)| = \frac{1}{\sqrt{1 + (f/f_3)^4}}$$

(a) What is the power spectral density of the output?
(b) Show that the autocorrelation function of the output is

$$R_0(\tau) = \frac{\pi f_3 N_0}{2} \exp\left(-\sqrt{2}\pi f_3 |\tau|\right) \cos\left[\sqrt{2}\pi f_3 |\tau| - \pi/4\right]$$

Plot as a function of $f_3\tau$. (*Hint:* Use the integral.)

$$\int_0^\infty \frac{\cos{(ax)}}{b^4 + x^4} \, dx = \frac{\sqrt{2}\pi}{4b^3} \exp{(-ab/\sqrt{2})}[\cos{(ab/\sqrt{2})}$$
$$+ \sin{(ab/\sqrt{2})}], \qquad a, b > 0$$

(c) Does the output power obtained by taking the proper limit of $R_0(\tau)$ check with that calculated using the equivalent noise bandwidth for a Butterworth filter as given by (5.101)?

5.21   A power spectral density given by

$$S_Y(f) = \frac{(2\pi f)^2}{(2\pi f)^4 + 10{,}000}$$

is desired. A white-noise source of two-sided power spectral density 1 W/Hz is available. What is the transfer function of the filter to be placed at the noise-source output to produce the desired power spectral density?

5.22   Obtain the autocorrelation functions and power spectral densities of the outputs of the following systems with the input autocorrelation functions or power spectral densities given.

(a) $H(f) = \Pi(f/2B)$
$$R_X(\tau) = \frac{N_0}{2} \delta(\tau)$$

N_0 and B are positive constants.

(b) $h(t) = A \exp{(-\alpha t)} u(t)$
$$S_X(f) = \frac{B}{1 + (2\pi\beta f)^2}$$

A, α, B, and β are positive constants.

5.23   The input to a lowpass filter with impulse response

$$h(t) = \exp{(-20\pi t)} u(t)$$

is white, Gaussian noise with single-sided power spectral density of 2 W/Hz. Obtain the following:

(a) The mean of the output.
(b) The power spectral density of the output.
(c) The autocorrelation function of the output.
(d) The probability density function of the output at an arbitrary time t_1.

(e) The joint probability density function of the output at time instants t_1 and $t_1 + 0.03$ s.

5.24 Find the noise-equivalent bandwidths for the following first- and second-order lowpass filters in terms of their 3-dB bandwidths. Refer to Chapter 2 to determine their transfer function magnitudes.

(a) Chebyshev
(b) Bessel

5.25 A second-order Butterworth filter has 3-dB bandwidth of 500 Hz. Determine the unit impulse response of the filter and use it to compute the noise-equivalent bandwidth of the filter. Check your result against the appropriate special case of Example 5.9.

5.26 Determine the noise-equivalent bandwidths for the two filters having transfer functions given below:

(a) $H_a(f) = \Pi(f/6) + \Pi(f/2)$
(b) $H_b(f) = 2\Lambda(f/100)$

5.27 A filter has transfer function

$$H(f) = H_0(f - 500) + H_0(f + 500)$$

where

$$H_0(f) = 2\Lambda(f/100)$$

Find its noise-equivalent bandwidth.

5.28 Determine the noise-equivalent bandwidths of the systems having the following transfer functions. (*Hint:* Use the time domain approach.)

(a) $H_a(f) = \dfrac{10}{(j2\pi f + 1)(j2\pi f + 20)}$

(b) $H_b(f) = \dfrac{25}{(j2\pi f + 5)^2}$

Section 5.5

5.29 Noise $n(t)$ has the power spectral density shown in Figure 5.18. We write

$$n(t) = n_c(t) \cos(2\pi f_0 t + \theta) - n_s(t) \sin(2\pi f_0 t + \theta)$$

FIGURE 5.18

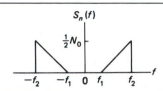

$S_n(f)$

$\frac{1}{2}N_0$

$-f_2 \quad -f_1 \quad 0 \quad f_1 \quad f_2$

Make plots of the power spectral densities of $n_c(t)$ and $n_s(t)$ for the
following cases:

(a) $f_0 = f_1$ (b) $f_0 = f_2$ (c) $f_0 = \frac{1}{2}(f_2 + f_1)$
(d) For which of these cases are $n_c(t)$ and $n_s(t)$ uncorrelated?

5.30   (a) If $S_n(f) = \alpha^2/(\alpha^2 + \omega^2)$, show that $R_n(\tau) = Ke^{-\alpha|\tau|}$. Find K.
       (b) Find $R_n(\tau)$ if

$$S_n(f) = \frac{\frac{1}{2}\alpha^2}{\alpha^2 + (\omega - \omega_0)^2} + \frac{\frac{1}{2}\alpha^2}{\alpha^2 + (\omega + \omega_0)^2}, \qquad \begin{array}{l} \omega = 2\pi f \\ \omega_0 = 2\pi f_0 \end{array}$$

   (c) If $n(t) = n_c(t) \cos(2\pi f_0 t + \theta) - n_s(t) \sin(2\pi f_0 t + \theta)$ find
       $S_{n_c}(f)$, and $s_{n_c n_c}(f)$, where $S_n(f)$ is as given in part (b). Sketch.

5.31   The double-sided power spectral density of noise $n(t)$ is shown in Fig-
       ure 5.19. If $n(t) = n_c(t) \cos(2\pi f_0 t + \theta) - n_s(t) \sin(2\pi f_0 t + \theta)$, find
       and plot $S_{n_c}(f)$, $S_{n_s}(f)$, and $S_{n_c n_s}(f)$ for the following cases:

(a) $f_0 = \frac{1}{2}(f_1 + f_2)$ (b) $f_0 = f_1$ (c) $f_0 = f_2$
(d) Find $R_{n_c n_s}(\tau)$ for each case for which $S_{n_c n_s}(f)$ is not zero. Plot.

5.32   A noise waveform $n_1(t)$ has the bandlimited power spectral density
       shown in Figure 5.20. Find and plot the power spectral density of
       $n_2(t) = n_1(t) \cos(\omega_0 t + \theta) - n_1(t) \sin(\omega_0 t + \theta)$, where θ is a uni-
       formly distributed random variable in $(0, 2\pi)$.

FIGURE 5.19

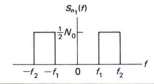

$S_{n_1}(f)$

$\frac{1}{2}N_0$

$-f_2 \quad -f_1 \quad 0 \quad f_1 \quad f_2$

FIGURE 5.20

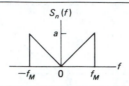

5.33   (Rice-Nakagami pdf) Consider the sum of a constant-amplitude sinu-soid and narrowband Gaussian noise:

$$y(t) = [A + n_c(t)] \cos (\omega_0 t + \theta) - n_s(t) \sin (\omega_0 t + \theta)$$

Let the variance of n_c and n_s be σ^2. Show that the pdf of

$$r = \sqrt{(A + n_c)^2 + n_s^2}$$

is

$$f_R(r) = \left(\frac{r}{\sigma^2}\right) \exp\left[-\frac{(r^2 + A^2)}{2\sigma^2}\right] I_0\left(\frac{rA}{\sigma^2}\right), \qquad r \geq 0$$

where $I_0(x) \triangleq (2\pi)^{-1} \int_0^{2\pi} e^{x\cos\theta} \, d\theta$ is the modified Bessel function of the first kind and of zero order.

Section 5.6

5.34   Consider a signal-plus-noise process of the form

$$z(t) = A \cos 2\pi (f_0 + f_d)t + n(t)$$

where $\omega_0 = 2\pi f_0$, with

$$n(t) = n_c(t) \cos \omega_0 t - n_s(t) \sin \omega_0 t$$

an ideal bandlimited white-noise process with double-sided power spectral density equal to $\frac{1}{2}N_0$, for $-\frac{1}{2}B \leq f \pm f_0 \leq \frac{1}{2}B$, and zero otherwise. Write $z(t)$ as

$$z(t) = A \cos 2\pi (f_0 + f_d)t + n_c'(t) \cos 2\pi (f_0 + f_d)t$$
$$- n_s'(t) \sin 2\pi (f_0 + f_d)t$$

   (a)   Express $n_c'(t)$ and $n_s'(t)$ in terms of $n_c(t)$ and $n_s(t)$. Using the tech-niques developed in Section 5.5, find the power spectral densities of $n_c'(t)$ and $n_s'(t)$, $S_{n_c'}(f)$ and $S_{n_s'}(f)$.
   (b)   Find the cross-spectral density of $n_c'(t)$ and $n_s'(t)$, $S_{n_c'n_s'}(f)$, and the cross-correlation function, $R_{n_c'n_s'}(\tau)$. Are $n_c'(t)$ and $n_s'(t)$

correlated? Are $n'_c(t)$ and $n'_s(t)$, sampled at the same instant, independent?

5.35 (a) Using the results of Problem (5.34), derive Equation (5.164) with $|\overline{\delta f}| = f_d$.

(b) Compare Equations (5.164) and (5.159) for a squarewave-modulated FM signal with deviation f_d by letting $|\overline{\delta f}| = f_d$ and $B = 2f_d$ for $f_d = 5$ and 10 for signal-to-noise ratios of $A^2/N_0B = 1$, 10, 100, 1000. Plot ν and $\overline{\delta \nu}$ versus A^2/N_0B.

Problems Extending Text Material

5.36 A random process is composed of sample functions of the form

$$x(t) = n(t) \sum_{k=-\infty}^{\infty} \delta(t - kT_s) = \sum_{k=-\infty}^{\infty} n_k \delta(t - kT_s)$$

where $n(t)$ is a wide-sense stationary random process with the auto-correlation function $R_n(\tau)$, and $n_k = n(kT_s)$. If T_s is chosen such that

$$R_n(kT_s) = 0, \qquad k = 1, 2, \dots$$

[that is, the samples $n_k = n(kT_s)$ are orthogonal], use Equation (5.28) to show that the power spectral density of $x(t)$ is

$$S_x(f) = \frac{R_n(0)}{T_s} = f_s R_n(0) = f_s \overline{n^2(t)}, \qquad -\infty < f < \infty$$

5.37 A random pulse-position-modulated waveform is as shown in Figure 5.21, where the delays Δt_i of the pulses are independent uniform random variables with identical pdf's of the form

$$f_{\Delta t_i}(x) = \begin{cases} \dfrac{4}{T}, & 0 \leq x \leq \tfrac{1}{4}T, \ \tau_0 < \tfrac{3}{4}T \\ \\ 0, & \text{otherwise} \end{cases}$$

Find the power spectral density of $y(t)$. (*Hint:* Use Equation (5.28).)

FIGURE 5.21

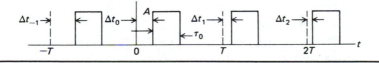

FIGURE 5.22

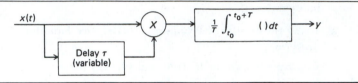

5.38 Consider the system shown in Figure 5.22 as a means of approximately measuring $R_x(\tau)$ where $x(t)$ is stationary.

(a) Show that $E\{y\} = R_x(\tau)$.

(b) Find an expression for $\sigma_y^{\,2}$ if $x(t)$ is Gaussian and has zero mean. *Hint:* If x_1, x_2, x_3, and x_4 are Gaussian with zero mean, it can be shown that

$$E\{x_1x_2x_3x_4\} = E\{x_1x_2\}E\{x_3x_4\} + E\{x_1x_3\}E\{x_2x_4\} + E\{x_1x_4\}E\{x_2x_3\}$$

5.39 A useful average in the consideration of noise in FM demodulation is the cross-correlation

$$R_{y\dot{y}}(\tau) \triangleq E\left\{ y(t) \frac{dy\,(t+\tau)}{dt} \right\}$$

where $y(t)$ is assumed stationary.

(a) Show that

$$R_{y\dot{y}}(\tau) = \frac{dR_y(\tau)}{d\tau}$$

where $R_y(\tau)$ is the autocorrelation function of $y(t)$. (*Hint:* The transfer function of a differentiator is $H(f) = j2\pi f$.)

(b) If $y(t)$ is Gaussian, write down the joint pdf of

$$Y \triangleq y(t) \qquad \text{and} \qquad Z \triangleq \frac{dy(t)}{dt}$$

at any time t, assuming the ideal lowpass power spectral density

$$S_y(f) = \tfrac{1}{2}N_0\Pi\left(\frac{f}{2B}\right)$$

Express your answer in terms of N_0 and B.

(c) Can one obtain a result for the joint pdf of y and $\dot{y}$ if $y(t)$ is obtained by passing white noise through a lowpass RC filter? Why or why not?

COMPUTER EXERCISES

5.1 Check the correlation between the random variable X and Y generated by the random number generator of Computer Exercise 4.2 by computing the sample correlation coefficient of 1000 pairs according to the definition

$$\rho(X,Y) = \frac{1}{(N-1)\hat{\sigma}_1\hat{\sigma}_2} \sum_{n=1}^{N} (X_n - \hat{\mu}_X)(Y_n - \hat{\mu}_Y)$$

where

$$\hat{\mu}_X = \frac{1}{N}\sum_{n=1}^{N} X_n$$

$$\hat{\mu}_Y = \frac{1}{N}\sum_{n=1}^{N} Y_n$$

$$\hat{\sigma}_X^2 = \frac{1}{N-1}\sum_{n=1}^{N} (X_n - \hat{\mu}_X)^2$$

and

$$\hat{\sigma}_Y^2 = \frac{1}{N-1}\sum_{n=1}^{N} (Y_n - \hat{\mu}_X)^2$$

5.2 Using the correlated Gaussian random-number generator designed in Computer Exercise 4.3, write a simulation for the origin encirclement problem described by Equations (5.151) and following. Note that the $\rho(\tau)$ given in that computer exercise corresponds to passing white noise through a first-order lowpass filter. Thus, you should be able to relate the filter bandwidth, and therefore of the noise, to α. You should design your simulation to check out the curve given in Figure 5.16.

6 NOISE IN MODULATION SYSTEMS

In Chapters 4 and 5 we investigated the subjects of probability and random processes. These concepts led to a representation for bandlimited noise, which will now be used in the analysis of analog communication systems operating in the presence of noise.

Noise is present in varying degrees in all electrical systems and in communication systems noise can be especially damaging. As illustrated in Appendix A, under ordinary conditions, whenever the temperature of a conductor is above 0 K, random motion of charge carriers results in thermal noise. This noise is often low-level and can usually be neglected in those portions of a system where the signal level is high. Often, however, systems are encountered in which the signal levels are low, and in such systems the effects of even low-level noise can seriously degrade overall system performance. Examples of this are found in the predetection portions of an AM or FM radio receiver that is tuned to a distant transmitter and in a radar receiver that is processing a return pulse from a distant target. There are many other examples.

As pointed out in Chapter 1, system noise results from sources external to the system as well as from sources internal to the system. Since noise is unavoidable in any practical system, techniques for minimizing the effects of noise must often be used if high-performance communications are desired. In the present chapter, appropriate performance criteria for system evaluation will be developed. After this, a number of systems will be analyzed to determine the effect of noise on system operation. It is especially important to note the differences between linear and nonlinear systems. We will find that the use of nonlinear modulation, such as FM, allows *improved performance* to be obtained at the expense of *increased transmission bandwidth*. Such tradeoffs do not exist when linear modulation is used.

Our concern in this chapter will be with systems using analog modulation, both continuous wave and pulse. For completeness, we briefly consider pulse code modulation in this chapter. In the following chapters, our study will be extended to include digital systems and the optimization of system performance in the presence of noise.

6.1 SIGNAL-TO-NOISE RATIOS

In Chapter 3, systems that involve the operations of modulation and demodulation were studied. In this section we extend that study to the performance of linear demodulators in the presence of noise. We concentrate our efforts on the calculation of signal-to-noise ratios. The signal-to-noise ratio is often a useful figure of merit for evaluating system performance.

Baseband Systems

In order to have a basis for comparing system performance, let us determine the signal-to-noise ratio at the output of a baseband system involving no modulation or demodulation. Such a system is illustrated in Figure 6.1(a). Assume that the signal power is finite at P_T W and that the noise added to

FIGURE 6.1   Baseband system. (a) Diagram. (b) Spectra at filter input. (c) Spectra at filter output.

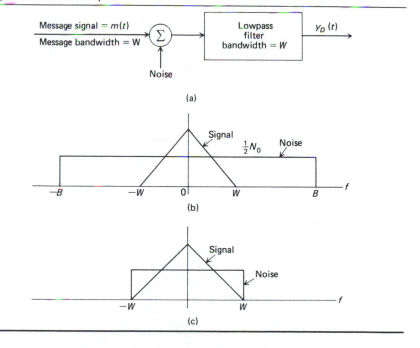

the signal has the double-sided power spectral density $\frac{1}{2}N_0$ W/Hz over a bandwidth B, which exceeds W. This is illustrated in Figure 6.1(b).

If the message signal $m(t)$ is assumed to be bandlimited, a lowpass filter can be used to enhance the signal-to-noise ratio, as shown in Figure 6.1(c). If the filter bandwidth equals W, the signal is totally passed and the signal power at the lowpass filter output is P_T. The power of the output noise is

$$\int_{-W}^{W} \tfrac{1}{2}N_0 \, df = N_0 W \tag{6.1}$$

Thus the signal-to-noise ratio (SNR) at the filter output is

$$\text{SNR} = \frac{P_T}{N_0 W} \tag{6.2}$$

The filter enhances the SNR by a factor of B/W. Since (6.2) describes the SNR achieved with a simple baseband system using no modulation, it is a reasonable standard for making comparisons of system performance.

DSB Systems

We can now compute the noise performance of a coherent double-sideband (DSB) demodulator. Consider the block diagram in Figure 6.2, which illustrates a coherent demodulator preceded by a predetection filter. Typically, the predetection filter is the intermediate-frequency (IF) filter discussed in Chapter 3. The input to this filter is the modulated signal plus white Gaussian noise of double-sided power spectral density $\frac{1}{2}N_0$ W/Hz. Since the transmitted signal $x_c(t)$ is assumed to be a DSB signal, the received signal $x_r(t)$ can be written as

$$x_r(t) = A_c m(t) \cos (\omega_c t + \theta) + n(t) \tag{6.3}$$

where $m(t)$ is the message and θ is used to denote our uncertainty of the carrier phase or, equivalently, the time origin. If the predetection filter bandwidth is $2W$, the DSB signal is completely passed by the filter. Using

FIGURE 6.2   DSB demodulator

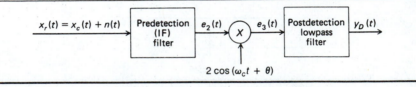

the technique developed in Chapter 5, the noise at the filter output $n_0(t)$ can be expanded into its direct and quadrature components as follows:

$$e_2(t) = A_c m(t) \cos(\omega_c t + \theta) + n_c(t) \cos(\omega_c t + \theta) - n_s(t) \sin(\omega_c t + \theta)$$
(6.4)

where the total noise power is $\overline{n_0^2(t)} = \frac{1}{2}\overline{n_c^2(t)} + \frac{1}{2}\overline{n_s^2(t)}$ and is equal to $2N_0W$.

The predetection SNR, measured at the input to the multiplier, is easily determined. The signal power is $\frac{1}{2}A_c^2\overline{m^2}$, where m is understood to be a function of t and the noise power is $2N_0W$ as shown in Figure 6.3(a). This yields the predetection SNR

$$(\text{SNR})_T = \frac{A_c^2\overline{m^2}}{4WN_0}$$
(6.5)

In order to compute the postdetection SNR, $e_3(t)$ is first computed:

$$e_3(t) = A_c m(t) + A_c m(t) \cos 2(\omega_c t + \theta) + n_c(t)$$
$$+ n_c(t) \cos 2(\omega_c t + \theta) - n_s(t) \sin 2(\omega_c t + \theta) \quad (6.6)$$

The double-frequency terms are removed by the postdetection filter so that

$$y_D(t) = A_c m(t) + n_c(t)$$
(6.7)

for the demodulated output. The postdetection signal power is $A_c^2\overline{m^2}$, and the postdetection noise power is $\overline{n_c^2}$ or $2N_0W$, as shown on Figure 6.3(b). The postdetection SNR is

$$(\text{SNR})_D = \frac{A_c^2\overline{m^2}}{2N_0W}$$

The ratio of $(\text{SNR})_D$ to $(\text{SNR})_T$ is referred to as the *detection gain* and is often used as a figure of merit for a demodulator. Thus, for the coherent DSB demodulator, the detection gain is

$$\frac{(\text{SNR})_D}{(\text{SNR})_T} = \frac{A_c^2\overline{m^2}}{2N_0W} \cdot \frac{4N_0W}{A_c^2\overline{m^2}} = 2$$
(6.8)

At first sight, this result is somewhat misleading, for it appears that we have gained 3 dB. This is true for the demodulator because it effectively suppresses the quadrature noise component. However, a comparison with the baseband system reveals that nothing is gained, insofar as the SNR at the system output is concerned, by the use of modulation, because the predetection filter bandwidth must be $2W$ if DSB modulation is used. Thus the channel bandwidth is doubled, resulting in double the noise bandwidth and,

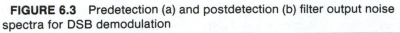

FIGURE 6.3 Predetection (a) and postdetection (b) filter output noise spectra for DSB demodulation

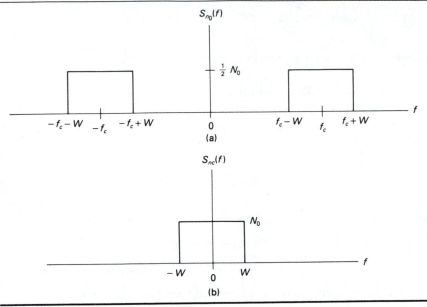

consequently, double the noise power at the demodulator input. The 3-dB detection gain is exactly enough to overcome this effect.

In order to see this mathematically, let us return to the expression for the postdetection SNR. The term $\frac{1}{2}A_c^2\overline{m^2}$ is the transmitted power P_T. Thus

$$(\text{SNR})_D = \frac{P_T}{N_0 W} \tag{6.9}$$

which is the output SNR of the equivalent baseband system.

SSB Systems

Similar calculations are easily carried out for SSB systems. For SSB, the predetection filter input can be written as

$$x_r(t) = A_c[m(t)\cos(\omega_c t + \theta) \pm \hat{m}(t)\sin(\omega_c t + \theta)] + n(t) \tag{6.10}$$

where $\hat{m}(t)$ denotes the Hilbert transform of $m(t)$. Recall from Chapter 3 that the plus sign is used for lower-sideband SSB and the minus sign is used for upper-sideband SSB. Since the minimum bandwidth of the predetection bandpass filter is W for single-sideband, the center frequency of the predetection filter is $\omega_x = \omega_c \pm \frac{1}{2}(2\pi W)$, where the sign depends on the choice of sideband.

We could expand the noise about the center frequency ω_x. However, as we saw in Chapter 5, we are free to expand the noise about any frequency we choose. It is slightly more convenient, however, to expand the noise about the carrier frequency ω_c. For this case, the predetection filter output can be written as $x_c(t) + n_0(t)$, which is

$$e_2(t) = A_c[m(t) \cos(\omega_c t + \theta) \pm \hat{m}(t) \sin(\omega_c t + \theta)]$$
$$+ n_c(t) \cos(\omega_c t + \theta) - n_s(t) \sin(\omega_c t + \theta) \quad (6.11)$$

where, as can be seen from Figure 6.4(a),

$$N_T = \overline{n_0^2} = \overline{n_c^2} = \overline{n_s^2} = N_0 W \quad (6.12)$$

Equation (6.11) can be written

$$e_2(t) = [A_c m(t) + n_c(t)] \cos(\omega_c t + \theta)$$
$$\pm [A_c \hat{m}(t) \mp n_s(t)] \sin(\omega_c t + \theta) \quad (6.13)$$

As discussed in Chapter 3, demodulation is accomplished by multiplying $e_2(t)$ by the demodulation carrier $2 \cos(\omega_c t + \theta)$ and lowpass filtering. Thus the DSB demodulator illustrated in Figure 6.2 also applies to the SSB demodulation problem. It follows that

$$y_D(t) = A_c m(t) + n_c(t) \quad (6.14)$$

The power spectral density of $n_c(t)$ is illustrated in Figure 6.4(b) for the case of lower-sideband SSB. Note that the postdetection filter passes $n_c(t)$, so that the postdetection noise power is

$$N_D = \overline{n_c^2} = N_0 W \quad (6.15)$$

From (6.14) it follows that the postdetection signal power is

$$S_D = A_c^2 \overline{m^2} \quad (6.16)$$

We now turn our attention to the predetection terms.

The predetection signal power is

$$S_T = \overline{\{A_c[m(t) \cos(\omega_c t + \theta) - \hat{m}(t) \sin(\omega_c t + \theta)]\}^2} \quad (6.17)$$

In Chapter 2 we pointed out that a function and its Hilbert transform are orthogonal. If $\overline{m(t)} = 0$, it follows that $\overline{m(t)\hat{m}(t)} = \langle m(t)\hat{m}(t)\rangle = 0$. Thus the preceding expression becomes

$$S_T = A_c^2 \{[\tfrac{1}{2}\overline{m^2(t)}] + [\tfrac{1}{2}\overline{\hat{m}^2(t)}]\} \quad (6.18)$$

It was also shown in Chapter 2 that a function and its Hilbert transform have equal power. Applying this to (6.18) yields

$$S_T = A_c^2 \overline{m^2} \quad (6.19)$$

FIGURE 6.4 Predetection (a) and postdetection (b) filter output spectra for lower-sideband SSB (+ and − signs denote spectral translation of positive and negative portions of $S_{n_0}(f)$ due to demodulation, respectively).

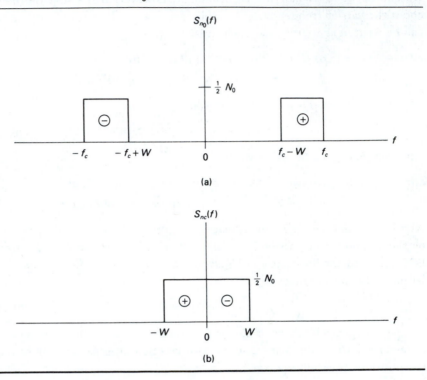

(a)

(b)

The same reasoning applied to the predetection and postdetection noise shows that they have equal power and that this power is

$$N_T = N_D = \tfrac{1}{2}\overline{n_c^2} + \tfrac{1}{2}\overline{n_s^2} = N_0 W \tag{6.20}$$

Therefore, the detection gain is

$$\frac{(\text{SNR})_D}{(\text{SNR})_T} = \frac{A_c^2 \overline{m^2}}{N_0 W} \cdot \frac{N_0 W}{A_c^2 \overline{m^2}} = 1 \tag{6.21}$$

The SSB system lacks the 3-dB detection gain of the DSB system. However, the predetection noise power of the SSB system is 3 dB less than that for the DSB system if the predetection filters have minimum bandwidth. This results in equal performance, given by

$$(\text{SNR})_D = \frac{A_c^2 \overline{m^2}}{N_0 W} = \frac{P_T}{N_0 W} \tag{6.22}$$

Thus coherent demodulation of both DSB and SSB results in performance equivalent to baseband.

AM Systems: Coherent Detection

We saw in Chapter 3 that the representation for an AM signal is

$$x_c(t) = A_c[1 + am_n(t)] \cos (\omega_c t + \theta) \tag{6.23}$$

where $m_n(t)$ is the modulation signal normalized so that the maximum value of $|m_n(t)|$ is unity and a is the modulation index. It is easily shown, by using a development parallel to that for DSB systems, that the demodulated output in the presence of noise is

$$y_D(t) = A_c am_n(t) + n_c(t) \tag{6.24}$$

The dc term that mathematically results in $y_D(t)$ is not included in (6.24) for two reasons. First, this term is not considered part of the signal since it contains no information. [Recall that we have assumed $\overline{m(t)} = 0$.] Second, most practical AM demodulators are not dc-coupled, so a dc term does not appear on the output of a practical system. In addition, the dc term is frequently used for AGC (automatic gain control) and has to be held constant at the transmitter.

From (6.24) it follows that the signal power in $y_D(t)$ is

$$S_D = A_c^2 a^2 \overline{m_n^2} \tag{6.25}$$

and the noise power is

$$N_D = \overline{n_c^2} = 2N_0 W \tag{6.26}$$

For the predetection case, the signal power is

$$S_T = \tfrac{1}{2}A_c^2 + \tfrac{1}{2}A_c^2 a^2 \overline{m_n^2} \tag{6.27}$$

and the noise power is

$$N_T = 2N_0 W \tag{6.28}$$

since the required transmission bandwidth for AM is $2W$. Thus the detection gain is

$$\frac{(\text{SNR})_D}{(\text{SNR})_T} = \frac{A_c^2 a^2 \overline{m_n^2}/2N_0 W}{(A_c^2 + A_c^2 a^2 \overline{m_n^2})/4N_0 W} = \frac{2a^2 \overline{m_n^2}}{1 + a^2 \overline{m_n^2}} \tag{6.29}$$

Recall that when we studied AM in Chapter 3 the efficiency of an AM transmission system was defined as the ratio of sideband power to total power in the transmitted signal $x_c(t)$. This resulted in the efficiency E being expressed as

$$E = \frac{a^2\overline{m_n^2}}{1 + a^2\overline{m_n^2}} \tag{6.30}$$

where the overbar, denoting a statistical average, has been substituted for the time-average notation $\langle \cdot \rangle$ used in Chapter 3. It follows from (6.29) and (6.30) that the detection gain can be expressed as

$$\frac{(SNR)_D}{(SNR)_T} = 2E \tag{6.31}$$

Since the predetection signal-to-noise ratio can be written as

$$(SNR)_T = \frac{S_T}{2N_0W} = \frac{P_T}{2N_0W} \tag{6.32}$$

it follows that the signal-to-noise ratio at the demodulator output can be written as

$$(SNR)_D = \frac{EP_T}{N_0W} \tag{6.33}$$

If the efficiency *could* be 100%, then AM would have the same postdetection signal-to-noise ratio as the ideal DSB and SSB systems. Of course, as we saw in Chapter 3, the efficiency of AM is typically much less than 100% and the postdetection signal-to-noise ratio is correspondingly lower.

▼ **EXAMPLE 6.1**   An AM system operates with a modulation index of 0.5 and the power in the normalized message signal is 0.1. The efficiency is

$$E = \frac{(0.5)^2(0.1)}{1 + (0.5)^2(0.1)} = 0.0244 \tag{6.34}$$

and the postdetection signal-to-noise ratio is

$$(SNR)_D = 0.0244 \frac{P_T}{N_0W} \tag{6.35}$$

The detection gain is

$$\frac{(SNR)_D}{(SNR)_T} = 2E = 0.0488 \tag{6.36}$$

This is more than 16 dB inferior to the ideal system requiring the same bandwidth. It should be remembered, however, that the motivation for using AM is not noise performance but rather that AM allows the use of simple envelope detectors for demodulation. The reason, of course, for the poor efficiency of AM is that a large fraction of the total transmitted power is devoted to the carrier component, which conveys no information since it is not a function of the message signal.

AM Systems: Envelope Detection

Since envelope detection is the usual method of demodulating an AM signal, it is important to understand how envelope demodulation differs from coherent demodulation in the presence of noise. The received signal, at the input to the envelope detector, is assumed to be $x_c(t)$ plus narrowband noise. Thus

$$x_r(t) = A_c[1 + am_n(t)] \cos(\omega_c t + \theta) + n_c(t) \cos(\omega_c t + \theta)$$
$$- n_s(t) \sin(\omega_c t + \theta) \quad (6.37)$$

where, as before, $\overline{n_c^2} = \overline{n_s^2} = 2N_0W$. The output of the envelope detector can be determined with the aid of the phasor diagram in Figure 6.5. The signal $x_r(t)$ can be written in terms of envelope and phase as

$$x_r(t) = r(t) \cos[\omega_c t + \theta + \phi(t)] \quad (6.38)$$

where

$$r(t) = \sqrt{\{A_c[1 + am_n(t)] + n_c(t)\}^2 + n_s^2(t)} \quad (6.39)$$

and

$$\phi(t) = \tan^{-1} \frac{n_s(t)}{A_c[1 + am_n(t)] + n_c(t)} \quad (6.40)$$

The output of an ideal envelope detector is independent of phase variations of the input. Thus the expression for $\phi(t)$ is of no interest, and we will

FIGURE 6.5   Phasor diagram of AM (drawn for $\theta = 0$)

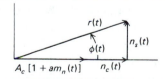

concentrate on $r(t)$. The envelope detector is assumed to be ac-coupled so that

$$y_D(t) = r(t) - \overline{r(t)} \tag{6.41}$$

where $\overline{r(t)}$ is the average value of the envelope amplitude.

Equation (6.41) will be evaluated for two cases. First, we consider the case in which the $(SNR)_T$ is large, and then we briefly consider the case in which the $(SNR)_T$ is small. For the first case, the solution is simple. From Figure 6.5 and Equation (6.39), we see that if

$$|A_c[1 + am_n(t)] + n_c(t)| \gg |n_s(t)| \tag{6.42}$$

then *most of the time*

$$r(t) \cong A_c[1 + am_n(t)] + n_c(t) \tag{6.43}$$

yielding

$$y_D(t) \cong A_c am_n(t) + n_c(t) \tag{6.44}$$

This is the final result for the case in which the SNR is large.

Comparing (6.44) and (6.24) illustrates that the output of the envelope detector is equivalent to the output of the coherent detector if $(SNR)_T$ is large. The detection gain for this case is given by (6.29).

For the case in which $(SNR)_T$ is small, the analysis is somewhat more complex. In order to analyze this case, we recall from Chapter 5 that $n_c(t) \cos(\omega_c t + \theta) - n_s(t) \sin(\omega_c t + \theta)$ can be written in terms of envelope and phase, so that the envelope detector input can be written as

$$e(t) = A_c[1 + am_n(t)] \cos(\omega_c t + \theta) + r_n(t) \cos[\omega_c t + \theta + \phi_n(t)] \tag{6.45}$$

For $(SNR)_T \ll 1$, the amplitude of $A_c[1 + am_n(t)]$ will usually be much smaller than $r_n(t)$. Thus the phasor diagram appears as illustrated in Figure 6.6. It can be seen that $r(t)$ is approximated by

$$r(t) \cong r_n(t) + A_c[1 + am_n(t)] \cos \phi_n(t) \tag{6.46}$$

yielding

$$y_D(t) \cong r_n(t) + A_c[1 + am_n(t)] \cos \phi_n(t) - \overline{r(t)} \tag{6.47}$$

The principal component of $y_D(t)$ is the Rayleigh-distributed noise envelope, and *no* component of $y_D(t)$ is proportional to the signal. Note that since $n_c(t)$ and $n_s(t)$ are random, $\cos \phi_n(t)$ is also random. Thus the *signal* is *multiplied* by a random quantity. This multiplication of the signal by a function of the noise has a significantly worse degrading effect than does additive noise.

This severe loss of signal at low-input SNR is known as the *threshold effect*

FIGURE 6.6   Phasor diagram for AM with $(SNR)_T \ll 1$ (drawn for $\theta = 0$)

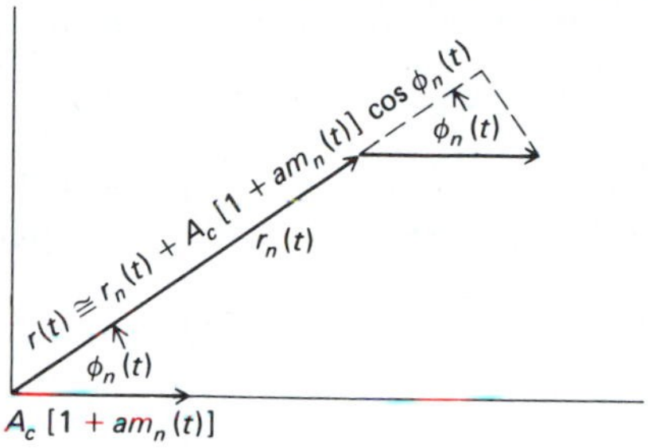

and results from the nonlinear action of the envelope detector. In coherent detectors, which are linear, the signal and noise are additive at the detector output *if* they are additive at the detector input. The result is that the signal retains its identity even when the input SNR is low. For this reason, coherent detection is often desirable when the noise is large.

Square-Law Detectors

The determination of the signal-to-noise ratio at the output of a nonlinear system is often a very difficult task. The *square-law detector*, however, is one system for which this is not the case. In this section, we conduct a simplified analysis to illustrate the phenomenon of thresholding, which is a character-istic of nonlinear systems. We will see that the detection gain is different for high and low signal-to-noise ratios. The intermediate region is known as the *threshold region*.

In the analysis that follows, the postdetection bandwidth will be assumed to be twice the message bandwidth W. As shown in the problem at the end of the chapter, this is not a necessary assumption, but it does result in a simplification of the analysis without having an impact on the threshold effect. We will also see that harmonic distortion is a problem with square-law detectors, an effect that often precludes their use.

Square-law detectors can be implemented as a squaring device followed by a lowpass filter. The response of a square-law detector to an AM signal is $r^2(t)$, where $r(t)$ is defined by (6.39). Thus the output can be written as

$$r^2(t) = \{A_c[1 + am_n(t)] + n_c(t)\}^2 + n_s^2(t) \tag{6.48}$$

We now determine the output SNR assuming sinusoidal modulation. Letting $m_n(t) = \cos \omega_m t$ and carrying out the indicated squaring operation, we obtain

$$r^2(t) = A_c^2 + 2A_c^2 a \cos \omega_m t + \tfrac{1}{2}A_c^2 a^2$$
$$+ \tfrac{1}{2}A_c^2 a^2 \cos 2\omega_m t + 2A_c n_c(t) \tag{6.49}$$
$$+ 2A_c a n_c(t) \cos \omega_m t + n_c^2(t) + n_s^2(t)$$

Assuming that the detector output is ac-coupled, so that the dc terms are blocked, the signal and noise components of the output are written as

$$s_D(t) = 2A_c^2 a \cos \omega_m t \tag{6.50}$$

and

$$n_D(t) = 2A_c n_c(t) + 2A_c a n_c(t) \cos \omega_m t + n_c^2(t) + n_s^2(t) \tag{6.51}$$

respectively. The power in the signal component is

$$S_D = 2A_c^4 a^2 \tag{6.52}$$

and the noise power is

$$N_D = 4A_c^2 \overline{n_c^2(t)} + 2A_c^2 a^2 \overline{n_c^2(t)} + 2\{\overline{n_c^4(t)} - \overline{n_c^2(t)^2}\} \tag{6.53}$$

where the dc noise power must be subtracted because of the assumed ac coupling. The total predetection noise power is

$$\sigma_n^2 = N_0 B_T = 2N_0 W = \overline{n_c^2(t)} = \overline{n_s^2(t)} \tag{6.54}$$

where W is the bandwidth of the message signal. In order to evaluate the quantity $\overline{n_c^4(t)}$ in (6.53) we write

$$\overline{n_c^4(t)} = \int_{-\infty}^{\infty} n_c^4 \frac{1}{\sqrt{2\pi}\sigma_n} e^{-n_c^2/2\sigma_n^2} \, dn_c = 3\sigma_n^4 \tag{6.55}$$

Thus (6.53) becomes $N_D = 2A_c^2(2 + a^2)\sigma_n^2 + 4\sigma_n^4$, and the output SNR is

$$(\text{SNR})_D = \frac{A_c^2 a^2 / \sigma_n^2}{(2 + a^2) + 2(\sigma_n^2/A_c^2)} \tag{6.56}$$

for the square-law detector.

It is interesting to compare (6.56) with baseband systems. The total transmitted power P_T of an AM system, with sinusoidal modulation, is

$$P_T = \tfrac{1}{2}A_c^2(1 + \tfrac{1}{2}a^2) \tag{6.57}$$

Substituting from (6.57) for A_c^2 and $2N_0 W$ for σ_n^2 yields

$$(\text{SNR})_D = 2\left(\frac{a}{2 + a^2}\right)^2 \frac{P_T/N_0 W}{1 + (N_0 W/P_T)} \tag{6.58}$$

This expression clearly illustrates the threshold effect. For large values of P_T/N_0W,

$$(\text{SNR})_D \cong 2 \left(\frac{a}{2 + a^2} \right)^2 \frac{P_T}{N_0W} \tag{6.59}$$

For small values of P_T/N_0W,

$$(\text{SNR})_D \cong 2 \left(\frac{a}{2 + a^2} \right)^2 \left(\frac{P_T}{N_0W} \right)^2 \tag{6.60}$$

Figure 6.7 illustrates (6.58) for several values of the modulation index a.

One term in the expression for $r^2(t)$ has been neglected. This term,

$$\tfrac{1}{2}A_c^2 a^2 \cos 2\omega_m t$$

represents *harmonic distortion* of the message due to the squaring operation. The power in this term is $\tfrac{1}{8}A_c^4 a^4$, and the ratio of the harmonic distortion power D_D to the signal power S_D is

$$\frac{D_D}{S_D} = \frac{1}{16} a^2 \tag{6.61}$$

FIGURE 6.7 Performance of a square-law detector

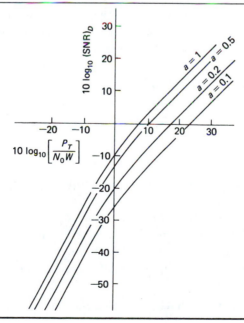

Thus harmonic distortion can be reduced by decreasing the modulation index. However, as illustrated in Figure 6.7, a reduction of the modulation index also results in a decrease in the output signal-to-noise ratio.

The *linear envelope detector* defined by (6.39) is much more difficult to analyze over a wide range of signal-to-noise ratios because of the square root. However, to a first approximation, the performance of a linear envelope detector and a square-law envelope detector are the same. Harmonic distortion is also present in linear envelope detectors, but the amplitude of the distortion component is significantly less than that observed for square-law detectors. In addition, it can be shown that for high signal-to-noise ratios and a modulation index of unity, the performance of a linear envelope detector is better by approximately 1.8 dB than the performance of a square-law detector. (See Problem 6.10.)*

6.2 NOISE AND PHASE ERRORS IN COHERENT SYSTEMS

In the preceding section we investigated the performance of various types of demodulators. Our main interests were detection gain and calculation of the demodulated output SNR. Where coherent demodulation was used, the demodulation carrier was assumed to have *perfect* phase coherence. In a practical system, this is often not a valid assumption, since in systems where noise is present, perfect phase coherence is not possible. In this section we will allow the demodulation carrier to have a time-varying phase error, and we will determine the combined effect of noise at the demodulator input and phase error in the demodulation carrier.

A General Analysis

The demodulator model is illustrated in Figure 6.8. The signal portion of $e(t)$ will be assumed to be the quadrature double-sideband (QDSB) signal

$$m_1(t) \cos (\omega_c t + \theta) + m_2(t) \sin (\omega_c t + \theta)$$

where any constant A_c is absorbed into $m_1(t)$ and $m_2(t)$ for notational simplicity.

Using this representation, we will obtain a general solution. After the analysis is complete, the DSB result is obtained by letting $m_1(t) = m(t)$ and $m_2(t) = 0$. The SSB result is obtained by letting $m_1(t) = m(t)$ and $m_2(t) = \pm \hat{m}(t)$, depending on the sideband of interest. For the QDSB system,

---

* For a detailed study of linear envelope detectors, see Bennett (1974).

FIGURE 6.8   Coherent demodulator with phase error

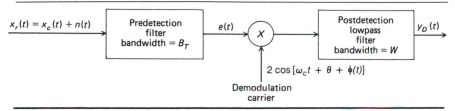

$y_D(t)$ is the demodulated output for the direct channel. The quadrature channel can be demodulated using a demodulation carrier of the form $2 \sin [\omega_c t + \theta + \phi(t)]$.

The noise portion of $e(t)$ is represented by

$$n_c(t) \cos (\omega_c t + \theta) - n_s(t) \sin (\omega_c t + \theta)$$

In the preceding expression,

$$\overline{n_c^{\,2}} = \overline{n_s^{\,2}} = N_0 B_T \qquad (6.62)$$

where B_T is the bandwidth of the predetection filter and $\frac{1}{2}N_0$ is the double-sided power spectral density of the noise at the filter input. The phase error of the demodulation carrier is assumed to be a sample function of a zero-mean Gaussian process of known variance $\sigma_\phi^{\,2}$. As before, the modulating signals are assumed to have zero mean.

With the preliminaries of defining the model and stating the assumptions disposed of, we now proceed with the analysis. The assumed performance criterion is mean-square error; that is, we will compute

$$\overline{\epsilon^2} = \overline{\{m_1(t) - y_D(t)\}^2} \qquad (6.63)$$

for the three cases of interest: DSB, SSB, and QDSB. The multiplier input signal $e(t)$ is

$$e(t) = m_1(t) \cos (\omega_c t + \theta) + m_2(t) \sin (\omega_c t + \theta)$$
$$+ n_c(t) \cos (\omega_c t + \theta) - n_s(t) \sin (\omega_c t + \theta) \qquad (6.64)$$

Multiplying by $2 \cos [\omega_c t + \theta + \phi(t)]$ and lowpass filtering gives us the generalized output

$$y_D(t) = [m_1(t) + n_c(t)] \cos \phi(t) - [m_2(t) - n_s(t)] \sin \phi(t) \qquad (6.65)$$

The error $m_1(t) - y_D(t)$ can be written as

$$\epsilon = m_1 - (m_1 + n_c) \cos \phi + (m_2 - n_s) \sin \phi \qquad (6.66)$$

where it is understood that ϵ, m_1, m_2, n_c, n_s, and ϕ are all functions of time. The mean-square error can be written as

$$\overline{\epsilon^2} = \overline{m_1^2} - 2\overline{m_1(m_1 + n_c)} \cos \phi$$
$$+ 2\overline{m_1(m_2 - n_s)} \sin \phi$$
$$+ \overline{(m_1 + n_c)^2} \cos^2 \phi$$
$$- 2\overline{(m_1 + n_c)(m_2 - n_s)} \sin \phi \cos \phi$$
$$+ \overline{(m_2 - n_s)^2} \sin^2 \phi \qquad (6.67)$$

The variables m_1, m_2, n_c, n_s, and ϕ are all assumed to be uncorrelated. It should be pointed out that for the SSB case, the power spectra of $n_c(t)$ and $n_s(t)$ will not be symmetrical about ω_c. However, as pointed out in Section 5.5, $n_c(t)$ and $n_s(t)$ are still uncorrelated, since there is no time displacement. Thus the mean-square error can be written as

$$\overline{\epsilon^2} = \overline{m_1^2} - 2\overline{m_1^2} \cos \phi + \overline{m_1^2} \cos^2 \phi + \overline{m_2^2} \sin^2 \phi + \overline{n^2} \qquad (6.68)$$

where

$$\overline{n_c^2} = \overline{n_s^2} = N_0 B_T = \sigma_n^2 = \overline{n^2}$$

At this point in our development we must define the type of system being analyzed. First, let us assume the system of interest in *QDSB with equal power* in each modulating signal. Under this assumption, $\overline{m_1^2} = \overline{m_2^2} = \sigma_m^2$, and the mean-square error is

$$\overline{\epsilon_Q^2} = 2\sigma_m^2 - 2\sigma_m^2 \overline{\cos \phi} + \sigma_n^2 \qquad (6.69)$$

This expression can be easily evaluated for the case in which the maximum value of $|\phi(t)| \ll 1$. For this case,

$$\overline{\cos \phi} \cong \overline{(1 - \tfrac{1}{2}\phi^2)} = 1 - \tfrac{1}{2}\sigma_\phi^2$$

and

$$\overline{\epsilon_Q^2} = \sigma_m^2 \sigma_\phi^2 + \sigma_n^2 \qquad (6.70)$$

In order to have an easily interpreted measure of system performance, the mean-square error is usually normalized to σ_m^2. This yields

$$\overline{\epsilon_{NQ}^2} = \sigma_\phi^2 + \frac{\sigma_n^2}{\sigma_m^2} \qquad (6.71)$$

The preceding expression is also valid for *the SSB case*, since an SSB signal is a QDSB signal with equal power in the direct and quadrature channels. The only possible point of confusion is that σ_n^2 may be different for the SSB

FIGURE 6.9   Mean-square error versus signal-to-noise ratio for QDSB system

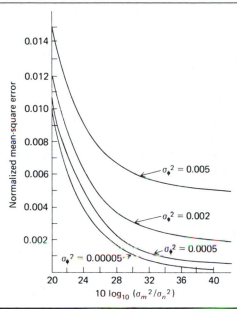

and QDSB cases, since the SSB predetection filter bandwidth need only be half the bandwidth of the predetection filter for the QDSB case. Equation (6.71) is of such general interest that it is illustrated in Figure 6.9.

In order to compute the mean-square error *for a DSB system*, we simply let $m_2 = 0$ and $m_1 = m$ in (6.68). This yields

$$\overline{\epsilon_D^2} = \overline{m^2} - 2\overline{m^2 \cos \phi} + \overline{m^2 \cos^2 \phi} + \overline{n^2} \tag{6.72}$$

or

$$\overline{\epsilon_D^2} = \sigma_m^2 \overline{(1 - \cos \phi)^2} + \overline{n^2} \tag{6.73}$$

which, for small ϕ, can be approximated as

$$\overline{\epsilon_D^2} \cong \sigma_m^2 (\tfrac{1}{4}) \overline{\phi^4} + \overline{n^2} \tag{6.74}$$

In the preceding section [see (6.55)], we pointed out that if ϕ is zero-mean Gaussian with variance σ_ϕ^2,

$$\overline{\phi^4} = \overline{(\phi^2)^2} = 3\sigma_\phi^4 \tag{6.75}$$

Thus

$$\overline{\epsilon_D^2} \cong \tfrac{3}{4}\sigma_m^2 \sigma_\phi^4 + \sigma_n^2 \tag{6.76}$$

and the normalized mean-square error becomes

$$\overline{\epsilon_{ND}^{2}} = \tfrac{3}{4}\sigma_{\phi}^{4} + \frac{\sigma_{n}^{2}}{\sigma_{m}^{2}} \tag{6.77}$$

Several items are of interest when comparing (6.77) and (6.71). First, equal *output* SNR's imply equal *normalized* mean-square errors for $\sigma_{\phi}^{2} = 0$. This is easy to understand, since the noise is additive at the output. The general expression for $y_{D}(t)$ is $y_{D}(t) = m(t) + n(t)$. The error is $n(t)$, and the normalized mean-square error is $\sigma_{n}^{2}/\sigma_{m}^{2}$. The analysis also illustrates that DSB systems are much less sensitive to phase errors of the demodulation carrier than are SSB or QDSB systems. This follows from the fact that if $\phi \ll 1$, the basic assumption made in the analysis, then $\sigma_{\phi}^{4} \ll \sigma_{\phi}^{2}$.

▼ **EXAMPLE 6.2**   Assume that the demodulation phase error variance of a coherent demodulator is described by $\sigma_{\phi}^{2} = 0.01$. The signal-to-noise ratio $\sigma_{m}^{2}/\sigma_{n}^{2}$ is 20 dB. If a DSB system is used, the normalized mean-square error is

$$\overline{\epsilon_{ND}^{2}} = \tfrac{3}{4}(0.01)^{2} + 10^{-20/10} = 0.010075 \quad \text{(DSB)} \tag{6.78}$$

while for the SSB case the normalized mean-square error is

$$\overline{\epsilon_{ND}^{2}} = (0.01) + 10^{-20/10} = 0.02 \quad \text{(SSB)} \tag{6.79}$$

Note that for the DSB demodulator, the demodulation phase error can probably be neglected for most applications. For the SSB case the demodulation phase error variance must clearly be considered. Another way to look at the results of this example is to recognize that the use of DSB gives a factor-of-two improvement over the performance of a SSB system, at the cost of twice ▲ the bandwidth.

Demodulation Phase Errors

At this point it is appropriate that we say a few words about the source of demodulation phase errors, the effects of which we have just analyzed. The demodulation carrier must be coherent with the original modulation carrier. Thus the modulation carrier must be the *source* of the demodulation carrier. This implies that the information necessary to generate the demodulation carrier passes through the channel and, consistent with the channel model, is perturbed by additive white noise.

One technique that can be used is illustrated in Figure 6.10(a). In this system, the demodulation carrier is derived from a *pilot signal,* which is har-

FIGURE 6.10   DSB system using pilot reconstruction. (a) System. (b) Single-sided spectra.

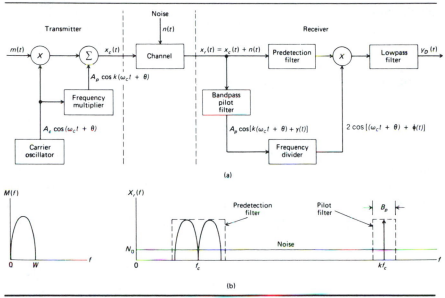

(a)

(b)

monically related to the carrier. The spectrum of the received signal is illustrated in Figure 6.10(b). As illustrated, the bandpass pilot filter extracts the pilot signal from the received signal. This pilot signal is then frequency-divided to yield the required demodulation carrier. Since the pilot is perturbed by channel noise, a pilot phase error $\gamma(t)$ is induced.

The statistics of $\gamma(t)$ are easily computed for the high SNR case. The pilot filter output can be expressed as

$$A_p \cos k(\omega_c t + \theta) + n_c(t) \cos k(\omega_c t + \theta) - n_s(t) \sin k(\omega_c t + \theta)$$

which can be written as

$$r(t) \cos [k(\omega_c t + \theta) + \gamma(t)]$$

where

$$\gamma(t) = \tan^{-1} \frac{n_s(t)}{A_p + n_c(t)} \tag{6.80}$$

In (6.80),

$$\overline{n_s^2} = \overline{n_c^2} = N_0 B_p \tag{6.81}$$

where B_p is the equivalent noise bandwidth of the pilot filter. For large signal-to-noise ratios, $A_p \gg n_c$ and the argument of the arctangent function is small. Thus

$$\gamma(t) \cong \frac{n_s(t)}{A_p} \tag{6.82}$$

and the phase error variance is

$$\sigma_\gamma^{\,2} = \frac{\overline{n_s^{\,2}}}{A_p^{\,2}} = \frac{1}{2\rho} \tag{6.83}$$

where ρ is the pilot SNR measured in the equivalent noise bandwidth of the pilot filter.

A frequency divider can be thought of as a device that divides the total instantaneous phase by a constant. Thus, if a frequency divider has input $2 \cos [k(\omega_c t + \theta) + \gamma(t)]$, the output is $2 \cos [\omega_c t + \theta + (1/k)\gamma(t)]$. Therefore, in Figure 6.9,

$$\sigma_\phi^{\,2} = \frac{1}{k^2} \sigma_\gamma^{\,2} \tag{6.84}$$

As pointed out in Chapter 3, a phase-lock loop, adjusted so that a harmonic of the voltage-controlled oscillator output is phase-locked to the input signal, is an effective frequency divider. The assumption has been made that the frequency divider contains a limiter, so that its output is insensitive to variations in input amplitude. It also should be mentioned that a phase-lock loop can be used as the bandpass pilot filter. In this application, the VCO output is the pilot signal, and B_p is the equivalent noise bandwidth of the phase-lock loop.

The technique just described is not usually used with single-channel systems because of the added bandwidth and power required for the pilot. However, in multichannel systems, a number of channels can use the same pilot. It is even possible to use a common pilot signal for a number of completely separate communication systems.

6.3 NOISE IN ANGLE MODULATION SYSTEMS

Now that we have investigated the effect of noise on a linear modulation system, we turn our attention to angle modulation. We will find that there are great differences between linear and angle modulation when noise effects are considered. We will even find significant differences between PM and FM. Finally, we will see that FM can offer greatly improved performance over both linear modulation and PM systems in noisy environments.

FIGURE 6.11   Angle demodulation system

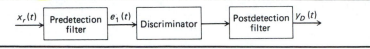

Demodulation in the Presence of Noise

Consider the system shown in Figure 6.11. The predetection filter bandwidth is B_T and can be determined by Carson's rule. In other words, B_T is approximately $2(D + 1)W$ Hz, where W is the bandwidth of the message signal and D is the deviation ratio, which is the peak frequency deviation divided by W. The input to the predetection filter is assumed to be the modulated carrier

$$x_c(t) = A_c \cos [\omega_c t + \theta + \phi(t)] \tag{6.85}$$

plus additive white noise that has the double-sided power spectral density $\frac{1}{2}N_0$ W/Hz.

At this point we recall that for PM,

$$\phi(t) = k_p m_n(t) \tag{6.86}$$

where k_p is the phase-deviation constant in radians per unit and $m_n(t)$ is the message signal normalized so that the peak value of $|m(t)|$ is unity. For the FM case,

$$\phi(t) = 2\pi f_d \int^t m_n(\alpha) \, d\alpha \tag{6.87}$$

where f_d is the deviation constant in hertz per unit. If the maximum value of $|m(t)|$ is not unity, as is usually the case, the scaling constant K, defined by $m(t) = Km_n(t)$, is contained in k_p or f_d. We shall analyze the PM and FM cases together and then specialize the results by replacing $\phi(t)$ by the proper function.

The output of the predetection filter can be written as

$$e_1(t) = A_c \cos [\omega_c t + \theta + \phi(t)]$$
$$+ n_c t \cos (\omega_c t + \theta) - n_s(t) \sin (\omega_c t + \theta) \tag{6.88}$$

where

$$\overline{n_c^2} = \overline{n_s^2} = N_0 B_T \tag{6.89}$$

Equation (6.88) can be written as

$$e_1(t) = A_c \cos [\omega_c t + \theta + \phi(t)] + r_n(t) \cos [\omega_c t + \theta + \phi_n(t)] \tag{6.90}$$

FIGURE 6.12 Phasor diagram for angle demodulation, assuming $(\text{SNR})_T > 1$ (drawn for $\theta = 0$)

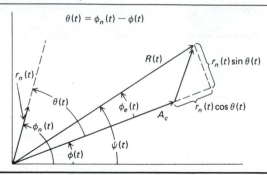

where $r_n(t)$ is the Rayleigh-distributed noise envelope and $\phi_n(t)$ is the uniformly distributed phase. By replacing $\omega_c t + \phi_n(t)$ with $\omega_c t + \phi(t) + \phi_n(t) - \phi(t)$, we can write (6.90) as

$$e_1(t) = A_c \cos [\omega_c t + \theta + \phi(t)]$$
$$+ \ r_n(t) \cos [\phi_n(t) - \phi(t)] \cos [\omega_c t + \theta + \phi(t)]$$
$$- \ r_n(t) \sin [\phi_n(t) - \phi(t)] \sin [\omega_c t + \theta + \phi(t)] \qquad (6.91)$$

which is

$$e_1(t) = \{A_c + r_n(t) \cos [\phi_n(t) - \phi(t)]\} \cos [\omega_c t + \theta + \phi(t)]$$
$$- \ r_n(t) \sin [\phi_n(t) - \phi(t)] \sin [\omega_c t + \theta + \phi(t)] \qquad (6.92)$$

The preceding expression can be written as

$$e_1(t) = R(t) \cos [\omega_c t + \theta + \phi(t) + \phi_e(t)] \qquad (6.93)$$

where $\phi_e(t)$ is the phase *deviation due to noise* and is given by

$$\phi_e(t) = \tan^{-1} \frac{r_n(t) \sin [\phi_n(t) - \phi(t)]}{A_c + r_n(t) \cos [\phi_n(t) - \phi(t)]} \qquad (6.94)$$

Since $\phi_e(t)$ adds to $\phi(t)$, *which conveys the message signal*, it is the noise component of interest.

A phasor diagram of the predetection filter output is given in Figure 6.12 for the case where $A_c > r_n(t)$. If $e_1(t)$ is expressed as

$$e_1(t) = R(t) \cos [\omega_c t + \theta + \psi(t)] \qquad (6.95)$$

the *phase deviation* of the discriminator input is

$$\psi(t) = \phi(t) + \phi_e(t) \qquad (6.96)$$

If the predetection signal-to-noise ratio $(\text{SNR})_T$ is large, $A_c \gg r_n(t)$ *most of the time. For this case,* (6.94) *becomes*

$$\phi_e(t) \cong \frac{r_n(t)}{A_c} \sin \left[\phi_n(t) - \phi(t)\right] \tag{6.97}$$

so that $\psi(t)$ is

$$\psi(t) = \phi(t) + \frac{r_n(t)}{A_c} \sin \left[\phi_n(t) - \phi(t)\right] \tag{6.98}$$

It is important to note that the effect of the noise $r_n(t)$ is reduced if the transmitted signal amplitude A_c is increased. Thus the output noise is affected by the transmitted signal amplitude even for above-threshold operation.

If the predetection SNR is small, $A_c \ll r_n(t)$ most of the time, and the phasor diagram appears as shown in Figure 6.13. For this case, the expression for $\psi(t)$ can be found from

$$r_n(t)[\phi_n(t) - \psi(t)] \cong A_c \sin \left[\phi_n(t) - \phi(t)\right] \tag{6.99}$$

or

$$\psi(t) = \phi_n(t) - \frac{A_c}{r_n(t)} \sin \left[\phi_n(t) - \phi(t)\right] \tag{6.100}$$

Several comments concerning (6.98) and (6.100) are in order. The signal $m(t)$ is contained in $\phi(t)$, and $r_n(t)$ contains only noise. For the case of a large

FIGURE 6.13   Phasor diagram for angle demodulation, assuming $(\text{SNR})_T < 1$ (drawn for $\theta = 0$)

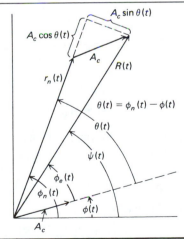

SNR, the phase of the predetection filter ouptut $\psi(t)$ is the sum of two terms. The first term is *signal alone,* and the second term contains components of both signal and noise. For the case of a small SNR, there is no term in $\psi(t)$ that can be identified as signal only. Actually, for small SNR, the only term containing the signal is *multiplied* by the noise. Thus the FM demodulator exhibits a threshold effect similar to that of the AM envelope detector. This effect will be examined in more detail in a later section.

The discriminator is assumed to contain a limiter and thus is insensitive to amplitude variations. Thus $R(t)$ can be assumed constant in (6.95). The discriminator output $y_D(t)$ for the PM case is given by

$$y_D(t) = K_D \psi(t) \tag{6.101}$$

and for the FM case is given by

$$y_D(t) = \frac{1}{2\pi} K_D \frac{d\psi}{dt} \tag{6.102}$$

where K_D is the discriminator constant. Substituting (6.96) into the preceding expression yields, *for the case of a large SNR,*

$$y_{DP}(t) = K_D \phi(t) + n_P(t) \tag{6.103}$$

for PM, and

$$y_{DF}(t) = \frac{1}{2\pi} K_D \frac{d\phi}{dt} + n_F(t) \tag{6.104}$$

for FM, where

$$n_P(t) = K_D \phi_e(t) = K_D \frac{r_n(t)}{A_c} \sin [\phi_n(t) - \phi(t)] \tag{6.105}$$

and

$$n_F(t) = \frac{K_D}{2\pi} \frac{d\phi_e}{dt} = \frac{K_D}{2\pi} \frac{d}{dt} \left\{ \frac{r_n(t)}{A_c} \sin [\phi_n(t) - \phi(t)] \right\} \tag{6.106}$$

are the output noise components for PM and FM, respectively. These outputs can be written in terms of the message signal as

$$y_{DP}(t) = K_D k_p m_n(t) + n_P(t) \tag{6.107}$$

and

$$y_{DF}(t) = K_D f_d m_n(t) + n_f(t) \tag{6.108}$$

The output signal power for PM is

$$S_{DP} = K_D^2 k_p^2 \overline{m_n^2} \tag{6.109}$$

and for FM is

$$S_{DF} = K_D^2 f_d^2 \overline{m_n^2} \tag{6.110}$$

Before the output signal-to-noise ratios can be calculated, the power spectral density of the output noise must be determined.

Output Noise Spectra

The power spectral density of the noise present at the discriminator output can, for the case of a large SNR, be computed from (6.105) and (6.106). As a first step in the analysis, we will set $\phi(t)$ equal to zero, so that $n_P(t)$ and $n_F(t)$ are functions of noise alone and contain no signal component. This assumption will greatly simplify our analysis, but it might seem that we are perhaps taking too many liberties. We could perform the analysis without assuming $\phi(t)$ equal to zero. The derivation would be much more complex and we would find that the effect of $\phi(t)$ is to produce frequency components in the demodulated output for $f > W$.* However, they can be removed by placing a lowpass filter in cascade with the discriminator output.

With $\phi(t) = 0$, the term $r_n(t) \sin [\phi_n(t) - \phi(t)]$ is equal to $n_s(t)$. Thus

$$n_P(t) = K_D \frac{n_s(t)}{A_c} \tag{6.111}$$

and

$$n_F(t) = \frac{K_D}{2\pi A_c} \frac{dn_s(t)}{dt} \tag{6.112}$$

From (6.111) it follows that the output noise power spectral density, for the PM case, is

$$S_{nP}(f) = \frac{K_D^2}{A_c^2} S_{ns}(f) = \frac{K_D^2}{A_c^2} N_0 \tag{6.113}$$

for $|f| < \frac{1}{2}B_T$ and zero otherwise, as illustrated in Figure 6.14(a). Thus the output noise power at the discriminator output is

$$N_{DP} = \frac{K_D^2}{A_c^2} \int_{-B_T/2}^{B_T/2} N_0 \, df = \frac{K_D^2}{A_c^2} N_0 B_T \tag{6.114}$$

The signal bandwidth is W. Since the predetection bandwidth B_T is greater than $2W$, the output SNR can be improved by following the discriminator

---

* An analysis that does not assume $\phi(t) = 0$ is contained in Downing (1964), pp. 96–98.

FIGURE 6.14   (a) Power spectral density for PM discriminator output, with portion for $|f| < W$ shaded. (b) Power spectral density for FM discriminator output, with portion for $|f| < W$ shaded.

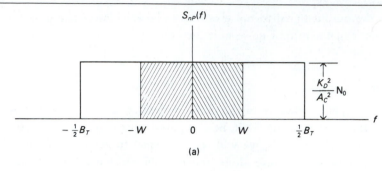

(a)

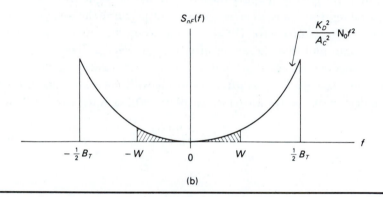

(b)

with a lowpass filter of bandwidth W. This filter has no effect on the signal but reduces the output noise power to

$$N_{DP} = \frac{K_D^{\,2}}{A_c^{\,2}} \int_{-W}^{W} N_0 \, df = 2 \, \frac{K_D^{\,2}}{A_c^{\,2}} N_0 W \qquad (6.115)$$

The output noise power is slightly more difficult to compute for the FM case because of the differentiation. It was shown in Chapter 5 that if $y(t) = dx/dt$, then $S_y(f) = (2\pi f)^2 S_x(f)$. Applying this result to (6.112) yields

$$S_{nF}(f) = \frac{K_D^{\,2}}{(2\pi)^2 A_c^{\,2}} (2\pi f)^2 N_0 = \frac{K_D^{\,2}}{A_c^{\,2}} N_0 f^2 \qquad (6.116)$$

for $|f| < \frac{1}{2} B_T$ and zero otherwise. This spectrum is illustrated in Figure 6.14(b). The parabolic shape of the noise spectrum results from the differentiating action of the FM discriminator and has a profound effect on the performance of FM systems operating in the presence of noise. It is clear from Figure 6.14(b) that low-frequency message-signal components are sub-jected to lower noise levels than are higher-frequency components. Once again, assuming that a lowpass filter having only sufficient bandwidth to pass the message follows the discriminator, the output noise power is

$$N_{DF} = \frac{K_D^{\,2}}{A_c^{\,2}}\, N_0 \int_{-W}^{W} f^2 \, df = \frac{2}{3}\frac{K_D^{\,2}}{A_c^{\,2}}\, N_0 W^3 \tag{6.117}$$

This quantity is indicated by the shaded area in Figure 6.14(b).

It is useful to write (6.115) for PM and (6.117) for FM in terms of P_T/N_0W. Since $P_T = A_c^{\,2}/2$ for both PM and FM, we have

$$\frac{P_T}{N_0W} = \frac{A_c^{\,2}}{2N_0W} \tag{6.118}$$

Thus, from (6.115) the discriminator output noise power for PM is

$$N_{DP} = K_D^{\,2} \left(\frac{P_T}{N_0W} \right)^{-1} \tag{6.119}$$

and from (6.117) the discriminator output noise power for FM is

$$N_{DF} = \tfrac{1}{3}K_D^{\,2}\, W^2 \left(\frac{P_T}{N_0W} \right)^{-1} \tag{6.120}$$

Note that for both PM and FM the noise power at the discriminator output is inversely proportional to P_T/N_0W.

Output Signal-to-Noise Ratios

The signal-to-noise ratio at the discriminator output is now easily deter-mined. For the PM system, the discriminator output SNR is the signal power, defined by (6.109), divided by the noise power, defined by (6.119). This gives

$$(\text{SNR})_{DP} = \frac{K_D^{\,2} k_p^{\,2} \overline{m_n^{\,2}}}{K_D^{\,2}(P_T/N_0W)^{-1}} \tag{6.121}$$

which gives

$$(\text{SNR})_{DP} = k_p^{\,2} \overline{m_n^{\,2}}\, \frac{P_T}{N_0W} \tag{6.122}$$

The above expression shows that the improvement of PM over linear mod-ulation depends on the phase-deviation constant and the power in the mod-ulating signal. It should be remembered that if the phase deviation of a PM signal exceeds π radians, unique demodulation cannot be accomplished. Thus the peak value of $|k_p m_n(t)|$ is π, and the maximum value of $k_p^2 m_n^2$ is π^2. This yields a *maximum* improvement of approximately 10 dB over baseband. In reality, the improvement is significantly less because $k_p^2 m_n^2$ is typically much less than the maximum value of π^2. It should be pointed out that if the constraint that the output of the phase demodulator is continuous is imposed, then it is possible for $|k_p m_n(t)|$ to exceed π.

We now turn our attention to the FM case. For FM the signal-to-noise ratio at the discriminator output is the signal power, defined by (6.110) divided by the noise power, defined by (6.120). This gives

$$(\text{SNR})_{DF} = \frac{K_D^2 f_d^2 \overline{m_n^2}}{\frac{1}{3}K_D^2 \left(\dfrac{P_T}{N_0 W}\right)^{-1}} \tag{6.123}$$

which can be expressed as

$$(\text{SNR})_{DF} = 3\left(\frac{f_d}{W}\right)^2 \overline{m_n^2} \frac{P_T}{N_0 W} \tag{6.124}$$

where P_T is the transmitted signal power $\frac{1}{2}A_c^2$. Since the ratio of peak deviation to W is the deviation ratio D, the output SNR is

$$(\text{SNR})_{DF} = 3D^2 \overline{m_n^2} \frac{P_T}{N_0 W} \tag{6.125}$$

where the maximum value of $|m_n(t)|$ is unity. Note that the maximum value of $m(t)$, together with f_d and W, determines D.

For the FM case, there is no restriction on the magnitude of the output SNR as there was for PM. As a matter of fact, it appears that we can increase D without bound, thereby increasing the output SNR to an arbitrarily large value. One price we pay for this enlarged SNR is excessive transmission bandwidth. For $D \gg 1$, the required bandwidth B_T is approximately $2DW$, which yields

$$(\text{SNR})_{DF} = \frac{3}{4}\left(\frac{B_T}{W}\right)^2 \overline{m_n^2} \left(\frac{P_T}{N_0 W}\right) \tag{6.126}$$

This expression illustrates the tradeoff that exists between bandwidth and output SNR. However, (6.126) is valid only if the discriminator input SNR is sufficiently large to result in operation *above threshold*. Thus the output SNR *cannot* be increased to any arbitrary value by simply increasing the deviation ratio and thus the transmission bandwidth. This effect will be studied in detail in a later section. First, however, we will study a simple technique for gaining additional improvement in the output SNR.

Performance Enhancement Through the Use of Deemphasis

In Chapter 3 we saw that preemphasis and deemphasis can be used to partially combat the effects of interference. These techniques can also be used to great advantage when noise is present in angle modulation systems.

As we saw in Chapter 3, the deemphasis filter is usually a first-order low-pass RC filter placed directly at the discriminator output. Prior to modulation, the signal is passed through a preemphasis filter with a transfer function so that the combination of the preemphasis and deemphasis filters has no net effect on the message signal. The deemphasis filter is followed by a low-pass filter, assumed to be ideal with bandwidth W, which eliminates the out-of-band noise. Assume the deemphasis filter to have the amplitude response

$$|H_{DE}(f)| = \frac{1}{\sqrt{1 + (f/f_3)^2}} \tag{6.127}$$

where f_3 is the 3-dB frequency $1/(2\pi RC)$ Hz. The total noise power output with deemphasis is

$$N_{DF} = \int_{-W}^{W} |H_{DE}(f)|^2 S_{nF}(f)\, df \tag{6.128}$$

Substituting $S_{nF}(f)$ from (6.116) and $|H_{DE}(f)|$ from (6.127) yields

$$N_{DF} = \frac{K_D^2}{A_c^2} N_0 f_3^2 \int_{-W}^{W} \frac{f^2}{f_3^2 + f^2}\, df \tag{6.129}$$

or

$$N_{DF} = 2\frac{K_D^2}{A_c^2} N_0 f_3^3 \left(\frac{W}{f_3} - \tan^{-1}\frac{W}{f_3}\right) \tag{6.130}$$

In a typical situation, $f_3 \ll W$, so $\tan^{-1}(W/f_3) \cong \frac{1}{2}\pi$, which is small compared to W/f_3. For this case,

$$N_{DF} = 2\left(\frac{K_D^2}{A_c^2}\right) N_0 f_3^2 W \tag{6.131}$$

and the output SNR becomes

$$(\text{SNR})_{DF} = \left(\frac{f_d}{f_3}\right)^2 \overline{m_n^2} \frac{P_T}{N_0 W} \tag{6.132}$$

A comparison of (6.132) with (6.124) illustrates that for $f_3 \ll W$, the improvement gained through the use of preemphasis and deemphasis can be very significant in noisy environments.

▼ EXAMPLE 6.3   Commerical FM operates with $f_d = 75$ kHz, $W = 15$ kHz, $D = 5$, and the standard value of 2.1 kHz for f_3. Assuming that $\overline{m_n}^2 = 0.1$ we have, for FM without deemphasis,

$$(\text{SNR})_{DF} = 7.5 \frac{P_T}{N_0 W} \qquad (6.133)$$

With deemphasis, the result is

$$(\text{SNR})_{DF,pe} = 128 \frac{P_T}{N_0 W} \qquad (6.134)$$

With the chosen values, FM without deemphasis is 8.75 dB superior to base-band, and FM with deemphasis is 21 dB superior. This improvement more ▲ than justifies the use of deemphasis.

As mentioned in Chapter 3, a price is paid for the SNR improvement gained by the use of preemphasis. The action of the preemphasis filter is to accentuate the high-frequency portion of the message signal. Thus preemphasis may increase the transmitter deviation and, consequently, the bandwidth required for signal transmission. Fortunately, many message signals of practical interest have relatively small energy in the high-frequency portion of their spectrum, so this effect is of little or no importance.

6.4 THRESHOLDS AND THRESHOLD EXTENSION IN FM

In Section 6.3 we observed that an FM discriminator exhibits a threshold effect. We now take a closer look at the threshold effect, and we examine techniques that can be used to improve the noise performance of FM systems by extending the threshold.

Threshold Effects in FM Discriminators

Significant insight into the mechanism by which threshold effects take place can be gained by performing a relatively simple laboratory experiment. On the input to an FM discriminator is placed an unmodulated sinusoid plus additive, bandlimited white noise having a power spectral density symmetrical about the frequency of the sinusoid. Starting out with a high SNR at the discriminator input, the noise power is gradually increased while continually observing the discriminator output on an oscilloscope. Initially, the discriminator output resembles bandlimited white noise. As the noise power spectral density is increased, thereby reducing the input SNR, a point is reached at which spikes or impulses appear in the discriminator output. The initial appearance of these spikes denotes that the discriminator is operating in the region of threshold.

The statistics for these spikes were examined in Section 5.6. In this section we review the phenomenon of spike noise with specific application to FM demodulation. The system under consideration is that of Figure 6.11. For this case,

$$e_1(t) = A_c \cos(\omega_c t + \theta) + n_c(t) \cos(\omega_c t + \theta) - n_s(t) \sin(\omega_c t + \theta)$$
$$(6.135)$$

which is

$$e_1(t) = A_c \cos(\omega_c t + \theta) + r_n(t) \cos[\omega_c t + \theta + \phi_n(t)] \quad (6.136)$$

or

$$e_1(t) = R(t) \cos[\omega_c t + \theta + \psi(t)] \quad (6.137)$$

The phasor diagram of this signal is given in Figure 6.15. Like Figure 5.15, it illustrates the mechanism by which spikes occur. The signal amplitude is A_c and the angle is θ, since the carrier is assumed unmodulated. The noise amplitude is $r_n(t)$. The angle difference between signal and noise is $\phi_n(t)$. As threshold is approached, the noise amplitude grows until, at least part of the time, $|r_n(t)| > A_c$. Also, since $\phi_n(t)$ is uniformally distributed, the phase of the noise is sometimes in the region of $-\pi$. Thus the resultant phasor $R(t)$ can occasionally encircle the origin. When $R(t)$ is in the region of the origin, a relatively small change in the amplitude or phase of the noise results in a rapid change in $\psi(t)$. Since the discriminator output is proportional to the time rate of change $\psi(t)$, the discriminator output is very large as the origin is encircled. This is essentially the same effect that was observed in Chapter

FIGURE 6.15   Phasor diagram near threshold (spike output) (drawn for $\theta = 0$)

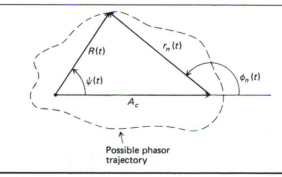

Possible phasor
trajectory

FIGURE 6.16   Phase deviation for a predetection signal-to-noise ratio of -4.0 dB

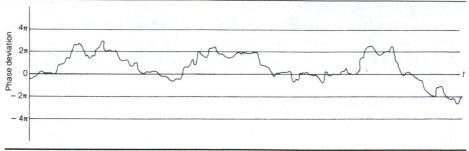

3 where the behavior of an FM discriminator operating in the presence of interference was studied.

The phase deviation $\psi(t)$ is illustrated in Figure 6.16 for the case in which the input signal-to-noise ratio is -4.0 dB. The origin encirclements can be observed by the 2π jumps in $\psi(t)$. The output of an FM discriminator for several predetection signal-to-noise ratios is shown in Figure 6.17. The decrease in spike noise as the signal-to-noise ratio is increased is clearly seen.

In Section 5.6 we showed that the power spectral density of spike noise is given by

$$S_{\dot\psi}(f) = (2\pi)^2(\nu + \overline{\delta\nu}) \qquad (6.138)$$

where ν is the average number of impulses per second of an unmodulated carrier plus noise and $\delta\nu$ is the net increase of the spike rate due to modulation. Since the discriminator output is given by

$$y_D(t) = \frac{1}{2\pi} K_D \frac{d\psi}{dt} \qquad (6.139)$$

The power spectral density due to spike noise at the discriminator output is

$$S_{D\delta} = K_D{}^2\nu + K_D{}^2\overline{\delta\nu} \qquad (6.140)$$

Using (5.159) for ν yields

$$K_D{}^2\nu = K_D{}^2 \frac{B_T}{\sqrt{3}} Q\left[\sqrt{\frac{A_c{}^2}{N_0 B_T}}\right] \qquad (6.141)$$

where $Q(x)$ is the probability Q-function defined in (4.186). Using (5.164) for $\overline{\delta\nu}$ yields

FIGURE 6.17   Output of FM discriminator due to input noise for various predetection signal-to-noise ratios. (a) Predetection SNR $= -10$ dB. (b) Predectection SNR $= -4$ dB. (c) Predetection SNR $= 0$ dB. (d) Predetection SNR $= 6$ dB. (e) Predetection SNR $= 10$ dB.

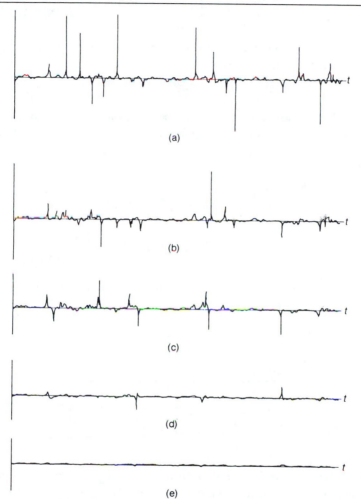

$$K_D{}^2 \delta \nu = K_D{}^2 |\overline{\delta f}| \exp \left[\frac{-A_c{}^2}{2N_0 B_T} \right] \tag{6.142}$$

Since the spike noise at the discriminator output is white, the spike noise power at the discriminator output is found by multiplying the power spectral density by the two-sided postdetection bandwidth, $2W$. Substituting (6.141) and (6.142) into (6.140) and multiplying by $2W$ yields

$$N_{D\delta} = K_D^2 \frac{2B_T W}{\sqrt{3}} Q \left[\sqrt{\frac{A_c^2}{N_0 B_T}} \right] + K_D^2 (2W) |\overline{\delta f}| \exp \left[\frac{-A_c^2}{2N_0 B_T} \right] \quad (6.143)$$

for the spike noise power. Now that the spike noise power is known we can determine the total noise power at the discriminator output. After this is accomplished the output signal-to-noise ratio at the discriminator output is easily determined.

The total noise power at the discriminator output is the sum of the Gaussian noise power and spike noise power. The total noise power is therefore found by adding (6.143) to (6.120). This gives

$$N_D = \tfrac{1}{3} K_D^2 W^2 \left(\frac{P_T}{N_0 W} \right)^{-1} + K_D^2 \frac{2B_T W}{\sqrt{3}} Q \left[\sqrt{\frac{A_c^2}{N_0 B_T}} \right]$$
$$+ K_D^2 (2W) |\overline{\delta f}| \exp \left[- \frac{A_c^2}{2N_0 B_T} \right] \quad (6.144)$$

The signal power at the discriminator output is given by (6.110). Dividing the signal power by the noise power given above yields, after canceling the K_D^2 terms,

$$(\text{SNR})_D =$$

$$\frac{f_d^2 \, \overline{m_n^2}}{\tfrac{1}{3} W^2 \left(\dfrac{P_T}{N_0 W} \right)^{-1} + \dfrac{2B_T W}{\sqrt{3}} Q \left[\sqrt{\dfrac{A_c^2}{N_0 B_T}} \right] + 2W |\overline{\delta f}| \exp \left[\dfrac{-A_c^2}{2N_0 B_T} \right]}$$

$$(6.145)$$

This result can be placed in standard form by dividing numerator and denominator by the leading term in the denominator. This gives the final result

$$(\text{SNR})_D =$$

$$\frac{3 \left(\dfrac{f_d}{W} \right)^2 \overline{m_n^2} \dfrac{P_T}{N_0 W}}{1 + 2\sqrt{3} \dfrac{B_T}{W} \left(\dfrac{P_T}{N_0 W} \right) Q \left[\sqrt{\dfrac{A_c^2}{N_0 B_T}} \right] + 6 \dfrac{|\overline{\delta f}|}{W} \left(\dfrac{P_T}{N_0 W} \right) \exp \left[\dfrac{-A_c^2}{2N_0 B_T} \right]}$$

$$(6.146)$$

For operation above threshold, the region of input signal-to-noise ratios where spike noise is negligible, the last two terms in the denominator of the preceding expression are much less than one and may therefore be neglected. For this case the postdetection signal-to-noise ratio defined by the preceding expression is equivalent to the above threshold result given by (6.124). It is worth noting that the quantity $A_c^2/(2N_0B_T)$ appearing in each of the two spike noise terms is the predetection signal-to-noise ratio. Note that the message signal explicitly affects two terms in the expression for the postdetection signal-to-noise ratio through $\overline{m_n^2}$ and $|\overline{\delta f}|$. Thus, before the detection gain can be determined, a message signal must be assumed. This is the subject of the following example.

▼ **EXAMPLE 6.4**   In this example the detection gain of an FM discriminator is determined assuming the sinusoidal message signal

$$m_n(t) = \sin 2\pi Wt \tag{6.147}$$

The instantaneous frequency deviation is given by

$$f_d m_n(t) = f_d \sin 2\pi Wt \tag{6.148}$$

and the average of the absolute value of the frequency deviation is therefore given by

$$|\overline{\delta f}| = 2W \int_0^{1/2W} f_d \sin 2\pi Wt \, dt \tag{6.149}$$

Carrying out the integration yields

$$|\overline{\sigma f}| = \frac{2}{\pi}(f_d) \tag{6.150}$$

Note that f_d is the peak frequency deviation which, by definition of the modulation index, β, is βW. Thus

$$|\overline{\delta f}| = \frac{2}{\pi} \beta W \tag{6.151}$$

From Carson's rule we have

$$\frac{B_T}{W} = 2(\beta + 1) \tag{6.152}$$

Since the message signal is sinusoidal

$$\left(\frac{f_d}{W}\right)^2 \overline{m_n^2} = \frac{1}{2}\left(\frac{f_d}{W}\right)^2 = \frac{1}{2}\beta^2 \tag{6.153}$$

Finally, the predetection signal-to-noise ratio can be written

$$\frac{A_c{}^2}{2N_0B_T} = \frac{1}{2(\beta + 1)} \frac{P_T}{N_0W} \tag{6.154}$$

Substituting (6.153) and (6.154) into (6.146) yields

$$(SNR)_D = \cfrac{1.5\beta^2 \, \dfrac{P_T}{N_0W}}{\begin{aligned}&1 + 4\sqrt{3}(\beta + 1)\left(\dfrac{P_T}{N_0W}\right)Q\left[\sqrt{\dfrac{1}{(\beta + 1)}\dfrac{P_T}{N_0W}}\right]\\[2mm]&+ \dfrac{12}{\pi}\beta\left(\dfrac{P_T}{N_0W}\right)\exp\left[\dfrac{-1}{2(\beta + 1)}\dfrac{P_T}{N_0W}\right]\end{aligned}} \tag{6.155}$$

for the postdetection signal-to-noise ratio.

The postdetection signal-to-noise ratio is illustrated in Figure 6.18 as a function of P_T/N_0W. The threshold value of P_T/N_0W is the value of P_T/N_0W at which the postdetection SNR is 3 dB below the value of the postdetection SNR given by the above threshold analysis. In other words, the threshold value of P_T/N_0W is the value of P_T/N_0W for which the denominator of (6.155) is equal to 2. It should be noted from Figure 6.18 that the threshold value of P_T/N_0W increases as the modulation index β increases. The study of this effect is the subject of one of the computer exercises at the end of this chapter.

In most communication systems it is important to maintain the operating point above threshold. There are two basic purposes served by Figure 6.18. First, we see the rapid convergence to the result of the above threshold analysis described by (6.124), with (6.153) used to allow (6.124) to be written in terms of the modulation index. The second important point emphasized by Figure 6.18 is the rapid deterioration of performance as the operating point moves into the below-threshold region.

Threshold Extension

There are several techniques that can be used for lowering the value of P_T/N_0W at which threshold occurs. Two systems, the frequency-compressive feedback loop and the phase-lock loop, have found widespread application in high-noise environments. Both these systems were presented briefly in Chapter 3.

FIGURE 6.18   FM system performance with sinusoidal modulation

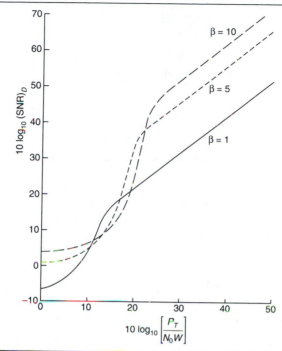

The block diagram of a frequency-compressive feedback loop is repeated in Figure 6.19. The input to the loop is assumed to be

$$x_r(t) = A_c \cos [\omega_c t + \psi(t)] \tag{6.156}$$

where $\psi(t)$ represents the phase deviation of the input resulting from both modulation and noise. Amplitude perturbations of $x_r(t)$ due to noise are neglected because they do not affect the discriminator output. In Chapter 3 we showed that the input to the discriminator can be written as

$$x(t) = K \cos \left[\omega_0 t + \frac{1}{1 + \alpha} \psi(t) \right] \tag{6.157}$$

where K is a constant related to the input amplitude and the VCO output amplitude and α is the loop gain. This is given, from (3.245), by

$$\alpha = \frac{1}{2\pi} K_D K_v \tag{6.158}$$

where $(1/2\pi)(K_D)$ is the discriminator gain and K_v is the gain of the VCO.

FIGURE 6.19   Frequency-compressive feedback loop

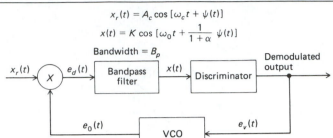

$$x_r(t) = A_c \cos\left[\omega_c t + \psi(t)\right]$$

$$x(t) = K \cos\left[\omega_0 t + \frac{1}{1+\alpha}\,\psi(t)\right]$$

In the frequency-compressive feedback loop, a spike occurs when $[1/(1+\alpha)]\psi(t)$ changes by 2π. The number of occurrences per second of this event, assuming sinusoidal modulation, is given by

$$\nu + \overline{\delta\nu} = \frac{B_p}{2\sqrt{3}}\,\text{erfc}\left(\sqrt{\frac{A_c^{\,2}}{2N_0 B_p}}\right) + \frac{2}{\pi}\,f_d\,\exp\left(-\frac{A_c^{\,2}}{2N_0 B_p}\right) \qquad (6.159)$$

where B_p is the bandwidth of the bandpass filter at the input to the discriminator. As shown in Chapter 3, B_p is smaller than the transmission bandwidth B_T due to bandwidth compression. Thus the rate of spikes is less with frequency-compressive feedback, and threshold is therefore extended.

Above threshold, (6.159) is negligible, and additive Gaussian noise is the only significant degradation of the output. For this case, both the signal and noise components are suppressed by $1 + \alpha$, as illustrated by (6.157). Thus, above threshold, the postdetection SNR is the same for the frequency-compressive feedback loop as for the ordinary discriminator.

The phase-lock loop also has been found to be useful for threshold extension. It is somewhat more difficult to analyze than the frequency-compressive feedback loop, and many developments have been published.* Thus we will not cover it here. We state, however, that the threshold extension obtained with the phase-lock loop is typically on the order of 2 to 3 dB. Even though this is a moderate extension, it can be important in high-noise environments.

6.5   NOISE IN PULSE MODULATION

We shall now investigate the performance of analog and digital pulse modulation systems in the presence of additive noise. We will see once again

---

* See Taub and Schilling (1986), pp. 419–422, for an introductory treatment.

that nonlinear systems allow a tradeoff between transmission bandwidth and postdetection signal-to-noise ratio.

Analog Pulse Modulation

For all analog pulse modulation techniques, demodulation involves the reconstruction of a continuous message signal from sample values that have been perturbed by noise. Assume that the received signal is of the form

$$x_r(t) = m_\delta(t) + n(t) \tag{6.160}$$

where $m_\delta(t)$ is the sampled message signal and $n(t)$ represents additive noise.

For pulse-amplitude modulation (PAM), the message signal is sampled, passed through the channel, and reconstructed using lowpass filtering. Thus, in effect, $m(t)$ is transmitted directly. Assuming that the output of the reconstruction filter is

$$y_D(t) = m(t) + n_0(t) \tag{6.161}$$

the maximum postdetection SNR is

$$(\text{SNR})_D = \frac{\overline{m^2(t)}}{N_0 W} \tag{6.162}$$

where W is the bandwidth of both $m(t)$ and the lowpass reconstruction filter. Thus PAM is essentially equivalent to continuous baseband transmission.

In both pulse-width modulation (PWM) and pulse-position modulation (PPM), the information resides in the position of pulse edges. Demodulation is accomplished by detecting pulse edges that have been perturbed by additive noise. If the channel bandwidth were infinite, perfectly rectangular pulses could be transmitted. For this case, additive noise would not cause a time perturbation of the pulse edge, and demodulation could be accomplished without error. Thus, for PWM and PPM, we can expect to see a tradeoff between postdetection SNR and transmission bandwidth.

If the channel bandwidth is finite, the pulses will have nonzero risetime, and noise will perturb pulse detection. This is illustrated in Figure 6.20. Edge detection is accomplished by detecting the time at which the pulse crosses a threshold. The error $\epsilon(t_i)$ is related to the noise $n(t_i)$, the pulse amplitude A, and the risetime T_R by

$$\frac{\epsilon(t_i)}{T_R} = \frac{n(t_i)}{A} \tag{6.163}$$

Thus the mean-square error is given by

$$\overline{\epsilon^2(t_i)} = \left(\frac{T_R}{A}\right)^2 \overline{n^2(t_i)} \tag{6.164}$$

FIGURE 6.20   PWM and PPM demodulation in the presence of noise

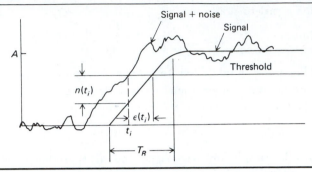

which, assuming the noise process is stationary, is

$$\overline{\epsilon^2} = \left(\frac{T_R}{A}\right)^2 \overline{n^2} = \left(\frac{T_R}{A}\right)^2 N_0 B_T \tag{6.165}$$

where B_T is the transmission bandwidth.

The preceding expression can be placed in more useful form by recalling from Chapter 2 that risetime is related to bandwidth by

$$T_R \cong \frac{1}{2B_T} \tag{6.166}$$

so that

$$\overline{\epsilon^2} = \frac{N_0}{4A^2 B_T} \tag{6.167}$$

Assume that the pulse position, or pulse width, is defined by

$$\tau(t) = \tau_0 + \tau_1 m_n(t) \tag{6.168}$$

so that τ_1 represents the maximum pulse *displacement* for PPM and the maximum *change* in pulse width for PWM. Thus the signal power is

$$S_D = \tau_1^2 \overline{m_n^2} \tag{6.169}$$

If the sampling period is T_s, the average transmitted power is

$$P_T = \frac{\tau_0}{T_s} A^2 \tag{6.170}$$

since the average pulse duration is τ_0 if $m(t)$ is assumed to be zero mean. Substituting (6.170) into (6.167) yields

$$\overline{\epsilon^2} = \frac{N_0 \tau_0}{4 T_s P_T B_T} \tag{6.171}$$

which represents the postdetection noise power. Using (6.169) for signal power and (6.171) for noise power, the postdetection SNR becomes

$$(\text{SNR})_D = \frac{4\tau_1^2 \overline{m_n^2} T_s P_T B_T}{\tau_0 N_0} \tag{6.172}$$

which is, in terms of $P_T/N_0 W$,

$$(\text{SNR})_D = \frac{4\tau_1^2 \overline{m_n^2} T_s B_T W}{\tau_0} \left(\frac{P_T}{N_0 W} \right) \tag{6.173}$$

The tradeoff between bandwidth and postdetection SNR is evident from (6.173). As with FM, however, the transmission bandwidth cannot be increased without bound. Both PPM and PWM are nonlinear schemes and have definite thresholds. As B_T is increased, the predetection SNR is increased, and large instantaneous noise amplitudes are detected as pulse edges. Thus the preceding analysis is valid only for large signal-to-noise ratios. These systems will be revisited in Chapter 9, where a more complete analysis will be presented.

Pulse Code Modulation

Pulse code modulation (PCM) was briefly presented in Chapter 3. We now wish to analyze the performance of such systems.

There are two error sources in PCM, the error resulting from quantization and the error resulting from additive channel noise. As we saw in Chapter 3, quantization involves representing each input sample by one of q quantization levels. Each quantization level is then transmitted using a sequence of binary symbols to uniquely represent each quantization level.

The sampled and quantized message waveform can be represented as

$$m_{\delta q}(t) = \Sigma m(t)\, \delta(t - iT_s) + \Sigma \epsilon(t)\, \delta(t - iT_s) \tag{6.174}$$

where the first term represents the sampling operation and the second term represents quantization. The ith sample of $m_{\delta q}(t)$ is

$$m(t_i) + \epsilon(t_i)$$

where $t_i = iT_s$. Thus the signal-to-noise ratio resulting from quantization is

$$(\text{SNR})_Q = \frac{\overline{m^2(t_i)}}{\overline{\epsilon^2(t_i)}} = \frac{\overline{m^2(t)}}{\overline{\epsilon^2(t)}} \tag{6.175}$$

assuming stationarity. The quantization error term is easily evaluated for the case in which the quantization levels have uniform spacing S. The quantization error is bounded by $\pm\frac{1}{2}S$. Thus, assuming that $\epsilon(t)$ is uniformally distributed in the range

$$-\tfrac{1}{2}S \leq \epsilon(t_i) \leq \tfrac{1}{2}S$$

the mean-square error due to quantization is

$$\overline{\epsilon^2} = \frac{1}{S} \int_{-S/2}^{S/2} x^2 \, dx = \tfrac{1}{12}S^2 \tag{6.176}$$

so that

$$(\text{SNR})_Q = 12 \, \frac{\overline{m^2(t)}}{S^2} \tag{6.177}$$

The next step is to express $\overline{m^2(t)}$ in terms of q and S. If there are q quantization levels, each of width S, it follows that the peak-to-peak value of $m(t)$ is qS. Assuming that $m(t)$ is *uniformally* distributed in this range,

$$\overline{m^2(t)} = \tfrac{1}{12}q^2 S^2 \tag{6.178}$$

Thus (6.177) becomes

$$(\text{SNR})_Q = q^2 = 2^{2n} \tag{6.179}$$

where n is the number of binary symbols used to represent each quantization level.

If the additive noise in the channel is so small that errors can be neglected, quantization is the only error source. Thus (6.179) becomes the postdetection SNR and is independent of $P_T/N_0 W$. If quantization is not the only error source, the postdetection SNR depends not only on $P_T/N_0 W$ but also on the signaling scheme. An analysis of these various signaling schemes is the subject of Chapter 7.

An *approximate* analysis of PCM is easily carried out for the case in which the signal-to-noise ratio is assumed not to be large. Each sample value is transmitted as a group of n pulses, and any of these n pulses can be in error at the receiver due to noise effects in the channel. Chapter 7 deals with the calculation of the error probability as a function of both the received signal-to-noise ratio and the modulation technique used for transmission of the pulses. The group of n pulses defines the quantizing level and is referred to as a *digital word*. Each individual pulse is a digital symbol, or *bit*. We assume that the bit-error probability P_b is known (it will be after the next chapter). Each of the n bits in the digital word is received correctly with probability

$1 - P_b$. Assuming independence, the probability that all n bits are received correctly is $(1 - P_b)^n$. The word-error probability P_w is therefore given by

$$P_w = 1 - (1 - P_b)^n \qquad (6.180)$$

The effect of a word error depends on which bit of the digital word is in error. If the least significant bit is in error, the resulting amplitude error is $\pm S$, or one quantizing level. If the next bit is in error, the resulting amplitude error is $\pm 2S$. The next bit results in an amplitude error of $\pm 4S$, and so forth. An error in the most significant bit results in an amplitude error of $\frac{1}{2}qS$. The effect of a word error is therefore an amplitude error in the range

$$-\tfrac{1}{2}qS \leq \epsilon_w \leq \tfrac{1}{2}qS$$

If we assume that ϵ_w is uniformally distributed in this range, then the mean-square word error is

$$\overline{\epsilon_w^2} = \tfrac{1}{12}q^2S^2 \qquad (6.181)$$

which is equal to the signal power.

The total noise power at the output of a PCM system is given by

$$N_D = \overline{\epsilon^2}(1 - P_w) + \overline{\epsilon_w^2}P_w \qquad (6.182)$$

The first term on the right-hand side of (6.182) is the contribution to the total noise due to quantizing error. It is equal to the quantizing error power weighted by the probability that all bits in the word are received correctly. The second term is the error power due to a word error weighted by the probability of word error. Using (6.182) for the signal power yields

$$(\text{SNR})_D = \frac{\tfrac{1}{12}q^2S^2}{\tfrac{1}{12}S^2(1 - P_w) + \tfrac{1}{12}S^2q^2P_w} \qquad (6.183)$$

which can be written as

$$(\text{SNR})_D = \frac{1}{q^{-2}(1 - P_w) + P_w} \qquad (6.184)$$

Equation (6.184) can be written in terms of the wordlength n using (6.179). The result is

$$(\text{SNR})_D = \frac{1}{2^{-2n} + P_w(1 - 2^{-2n})} \qquad (6.185)$$

The term 2^{-2n} is completely determined by the number of quantizing levels, while the word-error probability P_w is largely determined by the signal-to-noise ratio P_T/N_0W.

▼ **EXAMPLE 6.5**   The purpose of this example is to examine the postdetection signal-to-noise ratio for a PCM system. Before the postdetection signal-to-noise ratio, $(SNR)_D$, can be numerically evaluated, the word-error probability, P_w, must be known. As shown by (6.180) the word-error probability depends upon the bit-error probability, the calculation of which is the subject of the following chapter. Borrowing a result from Chapter 7, however, will allow us to illustrate the threshold effect of PCM. If we assume frequency-shift keying (FSK), which, as was discussed in Chapter 3, is transmission using one frequency to represent a binary zero and a second frequency to represent a binary one, and a noncoherent receiver, the probability of bit error is

$$P_b = \tfrac{1}{2} \exp\left[-\frac{P_T}{2N_0 B_p} \right] \tag{6.186}$$

In the preceding expression B_p is the bit-rate bandwidth, which is the reciprocal of the time required for transmission of a single bit in the n-symbol PCM digital word. Substitution of (6.186) into (6.180) and substitution of the result into (6.185) yields the postdetection signal-to-noise ratio, $(SNR)_D$. This result is shown in Figure 6.21. The threshold effect can be seen clearly.

Figure 6.21 shows that the postdetection signal-to-noise ratio approaches a value determined completely by the quantization noise. This can also be seen from (6.185) by letting the word-error probability tend to zero. The residual error, due to quantization, is defined by (6.179). Thus, above threshold,

$$(SNR)_D \cong 2^{2n} \tag{6.187}$$

which, expressed in decibels, is

$$10 \log_{10} (SNR)_D = 6.02n \tag{6.188}$$

We therefore gain 6 dB in signal-to-noise ratio for every bit added to the quantizer wordlength. However, the longer digital word means that more bits must be transmitted for each sample taken of the original time-domain signal, $m(t)$. This increases the bandwidth requirements of the system. Thus, the improved signal-to-noise ratio comes at the expense of a higher bit rate or system bandwidth. We therefore see again the tradeoff between signal-to-noise ratio and bandwidth. It is worth noting that, above threshold, the postdetection signal-to-noise ratio is independent of the modulation ▲ method.

Companding

As we saw in Chapter 3, a PCM signal is formed by modulating a digital signal. The digital signal is generated from an analog signal through the

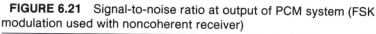

FIGURE 6.21   Signal-to-noise ratio at output of PCM system (FSK modulation used with noncoherent receiver)

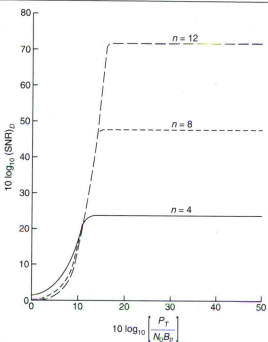

operations of sampling the analog waveform, quantizing the resulting samples, and encoding the quantized samples into digital words of finite length. These three operations are collectively referred to as analog-to-digital (A/D) conversion. The inverse process of forming an analog signal from a digital signal is known as digital-to-analog (D/A) conversion.

In the preceding section we saw that significant errors can result from the quantizing process if the wordlength n is chosen too small for a particular application. The result of these errors is expressed by the signal-to-quantization-noise ratio expressed by (6.179). Keep in mind that (6.179) was developed for the case of a uniformly distributed signal.

The level of quantizing noise added to a given sample, Equation (6.176), is independent of the signal amplitude, and small amplitude signals will therefore suffer more from quantization effects than large amplitude signals. This can be seen from (6.177). There are essentially two ways to combat this problem. First, the quantizing steps can be made small for small amplitudes and large for large amplitude portions of the signal. This scheme is known as nonuniform quantizing. An example of a nonuniform quantizer is the Max

FIGURE 6.22   Input-output compression characteristic

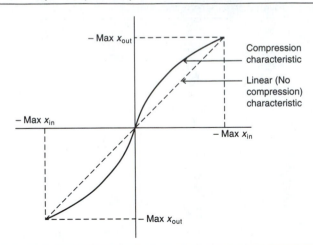

quantizer, in which the quantization steps are chosen so that the mean-square quantization error is minimized. The second technique, and the one of interest here, is to pass the analog signal through a nonlinear amplifier prior to the sampling process. The input-output characteristics of the amplifier are shown in Figure 6.22. For small values of the input x_{in}, the slope of the input-output characteristic is largest. A change in a low-amplitude signal will therefore force the signal through more quantization levels than the same change in a high-amplitude signal. This essentially yields smaller step sizes for small amplitude signals and therefore reduces the quantization error for small amplitude signals. It can be seen from Figure 6.22 that the peaks of the input signal are compressed. For this reason the characteristic shown in Figure 6.22 is known as a compressor.

The effect of the compressor must be compensated for when the signal is returned to analog form. This is accomplished by placing another nonlinear amplifier at the output of the D/A converter. This second nonlinear amplifier is known as an expander and is chosen so that the cascade combination of the compressor and expander yields a linear characteristic, as shown by the dashed line in Figure 6.22. The combination of a compressor and an expander is known as a compander. A companding system is shown in Figure 6.23.

The concept of predistorting a message signal in order to achieve better performance in the presence of noise, and then removing the effect of the predistortion, should remind us of the use of preemphasis and deemphasis filters in the implementation of FM systems.

FIGURE 6.23 Example of companding

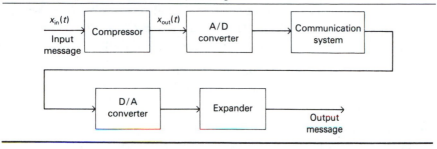

SUMMARY

1. The signal-to-noise ratio at the output of a baseband communication system operating in an additive Gaussian noise environment is P_T/N_0W, where P_T is the signal power, N_0 is the single-sided power spectral density of the noise ($\frac{1}{2}N_0$ is the two-sided power spectral density), and W is the signal bandwidth.

2. A double-sideband (DSB) system has an output signal-to-noise ratio of P_T/N_0W assuming perfect phase coherence of the demodulation carrier and a bandwidth of W.

3. A single-sideband (SSB) system also has an output signal-to-noise ratio of P_T/N_0W assuming perfect phase coherence of the demodulation carrier and a bandwidth of W. Thus, under ideal conditions, both SSB and DSB have performance equivalent to the baseband system.

4. An amplitude modulation (AM) system with coherent demodulation achieves an output signal-to-noise ratio of $E(P_T/N_0W)$, where E is the efficiency of the system. An AM system with envelope detection achieves the same output signal-to-noise ratio as an AM system with coherent demodulation for a high signal-to-noise ratio. If the predetection signal-to-noise ratio is small, the signal and noise at the demodulation output become multiplicative rather than additive. The output exhibits severe loss of signal for a small decrease in the input signal-to-noise ratio. This is known as the threshold effect.

5. The square-law detector is a nonlinear system that can be analyzed for all values of P_T/N_0W. Since the square-law detector is nonlinear, a threshold effect is observed.

6. Using a quadrature double-sideband (QDSB) signal model, a generalized analysis is easily carried out to determine the combined effect of both additive noise and demodulation phase errors on a communication system. The result shows that SSB and QDSB are equally sensitive to

demodulation phase errors if the power in the two QDSB signals is equal. DSB is much less sensitive to demodulation phase errors than SSB or QDSB because SSB and QDSB both exhibit crosstalk between the quadrature channels for nonzero demodulation phase errors.

7. The analysis of angle modulation systems shows that the output noise is suppressed as the signal carrier amplitude is increased. Thus the demodulator noise power output is a function of the input signal power.

8. The demodulator output power spectral density is constant over the range $|f| < W$ for PM and is parabolic over the range $|f| < W$ for FM. The parabolic power spectral density for an FM system is due to the fact that FM demodulation is essentially a differentiation process.

9. The demodulated output signal-to-noise ratio is proportional to k_p^2 for PM, where k_p is the phase-deviation constant. The output signal-to-noise ratio is proportional to D^2 for an FM system, where D is the deviation ratio.

10. The use of preemphasis and deemphasis can significantly improve the noise performance of an FM system. Typical values result in a better than 10-dB improvement in the signal-to-noise ratio of the demodulated output.

11. As the input signal-to-noise ratio of an FM system is reduced, spike noise appears. The spikes are due to origin encirclements of the total (Gaussian plus spike) noise phasor. The area of the spikes is constant at 2π, and the power spectral density is proportional to the spike frequency. Since the predetection bandwidth must be increased as the modulation index is increased, resulting in a decreased predetection signal-to-noise ratio, the threshold value of $P_T/N_0 W$ increases as the modulation index increases.

12. Threshold extension can be achieved using a phase-lock loop or a frequency-compressive feedback loop.

13. An analysis of nonlinear analog pulse modulation shows that, like FM, a tradeoff exists between bandwidth and output signal-to-noise ratio. With pulse code modulation, the signal-to-noise ratio and the system bandwidth are determined by the number of quantizing levels.

14. A most important result for this chapter is the postdetection signal-to-noise ratios for various modulation methods. A summary of these results is given in Table 6.1. Given in this table is the postdetection SNR for each technique as well as the required transmission bandwidth. The tradeoff between postdetection SNR and transmission bandwidth is evident.

TABLE 6.1 Summary of Noise Performance Characteristics

System	Postdetection SNR	Transmission Bandwidth
Baseband	$\dfrac{P_T}{N_0 W}$	W
DSB with coherent demodulation	$\dfrac{P_T}{N_0 W}$	$2W$
SSB with coherent demodulation	$\dfrac{P_T}{N_0 W}$	W
AM with envelope detection (above threshold) or AM with coherent demodulation. *Note:* E is efficiency	$\dfrac{EP_T}{N_0 W}$	$2W$
AM with square-law detection	$2\left(\dfrac{a^2}{2+a^2}\right)^2 \dfrac{P_T/N_0 W}{(N_0 W/P_T)}$	$2W$
PM above threshold	$k_p^2 \overline{m^2}\, \dfrac{P_T}{N_0 W}$	$2(D+1)W$
FM above threshold (without preemphasis)	$3D^2 \overline{m_n^2}\, \dfrac{P_T}{N_0 W}$	$2(D+1)W$
FM above threshold (with preemphasis)	$\left(\dfrac{f_d}{f_3}\right)^2 \overline{m^2}\, \dfrac{P_T}{N_0 W}$	$2(D+1)W$

FURTHER READING

All the books cited at the end of Chapter 3 contain material about noise effects in the systems studied in this chapter. The books by Lathi (1983), Haykin (1989), and Shanmugam (1979) are especially recommended for their completeness. The book by Taub and Schilling (1986) contains excellent sections on both PCM systems and threshold effects in FM systems.

PROBLEMS

Section 6.1

6.1 A baseband system, as shown in Figure 6.1, uses an nth-order Butterworth lowpass filter to enhance the signal-to-noise ratio. The 3-dB frequency of the Butterworth filter is $W \ll B$. Assuming that the signal is completely passed by the filter, determine the signal-to-noise ratio at the lowpass filter output as a function of the filter order, n. Write the result in the form

$$\text{SNR} = f(n)\,\frac{P_T}{N_0 W}$$

and plot $f(n)$ for values of n in the range $1 \leq n \leq 6$.

6.2   Derive the equation for $y_D(t)$ for an SSB system assuming that the noise is expanded about the frequency $\omega_x = \omega_c \pm \frac{1}{2}(2\pi W)$. Derive the detection gain and $(\text{SNR})_D$. Determine and plot $S_{n_c}(f)$ and $S_{n_s}(f)$.

6.3   Derive an expression for the detection gain of a DSB system for the case in which the bandwidth of the bandpass predetection filter is B_T and the bandwidth of the lowpass postdetection filter is B_D. Let $B_T > 2W$ and let $B_D > W$ simultaneously, where W is the bandwidth of the modulation. (There are two reasonable cases to consider.)

6.4   Repeat Problem 6.3 for an AM signal.

6.5   Assume that an AM system operates with an index of 0.7 and that the message signal is $15 \cos 8\pi t$. Compute the efficiency, the detection gain in decibels, and the output signal-to-noise ratio in decibels relative to the baseband performance $P_T/N_0 W$. Determine the improvement (in decibels) in the output signal-to-noise ratio that results if the modulation index is increased from 0.7 to 0.9.

6.6   An AM system has a message signal that has a zero-mean Gaussian amplitude distribution. The peak value of $m(t)$ is taken as that value that $|m(t)|$ exceeds 0.5% of the time. If the index is 0.5, what is the detection gain?

6.7   The threshold level for an envelope detector is sometimes defined as that value of $(\text{SNR})_T$ for which $A_c > r_n$ with probability 0.99. Assuming that $a^2 \overline{m_n^2} \cong 1$, derive the signal-to-noise ratio at threshold, expressed in decibels.

6.8   An envelope detector operates above threshold. The modulating signal is a sinusoid. Plot $(\text{SNR})_D$ in decibels as a function of $P_T/N_0 W$ for the modulation index equal to 0.5, 0.7, and 0.9.

6.9   A square-law demodulator for AM is illustrated in Figure 6.24. Assuming that

$$x_c(t) = A_c[1 + am_n(t)] \cos \omega_c t$$

FIGURE 6.24

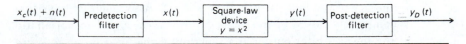

$x_c(t) + n(t) \longrightarrow$ Predetection filter $\xrightarrow{x(t)}$ Square-law device $y = x^2$ $\xrightarrow{y(t)}$ Post-detection filter $\xrightarrow{y_D(t)}$

FIGURE 6.25

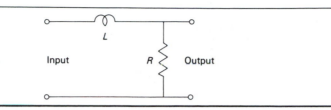

show that $y_D(t) = r^2(t)$, where $r^2(t)$ is defined by Equation (6.47). Let $m_n(t) = \cos \omega_m t$, and sketch the spectrum of each term that appears in $y_D(t)$. Do not neglect the noise power.

6.10 Compute $(\text{SNR})_D$ as a function of P_T/N_0W for a linear envelope detector assuming a high predetection SNR and a modulation index of unity. Compare this result to that for a square-law detector, and show that the square-law detector is inferior by approximately 1.8 dB.

6.11 The input to the filter shown in Figure 6.25 is $A \cos (2\pi f_c t)$ plus white noise with double-sided power spectral density $\frac{1}{2}N_0$. Compute the SNR at the filter output in terms of N_0, A, R, L, and f_c.

6.12 Repeat the preceding problem for the filter shown in Figure 6.26, writing the SNR in terms of N_0, A, R, C, and f_c.

6.13 Consider the system shown in Figure 6.27, in which an RC highpass filter is followed by an ideal lowpass filter having bandwidth W. Assume that the input to the system is $A \cos (2\pi f_c t)$, where $f_c < W$, plus white noise with double-sided power spectral density $\frac{1}{2}N_0$. Determine the SNR at the output of the ideal lowpass filter in terms of N_0, A, R, C, W, and f_c. What is the SNR in the limit as $W \to \infty$?

6.14 Compare the inprovement in $(\text{SNR})_D$ of a linear envelope detector over a square-law detector for a high-predetection SNR as the modulation index is varied. Assume a sinusoidal message signal.

FIGURE 6.26

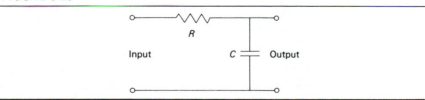

FIGURE 6.27

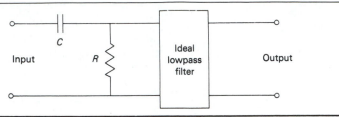

6.15   An AM system operates with a message signal that is a sample function of a zero-mean Gaussian process. The predetection signal-to-noise ratio is high, and envelope detection is used. The normalized message $m_n(t)$ is such that $|m_n(t)|$ exceeds 1 with probability 0.01. Determine the variance of $m_n(t)$, and plot $(SNR)_D$ as a function of P_T/N_0W for index a in the range $0 \le a \le 1$.

Section 6.2

6.16   Assume that a linear modulation system has negligible noise in the passband of the predetection filter. Sketch the normalized mean-square error as a function of the phase-error variance of the demodulation carrier for SSB, QDSB, and DSB. Comment on the significance of these curves.

6.17   Starting with (6.68) derive an expression for the mean-square error at the output of a coherent demodulator in the presence of noise and demodulation phase errors. Expand $\cos \phi$, $\cos^2 \phi$, and $\sin^2 \phi$ using power series expansions and discard all terms involving ϕ^k for $k > 4$. Show that the result is

$$\overline{\epsilon^2} = \tfrac{3}{4}\sigma_{m_1}^2\sigma_\phi^2 + \sigma_{m_2}^2\sigma_\phi^2 - \sigma_{m_2}^2\sigma_\phi^4 + \sigma_n^4$$

Show that, under appropriate assumptions, this result implies (6.71) and (6.77).

6.18   An SSB system is to be operated with a normalized mean-square error of 0.1 or less. By making a plot of output SNR versus demodulation phase-error variance for the case in which normalized mean-square error is 0.1%, show the region of satisfactory system performance. Repeat for a DSB system. Plot both curves on the same set of axes.

6.19   A DSB communication system is implemented as shown in Figure 6.10. The pilot frequency is k times the carrier frequency. Let the bandwidth of the predetection filter be α times the bandwidth of the pilot filter, and let ρ equal the pilot signal-to-noise ratio at the pilot filter output. Assume the pilot power is equal to the power in the modulating signal. Derive an expression for the normalized mean-square error of the system in terms of k, α, and ρ. Plot the normalized mean-square error as a function of α for ρ equal to 15 dB and k equal to 4.

6.20 It was shown in Chapter 2 that the output of a distortionless linear system is given by

$$y(t) = Ax(t - \tau)$$

where A is the gain of the system, τ is the system time delay, and $x(t)$ is the system input. It is often convenient to evaluate the performance of a linear system by comparing the system output with an amplitude-scaled and time-delayed version of the input. The mean-square error is then

$$\overline{\epsilon^2(A, \tau)} = \overline{[y(t) - Ax(t - \tau)]^2}$$

The values of A and τ that minimize this expression, denoted A_m and τ_m, respectively, are defined as the system gain and the system time delay. Show that, with these definitions, the system gain is

$$A_m = \frac{R_{xy}(\tau_m)}{R_x(0)}$$

and the resulting system mean-square error is

$$\overline{\epsilon^2(A_m, \tau_m)} = R_y(0) - \frac{R_{xy}^{\,2}(\tau_m)}{R_x(0)}$$

Also show that the signal power at the system output is

$$S_D = A_m^{\,2} R_x(0) = \frac{R_{xy}^{\,2}(\tau_m)}{R_x(0)}$$

and the output signal-to-noise ratio is

$$\frac{S_D}{N_D} = \frac{R_{xy}^{\,2}(\tau_m)}{R_x(0)R_y(0) - R_{xy}^{\,2}(\tau_m)}$$

in which N_D is the mean-square error.*

---

* For a discussion of these techniques, see Houts and Simpson (1968).

6.21   A linear system has the input

$$x(t) = 6 \cos (20\pi t) + 3 \cos (40\pi t)$$

and the output

$$y(t) = 3 \cos \left(20\pi t - \frac{\pi}{2} \right) + 2 \cos \left(40\pi t - \frac{3\pi}{4} \right)$$

Using the definitions of system gain and system time delay given in the previous problem, determine the gain, A_m, and time delay, τ_m, of this system. Also, determine the signal-to-noise ratio, in decibels, at the system output using the system input, $x(t)$, as the reference.

Section 6.3

6.22   The process of stereophonic broadcasting was illustrated in Chapter 3. By comparing the noise power in the $l(t) - r(t)$ channel to the noise power in the $l(t) + r(t)$ channel, explain why stereophonic broadcasting is more sensitive to noise than nonstereophonic broadcasting.

6.23   An FDM communication system uses DSB modulation to form the baseband and FM modulation for transmission of the baseband. Assume that there are 8 channels and that all 8 message signals have equal power P_0 and equal bandwidth W. One channel does *not* use subcarrier modulation. The other channels use subcarriers of the form

$$A_k \cos k\omega_1 t, \qquad 1 \le k \le 7$$

The width of the guardbands is $4W$. Sketch the power spectrum of the received *baseband* signal showing both the signal and noise components. Calculate the relationship between the values of A_k if the channels are to have equal signal-to-noise ratios.

6.24   Using Equation (6.130), derive an expression for the ratio of the noise power in $y_D(t)$ with deemphasis to the noise power in $y_D(t)$ without deemphasis. Plot this ratio as a funtion of W/f_3. Evaluate the ratio for the standard values of $f_3 = 2.1$ kHz and $W = 15$ kHz, and use the result to determine the improvement, in decibels, that results through the use of deemphasis. Compare the result with that found in Example 6.3.

6.25   White noise with two-sided power spectral density $\frac{1}{2}N_0$ is added to a signal having the power spectral density shown in Figure 6.28. The sum (signal plus noise) is filtered with an ideal lowpass filter with

FIGURE 6.28

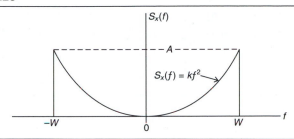

unity passband gain and bandwidth $B > W$. Determine the signal-to-noise ratio at the filter output. By what factor will the signal-to-noise ratio increase if B is reduced to W?

6.26 Consider the system shown in Figure 6.29. The signal $x(t)$ is defined by

$$x(t) = A \cos 2\pi f_c t$$

The lowpass filter has unity gain in the passband and bandwidth W, where $f_c < W$. The noise $n(t)$ is white with two-sided power spectral density $\frac{1}{2}N_0$. The signal component of $y(t)$ is defined to be the component at frequency f_c. Determine the signal-to-noise ratio of $y(t)$.

6.27 Repeat the preceding problem for the system shown in Figure 6.30.

6.28 Consider the system shown in Figure 6.31. The noise is white with two-sided power spectral density $\frac{1}{2}N_0$. The power spectral density of the signal is

$$S_x(f) = \frac{A}{1 + (f/f_3)^2}, \qquad -\infty < f < \infty$$

FIGURE 6.29

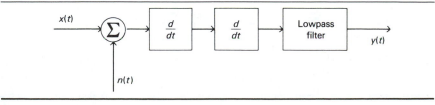

FIGURE 6.30

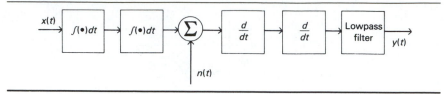

The parameter f_3 is the 3-dB frequency of the signal. The bandwidth of the lowpass filter is W. Determine the signal-to-noise ratio of $y(t)$. Plot the signal-to-noise ratio as a function of W/f_3.

Section 6.4

6.29   Derive an expression that illustrates the ratio of spike noise to Gaussian noise power at the output of an FM discriminator. Your answer should be a function of the deviation ratio D and P_T/N_0W. Assume an unmodulated carrier. Plot your answer as a function of P_T/N_0W for $D = 5$. At what value of P_T/N_0W does the spike noise power equal the Gaussian noise power? What happens to this value as D is changed?

6.30   Derive an expression, similar to Equation (6.155), that gives the output signal-to-noise ratio of an FM discriminator for the case in which the message signal is random with a Gaussian amplitude probability density function.

Section 6.5

6.31   Assume that a PPM system uses Nyquist rate sampling and that the minimum channel bandwidth is used for a given pulse duration. Show that the postdetection SNR can be written as

FIGURE 6.31

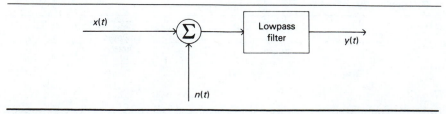

$$(\text{SNR})_D = K \left(\frac{B_T}{W} \right)^2 \frac{P_T}{N_0 W}$$

and evaluate K.

6.32 The message signal on the input to an analog-to-digital converter is a sinusoid of 15 V peak-to-peak. Compute the signal-to-quantization-noise power ratio as a function of the wordlength of the analog-to-digital converter. State any assumptions you make.

COMPUTER EXERCISES

6.1 Develop a set of performance curves, similar to those shown in Figure 6.9, that illustrate the performance of a coherent demodulator as a function of the phase error variance. Let the signal-to-noise ratio (SNR) be a parameter and express the SNR in decibels. As in Figure 6.9, assume a QDSB system. Repeat this exercise for a DSB system.

6.2 In many problems of practical interest it is important to know the total noise power at the output of an FM discriminator in which the discriminator input consists of an unmodulated sinusoid plus bandlimited white Gaussian noise. Give the expression for the noise power at the output of the postdetection filter as a function of B_T, W, P_T, and N_0, where each of these terms is defined in Section 6.3. Assume that the discriminator constant K_D is 1. Plot, on a log scale, the normalized noise power, N_D/B_T^2, as a function of the predetection SNR, $A_c^2/2N_0 B_T$, expressed in decibels. Develop a second set of curves in which N_D/B_T^2 is shown as a function of $P_T/N_0 W$, expressed in decibels. What do you conclude from the result of this exercise?

6.3 Develop a computer program to verify the FM discriminator performance characteristics illustrated in Figure 6.18. Use the resulting program to generate a set of performance curves for $\beta = 1, 3, 5, 10$, and 20. On a second set of axes plot the performance characteristic for $\beta = 0.01$. Is the threshold effect more or less pronounced? Why?

6.4 The value of the input signal-to-noise ratio at threshold is often defined as the value of $P_T/N_0 W$ at which the denominator of (6.155) is equal to 2. Note that this value yields a postdetection signal-to-noise ratio, $(\text{SNR})_D$, that is 3 dB below the value of $(\text{SNR})_D$ predicted by the above threshold (linear) analysis. Using this definition of threshold, plot the threshold value of $P_T/N_0 W$, in decibels, as a function of β. What do you conclude?

6.5   In analyzing the performance of an FM discriminator, operating in the presence of noise, the postdetection signal-to-noise ratio, $(SNR)_D$, is often determined using the approximation that the effect of modulation on $(SNR)_D$ is negligible. In other words, $|\overline{\delta f}|$ is set equal to zero. Assuming sinusoidal modulation, investigate the error induced by making this approximation.

6.6   Develop a computer program to verify the performance curves shown in Figure 6.21. Compare the performance of the noncoherent FSK system to the performance of both coherent FSK and coherent PSK with a modulation index of 1. We will show in the following chapter that the bit-error probability for coherent FSK is

$$P_b = Q\left(\sqrt{\frac{P_T}{N_0 B_p}}\right)$$

and that the bit-error probability for coherent BPSK with a unity modulation index is

$$P_b = Q\left(\sqrt{\frac{2P_T}{N_0 B_p}}\right)$$

where B_p is the system bit-rate bandwidth. Compare the results of the three systems studied in this example for $n = 8$.

BINARY DATA TRANSMISSION 7

In Chapter 6 we studied the effects of noise in analog communication systems. We now consider a distinctly different situation. Instead of being concerned with continuous-time, continuous-level message signals, we are concerned with the transmission of information from sources that produce discrete symbols. That is, the input signal to the transmitter block of Figure 1.1 would be a signal that assumes only discrete values.

The purpose of this chapter is to consider various systems for the transmission of digital data and their relative performances. Before beginning, however, let us consider the block diagram of a digital data transmission system, shown in Figure 7.1, which is somewhat more detailed than Figure 1.1. The focus of our attention will be on the portion of the system between the optional blocks labeled *Encoder* and *Decoder*. In order to gain a better perspective of the overall problem of digital data transmission, we will briefly discuss the operations performed by the blocks shown as dashed lines.

While many sources result in message signals that are inherently digital, such as from computers, it is often advantageous to represent analog signals in digital form (referred to as *analog-to-digital conversion*) for transmission and then convert them back to analog form upon reception (referred to as *digital-to-analog conversion*), as discussed in the preceding chapter. Pulse code modulation (PCM), introduced in Chapter 3, is an example of a modulation technique that can be employed to transmit analog messages in digital form. The signal-to-noise ratio performance characteristics of a PCM system, which were presented in Chapter 6, show one advantage of this system to be the option of exchanging bandwidth for signal-to-noise ratio improvement.*

Regardless of whether a source is purely digital or an analog source that

---

* A device for converting voice signals from analog to digital and from digital to analog form is known as a *vocoder*. See, for example, Qualcomm's technical data sheet for the Q4400 variable rate vocoder. Qualcomm, Inc., VLSI Products, 10555 Sorrento Valley Road, San Diego, CA 92121-1617.

FIGURE 7.1   Block diagram of a digital data system. (a) Transmitter.
(b) Receiver.

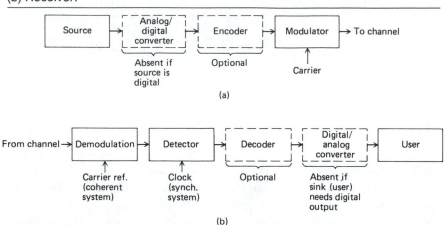

(a)

(b)

has been converted to digital, it may be advantageous to add or remove redundant digits to the digital signal. Such procedures, referred to as *coding*, are performed by the encoder-decoder blocks of Figure 7.1 and will be considered in Chapter 10.

We now consider the basic system in Figure 7.1, shown as the blocks with solid lines. If the digital signals at the modular input take on one of only two possible values, the communication system is referred to as *binary*. If one of $M > 2$ possible values is available, the system is referred to as *M-ary*. For long-distance transmission, these digital baseband signals from the source may modulate a carrier before transmission, as briefly mentioned in Chapter 3. The result is referred to as *amplitude-shift keying* (ASK), *phase-shift keying* (PSK), or *frequency-shift keying* (FSK) if it is amplitude, phase, or frequency, respectively, that is varied in accordance with the baseband signal. An important *M*-ary modulation scheme, *quadriphase-shift keying* (QPSK), is often employed in situations in which bandwidth efficiency is a consideration. Other schemes related to QPSK include offset QPSK and minimum-shift keying (MSK). These schemes will be discussed in Chapter 8.

A digital communication system is referred to as *coherent* if a local reference is available for demodulation that is in phase with the transmitted carrier (with fixed phase shifts due to transmission delays accounted for). Otherwise, it is referred to as *noncoherent*. Likewise, if a periodic signal is available at the receiver that is in synchronism with the transmitted sequence of digital signals (referred to as a *clock*), the system is referred to as *synchronous;* if a signaling technique is employed in which such a clock is unnecessary, the system is called *asynchronous*.

The primary measure of system performance for digital data communication systems is the probability of error P_E. In this chapter we will obtain expressions for P_E for various types of digital communication systems. We are, of course, interested in receiver structures that give minimum P_E for given conditions. Synchronous detection in a white Gaussian-noise background requires a correlation or a *matched filter* detector to give minimum P_E for fixed signal and noise conditions.

We begin our consideration of digital data transmission systems in Section 7.1 with the analysis of a simple, synchronous baseband system that employs a special case of the matched filter detector known as an *integrate-and-dump detector*. This analysis is then generalized in Section 7.2 to the matched-filter receiver. Section 7.3 contains several specializations of the results of Section 7.2 to common coherent digital signaling schemes, while noncoherent schemes are analyzed in Section 7.4. After analyzing these modulation schemes, which operate in an ideal environment in the sense that infinite bandwidth is available, we look at optimum signaling through bandlimited baseband channels in Section 7.5. In Sections 7.6 and 7.7, the effect of multipath interference and signal fading on data transmission is analyzed, and in Section 7.8, the use of equalizing filters to mitigate the effects of channel distortion is examined.

7.1   BASEBAND DATA TRANSMISSION IN WHITE GAUSSIAN NOISE

Consider the binary digital data communication system illustrated in Figure 7.2(a), in which the transmitted signal consists of a sequence of constant-amplitude pulses of either A or $-A$ units in amplitude and T seconds in duration. A typical transmitted sequence is shown in Figure 7.2(b). We may think of a positive pulse as representing a logic 1 and a negative pulse as representing a logic 0 from the data source. Each T-second pulse is called a *binit* for binary digit or, more simply, a *bit*. (In Chapter 10, the term *bit* will take on a new meaning.)

As in Chapter 6, the channel is idealized as simply adding white Gaussian noise with double-sided power spectral density $\frac{1}{2}N_0$ to the signal. A typical sample function of the received signal plus noise is shown in Figure 7.2(c). For sketching purposes, it is assumed that the noise is bandlimited, although it is modeled as white noise later when the performance of the receiver is analyzed. It is assumed that the starting and ending times of each pulse are known by the receiver. The problem of acquiring this information, referred to as *synchronization*, will not be considered at this time.

The function of the receiver is to decide whether the transmitted signal

FIGURE 7.2 System model and waveforms for synchronous baseband digital data transmission. (a) Baseband digital data communication system. (b) Typical transmitted sequence. (c) Received sequence plus noise.

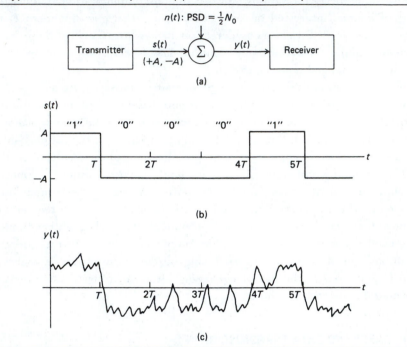

(a)

(b)

(c)

was A or $-A$ during each bit period. A straightforward way of accomplishing this is to pass the signal plus noise through a lowpass predetection filter, sample its output at some time within each T-second interval, and determine the sign of the sample. If the sample is greater than zero, the decision is made that $+A$ was transmitted. If the sample is less than zero, the decision is that $-A$ was transmitted. With such a receiver structure, however, we do not take advantage of everything known about the signal. Since the starting and ending times of the pulses are known, a better procedure is to compare the area of the received signal-plus-noise waveform (data) with zero at the end of each signaling interval by integrating the received data over the T-second signaling interval. Of course, a noise component is present at the output of the integrator, but since the input noise has zero mean, it takes on positive and negative values with equal probability. Thus the output noise component has zero mean. The proposed receiver structure and a typical

FIGURE 7.3   Receiver structure and integrator output. (a) Integrate-and-dump receiver. (b) Output from the integrator.

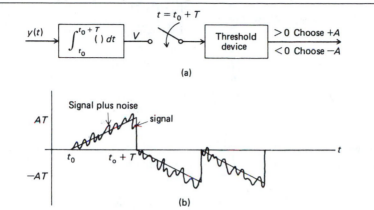

(a)

(b)

waveform at the output of the integrator are shown in Figure 7.3 where t_0 is the start of an arbitrary signaling interval. For obvious reasons, this receiver is referred to as an integrate-and-dump detector.

The question to be answered is: How well does this receiver perform, and on what parameters does its performance depend? As mentioned previously, a useful criterion of performance is probability of error, and it is this we now compute. The output of the integrator at the end of a signaling interval is

$$V = \int_{t_0}^{t_0+T} [s(t) + n(t)]\, dt$$

$$= \begin{cases} + AT + N & \text{if } +A \text{ is sent} \\ -AT + N & \text{if } -A \text{ is sent} \end{cases} \tag{7.1}$$

where N is a random variable defined as

$$N = \int_{t_0}^{t_0+T} n(t)\, dt \tag{7.2}$$

Since N results from a linear operation on a sample function from a Gaussian process, it is a Gaussian random variable. It has mean

$$E\{N\} = E\left\{ \int_{t_0}^{t_0+T} n(t)\, dt \right\} = \int_{t_0}^{t_0+T} E\{n(t)\}\, dt = 0 \tag{7.3}$$

since $n(t)$ has zero mean. Its variance is therefore

$$\text{var}\{N\} = E\{N^2\} = E\left\{\left[\int_{t_0}^{t_0+T} n(t)\,dt\right]^2\right\}$$

$$= \int_{t_0}^{t_0+T}\int_{t_0}^{t_0+T} E\{n(t)n(\sigma)\}\,dt\,d\sigma \tag{7.4}$$

$$= \int_{t_0}^{t_0+T}\int_{t_0}^{t_0+T} \tfrac{1}{2}N_0\,\delta(t-\sigma)\,dt\,d\sigma$$

where we have made the substitution $E\{n(t)n(\sigma)\} = \tfrac{1}{2}N_0\,\delta(t-\sigma)$. Using the sifting property of the delta function, we obtain

$$\text{var}\{N\} = \int_{t_0}^{t_0+T} \tfrac{1}{2}\,N_0\,d\sigma$$

$$= \tfrac{1}{2}N_0 T \tag{7.5}$$

Thus the pdf of N is

$$f_N(\eta) = \frac{e^{-\eta^2/N_0 T}}{\sqrt{\pi N_0 T}} \tag{7.6}$$

where η is used as the dummy variable for N to avoid confusion with $n(t)$.

There are two ways in which errors occur. If $+A$ is transmitted, an error occurs if $AT + N < 0$, that is, if $N < -AT$. From (7.6), the probability of this event is

$$P(\text{error}\,|\,A\text{ sent}) = P(E\,|\,A) = \int_{-\infty}^{-AT} \frac{e^{-\eta^2/N_0 T}}{\sqrt{\pi N_0 T}}\,d\eta \tag{7.7}$$

which is the area to the left of $\eta = -AT$ in Figure 7.4. Letting $u = -\sqrt{2}\eta/\sqrt{N_0 T}$, we can write this as*

$$P(E\,|\,A) = \int_{\sqrt{2A^2 T/N_0}}^{\infty} \frac{e^{-u^2/2}}{\sqrt{2\pi}}\,du \triangleq Q\left(\sqrt{\frac{2A^2 T}{N_0}}\right) \tag{7.8}$$

The other way in which an error can occur is if $-A$ is transmitted and $-AT + N > 0$. The probability of this event is the same as the probability that $N > AT$, which can be written as

$$P(E\,|-A) = \int_{AT}^{\infty} \frac{e^{-\eta^2/N_0 T}}{\sqrt{\pi N_0 T}}\,d\eta \triangleq Q\left(\sqrt{\frac{2A^2 T}{N_0}}\right) \tag{7.9}$$

* See Appendix C.6 for a discussion and tabulation of the Q-function.

FIGURE 7.4   Illustration of error probabilities for binary signaling

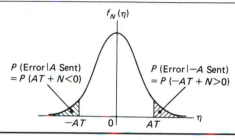

which is the area to the right of $\eta = AT$ in Figure 7.4. The average probability of error is

$$P_E = P(E\,|\,+A)P(+A) + P(E\,|\,-A)P(-A) \tag{7.10}$$

Substituting (7.8) and (7.9) into (7.10) and noting that $P(+A) + P(-A) = 1$, we obtain

$$P_E = Q\left(\sqrt{\frac{2A^2T}{N_0}}\right) \tag{7.11}$$

Thus the important parameter is A^2T/N_0. We can interpret this ratio in two ways. First, since the energy in each signal pulse is

$$E_b = \int_{t_0}^{t_0+T} A^2\,dt = A^2T \tag{7.12}$$

we see that the ratio of signal energy per pulse to noise power spectral density is

$$z = \frac{A^2T}{N_0} = \frac{E_b}{N_0} \tag{7.13}$$

where E_b is called the *energy per bit* because each signal pulse ($+A$ or $-A$) carries one *bit* of information. Second, we recall that a rectangular pulse of duration T seconds has amplitude spectrum AT sinc Tf and that $B_p = 1/T$ is a rough measure of its bandwidth. Thus

$$z = \frac{A^2}{N_0(1/T)} = \frac{A^2}{N_0 B_p} \tag{7.14}$$

can be interpreted as the ratio of signal power to noise power in the signal bandwidth. The bandwidth B_p is sometimes referred to as the *bit-rate band-*

FIGURE 7.5 P_E for baseband signaling

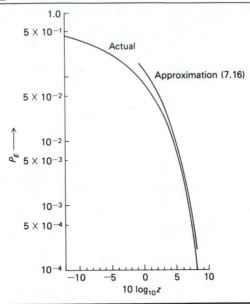

width. We will refer to z as the signal-to-noise ratio (SNR). An often-used reference to this signal-to-noise ratio in the digital communications industry is "E_b-over-N_0."

A plot of P_E versus z is shown in Figure 7.5, where z is given in decibels. Also shown is an approximation for P_E using the asymptotic expansion for the Q-function [Equation (4.187)]:

$$Q(u) \cong \frac{e^{-u^2/2}}{u\sqrt{2\pi}}, \qquad u \gg 1 \qquad (7.15)$$

Using this approximation,

$$P_E \cong \frac{e^{-z}}{2\sqrt{\pi z}}, \qquad z \gg 1 \qquad (7.16)$$

which shows that P_E essentially decreases exponentially with increasing z. Figure 7.5 shows that the approximation of (7.16) is close to the true result of (7.11) for $z \gtrsim 3$ dB.

▼ **EXAMPLE 7.1** Digital data is to be transmitted through a baseband system with $N_0 = 10^{-7}$ W/Hz and the received signal amplitude $A = 20$ mV. (a) If 10^3 bits per second (bps) are transmitted, what is P_E? (b) If 10^4 bps are transmitted, to what value must A be adjusted in order to attain the same P_E as in part (a)?

SOLUTION To solve part (a), note that

$$z = \frac{A^2 T}{N_0} = \frac{(0.02)^2(10^{-3})}{10^{-7}} = 4 \tag{7.17}$$

Using (7.16), $P_E \cong e^{-4}/2\sqrt{4\pi} = 2.58 \times 10^{-3}$. Part (b) is solved by finding
▲ A such that $A^2(10^{-4})/(10^{-7}) = 4$, which gives $A = 63.2$ mV.

▼ **EXAMPLE 7.2** The conditions are the same as in the preceding example, but a bandwidth of 5000 Hz is available. (a) What is the maximum data rate that can be supported by the channel? (b) Find the transmitter power required to give a probability of error of 10^{-6} at the data rate found in part (a).

SOLUTION (a) Since a rectangular pulse has Fourier transform

$$\Pi(t/T) \leftrightarrow T \text{ sinc } (fT)$$

we take the signal bandwidth to be that of the first null of the sinc function. Therefore, $1/T = 5000$ Hz, which implies a maximum data rate of $R = 5000$ bps. (b) To find the transmitter power to give $P_E = 10^{-6}$, we solve

$$10^{-6} = Q\left[\sqrt{2A^2 T/N_0}\right] = Q\left[\sqrt{2z}\right] \tag{7.18}$$

Using the approximation (7.15) for the error function, we need to solve

$$10^{-6} = \frac{e^{-z}}{2\sqrt{\pi z}}$$

iteratively. This gives the result

$$z \cong 10.54 \text{ dB} = 11.31 \text{ (ratio)}$$

Thus, $A^2 T/N_0 = 11.31$, or

$$A^2 = (11.31)N_0/T = 5.65 \text{ mW (actually V}^2 \times 10^{-3})$$

▲ This corresponds to a signal amplitude of approximately 75.2 mV.

7.2 BINARY SYNCHRONOUS DATA TRANSMISSION WITH ARBITRARY SIGNAL SHAPES

In Section 7.1 we analyzed a simple baseband digital communication system. As in the case of analog transmission, it is often necessary to utilize modulation to condition a digital message signal so that it is suitable for transmission through a channel. Thus, instead of the constant-level signals considered in Section 7.1, we will let a logic 1 be represented by $s_1(t)$ and a logic 0 by $s_2(t)$. The only restriction on $s_1(t)$ and $s_2(t)$ is that they must have finite energy in a T-second interval. The energies of $s_1(t)$ and $s_2(t)$ are denoted by

$$E_1 \triangleq \int_{t_0}^{t_0+T} s_1^2(t) \, dt \qquad (7.19a)$$

and

$$E_2 \triangleq \int_{t_0}^{t_0+T} s_2^2(t) \, dt \qquad (7.19b)$$

respectively. In Table 7.1, several commonly used choices for $s_1(t)$ and $s_2(t)$ are given.

Receiver Structure and Error Probability

A possible receiver structure for detecting $s_1(t)$ or $s_2(t)$ in additive white Gaussian noise is shown in Figure 7.6. Since the signals chosen may have zero average value over a T-second interval (see the examples in Table 7.1), we can no longer employ an integrator followed by a threshold device as in the case of constant-amplitude signals. Instead of the integrator, we employ a filter with, as yet, unspecified impulse response $h(t)$ and corresponding transfer function $H(f)$. The received signal plus noise is either

$$y(t) = s_1(t) + n(t), \qquad t_0 \le t \le t_0 + T \qquad (7.20a)$$

or

$$y(t) = s_2(t) + n(t), \qquad t_0 \le t \le t_0 + T \qquad (7.20b)$$

where the noise, as before, is assumed to be white with power spectral density $\frac{1}{2}N_0$. We can assume that $t_0 \doteq 0$ without loss of generality; that is, the signaling interval under consideration is $0 \le t \le T$.

To find P_E, we again note that an error can occur in either one of two ways. Assume that $s_1(t)$ and $s_2(t)$ were chosen such that $s_{01}(T) < s_{02}(T)$, where $s_{01}(t)$ and $s_{02}(t)$ are the outputs of the filter due to $s_1(t)$ and $s_2(t)$,

TABLE 7.1   Possible Signal Choices for Binary Digital Signaling

Case	$s_1(t)$	$s_2(t)$	Type of Signaling
1	0	$A \cos \omega_c t$	Amplitude-shift keying
2	$A \sin (\omega_c t + \cos^{-1} m)$	$A \sin (\omega_c t - \cos^{-1} m)$	Phase-shift keying with carrier ($\cos^{-1} m \triangleq$ modulation index)
3	$A \cos \omega_c t$	$A \cos (\omega_c + \Delta\omega)t$	Frequency-shift keying

respectively, at the input. If not, the roles of $s_1(t)$ and $s_2(t)$ at the input can be reversed to ensure this. Referring to Figure 7.6, if $v(T) > k$, where k is the threshold, we decide that $s_2(t)$ was sent; if $v(T) < k$, we decide that $s_1(t)$ was sent. Letting $n_0(t)$ be the noise component at the filter output, an error is made if $s_1(t)$ is sent and $v(T) = s_{01}(T) + n_0(T) > k$; if $s_2(t)$ is sent, an error occurs if $v(T) = s_{02}(T) + n_0(T) < k$. Since $n_0(t)$ is the result of passing white Gaussian noise through a fixed linear filter, it is a Gaussian process. Its power spectral density is

$$S_{n_0}(f) = \tfrac{1}{2}N_0 |H(f)|^2 \tag{7.21}$$

Because the filter is fixed, $n_0(t)$ is a stationary process with mean zero and variance

$$\sigma_0{}^2 = \int_{-\infty}^{\infty} \tfrac{1}{2}N_0 |H(f)|^2 \, df \tag{7.22}$$

Since $n_0(t)$ is stationary, $N = n_0(T)$ is a random variable with mean zero and variance $\sigma_0{}^2$. Its pdf is

$$f_N(\eta) = \frac{e^{-\eta^2/2\sigma_0{}^2}}{\sqrt{2\pi\sigma_0{}^2}} \tag{7.23}$$

Given that $s_1(t)$ is transmitted, the sampler output is

$$V \triangleq v(T) = s_{01}(T) + N \tag{7.24}$$

FIGURE 7.6   A possible receiver structure for detecting binary signals in white Gaussian noise

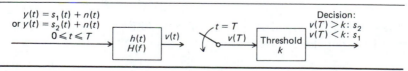

FIGURE 7.7 Conditional probability density functions of the filter output at time $t = T$

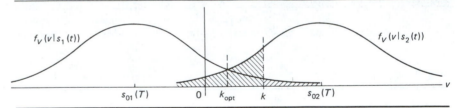

and if $s_2(t)$ is transmitted, the sampler output is

$$V \triangleq v(T) = s_{02}(T) + N \qquad (7.25)$$

These are also Gaussian random variables, since they result from linear operations on Gaussian random variables. They have means $s_{01}(T)$ and $s_{02}(T)$, respectively, and the same variance as N, that is, $\sigma_0{}^2$. Thus the conditional pdf's of V given $s_1(t)$ is transmitted, $f_V(v \,|\, s_1(t))$, and given $s_2(t)$ is transmitted, $f_V(v \,|\, s_2(t))$, are as shown in Figure 7.7. Also illustrated is a decision threshold k.

From Figure 7.7, we see that the probability of error, given $s_1(t)$ is transmitted, is

$$
\begin{aligned}
P(E \,|\, s_1(t)) &= \int_k^\infty f_V(v \,|\, s_1(t)) \, dv \\
&= \int_k^\infty \frac{e^{-[v - s_{01}(T)]^2 / 2\sigma_0{}^2}}{\sqrt{2\pi\sigma_0{}^2}} \, dv
\end{aligned}
\qquad (7.26)
$$

which is the area under $f_V(v \,|\, s_1(t))$ to the right of $v = k$. Similarly, the probability of error, given $s_2(t)$ is transmitted, which is the area under $f_V(v \,|\, s_2(t))$ to the left of $v = k$, is given by

$$P(E \,|\, s_2(t)) = \int_{-\infty}^k \frac{e^{-[v - s_{02}(T)]^2 / 2\sigma_0{}^2}}{\sqrt{2\pi\sigma_0{}^2}} \, dv \qquad (7.27)$$

Assuming that $s_1(t)$ and $s_2(t)$ are *a priori* equally probable,* the average probability error is

$$P_E = \tfrac{1}{2} P[E \,|\, s_1(t)] + \tfrac{1}{2} P[E \,|\, s_2(t)] \qquad (7.28)$$

The task now is to minimize this error probability by adjusting the threshold k and the impulse response $h(t)$.

---

* See Problem 7.8 for the case of unequal *a priori* probabilities.

Because of the equal *a priori* probabilities for $s_1(t)$ and $s_2(t)$ and the symmetrical shapes of $f_V(v \mid s_1(t))$ and $f_V(v \mid s_2(t))$, it is reasonable that the optimum choice for k is the intersection of the conditional pdf's, which is

$$k_{opt} = \tfrac{1}{2}[s_{01}(T) + s_{02}(T)] \tag{7.29}$$

The optimum threshold is illustrated in Figure 7.7 and can be derived by differentiating (7.28) with respect to k after substitution of (7.26) and (7.27). Because of the symmetry of the pdf's, the probabilities of either type of error, (7.26) or (7.27), are equal for this choice of k.

With this choice of k, the probability of error given by (7.28) reduces to

$$P_E = Q\left[\frac{s_{02}(T) - s_{01}(T)}{2\sigma_0}\right] \tag{7.30}$$

Thus we see that P_E is a function of the difference between the two output signals at $t = T$. Remembering that the Q-function decreases monotonically with u, we see that P_E decreases with increasing *distance* between the two output signals, a reasonable result. We will encounter this interpretation again in Chapter 9, where we discuss concepts of signal space.

We now consider the minimization of P_E by proper choice of $h(t)$. This will lead us to the matched filter.

The Matched Filter

For a given choice of $s_1(t)$ and $s_2(t)$, we wish to determine an $H(f)$, or equivalently, an $h(t)$, that maximizes

$$\zeta = \frac{s_{02}(T) - s_{01}(T)}{\sigma_0} \tag{7.31}$$

Letting $g(t) = s_2(t) - s_1(t)$, the problem is to find the $H(f)$ that maximizes $\zeta = g_0(T)/\sigma_0$, where $g_0(t)$ is the signal portion of the output due to the input $g(t)$. This situation is illustrated in Figure 7.8. We can equally well consider the maximization of

$$\zeta^2 = \frac{g_0{}^2(T)}{\sigma_0{}^2} = \frac{g_0{}^2(t)}{E\{n_0{}^2(t)\}}\bigg|_{t=T} \tag{7.32}$$

FIGURE 7.8   Choosing $H(f)$ to minimize P_E

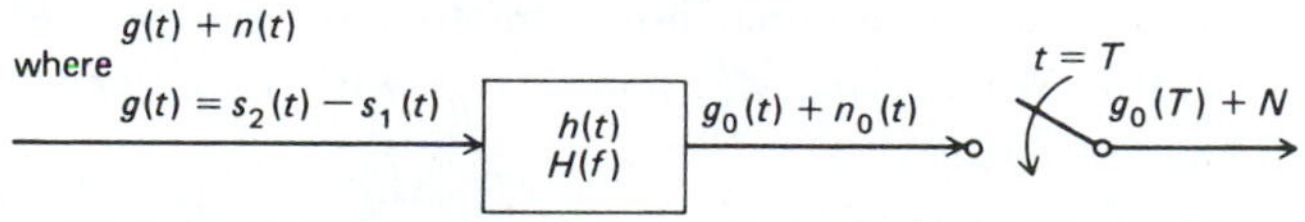

Since the input noise is stationary,

$$E\{n_0{}^2(t)\} = E\{n_0{}^2(T)\} = \tfrac{1}{2}N_0 \int_{-\infty}^{\infty} |H(f)|^2 \, df \qquad (7.33)$$

We can write $g_0(t)$ in terms of $H(f)$ and the Fourier transform of $g(t)$, $G(f)$, as

$$g_0(t) = \mathcal{F}^{-1}[G(f)H(f)] = \int_{-\infty}^{\infty} H(f)G(f)e^{j2\pi ft} \, df \qquad (7.34)$$

Setting $t = T$ in (7.34) and using this result along with (7.33) in (7.32), we obtain

$$\zeta^2 = \frac{\left| \int_{-\infty}^{\infty} H(f)G(f)e^{j2\pi fT} \, df \right|^2}{\tfrac{1}{2}N_0 \int_{-\infty}^{\infty} |H(f)|^2 \, df} \qquad (7.35)$$

To maximize this equation with respect to $H(f)$, we employ *Schwarz's inequality*. Schwarz's inequality is a generalization of the inequality

$$|\mathbf{A} \cdot \mathbf{B}| = |AB \cos \theta| \leq |\mathbf{A}||\mathbf{B}| \qquad (7.36)$$

where $\mathbf{A}$ and $\mathbf{B}$ are ordinary vectors, with θ the angle between them, and $\mathbf{A} \cdot \mathbf{B}$ denotes their inner, or dot, product. Since $|\cos \theta|$ equals unity if and only if θ equals zero or an integer multiple of π, equality holds if and only if $\mathbf{A}$ equals $k\mathbf{B}$, where k is a constant. Considering the case of two complex functions $X(f)$ and $Y(f)$, and defining the inner product as

$$\int_{-\infty}^{\infty} X(f)Y^*(f) \, df$$

Schwarz's inequality assumes the form

$$\left| \int_{-\infty}^{\infty} X(f)Y^*(f) \, df \right|^2 \leq \int_{-\infty}^{\infty} |X(f)|^2 \, df \int_{-\infty}^{\infty} |Y(f)|^2 \, df \quad (7.37)$$

Equality holds if and only if $X(f) = kY(f)$ where k is, in general, complex. We will prove Schwarz's inequality in Chapter 9 with the aid of signal-space notation.

We now return to our original problem, that of finding the $H(f)$ that maximizes (7.35). We replace $X(f)$ in (7.37) with $H(f)$ and $Y^*(f)$ with $G(f)e^{j2\pi Tf}$. Thus

$$\zeta^2 = \frac{2}{N_0} \frac{\left| \int_{-\infty}^{\infty} X(f)Y^*(f)\, df \right|^2}{\int_{-\infty}^{\infty} |H(f)|^2\, df} \leq \frac{2}{N_0} \frac{\int_{-\infty}^{\infty} |H(f)|^2\, df \int_{-\infty}^{\infty} |G(f)|^2\, df}{\int_{-\infty}^{\infty} |H(f)|^2\, df}$$

$$(7.38)$$

Canceling the integral over $|H(f)|^2$ in the numerator and denominator, we find the maximum value of ζ^2 to be

$$\zeta_{max}^2 = \frac{2}{N_0} \int_{-\infty}^{\infty} |G(f)|^2\, df = \frac{2E_g}{N_0} \qquad (7.39)$$

where $E_g = \int_{-\infty}^{\infty} |G(f)|^2\, df$ is the energy contained in $g(t)$, which follows by Rayleigh's energy theorem. Equality holds in (7.38) if and only if

$$H(f) = kG^*(f)e^{-j2\pi Tf} \qquad (7.40)$$

where k is an arbitrary constant. Since k just fixes the gain of the filter, we can set it to unity. Thus the optimum choice for $H(f)$, $H_0(f)$, is

$$H_0(f) = G^*(f)e^{-j2\pi Tf} \qquad (7.41)$$

The impulse response corresponding to this choice of $H_0(f)$ is

$$\begin{aligned}
h_0(t) &= \mathscr{F}^{-1}[H_0(f)] \\
&= \int_{-\infty}^{\infty} G^*(f)e^{-j2\pi Tf}e^{j2\pi ft}\, df \\
&= \int_{-\infty}^{\infty} G(-f)e^{-j2\pi f(T-t)}\, df \\
&= \int_{-\infty}^{\infty} G(f')e^{j2\pi f'(T-t)}\, df'
\end{aligned} \qquad (7.42)$$

Recognizing this as the inverse Fourier transform of $g(t)$ with t replaced by $T - t$, we obtain

$$h_0(t) = s_2(T - t) - s_1(T - t) = g\,(T - t) \qquad (7.43)$$

Thus, in terms of the original signals, the optimum receiver corresponds to passing the received signal plus noise through two parallel filters whose impulse responses are the time reverses of $s_1(t)$ and $s_2(t)$, respectively, and comparing the difference of their outputs at time T with the threshold given by (7.29). This operation is illustrated in Figure 7.9.

FIGURE 7.9 Matched filter receiver for binary signaling in white Gaussian noise

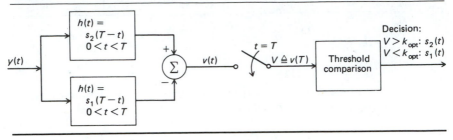

▼ EXAMPLE 7.3   Consider the pulse signal

$$s(t) = \begin{cases} A, & 0 \le t \le T \\ 0, & \text{otherwise} \end{cases} \tag{7.44}$$

A filter matched to this signal has the impulse response

$$h_0(t) = s(t_0 - t) \tag{7.45}$$

where the parameter t_0 will be left to be fixed later. We note that if $t_0 < T$, the filter will be unrealizable, since it will have nonzero impulse response for $t < 0$. The response of the filter to $s(t)$ is

$$y(t) = h_0(t) * s(t) = \int_{-\infty}^{\infty} h_0(\tau)s(t - \tau) \, d\tau \tag{7.46}$$

The factors in the integrand are shown in Figure 7.10(a). The resulting integrations are familiar from our previous considerations of linear systems, and the filter output is easily found to be as shown in Figure 7.10(b). We note that the peak output signal occurs at $t = t_0$. This is also the time of peak-signal-to-rms-noise ratio, since the noise is stationary. Clearly, in digital sig-
▲ naling, we want $t_0 = T$.

FIGURE 7.10 Signals pertinent to finding the matched-filter response of Example 7.3

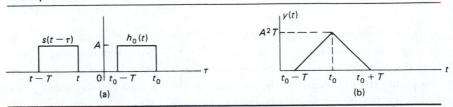

▼ EXAMPLE 7.4   For a given value of N_0, consider the peak-signal-to-rms-noise ratio at the output of a matched filter for the two pulses

$$g_1(t) = A\Pi\left(\frac{t - t_0}{T}\right) \tag{7.47}$$

and

$$g_2(t) = B\cos\left[\frac{2\pi(t - t_0)}{T}\right]\Pi\left(\frac{t - t_0}{T}\right) \tag{7.48}$$

Relate A and B such that both pulses provide the same signal-to-noise ratio at the matched filter output.

SOLUTION   Since the signal-to-noise ratio at the matched filter output is $2E_g/N_0$ and N_0 is the same for both cases, we can obtain equal signal-to-noise ratios for both cases by computing the energy of each pulse and setting the two energies equal. The results are

$$E_{g_1} = \int_{t_0 - T/2}^{t_0 + T/2} A^2\, dt = A^2T \tag{7.49}$$

$$E_{g_2} = \int_{t_0 - T/2}^{t_0 + T/2} B^2\cos^2\left[\frac{2\pi(t - t_0)}{T}\right] dt = \frac{B^2T}{2} \tag{7.50}$$

Setting these equal, we have that $A = B/\sqrt{2}$ to give equal signal-to-noise ratios. The peak signal-squared-to-mean-square-noise ratio is

$$\frac{E_g}{N_0} = \frac{A^2T}{N_0} = \frac{B^2T}{2N_0} \tag{7.51}$$

▲

Error Probability for the Matched-Filter Receiver

From (7.30), the error probability for the matched-filter receiver of Figure 7.9 is

$$P_E = Q\left(\frac{\zeta}{2}\right) \tag{7.52}$$

where ζ is the maximum value,

$$\zeta = \left[\frac{2}{N_0}\int_{-\infty}^{\infty} |G(f)|^2\, df\right]^{1/2} = \left[\frac{2}{N_0}\int_{-\infty}^{\infty} |S_2(f) - S_1(f)|^2\, df\right]^{1/2} \tag{7.53}$$

given by (7.39). Using Parseval's theorem, we can write ζ^2 in terms of $g(t) = s_2(t) - s_1(t)$ as

$$
\begin{aligned}
\zeta^2 &= \frac{2}{N_0} \int_{-\infty}^{\infty} [s_2(t) - s_1(t)]^2 \, dt \\
&= \frac{2}{N_0} \left\{ \int_{-\infty}^{\infty} s_2^2(t) \, dt + \int_{-\infty}^{\infty} s_1^2(t) \, dt - 2 \int_{-\infty}^{\infty} s_1(t)s_2(t) \, dt \right\}
\end{aligned}
\tag{7.54}
$$

From (7.19), we see that the first two terms inside the braces are E_1 and E_2, respectively. We define

$$
\rho_{12} = \frac{1}{\sqrt{E_1 E_2}} \int_{-\infty}^{\infty} s_1(t)s_2(t) \, dt
\tag{7.55}
$$

as the correlation coefficient of $s_1(t)$ and $s_2(t)$. Just as for random variables, ρ_{12} is a measure of the similarity between $s_1(t)$ and $s_2(t)$ and is normalized such that $-1 \leq \rho_{12} \leq 1$ (ρ_{12} achieves the end points for $s_1(t) = \pm k s_2(t)$, where k is a constant). Thus

$$
\zeta^2 = \frac{2}{N_0} (E_1 + E_2 - 2\sqrt{E_1 E_2}\rho_{12})
\tag{7.56}
$$

and the error probability is

$$
P_E = Q \left[\sqrt{2} \left(\frac{E_1 + E_2 - 2\sqrt{E_1 E_2}\rho_{12}}{N_0} \right)^{1/2} \right]
\tag{7.57}
$$

It is apparent from (7.57) that in addition to depending on the signal energies, as in the constant-signal case, P_E also depends on the similarity between the signals through ρ_{12}. We note that (7.56) takes on its maximum value of $(2/N_0)(\sqrt{E_1} + \sqrt{E_2})^2$ for $\rho_{12} = -1$, which gives the minimum value of P_E possible as $s_1(t)$ and $s_2(t)$ vary. This is reasonable, for then the transmitted signals are as dissimilar as possible.

By noting that $E = \frac{1}{2}(E_1 + E_2)$ is the average received signal energy, since $s_1(t)$ and $s_2(t)$ are transmitted with equal *a priori* probability, we can write (7.57) as

$$
P_E = Q[\sqrt{z(1 - R_{12})}]
\tag{7.58}
$$

where z is the average energy per bit divided by noise power spectral density E_b/N_0, as it was for baseband systems. The parameter R_{12} is defined as

$$
R_{12} = \frac{2\sqrt{E_1 E_2}}{E_1 + E_2} \rho_{12} = \frac{\sqrt{E_1 E_2}}{E} \rho_{12}
\tag{7.59}
$$

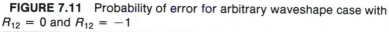

FIGURE 7.11   Probability of error for arbitrary waveshape case with $R_{12} = 0$ and $R_{12} = -1$

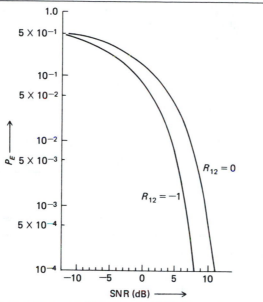

and is a convenient parameter related to the correlation coefficient, but which should *not* be confused with a correlation function. The minimum value of R_{12} is -1, which is attained for $E_1 = E_2$ and $\rho_{12} = -1$. For this value of R_{12},

$$P_E = Q(\sqrt{2z}) \tag{7.60}$$

which is identical to (7.11), the result for the baseband system.

The probability of error versus the signal-to-noise ratio is compared in Figure 7.11 for $R_{12} = 0$ (orthogonal signals) and $R_{12} = -1$ (antipodal signals).

Correlator Implementation of the Matched-Filter Receiver

In Figure 7.9, the optimum receiver involves two filters with impulse responses equal to the time reverse of the respective signals being detected. An alternative receiver structure can be obtained by noting that the matched filter in Figure 7.12(a) can be replaced by a multiplier-integrator cascade as shown in Figure 7.12(b). Such a series of operations is referred to as *correlation detection*.

FIGURE 7.12 Equivalence of the matched-filter and correlator receivers.
(a) Matched-filter sampler. (b) Correlator sampler.

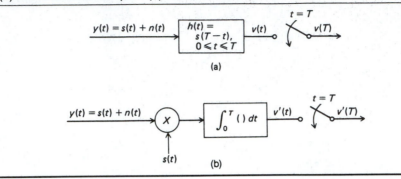

(a)

(b)

To show that the operations given in Figure 7.12 are equivalent, we will show that $v(T)$ in Figure 7.12(a) is equal to $v'(T)$ in Figure 7.12(b). The output of the matched filter in Figure 7.12(a) is

$$v(t) = h(t) * y(t) = \int_0^T s(T - \tau)\, y(t - \tau)\, d\tau \qquad (7.61)$$

which follows because $h(t) = s(T - t)$, $0 \le t \le T$, and zero otherwise. Letting $t = T$ and changing variables in the integrand to $\alpha = T - \tau$, we obtain

$$v(T) = \int_0^T s(\alpha)\, y(\alpha)\, d\alpha \qquad (7.62)$$

Considering next the output of the correlator configuration in Figure 7.12(b), we obtain

$$v'(T) = \int_0^T y(t)s(t)\, dt \qquad (7.63)$$

which is identical to (7.62). Thus the matched filters for $s_1(t)$ and $s_2(t)$ in Figure 7.9 can be replaced by correlation operations with $s_1(t)$ and $s_2(t)$, respectively, and the receiver operation will not be changed. We note that the integrate-and-dump receiver for the constant signal case of Section 7.1 is actually a correlation or, equivalently, a matched-filter receiver.

Optimum Threshold

The optimum threshold for binary signal detection is given by (7.29), where $s_{01}(T)$ and $s_{02}(T)$ are the outputs of the detection filter in Figure 7.6 at time

T due to the input signals $s_1(t)$ and $s_2(t)$, respectively. We now know that the optimum detection filter is a matched filter, matched to the difference of the input signals, and has impulse response given by (7.43). From the superposition integral, we have

$$
\begin{aligned}
s_{01}(T) &= \int_{-\infty}^{\infty} h(\lambda) s_1(T - \lambda) \, d\lambda \\
&= \int_{-\infty}^{\infty} [s_2(T - \lambda) - s_1(T - \lambda)] s_1(T - \lambda) \, d\lambda \\
&= \int_{-\infty}^{\infty} s_2(u) s_1(u) \, du - \int_{-\infty}^{\infty} [s_1(u)]^2 \, du \\
&= \sqrt{E_1 E_2} \, \rho_{12} - E_1
\end{aligned}
\tag{7.64}
$$

where the substitution $u = T - \lambda$ has been used to go from the second equation to the third, and the definition of the correlation coefficient (7.55) has been used to get the last equation along with the definition of energy of a signal. Similarly, it follows that

$$
\begin{aligned}
s_{02}(T) &= \int_{-\infty}^{\infty} [s_2(T - \lambda) - s_1(T - \lambda)] s_2(T - \lambda) \, d\lambda \\
&= \int_{-\infty}^{\infty} [s_2(u)]^2 \, du - \int_{-\infty}^{\infty} s_2(u) s_1(u) \, du \\
&= E_2 - \sqrt{E_1 E_2} \, \rho_{12}
\end{aligned}
\tag{7.65}
$$

Substituting (7.64) and (7.65) into (7.29), we find the optimum threshold to be

$$
k_{opt} = \tfrac{1}{2}(E_2 - E_1)
\tag{7.66}
$$

Note that equal energy signals will always result in an optimum threshold of zero. Also note that the waveshape of the signals, as manifested through the correlation coefficient, has no effect on the optimum threshold, but only the signal energies do.

Nonwhite (Colored) Noise Backgrounds

The question naturally arises about the optimum receiver for nonwhite noise backgrounds. Usually, the noise in a receiver system is generated primarily in the front-end stages and is due to thermal agitation of electrons in the electronic components (see Appendix A). This type of noise is well approximated as white. If a bandlimited channel precedes the introduction of the white noise, then we need only work with modified transmitted signals. If, for some reason, a bandlimiting filter follows the introduction of the white noise (for example, an intermediate-frequency amplifier following the radio-frequency amplifier and mixers where most of the noise is generated in a heterodyne receiver), we can use a simple artifice to approximate the matched filter receiver. The colored noise plus signal is passed through a "whitening filter" with a transfer function that is the inverse square root of the noise spectral density. Thus, the output of this whitening filter is white noise plus a signal component that has been transformed by the whitening filter. We then build a matched filter receiver with impulse response that is the difference of the time-reverse of the "whitened" signals. The cascade of a whitening filter and matched filter (matched to the whitened signals) is called a *whitened matched filter*. This combination provides only an approximately optimum receiver for two reasons. Since the whitening filters will spread the received signals beyond the T-second signaling interval, two types of degradation will result: (1) the signal energy spread beyond the interval under consideration is not used by the matched filter in making a decision; (2) previous signals spread out by the whitening filter will interfere with the matched filtering operation on the signal on which a decision is being made. The latter is referred to as *intersymbol interference* and is explored further in Sections 7.6 and 7.8. It is apparent that degradation due to these effects is minimized if the signal *duration* is short compared with T, such as in a pulsed radar system. Finally, signal intervals adjacent to the interval being used in the decision process contain information that is relevant to making a decision on the basis of the correlation of the noise. In short, the whitened matched-filter receiver is nearly optimum if the signaling interval is large compared with the inverse bandwidth of the whitening filter. The question of bandlimited channels, and nonwhite background noise, is explored further in Section 7.5.

Receiver Implementation Imperfections

In the theory developed in this section, it is assumed that the signals are known *exactly* at the receiver. This is, of course, an idealized situation. Two possible deviations from this assumption are: (1) the phase of the receiver's replica of the transmitted signal may be in error, and (2) the exact arrival time of the received signal may be in error. These are called *synchronization* errors. The first case is explored in Section 7.3, and the latter is explored in the problems. Methods of synchronization are discussed in Chapter 8.

FIGURE 7.13   Waveforms for ASK, PSK, and FSK modulation

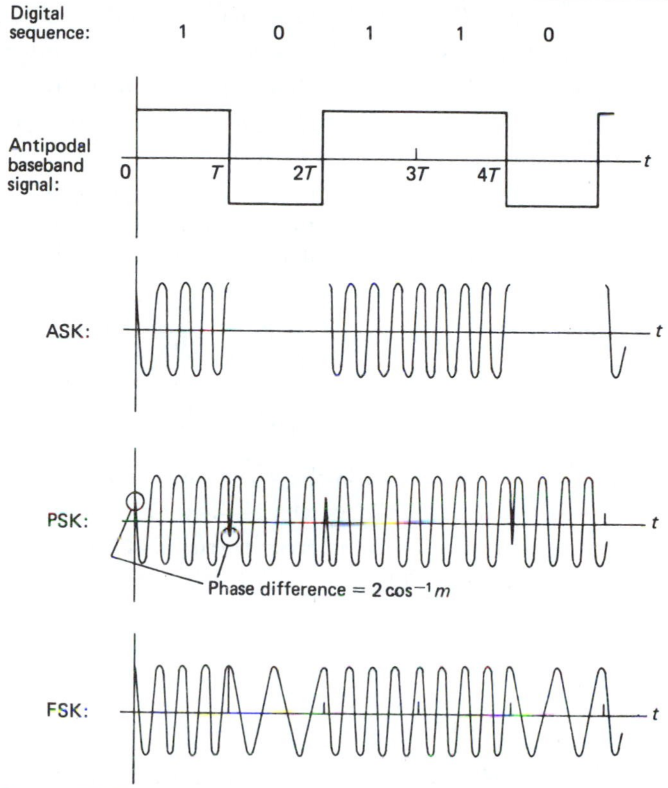

7.3   ERROR PROBABILITIES FOR COHERENT BINARY SIGNALING SCHEMES

We now compare the performance of several commonly used coherent binary signaling schemes. Then we will examine noncoherent systems. To obtain the error probability for coherent systems, the results of Section 7.2 will be applied directly. The three types of coherent systems to be considered in this section are amplitude-shift keyed (ASK), phase-shift keyed (PSK), and frequency-shift keyed (FSK). Typical transmitted waveforms for these three types of digital modulation are shown in Figure 7.13. We also will consider the effect of an imperfect phase reference on the performance of a coherent PSK system. Such systems are often referred to as partially coherent.

Amplitude-Shift Keying (ASK)

In Table 7.1, $s_1(t)$ and $s_2(t)$ for ASK are given as 0 and $A \cos \omega_c t$, where $f_c = \omega_c/2\pi$ is the carrier frequency. We note that the transmitter for such a system simply consists of an oscillator that is gated on and off; accordingly, ASK is often referred to as *on-off keying*. It is important to note that the oscillator runs continuously as the on-off gating is carried out.

The correlator realization for the optimum receiver consists of multiplication of the received signal plus noise by $A \cos \omega_c t$, integration over $(0, T)$, and comparison of the integrator output with the threshold $\frac{1}{4}A^2 T$ as calculated from (7.66).

From (7.55) and (7.59), $R_{12} = \rho_{12} = 0$, and the probability of error, from (7.58), is

$$P_E = Q(\sqrt{z})$$ (7.67)

Because of the lack of a factor $\sqrt{2}$ in the argument of the Q function, ASK is seen to be 3 dB worse in terms of signal-to-noise ratio than antipodal baseband signaling. The probability of error versus SNR corresponds to the curve for $R_{12} = 0$ in Figure 7.11.

Phase-Shift Keying (PSK)

From Table 7.1, the signals for PSK are

$$s_k(t) = A \sin [\omega_c t - (-1)^k \cos^{-1} m], \qquad 0 \le t \le T, k = 1, 2 \quad (7.68)$$

where $\cos^{-1} m$, the modulation index, is written in this fashion for future convenience. For simplicity, we assume that $\omega_c = 2\pi n/T$, where n is an integer. Using $\sin(-x) = -\sin x$ and $\cos(-x) = \cos(x)$, we can write (7.68) as

$$s_k(t) = Am \sin \omega_c t - (-1)^k A \sqrt{1 - m^2} \cos \omega_c t,$$
$$0 \le t \le T, k = 1, 2 \quad (7.69)$$

where we note that $\cos(\cos^{-1} m) = m$ and $\sin(\cos^{-1} m) = \sqrt{1 - m^2}$.

The first term on the right-hand side of (7.69) represents a carrier component included in some systems for synchronization of the local carrier reference at the receiver to the transmitted carrier. The power in the carrier component is $\frac{1}{2}(Am)^2$, and the power in the modulation component is $\frac{1}{2}A^2(1 - m^2)$. Thus m^2 is the fraction of the total power in the carrier. The correlator receiver is shown in Figure 7.14, where, instead of two correlators, only a single correlation with $s_2(t) - s_1(t)$ is used. The threshold, calculated from (7.66), is zero. We note that the carrier component of $s_k(t)$ is

FIGURE 7.14   Correlator realization of optimum receiver for PSK

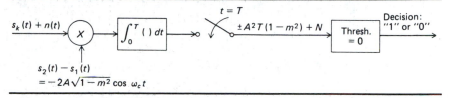

of no consequence in the correlation operation because it is orthogonal to the modulation component over the bit interval. For PSK, $E_1 = E_2 = \frac{1}{2}A^2T$ and

$$\sqrt{E_1 E_2}\,\rho_{12} = \int_0^T s_1(t)s_2(t)\,dt$$

$$= \int_0^T (Am \sin \omega_c t + A\sqrt{1 - m^2}\cos \omega_c t)$$

$$\cdot\,(Am \sin \omega_c t - A\sqrt{1 - m^2}\cos \omega_c t)\,dt$$

$$= \tfrac{1}{2}A^2 Tm^2 - \tfrac{1}{2}A^2 T(1 - m^2)$$

$$= \tfrac{1}{2}A^2 T(2m^2 - 1) \tag{7.70}$$

Thus R_{12}, from (7.59), is

$$R_{12} = \frac{2\sqrt{E_1 E_2}}{E_1 + E_2}\rho_{12}$$

$$= 2m^2 - 1 \tag{7.71}$$

and the probability of error for PSK is

$$P_E = Q[\sqrt{2(1 - m^2)z}] \tag{7.72}$$

The effect of allocating a fraction m^2 of the total transmitted power to a carrier component is to degrade P_E by $10 \log_{10}(1 - m^2)$ decibels from the ideal $R_{12} = -1$ curve of Figure 7.11.

For $m = 0$, the resultant error probability is 3 dB better than ASK and corresponds to the $R_{12} = -1$ curve in Figure 7.11. We will refer to the case for which $m = 0$ as *biphase-shift keying* (BPSK) to avoid confusion with the case for which $m \neq 0$.

▼ | **EXAMPLE 7.5**   Consider PSK with $m = 1/\sqrt{2}$. (a) By how many degrees does the modulated carrier shift in phase each time the binary data changes? (b) What percent of the total power is in the carrier, and what percent is in the modulation component? (c) What value of $z = E_b/N_0$ is required to give $P_E = 10^{-6}$?

SOLUTION   (a) Since the change in phase is from $-\cos^{-1} m$ to $\cos^{-1} m$ whenever the phase switches, the phase change of the modulated carrier is

$$2 \cos^{-1} m = 2 \cos^{-1} (1/\sqrt{2}) = 2(45°) = 90° \qquad (7.73)$$

(b) The carrier and modulation components are

$$\text{carrier} = Am \sin (\omega_c t) \qquad (7.74a)$$

and

$$\text{modulation} = \pm A\sqrt{1 - m^2} \cos (\omega_c t) \qquad (7.74b)$$

respectively. Therefore, the power in the carrier component is

$$P_c = \frac{A^2 m^2}{2} \qquad (7.75a)$$

and the power in the modulation component is

$$P_m = \frac{A^2(1 - m^2)}{2} \qquad (7.75b)$$

Since the total power is $A^2/2$, the percent power in each of these components is

$$\%P_c = m^2 \times 100 = 50\% \qquad \text{and} \qquad \%P_m = (1 - m^2) \times 100 = 50\%$$

respectively.

(c) We have, for the probability of error,

$$P_E = Q[\sqrt{2(1 - m^2)z}] \cong \frac{e^{-(1 - m^2)z}}{2\sqrt{\pi(1 - m^2)z}} \qquad (7.76)$$

Solving this iteratively, we obtain, for $m = 0.5$, $z = 22.6$ or $E_b/N_0 = 13.54$ dB. Actually, we do not have to solve the error probability relationship iteratively again. From Example 7.2 we already know that $z = 10.54$ dB gives $P_E = 10^{-6}$ for PSK. In this example we simply note that the required power is twice as much as for BPSK, which is equivalent to adding 3 dB on to the 10.54 dB required in Example 7.2.

Biphase-Shift Keying with Imperfect Reference

The results obtained earlier for PSK are for the case of a perfect reference at the receiver. If $m = 0$, it is simple to consider the case of an imperfect reference at the receiver as represented by an input of the form $\pm A \cos (\omega_c t + \theta) + n(t)$ and the reference by $A \cos (\omega_c t + \hat{\theta})$, where θ is an unknown carrier phase and $\hat{\theta}$ is the phase estimate at the receiver.

FIGURE 7.15   Effect of phase error in reference signal for correlation detection of BPSK

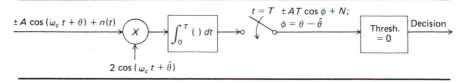

The correlator implementation for the receiver is shown in Figure 7.15. Using appropriate trigonometric identities, we find that the signal component of the correlator output at the sampling instant is $\pm AT \cos \phi$, where $\phi = \theta - \hat{\theta}$ is the phase error. It follows that the error probability *given* the phase error ϕ is

$$P_E(\phi) = Q(\sqrt{2z \cos^2 \phi}) \qquad (7.77)$$

We note that the performance is degraded by $20 \log_{10} \cos \phi$ decibels compared with the perfect reference case.

If we assume ϕ to be fixed at some maximum value, we may obtain an upper bound on P_E due to phase error in the reference. However, a more exact model is often provided by approximating ϕ as a Gaussian random variable with the pdf

$$p(\phi) = \frac{e^{-\phi^2/2\sigma_\phi^2}}{\sqrt{2\pi\sigma_\phi^2}} \qquad (7.78)$$

This is an especially appropriate model if the phase reference at the receiver is derived by means of a phase-locked loop operating with high signal-to-noise ratio at its input. If this is the case, σ_ϕ^2 is related to the signal-to-noise ratio at the input of the phase estimation device, whether it is a phase-locked loop or a bandpass-filter-limiter combination.

To find the error probability averaged over all possible phase errors, we simply find the expectation of $P(E|\phi)$ with respect to the phase-error pdf, $p(\phi)$, that is,

$$P_E = \int_{-\pi}^{\pi} P_E(\phi)p(\phi) \, d\phi \qquad (7.79)$$

The resulting integral must be evaluated numerically for typical phase error pdf's.* Typical results are given in Table 7.2 for $p(\phi)$ Gaussian.

---

* See, for example, Van Trees (1968), Chapter 4.

TABLE 7.2   Effect of Gaussian Phase Reference Jitter on the Detection of BPSK

E/N_0, dB	$P_E, \sigma_\phi^2 = 0.01 \text{ rad}^2$	$P_E, \sigma_\phi^2 = 0.05 \text{ rad}^2$	$P_E, \sigma_\phi^2 = 0.1 \text{ rad}^2$
9	3.68×10^{-5}	6.54×10^{-5}	2.42×10^{-4}
10	4.55×10^{-6}	1.08×10^{-5}	8.96×10^{-5}
11	3.18×10^{-7}	1.36×10^{-6}	3.76×10^{-5}
12	1.02×10^{-8}	1.61×10^{-7}	1.83×10^{-5}

Frequency-Shift Keying (FSK)

In Table 7.1, the signals for FSK are given as

$$\left.\begin{array}{ll} & s_1(t) = A \cos \omega_c t \\ \text{and} & s_2(t) = A \cos (\omega_c + \Delta\omega)t \end{array}\right\} \quad 0 \le t \le T \qquad (7.80)$$

For simplification, we assume that

$$\omega_c = \frac{2\pi n}{T} \qquad (7.81a)$$

and

$$\Delta\omega = \frac{2\pi m}{T} \qquad (7.81b)$$

where m and n are integers. This ensures that both $s_1(t)$ and $s_2(t)$ will go through an integer number of cycles in T seconds. As a result,

$$\begin{aligned} \sqrt{E_1 E_2} \rho_{12} &= \int_0^T A^2 \cos \omega_c t \cos (\omega_c + \Delta\omega)t \; dt \\ &= \tfrac{1}{2}A^2 \int_0^T [\cos \Delta\omega t + \cos (2\omega_c + \Delta\omega)t] \; dt \\ &= 0 \qquad\qquad\qquad\qquad\qquad\qquad\qquad (7.82) \end{aligned}$$

and $R_{12} = 0$. Thus

$$P_E = Q(\sqrt{z}) \qquad (7.83)$$

which is the same as for ASK. The error probability versus signal-to-noise ratio therefore corresponds to the curve $R_{12} = 0$ in Figure 7.11.

We note that the reason ASK and FSK have the same P_E versus SNR characteristics is that we are making our comparison on the basis of *average* signal power. If we constrain *peak* signal power to be equal, ASK will be 3 dB worse than FSK.

   We denote the three schemes just considered as coherent, binary ASK, PSK, and FSK to indicate the fact that they are binary. We will consider *M*-ary ($M > 2$) schemes in Chapter 8.

▼ **EXAMPLE 7.6**   Compare binary ASK, PSK, and FSK on the basis of E_b/N_0 required for $P_E = 10^{-6}$ and on the basis of transmission bandwidth for a constant data rate. Take the required bandwidth as the null-to-null bandwidth of the square-pulse modulated carrier. Assume the minimum bandwidth possible for BFSK.

SOLUTION   From before, we know that to give $P_{E,\text{BPSK}} = 10^{-6}$, the required E_b/N_0 is 10.54 dB. ASK, on an average basis, and FSK require 3 dB more, or 13.54 dB, to give $P_E = 10^{-6}$. The Fourier transform of a square-pulse modulated carrier is

$$\Pi(t/T) \cos (2\pi f_c t) \leftrightarrow (T/2) \left\{ \text{sinc} \left[T(f - f_c) \right] + \text{sinc} \left[T(f + f_c) \right] \right\}$$

The null-to-null bandwidth of the positive-frequency portion of this spectrum is

$$B_{RF} = \frac{2}{T} \text{ Hz} \tag{7.84}$$

For binary ASK and PSK, the required bandwidth is

$$B_{PSK} = B_{ASK} = \frac{2}{T} = 2R \text{ Hz} \tag{7.85}$$

where R is the data rate in bits per second. For FSK, the spectra for

$$s_1(t) = A \cos \omega_c t, \qquad 0 \le t \le T, \qquad \omega_c = 2\pi f_c$$

and

$$s_2(t) = A \cos (\omega_c t + \Delta\omega)t, \qquad 0 \le t \le T, \qquad \Delta\omega = 2\pi \, \Delta f$$

are assumed to be separated by $1/2T$ Hz, which is the minimum spacing for orthogonality of the signals. The required bandwidth is therefore

$$B_{FSK} = \frac{1}{T} + \frac{1}{2T} + \frac{1}{T} = \frac{2.5}{T} = 2.5R \text{ Hz}$$

$$\underbrace{}$$

$$f_c \text{ burst} \tag{7.86}$$

$$\underbrace{}$$

$$f_c + \Delta f \text{ burst}$$

We often specify bandwidth efficiency R/B in terms of bits per second per hertz. For binary ASK and PSK the bandwidth efficiency is 0.5 bits/s/Hz, while for binary FSK it is 0.4 bits/s/Hz.

7.4 MODULATION SCHEMES NOT REQUIRING COHERENT REFERENCES

We now consider several modulation schemes that do not require the acquisition of a local reference signal in phase coherence with the received carrier. The first scheme to be considered is referred to as *differentially coherent phase-shift keying* (DPSK) and may be thought of as the noncoherent version of BPSK considered in Section 7.3. Also considered in this section will be noncoherent, binary FSK.

Differential Phase-Shift Keying (DPSK)

One way of obtaining a phase reference for the demodulation of BPSK is to use the carrier phase of the preceding signaling interval. The implementation of such a scheme presupposes two things: (1) The mechanism causing the unknown phase perturbation on the signal varies so slowly that the phase is essentially constant from one signaling interval to the next. (2) The phase during a given signaling interval bears a known relationship to the phase during the preceding signaling interval. The former is determined by the stability of the transmitter oscillator, time-varying changes in the channel, and so on. The latter requirement can be met by employing what is referred to as *differential encoding* of the message sequence at the transmitter.

Differential encoding of a message sequence is illustrated in Table 7.3. An arbitrary reference binary digit is assumed for the initial digit of the encoded sequence. In the example shown in Table 7.3, a 1 has been chosen. For each digit of the encoded sequence, the present digit is used as a reference for the following digit in the sequence. A 0 in the message sequence is encoded as a transition from the state of the reference digit to the opposite state in the encoded message sequence; a 1 is encoded as no change of state. In the example shown, the first digit in the message sequence is a 1,

TABLE 7.3 Differential Encoding Example

Message sequence:		1	0	0	1	1	1	0	0	0
Encoded sequence:	1	1	0	1	1	1	1	0	1	0
Reference digit	⟶	↑								
Transmitted phase:		0	0	π	0	0	0	0	π	0	π

FIGURE 7.16 Block diagram of a DPSK modulator

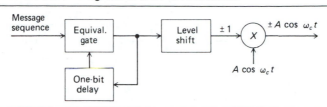

so no change in state is made in the encoded sequence, and a 1 appears as the next digit in the encoded sequence. This serves as the reference for the next digit to be encoded. Since the next digit appearing in the message sequence is a 0, the next encoded digit is the opposite of the reference digit, or a 0. The encoded message sequence then phase-shift keys a carrier with the phases 0 and π as shown in the table.

The block diagram in Figure 7.16 illustrates the generation of DPSK. The equivalence gate, which is the negation of an EXCLUSIVE-OR, is a logic circuit that performs the operations listed in Table 7.4. By a simple level shift at the output of the logic circuit, so that the encoded message is bipolar, the DPSK signal is produced by multiplication by the carrier, or double-sideband modulation.

A possible implementation of a differentially coherent demodulator for DPSK is shown in Figure 7.17. The received signal plus noise is correlated bit by bit with a one-bit delayed version of the signal plus noise. The output of the correlator is then compared with a threshold set at zero, a decision being made in favor of a 1 or a 0, depending on whether the correlator output is positive or negative, respectively.

To illustrate that the received sequence will be correctly demodulated, consider the example given in Table 7.3, assuming no noise is present. After the first two bits have been received (the reference bit plus the first encoded

TABLE 7.4 Truth Table for the Equivalence Operation

Input 1 (Message)	Input 2 (Reference)	Output
0	0	1
0	1	0
1	0	0
1	1	1

FIGURE 7.17 Demodulation of DPSK

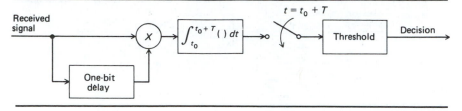

bit), the signal input to the correlator is $S_1 = A \cos \omega_c t$, and the reference, or delayed, input is $R_1 = A \cos \omega_c t$. The output of the correlator is

$$v_1 = \int_0^T A^2 \cos^2 \omega_c t \, dt = \tfrac{1}{2} A^2 T \qquad (7.87a)$$

and the decision is that a 1 was transmitted. For the next bit interval, the inputs are $S_2 = -A \cos \omega_c t$ and $R_2 = S_1 = A \cos \omega_c t$, resulting in a correlator output of

$$v_2 = \int_0^T (-A^2 \cos^2 \omega_c t) \, dt = -\tfrac{1}{2} A^2 T \qquad (7.87b)$$

and a decision that a 0 was transmitted is made. Continuing in this fashion, we see that the original message sequence is obtained if there is no noise at the input.

This detector, while simple to implement, is actually not optimum. The optimum detector for binary DPSK is shown in Figure 7.18. The test statistic for this detector is

$$\ell = x_k x_{k-1} + y_k y_{k-1} \qquad (7.88)$$

If $\ell > 0$, the receiver chooses the signal sequence

$$s_1(t) = \begin{cases} A \cos (\omega_c t + \theta), & -T \le t < 0 \\ A \cos (\omega_c t + \theta), & 0 \le t < T \end{cases} \qquad (7.89)$$

as having been sent. If $\ell < 0$, the receiver chooses the signal sequence

$$s_2(t) = \begin{cases} A \cos (\omega_c t + \theta), & -T \le t < 0 \\ -A \cos (\omega_c t + \theta), & 0 \le t < T \end{cases} \qquad (7.90)$$

as having been sent.

Without loss of generality, we can choose $\theta = 0$ (the noise and signal orientations with respect to the sine and cosine mixers in Figure 7.18 are completely random). The probability of error can then be computed from

$$P_E = Pr(x_k x_{k-1} + y_k y_{k-1} < 0 \mid s_1 \text{ sent}, \theta = 0)$$

FIGURE 7.18 Optimum receiver for binary differential phase-shift keying

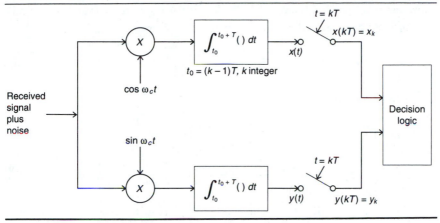

(it is assumed that s_1 and s_2 are equally likely). Assuming that the $\omega_c T$ is an integer multiple of 2π, we find the outputs of the integrators at time $t = 0$ to be

$$x_0 = \frac{AT}{2} + n_1 \qquad \text{and} \qquad y_0 = n_3 \qquad (7.91)$$

where

$$n_1 = \int_{-T}^{0} n(t) \cos(\omega_c t) \, dt \qquad (7.92a)$$

and

$$n_3 = \int_{-T}^{0} n(t) \sin(\omega_c t) \, dt \qquad (7.92b)$$

Similarly, at time $t = T$, the outputs are

$$x_1 = \frac{AT}{2} + n_2 \qquad \text{and} \qquad y_1 = n_4 \qquad (7.93)$$

where

$$n_2 = \int_{0}^{T} n(t) \cos(\omega_c t) \, dt \qquad (7.94a)$$

and

$$n_4 = \int_{0}^{T} n(t) \sin(\omega_c t) \, dt \qquad (7.94b)$$

It follows that n_1, n_2, n_3, and n_4 are uncorrelated, zero-mean Gaussian random variables with variances $N_0 T/4$. Since they are uncorrelated, they are also independent, and the expression for P_E becomes

$$P_E = Pr\left[\left(\frac{AT}{2} + n_1\right)\left(\frac{AT}{2} + n_2\right) + n_3 n_4 < 0\right] \tag{7.95}$$

This can be rewritten as

$$P_E = Pr\left[\left(\frac{AT}{2} + \frac{n_1}{2} + \frac{n_2}{2}\right)^2 - \left(\frac{n_1}{2} - \frac{n_2}{2}\right)^2\right.$$
$$\left. + \left(\frac{n_3}{2} + \frac{n_4}{2}\right)^2 - \left(\frac{n_3}{2} - \frac{n_4}{2}\right)^2 < 0\right] \tag{7.96}$$

[To check this, simply square the separate terms in the argument of (7.96), collect like terms, and compare with the argument of (7.95).] Defining new Gaussian random variables as

$$w_1 = \frac{n_1}{2} + \frac{n_2}{2}$$

$$w_2 = \frac{n_1}{2} - \frac{n_2}{2}$$

$$w_3 = \frac{n_3}{2} + \frac{n_4}{2} \tag{7.97}$$

$$w_4 = \frac{n_3}{2} - \frac{n_4}{2}$$

the probability of error can be written as

$$P_E = Pr\left[\left(\frac{AT}{2} + w_1\right)^2 + w_3^2 < w_2^2 + w_4^2\right] \tag{7.98}$$

The positive square roots of the quantities on either side of the inequality sign inside the brackets can be compared just as well as the quantities themselves. From the definitions of w_1, w_2, w_3, and w_4, it can be shown that they are uncorrelated with each other and all are zero mean with variances $N_0 T/8$. Since they are uncorrelated and Gaussian, they are also independent. It follows that

$$R_1 = \sqrt{\left(\frac{AT}{2} + w_1\right)^2 + w_3^2} \tag{7.99}$$

is a Rician random variable (see Problem 5.33). It is also true that

$$R_2 = \sqrt{w_2^2 + w_4^2} \tag{7.100}$$

is a Rayleigh random variable. It follows that the probability of error can be written as the double integral

$$P_E = \int_0^\infty \left[\int_{r_2}^\infty f_{R_2}(r_2) \right] f_{R_1}(r_1)\, dr_1 \tag{7.101}$$

where $f_{R_1}(r_1)$ is a Rayleigh pdf and $f_{R_2}(r_2)$ is a Rician pdf. This double integral will arise later when we consider noncoherent FSK modulation. It is rather amazing that it evaluates to the simple result

$$P_E = \tfrac{1}{2}e^{-E_b N_0} = \tfrac{1}{2}e^{-z} \tag{7.102}$$

It has been shown in the literature that the suboptimum integrate-and-dump detector of Figure 7.12 gives an approximate probability of error*

$$P_E \cong Q[\sqrt{E_b/N_0}] = Q[\sqrt{z}] \tag{7.103}$$

The result is about a 2-dB degradation in signal-to-noise ratio for a specified probability of error from that of the optimum detector.

Recalling the result for BPSK, (7.72) with $m = 0$, and using the asymptotic expression $Q(u) \cong e^{-u^2/2}/[(2\pi)^{1/2}u]$, we obtain the following result for BPSK valid for large z:

$$P_E \cong \frac{e^{-z}}{2\sqrt{\pi z}}, \qquad (BPSK;\ z \gg 1) \tag{7.104}$$

For large z, DPSK and BPSK differ only by the factor $(\pi z)^{1/2}$, or roughly a 1-dB degradation in signal-to-noise ratio of DPSK with respect to BPSK at low probability of error. This makes DPSK an extremely attractive solution to the problem of carrier reference acquisition required for demodulation of BPSK. The only significant disadvantages of DPSK are that the signaling rate is locked to the specific value dictated by the delay elements in the transmitter and receiver and that errors tend to occur in groups of two because of the correlation imposed between successive bits by the differential encoding process (the latter is the main reason for the 1-dB degradation in performance of DPSK over BPSK at high signal-to-noise ratios).

Noncoherent FSK

The computation of error probabilities for noncoherent systems is somewhat more difficult than it is for coherent systems. Since more is known about the received signal in a coherent system than in a noncoherent system, it is not surprising that the performance of the latter is worse than the corre-

---

* See J. H. Park, "On Binary DPSK Reception," *IEEE Transactions on Communications*, Vol. COM-26, April 1978, pp. 484–486.

FIGURE 7.19 Receiver for noncoherent FSK

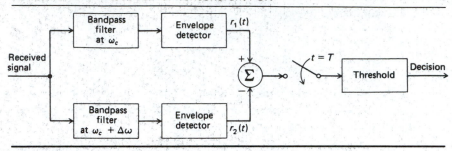

sponding coherent system. Even with this loss in performance, noncoherent systems are often used when simplicity of implementation is a predominent consideration.

Only noncoherent FSK will be discussed here.* A truly noncoherent PSK system does not exist, but DPSK can be viewed as such.

For noncoherent FSK, the transmitted signals are

$$s_1(t) = A \cos (\omega_c t + \theta), \qquad 0 \le t \le T \qquad (7.105)$$

and

$$s_2(t) = A \cos [(\omega_c + \Delta\omega)t + \theta], \qquad 0 \le t \le T$$

where $\Delta\omega$ is sufficiently large that $s_1(t)$ and $s_2(t)$ occupy different spectral regions. The receiver for FSK is shown in Figure 7.19; note that it consists of two receivers for noncoherent ASK in parallel. As such, calculation of the probability of error for FSK proceeds much the same way as for ASK, although we are not faced with the dilemma of a threshold that must change with signal-to-noise ratio. Indeed, because of the symmetries involved, an exact result for P_E can be obtained. Assuming $s_1(t)$ has been transmitted, the output of the upper detector at time T, $R_1 \triangleq r_1(T)$, has the Rician pdf

$$f_{R_1}(r_1) = \frac{r_1}{N} e^{-(r_1^2 + A^2)/2N} I_0 \left(\frac{Ar_1}{N} \right), \qquad r_1 > 0 \qquad (7.106)$$

where $I_0(\cdot)$ is the modified Bessel function of the first kind of order zero and we have made use of Problem 5.33. The noise power is $N = N_0 B_T$. The output of the lower filter at time T, $R_2 \triangleq r_2(T)$, results from noise alone; its pdf is therefore Rayleigh:

$$f_{R_2}(r_2) = \frac{r_2}{N} e^{-r_2^2/2N}, \qquad r_2 > 0 \qquad (7.107)$$

---

* See Problem 7.27 for a sketch of the derivation of P_E for noncoherent ASK.

It is a special case of the Rician pdf with $A = 0$. An error occurs if $R_2 > R_1$, which can be written as

$$P(E \mid s_1(t)) = \int_0^\infty f_{R_1}(r_1) \left[\int_{r_1}^\infty f_{R_2}(r_2) \, dr_2 \right] dr_1 \qquad (7.108)$$

By symmetry, it follows that $P(E \mid s_1(t)) = P(E \mid s_2(t))$, so that (7.108) is the average probability of error. The inner integral in (7.108) integrates to $e^{-r_1^2/2N}$, which results in the expression

$$P_E = e^{-z} \int_0^\infty \frac{r_1}{N} I_0 \left(\frac{Ar_1}{N} \right) e^{-r_1^2/N} \, dr_1 \qquad (7.109)$$

where $z = A^2/2N$ as before. If we use a table of definite integrals, we can reduce (7.109) to

$$P_E = \tfrac{1}{2} e^{-z/2} \qquad (7.110)$$

For coherent, binary FSK, the error probability for large signal-to-noise ratios, using the asymptotic expansion for the Q-function, is

$$P_E \cong \exp\left(-\tfrac{1}{2}z\right) / \sqrt{2\pi z} \qquad \text{for } z \gg 1 \qquad (7.111)$$

which indicates that *the additional power required for noncoherent detection of FSK over coherent detection at large signal-to-noise ratios is inconsequential.* Thus, because of the comparable performance and the added simplicity of noncoherent FSK, it is employed almost exclusively in practice, instead of coherent FSK.

Comparison of Digital Modulation Systems

The binary digital modulation systems analyzed in this section and the preceding one are compared in Figure 7.20. We see that there is less than a 4-dB difference between the best (BPSK) and worst (noncoherent ASK and FSK) at high signal-to-noise ratios. This may seem to be a small price to pay for the simplicity in going from a coherent PSK system to a noncoherent FSK system. However, in some applications, even a 1-dB savings is worthwhile.

In addition to cost and complexity of implementation, there are many other considerations in choosing one type of digital data system over another. For some channels, where the channel gain or phase characteristics (or both) are perturbed by randomly varying propagation conditions, use of a noncoherent system may be dictated because of the near impossibility of establishing a coherent reference at the receiver under such conditions. Such channels are referred to as *fading*. In other situations, the choice of a coherent system may be demanded by considerations other than power savings.

FIGURE 7.20   Error probabilities for several binary digital signaling schemes

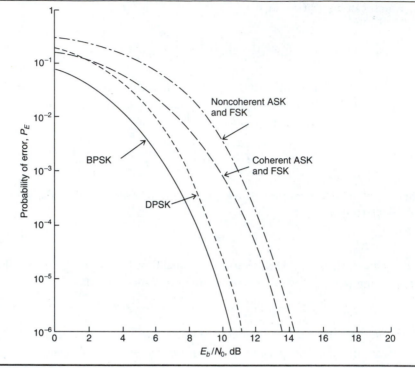

The following example illustrates some typical signal-to-noise ratio calculations for these digital modulation techniques.

▼ **EXAMPLE 7.7**   Suppose a P_E of 10^{-4} is desired for a certain digital data
transmission system. (a) Compare the necessary SNR's for BPSK, DPSK, noncoherent ASK, and noncoherent FSK. (b) If a PSK system is used with 10%
of the transmitted signal power allocated to the carrier, how much must the
SNR be increased to maintain $P_E = 10^{-4}$? (c) If BPSK is used and a maximum
demodulator phase error of 10% is anticipated, what must the SNR power
margin be to ensure that $P_E \leq 10^{-4}$?

SOLUTION   For part (a), we use asymptotic expressions for P_E and neglect
factors of $z^{-1/2}$. The error in doing so is less than 1 dB, as shown in
Table 7.5.

TABLE 7.5   Comparison of Binary Modulation Schemes at $P_E = 10^{-4}$

Modulation Method	P_E A = Asymptotic E = Exact	Required SNR (Approx.) (dB)	Actual SNR (dB)
BPSK	$e^{-z/2}\sqrt{\pi}$ (A)	9.0	8.4
DPSK	$e^{-z/2}$ (E)	9.3	9.3
Noncoherent ASK	$e^{-z/2}/2$ (A)	12.3	—
Noncoherent FSK	$e^{-z/2}/2$ (E)	12.3	12.3

For part (b), $m^2 = 0.1$ and the degradation in SNR over a BPSK system is $10 \log_{10} (1 - m^2) = -0.46$ dB. Thus an SNR of $8.4 + 0.46 = 8.86$ dB is required to maintain $P_E = 10^{-4}$.

In part (c), the degradation is $20 \log_{10} \cos 10° = -0.133$ dB. Thus an SNR of $8.4 + 0.133 = 8.533$ dB will ensure that $P_E \leq 10^{-4}$.

7.5   DIGITAL SIGNALING THROUGH BANDLIMITED CHANNELS

The analyses of the digital modulation techniques considered so far assumed the availability of an infinite-bandwidth channel. In practical applications, the channel bandwidth is never infinite. Two approaches to the finite-band-width channel case are the following:

1. Make use of one of the standard digital modulation schemes considered previously and simply transmit a higher-power signal than ideally required to compensate for the degradation imposed by the finite-band-width channel.
2. Choose designs for transmitter and receiver filters that shape the signal pulse-shape function so as to ideally eliminate interference between adjacent pulses.

To see how one might make use of the latter approach, we recall the sampling theorem, which gives a theoretical minimum spacing between samples to be taken from a signal with an ideal lowpass spectrum in order that the signal can be reconstructed exactly from the sample values. In particular, the transmission of a lowpass signal with bandwidth W can be viewed as sending $2W$ independent samples per second. If these $2W$ samples per second represent $2W$ independent pieces of data, this transmission can be viewed as sending $2W$ pulses per second through a channel represented by an ideal lowpass filter of bandwidth W. The transmission of the nth piece of information through the channel at time $t = nT = n/2W$ is accomplished by sending an impulse of amplitude a_n. The output of the channel due to this impulse at the input is

$$y_n(t) = a_n \text{ sinc} \left[2W \left(t - \frac{n}{2W} \right) \right] \qquad (7.112)$$

For an input consisting of a train of impulses spaced by $T = 1/2W$, the channel output is

$$y(t) = \sum_n y_n(t) = \sum_n a_n \text{ sinc} \left[2W \left(t - \frac{n}{2W} \right) \right] \qquad (7.113)$$

If the channel output is sampled at time $t_m = m/2W$, the sample value is a_m because

$$\text{sinc } (m - n) = \begin{cases} 1, & m = n \\ 0, & m \neq n \end{cases} \qquad (7.114)$$

which results in all terms in (7.113) except the mth being zero. In other words, the mth sample value at the output is not affected by preceding or succeeding sample values; it represents an independent piece of information.

Note that the bandlimited channel implies that the time response due to the nth impulse at the input is infinite in extent; a waveform cannot be simultaneously bandlimited and time-limited. It is of interest to inquire if there are any other bandlimited waveforms other than sinc $(2Wt)$ that have the property of (7.114), that is, that their zero crossings are spaced by T seconds. One such family of pulses are those having *raised cosine spectra*. Their time response is given by

$$p(t) = \frac{\cos (2\pi\beta t)}{1 - (4\beta t)^2} \text{ sinc} \left(\frac{t}{T} \right) \qquad (7.115)$$

and their spectra by

$$P(f) = \begin{cases} T, & |f| \leq 1/2T - \beta \\ \dfrac{T}{2} \left\{ 1 + \cos \left[\dfrac{\pi(|f| - 1/2T + \beta)}{2\beta} \right] \right\}, & 1/2T - \beta < |f| \\ & \qquad \leq 1/2T + \beta \\ 0, & |f| > 1/2T + \beta \end{cases} \qquad (7.116)$$

Figure 7.21 shows this waveform and its spectra for several values of β. Note that zero crossings for $p(t)$ occur at least every T seconds. If $\beta = 1/2T$, the bandwidth of $P(f)$ is $1/T$ Hz, which is twice that of the sinc (t/T) pulse ($\beta = 0$). The increased bandwidth, however, results from a cosine rolloff with

FIGURE 7.21   (a) Pulse response for the raised cosine spectra shown in (b)

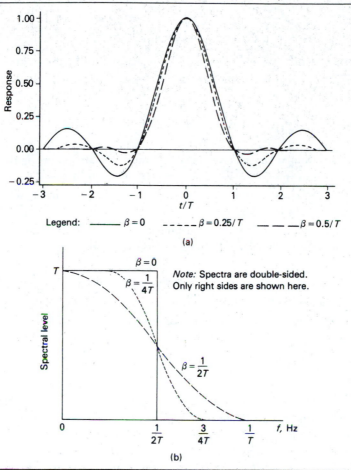

(a)

(b)

frequency of $P(f)$, which may be easier to realize in a practical filter. Also, $p(t)$ for $\beta = 1/2T$ has a narrow main lobe with very low side lobes. This is advantageous in that interference with neighboring pulses is minimized. Pulses with raised cosine spectra are used extensively in the design of digital communication systems.

Given a pulse-shape function $p(t)$, which has the Fourier transform $P(f)$ that satisfies the criterion

$$\sum_{k=-\infty}^{\infty} P\left(f + \frac{k}{T}\right) = T, \qquad |f| \leq \frac{1}{2T} \tag{7.117}$$

then

$$p(nT) = \begin{cases} 1, & n = 0 \\ 0, & n \neq 0 \end{cases} \qquad (7.118)$$

Using this result, we can see that no adjacent pulse interference will be obtained if the received data stream is represented as

$$y(t) = \sum_{k=-\infty}^{\infty} a_k p(t - kT) \qquad (7.119)$$

and the sampling at the receiver occurs at multiples of T seconds at the pulse epochs. For example, to obtain the $k = 10$ sample, one simply sets $t = 10T$ in (7.119), and the resulting sample is a_{10}, given that Nyquist's pulse-shaping criterion of (7.117) holds.

The proof of Nyquist's pulse-shaping criterion follows easily by making use of the inverse Fourier representation for $p(t)$, which is

$$p(t) = \int_{-\infty}^{\infty} P(f) \exp(j2\pi ft) \, df \qquad (7.120)$$

For the nth sample value, this expression can be written as

$$p(nT) = \sum_{k=-\infty}^{\infty} \int_{(2k-1)/2T}^{(2k+1)/2T} P(f) \exp(j2\pi fnT) \, df \qquad (7.121)$$

where the inverse Fourier transform integral for $p(t)$ has been broken up into contiguous frequency intervals of length $1/T$ Hz. By the change of variables $u = f - k/T$, (7.121) becomes

$$\begin{aligned} p(nT) &= \sum_{k=-\infty}^{\infty} \int_{-1/2T}^{1/2T} P\left(u + \frac{k}{T}\right) \exp(j2\pi nTu) \, du \\ &= \int_{-1/2T}^{1/2T} \sum_{k=-\infty}^{\infty} P\left(u + \frac{k}{T}\right) \exp(j2\pi nTu) \, du \qquad (7.122) \end{aligned}$$

where the order of integration and summation has been reversed. By hypothesis, $\sum_{k=-\infty}^{\infty} P(u + k/T) = T$ between the limits of integration, so (7.122) becomes

$$\begin{aligned} p(nT) &= \int_{-1/2T}^{1/2T} T \exp(j2\pi nTu) \, du = \operatorname{sinc} n \\ &= \begin{cases} 1, & n = 0 \\ 0, & n \neq 0 \end{cases} \qquad (7.123) \end{aligned}$$

FIGURE 7.22   Baseband system for signaling through a bandlimited channel

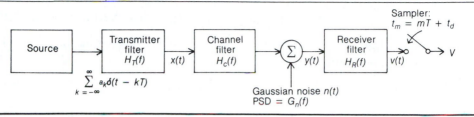

which completes the proof of Nyquist's pulse-shaping criterion. With the aid of this result, it is now apparent why the raised-cosine pulse family is free of interference, even though the family is by no means unique.

To use this result in designing digital communication systems for bandlimited channels, we consider a model for a baseband system such as is depicted in Figure 7.22. The transmitted signal is

$$
x(t) = \sum_{k=-\infty}^{\infty} a_k \delta(t - kT) * h_T(t)
$$

$$
= \sum_{k=-\infty}^{\infty} a_k h_T(t - kT) \tag{7.124}
$$

where $h_T(t)$ is the impulse response of the transmitter filter that has the lowpass transfer function $H_T(f) = \mathcal{F}[h_T(t)]$. This signal passes through a bandlimited channel filter, after which Gaussian noise with power spectral density $G_n(f)$ is added to give the received signal

$$
y(t) = x(t) * h_C(t) + n(t) \tag{7.125}
$$

where $h_C(t) = \mathcal{F}^{-1}[H_C(f)]$ is the impulse response of the channel. Detection at the receiver is accomplished by passing $y(t)$ through a filter with impulse response $h_R(t)$ and sampling its output at intervals of T seconds. If we require that the cascade of transmitter, channel, and receiver filters satisfies Nyquist's pulse-shaping criterion, it then follows that the output sample at time $t = t_d$, where t_d is the delay imposed by the channel for the receiver filter, is

$$
V = Aa_0 p(0) + N
$$

$$
= Aa_0 + N, \tag{7.126}
$$

where

$$
Ap(t - t_d) = h_T(t) * h_C(t) * h_R(t) \tag{7.127}
$$

or, by Fourier transforming both sides, we have

$$AP(f) \exp(-j2\pi f t_d) = H_T(f) H_C(f) H_R(f) \tag{7.128}$$

In (7.126), A is a scale factor, t_d is a time delay accounting for all delays in the system, and

$$N = n(t) * h_R(t)|_{t=t_d} \tag{7.129}$$

is the Gaussian noise component at the output of the detection filter at time $t = t_d$.

To analyze the probability of error for the performance of the system, we assume binary signaling ($a_m = +1$ or -1) so that the average probability of error is

$$
\begin{aligned}
P_E &= Pr(a_m = 1)Pr(Aa_m + N \leq 0 \quad \text{given } a_m = 1) \\
&\quad + Pr(a_m = -1)Pr(Aa_m + N \geq 0 \quad \text{given } a_m = -1) \quad (7.130) \\
&= Pr(Aa_m + N \leq 0 \quad \text{given } a_m = 1) \\
&= Pr(Aa_m + N \geq 0 \quad \text{given } a_m = -1) \tag{7.131}
\end{aligned}
$$

where the latter two equations result by assuming $a_m = 1$ and $a_m = -1$ are equally likely. Taking the last equation of (7.131), it follows that

$$P_E = Pr(N \geq A) = \int_A^\infty \frac{\exp(-u^2/2\sigma^2)}{\sqrt{2\pi\sigma^2}} \, du = Q\left(\frac{A}{\sigma}\right) \tag{7.132}$$

where

$$\sigma^2 = \text{var}(N) = \int_{-\infty}^\infty G_n(f) |H_R(f)|^2 df \tag{7.133}$$

Because the Q-function is a monotonically decreasing function of its argument, it follows that the average probability of error can be minimized through proper choice of $H_T(f)$ and $H_R(f)$ [$H_C(f)$ is assumed to be fixed], by maximizing A/σ or by minimizing σ^2/A^2. The minimization can be carried out, subject to the constraint in (7.128), by applying Schwarz's inequality. The result* is

$$|H_R(f)|_{\text{opt}} = \frac{K|P(f)|^{1/2}}{G_n^{1/4}(f)|H_C(f)|^{1/2}} \tag{7.134}$$

---

* For a derivation, see Ziemer and Peterson (1992).

and

$$|H_T(f)|_{\text{opt}} = \frac{A}{K}|P(f)|^{1/2}G_n^{1/4}(f)/|H_c(f)|^{1/2} \qquad (7.135)$$

where K is an arbitrary constant and any appropriate phase response can be used.

A special case of interest occurs when

$$G_n(f) = \frac{N_0}{2}, \qquad \text{all } f \quad \text{(white noise)} \qquad (7.136)$$

and

$$H_c(f) = H_0, \qquad |f| \le \frac{1}{2T} \qquad (7.137)$$

The minimum probability of error then simplifies to

$$P_{E,\text{min}} = Q(\sqrt{2E_T/N_0}) \qquad (7.138)$$

where

$$E_T = \int_{-\infty}^{\infty} |H_T(f)|^2 df \qquad (7.139)$$

is the average transmitted signal energy. This result is identical to that obtained previously for binary signaling in an infinite bandwidth baseband channel.

7.6   MULTIPATH INTERFERENCE

The channel models that we have assumed so far have been rather idealistic in that the only signal perturbation considered was due to additive Gaussian noise. Although realistic for many situations, additive Gaussian-noise channel models do not accurately represent many transmission phenomena. Other important sources of degradation in many digital data systems are bandlimiting of the signal by the channel, as examined in the previous section; non-Gaussian noise, such as impulse noise due to lightning discharges or switches; radio frequency interference due to other transmitters; and multiple transmission paths due to stratifications in the transmission medium or reflecting objects.

In this section we consider the effects of multipath transmission because it is a fairly common transmission perturbation and its effects on digital data

FIGURE 7.23 Channel model for multipath transmission

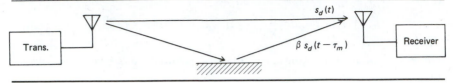

transmission can, in the simplest form, be analyzed in a straightforward fashion.

We will limit our discussion to a two-ray multipath model as illustrated in Figure 7.23. In addition to the multiple transmission, the channel perturbs the signal with white Gaussian noise with double-sided power spectral density $\frac{1}{2}N_0$. Thus the received signal plus noise is given by

$$y(t) = s_d(t) + \beta s_d(t - \tau_m) + n(t) \tag{7.140}$$

where $s_d(t)$ is the received direct-path signal, β is the attenuation of the multipath component, and τ_m is its delay. For simplicity, only binary BPSK will be considered in the following analysis. Thus the direct-path signal can be represented as

$$s_d(t) = Ad(t) \cos \omega_c t \tag{7.141}$$

where $d(t)$, the data stream, is a sequence of plus or minus 1's, each one of which is T seconds in duration. Because of the multipath component, we must consider a sequence of bits at the receiver input. We will analyze the effect of the multipath component and noise on a correlation receiver as shown in Figure 7.24, which, we recall, detects the data in the presence of Gaussian noise alone with minimum probability of error. Writing the noise in terms of quadrature components $n_c(t)$ and $n_s(t)$, we find that the input to the integrator, ignoring double frequency terms, is

$$x(t) = Lp[2y(t) \cos \omega_c t]$$
$$= Ad(t) + \beta Ad(t - \tau_m) \cos \omega_c \tau_m + n_c(t) \tag{7.142}$$

where $Lp[\ ]$ stands for the lowpass part of the bracketed quantity.

The second term in (7.142) represents interference due to the multipath. It is useful to consider two special cases:

1. $\tau_m/T \cong 0$, so that $d(t - \tau_m) \cong d(t)$. For this case, we assume $\omega_0\tau_m$ is a uniformly distributed random variable in $(-\pi, \pi)$. This will be true if $\omega_c\tau_m$ fluctuates much faster than the carrier tracking loop time constant. The predominant effect of the multipath for this case will be fading of the signal component. This case will be analyzed in the next section.

FIGURE 7.24   Correlation receiver for BPSK with signal plus multipath at its input

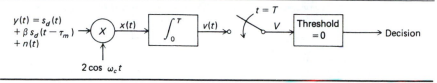

2. $0 < \tau_m/T \leq 1$, so that successive bits of $d(t)$ and $d(t - \tau_m)$ overlap; in other words, we will have intersymbol interference. For this case, we will let $\delta = \beta \cos \omega_c \tau_m$ be a parameter in our analysis.

We now analyze the receiver performance for case 2, for which the effect of intersymbol interference is nonnegligible. To simplify notation, let

$$\delta = \beta \cos \omega_c \tau_m \tag{7.143}$$

so that (7.142) becomes

$$x(t) = Ad(t) + A\delta d(t - \tau_m) + n_c(t) \tag{7.144}$$

If $\tau_m/T \leq 1$, only adjacent bits of $Ad(t)$ and $A\delta d(t - \tau_m)$ will overlap. Thus we can compute the signal component of the integrator output in Figure 7.24 by considering the four combinations shown in Figure 7.25. Assuming 1's and 0's are equally probable, the four combinations shown in Figure 7.25 will occur with equal probabilities. Thus the average probability of error is

$$P_E = \tfrac{1}{4}[P(E|++) + P(E|-+) + P(E|+-) + P(E|--)] \tag{7.145}$$

where $P(E|++)$ is the probability of error given two 1's were sent, and so on. The noise component of the integrator output, namely

$$N = \int_0^T 2n(t) \cos \omega_c t \; dt \tag{7.146}$$

is Gaussian with zero mean and variance

$$
\begin{aligned}
\sigma_n^{\,2} &= E\left\{ 4 \int_0^T \int_0^T n(t)n(\sigma) \cos \omega_c t \cos \omega_c \sigma \; dt \; d\sigma \right\} \\
&= 4 \int_0^T \int_0^T \frac{N_0}{2} \delta(t - \sigma) \cos \omega_c t \cos \omega_c \sigma \; d\sigma \; dt \\
&= 2N_0 \int_0^T \cos^2 \omega_c t \; dt \\
&= N_0 T \qquad (\omega_c T \text{ an integer multiple of } 2\pi) \tag{7.147}
\end{aligned}
$$

FIGURE 7.25 The various possible cases for intersymbol interference in multipath transmission

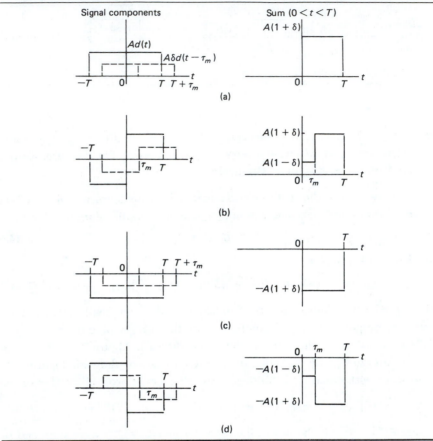

Because of the symmetry of the noise pdf and the symmetry of the signals in Figure 7.25, it follows that

$$P(E|++) = P(E|--) \quad \text{and} \quad P(E|-+) = P(E|+-) \quad (7.148)$$

so that we have to compute only two probabilities instead of four. From Figure 7.25, it follows that the signal component at the integrator output, given a 1-1 was transmitted, is

$$V_{++} = AT(1 + \delta) \quad (7.149)$$

and if a 0-1 was transmitted, it is

$$V_{-+} = AT(1 + \delta) - 2A\delta\tau_m$$

$$= AT\left[(1 + \delta) - \frac{2\delta\tau_m}{T}\right] \tag{7.150}$$

The conditional error probability $P(E \mid + +)$ is therefore

$$P(E \mid + +) = P[AT(1 + \delta) + N < 0] = \int_{-\infty}^{-AT(1+\delta)} \frac{e^{-u^2/2N_0T}}{\sqrt{2\pi N_0 T}} \, du$$

$$= Q\left[\sqrt{\frac{2E}{N_0}}(1 + \delta)\right] \tag{7.151}$$

where $E = \frac{1}{2}A^2T$ is the energy of the direct signal component. Similarly, $P(E \mid - +)$ is given by

$$P(E \mid - +) = P\left\{AT\left[(1 + \delta) - \frac{2\delta\tau_m}{T}\right] + N < 0\right\}$$

$$= \int_{-\infty}^{-AT[(1+\delta)-2\delta\tau_m/T]} \frac{e^{-u^2/2N_0T}}{\sqrt{2\pi N_0 T}} \, du$$

$$= Q\left\{\sqrt{\frac{2E}{N_0}}\left[1 + \delta - \frac{2\delta\tau_m}{T}\right]\right\} \tag{7.152}$$

Substituting these results into (7.145) and using the symmetry properties for the other conditional probabilities, we have the average probability of error

$$P_E = \tfrac{1}{2}Q\sqrt{2z_0}\,[(1 + \delta)] + \tfrac{1}{2}Q\{\sqrt{2z_0}\,[(1 + \delta) - 2\delta\tau_m/T]\} \tag{7.153}$$

where $z_0 \triangleq E/N_0 = A^2T/2N_0$ as before.

A plot of P_E versus z_0 for various values of δ and τ_m/T, as shown in Figure 7.26, gives an indication of the effect of multipath on signal transmission. A question arises as to which curve in Figure 7.26 should be used as a basis of comparison. The one for $\delta = \tau_m/T = 0$ corresponds to the error probability for BPSK signaling in a nonfading channel. However, we note that

$$z_m = \frac{E(1 + \delta)}{N_0} = z_0(1 + \delta) \tag{7.154}$$

FIGURE 7.26 P_E versus z for various conditions of fading and intersymbol interference due to multipath

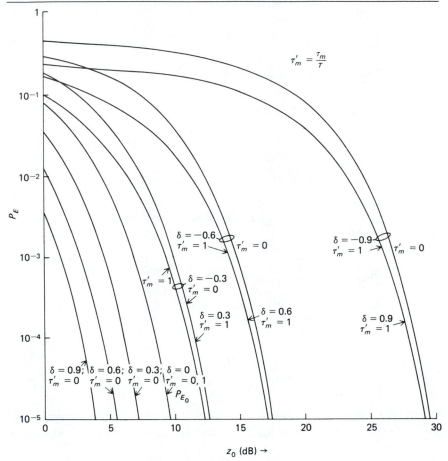

is the signal-to-noise ratio that results if the total effective received signal energy, including that of the indirect component, is used. Indeed, from (7.153) it follows that this is the curve for $\tau_m/T = 0$ for a given value of δ. Thus, if we use this curve for P_E as a basis of comparison for P_E with τ_m/T nonzero for each δ, we will be able to obtain the increase in P_E due to intersymbol interference alone. However, it is more useful for system design purposes to have degradation in SNR instead. That is, we want the increase in signal-to-noise ratio (or signal energy) necessary to maintain a given P_E in the presence of multipath relative to a channel with $\tau_m = 0$. Figure 7.27 shows typical results for $P_E = 10^{-4}$.

FIGURE 7.27   Degradation versus δ for correlation detection of BPSK in specular multipath for $P_E = 10^{-4}$

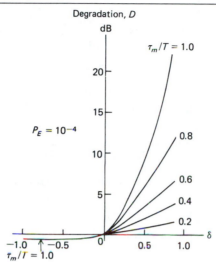

We note that the degradation is actually negative for $\delta < 0$; that is, the performance with intersymbol interference is better than for no intersymbol interference, provided the indirect received signal fades out of phase with respect to the direct component. This seemingly contradictory result is explained by consulting Figure 7.25, which shows that the direct and indirect received signal components being out of phase, as implied by $\delta < 0$, results in additional signal energy being received for cases (b) and (d) with $\tau_m/T > 0$ over what would be received if $\tau_m/T = 0$. On the other hand, the received signal energy for cases (a) and (c) is independent of τ_m/T.

Two interesting conclusions may be drawn from Figure 7.27. First, we note that when $\delta < 0$, the effect of *intersymbol interference* is negligible, since variation of τ_m/T has no significant effect on the degradation. The degradation is due primarily to the decrease in signal amplitude owing to the destructive interference because of the phase difference of the direct and indirect signal components. Second, when $\delta > 0$, the degradation shows a strong dependence on τ_m/T, indicating that intersymbol interference is the primary source of the degradation.

The adverse effects of intersymbol interference due to multipath can be combated by using an *equalization filter* that precedes detection of the received data.* To illustrate the basic idea of such a filter, we take the Fou-

---

* Equalization can be used to improve performance whenever intersymbol interference is a problem, for example, due to filtering.

rier transform of (7.140) with $n(t) = 0$ to obtain the transfer function of the channel, $H_C(f)$:

$$H_C(f) = \frac{\mathcal{F}[y(t)]}{\mathcal{F}[s_d(t)]}$$

$$= 1 + \beta e^{-j2\pi\tau_m f} \tag{7.155}$$

If β and τ_m are known, the correlation receiver of Figure 7.24 can be preceded by a filter, referred to as an equalizer, with the transfer function

$$H_{eq}(f) = \frac{1}{H_C(f)} = \frac{1}{1 + \beta e^{-j2\pi\tau_m f}} \tag{7.156}$$

to fully compensate for the signal distortion introduced by the multipath. Since β and τ_m will not be known exactly, or may even change with time, provision must be made for adjusting the parameters of the equalization filter. Noise, although important, is neglected for simplicity.

7.7   FLAT FADING CHANNELS

Returning to (7.140), we assume that the delay of the multipath portion is small compared with a bit period so that we have as the approximate integrate-and-dump detector output

$$x(t) = A(1 + \beta) d(t) \cos \omega_c \tau_m + n_c(t) \tag{7.157}$$

Since the multipath channel is usually unknown, it is sensible to model β and $\omega_c\tau_m$ as random variables, assuming that the channel varies slowly relative to the bit duration. This is known as a slowly fading channel. If β and $\omega_c\tau_m$ vary quickly with respect to $d(t)$, the channel is said to be fast fading. This is a more difficult case to analyze than the slowly fading case, and we do not consider it here. A common model for the envelope of the received signal in the slowly fading case is a Rayleigh random variable, which is also the simplest case to analyze. Somewhat more general, but more difficult to analyze, is to model the envelope of the received signal as a Rician random variable. We illustrate the consideration of a BPSK signal received from a Rayleigh slowly fading channel as follows. Let the demodulated signal (7.157) be written in the simpler form

$$x(t) = R d(t) + n_c(t) \tag{7.158}$$

where R is a Rayleigh random variable with pdf given by (7.107). If R were a constant, we know that the probability of error is given by (7.72) with $m = 0$. In other words, *given* R, we have for the probability of error

$$P_E(R) = Q(\sqrt{2z}) \tag{7.159}$$

In order to find the average probability of error over the amplitude R, we average (7.159) with respect to the pdf of R, which is assumed to be Rayleigh in this case. However, R is not explicitly present in (7.159) because it is buried in z:

$$z = \frac{E_b}{N_0} = \frac{R^2 T_b}{2N_0} \tag{7.160}$$

Now if R is Rayleigh-distributed, it can be shown by transformation of pdf's that R^2, and therefore z, is exponentially distributed. Thus, the average of (7.159) is

$$P_E = \int_0^\infty Q(\sqrt{2z}) \frac{1}{\bar{z}} e^{-z/\bar{z}} \, dz \tag{7.161}$$

where $\bar{z}$ is the average signal-to-noise ratio. This integration can be carried out with the result that

$$P_E = \frac{1}{2} \left[1 - \sqrt{\frac{\bar{z}}{1 + \bar{z}}} \right] \qquad \text{BPSK} \tag{7.162}$$

A similar analysis for binary, coherent FSK results in the expression

$$P_E = \frac{1}{2} \left[1 - \sqrt{\frac{\bar{z}}{2 + \bar{z}}} \right] \qquad \text{coherent FSK} \tag{7.163}$$

Other modulation techniques that we can consider in a similar fashion are DPSK and noncoherent FSK. For these modulation schemes, we obtain

$$P_E = \frac{1}{2(1 + \bar{z})} \qquad \text{DPSK} \tag{7.164}$$

and

$$P_E = \frac{1}{2 + \bar{z}} \qquad \text{noncoherent FSK} \tag{7.165}$$

respectively. The derivations are left to the problems. These results are plotted in Fig. 7.28 and compared with the corresponding results for nonfading channels. Note that the penalty imposed by the fading is severe.

What can be done to combat the adverse effects of fading? We note that the degradation in performance due to fading results from the received signal amplitude being less than what it would be for a nonfading channel because of the factor $\beta \cos \phi$. If the transmitted signal power can be divided between two or more subchannels that fade independently of one another, then the degradation will most likely not be severe in all subchannels for

FIGURE 7.28   Error probabilities for various modulation schemes in flat fading Rayleigh channels. (a) Coherent and noncoherent FSK. (b) BPSK and DPSK.

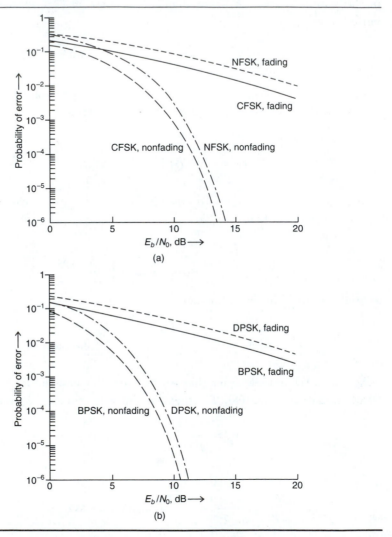

the same binary digit. Then, if the outputs of these subchannels are recombined in the proper fashion, it seems reasonable that better performance can be obtained than that from using a single transmission path. The use of such multiple transmission paths to combat fading is referred to as *diversity transmission*. There are various ways to obtain the independent transmission paths; chief ones are by transmitting over spatially different paths (space diversity),

at different times (time diversity), with different carrier frequencies (frequency diversity), or with different polarizations of the propagating wave
(polarization diversity).

In any case, an optimum number of subpaths exist that give the maximum
improvement. The number of subpaths D employed is referred to as the *order
of diversity*.

It is apparent that an optimum value of D exists. Increasing D provides
additional diversity and decreases the probability that most of the subchannel outputs are badly faded. On the other hand, as D increases with total
signal energy held fixed, the average signal-to-noise ratio per subchannel
decreases, thereby resulting in a larger probability of error per subchannel.
Clearly, therefore, a compromise between these two situations must be
made. The problem of fading is again reexamined in Chapter 9 (Section 9.3),
and the optimum selection of D is considered in Problem 9.17.

7.8 EQUALIZATION

As explained in Section 7.6, an equalization filter can be used to combat
channel-induced distortion caused by perturbations such as multipath propagation or bandlimiting due to filters. According to (7.156), a simple
approach to the idea of equalization leads to the concept of an inverse filter.
In this section, we specialize our considerations of an equalization filter to
a particular form—a transversal or tapped-delay-line filter. Figure 7.29 is a
block diagram of such a filter.

There are at least two reasons for considering a transversal structure for
the purpose of equalization. First, it is simple to analyze. Second, it is easy

FIGURE 7.29 Transversal filter implementation for equalization of
intersymbol interference

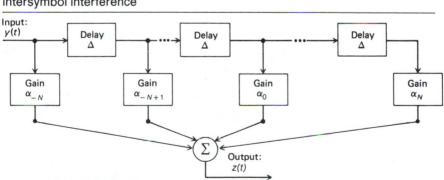

to mechanize by means of *surface acoustic wave* (SAW) filters* at high frequencies and by digital signal processors and *charge-transfer devices*† (CTDs) at
lower frequencies.

We can take two approaches to determining the tap weights, $\alpha_{-N}, \ldots,$
$\alpha_0, \ldots, \alpha_N$, in Figure 7.29 for given channel conditions. One is *zero-forcing*,
and the other is *minimization of mean-square error* (MMSE). Let us briefly discuss
these two techniques.

Equalization by Zero-Forcing

Let the pulse response of the channel output be $p_c(t)$. The output of the
equalizer in response to $p_c(t)$ is

$$p_{eq}(t) = \sum_{n=-N}^{N} \alpha_n p_c(t - n\Delta) \tag{7.166}$$

We want $p_{eq}(t)$ to satisfy Nyquist's pulse-shaping criterion, which we will
call the *zero-ISI condition*. Since the output of the equalizer is sampled every
T seconds, it is reasonable that the tap spacing be $\Delta = T$. The zero-ISI
condition therefore becomes

$$p_{eq}(mT) = \sum_{n=-N}^{N} \alpha_n p_c[(m - n)T]$$

$$= \begin{cases} 1, & m = 0 \\ 0, & m \neq 0 \end{cases} \quad m = 0, \pm 1, \pm 2, \ldots, \pm N \tag{7.167}$$

Note that the zero-ISI condition can be satisfied at only $2N + 1$ time
instants because there are only $2N + 1$ coefficients to be selected in (7.167).
Defining the matrices

$$[P_{eq}] = \begin{bmatrix} 0 \\ 0 \\ \vdots \\ 0 \\ 1 \\ 0 \\ 0 \\ \vdots \\ 0 \end{bmatrix} \begin{matrix} \left.\vphantom{\begin{matrix}0\\0\\\vdots\\0\end{matrix}}\right\} N \text{ zeros} \\ \\ \\ \left.\vphantom{\begin{matrix}0\\0\\\vdots\\0\end{matrix}}\right\} N \text{ zeros} \end{matrix} \tag{7.168}$$

---

* See "SAW Bandpass Filters," Anderson Laboratories (1980), 1280 Blue Hill Avenue,
Bloomfield, CT 06002.
† Allen Gersho, "Charge-Coupled Devices: The Analog Shift Register Comes of Age," *IEEE
Communications Magazine*, Vol. 13, November 1975, pp. 27–32.

$$[A] = \begin{bmatrix} \alpha_{-N} \\ \alpha_{-N+1} \\ \vdots \\ \alpha_N \end{bmatrix} \tag{7.169}$$

and

$$[P_c] = \begin{bmatrix} p_c(0) & p_c(-T) & \cdots & p_c(-2NT) \\ p_c(T) & p_c(0) & \cdots & p_c[(-2N+1)T] \\ \vdots & & & \vdots \\ p_c(2NT) & & & p_c(0) \end{bmatrix} \tag{7.170}$$

it follows that (7.167) can be written in matrix form as

$$[P_{eq}] = [P_c][A] \tag{7.171}$$

The method of solution of the zero-forcing coefficients is now clear. Since $[P_{eq}]$ is specified by the zero-ISI condition, all we must do is find the inverse of $[P_c]$. The desired coefficient matrix $[A]$ is then the middle column of $[P_c]^{-1}$, which follows by multiplying $[P_c]^{-1}$ and $[P_{eq}]$.

▼ **EXAMPLE 7.8**   Consider a channel for which the following sample values of the channel pulse response are obtained:

$$p_c(-2T) = -0.05 \qquad p_c(-T) = 0.2 \qquad p_c(0) = 1.0$$
$$p_c(T) \quad = 0.3 \qquad p_c(2T) \quad = -0.07$$

The matrix $[P_c]$ is

$$[P_c] = \begin{bmatrix} 1.0 & 0.2 & -0.05 \\ 0.3 & 1.0 & 0.2 \\ -0.07 & 0.3 & 1.0 \end{bmatrix} \tag{7.172}$$

and the inverse matrix is

$$[P_c]^{-1} = \begin{bmatrix} 1.0815 & -0.2474 & 0.1035 \\ -0.3613 & 1.1465 & -0.2474 \\ 0.1841 & -0.3613 & 1.0815 \end{bmatrix} \tag{7.173}$$

The zero-forcing tap coefficients are given by the middle column of this inverse matrix. Using these coefficients, the equalizer output is

$$p_{eq}(mT) = -0.2474p_c[(m+1)T] + 1.1465p_c(mT)$$
$$- 0.3613p_c[(m-1)T] \tag{7.174}$$

▲ Note that $p_{eq}(0) = 1.0$ and that samples on either side of $p_{eq}(0)$ are zero.

Equalization by MMSE

Suppose that the desired output from the transversal filter equalizer of Figure 7.29 is $d(t)$. An MMSE criterion then seeks the tap weights that minimize the mean-square error between the desired output from the equalizer and its actual output. Since this output includes noise, we denote it by $z(t)$ to distinguish it from the pulse response of the equalizer. The MMSE criterion is therefore expressed as

$$\mathcal{E} = E\{[z(t) - d(t)]^2\} = \text{minimum} \tag{7.175}$$

where, if $y(t)$ is the equalizer input including noise, the equalizer output is

$$z(t) = \sum_{n=-N}^{N} \alpha_n y(t - n\Delta) \tag{7.176}$$

Since $\mathcal{E}$ is a concave function of the tap weights, a set of sufficient conditions for minimizing the tap weights is

$$\frac{\partial \mathcal{E}}{\partial \alpha_m} = 0 = 2E\left\{[z(t) - d(t)]\frac{\partial z(t)}{\partial \alpha_m}\right\}, \qquad m = 0, \pm 1, \ldots, \pm N \tag{7.177}$$

Substituting (7.176) in (7.177) and carrying out the differentiation, we obtain the conditions

$$E\{[z(t) - d(t)]y(t - m\Delta)\} = 0, \qquad m = 0, \pm 1, \pm 2, \ldots, \pm N \tag{7.178}$$

or

$$R_{yz}(m\Delta) = R_{yd}(m\Delta), \qquad m = 0, \pm 1, \pm 2, \ldots, \pm N \tag{7.179}$$

where

$$R_{yz}(\tau) = E[y(t)z(t + \tau)] \tag{7.180}$$

and

$$R_{yd}(\tau) = E[y(t)d(t + \tau)] \tag{7.181}$$

are the cross-correlations of the received signal with the equalizer output and with the data, respectively.

Using the expression in (7.176) for $z(t)$ in (7.179), these conditions can be expressed as the matrix equation

$$[R_{yy}][A] = [R_{yd}] \tag{7.182}$$

where

$$[R_{yy}] = \begin{bmatrix} R_{yy}(0) & R_{yy}(\Delta) & \cdots & R_{yy}(2N\Delta) \\ R_{yy}(-\Delta) & R_{yy}(0) & \cdots & R_{yy}[2(N - 1)\Delta] \\ \vdots & & & \vdots \\ R_{yy}(-2N\Delta) & & \cdots & R_{yy}(0) \end{bmatrix} \tag{7.183}$$

and

$$[R_{yd}] = \begin{bmatrix} R_{yd}(-N\Delta) \\ R_{yd}[-(N-1)\Delta] \\ \vdots \\ R_{yd}(N\Delta) \end{bmatrix} \qquad (7.184)$$

and $[A]$ is defined by (7.169). Note that these conditions for the optimum tap weights using the MMSE criterion are similar to the conditions for the zero-forcing weights, except correlation-function samples are used instead of pulse-response samples.

Two questions remain with regard to setting the tap weights. The first is what should be used for the desired response $d(t)$. In the case of digital signaling, one could use the detected data, even if the modem performance is very poor, since an error probability of only 10^{-2} still implies that $d(t)$ is correct for 99 out of 100 bits. Algorithms using the detected data as $d(t)$, the desired output, are called *decision-directed*.

The second question is what procedure should be followed if the sample values of the pulse needed in the zero-forcing criterion or the samples of the correlation function required for the MMSE criterion are not available. A useful strategy to follow in such cases is referred to as *adaptive equalization*. In this approach, known as the LMS algorithm, an initial guess for the coefficients is corrected according to the recursive relationship

$$A(n+1) = A(n) - K\epsilon(n)Y(n) \qquad (7.185)$$

where K is an adjustment parameter, $Y(n)$ is a vector of received data samples,

$$Y(n) = \begin{bmatrix} y(n\Delta) \\ y[(n-1)\Delta] \\ \vdots \\ y(2N\Delta) \end{bmatrix} \qquad (7.185)$$

and $\epsilon(n)$ is the error $y(n) - d(n)$.

Much more could be said about equalization, but we will close here.*

SUMMARY

1. Binary baseband data transmission in additive white Gaussian noise with equally likely signals having constant amplitudes of $\pm A$ and of duration T results in an average error probability of

---

* For an excellent survey article dealing with equalization methods in part, see S. Qureshi "Adaptive Equalization," *IEEE Communications Magazine*, Vol. 20, March 1982, pp. 9–16.

$$P_E = Q\left(\sqrt{\frac{2A^2T}{N_0}}\right)$$

where N_0 is the single-sided power spectral density of the noise. The hypothesized receiver was the integrate-and-dump receiver, which turns out to be the optimum receiver in terms of minimizing the probability of error.

2. An important parameter in binary data transmission is $z = E_b/N_0$, the energy per bit divided by the noise power spectral density (single sided). For binary baseband signaling, it can be expressed in the following equivalent forms:

$$z = \frac{E_b}{N_0} = \frac{A^2T}{N_0} = \frac{A^2}{N_0(1/T)} = \frac{A^2}{N_0 B_p}$$

where B_p is the "pulse" bandwidth, or roughly the bandwidth required to pass the baseband pulses. The latter expression then allows the interpretation that z is the signal power divided by the noise power in a pulse, or bit-rate, bandwidth.

3. For binary data transmission with arbitrary (finite energy) signal shapes, $s_1(t)$ and $s_2(t)$, the error probability for equally probable signals, was found to be

$$P_E = Q(\sqrt{w})$$

where

$$w = \frac{1}{2N_0} \int_{-\infty}^{\infty} |\mathscr{S}_1(f) - \mathscr{S}_2(f)|^2 \, df = \frac{1}{2N_0} \int_{-\infty}^{\infty} |s_1(t) - s_2(t)|^2 \, dt$$

in which $\mathscr{S}_1(f)$ and $\mathscr{S}_2(f)$ are the Fourier transforms of $s_1(t)$ and $s_2(t)$, respectively. This expression resulted from minimizing the average probability of error, assuming a linear-filter/threshold-comparison type of receiver. The receiver involves the concept of a matched filter; such a filter is matched to a specific signal pulse and maximizes peak signal divided by rms noise ratio at its output. In a matched filter receiver for binary signaling, two matched filters are used in parallel, each matched to one of the two signals, and their outputs are compared at the end of each signaling interval. The matched filters also can be realized as correlators.

4. The expression for the error probability of a matched-filter receiver also can be written as

$$P_E = Q\{[z(1 - R_{12})]^{1/2}\}$$

where $z = E/N_0$, E being the *average* signal energy given by $E = \frac{1}{2}(E_1 + E_2)$. R_{12} is a parameter that is a measure of the similarity of the two signals; it is given by

$$R_{12} = \frac{2}{E_1 + E_2} \int_{-\infty}^{\infty} s_1(t)s_2(t)\, dt$$

If $R_{12} = -1$, the signaling is termed *antipodal*, while if $R_{12} = 0$, the signaling is termed *orthogonal*.

5. Examples of coherent (that is, the signal arrival time and carrier phase are known at the receiver) signaling techniques at a carrier frequency ω_c rad/s are the following:

$$\text{PSK: } s_k(t) = A \sin [\omega_c t - (-1)^k \cos^{-1} m], \qquad nt_0 \leq t \leq nt_0 + T,$$
$$k = 1, 2$$

$$(\cos^{-1} m \text{ is called the } \textit{modulation index})$$

$$\text{ASK: } s_1(t) = 0, \qquad\qquad nt_0 \leq t \leq nt_0 + T$$
$$s_2(t) = A \cos \omega_c t, \qquad nt_0 \leq t \leq nt_0 + T$$

$$\text{FSK: } s_1(t) = A \cos \omega_c t, \qquad\qquad\quad nt_0 \leq t \leq nt_0 + T$$
$$s_2(t) = A \cos [(\omega_c + \Delta\omega)t], \qquad nt_0 \leq t \leq nt_0 + T$$

If $\Delta\omega = 2\pi \ell/T$ for FSK, where ℓ is an integer, it is an example of an orthogonal signaling technique. If $m = 0$ for PSK, it is an example of an antipodal signaling scheme. A value of E_b/N_0 of approximately 10.54 dB is required to achieve an error probability of 10^{-6} for PSK with $m = 0$; 3 dB more than this is required to achieve the same error probability for ASK and FSK.

6. Examples of signaling schemes not requiring coherent carrier references at the receiver are differential phase-shift keying (DPSK) and noncoherent FSK. Using ideal minimum-error-probability receivers, DPSK yields the error probability

$$P_E = \tfrac{1}{2} \exp (-E_b/N_0)$$

while noncoherent FSK gives the error probability

$$P_E = \tfrac{1}{2} \exp\left(-E_b/2N_0\right)$$

Noncoherent ASK is another possible signaling scheme with about the same error probability performance as noncoherent FSK.

7. In general, if a sequence of signals is transmitted through a bandlimited channel, adjacent signal pulses are smeared into each other by the transient response of the channel. Such interference between signals is termed *intersymbol interference*. By appropriately choosing transmitting and receiving filters, it is possible to signal through bandlimited channels while eliminating intersymbol interference. This signaling technique was examined by using Nyquist's pulse-shaping criterion and Schwarz's inequality. A useful family of pulse shapes for such signaling are those having raised cosine spectra.

8. One form of channel distortion is multipath interference. The effect of a simple two-ray multipath channel on binary data transmission is examined. Half of the time the received signal pulses interfere destructively, and the rest of the time they interfere constructively. The interference can be separated into intersymbol interference of the signaling pulses and cancelation due to the carriers of the direct and multipath components arriving out of phase.

9. Fading results from channel variations caused by propagation conditions. One of these conditions is multipath if the differential delay is short compared with the bit period but made up of many wavelengths. A commonly used model for a fading channel is one where the envelope of the received signal has a Rayleigh pdf. In this case, the signal power or energy has an exponential pdf, and the probability of error can be found by using the previously obtained error probability expressions for nonfading channels and averaging over the signal energy with respect to the assumed exponential pdf of the energy. Figure 7.28 compares the error probability for fading and nonfading cases for various modulation schemes. Fading results in severe degradation of the performance of a given modulation scheme. A way to combat fading is to use *diversity*.

10. Equalization can be used to remove a large part of the intersymbol interference introduced by channel filtering. Two techniques were briefly examined: zero-forcing and minimum-mean-squared error. Both can be realized by tapped delay-line filters. In the former technique, zero intersymbol interference is forced at sampling instants separated by multiples of a symbol period. If the tapped delay line is of length

$(2N + 1)T$, then N zeros can be forced on either side of the desired pulse. In a minimum mean-square-error equalizer, the tap weights are sought that give minimum mean square error between the desired output from the equalizer and the actual output. The resulting weights for either case can be precalculated and preset, or adaptive circuitry can be implemented to automatically adjust the weights. The latter technique can make use of a training sequence periodically sent through the channel, or it can make use of the received data itself in order to carry out the minimizing adjustment.

FURTHER READING

A number of excellent recently published textbooks deal exclusively with the subject of digital communications and treat the binary case in detail.

A very complete treatment of digital communications is contained in the book by Sklar (1989). Covered in this one volume are the concepts of basic modulation theory, error-correcting codes, link-power budget analysis, synchronization, and encryption. A very readable treatment of digital communications, although not as complete as the book by Sklar, is the volume by Haykin (1989).

The books by Proakis (1989) and Lee and Messerschmitt (1988) are other good treatments of digital communication systems. Proakis includes excellent treatments of the bandlimited channel problem and the problem of combating fading and multipath, which will be treated in Chapter 9. Proakis also includes a good introductory chapter on spread-spectrum communications, which will be treated in Chapter 8.

The book by Benedetto, Biglieri, and Castellani (1987) also provides an excellent overall treatment of the digital communication problem. They include a chapter on transmission over "real" channels and have good treatments of intersymbol interference, equalization, coding, and nonlinear channels.

The book by Ziemer and Peterson (1992) develops digital communication systems in more depth than in this chapter.

PROBLEMS

Section 7.1

7.1 A baseband digital transmission system that sends $\pm A$-valued rectangular pulses through a channel at a rate of 10,000 bps is to achieve an error probability of 10^{-5}. If the noise power spectral density is

$N_0 = 10^{-7}$ W/Hz, what is the required value of A? What is a rough estimate of the bandwidth required?

7.2 (a) Consider a baseband digital transmission system with noise level of $N_0 = 10^{-5}$ W/Hz. The signal bandwidth is defined to be that required to pass the main lobe of the signal spectrum. Fill in the following table with the required signal power and bandwidth to achieve the error probability/data rate combinations given.

Required Signal Power A^2 and Bandwidth

R, bps	$P_E = 10^{-3}$	$P_E = 10^{-4}$	$P_E = 10^{-5}$	$P_E = 10^{-6}$
1,000				
10,000				
100,000				

(b) Assume the noise level of part (a), but assume that the bandwidth is that required to pass the main lobe and the first side lobe of the signal spectrum. Recompute the preceding table.

7.3 A receiver for baseband digital data has a threshold set at ϵ instead of zero. Rederive (7.8), (7.9), and (7.10) taking this into account. If $P(+A) = P(-A) = \frac{1}{2}$, find E_b/N_0 in decibels as a function of ϵ for $0 \le \epsilon/\sigma \le 1$ to give $P_E = 10^{-6}$, where σ^2 is the variance of N.

7.4 With $N_0 = 10^{-6}$ W/Hz and $A = 20$ mV in a baseband data transmission system, what is the maximum data rate that will allow a P_E of 10^{-4} or less? 10^{-5}? 10^{-6}?

7.5 In a practical implementation of a baseband data transmission system, the sampler at the output of the integrate-and-dump detector requires 1 μs to sample the output. How much additional E_b/N_0, in decibels, is required to achieve a given P_E for a practical system over an ideal system for the following data rates? (a) 10 kbps; (b) 50 kbps; (c) 100 kbps.

7.6 (a) The received signal in a digital baseband system is either $+A$ or $-A$, equally likely, for T-second contiguous intervals. However, the timing is off at the receiver so that the integration starts ΔT seconds late (positive) or early (negative). Assume that the timing error is less than one signaling interval. By assuming a zero threshold and considering two successive intervals [i.e., $(+A,$

$+A$), $(+A, -A)$, $(-A, +A)$, $(-A, -A)]$ obtain an expression for the probability of error as a function of ΔT. Show that it is

$$P_E = \tfrac{1}{2}Q\left(\sqrt{\frac{2E_b}{N_0}}\right) + \tfrac{1}{2}Q\left[\sqrt{\frac{2E_b}{N_0}\left(1 - \frac{2|\Delta T|}{T}\right)}\right]$$

Plot for $|\Delta T|/T = 0, 0.1, 0.2,$ and 0.3.

(b) The degradation due to timing error in E_b/N_0 at a particular probability of error is the decibel difference between the curve for no timing error and that for a given timing error. Plot the degradation due to timing error at $P_E = 10^{-6}$ as a function of $|\Delta T|/T$. What is the additional amount of signal power (or energy) required to maintain an error probability of 10^{-6} for a timing error of 15%?

7.7 As an approximation to the integrate-and-dump detector in Figure 7.3(a), we replace the integrator with a lowpass RC filter with transfer function

$$H(f) = \frac{1}{1 + j(f/f_3)}$$

where f_3 is the 3-dB cutoff frequency.

(a) Find $s_0^2(T)/E\{n_0^2(t)\}$, where $s_0(T)$ is the value of the output signal at $t = T$ due to $+A$ being applied at $t = 0$, and $n_0(t)$ is the output noise. (Assume that the filter initial conditions are zero.)

(b) Find the relationship between T and f_3 such that the signal-to-noise ratio found in part (a) is maximized.

Section 7.2

7.8 Assume that the probabilities of sending the signals $s_1(t)$ and $s_2(t)$ are not equal, but are given by p and $q = 1 - p$, respectively. Derive an expression for P_E that replaces (7.30) that takes this into account. Show that the error probability is minimized by choosing the threshold to be

$$k_{opt} = \frac{\sigma_0^2}{s_{0_1}(T) - s_{0_2}(T)} \ln(p/q) + \frac{s_{0_1}(T) + s_{0_2}(T)}{2}$$

7.9 The general definition of a matched filter is a filter that maximizes *peak* signal-to-rms noise at some prechosen instant of time t_0.

(a) Assuming white noise at the input, use Schwarz's inequality to show that the transfer function of the matched filter is

$$H_m(f) = S^*(f) \exp(-j2\pi f t_0)$$

where $S(f) = \mathcal{F}[s(t)]$ and $s(t)$ is the signal to which the filter is matched.

(b) Show that the impulse response for the matched-filter transfer function found in part (a) is

$$h_m(t) = s(t_0 - t)$$

(c) If $s(t)$ is not zero for $t > t_0$, the matched-filter impulse response is nonzero for $t < 0$; that is, the filter is noncausal and cannot be physically realized because it responds before the signal is applied. If we want a realizable filter, we use

$$h_{mr}(t) = \begin{cases} s(t_0 - t), & t \geq 0 \\ 0, & t < 0 \end{cases}$$

Find the realizable matched-filter impulse response corresponding to the signal

$$s(t) = A\Pi\left[(t - T/2)/T\right] \text{ and } t_0 \text{ equal to } 0, T/2, T, \text{ and } 2T.$$

(d) Find the peak output signal for all cases in part (c). Plot them versus t_0. What do you conclude about the relation between t_0 and the causality condition?

7.10 Referring to Problem 7.9 for the general definition of a matched filter, find the following in relation to the two signals shown in Figure 7.30.

(a) The causal matched-filter impulse responses. Sketch them.
(b) Relate the constants A and B so that both cases give the same peak-signal-to-rms noise ratio at the matched-filter output.

FIGURE 7.30

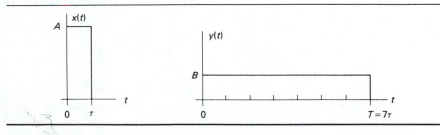

FIGURE 7.31

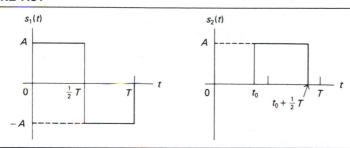

(c) Sketch the output of the matched filters as a function of time with signal only at the input.

(d) Comment on the ability of the two matched filters for these signals to provide an accurate measurement of time delay. What do you estimate the maximum error to be in so doing in each case?

(e) If *peak* transmitted power is a consideration, which waveform (and matched filter) is preferable?

7.11 (a) Find the optimum (matched) filter impulse response $h_0(t)$, as given by (7.43) for $s_1(t)$ and $s_2(t)$, shown in Figure 7.31.

    (b) Find ζ^2 as given by (7.54). Plot ζ^2 versus t_0.

    (c) What is the best choice for t_0 such that the error probability is minimized?

    (d) Sketch a correlator receiver structure for these signals.

7.12 Find the peak-signal-squared-to-mean-squared-noise ratio for the output of a matched filter for each of the following signals in terms of A and T. Take the noise spectral density (single-sided) as N_0.

(a) $s_1(t) = A\Pi\left[\dfrac{(t - T/2)}{T}\right]$

(b) $s_2(t) = \dfrac{A}{2}\left\{1 + \cos\left[\dfrac{2\pi(t - T/2)}{T}\right]\right\}\Pi\left[\dfrac{(t - T/2)}{T}\right]$

(c) $s_3(t) = A\cos\left[\dfrac{\pi(t - T/2)}{T}\right]\Pi\left[\dfrac{(t - T/2)}{T}\right]$

(d) $s_4(t) = A\Lambda\left[\dfrac{(t - T/2)}{T}\right]$

The signals $\Pi(t)$ and $\Lambda(t)$ are the unit-rectangular and unit-triangular functions defined in Chapter 2.

7.13   Given these signals:

$$s_A(t) = A\Pi\left[\frac{(t - T/2)}{T}\right]$$

$$s_B(t) = B\cos\left[\frac{\pi(t - T/2)}{T}\right]\Pi\left[\frac{(t - T/2)}{T}\right]$$

$$s_C(t) = \frac{C}{2}\left\{1 + \cos\left[\frac{2\pi(t - T/2)}{T}\right]\right\}\Pi\left[\frac{(t - T/2)}{T}\right]$$

Assume that these signals are used in a binary digital data transmission system in the following combinations. In each case, calculate R_{12} in (7.59) in terms of A, B, C, and T. Write down an expression for P_E according to (7.58). What is the optimum threshold in each case?

(a) $s_1(t) = s_A(t)$; $s_2(t) = s_B(t)$ (d) $s_1(t) = s_B(t)$; $s_2(t) = -s_B(t)$

(b) $s_1(t) = s_A(t)$; $s_2(t) = s_C(t)$ (e) $s_1(t) = s_C(t)$; $s_2(t) = -s_C(t)$

(c) $s_1(t) = s_B(t)$; $s_2(t) = s_C(t)$

Section 7.3

7.14   Verify the numbers given in Table 7.2 by numerical integration. To evaluate $Q(x)$, use the rational approximation given in Appendix C.6.

7.15   Plot the results for P_E given in Table 7.2 versus $z = E_b/N_0$ in decibels with P_E plotted on a semilog axis. Estimate the additional E_b/N_0 at $P_E = 10^{-5}$ in decibels over the case for no phase error. Compare these results with that for constant phase error, as given by (7.77), of the same magnitude (ϕ for constant phase error equals σ_ϕ for the Gaussian phase-error case).

7.16   Find $z = E_b/N_0$ required to give $P_E = 10^{-7}$ for the following coherent digital modulation techniques: (a) binary ASK; (b) BPSK; (c) binary FSK; (d) BPSK with no carrier component but with a phase error of 5 degrees in the demodulator; (e) PSK with no phase error in demodulation, but with $m = 1/\sqrt{2}$; (f) PSK with $m = 1/\sqrt{2}$ and with a phase error of 5 degrees in the demodulator.

7.17   (a) Make a plot of degradation in decibels versus ϕ, the phase error in demodulation, for BPSK. Assume ϕ constant.

(b) Given that $z = 10.54$ dB is required to give $P_E = 10^{-6}$ for BPSK with no phase error in demodulation, what values of z in decibels are required to give $P_E = 10^{-6}$ for the following static phase errors in demodulation? (i) $\phi = 3$ degrees; (ii) $\phi = 5$ degrees; (iii) $\phi = 10$ degrees; (iv) $\phi = 15$ degrees.

7.18 (a) Plot the required signal-to-noise ratio $z = E_b/N_0$, in decibels, to give $P_E = 10^{-4}$ versus m for PSK with a carrier component for $0 \le m \le 1$.

      (b) Repeat part (a) for $P_E = 10^{-5}$.

7.19 (a) Consider the transmission of digital data at a rate of $R = 50$ kbps and at an error probability of $P_E = 10^{-4}$. Using the bandwidth of the main lobe as a bandwidth measure, give an estimate of the required transmission bandwidth and signal-to-noise ratio in decibels required for the following coherent modulation schemes: (i) binary ASK; (ii) BPSK; (iii) binary coherent FSK (take the minimum spacing possible between the signal representing the logic 1 and that representing the logic 0).

      (b) Consider the same question as in part (a), but with $R = 100$ kbps and $P_E = 10^{-5}$.

      (c) Repeat part (a), but with $R = 25$ kbps and $P_E = 10^{-6}$.

7.20 Derive an expression for P_E for binary coherent FSK if the frequency separation of the two transmitted signals is chosen to give a *minimum* correlation coefficient between the two signals. How much improvement in SNR over the orthogonal-signal case is obtained?

Section 7.4

7.21 Differentially encode the following binary sequences. Arbitrarily choose a 1 as the reference bit to begin the encoding process. (*Note:* Spaces are used to add clarity.)

    (a) 100 111 000 101   (d) 000 000 000 000

    (b) 101 010 101 010   (e) 111 111 000 000

    (c) 111 111 111 111

7.22 Consider the sequence to 110 111 001 010. Differentially encode it and assume that the differentially encoded sequence is used to biphase modulate a sinusoidal carrier of arbitrary phase. Prove that the demodulator of Figure 7.17 properly gives back the original sequence.

7.23 (a) In the analysis of the optimum detector for DPSK, show that the random variables n_1, n_2, n_3, and n_4 have zero means and variances $N_0 T/4$.

      (b) Show that w_1, w_2, w_3, and w_4 have zero means and variances $n_0 T/8$.

7.24 Compare (7.102) and (7.103) to show that for large z, nonoptimum detection and optimum detection of DPSK differ by approximately 2 dB.

7.25 (a) Compute z in decibels required to give $P_E = 10^{-6}$ for noncoherent, binary FSK and DPSK. For the latter, carry out the computation for both the optimum and suboptimum detectors.
   (b) Repeat part (a) for $P_E = 10^{-5}$.
   (c) Repeat part (a) for $P_E = 10^{-4}$.

7.26 Compare bandwidth requirements for noncoherent FSK and for DPSK for the following data rates R. Use null-to-null RF bandwidths. Assume that the FSK "tones" must be separated by $2/T$ Hz.

   (a) $R = 10$ kbps   (d) $R = 500$ kbps
   (b) $R = 50$ kbps   (e) $R = 1$ Mbps
   (c) $R = 100$ kbps

7.27 Find the probability of error for noncoherent ASK, with signal set

$$s_i(t) = \begin{cases} 0, & 0 \le t \le T, i = 1 \\ A \cos(\omega_c t + \theta), & 0 \le t \le T, i = 2 \end{cases}$$

where θ is a uniformly distributed random variable in $(0, 2\pi)$. White Gaussian noise is added to this signal in the channel. The receiver is a bandpass filter followed by an envelope detector followed by a sampler and threshold comparison. Show that the envelope detector output with signal 1 present (i.e., zero signal) is Rayleigh-distributed, and that the envelope detector output with signal 2 present is Rician-distributed. Assuming that the threshold is set at $A/2$, find an expression for the probability of error. You will not be able to integrate this expression. However, by making use of the approximation

$$I_0(v) \approx \frac{e^v}{\sqrt{2\pi v}}, \qquad v \gg 1$$

you will be able to approximate the pdf of the sampler output for large signal-to-noise ratio as Gaussian and express the probability of error in terms of a Q-function. (*Hint:* Neglect the $v^{-1/2}$ in the above approximation.) Plot the error probability versus the signal-to-noise ratio.

Section 7.5

7.28 Show that (7.115) and (7.116) are Fourier transform pairs.

7.29 Assume a raised-cosine pulse with $\beta = 1/2T$, additive noise with power spectral density

$$G_n(f) = \frac{2\sqrt{2}\sigma_n^2/\pi f_3}{1 + (f/f_3)^4}$$

and a channel filter with transfer-function-squared magnitude given by

$$|H_C(f)|^2 = \frac{1}{1 + (f/f_c)^2}$$

Find and sketch the optimum transmitter and receiver filters for binary signaling for the following cases:

(a) $f_3 = f_c = \dfrac{1}{2T}$ (b) $f_c = 2f_3 = \dfrac{1}{T}$ (c) $f_3 = 2f_c = \dfrac{1}{T}$

(d) $\dfrac{1}{2T} = f_3 = 2f_c$ (e) $\dfrac{1}{2T} = f_c = 2f_3$

7.30 By appropriate sketches, show that the raised-cosine spectra satisfy Nyquist's pulse-shaping criterion.

7.31 (a) Using (7.128), and (7.139), show that the transmitted signal energy is

$$E_T = A^2 \int_{-\infty}^{\infty} \frac{|P(f)|^2 \, df}{|H_C(f)|^2 |H_R(f)|^2}$$

(b) Solving for A^2 in the result obtained for E_T above and using (7.129) to find var $(N) = \sigma^2$, show that

$$\frac{\sigma^2}{A^2} = \frac{1}{E_T} \int_{-\infty}^{\infty} G_n(f) |H_R(f)|^2 \, df \int_{-\infty}^{\infty} \frac{|P(f)|^2 \, df}{|H_C(f)|^2 |H_R(f)|^2}$$

(c) Use Schwarz's inequality to show that the minimum for σ^2/A^2 is

$$\left(\frac{\sigma}{A}\right)^2_{\min} = E_T^{-1} \left[\int_{-\infty}^{\infty} \frac{G_n^{1/2}(f)|P(f)|}{|H_C(f)|} \, df \right]^2$$

which is achieved for $|H_R(f)|_{opt}$ and $|H_T(f)|_{opt}$ given when (7.134) and (7.135) are used.

(d) Show the minimum error probability is given by

$$P_{E,min} = Q\left\{ \sqrt{E_T} \left[\int_{-\infty}^{\infty} \frac{G_n^{1/2}(f)|P(f)|}{|H_C(f)|} df \right]^{-1/2} \right\}$$

7.32 Data are to be transmitted through a bandlimited channel at a rate $R = 1/T = 9600$ bps. The channel filter has transfer function

$$H_C(f) = \frac{1}{1 + j(f/4800)}$$

The noise is white with power spectral density

$$\frac{N_0}{2} = 10^{-12} \text{ W/Hz}$$

Assume a received pulse with raised-cosine spectrum given by (7.116) with

$$\beta = \frac{1}{2T} = 4800 \text{ Hz}$$

is desired.

(a) Find the magnitudes of the transmitter and receiver filter transfer functions that give zero intersymbol interference and optimum detection.

(b) Using a table or the asymptotic approximation for the Q-function, find the value of A/σ required to give $P_{E,min} = 10^{-4}$.

(c) Referring to Problem 7.31, find E_T to give this value of A/σ for the N_0, $G_n(f)$, $P(f)$, and $H_C(f)$ given above. (Numerical integration required.)

Section 7.6

7.33 Plot P_E from Equation (7.153) versus z_0 for $\delta = 0.5$ and $\tau_m/T = 0.2$, 0.6, and 1.0. Note that effective use can be made of Figure 7.26 by appropriately shifting the abscissa.

7.34 Redraw Figure 7.27 for $P_E = 10^{-5}$. For a rough plot, make use of Figure 7.26. For a more refined plot, write a Mathcad program using solve blocks to obtain the degradation for various values of δ and τ_m/T.

FIGURE 7.32

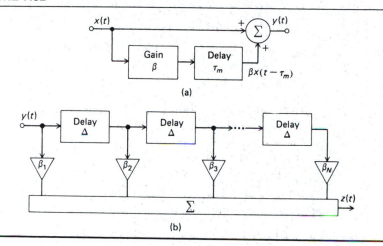

(a)

(b)

7.35 A simple model for a multipath communications channel is shown in Figure 7.32(a).

(a) Find $H_c(f) = Y(f)/X(f)$ for this channel and plot $|H_c(f)|$ for $\beta = 1$ and 0.5.

(b) In order to equalize, or undo, the channel-induced distortion, an equalization filter is used. Ideally, its transfer function should be

$$H_{eq}(f) = \frac{1}{H_c(f)}$$

if the effects of noise are ignored and only distortion caused by the channel is considered. A tapped delay-line or transversal filter, as shown in Figure 7.32(b), is commonly used to approximate $H_{eq}(f)$. Write down a series expression for $H'_{eq}(f) = Z(f)/Y(f)$.

(c) Using $(1 + x)^{-1} = 1 - x + x^2 - x^3 + \ldots$, $|x| < 1$, find a series expression for $1/H_c(f)$. Equating this with $H_{eq}(f)$ found in part (b), find the values for $\beta_1, \beta_2, \ldots, \beta_N$, assuming $\tau_m = \Delta$.

Section 7.7

7.36 Fading margin can be defined as the incremental E_b/N_0, in decibels, required to provide a certain desired error probability in a fading channel as could be achieved with the same modulation technique in a nonfading channel. Assume that a bit error probability of 10^{-2}

is specified. Find the fading margin required for the following cases:
(a) BPSK; (b) DPSK; (c) coherent FSK; (d) noncoherent FSK.

7.37   Repeat Problem 7.36 for a specified error probability of 10^{-3}.

7.38   Carry out the integrations leading to (7.162), (7.163), (7.164), and
(7.165) given that the signal-to-noise ratio pdf is given by

$$f_z(z) = \frac{1}{\bar{z}} e^{-z/\bar{z}}, \qquad z \geq 0$$

Section 7.8

7.39   Given the following channel pulse-response samples:

$p_c(-3T) = 0.001$ $p_c(-2T) = -0.01$ $p_c(-T) = 0.1$ $p_c(0) = 1.0$

$p_c(T) \quad = 0.2$ $p_c(2T) \quad = -0.02$ $p_c(3T) \quad = 0.005$

(a) Find the tap coefficients for a three-tap zero-forcing equalizer.
(b) Find the output samples for $mT = -2T, -T, 0, T,$ and $2T$.

7.40   (a) Consider the design of an MMSE equalizer for a multipath channel whose output is of the form

$$y(t) = A\, d(t) + bA\, d(t - T_m) + n(t)$$

where the second term is a multipath component and the third
term is noise independent of the data, $d(t)$. Assume $d(t)$ is a random (coin-toss) binary sequence with autocorrelation function
$R_{dd}(\tau) = \Lambda(\tau/T)$. Let the noise have a lowpass-RC-filtered spectrum with 3-dB cutoff frequency $f_3 = 1/T$. Let the tap spacing be
$\Delta = T_m = T$. Express the matrix $[R_{yy}]$ in terms of the signal-to-
noise ratio $E_b/N_0 = A^2 T/N_0$, where N_0 is the single-sided power
spectral density of the noise into the lowpass RC filter that produced the noise into the equalizer.

(b) Obtain the optimum tap weights for a three-tap MMSE equalizer
and at a signal-to-noise ratio of 10 dB.

COMPUTER EXERCISES*

7.1   Develop a computer simulation of an integrate-and-dump detector for
antipodal baseband signaling. Generate Gaussian noise samples so that

---

* Computer exercises requiring a computer simulation, such as 7.1 and 7.4, are best done
with a high-level programming language such as C or Fortran. Matlab may also suffice in these
exercises.

white noise of two-sided spectral density $N_0/2$ is simulated. You can do this by drawing independent Gaussian random numbers with zero mean with variance σ^2 obeying*

$$\sigma^2 T_s = \frac{N_0}{2}$$

where T_s is the sample interval (this should be one-tenth the bit period or less). Obtain an estimate of the probability of error versus E_b/N_0 by stimulating the transmission of many bits into the receiver, counting the number of errors, and dividing by the total number of bits received for a number of values of E_b/N_0. Do your results check with Figure 7.5?

7.2 Write a computer program to evaluate the degradation imposed by bit timing error at a desired error probability as discussed in Problem 7.6.

7.3 Write a computer program to evaluate the degradation imposed by Gaussian phase jitter at a desired error probability as discussed in connection with the data presented in Table 7.2. This will require numerical integration.

7.4 Referring to Computer Exercise 7.1 and Problem 7.7, write a simulation of a receiver in which the integrate-and-dump detector is replaced by a first-order filter, thus verifying the theoretical result obtained in Problem 7.7. You will have to simulate the filter digitally. It is suggested that you use the bilinear-z transform realization method for the digital filter.†

7.5 Write a computer program to evaluate various digital modulation techniques:

(a) For a specified data rate and error probability, find the required bandwidth and E_b/N_0 in decibels. Corresponding to the data rate and required E_b/N_0, find the required received signal power for $N_0 = 1$ W/Hz.

(b) For a specified bandwidth and error probability find the allowed data rate and required E_b/N_0 in decibels. Corresponding to the data rate and required E_b/N_0, find the required received signal power for $N_0 = 1$ W/Hz.

7.6 Write a computer program to compute and plot the transfer functions of the optimum filters as specified by (7.134) and (7.135) for specified

---

* See Ziemer and Peterson, pp. 121–122.
† See, for example, Ziemer, Tranter, and Fannin, Chapter 9.

noise power spectral density and channel transfer function for raised cosine spectra.

7.7   Write a computer program to verify Figures 7.26 and 7.27.

7.8   Write a computer program to evaluate degradation due to fading at a specified error probability.

7.9   Write a computer program to design equalizers for specified channel conditions for: (a) the zero-forcing criterion; (b) the MMSE criterion.

ADVANCED DATA COMMUNICATIONS TOPICS 8

In this chapter we consider some topics on data transmission that are beyond the fundamental ones considered in Chapter 7. The first topic considered is that of *M*-ary digital modulation systems, where $M > 2$. We next examine bandwidth requirements for data transmission systems. An important consideration in any communications system is synchronization including carrier, symbol, and word. This is considered next. Following this, we look at modulation techniques that utilize bandwidths much larger than that required for data modulation itself. Such schemes are referred to as *spread-spectrum systems*. The final section deals with satellite communications links, which provides a specific example of the application of some of the digital communications topics considered in Chapters 7 and 8.

8.1 *M*-ARY DATA COMMUNICATIONS SYSTEMS

With the binary digital communications systems we have considered so far, one of only two possible signals can be transmitted during each signaling interval. In an *M*-ary system, one of *M* possible signals may be transmitted during each T_s-second signaling interval, where $M \geq 2$ (we now place a subscript *s* on the signaling interval *T* to denote "symbol"; we will place the subscript *b* on *T* to denote "bit" when $M = 2$). Thus binary data transmission is a special case of *M*-ary data transmission. We refer to each possible transmitted signal of an *M*-ary message sequence as a *character* or *symbol*. The rate at which *M*-ary symbols are transmitted through the channel is called the *baud rate* in bauds.

M-ary Schemes Based on Quadrature Multiplexing

In Section 3.6 we demonstrated that two different messages can be sent through the same channel by means of quadrature multiplexing. In a quadrature-multiplexed system, the messages $m_1(t)$ and $m_2(t)$ are used to double-sideband modulate two carrier signals, which are in phase quadrature, to produce the modulated signal

$$x_c(t) = A[m_1(t) \cos \omega_c t - m_2(t) \sin \omega_c t]$$

$$\triangleq R(t) \cos [\omega_c t + \theta(t)] \tag{8.1}$$

Demodulation at the receiver is accomplished by coherent demodulation with two reference sinusoids that are ideally phase and frequency coherent with the quadrature carriers. This same principle can be applied to transmission of digital data and results in several modulation schemes, three of which will be described here: (1) quadriphase-shift keying (QPSK), (2) offset quadriphase-shift keying (OQPSK), and (3) minimum-shift keying (MSK).

In the analysis of these systems, we make use of the fact that coherent demodulation ideally results in the two messages $m_1(t)$ and $m_2(t)$ being separate at the outputs of the quadrature mixers. Thus these quadrature-multiplexed schemes can be viewed as two separate digital modulation schemes operating in parallel.

The block diagram of a parallel realization for a QPSK transmitter is shown in Figure 8.1, along with typical signal waveforms. In the case

FIGURE 8.1   Modulator and typical waveforms for QPSK

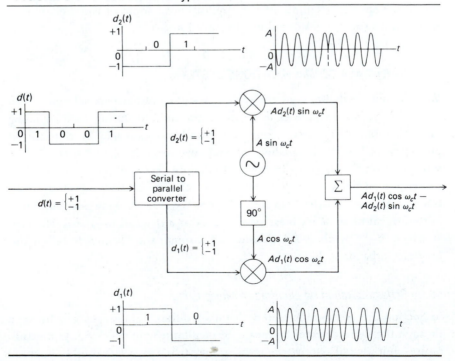

FIGURE 8.2   Detection of QPSK

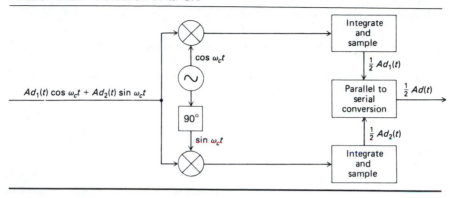

of QPSK, we set $m_1(t) = d_1(t)$ and $m_2(t) = d_2(t)$, where d_1 and d_2 are ± 1-valued waveforms that have possible transitions each T_s seconds. Transitions are aligned. Note that we may think of $d_1(t)$ and $d_2(t)$, the bit streams that modulate the quadrature carriers, as being obtained by grouping the bits of a binary signal $d(t)$ with a bit period half the symbol period of $d_1(t)$ and $d_2(t)$, two bits at a time. Since the phase of the transmitted signal is

$$\theta_i = \tan^{-1}\left[\frac{-d_2(t)}{d_1(t)}\right] \tag{8.2}$$

we see that θ_i takes on the four possible values $\pm 45°$ and $\pm 135°$. Consequently, a QPSK transmitter can be realized in a serial fashion where $d_1(t)$ and $d_2(t)$ impose phase shifts on the carrier that are integer multiples of $90°$.

Because the transmitted signal for a QPSK system can be viewed as two binary PSK signals summed as shown in Figure 8.1, it is reasonable that demodulation and detection involve two binary receivers in parallel, one for each quadrature carrier. The block diagram of such a system is shown in Figure 8.2. We note that a symbol in $d(t)$ will be correct only if the corresponding bits in both $d_1(t)$ and $d_2(t)$ are correct. Thus the probability of correct reception P_c for each bit of $d(t)$ is given by

$$P_c = (1 - P_{E_1})(1 - P_{E_2}) \tag{8.3}$$

where P_{E_j} is the probability of error for quadrature channel j, $j = 1$ or 2. In writing (8.3), it has been assumed that errors in the quadrature channels are independent. We will discuss this assumption shortly.

Turning now to the calculation of P_{E_j}, $j = 1, 2$, we note that because of symmetry, $P_{E_1} = P_{E_2}$. Assuming that the input to the receiver is signal plus white Gaussian noise with double-sided power spectral density $\frac{1}{2}N_0$, that is,

$$y(t) = Ad_1(t) \cos \omega_c t - Ad_2(t) \sin \omega_c t + n(t) \tag{8.4}$$

we find that the output of the upper correlator in Figure 8.2 at the end of a signaling interval T_s is

$$V_1 = v_1(T_s) = \pm \tfrac{1}{2} A T_s + N_1 \tag{8.5}$$

where

$$N_1 = \int_0^{T_s} n(t) \cos \omega_c t \, dt \tag{8.6}$$

Similarly, the output of the lower correlator at $t = T_s$ is

$$V_2 = v_2(T_s) = \pm \tfrac{1}{2} A T_s + N_2 \tag{8.7}$$

where

$$N_2 = \int_0^{T_s} n(t) \sin \omega_c t \, dt \tag{8.8}$$

Errors at either correlator output will be independent if V_1 and V_2 are independent, which requires that N_1 and N_2 be independent. We can show that N_1 and N_2 are uncorrelated (Problem 8.4), and since they are Gaussian (why?), they are independent.

Returning to the calculation of P_{E_1}, we note that the problem is similar to the antipodal baseband case. The mean of N_1 is zero, and its variance is

$$\sigma_1^2 = E\{N_1^2\} = E\left\{ \left[\int_0^{T_s} n(t) \cos \omega_c t \, dt \right]^2 \right\}$$

$$= \int_0^{T_s} \int_0^{T_s} E\{n(t)n(\alpha)\} \cos \omega_c t \cos \omega_c \alpha \, dt \, d\alpha$$

$$= \int_0^{T_s} \int_0^{T_s} \frac{N_0}{2} \delta(t - \alpha) \cos \omega_c t \cos \omega_c \alpha \, d\alpha \, dt$$

$$= \frac{N_0}{2} \int_0^{T_s} \cos^2 \omega_c t \, dt$$

$$= \frac{N_0 T_s}{4} \tag{8.9}$$

Thus, following a series of steps similar to the case of binary antipodal signaling, we find that

$$P_{E_1} = P(d_1 = +1)P(E_1|d_1 = +1) + P(d_1 = -1)P(E_1|d_1 = -1)$$
$$= P(E_1|d_1 = +1) = P(E_1|d_1 = -1) \tag{8.10}$$

where the latter equation follows by noting the symmetry of the pdf of V_1. But

$$P(E|d_1 = +1) = P(\tfrac{1}{2}AT_s + N_1 < 0) = P(N_1 < -\tfrac{1}{2}AT_s)$$
$$= \int_{-\infty}^{-AT_s/2} \frac{e^{-n_1^2/2\sigma_1^2}}{\sqrt{2\pi\sigma_1^2}} \, dn_1$$
$$= Q\left(\sqrt{\frac{A^2 T_s}{N_0}}\right) \tag{8.11}$$

Thus the probability of error for the upper channel in Figure 8.2 is

$$P_{E_1} = Q\left(\sqrt{\frac{A^2 T_s}{N_0}}\right) \tag{8.12}$$

the same as P_{E_2}. Noting that $\tfrac{1}{2}A^2 T_s$ is the average energy for *one* quadrature channel, we see that (8.12) is identical to binary PSK. Thus, considered on a *per channel* basis, QPSK performs identically to binary PSK.

However, if we consider the probability of error for a single phase of a QPSK system we obtain, from (8.3), the result

$$P_E = 1 - P_c = 1 - (1 - P_{E_1})^2 \cong 2P_{E_1}, \qquad P_{E_1} \ll 1$$
$$= 2Q\left(\sqrt{\frac{A^2 T_s}{N_0}}\right) \tag{8.13}$$

Noting that the energy per symbol, or character, is $A^2 T_s \triangleq E_s$ for the quadriphase signal, we may write (8.13) as

$$P_E = 2Q\left(\sqrt{\frac{E_s}{N_0}}\right) \tag{8.14}$$

If we compare QPSK and binary PSK on the basis of average energy-per-symbol-to-noise-spectral-density ratio, *QPSK is approximately 3 dB worse than binary PSK.* However, we are transmitting twice as many bits with the QPSK system,

FIGURE 8.3 Error probability for QPSK

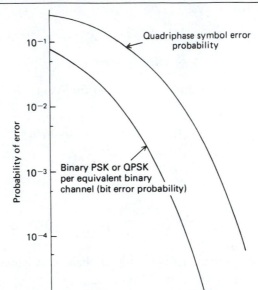

assuming T_s the same. Comparing QPSK and binary PSK on the basis of the systems transmitting equal numbers of bits per second (two bits per QPSK phase), we find that the performances are the same. Binary PSK and QPSK are compared in Figure 8.3 on the basis of signal-to-noise ratio $z = E_s/N_0$, where E_s is the average energy per symbol. Note that the curve for QPSK approaches $\frac{3}{4}$ as the SNR approaches zero ($-\infty$ dB). This is reasonable because the receiver will, on average, make only one correct decision for every four signaling intervals (one of four possible phases) if the input is noise alone.

OQPSK Systems

Because the quadrature data streams $d_1(t)$ and $d_2(t)$ can switch signs simultaneously in a QPSK system, it follows that the data-bearing phase θ_i of the modulated signal can change by 180°. This can have an undesirable effect in terms of envelope deviation if the modulated signal is filtered, which is invariably the case in a practical system. To avoid the possibility of 180° phase switching, the switching instants of the quadrature-channel data signals $d_1(t)$ and $d_2(t)$ of a QPSK system can be offset by $T_s/2$ relative to each

other, where T_s is the signaling interval in either channel. The resulting modulation scheme is referred to as *offset QPSK*, which is abbreviated OQPSK; it
is also sometimes called *staggered QPSK*. With the staggering or offsetting of
quadrature data streams by $T_s/2$, the maximum phase change due to data
modulation of the transmitted carrier is 90°. For an ideal system, the error
probability performance of OQPSK is identical to that of QPSK. One limi-
tation of an OQPSK system is that the data streams $d_1(t)$ and $d_2(t)$ *must have
the same symbol durations*, whereas for QPSK they need not.

MSK Systems

In (8.1), suppose that message $m_1(t)$ is of the form

$$m_1(t) = d_1(t) \cos \omega_1 t \tag{8.15}$$

and message $m_2(t)$ is given by

$$m_2(t) = d_2(t) \sin \omega_1 t \tag{8.16}$$

where $d_1(t)$ and $d_2(t)$ are binary data signals taking on the value $+1$ or
-1 in symbol intervals of length $T_s = 2T_b$ seconds, and ω_1 is the radian
frequency of the weighting functions, $\cos \omega_1 t$ and $\sin \omega_1 t$, to be specified
later. As in the case of QPSK, these data signals can be thought of as having
been derived from a serial binary data stream whose bits occur each T_b seconds, with even-indexed bits used to produce $d_1(t)$ and odd-indexed bits
used to produce $d_2(t)$, or vice versa. These binary data streams are weighted
by a cosine or sine waveform as shown in Figure 8.4. If we substitute (8.15)
and (8.16) into (8.1) and keep in mind that $d_1(t)$ and $d_2(t)$ are either $+1$ or
-1, then, through the use of appropriate trigonometric identities, it follows
that the modulated signal can be written as

$$x_c(t) = A \cos [\omega_c t + \theta_i(t)] \tag{8.17}$$

where

$$\theta_i(t) = -\tan^{-1}\left\{ \left[\frac{d_2(t)}{d_1(t)}\right] \tan (\omega_1 t) \right\} \tag{8.18}$$

If $d_2(t) = d_1(t)$ (i.e., successive bits in the serial data stream are the same,
either both logic 1's or both logic 0's), then

$$\theta_i(t) = -\omega_1 t \tag{8.19}$$

whereas, if $d_2(t) = -d_1(t)$ (i.e., successive bits in the serial data stream are
different), then

$$\theta_i(t) = \omega_1 t \tag{8.20}$$

FIGURE 8.4 Block diagrams for parallel MSK modulator and demodulator. (a) Modulator. (b) Demodulator. (From R. E. Ziemer and C. R. Ryan, "Minimum-Shift Keyed Modem Implementations for High Data Rates," *IEEE Communications Magazine*, Vol. 21, October 1983. Copyright © 1983 IEEE. Reprinted with permission.)

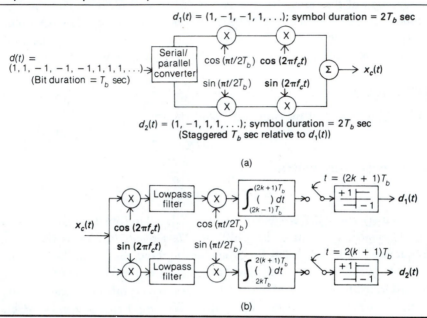

One form of minimum-shift keying (MSK) results if $\omega_1 = \pi/T_s = \pi/2T_b$. In this case, each symbol of the data signal $d_1(t)$ is multiplied or weighted by one-half cycle of a sine waveform, and each symbol of the data signal $d_2(t)$ is weighted by one-half cycle of a cosine waveform, as shown in Figure 8.5. This form of MSK, wherein the weighting functions for each symbol are alternating half cycles of sine or cosine waveforms, is referred to as *MSK type I*. *MSK type II* modulation results if the weighting is always a positive half-cosinusoid or half-sinusoid, depending on whether it is the upper or lower arm in Figure 8.4 being referred to. This type of MSK modulation, which is also illustrated in Figure 8.5, bears a closer relationship to OQPSK than to MSK type I.

Using $\omega_1 = \pi/2T_b$ in (8.18) and substituting the result into (8.19) gives

$$x_c(t) = A \cos\left[2\pi\left(f_c \pm \frac{1}{4T_b}\right)t + u_k\right] \qquad (8.21)$$

FIGURE 8.5   (a) MSK type I modulation. (b) MSK type II modulation. (Adapted from Figure 3.6 of *Digital Communications by Satellite* by V. K. Bhargava et al. Copyright © 1981 John Wiley & Sons, Inc. Used with permission.)

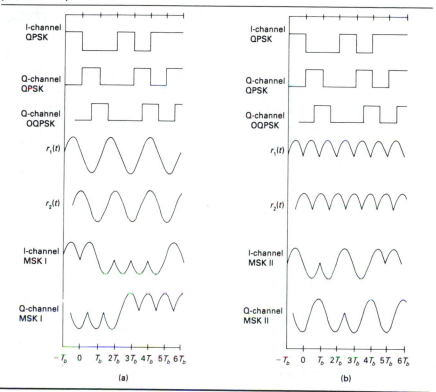

where $f_c = \omega_c/2\pi$ and $u_k = 0$ or $u_k = k\pi$ modulo · 2π, according to whether d_2/d_1 equals $+1$ or -1, respectively. From this form of an MSK-modulated signal, we can see that MSK can be viewed as frequency modulation in which the transmitted tones* are either one-quarter data rate $(1/4T_b)$ above or one-quarter data rate below the carrier f_c in instantaneous frequency (since the carrier is not actually transmitted, f_c is sometimes referred to as the *apparent carrier*). Note that the frequency spacing between the tones is $\Delta f = 1/2T_b$, which is the minimum frequency spacing required for the tones to be coherently orthogonal [that is, ρ_{12} of (7.57) is zero].

---

* One should not infer from this that the spectrum of the transmitted signal consists of impulses at frequencies $f_c \pm 1/4T_b$.

In neither MSK type I nor MSK type II modulation formats is there a one-to-one correspondence between the data bits of the serial bit stream and the instantaneous frequency of the transmitted signal. A modulation format in which this is the case, referred to as *fast frequency-shift keying (FFSK)*, can be obtained by differentially encoding the serial bit stream before modulation by means of an MSK type I modulator.

Viewing (8.21) as a phase-modulated signal, we note that the argument of the cosine can be separated into two phase terms, one due solely to the carrier frequency, or $2\pi f_c t$, and the other due to the modulation, or $\pm \pi(t/2T_b) + u_k$. The latter term is referred to as the *excess phase* and is conveniently portrayed by a trellis diagram, such as that in Figure 8.6, which shows the excess phase trajectories for all possible serial data sequences. Note that the excess phase changes by exactly $\pi/2$ radians each T_b seconds and that it is a continuous function of time. This results in even better envelope deviation characteristics than OQPSK when filtered. In the excess-phase trellis diagram, straight lines with positive slope correspond to alternating 1's and 0's in serial-data sequences, and straight lines with negative slope correspond to all 1's or all 0's in serial-data sequences.

The detector for MSK signals can be realized in parallel form in analogous fashion to QPSK or OQPSK, as shown in Figure 8.2, except that multiplication by $\cos(\pi t/2T_b)$ is required in the upper arm and multiplication by $\sin(\pi t/2T_b)$ is required in the lower arm in order to realize the optimum

FIGURE 8.6   MSK trellis diagram (From R. E. Ziemer and C. R. Ryan, "Minimum-Shift Keyed Modem Implementations for High Data Rates," *IEEE Communications Magazine*, Vol. 21, October 1983. Copyright © 1983 IEEE. Reprinted with permission.)

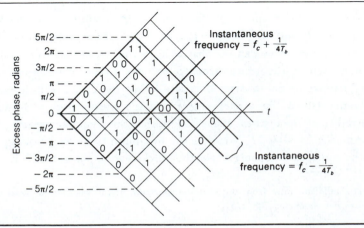

FIGURE 8.7   Block diagrams of serial MSK modulator and demodulator.
(a) Serial MSK modulator. (b) Serial MSK demodulator. (From R. E. Ziemer
and C. R. Ryan, "Minimum-Shift Keyed Modem Implementations for High Data
Rates," *IEEE Communications Magazine,* Vol. 21, October 1983. Copyright ©
1983 IEEE. Reprinted with permission.)

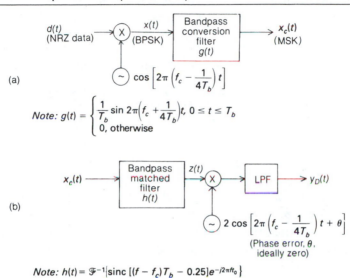

correlation detector for the two data signals $d_1(t)$ and $d_2(t)$. As in the case
of QPSK (or OQPSK), it can be shown that the noise components at the
integrator outputs of the upper and lower arms are uncorrelated. Except for
a different scaling factor, the error probability analysis for MSK is identical
to that for QPSK, and consequently, the error probability performance of
MSK is identical to that of QPSK or OQPSK.

   In the discussion of MSK so far, we have viewed the modulation and
detection processes as being accomplished by parallel structures like those
shown in Figures 8.1 and 8.2 for QPSK. It turns out that MSK can be pro-
cessed in a serial fashion as well. The serial modulator structure, as shown
in Figure 8.7(a), consists of a BPSK modulator with a conversion filter at its
output with the transfer function

$$G(f) = \{\text{sinc } [(f - f_c)T_b - 0.25] + \text{sinc } [(f + f_c)T_b + 0.25]\}e^{-j2\pi f t_0}$$

$$(8.22)$$

where t_0 is an arbitrary filter delay and f_c is the apparent carrier frequency
of the MSK signal. Note that the peak of the frequency response of the
conversion filter is offset in frequency one-quarter data rate above the appar-

ent carrier. The BPSK signal, on the other hand, is offset one-quarter data rate below the desired apparent carrier of the MSK signal. Its power spectrum can be written as

$$S_{\text{BPSK}}(f) = (A^2 T_b/2)\{\text{sinc}^2\ [(f - f_c)T_b + 0.25]$$
$$+ \text{sinc}^2\ [(f + f_c)T_b - 0.25]\} \quad (8.23)$$

The product of $|G(f)|^2$ and $S_{\text{BPSK}}(f)$ gives the power spectrum of the conversion filter output, which, after some simplification, can be shown to be

$$S_{\text{MSK}}(f) = \frac{32A^2 T_b}{\pi^4} \left\{ \frac{\cos^2\ 2\pi T_b(f - f_c)}{[1 - 16T_b^2(f - f_c)^2]^2} + \frac{\cos^2\ 2\pi T_b(f + f_c)}{[1 - 16T_b^2(f + f_c)^2]^2} \right\}$$
$$(8.24)$$

as illustrated graphically in Figure 8.8. This is the double-sided power spectrum of an MSK-modulated signal,* which demonstrates in the frequency domain the validity of the serial approach to the generation of MSK. Thus the parallel modulator structure can be replaced by a serial modulator structure, which means that the difficult task of producing amplitude-matched phase-quadrature signals in the parallel structure can be replaced by the perhaps easier task of generation of BPSK signals and synthesis of conversion filters.

At the receiver, essentially the reverse of the signal-processing procedure at the transmitter is carried out. The received signal is passed through a filter whose frequency response is proportional to the square root of the MSK spectrum. As shown in Figure 8.7(b), the filter is followed by a coherent demodulator, a sampler, and threshold comparison to carry out the data detection. Although the details will not be given here,† it can be shown that each symbol is sampled independently of those preceding or following it at the proper sampling instants.

M-ary Data Transmission in Terms of Signal Space

A convenient framework for discussing *M*-ary data transmission systems is that of signal space. The approach used here in terms of justifying the receiver structure is heuristic. It is placed on a firm theoretical basis in Chapter 9, where optimum signal detection principles are discussed.

---

* This will be shown in Section 8.2.
† See F. Amoroso and J. A. Kivett, "Simplified MSK Signaling Technique," *IEEE Transactions on Communication*, Vol. COM-25, April 1977, pp. 433–441.

FIGURE 8.8 Spectra pertinent to producing serial MSK. (a) PRK power spectrum. (b) Frequency response of conversion filter magnitude squared. (c) MSK power spectrum. Only positive-frequency halves of double-sided spectra are shown. (From R. E. Ziemer and C. R. Ryan, "Minimum-Shift Keyed Modem Implementations for High Data Rates," *IEEE Communications Magazine,* Vol. 21, October 1983. Copyright © 1983 IEEE. Reprinted with permission.)

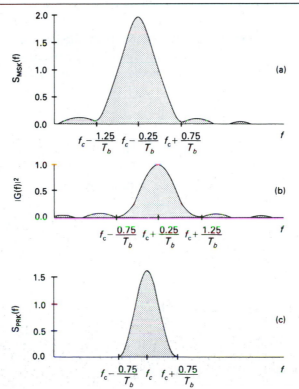

We consider coherent communication systems with signal sets of the form

$$s_i(t) = \sum_{j=1}^{K} a_{ij}\phi_j(t), \qquad 0 \le t \le T_s, \quad K \le M, \quad i = 1, 2, \ldots, M \quad (8.25)$$

where the functions $\phi_j(t)$ are orthonormal over the symbol interval. That is,

$$\int_0^{T_s} \phi_m(t)\phi_n(t)\, dt = \begin{cases} 1, & m = n \\ 0, & m \ne n \end{cases} \qquad (8.26)$$

Based on (8.25), we can visualize the possible transmitted signals as points in a space with coordinate axes $\phi_1(t)$, $\phi_2(t)$, $\phi_3(t)$, ..., $\phi_K(t)$, much as illustrated in Figure 2.5.

At the output of the channel it is assumed that signal plus additive white Gaussian noise is received; that is,

$$y(t) = s_i(t) + n(t), \qquad t_0 \le t \le t_0 + T_s, \, i = 1, \ldots, M \qquad (8.27)$$

where t_0 is an arbitrary starting time equal to an integer times T_s. As shown in Figure 8.9, the receiver consists of a bank of K correlators, one for each orthonormal function. The output of the jth correlator is

$$Z_j = a_{ij} + N_j \qquad (8.28)$$

where the noise component N_j is given by ($t_0 = 0$ for notational ease)

$$N_j = \int_0^{T_s} n(t)\phi_j(t) \, dt \qquad (8.29)$$

FIGURE 8.9 Computation of signal-space coordinates

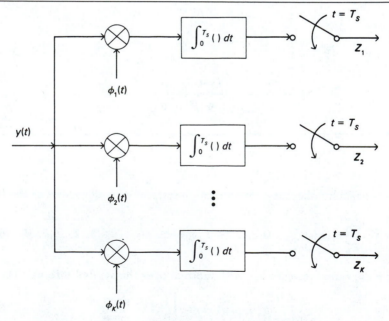

Note: $y(t) = s_i(t) + n(t)$ where $n(t)$ is white Gaussian noise.

Since $n(t)$ is Gaussian and white, the random variables $N_1, N_2, \ldots, N_K$ can be shown to be independent, zero-mean, Gaussian random variables with variances $N_0/2$, which is the two-sided spectral density of the noise.

It can be shown that this process preserves all the information required to make a minimum error-probability decision regarding which signal was transmitted. The next operation in the receiver is a decision box that performs the following function:

Compare the received signal plus noise coordinates with the stored signal coordinates, a_{ij}. Choose as the transmitted signal that one closest to the received signal plus noise point with distance measured in the Euclidian sense; i.e., choose the transmitted signal as the one whose a_{ij}'s minimize $\sum_{j=1}^{K} [Z_j - a_{ij}]^2$.

This decision procedure will be shown in Chapter 9 to result in the minimum error probability possible.

▼ **EXAMPLE 8.1**   Consider BPSK. Only one orthonormal function is required in this case, and it is

$$\phi(t) = \sqrt{\frac{2}{T_b}} \cos \omega_c t, \qquad 0 \le t \le T_b \qquad (8.30)$$

The possible transmitted signals can be represented as

$$s_1(t) = \sqrt{E_b}\, \phi(t) \qquad \text{and} \qquad s_2(t) = -\sqrt{E_b}\, \phi(t) \qquad (8.31)$$

▲ where E_b is the bit energy.

QPSK in Terms of Signal Space

From Figures 8.9 and 8.2 we see that the receiver for QPSK consists of a bank of two correlators. Thus the received data can be represented in a two-dimensional signal space as shown in Figure 8.10. The transmitted signals can be represented in terms of two orthonormal functions $\phi_1(t)$ and $\phi_2(t)$ as

$$x_c(t) = s_i(t) = \sqrt{E_s}\,[d_1(t)\phi_1(t) - d_2(t)\phi_2(t)] = \sqrt{E_s}\,[\pm\phi_1(t) \pm \phi_2(t)] \qquad (8.32)$$

where

$$\phi_1(t) = \sqrt{\frac{2}{T_s}} \cos \omega_c t, \qquad 0 \le t \le T_s \qquad (8.33)$$

FIGURE 8.10 Signal space for QPSK

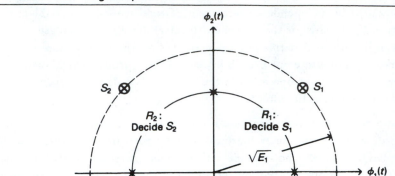

and

$$\phi_2(t) = \sqrt{\frac{2}{T_s}} \sin \omega_c t, \qquad 0 \le t \le T_s \qquad (8.34)$$

E_s is the energy contained in $x_c(t)$ in one symbol interval. The resulting regions for associating a received data point with a possible signal point are also illustrated in Figure 8.10. It can be seen that the coordinate axes provide the boundaries of the regions that determine the signal point to be associated with a received data point. For example, if the received data point is in the first quadrant (region R_1), the decision is made that $d_1 = 1$ and $d_2 = -1$ (this will be denoted as signal point S_1 in the signal space). A simple bound on symbol error probability can be obtained by noting that

$$P_E = Pr(Z \in R_2 \text{ or } R_3 \text{ or } R_4 | S_1 \text{ sent})$$

$$\le Pr(Z \in R_2 \text{ or } R_3 | S_1 \text{ sent}) + Pr(Z \in R_3 \text{ or } R_4 | S_1 \text{ sent}) \qquad (8.35)$$

The two probabilities on the right-hand side of (8.35) can be shown to be equal. Thus,

$$P_E \le 2Pr(Z \in R_2 \text{ or } R_3) = 2Pr(\sqrt{E_s/2} + N_\perp < 0) \qquad (8.36)$$

$$= 2Pr(N_\perp < -\sqrt{E_s/2})$$

where $N_\perp$, as shown in Figure 8.11, is the noise component perpendicular to the decision boundary between R_1 and R_2. It can be shown that it has zero mean and variance $N_0/2$. Thus,

$$P_E \le 2 \int_{-\infty}^{-\sqrt{E_s/2}} \frac{e^{-u^2/N_0}}{\sqrt{\pi N_0}} \, du = 2 \int_{\sqrt{E_s/2}}^{\infty} \frac{e^{-u^2/N_0}}{\sqrt{\pi N_0}} \, du \qquad (8.37)$$

Making the change of variables $u = v/\sqrt{N_0/2}$, we can reduce this to the form

$$P_E = 2Q\left(\sqrt{E_s/N_0}\right) \qquad (8.38)$$

This is identical to (8.14), which resulted in neglecting the square of P_{E_1} in (8.13).

M-ary Phase-Shift Keying

The signal set for QPSK can be generalized to an arbitrary number of phases. The modulated signal takes the form

$$s_i(t) = \sqrt{\frac{2E_s}{T_s}} \cos\left[\omega_c t + \frac{2\pi(i-1)}{M}\right],$$

$$0 \le t \le T_s, \quad i = 1, 2, \ldots, M \qquad (8.39)$$

FIGURE 8.11   Representation of signal plus noise in signal space, showing $N_\perp$, the noise component that can cause the received data vector to land in R_2

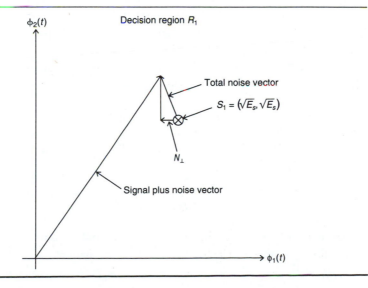

Using trigonometric identities, this can be expanded as

$$s_i(t) = \sqrt{E_s}\left[\cos\left(\frac{2\pi(i-1)}{M}\right)\sqrt{\frac{2}{T_s}}\cos\omega_c t\right.$$

$$\left. -\sin\left(\frac{2\pi(i-1)}{M}\right)\sqrt{\frac{2}{T_s}}\sin\omega_c t\right]$$

$$= \sqrt{E_s}\left[\cos\left(\frac{2\pi(i-1)}{M}\right)\phi_1(t) - \sin\left(\frac{2\pi(i-1)}{M}\right)\phi_2(t)\right] \qquad (8.40)$$

where $\phi_1(t)$ and $\phi_2(t)$ are the orthonormal functions defined by (8.33) and (8.34).

A plot of the signal points S_i, $i = 1, 2, \ldots, M$ along with the optimum decision regions is shown in Figure 8.12(a). The probability of error can be overbounded by noting from Figure 8.12(b) that the total area represented by the two half planes D_1 and D_2 is greater than the total shaded area in Figure 8.12(b), and thus the probability of symbol error is overbounded by the probability that the received data point Z lies in either half plane. Because of symmetry of the noise distribution, both probabilities are equal. Consider a single half plane along with a single signal point, which is at a minimum distance of

$$d = \sqrt{E_s}\sin(\pi/M) \qquad (8.41)$$

away from the boundary of the half plane. As in Figure 8.11, consider the noise component $N_\perp$, which is perpendicular to the boundary of the half plane. It is the only noise component that can possibly put the received data point on the wrong side of the decision boundary; it has zero mean and a variance $N_0/2$. From this discussion and referring to Figure 8.12(b), it follows that the probability of error is overbounded by

$$P_E < Pr(Z \in D_1 \text{ or } D_2) = 2Pr(Z \in D_1)$$

$$= 2Pr(d + N_\perp < 0) = 2Pr(N_\perp < -d)$$

$$= 2\int_{-\infty}^{-d}\frac{e^{-u^2/N_0}}{\sqrt{\pi N_0}}\,du = 2Q\left(\sqrt{\frac{2E_s}{N_0}}\sin(\pi/M)\right) \qquad (8.42)$$

From Figure 8.12(b) it can be seen that the bound becomes tighter as M gets larger (because the overlap of D_1 and D_2 gets smaller).

FIGURE 8.12   (a) Signal space for *M*-ary PSK with *M* = 8. (b) Signal space for *M*-ary PSK showing two half-planes that can be used to overbound P_E.

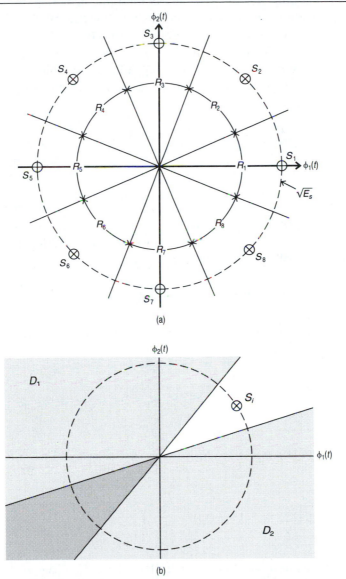

(a)

(b)

Quadrature-Amplitude-Shift Keying (QASK)

Another signaling scheme that allows multiple signals to be transmitted using quadrature carriers is *quadrature-amplitude-shift keying* (QASK), which is also referred to as *quadrature-amplitude modulation* (QAM). For example, *M*-QASK utilizes a signal structure similar to that of (8.32) but with the data sequences d_1 and d_2 each taking on $\sqrt{M}$ different possible levels. Thus, for example, we represent the transmitted signal in an arbitrary signaling interval for 16-QASK as

$$s_i(t) = \sqrt{\frac{2E_s}{T_s}}\,(A_i \cos \omega_c t + B_i \sin \omega_c t), \qquad 0 \le t \le T_s \quad (8.43)$$

where A_i and B_i take on the possible values $\pm a$ and $\pm 3a$ with equal probability. A signal-space representation for 16-QASK is shown in Figure 8.13(a), and the receiver structure is shown in Figure 8.13(b). The probability of symbol error for 16-QASK can be shown to be

$$P_E = 1 - [\tfrac{1}{4}P(C\,|\,\text{I}) + \tfrac{1}{2}P(C\,|\,\text{II}) + \tfrac{1}{4}P(C\,|\,\text{III})] \qquad (8.44)$$

where the probabilities $P(C\,|\,\text{I})$, $P(C\,|\,\text{II})$, and $P(C\,|\,\text{III})$ are given by

$$P(C\,|\,\text{I}) = \left[1 - 2Q\left(\sqrt{\frac{2a^2}{N_0}}\right) \right]^2 \qquad (8.45a)$$

$$P(C\,|\,\text{II}) = \left[1 - 2Q\left(\sqrt{\frac{2a^2}{N_0}}\right) \right]\left[1 - Q\left(\sqrt{\frac{2a^2}{N_0}}\right) \right] \qquad (8.45b)$$

$$P(C\,|\,\text{III}) = \left[1 - Q\left(\sqrt{\frac{2a^2}{N_0}}\right) \right]^2 \qquad (8.45c)$$

In the preceding equations $a^2 = E_s/10$, where E_s is the average energy per symbol. The notation I, II, or III denotes that the particular probability refers to the probability of correct reception for the three types of decision regions shown in Figure 8.13(a). This error probability will be compared with that for *M*-ary PSK later. The error probability for 64-QASK or 256-QASK also can be obtained in a straightforward, but tedious, manner.

FIGURE 8.13 Signal space and detector structure for 16-QASK. (a) Signal constellation and decision regions for 16-QASK. (b) Detector structure for 16-QASK. (Binary representations for signal points are Gray encoded.)

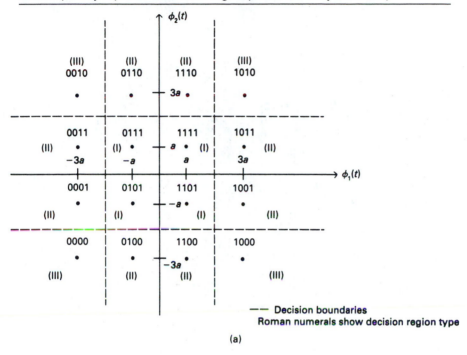

(a)

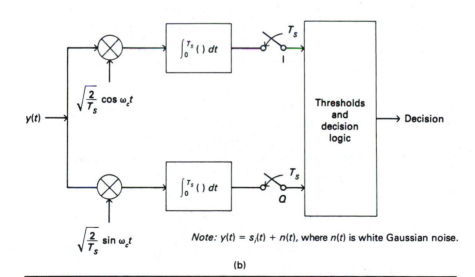

Note: $y(t) = s_i(t) + n(t)$, where $n(t)$ is white Gaussian noise.

(b)

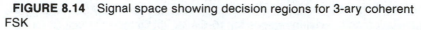

FIGURE 8.14   Signal space showing decision regions for 3-ary coherent FSK

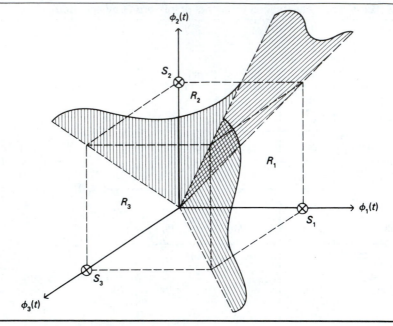

Coherent Frequency-Shift Keying

The error probability for M-ary FSK will be derived in Chapter 9. The transmitted signals have the form

$$s_i(t) = \sqrt{\frac{2E_s}{T_s}} \cos\{2\pi[f_c + (i-1)\Delta f]t\},$$

$$0 \le t \le T_s, \quad i = 1, 2, \ldots, M \quad (8.46)$$

where Δf is a frequency separation large enough to make the signals represented by (8.46) coherently orthogonal (the minimum separation is $\Delta f = 1/2T_s$). Since each of the possible M transmitted signals is orthogonal to the rest, it follows that the signal space must be M-dimensional. It is shown in Figure 8.14 for $M = 3$. An upper bound for the probability of error that becomes tighter as M gets larger is given by[*]

$$P_E \le M\, Q\!\left(\sqrt{\frac{E_s}{N_0}}\right) \qquad (8.47)$$

[*]See Ziemer and Peterson (1992), Chapter 4.

FIGURE 8.15 Receiver for noncoherent FSK

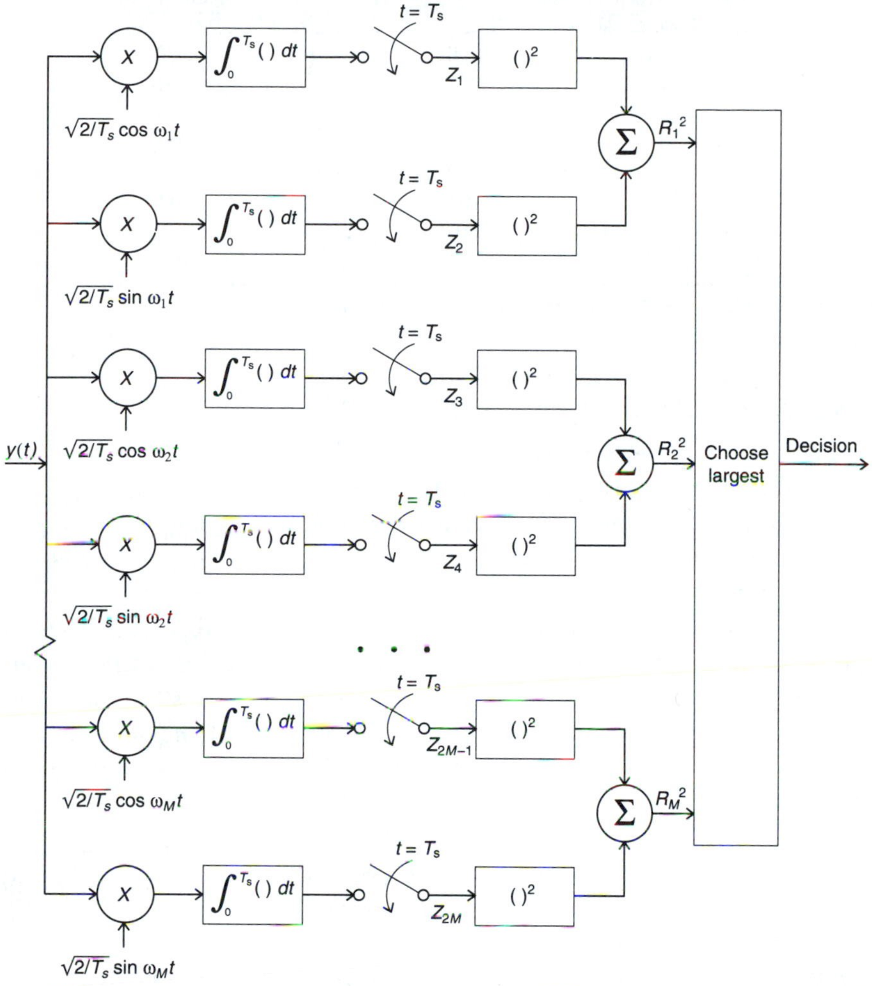

Note: $y(t) = s_1(t) + n(t)$, where $n(t)$ is white Gaussian noise.

Noncoherent Frequency-Shift Keying

Noncoherent *M*-ary FSK employs the same signal set as coherent FSK; however, a receiver structure is used that does not require the acquisition of a coherent carrier reference. A block diagram of a suitable receiver structure is shown in Figure 8.15. The symbol error probability can be shown to be*

$$P_E = \sum_{k=1}^{M-1} \binom{M-1}{k} \frac{(-1)^{k+1}}{k+1} \exp\left(-\frac{k}{k+1}\frac{E_s}{N_0}\right) \tag{8.48}$$

*See Ziemer and Peterson (1992), Chapter 4.

TABLE 8.1 Gray Code Illustration

Digit	Binary Code	Gray Code
0	000	000
1	001	001
2	010	011
3	011	010
4	100	110
5	101	111
6	110	101
7	111	100

Note: See problem 8.14 for the encoding algorithm.

Bit Error Probability from Symbol Error Probability

If one of M possible symbols is transmitted, the number of bits required to specify this symbol is $\log_2 M$. It is possible to number the signal points using a binary code such that only one bit changes in going from a signal point to an adjacent signal point. Such a code is a *Gray code* and is given in Table 8.1 for $M = 8$.

Since mistaking an adjacent signal point for a given signal point is the most probable error, we assume that only such errors occur and that Gray encoding has been used so that a symbol error corresponds to a single bit error. We may then write the bit error probability in terms of the symbol error probability for an M-ary communications system for which these assumptions are valid as

$$P_{E,\text{bit}} = \frac{P_{E,\text{symbol}}}{\log_2 M} \tag{8.49}$$

Since we neglect probabilities of symbol errors where the symbols are further away than adjacent symbols, this result gives a *lower bound* for the bit error probability.

A second way that we can relate bit error probability to symbol error probability is as follows. Consider an M-ary modulation scheme for which $M = 2^n$. Then each symbol (M-ary signal) can be represented by an n-bit binary number, for example, the binary representation of the signal's index minus one. Such a representation is given in Table 8.2 for $M = 8$.

Take any column, say the last, which is enclosed by a box. In this column, there are $M/2$ zeros and $M/2$ ones. If a symbol (M-ary signal) is received in error, then for any given bit position of the binary representation (in this example, we are considering the right-most bit), there are $M/2$ of a possible

TABLE 8.2   Computation of Bit Error Probability for Orthogonal Signaling

M-ary Signal	Binary Representation
1 (0)	0 0 0
2 (1)	0 0 1
3 (2)	0 1 0
4 (3)	0 1 1
5 (4)	1 0 0
6 (5)	1 0 1
7 (6)	1 1 0
8 (7)	1 1 1

$M - 1$ ways that the chosen bit can be in error (one of the M total possibilities is correct). Therefore, the probability of a given data bit being in error, given that a signal (symbol) was received in error, is

$$P(B \mid S) = \frac{M/2}{M - 1} \tag{8.50}$$

Since a symbol is in error if a bit in the binary representation of it is in error, it follows that the probability $P(S \mid B)$ of a symbol error given a bit error is unity. Employing Bayes' rule, we find the equivalent bit error probability of an M-ary system can be approximated by

$$P_{E,\text{bit}} = \frac{P(B \mid S) P_{E,\text{symbol}}}{P(S \mid B)} = \frac{M}{2(M - 1)} P_{E,\text{symbol}} \tag{8.51}$$

This result is especially useful for orthogonal signaling schemes such as FSK.

One more thing must be accomplished before we can compare communications systems using a different number of symbols on an equivalent basis. This is that the energies must be equivalent. For this purpose, we note that the energy per bit E_b in terms of the energy per symbol E_s is

$$E_b = \frac{E_s}{\log_2 M} \tag{8.52}$$

▼| **EXAMPLE 8.2**   Compare the performances of noncoherent and coherent FSK on the basis of E_b/N_0 required to provide a bit error probability of 10^{-6} for various values of M.

TABLE 8.3 Power Efficiencies for Noncoherent and Coherent FSK

	E_b/N_0 for $P_{E,\text{bit}} = 10^{-6}$	
M	Noncoherent	Coherent
2	14.2	13.6
4	11.3	10.8
8	9.7	9.3
16	8.7	8.2

SOLUTION   Using (8.47), (8.48), (8.51), and (8.52), the results in Table 8.3 can be obtained with the aid of the asymptotic approximation for the ▲ Q-function:

Comparison of M-ary Communications Systems

Figure 8.16 compares different *M*-ary PSK systems on the basis of bit error probability versus E_b/N_0. This figure shows that the bit error probability gets

FIGURE 8.16 Bit error probability versus E_b/N_0 for *M*-ary PSK and 16-QASK

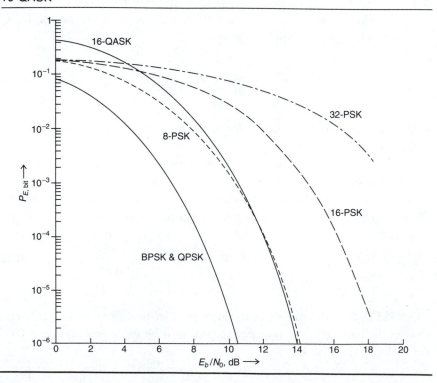

FIGURE 8.17 Bit error probability versus E_b/N_0 for coherent *M-ary* FSK

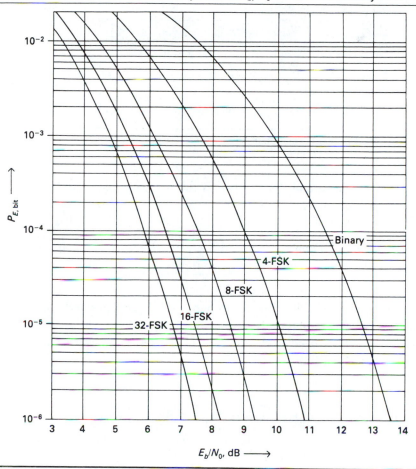

worse as *M* gets larger. This can be attributed to the signal points being crowded closer together in the two-dimensional signal space as *M* gets larger. Not all *M-ary* digital modulation schemes exhibit this undesirable behavior. We have seen that *M-ary* FSK is a signaling scheme in which the number of dimensions in the signal space grows directly with *M*. This means that the bit error probability for *M-ary* FSK *decreases* as *M* increases. This is illustrated in Figure 8.17, which compares bit error probability for coherent FSK for various values of *M*. Unfortunately, the bandwidth required for this signaling scheme grows larger directly with *M*, whereas this is not the case for *M-ary* PSK.

If one considers the required bandwidth to be that required to pass the main lobe of the signal spectrum (null to null), it follows that the *bandwidth*

TABLE 8.4   Bandwidth Efficiencies of Various M-ary Digital Modulation Schemes[a]

M-ary Scheme	Bandwidth Efficiency (bits/s/Hz)
PSK, QASK	$0.5 \log_2 M$
Coherent FSK	$\dfrac{2 \log_2 M}{M + 3}$ (Assumes tone spacing of $1/2T_s$ Hz)
Noncoherent FSK	$\dfrac{\log_2 M}{2M}$ (Assumes tone spacing of $2/T_s$ Hz)

[a] For justification, see Ziemer and Peterson (1992), Chapter 4.

efficiencies of the various *M*-ary schemes that we have just considered are as given in Table 8.4.

▼ **EXAMPLE 8.3**   Compare bandwidth efficiencies on a mainlobe spectrum basis for PSK, QASK, and FSK for various *M*.

SOLUTION   Bandwidth efficiencies in bits per second per hertz for various values of *M* are as follows. Note that for QASK, *M* must be a power of 4. Also note that the bandwidth efficiency of *M*-ary PSK *goes up* with increasing *M* while that for FSK *goes down*.

M	QASK	PSK	Coh. FSK	Noncoh. FSK
2		0.5	0.4	0.25
4	1	1.0	0.57	0.25
8		1.5	0.55	0.19
16	2	2.0	0.42	0.13
32		2.5	0.29	0.08

▲

8.2 BANDWIDTH EFFICIENCIES OF DIGITAL MODULATION FORMATS

Our principal measure of performance for the various modulation schemes considered so far has been probability of error. Actually, this measure tells only half the story of how well a given modulation scheme performs, the other half being given by the bandwidth occupancy of the modulated signal. Specifically, we are interested in the data rate in bits per second that can be achieved by a given modulation scheme through a channel of a given bandwidth in hertz. Theoretically, of course, an infinite bandwidth may be required, depending on the signal pulse shape used to represent a symbol. Consequently, a definition of bandwidth is required that, for purposes of this discussion, is based on the criterion of fractional power of the signal within a certain specified bandwidth. That is, if $S(f)$ is the double-sided power spectrum of a given modulation format, the fraction of total power in a bandwidth B is given by

$$\Delta P_{IB} = 2P_T^{-1} \int_{f_c-B/2}^{f_c+B/2} S(f) \, df \tag{8.53}$$

where the factor of 2 is used since we only integrate over positive frequencies,

$$P_T = \int_{-\infty}^{\infty} S(f) \, df \tag{8.54}$$

is the total power, and f_c is the "center" frequency of the spectrum (usually the carrier frequency, apparent or otherwise). The percent out-of-band power ΔP_{OB} is defined as

$$\Delta P_{OB} = (1 - \Delta P_{IB}) \times 100\% \tag{8.55}$$

The definition of modulated signal bandwidth is conveniently given by setting ΔP_{OB} equal to some acceptable value, say 0.01 or 1%, and solving for the corresponding bandwidth. A curve showing ΔP_{OB}, in decibels, versus bandwidth is a convenient tool for carrying out this procedure, since the 1% out-of-band power criterion for bandwidth corresponds to the bandwidth at which the out-of-band power curve has a value of -20 dB. Later we will present several examples to illustrate this procedure. First, however, we will discuss the spectra for various baseband data formats and for the digital modulation schemes considered so far.

The spectrum of a digitally modulated signal is influenced both by the particular baseband data format used to represent the digital data and by the type of modulation scheme used to prepare the signal for transmission. Several commonly used baseband data formats are illustrated in Figure 8.18. These data formats are referred to as *return-to-zero* (RZ), *non-return-to-zero* (NRZ), and *split-phase* (or *Manchester*). Split-phase, we note, is obtained from NRZ by multiplication by a squarewave clock waveform with a period equal to the bit duration. Typical power spectra are also shown in Figure 8.18 for these three modulation formats, assuming a random (coin toss) bit sequence.

FIGURE 8.18 Typical digital data waveforms and corresponding power spectra. (a) Return-to-zero (RZ) waveform. (b) Non-return-to-zero (NRZ) waveform. (c) Split-phase (Manchester) waveform.

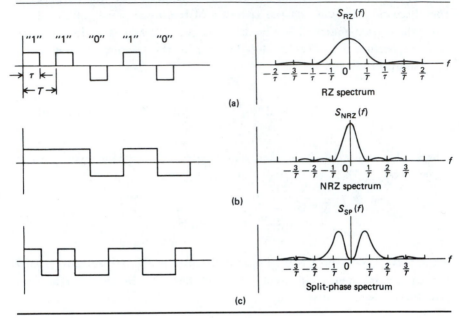

The spectra are derived using the techniques of Chapter 5, in particular, of Example 5.6, and using appropriate expressions for the pulse-shape function $p(t)$. None of these formats results in power spectra with significant frequency content at multiples of the bit rate $1/T_b$. Thus, for the purpose of bit synchronization, nonlinear operations are required to generate power at a frequency of $1/T_b$ Hz or multiples thereof. Both the RZ and split-phase formats guarantee at least one zero crossing per bit interval, but both require twice the transmission bandwidth of NRZ. Around 0 Hz, NRZ and RZ possess significant energy content and therefore may have an adverse effect on carrier-acquisition devices.

These data formats have no memory between symbols. Other data formats are possible in which memory is introduced between symbols for various reasons, including spectral shaping (refer to Example 5.6, Case 2).

In order to obtain the spectrum of a BPSK-modulated signal using any of these data formats, the appropriate spectrum shown in Figure 8.18 is simply shifted up in frequency and centered around the carrier (assuming a single-sided spectrum). An analysis similar to that carried out in Example 5.6 can be used to obtain the spectrum of a quadrature-modulated waveform of the form given by (8.1), where $m_1(t) = d_1(t)$ and $m_2(t) = d_2(t)$ are random (coin toss) waveforms of the forms

$$d_1(t) = \sum_{k=-\infty}^{\infty} a_k p(t - kT_s - \Delta_1) \tag{8.56}$$

and

$$d_2(t) = \sum_{k=-\infty}^{\infty} b_k q(t - kT_s - \Delta_2) \tag{8.57}$$

with

$$E\{a_k\} = E\{b_k\} = 0, \qquad E\{a_k^2\} = A^2, \qquad E\{b_k^2\} = B^2$$

The pulse-shape functions $p(t)$ and $q(t)$ in (8.56) and (8.57) may be the same, or one of them may be zero. It is left to the problems to show that the double-sided spectrum of (8.1), with (8.56) and (8.57) substituted, is

$$S(f) = G(f - f_c) + G(f + f_c) \tag{8.58}$$

where

$$G(f) = \frac{A^2 |P(f)|^2 + B^2 |Q(f)|^2}{T_s} \tag{8.59}$$

in which $P(f)$ and $Q(f)$ are the Fourier transforms of $p(t)$ and $q(t)$, respectively.

This result can be applied to BPSK, for example, by letting $q(t) = 0$ and $p(t) = \Pi(t/T_b)$. The resulting baseband spectrum is

$$G_{PRK}(f) = A^2 T_b \, \text{sinc}^2 \, (T_b f) \tag{8.60}$$

The spectrum for QPSK follows by letting

$$p(t) = q(t) = \frac{1}{\sqrt{2}} \, \Pi \left(\frac{t}{2T_b} \right) \tag{8.61}$$

and $A = B$. The result for $G(f)$ is

$$G_{QPSK}(f) = 2A^2 T_b \, \text{sinc}^2 \, (2T_b f) \tag{8.62}$$

This result also holds for OQPSK because the pulse-shape function $q(t)$ differs from $p(t)$ only by a time shift that results in a factor of $\exp(-j2\pi T_b f)$ in the amplitude spectrum $Q(f)$ (its magnitude is unity).

The baseband spectrum for MSK is found by choosing the pulse-shape functions

$$p(t) = q(t - T_b) = \cos\left(\frac{\pi t}{2T_b}\right) \Pi\left(\frac{t}{2T_b}\right) \tag{8.63}$$

and by letting $A = B$. It can be shown (see Problem 8.25) that

$$\mathcal{F}\left\{\cos\left(\frac{\pi t}{2T_b}\right) \Pi\left(\frac{t}{2T_b}\right)\right\} = \frac{4T_b \cos(2\pi T_b f)}{\pi[1 - (4T_b f)^2]} \tag{8.64}$$

which results in the baseband spectrum for MSK,

$$G_{MSK}(f) = \frac{16A^2 T_b \cos^2(2\pi T_b f)}{\pi^2[1 - (4T_b f)^2]^2} \tag{8.65}$$

Using these results for the baseband spectra of BPSK, QPSK (or OQPSK), and MSK in the definition of percent out-of-band power, Equation (8.55), results in the set of plots shown in Figure 8.19. From this figure, it follows that the bandwidths containing 90% of the power for these modulation formats are approximately

$$B_{90\%} = \frac{1}{T_b} \text{ Hz} \qquad \text{(QPSK, OQPSK, MSK)} \tag{8.66}$$

$$B_{90\%} = \frac{2}{T_b} \text{ Hz} \qquad \text{(BPSK)}$$

These are obtained by noting the bandwidths on the curves corresponding to $\Delta P_{OB} = -10$ dB and doubling these values, since the plots are for baseband bandwidths.

Because the MSK out-of-band power curve rolls off at a much faster rate than do the curves for BPSK or QPSK, a more stringent in-band power specification, such as 99%, results in a much smaller containment bandwidth for MSK than for BPSK or QPSK. The bandwidths containing 99% of the power are

$$B_{99\%} = \frac{1.2}{T_b} \qquad \text{(MSK)} \tag{8.67}$$

$$B_{99\%} = \frac{8}{T_b} \qquad \text{(QPSK or OQPSK)}$$

FIGURE 8.19   Fractional out-of-band power for BPSK, QPSK or OQPSK, and MSK

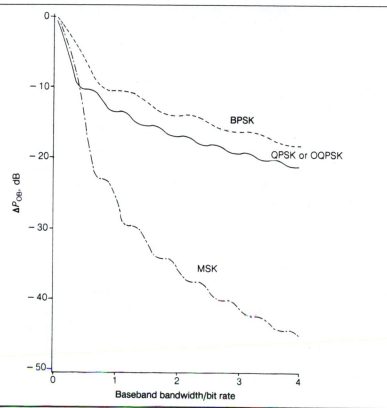

For binary FSK, the following formula can be used to compute the power spectrum if the phase is continuous:*

$$G(f) = G_+(f) + G_-(f) \tag{8.68a}$$

where

$$
G_\pm(f) = \frac{A^2 \sin^2 [\pi(f \pm f_1)T_b \, \sin^2 [\pi(f \pm f_2)T_b]}{2\pi^2 T_b \{1 - 2 \cos [2\pi(f \pm \alpha)T_b] \cos (2\pi\beta T_b) + \cos^2 (2\pi\beta T_b)\}}
$$

$$
\times \left[\frac{1}{f \pm f_1} - \frac{1}{f \pm f_2} \right]^2 \tag{8.68b}
$$

* See W. R. Bennett and J. R. Davey, *Data Transmission* (New York: McGraw-Hill, 1965).

In (8.68b), the following definitions are used:

$$f_1, f_2 = \text{the signaling frequencies in hertz (that is, } f_c \text{ and } f_c + \Delta f)$$
$$\alpha = \tfrac{1}{2}(f_1 + f_2)$$
$$\beta = \tfrac{1}{2}(f_2 - f_1)$$

Equation (8.58) is used to get the bandpass (modulated) signal spectrum.

The preceding approach to determining bandwidth occupancy of digitally modulated signals provides one criterion for selecting modulation schemes based on bandwidth considerations. It is not the only approach by any means. Another important criterion is adjacent channel interference. In other words, what is the degradation imposed on a given modulation scheme by channels adjacent to the channel of interest? In general, this is a difficult problem. For one approach, the reader is referred to a series of papers on the concept of crosstalk.*

8.3 SYNCHRONIZATION

We have seen that at least two levels of synchronization are necessary in a coherent communication system. For the known-signal-shape receiver considered in Section 7.2, the beginning and ending times of the signals had to be known. When specialized to the case of ASK, PSK, or FSK, knowledge was needed not only of the bit timing but of carrier phase as well. In addition, if the bits are grouped into words, the starting and ending times of the words are also required. In this section we will look at methods for achieving synchronization at these three levels. In order of consideration, we will look at methods for (1) carrier synchronization, (2) bit synchronization, and (3) word synchronization. There are also other levels of synchronization in some communication systems that will not be considered here.

Carrier Synchronization

The three main types of digital modulation methods considered were ASK, PSK, and FSK. Typical power spectra for ASK, FSK, and PSK, assuming a random NRZ data sequence, are illustrated in Figure 8.20. In each case, impulses are shown at $f = \pm f_c$ Hz corresponding to the carrier. For PSK, the impulses are dashed, signifying that they are absent for BPSK. Assuming the presence of a carrier component in the modulated signal spectrum, car-

---

* See I. Kalet, "A Look at Crosstalk in Quadrature-Carrier Modulation Systems," *IEEE Transactions on Communications*, Vol. COM-25, September 1977, pp. 884–892.

FIGURE 8.20   Power spectra. (a) ASK. (b) PSK. (c) FSK.

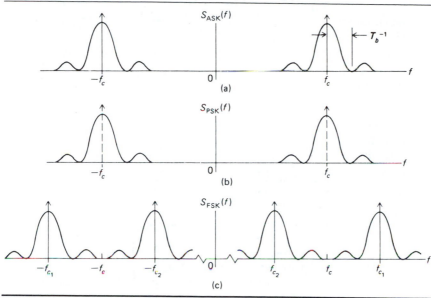

rier coherence is very simply obtained by locking onto the carrier component with a phase-locked loop. To minimize the adverse effects of
modulation power on the loop, a split-phase modulation format is typically
used. (See Problem 8.24 for derivations of these spectra.)

If a carrier component is not present, as in BPSK, two alternatives that
may be employed are squaring and Costas loops, as discussed in Chapter 3
for double-sideband modulation.* When used for digital data demodulation,
however, these loop mechanizations introduce a problem that was not present for demodulation of analog message signals. We note that either loop
will lock if we assume $\pm d(t) \cos \omega_c t$ at the loop input. Some method is
usually required to resolve this sign ambiguity at the demodulator output.
One method of doing so is to differentially encode the data stream before
modulation and differentially decode it at the detector output with a resultant small loss in signal-to-noise ratio. This is referred to as *coherent detection
of differentially encoded BPSK* and is different from differentially coherent
detection of BPSK.

Circuits similar to the Costas and squaring loops may be constructed for
four-phase PSK or QPSK.

---

* The decision-feedback loop is yet another alternative. See Lindsey and Simon (1973).

TABLE 8.5 Tracking Loop Error Variances

No modulation (PLL)	$\sigma_\phi^2 = \dfrac{N_0 B_L}{P_c}$
PSK (squaring or Costas loop)	$\sigma_\phi^2 = \dfrac{1}{L}\left(\dfrac{1}{z} + \dfrac{1}{2z^2}\right); \ z = E_s/N_0; \ L = \dfrac{1}{B_L T_s} \propto \dfrac{B_s}{B_L} = \dfrac{T_L}{T_s}$
QPSK (quadrupling or data estimation loop)	$\sigma_\phi^2 = \dfrac{1}{L}\left(\dfrac{1}{z} + \dfrac{9}{2}\dfrac{1}{z^2} + \dfrac{6}{z^3} + \dfrac{3}{2}\dfrac{1}{z^4}\right)$

Definitions: T_s = symbol duration
B_L = single-sided loop bandwidth
N_0 = single-sided noise spectral density
$T_L = B_L^{-1}$ = loop memory time
L = effective number of symbols used by loop in making phase estimate
P_c = signal power (carrier component only)
$B_s = T_s^{-1}$ = symbol bandwidth (bit-rate bandwidth for binary case)
E_s = symbol energy

The question naturally arises as to the effect of noise on these phase-tracking devices. The *phase error,* that is, the difference between the input signal phase and the VCO phase, can be shown to be approximately Gaussian with zero mean at high signal-to-noise ratios at the loop input. Table 8.5 summarizes the phase-error variance for these various cases.[*] When used with equations such as (7.79), these results provide a measure for the average performance degradation due to an imperfect phase reference. Note that in all cases, σ_ϕ^2 is inversely proportional to the signal-to-noise ratio raised to integer powers and to the effective number L of symbols remembered by the loop in making the phase estimate. (See Problem 8.28.)

We next look at methods for acquiring bit synchronization.

Bit Synchronization[†]

Three general methods by which bit synchronization can be obtained are (1) derivation from a primary or secondary standard (for example, transmitter and receiver slaved to a master timing source), (2) utilization of a separate synchronization signal (pilot clock), and (3) derivation from the modulation itself, referred to as *self-synchronization,* as illustrated in Figure 8.21. Because

---

[*] Stiffler (1971), Equation (8.3.13).
[†] See Stiffler (1971) or Lindsey and Simon (1973) for a more extensive discussion.

FIGURE 8.21   Block diagram of a system for deriving a clock that is coherent with a random-bit stream

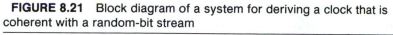

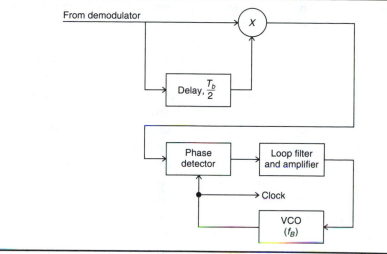

of the nonlinear operation provided by the delay-and-multiply operation (almost any even-order nonlinearity will do), energy is produced at twice the bit frequency, which permits the acquisition of a clock by locking a phase-locked loop to this frequency component and dividing it by two. This loop configuration is analogous to the squaring loop for carrier acquisition from a BPSK waveform. Loop configurations for acquiring bit synchronization that are similar in form to the Costas loop are also possible but will not be discussed here.*

The analysis of bit synchronizer performance in noise proceeds similarly to the analysis of carrier acquisition devices, and expressions may be derived for the timing-error variance similar to the expressions given in Table 8.5 for phase-error variance.†

Word Synchronization

The same principles used for bit synchronization may be applied to word synchronization. These are (1) derivation from a primary or secondary standard, (2) utilization of a separate synchronization signal, and (3) self-synchronization.

Only the second method will be discussed here. The third method involves the utilization of self-synchronizing codes. It is clear that such

---

* Again, see Stiffler (1971) or Lindsey and Simon (1973).
† See Stiffler (1971) or Lindsey and Simon (1973).

codes, which consist of sequences of 1's and 0's in the binary case, must be such that no shift of an arbitrary sequence of code words produces another code word. If this is the case, proper alignment of the code words at the receiver is accomplished simply by comparing all possible time shifts of a received digital sequence with all code words in the code dictionary (assumed available at the receiver) and choosing the shift and code word having maximum correlation. For long code words, this could be very time-consuming. Furthermore, the construction of good codes is not a simple task and requires computer search procedures in some cases.*

When a separate synchronization signal is employed, this signal may be transmitted over a channel separate from the one being employed for data transmission, or over the data channel by inserting the synchronization signal periodically between data words. If the first alternative is employed, the synchronizing waveform should have small cross correlation with its own cyclic permutations. The second alternative is often accomplished by inserting a known sequence, referred to as a *prefix*, in front of each data word. In this case, the synchronization waveform employed should have small aperiodic cross correlation with its cyclic permutations. Waveforms having such properties will be discussed shortly. In both the case of a separate channel and the case of employment of a prefix, it should be noted that power is wasted in achieving synchronization. Thus the self-synchronization approach is preferable in cases in which power is at a premium. For example, self-synchronizing codes were used in the deep space missions by NASA.

To see how the second method can be implemented, we digress briefly to discuss pseudo-noise sequences.

Pseudo-Noise (PN) Sequences

Pseudo-noise (PN) codes are binary-valued, noiselike sequences; they approximate a sequence of coin tossings for which a 1 represents a head and a 0 represents a tail. However, their primary advantages are that they are deterministic, being easily generated by feedback shift registers, and they have a correlation function that is highly peaked for zero delay and approximately zero for other delays. Thus they find application wherever waveforms at remote locations must be synchronized. These applications include not only word synchronization but also the determination of range between two points and the measurement of the impulse response of a system by cross-correlation of input with output, as discussed in Chapter 5 (Example 5.7).

Figure 8.22 illustrates the generation of a PN code of length $2^3 - 1 =$

---

* See Stiffler (1971) or Lindsey and Simon (1973).

FIGURE 8.22   Generation of a 7-bit PN sequence. (a) Generation. (b) Shift register contents.

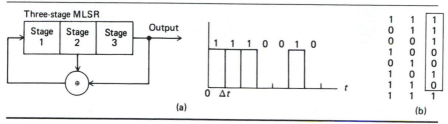

(a)

(b)

7, which is accomplished with the use of a shift register three stages in length. After each shift of the contents of the shift register to the right, the contents of the second and third stages are used to produce an input to the first stage through an EXCLUSIVE-OR (XOR) operation (that is, a binary add without carry). The logical operation performed by the XOR circuit is given in Table 8.6. Thus, if the initial contents of the shift register are 1 1 1, as shown in the first row of Figure 8.22(b), the contents for seven more successive shifts are given by the remaining rows of this table. Therefore, the shift register again returns to the 1 1 1 state after $2^3 - 1 = 7$ more shifts, which is also the length of the output sequence taken at the third stage. By using an n-stage shift register with proper feedback connections, PN sequences of length $2^n - 1$ may be obtained. Proper feedback connections for several values of n are given in Table 8.7.

Considering next the autocorrelation function of the periodic waveform (normalized to a peak value of unity) obtained by letting the shift register in Figure 8.22(a) run indefinitely, we see that its values for integer multiples of the output pulse width $\Delta = n\Delta t$ are given by

$$R(\Delta) = \frac{N_A - N_U}{\text{sequence length}} \tag{8.69}$$

where N_A is the number of like digits of the sequence and a sequence shifted by n pulses and N_U is the number of unlike digits of the sequence and a sequence shifted by n pulses. This equation is a direct result of the definition of the autocorrelation function for a periodic waveform, given by Equation

TABLE 8.6   Truth Table for the XOR Operation

Input 1	Input 2	Output
1	1	0
1	0	1
0	1	1
0	0	0

TABLE 8.7 Feedback Connections for Generation of PN Codes[a]

n	Sequence Length	Sequence (Initial State: All Ones)		Feedback Digit
2	3	110		$x_1 \oplus x_2$
3	7	11100	10	$x_2 \oplus x_3$
4	15	11110	00100	$x_3 \oplus x_4$
		11010		
5	31	11111	00110	$x_2 \oplus x_5$
		10010	00010	
		10111	01100 0	

[a] See R. E. Ziemer and R. L. Peterson (1992), Chapter 8, for additional sequences and proper feedback connections.

(2.150), and the binary-valued nature of the shift register output if the peak value is normalized to unity. Thus the autocorrelation function for the sequence generated by the feedback shift register of Figure 8.22(a) is as shown in Figure 8.23, as you may readily verify. Applying the definition of the autocorrelation function, given by Equation (2.150), we could also easily show that the shape for noninteger values of delay is as shown in Figure 8.23. In general, for a sequence of length N, the minimum correlation is $-1/N$. Because the correlation function of a PN sequence consists of a narrow triangle around zero delay and is essentially zero otherwise, it resembles white noise when used to drive any system whose bandwidth is small compared with the inverse pulse width. This explains the reason for the name "pseudo-noise."

Returning to our original problem, that of synchronization of waveforms

FIGURE 8.23 Correlation function of a PN code

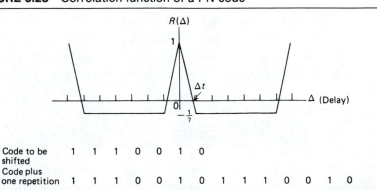

FIGURE 8.24   Synchronization by PN code. (a) PN transmitter-receiver
portion for synchronization. (b) Error signal at VCO.

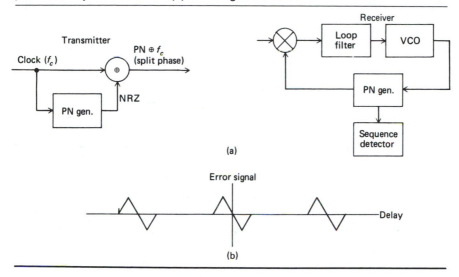

(a)

(b)

at remotely located points, consider the system illustrated in Figure 8.24(a).
Assume that the transmitter transmits a PN sequence (with levels $+1$ and
-1 instead of 1 and 0) that has been split-phase encoded. We recall that
split-phase encoding can be thought of as multiplication of the sequence by
a squarewave clock with period equal to the pulse width of the PN sequence.
For simplicity, the operations of modulation and coherent demodulation,
which would be performed in most systems, are not shown, and we view the
system as operating at baseband. Initially, suppose that the transmitter and
receiver clocks are unsynchronized but operating near the same frequency.
Thus the PN sequence generated at the receiver will slowly slide by the
received split-phase-encoded PN sequence. If the received sequence were
not split-phase encoded, the output of the loop filter would be essentially
the autocorrelation function of the sequence with the delay proportional to
time because of the sequences sliding by each other. However, because of
the multiplication of the transmitted sequence by the clock, an error signal
at the loop filter output, as shown in Figure 8.24(b), will result. Eventually,
the input to the voltage-controlled oscillator (VCO) will be the negative-
slope portion of this error signal, and the frequency of the VCO will be
driven in the proper direction so that the loop will lock at zero delay (with
the delay due to the transmission path excluded). At this point, the VCO
output will be locked to the transmitter clock, and the PN sequences at
transmitter and receiver will be time-coincident if the channel delay is

TABLE 8.8 The Barker Sequences

$+-$
$++-$
$++-+$
$+++-+$
$+++--+-$
$+++---+--+-$
$+++++--++-+-+$

excluded. More general schemes exist for the synchronization of PN sequences.*

It is not difficult to see how the system just described could be used for measuring the range between two points if the transmitter and receiver were colocated and a transponder at a remote location simply retransmitted whatever it received. This is, in fact, the technique used for the Global Positioning System (GPS).

To use this system for word synchronization, all that would be necessary would be to insert the split-phase-encoded PN sequence at the beginning of each word. The PN sequence generated at the receiver would, on average, have low correlation with the data. It is clear that if the PN sequence is inserted as a prefix before each data word, it is not the periodic correlation function that is important, but the *aperiodic* correlation function obtained by sliding the sequence past itself rather than past its periodic extension as in Figure 8.23. Sequences with good aperiodic correlation properties are the Barker codes.† Unfortunately the longest known Barker code is of length 13. Table 8.8 lists all known Barker sequences. (See Problem 8.31.) Other digital sequences with good correlation properties can be constructed from Hadamard matrices.‡

8.4 SPREAD-SPECTRUM COMMUNICATION SYSTEMS

We next consider a special class of modulation technique referred to as *spread-spectrum modulation*. In general, spread-spectrum modulation refers to any modulation technique in which the bandwidth of the modulated signal

---

* For techniques that will work with data present, see Ziemer and Peterson (1992), Chapter 8.

† See Skolnik (1970), Chapter 20.

‡ See Lindsey and Simon (1973). Also, see *IEEE Transactions on Communications,* August 1980 (special issue on synchronization).

is spread well beyond the bandwidth of the modulating signal, independently of the modulating signal bandwidth. The following are reasons for employing spread-spectrum modulation:

1. To provide resistance to intentional jamming by another source
2. To provide a means for masking the transmitted signal in the background noise and prevent another party from eavesdropping
3. To provide resistance to the degrading effects of multipath transmission
4. To provide a means for more than one user to use the same transmission channel
5. To provide range-measuring capability

The two most common techniques for effecting spread-spectrum modulation are referred to as *direct sequence* (DS) and *frequency hopping* (FH). Figures 8.25 and 8.26 are block diagrams of these generic systems. Variations and combinations of these two basic systems are also employed.

FIGURE 8.25   Block diagram of a DS spread-spectrum communication system. (a) Transmitter. (b) Receiver.

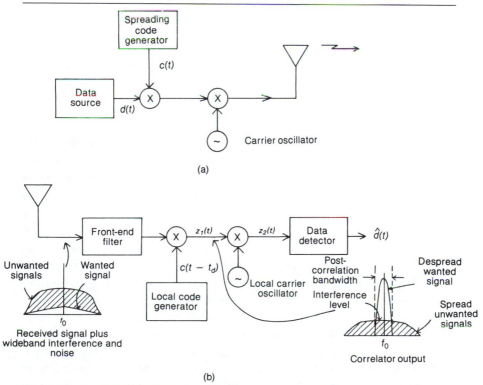

FIGURE 8.26   Block diagram of an FH spread-spectrum communication system. (a) Transmitter. (b) Receiver.

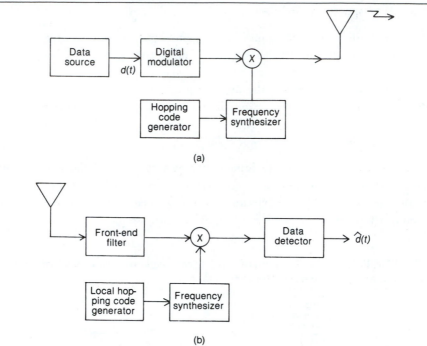

(a)

(b)

Direct-Sequence Spread Spectrum

In a *direct-sequence spread-spectrum* (DSSS) communication system, the modulation format may be almost any of the coherent digital techniques discussed previously, although BPSK, QPSK, and MSK are the most common. Figure 8.25 illustrates the use of BPSK. The spectrum spreading is effected by multiplying the data $d(t)$ by the spreading code $c(t)$. In this case, both are assumed to be binary sequences taking on the values $+1$ and -1. The duration of a data symbol is T_b, and the duration of a spreading-code symbol, called a *chip period*, is T_c. There are usually many chips per bit, so that $T_c \ll T_b$. In this case, it follows that the spectrum of the modulated signal is essentially dependent only on the inverse chip period. The spreading code is chosen to have the properties of a random binary sequence; an often-used choice for $c(t)$ is a PN sequence, as described in the previous section. Often, however, a sequence generated using nonlinear generation techniques is used for security reasons. It is also advantageous, from the standpoint of security, to use the same clock for both the data and spreading code so that the data changes sign coincident with a sign change for the spreading code. This is not necessary for proper operation of the system, however.

In the system illustrated in Figure 8.25, BPSK data modulation is used. Typical spectra are shown directly below the corresponding blocks. At the receiver, it is assumed that a replica of the spreading code is available and is time-synchronized with the incoming code used to multiply the BPSK-modulated carrier. This synchronization procedure is composed of two steps, called *acquisition* and *tracking*. A very brief discussion of methods for acquisition will be given later. For a fuller discussion and analyses of both procedures, the student is referred to Peterson, Ziemer, and Borth (1995).

A rough approximation to the spectrum of a DSSS signal employing BPSK data modulation can be obtained by representing the modulated carrier as

$$x_c(t) = A \ d(t)c(t) \ \cos{(\omega_c t + \theta)} \tag{8.70}$$

where it is assumed that θ is a random phase uniformly distributed in $(0, 2\pi)$ and $d(t)$ and $c(t)$ are independent, random binary sequences (if derived from a common clock, the independence assumption for $d(t)$ and $c(t)$ is not strictly valid). With these assumptions, the autocorrelation function for $x_c(t)$ is

$$R_{x_c}(\tau) = \frac{A^2}{2} R_d(\tau)R_c(\tau) \ \cos{(\omega_c \tau)} \tag{8.71}$$

where $R_d(\tau)$ and $R_c(\tau)$ are the autocorrelation functions of the data and spreading code, respectively. If they are modeled as random "coin toss" sequences as considered in Example 5.6, their autocorrelation functions are given by

$$R_d(\tau) = \Lambda(\tau/T_b) \tag{8.72a}$$

and

$$R_c(\tau) = \Lambda(\tau/T_c) \tag{8.72b}$$

respectively. Their corresponding power spectral densities are

$$S_d(f) = T_b \ \text{sinc}^2 \ (T_b f) \tag{8.73}$$

and

$$S_c(f) = T_c \ \text{sinc}^2 \ (T_c f) \tag{8.74}$$

respectively, where the width of the main lobe of (8.73) is T_b^{-1} and that for (8.74) is T_c^{-1}.

The power spectral density of $x_c(t)$ can be obtained by taking the Fourier transform of (8.71):

$$S_{x_c}(f) = \frac{A^2}{2} S_d(f) * S_c(f) * \mathscr{F} [\cos (\omega_c \tau)] \tag{8.75}$$

Since the spectral width of $S_d(f)$ is much less than that for $S_c(f)$, the convolution of these two spectra is approximately $S_c(f)$. Thus the spectrum of the DSSS modulated signal is very closely approximated by

$$S_{x_c}(f) = \frac{A^2}{4} [S_c(f - f_c) + S_c(f + f_c)]$$

$$= \frac{A^2 T_c}{4} \{\text{sinc}^2 [T_c(f - f_c)] + \text{sinc}^2 [T_c(f + f_c)]\} \tag{8.76}$$

The spectrum, as stated above, is approximately independent of the data spectrum and has a null-to-null bandwidth around the carrier of $2/T_c$ Hz.

We next look at the error probability performance. First, assume a DSSS signal plus additive white Gaussian noise is present at the receiver. Ignoring propagation delays, the output of the local code multiplier at the receiver (see Fig. 8.25) is

$$z_1(t) = A \, d(t)c(t)c(t - \Delta) \cos (\omega_c t + \theta) + n(t)c(t - \Delta) \tag{8.77}$$

where Δ is the misalignment of the locally generated code at the receiver with the code on the received signal. Assuming perfect code synchronization ($\Delta = 0$), the output of the coherent demodulator is

$$z_2(t) = A \, d(t) + n'(t) + \text{double frequency terms} \tag{8.78}$$

where

$$n'(t) = 2n(t)c(t) \cos (\omega_c t + \theta) \tag{8.79}$$

is a new Gaussian random process with zero mean. Passing $z_2(t)$ through an integrate-and-dump circuit, we have for the signal component at the output

$$V_0 = \pm AT_b \tag{8.80}$$

where the sign depends on the sign of the bit at the input. The noise component at the integrator output is

$$N_g = \int_0^{T_b} 2n(t)c(t) \cos (\omega_c t + \theta) \, dt \tag{8.81}$$

Since $n(t)$ has zero mean, N_g has zero mean. Its variance, which is the same as its second moment, can be found by squaring the integral, writing it as an iterated integral, and taking the expectation inside the double integral—a procedure that has been used several times before in this chapter and the previous one. The result is

$$\text{var}\,(N_g) \;=\; E(N_g^2) \;=\; N_0 T_b \tag{8.82}$$

where N_0 is the single-sided power spectral density of the input noise. This, together with the signal component of the integrator output, allows us to write down an expression similar to the one obtained for the baseband receiver analysis carried out in Section 7.1 (the only difference is that the signal power is $A^2/2$ here, whereas it was A^2 for the baseband signal considered there). The result for the probability of error is

$$P_E \;=\; Q\,(\sqrt{A^2 T_b/N_0}) \;=\; Q\,(\sqrt{2E_b/N_0}) \tag{8.83}$$

With Gaussian noise alone, DSSS performs the same as BPSK without the spread-spectrum modulation.

Consider next a cw interference component. Now, the input to the integrate-and-dump detector, excluding double frequencies, is

$$z_2'(t) = A\,d(t) + n'(t) + A_t\,c(t)\,\cos\,[(\omega_c + \Delta\omega)t + \phi][2\,\cos\,(\omega_c t + \theta)] \tag{8.84}$$

where A_I is the amplitude of the interference component and $\Delta\omega$ is its offset frequency from the carrier frequency. It is assumed that $\Delta\omega < 2\pi/T_c$. The output of the integrate-and-dump detector is

$$V_0' \;=\; \pm A T_b + N_g + N_I \tag{8.85}$$

The first two terms are the same as obtained before. The last term is the result of interference and is given by

$$\begin{aligned} N_I &= \int_0^{T_b} c(t)A_I\,\cos\,[(\omega_c + \Delta\omega(t + \phi][2\,\cos\,(\omega_c t + \theta)]\,dt \\[4pt] &= \int_0^{T_b} A_I c(t)\,\cos\,(\Delta\omega t + \theta - \phi)\,dt \end{aligned} \tag{8.86}$$

Because of the multiplication by the wideband spreading code $c(t)$ and the subsequent integration, we approximate this term by an equivalent Gaussian random variable (the integral is a sum of a large number of random variables). Its mean is zero, and for $\Delta\omega \ll 2\pi/T_c$, its variance can be shown to be

$$\text{var}\,(N_I) \;=\; \frac{T_c T_b A_I^{\,2}}{2} \tag{8.87}$$

With this Gaussian approximation for N_I, the probability of error can be shown to be

$$P_E = Q\left(\sqrt{\frac{A^2 T_b^2}{\sigma_T^2}}\right) \tag{8.88}$$

where

$$\sigma_T^2 = N_0 T_b + \frac{T_c T_b A_I^2}{2} \tag{8.89}$$

is the total variance of the noise plus interference components at the integrator output. The quantity under the square root can be further manipulated as

$$\frac{A^2 T_b^2}{2\sigma_T^2} = \frac{A^2/2}{N_0/T_b + (T_c/T_b)(A_I^2/2)}$$

$$= \frac{P_s}{P_n + P_I/G_p} \tag{8.90a}$$

where

$P_s = A^2/2$ is the signal power at the input

$P_n = N_0/T_b$ is the Gaussian noise power in the bit-rate bandwidth

$P_I = A_I^2/2$ is the power of the interfering component at the input

$G_p = T_b/T_c$ is the processing gain of the DSSS system.

It is seen that the effects of the interfering component is decreased by the processing gain G_p. Equation (8.90a) can be rearranged as

$$\frac{A^2 T_b^2}{2\sigma_T^2} = \frac{\text{SNR}}{1 + \dfrac{(\text{SNR})(\text{JSR})}{G_p}} \tag{8.90b}$$

where

$$\text{SNR} = P_s/P_n \text{ is the signal-to-noise ratio}$$

$$\text{JSR} = P_I/P_s \text{ is the jamming-to-signal ratio}$$

Figure 8.27 shows P_E versus the SNR for several values of JSR/G_p.

FIGURE 8.27 P_E versus SNR for DSSS with $G_p = 30$ dB

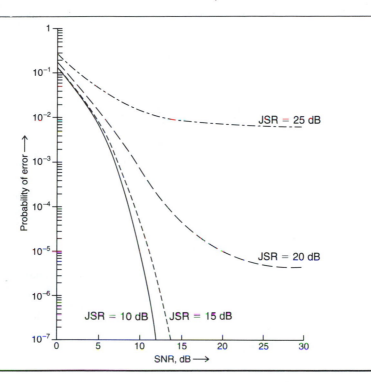

Frequency-Hop Spread Spectrum

In the case of *frequency-hop spread spectrum* (FHSS), the modulated signal is hopped in a pseudorandom fashion among a set of frequencies so that a potential eavesdropper does not know in what band to listen or jam. Consequently, the eavesdropper must listen or jam in the full bandwidth in which the signal is hopped. Current FHSS systems may be classified as *fast hop* or *slow hop*, depending on whether one (or less) or several data bits are included in a hop. The data modulator for either is usually a noncoherent type such as FSK or DPSK, since frequency synthesizers are not usually coherent from hop to hop. Even if one goes to the expense of building a coherent frequency synthesizer, the channel may not preserve the coherency property of the synthesizer output. At the receiver, as shown in Figure 8.26, a replica of the hopping code is produced and synchronized with the hopping pattern of the received signal. One possible circuit for achieving this synchronization process is shown later.

FIGURE 8.28   Code acquisition circuits for (a) DSSS and (b) FHSS using serial search

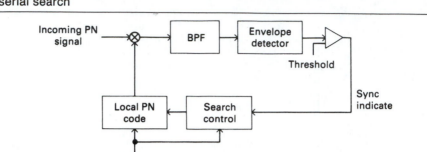

(a)

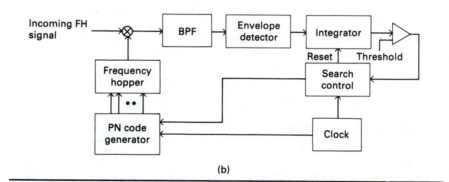

(b)

Code Synchronization

Only a brief discussion of code synchronization will be given here. For detailed discussions and analyses of such systems, the reader is referred to Ziemer and Petersen (1985).*

Figure 8.28(a) shows a serial-search acquisition circuit for DSSS. A replica of the spreading code is generated at the receiver and multiplied by the incoming spread-spectrum signal (the carrier is assumed absent in Figure 8.28 for simplicity). Of course, the code epoch is unknown, so an arbitrary local code delay relative to the incoming code is tried. If it is within $\pm\frac{1}{2}$

---

* For an excellent tutorial paper on acquisition and tracking, see S. S. Rappaport and D. M. Grieco, "Spread-Spectrum Signal Acquisition: Methods and Technology," *IEEE Communications Magazine*, Vol. 22, No. 6, June 1984, pp. 6–21.

chip of the correct code epoch, the output of the multiplier will be despread data and its spectrum will pass through the bandpass filter whose bandwidth is of the order of the data bandwidth. If the code delay is not correct, the output of the multiplier remains spread and little power passes through the bandpass filter. The envelope of the bandpass filter output is compared with a threshold—a value below threshold denotes an unspread condition at the multiplier output and, hence, a delay that does not match the delay of the spreading code at the receiver input, while a value above threshold indicates that the codes are approximately aligned. If the latter condition holds, the synchronization indicator stops the code search and a tracking mode is entered. If the below-threshold condition holds, the codes are assumed to be not aligned, so the search control steps to the next code delay (usually a half chip) and the process is repeated. It is apparent that such a process can take a relatively long time to achieve lock. The mean time to acquisition is given by

$$T_{acq} = \frac{nN_cT_e}{2(1 - P_{FH})} \frac{2 - P_H}{P_H} \tag{8.91}$$

where

T_e = examination time for each cell

n = number of cells examined per chip

N_c = initial number of chips to be searched

P_{FH} = probability of false threshold crossing at the end of an examination interval

P_H = probability of threshold crossing when correct alignment exists

Other techniques are available that speed up the acquisition, but at the expense of more hardware or special code structures.

A synchronization scheme for FHSS is shown in Figure 8.28(b). The discussion of its operation would be similar to that for acquisition in DSSS except that the correct frequency pattern for despreading is sought.

Conclusion

From the preceding discussions and the block diagrams of the DS and FH spread-spectrum systems, you can see that *nothing is gained by using a spread-spectrum system in terms of performance in an additive white Gaussian noise channel.* Indeed, using such a system may result in more degradation than using a conventional system, owing to the additional operations required. The advantages of spread-spectrum systems accrue in environments that are hostile to digital communications—environments such as those in which multipath transmission or intentional jamming of channels exist. In addition, since the signal power is spread over a much wider bandwidth than it is in an ordinary system, it follows that the average power density of the transmitted spread-spectrum signal is much lower than the power density when the spectrum is *not* spread. This lower power density gives the sender of the signal a chance to mask the transmitted signal by the background noise and thereby lower the probability that anyone may intercept the signal.

One last point is perhaps worth making: It is knowledge of the structure of the signal that allows the intended receiver to pull the received signal out of the noise. The use of correlation techniques is indeed powerful.

8.5 SATELLITE COMMUNICATIONS

In this section we look at the special application area of satellite communications to illustrate the use of some of the error probability results derived in this chapter. Satellite communications were first conceived in the 1950s. The first communications satellite was Echo I, a passive reflecting sphere, which was launched in 1956. The first active satellite, Courier, where "active" refers to the satellite's ability to receive, amplify, and retransmit signals, was launched in 1960. It had only two transmitters and had a launch weight of only 500 pounds. (Score was launched in 1958, but transmitted a prerecorded message.) In contrast, Intelsat VI, launched in 1986, has 77 transmitters and weighs 3600 pounds. Figure 8.29(a) shows a typical satellite repeater link, and Figure 8.29(b) shows a frequency-translating "bent-pipe" satellite communications system. Frequency translation is necessary to separate the receive and transmit frequencies and thus prevent "ring-around." Another type of satellite communication system, known as a *demod/remod system* (also referred to as on board processing, or OBP), is shown in Figure 8.29(c). In such a satellite repeater, the data are actually demodulated and subsequently remodulated onto the downlink carrier. In addition to the relay communications system on board the satellite, other communications systems include ranging (to provide a range measurement to the satellite), command (to receive commands from an earth station to control the satellite), and telemetry (to relay data about the satellite's condition back to the earth).

Early satellite transmissions took place in the UHF, C, or X bands. Because of the subsequent crowding of these bands, additional frequency allocations were added at K, V, and Q bands. Services are classified as *fixed-point* (communications between a satellite and fixed ground station), *broadcast* (trans-

FIGURE 8.29   Various satellite relay link communications configurations. (a) Satellite repeater link. (b) Frequency-translation satellite communications relay. (c) Demod/remod satellite communications relay.

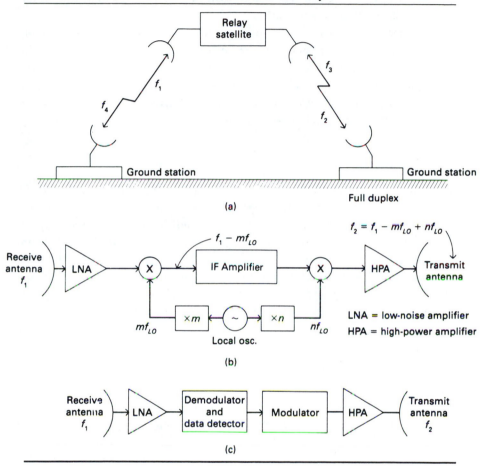

(a)

(b)

(c)

mission to many receivers), and *mobile* (e.g., communications to aircraft, ships, and land vehicles). *Intersatellite* refers to communications between satellites.

It is important that satellites be stabilized so that the antennas can be pointed to predetermined points on the earth's surface. Early satellites were *spin-stabilized,* which means that the satellites were physically spun about an axis that kept them oriented in a particular relationship to the earth as a result of the gyroscopic effect. Because of the difficulty in despinning ever-more-complicated antenna systems, present-day satellites are almost all *three-axis stabilized.* This means that a three-axis gyroscope system is on board to sense deviations from the desired orientation, and the resulting control signals derived from them are used to turn thruster jets on and off in order to maintain the desired orientation.

Satellites can be in low-earth orbits (LEO), medium-earth orbits (MEO), geostationary orbits, or interplanetary orbits. A geostationary orbit is one such that the satellite is at an altitude over the equator so that its angular rotation rate exactly matches that of the earth's, and it therefore appears to be stationary with respect to the earth. Geostationary altitude is 35,784 km or 22,235 statute miles (1 mile is approximately 1.6 km).

Antenna Coverage

Coverage of the earth by an antenna mounted on a satellite can be hemispherical, continental, or zonal depending on the antenna design. Antenna designs are now possible that cover several zones or spots simultaneously on the earth's surface. Such designs allow *frequency reuse,* in that the same band of frequencies can be reused in separate beams, which effectively multiplies the bandwidth of the satellite transponder available for communications by the reuse factor. Figure 8.30 shows a typical antenna gain pattern in polar coordinates. The maximum gain can be roughly calculated from

$$G_0 = \rho_a \left(\frac{4\pi}{\lambda^2} \right) A \tag{8.92a}$$

FIGURE 8.30   Polar representation of a general antenna gain function

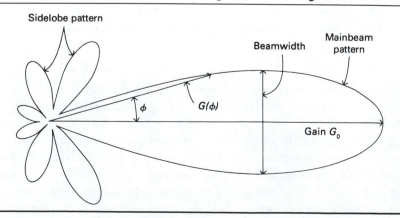

where

$$\rho_a = \text{antenna efficiency } (\leq 1)$$
$$\lambda = \text{wavelength}$$
$$A = \text{aperture area}$$

For a circular aperture of diameter d, this expression becomes

$$G_0 = \rho_a \left(\frac{\pi d}{\lambda}\right)^2 \tag{8.92b}$$

The half-power beamwidth in radians can be approximated as

$$\phi_{3\text{dB}} = \frac{\lambda}{d \sqrt{\rho_a}} \tag{8.93}$$

A convenient approximation for the antenna pattern of a parabolic reflector antenna for small angles off boresight (such that the gain is within 6 dB of the maximum value) is

$$G(\phi) = \rho_a \left(\frac{\pi d}{\lambda}\right)^2 \exp\left[-2.76(\phi/\phi_{3\text{dB}})^2\right] \tag{8.94}$$

▼ **EXAMPLE 8.4** Find the aperture diameter and maximum gain for a transmit frequency of 10 GHz and $\rho_a = 0.8$ if, from geosynchronous altitude, the following coverages are desired: (a) hemispherical, (b) continental United States (CONUS), and (c) a 150-mile-diameter spot.

SOLUTION The wavelength at 10 GHz is

$$\lambda = \frac{3 \times 10^8 \text{ m/s}}{10 \times 10^9 \text{ Hz}} = 0.03 \text{ m}$$

$$= \frac{0.03 \text{ m}}{0.3048 \text{ m/ft}} = 0.0984 \text{ ft}$$

(a) Geosynchronous altitude is 22,235 statute miles, and the earth's radius is 3963 miles. The angle subtended by the earth from geosynchronous altitude is

$$\phi_{\text{hemis}} = \frac{2(3963)}{22,235} = 0.356 \text{ rad}$$

Equating this to ϕ_{3dB} in (8.93) and solving for d, we have

$$d = \frac{0.0984}{(0.356)\sqrt{0.8}} = 0.31 \text{ ft}$$

(b) The angle subtended by CONUS from geosynchronous altitude is

$$\phi_{CONUS} = \frac{4000}{22,235} = 0.18 \text{ rad}$$

Thus,

$$d = \frac{0.0984}{(0.18)\sqrt{0.8}} = 0.61 \text{ ft}$$

(c) A 150-mile-diameter spot on the earth's surface directly below the satellite subtends an angle of

$$\phi_{150} = \frac{150}{22,235} = 0.0067 \text{ rad}$$

from geosynchronous orbit. The diameter of an antenna with this beamwidth is

$$d = \frac{0.0984}{0.0067\sqrt{0.8}} = 16.3 \text{ ft}$$

▲ Note that doubling the frequency to 20 GHz would halve these diameters.

Earth Stations and Transmission Methods

Figure 8.31 shows a block diagram of the transmitter and receiving end of an earth station. Signals from several sources enter the earth station (e.g., telephone, television, etc.), whereupon two transmission options are available. First, the information from a single source can be placed on a single carrier. This is referred to as *single-channel-per-carrier* (SCPC). Second, information from several sources can be multiplexed together and placed onto contiguous intermediate frequency carriers, the sum translated to radio frequency, power amplified, and transmitted. At the receiving end, the reverse process takes place.

At this point, it is useful to draw a distinction between *multiplexing* and *multiple access*. Multiple access (MA), like multiplexing, involves sharing of a common communications resource between several users. However, whereas multiplexing involves a fixed assignment of this resource at a local level, MA involves the remote sharing of a resource, and this sharing may under certain

FIGURE 8.31   Satellite ground station receiver/transmitter configuration.
The command transmitter and telemetry receivers are not shown.

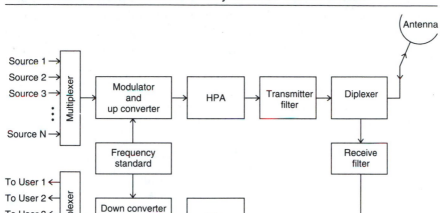

circumstances change dynamically under the control of a system controller.
There are three main techniques in use for utilizing the bandwidth resources
of a remote resource, such as a relay satellite. These are

1. Frequency division multiple access (FDMA), wherein the resource is
   divided up in frequency
2. Time division multiple access (TDMA), wherein the resource is divided
   up in time
3. Code division multiple access (CDMA), wherein a unique code is assigned
   to each intended user and the separate transmissions are separated by
   correlation with the code of the desired transmitting party

Figure 8.32 illustrates these three accessing schemes. In FDMA, signals
from various users are stacked up in frequency, just as for frequency division
multiplexing, as shown in Figure 8.32(a). Guard bands are maintained
between adjacent signal spectra to minimize crosstalk between channels. If
frequency slots are assigned permanently to the users, the system is referred
to as *fixed-assigned multiple-access* (FAMA). If some type of dynamic allocation
scheme is used to assign frequency slots, it is referred to as a *demand-assigned
multiple-access* (DAMA) system.

   In TDMA, the messages from various users are interlaced in time, just as
for TDM, as shown in Figure 8.32(b). As illustrated in Figure 8.33, the data

from each user are conveyed in time intervals called *slots*. A number of slots make up a frame. Each slot is made up of a preamble plus information bits. The functions of the preamble are to provide identification and allow synchronization of the slot at the intended receiver. Guard times are utilized between each user's transmission to minimize crosstalk between channels.

FIGURE 8.32 Illustration of multiple-access techniques. (a) FDMA. (b) TDMA. (c) CDMA using frequency-hop modulation (numbers denote hopping sequences for channels 1, 2, and 3).

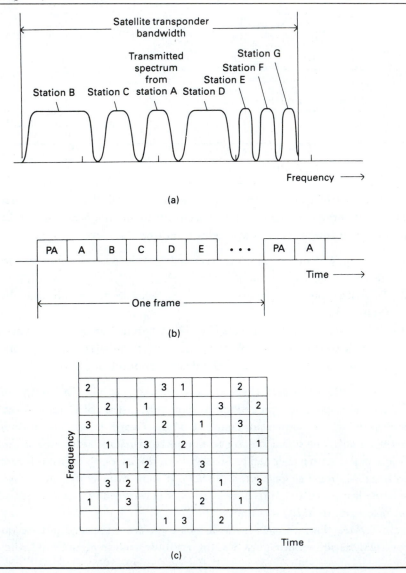

FIGURE 8.33   Details of a TDMA frame format

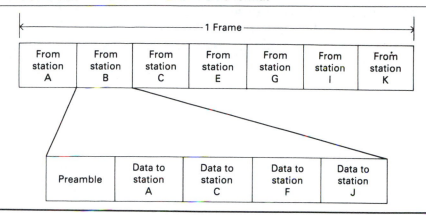

It is necessary to maintain overall network synchronization in TDMA, unlike FDMA. If, in a TDMA system, the time slots that make up each frame are preassigned to specific sources, it is referred to as FAMA; if time slots are not preassigned, but assigned on a dynamic basis, the technique is referred to as DAMA. DAMA schemes require a central network controller and a separate low-information-rate channel between each user and the controller to carry out the assignments. A DAMA TDMA system is more efficient in the face of bursty traffic than a FAMA system.

In CDMA, each user is assigned a code that ideally does not correlate with the codes assigned to other users, and the transmissions of a desired user are separated from those of all other users at a given receiving site through correlation with a locally generated replica of the desired user's code. Two ways that the messages can be modulated with the code for a given user is through direct-sequence spread spectrum or frequency-hop spread spectrum (see Section 8.4). Although CDMA schemes can be operated with network synchronization, it is obviously more difficult to do this than to operate the system asynchronously, and therefore asynchronous operation is the preferred mode. When operated asynchronously, one must account for multiple access noise, which is a manifestation of the partial correlation of a desired user's code with all other users' codes present on the system.

Link Analysis: Bent-Pipe Relay

In Appendix A, a single one-way link budget is considered for a satellite communications system. Consider now the situation depicted in Figure 8.34. A transmitted signal from a ground station is broadcast to a satellite with power P_{us}, where the subscript u stands for "uplink." Noise referred to

FIGURE 8.34 Signal and noise powers in the uplink and downlink portions of a bent-pipe satellite relay system

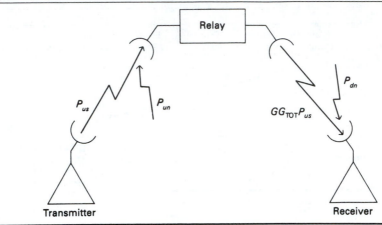

the satellite input has power P_{un}. The sum of the signal and noise is amplified by the satellite repeater to give a transmitted power from the satellite of

$$P_T = G(P_{us} + P_{un}) \tag{8.95}$$

where G is the gain of the satellite repeater. The received signal power from the satellite at the receiving ground station is

$$P_{rs} = GG_{TOT}P_{us} \tag{8.96}$$

where G_{TOT} represents total system losses and gains on the downlink. It can be expressed as

$$G_{TOT} = \frac{G_t G_r}{L_a L_p} \tag{8.97}$$

where

G_t = gain of the satellite transmitter antenna

L_a = atmospheric losses on the downlink

L_p = propagation losses on the downlink

G_r = gain of the ground station receive antenna

The uplink noise power transmitted by the satellite repeater and appearing at the ground station input is

$$P_{ru} = GP_{un}G_{TOT} \tag{8.98}$$

Additional noise generated by the ground station itself is added to this noise at the ground station. The ratio of P_{rs} to total noise is the downlink carrier-to-noise power ratio. It is given by

$$(\text{CNR})_r = \frac{P_{rs}}{P_{ru} + P_{rd}} \tag{8.99}$$

Substituting previously derived expressions for each of the powers appearing on the right side of (8.99), we obtain

$$(\text{CNR})_r = \frac{GG_{\text{TOT}}P_{us}}{GP_{un}G_{\text{TOT}} + P_{dn}}$$

$$= \frac{1}{\dfrac{P_{un}}{P_{us}} + \dfrac{P_{dn}}{GG_{\text{TOT}}P_{us}}} \tag{8.100}$$

$$= \frac{1}{(\text{CNR})_u^{-1} + (\text{CNR})_d^{-1}}$$

where

$$(\text{CNR})_u = \frac{P_{us}}{P_{un}} = \text{carrier-to-noise power ratio on the uplink}$$

$$(\text{CNR})_d = \frac{GG_{\text{TOT}}P_{us}}{P_{dn}} = \text{carrier-to-noise power ratio on the downlink}$$

Note that the weakest of the two CNRs affects the overall carrier-to-noise power ratio the most. The overall carrier-to-noise power ratio cannot be better than the worse of two carrier-to-noise ratios that make it up. To obtain $(\text{CNR})_u$ and $(\text{CNR})_d$, we use the link equations developed in Appendix A.

To relate carrier-to-noise ratio to E_s/N_0 in order to calculate the error probability, we note that

$$\text{CNR} = \frac{P_c}{N_0 B_{RF}} \tag{8.101}$$

where

$$P_c = \text{average carrier power}$$

$$N_0 = \text{noise power spectral density}$$

$$B_{RF} = \text{modulated signal (radio frequency) bandwidth}$$

Multiplying numerator and denominator by the symbol duration T_s, we note that $P_c T_s = E_s$ is the symbol energy and obtain

$$\text{CNR} = \frac{E_s}{N_0 B_{RF} T_s}$$

or, solving for E_s/N_0,

$$\frac{E_s}{N_0} = (\text{CNR}) B_{RF} T_s \tag{8.102}$$

Given a modulation scheme, we can use a suitable bandwidth criterion to determine $B_{RF} T_s$. For example, using the null-to-null bandwidth for BPSK as B_{RF}, we have $B_{RF} = 2/T_b$ or $T_b B_{RF} = 2$, where $T_s = T_b$, since we are considering binary signaling.

Because the CNR is related to E_s/N_0 by the constant $B_{RF} T_s$, we can write (8.100) as

$$\left(\frac{E_s}{N_0}\right)_r = \frac{1}{(E_s/N_0)_u^{-1} + (E_s/N_0)_d^{-1}} \tag{8.103}$$

where

$(E_s/N_0)_u = $ symbol-energy-to-noise-spectral-density ratio on the uplink

$(E_s/N_0)_d = $ symbol-energy-to-noise-spectral-density ratio on the downlink

▼ **EXAMPLE 8.5**   Compute the relationship between $(E_s/N_0)_u$ and $(E_s/N_0)_d$ required to yield an error probability of $P_E = 10^{-6}$ on a bent-pipe satellite relay communications link if BPSK modulation is used.

SOLUTION   For BPSK, $(E_b/N_0)_r \cong 10.54$ dB gives $P_E = 10^{-6}$. Thus (8.103) becomes

$$\frac{1}{(E_b/N_0)_u^{-1} + (E_b/N_0)_d^{-1}} = 10^{1.054} \cong 11.324 \tag{8.104}$$

Solving for $(E_b/N_0)_d$ in terms of $(E_b/N_0)_u$, we have the relationship

$$\left(\frac{E_b}{N_0}\right)_d = \frac{1}{0.0883 - (E_b/N_0)_u^{-1}} \tag{8.105}$$

▲ Several values are given in Table 8.9.

A curve showing the graphical relationship between the uplink and downlink values of E_b/N_0 will be shown later in conjunction with another exam-

TABLE 8.9 Uplink and Downlink Values of E_b/N_0 Required for $P_E = 10^{-6}$

$(E_s/N_0)_u$, dB	$(E_s/N_0)_d$, dB
20.0	11.06
15.0	12.47
14.0	13.14
13.55	13.55
12.0	15.98
11.0	20.52

ple. Note that the received E_b/N_0 is never better than the uplink or downlink values of E_b/N_0. For $(E_b/N_0)_u = (E_b/N_0)_d \cong 13.55$ dB, the value of $(E_b/N_0)_r$ is 10.54, which is that value required to give $P_E = 10^{-6}$. Note that as either $(E_b/N_0)_u$ or $(E_b/N_0)_d$ approaches infinity, the other energy-to-noise-spectral-density ratio approaches 10.54 dB.

Link Analysis: OBP Digital Transponder

Consider a satellite relay link in which the modulation is digital binary and detection takes place on board the satellite with subsequent remodulation of the detected bits on the downlink carrier and subsequent demodulation and detection at the receiving ground station. This situation can be illustrated in terms of bit errors as shown in Figure 8.35.

The channel is considered symmetrical in that errors for 1's and 0's are equally likely. It is also assumed that errors on the downlink are statistically independent of errors on the uplink and that errors in both links are independent of each other. From Figure 8.35, it follows that the overall probability of no error given that a 1 is transmitted is

$$P(C \mid 1) = q_u q_d + p_u p_d \tag{8.106}$$

where

$$q_u = 1 - p_u \text{ is the probability of no error on the uplink}$$

$$q_d = 1 - p_d \text{ is the probability of no error on the downlink}$$

A similar expression holds for the probability $P(C \mid 0)$ of correct transmission through the channel given that a 0 is transmitted, and it therefore follows that the probability of correct reception averaged over both 1's and 0's is

$$P(C) = P(C \mid 1) = P(C \mid 0) \tag{8.107}$$

FIGURE 8.35   Transition probability diagram for uplink and downlink errors on a demod/remod satellite relay

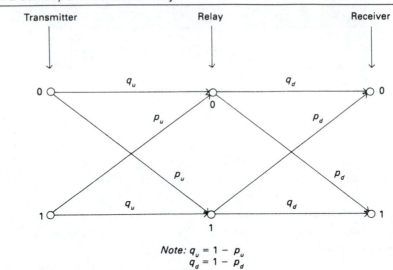

Note: $q_u = 1 - p_u$
$q_d = 1 - p_d$

The average probability of error is

$$P_E = 1 - P(C)$$
$$= 1 - (q_u q_d + p_u p_d)$$
$$= 1 - (1 - p_u)(1 - p_d) - p_u p_d$$
$$= p_u + p_d - 2p_u p_d \qquad (8.108)$$

The following example illustrates how to obtain the uplink and downlink signal energy-to-noise-spectral-density ratios required for an overall desired P_E for a given modulation technique.

▼ **EXAMPLE 8.6**   Consider an OBP satellite communications link where BPSK is used on the uplink and the downlink. For this modulation technique,

$$p_u = Q(\sqrt{2(E_b/N_0)_u}) \cong \frac{e^{-(E_b/N_0)_u}}{2\sqrt{\pi(E_b/N_0)_u}}, \qquad (E_b/N_0)_u > 3 \quad (8.109)$$

with a similar expression for p_d. Say we want the error probability for the overall link to be 10^{-6}. Thus, from (8.108), we have

$$10^{-6} = p_u + p_d - 2p_u p_d \qquad (8.110)$$

TABLE 8.10 E_b/N_0 Values for the Uplink and Downlink Required in an OBP Satellite Communications Link to Give $P_E = 10^{-6}$

p_u	$(E_b/N_0)_u$, dB	p_d	$(E_b/N_0)_d$, dB
10^{-7}	11.31	9×10^{-7}	10.58
5×10^{-7}	10.79	5×10^{-7}	10.79
6×10^{-7}	10.73	4×10^{-7}	10.87
7×10^{-7}	10.67	3×10^{-7}	10.96

Solving for p_d in terms of p_u, we have

$$p_d = \frac{10^{-6} - p_u}{1 - 2p_u} \tag{8.111}$$

A table of values of p_d can be made up versus values of p_u and the corresponding required values of $(E_b/N_0)_u$ can then be calculated from (8.109) with a similar procedure followed for $(E_b/N_0)_d$. Such a table of values is presented as Table 8.10.

Figure 8.36 shows $(E_b/N_0)_u$ versus $(E_b/N_0)_d$ for both the bent-pipe and

FIGURE 8.36   Comparison of bent-pipe and OBP relay characteristics

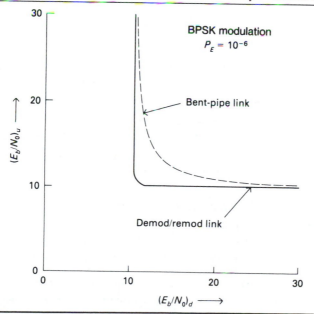

demod/remod satellite relays for an overall bit error probability of 10^{-6}. Curves for other values of P_E or other digital modulation schemes can be obtained in a similar manner. This is left to the problems.

SUMMARY

1. When dealing with M-ary digital communications systems, with $M \geq 2$, it is important to distinguish between a bit and a symbol or character. A symbol conveys $\log_2(M)$ bits. The symbol rate in symbols per second is called the band rate in bauds. We must also distinguish between bit error probability and symbol error probability.

2. M-ary schemes based on quadrature multiplexing include quadrature phase-shift keying (QPSK), offset QPSK (OQPSK), and minimum-shift keying (MSK). All have a bit error rate performance that is essentially the same as binary BPSK if precoding is used to ensure that only one bit error results from mistaking one phase for an adjacent phase.

3. MSK can be produced by quadrature modulation or by serial modulation. In the latter case, MSK is produced by filtering BPSK with a properly designed conversion filter. At the receiver, serial MSK can be recovered by first filtering it with a bandpass matched filter and performing coherent demodulation with a carrier at $f_c + 1/4T_b$ (i.e., at the carrier plus a quarter data rate). Serial MSK performs identically to quadrature-modulated MSK and has advantageous implementation features at high data rates.

4. It is convenient to view M-ary data modulation in terms of signal space. Examples of data formats that can be considered in this way are M-ary PSK, quadrature-amplitude-shift keying (QASK), and M-ary FSK. For the former two modulation schemes, the dimensionality of the signal space stays constant as more signals are added; for the latter, it increases directly as the number of signals added. A constant-dimensional signal space means signal points are packed closer as the number of signal points is increased, thus degrading the error probability; the bandwidth remains essentially constant. In the case of FSK, with increasing dimensionality as more signals are added, the signal points are not compacted, and the error probability decreases for a constant signal-to-noise ratio; the bandwidth increases with an increasing number of signals, however.

5. A convenient measure of bandwidth occupancy for digital modulation is in terms of out-of-band power or power-containment bandwidth. An ideal brickwall containment bandwidth that passes 90% of the signal

power is approximately $1/T_b$ Hz for QPSK, OQPSK, and MSK, and about $2/T_b$ Hz for BPSK.

6. The different types of synchronization that may be necessary in a digital modulation system are carrier (only for coherent systems), symbol or bit, and possibly word. Carrier and symbol synchronization can be carried out by an appropriate nonlinearity followed by a narrowband filter or phase-lock loop. Alternatively, appropriate feedback structures may be used.

7. A pseudo-noise sequence resembles a random "coin toss" sequence but can be generated easily with linear feedback shift-register circuits. They have a narrow correlation peak for zero delay and low sidelobes for nonzero delay, a property that makes them ideal for synchronization of words or measurement of range.

8. Spread-spectrum communications systems are useful for providing resistance to jamming, to provide a means for masking the transmitted signal from unwanted interceptors, to provide resistance to multipath, to provide a way for more than one user to use the same time-frequency allocation, and to provide range-measuring capability.

9. The two major types of spread-spectrum systems are direct-sequence spread-spectrum (DSSS) and frequency-hop spread-spectrum (FHSS). In the former, a spreading code with rate much higher than the data rate multiplies the data sequence, thus spreading the spectrum, while for FHSS, a synthesizer driven by a pseudorandom code generator provides a carrier that hops around in a pseudorandom fashion. A combination of these two schemes, referred to as *hybrid spread spectrum,* is also another possibility.

10. Spread spectrum performs identically to whatever data-modulation scheme is employed without the spectrum spreading as long as the background is additive white Gaussian noise and synchronization is perfect.

11. The performance of a spread-spectrum system in interference is determined in part by its processing gain, which can be defined as the ratio of bandwidth of the spread system to that for an ordinary system employing the same type of data modulation as the spread-spectrum system. For DSSS the processing gain is the ratio of the data bit duration to the spreading code bit (or chip) duration.

12. An additional level of synchronization, referred to as *code synchronization,* is required in a spread-spectrum system. The serial search method is

perhaps the simplest in terms of hardware and to explain, but it is relatively slow in achieving synchronization.

13. Satellite communications systems provide a realistic example to which the digital modulation schemes considered in this chapter can be applied. Two general types of relay satellite configurations were considered in the last section of this chapter: bent-pipe and OBP. In the OBP system, the data on the uplink are demodulated and detected and then used to remodulate the downlink carrier. In the bent-pipe relay, the uplink transmissions are translated in frequency, amplified, and retransmitted on the downlink. Performance characteristics of both types of links were considered by example.

FURTHER READING

In addition to the references given in Chapter 7, for a fuller discussion and in-depth treatment of the digital modulation techniques presented here, see Ziemer and Peterson (1992), and Peterson, Ziemer, and Borth (1995).

PROBLEMS

Section 8.1

8.1 An *M*-ary communication system transmits at a rate of 1000 symbols per second. What is the equivalent bit rate in bits per second for $M = 4$? $M = 8$? $M = 64$?

8.2 A serial bit stream, proceeding at a rate of 10 kbps from a source, is given as

$$110\ 011\ 100\ 101 \qquad \text{(spacing for clarity)}$$

Number the bits from left to right starting with 1 and going through 12 for the right-most bit. Associate the odd-indexed bits with $d_1(t)$ and the even-indexed bits with $d_2(t)$ in Figure 8.1.

(a) What is the symbol rate for d_1 and d_2?
(b) What are the successive values of θ_i given by (8.2) assuming QPSK modulation? At what time intervals may θ_i switch?
(c) What are the successive values of θ_i given by (8.2) assuming OQPSK modulation? At what time intervals may θ_i switch values?

8.3 QPSK is used to transmit data through a channel that adds Gaussian noise with power spectral density $N_0 = 10^{-12}$ V^2/Hz. What are the values of the quadrature-modulated carrier amplitudes required to give $P_{E,\text{symbol}} = 10^{-4}$ for the following data rates?

(a) 10 kbps   (c) 1 Mbps
(b) 100 kbps   (d) 1 Gbps

8.4 Show that the noise components N_1 and N_2 for QPSK, given by Equations (8.6) and (8.8), are uncorrelated.

8.5 (a) Given the following values for the transmitted signal phase, θ, in QPSK, from Figure 8.1 and the text, tell whether $d_1(t)$ and $d_2(t)$ are $+1$ or -1 for the following cases:
     (1) $\theta_i = 45°$ (2) $\theta_i = 135°$ (3) $\theta_i = -45°$ (4) $\theta_i = -135°$
  (b) For the results obtained in part (a) assume that an error is made in detecting $d_2(t)$ What is the corresponding $\hat{\theta}$?
  (c) Same question as part (b), but an error is made in $d_1(t)$.

8.6 (a) A BPSK system and a QPSK system are designed to transmit at equal rates; that is, two bits are transmitted with the BPSK system for each phase in the QPSK system. Compare their *symbol* error probabilities versus SNR.
  (b) A BPSK system and a QPSK system are designed to have equal transmission bandwidths. Compare their symbol error probabilities versus SNR.
  (c) On the basis of parts (a) and (b), what do you conclude about the deciding factor(s) in choosing BPSK versus QPSK?

8.7 Compare P_E as computed from Equation (8.14) with the exact result for SNR's of $E_s/N_0 = 0, 3, 6,$ and 9 dB.

8.8 Given the serial data sequence

$$1\ 1\ 1\ 0\ 0 \quad 1\ 0\ 1\ 1\ 1 \quad 0\ 0\ 1\ 0\ 0 \quad 0\ 0\ 0\ 1\ 1$$

associate every other bit with the upper and lower data streams of the block diagrams of Figures 8.2 and 8.4. Draw on the same time scale (one below the other) the quadrature waveforms for the following data modulation schemes: QPSK, OQPSK, MSK type I, and MSK type II.

8.9 Sketch an excess phase trellis diagram for each of the cases of Problem 8.8. Show as a heavy line the actual path through the trellis represented by the data sequence given.

8.10 Derive Equation (8.24) for the spectrum of an MSK signal, assuming that the block diagram of Figure 8.7(a) is correct. That is, show that serial modulation of MSK works from the standpoint of spectral arguments. (*Hint:* Work only with the positive-frequency portions of the spectrum.)

8.11 An MSK system has a carrier frequency of 100 MHz and transmits data at a rate of 100 kbps.

(a) For the data sequence 1010101010 . . . , what is the instantaneous frequency?

(b) For the data sequence 000000000 . . . , what is the instantaneous frequency?

8.12 Sketch the signal space with decision regions for 16-ary PSK [see (8.39)].

8.13 Using (8.49) and appropriate bounds for $P_{E,\text{symbol}}$, obtain E_b/N_0 required for achieving $P_{E,\text{bit}} = 10^{-4}$ for *M*-ary PSK with $M = 8, 16, 12, 64$. Repeat for 16-QASK.

8.14 Gray encoding of decimal numbers ensures that only one bit changes when the decimal number changes by one unit. Let $b_1 b_2 b_3 \ldots b_n$ represent an ordinary binary representation of a number, with b_1 being the most significant bit. Let the corresponding Gray code digits be $g_1 g_2 g_3 \ldots g_n$. Then the Gray code representation is obtained by the algorithm

$$g_i = b_1$$
$$g_n = b_n \oplus b_{n-1}$$

where $\oplus$ denotes modulo-2 addition (i.e., $0 \oplus 0 = 0$, $0 \oplus 1 = 1$, $1 \oplus 0 = 1$, and $1 \oplus 1 = 0$). Find the Gray code representation for the decimal numbers 0 through 16.

8.15 Derive the three equations numbered (8.45) for 16-QASK.

8.16 Using (8.51) and (8.47), obtain E_b/N_0 required for achieving $P_{E,\text{bit}} = 10^{-4}$ for *M*-ary coherent FSK for $M = 8, 16, 32, 64$.

8.17 Repeat Problem 8.16 for noncoherent *M*-ary FSK for $M = 2, 4, 8$.

8.18 Sketch the signal space for binary coherent FSK showing the decision regions.

8.19 On the basis of null-to-null bandwidths, give the required transmission bandwidth to achieve a bit rate of 10 kbps for the following:

(a) 16-QASK or 16-PSK   (d) 8-FSK, coherent
(b) 8-PSK   (e) 16-FSK, coherent
(c) 32-PSK   (f) 32-FSK, coherent

Section 8.2

8.20 On the basis of 90% power-containment bandwidth, give the required transmission bandwidth to achieve a bit rate of 10 kbps for

(a) BPSK
(b) QPSK or OQPSK
(c) MSK

8.21 Repeat Problem 8.20, but use a 99% power-containment bandwidth. (Hint: For BPSK, use an appropriate reinterpretation of the abscissa of the curve for QPSK to extend the curve.)

8.22 Generalize the results for power-containment bandwidth for quadrature-modulation schemes given in Section 8.2 to M-ary PSK. With appropriate reinterpretation of the abscissa of Figure 8.19 and using the 90% power containment bandwidth, obtain the required transmission bandwidth to support a bit rate of 10 kbps for

(a) 8-PSK
(b) 16-PSK
(c) 32-PSK

8.23 Note that an NRZ waveform can be converted to split phase by multiplying the NRZ waveform by a squarewave with period equal to the bit period. Assume a random (coin-toss) NRZ bit stream with the autocorrelation function

$$R(\tau) = \begin{cases} 1 - \left| \dfrac{\tau}{T} \right|, & \left| \dfrac{\tau}{T} \right| \le 1 . \\[2mm] 0, & \text{otherwise} \end{cases} = \Lambda\left(\dfrac{\tau}{T}\right)$$

Obtain a series expression for the power spectral density using the Fourier series expansion for a squarewave and the modulation theorem. Sketch the result, using the first few terms of the series.

8.24 Assume that a data stream $d(t)$ consists of a random (coin toss) sequence of $+1$ and -1 that is T seconds in duration. The autocorrelation function for such a sequence is

$$R_d(\tau) = \begin{cases} 1 - \left| \dfrac{\tau}{T} \right|, & \left| \dfrac{\tau}{T} \right| \le 1 \\[2mm] 0, & \text{otherwise} \end{cases}$$

(a) Find and sketch the power spectral density for an ASK-modulated signal given by

$$s_{ASK}(t) = \tfrac{1}{2}A[1 + d(t)] \cos (\omega_c t + \theta)$$

where θ is a uniform random variable in $(0, 2\pi)$.

(b) Compute and sketch the power spectral density of a PSK-modulated signal given by

$$s_{PSK}(t) = A \sin [\omega_c t + \cos^{-1} m \, d(t) + \theta]$$

for the three cases $m = 0, 0.5$, and 1. (*Hint:* Expand $s_{PSK}(t)$ into its carrier and modulation components.)

8.25 Derive the Fourier transform pair in Equation (8.64).

8.26 Derive Equation (8.59), given Equations (8.56) and (8.57) and assuming that Δ_1 and Δ_2 are independent and uniformly distributed in the interval $(0, T)$.

8.27 Accurately plot spectra for BPSK, QPSK, and MSK. Use the independent variable $T_b f$.

Section 8.3

8.28 Plot σ_ϕ^2 versus z for the various cases given in Table 8.5. Assume 10% of the signal power is in the carrier for the PLL and all signal power is in the modulation for the Costas and data estimation loops. Assume values of $L = 100, 10, 5$.

8.29 Consider a 15-bit, maximal-length PN code. It is generated by feeding back the last two stages of a four-stage shift register. Assuming a 1 1 1 1 initial state, find all the other possible states of the shift register. What is the sequence? Find and plot its periodic autocorrelation function.

8.30 Repeat Problem 8.29 for a 31-bit maximal-length PN code.

8.31 The aperiodic autocorrelation function of a binary code is important for word synchronization. In computing it, the code is not assumed to periodically repeat itself, but each end is padded with zeros.

(a) Find the aperiodic autocorrelation function for the 7-bit sequence of Figure 8.22. What is the maximum of the absolute value of the autocorrelation function for nonzero delay? This is a Barker sequence.

(b) Compute the aperiodic autocorrelation function of the 15-bit PN sequence found in Problem 8.29 and compare with the result of part (a). Note from Table 8.4 that this is *not* a Barker sequence.

Section 8.4

8.32 Show that the variance of N_g as given by (8.81) is $N_0 T_b$.

8.33 Show that the variance of N_I as given by (8.86) is approximated by the result given by (8.87). (*Hint:* You will have to make use of the fact that $T_c^{-1} \Lambda(\tau/T_c)$ is approximately a delta function for small T_c.)

8.34 A DSSS system employing BPSK data modulation operates with a data rate of 1 Mbps. A processing gain of 100 (20 dB) is desired.

(a) Find the required chip rate.
(b) What is the RF transmission bandwidth required (null-to-null)?
(c) An SNR of 10 dB is employed. What is P_E for the following JSRs? 5 dB; 10 dB; 15 dB; 30 dB.

8.35 Repeat Problem 8.34 for $G_p = 1000$ (30 dB).

8.36 Given a sequential search code acquisition system with $n = 2$ cells per chip, $N_c = 1000$ chips, $T_e = 1$ ms, $P_{FH} = 10^{-3}$, and $P_H = 0.9$. Find T_{acq}.

Section 8.5

8.37 Rederive the curves shown in Figure 8.36, assuming BPSK modulation, for an overall P_E of (a) 10^{-5}; (b) 10^{-4}.

8.38 Rederive the curves shown in Figure 8.36, assuming $P_E = 10^{-6}$, if the modulation technique used on the uplink and downlink is (a) binary noncoherent FSK; (b) binary DPSK.

8.39 Rederive the curves shown in Figure 8.36, assuming $P_{E,bit} = 10^{-6}$, if the modulation technique is changed to M-ary coherent FSK for (a) $M = 4$; (b) $M = 8$; (c) $M = 16$.

COMPUTER EXERCISES

8.1 Use a computer mathematics package to plot curves of P_b versus E_b/N_0 for

(a) M-ary PSK (use the upper-bound expression as an approximation to the actual error probability)

(b) Coherent *M*-ary FSK (again, use the upper bound as an approximation)

(c) Noncoherent *M*-ary FSK

Compare your results with Figures 8.16 and 8.17.

8.2 Use a computer mathematics package to plot out-of-band power for *M*-ary PSK, QPSK (or OQPSK), and MSK. Compare with Figure 8.19.

8.3 Approximate the power spectrum of coherent *M*-ary FSK by adding voltage spectra of sinusoidal bursts of duration T_b and of the appropriate frequency coherently, and then plotting the magnitude squared. What is the minimum spacing of the "tones" in order to maintain them coherently orthogonal? Write a program to evaluate and plot (8.68b). Compare with the spectrum found in the first part of this exercise.

8.4 Use a computer mathematics package to plot curves like those shown in Figure 8.27. Use a solve routine to find the processing gain required to give a desired probability of bit error for a given JSR and SNR. Note that your program should check to see if the desired bit error probability is possible for the given JSR and SNR.

8.5 Given a satellite altitude and desired illumination spot diameter on the earth's surface, use a computer mathematics package to determine the antenna aperture diameter and maximum gain to give the desired spot diameter.

8.6 Use a computer mathematics package to plot Figure 8.36 for a given probability of bit error, and

(a) *M*-ary PSK
(b) Coherent *M*-ary FSK
(c) Noncoherent *M*-ary FSK

OPTIMUM RECEIVERS AND SIGNAL-SPACE CONCEPTS 9

For the most part, this book has been concerned with the *analysis* of communication systems. An exception occurred in Chapter 7, where we sought the best receiver in terms of minimum probability of error for binary digital signals of known shape. In this chapter we deal with the *optimization* problem; that is, we wish to find the communication system for a given task that performs the *best,* within a certain class, of all possible systems. In taking this approach, we are faced with three basic problems:

1. What is the optimization criterion to be used?
2. What is the optimum structure for a given problem under this optimization criterion?
3. What is the performance of the optimum receiver?

We will consider the simplest type of problem of this nature possible—that of fixed transmitter and channel structure with only the receiver to be optimized.

We have two purposes for including this subject in our study of information transmission systems. First, in Chapter 1 we stated that the application of probabilistic systems analysis techniques coupled with statistical optimization procedures has led to communication systems distinctly different in character from those of the early days of communications. The material in this chapter will, we hope, give you an indication of the truth of this statement, particularly when you see that some of the optimum structures considered here are building blocks of systems analyzed in earlier chapters. Additionally, the signal-space techniques to be further developed later in this chapter provide a unification of the performance results for the analog and digital communication systems that we have obtained so far.

9.1 BAYES OPTIMIZATION

Signal Detection Versus Estimation

Based on our considerations in Chapters 7 and 8, we see that it is perhaps advantageous to separate the signal-reception problem into two domains.

The first of these we shall refer to as *detection,* for we are interested merely in detecting the presence of a particular signal, among other candidate signals, in a noisy background. The second is referred to as *estimation,* in which we are interested in estimating some characteristic of a signal that is assumed to be present in a noisy environment. The signal characteristic of interest may be a time-independent parameter such as a constant (random or nonrandom) amplitude or phase or an estimate (past, present, or future value) of the waveform itself (or a functional of the waveform). The former problem is usually referred to as *parameter estimation.* The latter is referred to as *filtering.* We see that demodulation of analog signals (AM, DSB, and so on), if approached in this fashion, would be a signal-filtering problem.*

While it is often advantageous to categorize signal-reception problems as either detection or estimation, both are usually present in practical cases of interest. For example, in the detection of phase-shift-keyed signals, it is necessary to have an estimate of the signal phase available to perform coherent demodulation. In some cases, we may be able to ignore one of these aspects, as in the case of noncoherent digital signaling, in which signal phase was of no consequence. In other cases, the detection and estimation operations may be inseparable. However, we will look at signal detection and estimation as separate problems in this chapter.

Optimization Criteria

In Chapter 7, the optimization criterion that was employed to find the matched filter receiver for binary signals was *minimum average probability of error.* In this chapter we will generalize this idea somewhat and seek signal detectors or estimators that *minimize average cost.* Such devices will be referred to as *Bayes* receivers for reasons that will become apparent later.

Bayes Detectors

To illustrate the use of minimum average cost optimization criteria to find optimum receiver structures, we will first consider detection. For example, suppose we are faced with a situation in which the presence or absence of a constant signal of value k is to be detected in the presence of an additive Gaussian noise component N (for example, as would result by taking a single

---

* See Van Trees (1968), Vol. 1, for a consideration of filtering theory applied to optimal demodulation.

sample of a signal plus noise waveform). Thus we may hypothesize two situations for the observed data Z:

Hypothesis 1 (H_1): $Z = N$ (noise alone) $P(H_1 \text{ true}) = p_0$
Hypothesis 2 (H_2): $Z = k + N$ (signal plus noise) $P(H_2 \text{ true}) = 1 - p_0$

Assuming the noise to have zero mean and variance σ_n^2, we may write down the pdf's of Z given hypotheses H_1 and H_2, respectively. Under hypothesis H_1, Z is Gaussian with mean zero and variance σ_n^2. Thus

$$f_Z(z|H_1) = \frac{e^{-z^2/2\sigma_n^2}}{\sqrt{2\pi\sigma_n^2}} \tag{9.1}$$

Under hypothesis H_2, since the mean is k,

$$f_Z(z|H_2) = \frac{e^{-(z-k)^2/2\sigma_n^2}}{\sqrt{2\pi\sigma_n^2}} \tag{9.2}$$

These conditional pdf's are illustrated in Figure 9.1. We note in this example that Z, the observed data, can range over the real line $-\infty < Z < \infty$. Our objective is to partition this one-dimensional observation space into two regions R_1 and R_2 such that if Z falls into R_1, we decide hypothesis H_1 is true, while if Z is in R_2, we decide H_2 is true. We wish to accomplish this in such a manner that the average cost of making a decision is minimized. It may happen, in some cases, that R_1 or R_2 or both will consist of multiple segments of the real line. (See Problem 9.2.)

Taking a general approach to the problem, we note that four *a priori* costs are required, since there are four types of decisions that we can make. These costs are

c_{11}—the cost of deciding in favor of H_1 when H_1 is actually true
c_{12}—the cost of deciding in favor of H_1 when H_2 is actually true

FIGURE 9.1   Conditional pdf's for a two-hypothesis detection problem

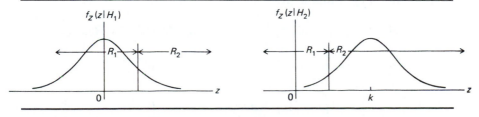

c_{21}—the cost of deciding in favor of H_2 when H_1 is actually true
c_{22}—the cost of deciding in favor of H_2 when H_2 is actually true

Given that H_1 was actually true, the conditional average cost of making a decision, $C(D|H_1)$, is

$$C(D|H_1) = c_{11}P[\text{decide } H_1|H_1 \text{ true}] + c_{21}P[\text{decide } H_2|H_1 \text{ true}] \quad (9.3)$$

In terms of the conditional pdf of Z given H_1, we may write

$$P[\text{decide } H_1|H_1 \text{ true}] = \int_{R_1} f_Z(z|H_1)\, dz \quad (9.4)$$

and

$$P[\text{decide } H_2|H_1 \text{ true}] = \int_{R_2} f_Z(z|H_1)\, dz \quad (9.5)$$

where the one-dimensional regions of integration are as yet unspecified.

We note that Z must lie in either R_1 or R_2, since we are forced to make a decision. Thus

$$P[\text{decide } H_1|H_1 \text{ true}] + P[\text{decide } H_2|H_1 \text{ true}] = 1$$

or if expressed in terms of the conditional pdf $f_Z(z|H_1)$, we obtain

$$\int_{R_2} f_Z(z|H_1)\, dz = 1 - \int_{R_1} f_Z(z|H_1)\, dz \quad (9.6)$$

Thus, combining (9.3) through (9.6), the conditional average cost given H_1, $C(D|H_1)$, becomes

$$C(D|H_1) = c_{11} \int_{R_1} f_Z(z|H_1)\, dz + c_{21} \left[1 - \int_{R_1} f_Z(z|H_1)\, dz \right] \quad (9.7)$$

In a similar manner, the average cost of making a decision given that H_2 is true, $C(D|H_2)$, can be written as

$$C(D|H_2) = c_{12}P[\text{decide } H_1|H_2 \text{ true}] + c_{22}P[\text{decide } H_2|H_2 \text{ true}]$$

$$= c_{12} \int_{R_1} f_Z(z|H_2)\, dz + c_{22} \int_{R_2} f_Z(z|H_2)\, dz$$

$$= c_{12} \int_{R_1} f_Z(z|H_2)\, dz + c_{22} \left[1 - \int_{R_1} f_Z(z|H_2)\, dz \right] \quad (9.8)$$

To find the average cost without regard to which hypothesis is actually true, we must average (9.7) and (9.8) with respect to the prior probabilities of hypotheses H_1 and H_2, $p_0 = P[H_1 \text{ true}]$ and $q_0 = 1 - p_0 = P[H_2 \text{ true}]$. The average cost of making a decision is then

$$C(D) = p_0 C(D|H_1) + q_0 C(D|H_2) \tag{9.9}$$

Substituting (9.7) and (9.8) into (9.9) and collecting terms, we obtain

$$C(D) = p_0 \left\{ c_{11} \int_{R_1} f_Z(z|H_1)\, dz + c_{21} \left[1 - \int_{R_1} f_Z(z|H_1)\, dz \right] \right\}$$

$$+ q_0 \left\{ c_{12} \int_{R_1} f_Z(z|H_2)\, dz + c_{22} \left[1 - \int_{R_1} f_Z(z|H_2)\, dz \right] \right\} \tag{9.10}$$

for the average cost, or risk, in making a decision. Collection of all terms under a common integral that involves integration over R_1 results in

$$C(D) = [p_0 c_{21} + q_0 c_{22}]$$

$$+ \int_{R_1} \{[q_0(c_{12} - c_{22})f_Z(z|H_2)] - [p_0(c_{21} - c_{11})f_Z(z|H_1)]\}\, dz \tag{9.11}$$

The first term in brackets represents a fixed cost once p_0, q_0, c_{21}, and c_{22} are specified. The value of the integral is determined by those points which are assigned to R_1. Since wrong decisions should be more costly than right decisions, it is reasonable to assume that $c_{12} > c_{22}$ and $c_{21} > c_{11}$. Thus the two bracketed terms within the integral are positive because q_0, p_0, $f_Z(z|H_2)$, and $f_Z(z|H_1)$ are probabilities. Hence all values of z that give a larger value for the second term in brackets within the integral than for the first term in brackets should be assigned to R_1 because they contribute a negative amount to the integral. Values of z that give a larger value for the first bracketed term than for the second should be assigned to R_2. In this manner, $C(D)$ will be minimized. Mathematically, the preceding discussion can be summarized by the pair of inequalities

$$q_0(c_{12} - c_{22})f_Z(Z|H_2) \underset{H_1}{\overset{H_2}{\gtrless}} p_0(c_{21} - c_{11})f_Z(Z|H_1)$$

or

$$\frac{f_Z(Z|H_2)}{f_Z(Z|H_1)} \underset{H_1}{\overset{H_2}{\gtrless}} \frac{p_0(c_{21} - c_{11})}{q_0(c_{12} - c_{22})} \tag{9.12}$$

which are interpreted as follows: If an observed value for Z results in the left-hand ratio of pdf's being greater than the right-hand ratio of constants, choose H_2; if not, choose H_1. The left-hand side of (9.12), denoted by $\Lambda(Z)$,

$$\Lambda(Z) \triangleq \frac{f_Z(Z \mid H_2)}{f_Z(Z \mid H_1)} \tag{9.13}$$

is called the *likelihood ratio*. The right-hand side of (9.12),

$$\eta \triangleq \frac{p_0(c_{21} - c_{11})}{q_0(c_{12} - c_{22})} \tag{9.14}$$

is called the *threshold* of the test. Thus the Bayes criterion of minimum average cost has resulted in a test of the likelihood ratio, which is a random variable, against the threshold value η. Note that the development has been general, in that no reference has been made to the particular form of the conditional pdf's in obtaining (9.12). We now return to the specific example that resulted in the conditional pdf's of (9.1) and (9.2).

▼ **EXAMPLE 9.1**   Consider the pdf's of (9.1) and (9.2). Let the costs for a Bayes test be $c_{11} = c_{22} = 0$ and $c_{21} = c_{12}$.

(a) Find $\Lambda(Z)$.
(b) Write down the likelihood ratio test for $p_0 = q_0 = \frac{1}{2}$.
(c) Compare the result of part (b) with the case $p_0 = \frac{1}{4}$ and $q_0 = \frac{3}{4}$.

SOLUTION

(a) $\Lambda(Z) = \dfrac{\exp\left[-(Z - k)^2 / 2\sigma_n^2\right]}{\exp\left(-Z^2 / 2\sigma_n^2\right)} = \exp\left[\dfrac{2kZ - k^2}{2\sigma_n}\right] \tag{9.15}$

(b) For this case $\eta = 1$, which results in the test

$$\exp\left[\frac{2kZ - k^2}{2\sigma_n^2}\right] \underset{H_1}{\overset{H_2}{\gtrless}} 1 \tag{9.16a}$$

Taking the natural logarithm of both sides [this is permissible because $\ln(x)$ is a monotonic function of x] and simplifying, we obtain

$$Z \underset{H_1}{\overset{H_2}{\gtrless}} \tfrac{1}{2}k \tag{9.16b}$$

which states that if the noisy received data are less than half the signal amplitude, the decision that minimizes risk is that the signal was absent, which is reasonable.

(c) For this situation, $\eta = \frac{1}{3}$, and the likelihood ratio test is

$$\exp\left[(2kZ - k^2)/2\sigma_n^2\right] \overset{H_2}{\underset{H_1}{\gtrless}} \frac{1}{3} \tag{9.17a}$$

or, simplifying,

$$Z \overset{H_2}{\underset{H_1}{\gtrless}} \frac{k}{2} - \frac{\sigma_n^2}{k} \ln 3 < \frac{k}{2} \tag{9.17b}$$

Thus, if the prior probability of a signal being present in the noise is increased, the optimum threshold is decreased so that the signal-present hypothesis (H_2) will be chosen with higher probability.

Performance of Bayes Detectors

Since the likelihood ratio is a function of a random variable, it is itself a random variable. Thus, whether we compare the likelihood ratio $\Lambda(Z)$ with the threshold η or we simplify the test to a comparison of Z with a modified threshold as in Example 9.1, we are faced with the prospect of making wrong decisions. The average cost of making a decision, given by (9.11), can be written in terms of the conditional probabilities of making wrong decisions, of which there are two.* These are given by

$$P_F = \int_{R_2} f_Z(z \mid H_1) \, dz \tag{9.18}$$

and

$$P_M = \int_{R_1} f_Z(z \mid H_2) \, dz$$

$$= 1 - \int_{R_2} f_Z(z \mid H_2) \, dz = 1 - P_D \tag{9.19}$$

The subscripts F, M, and D stand for "false alarm," "miss," and "detection," respectively, a terminology that grew out of the application of detection theory to radar. (It is implicitly assumed that hypothesis H_2 corresponds to the signal-present hypothesis and that hypothesis H_1 corresponds to noise-

---

* As will be apparent soon, the probability of error introduced in Chapter 7 can be expressed in terms of P_M and P_F. Thus these conditional probabilities provide a complete performance characterization of the detector.

alone when this terminology is used.) When (9.18) and (9.19) are substituted into (9.11), the risk per decision becomes

$$C(D) = p_0 c_{21} + q_0 c_{22} + q_0(c_{12} - c_{22})P_M - p_0(c_{21} - c_{11})(1 - P_F) \quad (9.20)$$

Thus it is seen that if the probabilities P_F and P_M (or P_D) are available, the Bayes risk can be computed.

Alternative expressions for P_F and P_M can be written in terms of the conditional pdf's of the likelihood ratio given H_1 and H_2 as follows: Given that H_2 is true, an erroneous decision is made if

$$\Lambda(Z) < \eta \quad (9.21)$$

for the decision, according to (9.12), is in favor of H_1. The probability of inequality (9.21) being satisfied, given H_2 is true, is

$$P_M = \int_0^\eta f_\Lambda(\lambda \mid H_2) \, d\lambda \quad (9.22)$$

where $f_\Lambda(\lambda \mid H_2)$ is the conditional pdf of $\Lambda(Z)$ given that H_2 is true. The lower limit of the integral in (9.22) is $\eta = 0$ since $\Lambda(Z)$ is nonnegative, being the ratio of pdfs. Similarly,

$$P_F = \int_\eta^\infty f_\Lambda(\lambda \mid H_1) \, d\lambda \quad (9.23)$$

because, given H_1, an error occurs if

$$\Lambda(Z) > \eta \quad (9.24)$$

[The decision is in favor of H_2 according to (9.12).] The conditional probabilities $f_\Lambda(\lambda \mid H_2)$ and $f_\Lambda(\lambda \mid H_1)$ can be found, in principle at least, by transforming the pdf's $f_Z(z \mid H_2)$ and $f_Z(z \mid H_1)$ in accordance with the transformation of random variables defined by (9.13). Thus two ways of computing P_M and P_F are given by using either (9.18) and (9.19) or (9.22) and (9.23). Often, however, P_M and P_F are computed by using a monotonic function of $\Lambda(Z)$ that is convenient for the particular situation being considered, as in Example 9.2.

A plot of $P_D = 1 - P_M$ versus P_F is called the *operating characteristic* of the likelihood ratio test, or the *receiver operating characteristic* (ROC). It provides all the information necessary to evaluate the risk through (9.20), provided the costs c_{11}, c_{12}, c_{21}, and c_{22} are known. To illustrate the calculation of an ROC, we return to the example involving detection of a constant in Gaussian noise.

▼ EXAMPLE 9.2   Consider the conditional pdf's of (9.1) and (9.2). For an arbitrary threshold η, the likelihood ratio test of (9.12), after taking the natural logarithm of both sides, reduces to

$$\frac{2kZ - k^2}{2\sigma_n^2} \overset{H_2}{\underset{H_1}{\gtrless}} \ln \eta \qquad \text{or} \qquad \frac{Z}{\sigma_n} \overset{H_2}{\underset{H_1}{\gtrless}} \left(\frac{\sigma_n}{k}\right) \ln \eta + \frac{k}{2\sigma_n} \tag{9.25}$$

Defining the new random variable $X \triangleq Z/\sigma_n$ and the parameter $d \triangleq k/\sigma_n$, we can further simplify the likelihood ratio test to

$$X \overset{H_2}{\underset{H_1}{\gtrless}} d^{-1} \ln \eta + \tfrac{1}{2}d \tag{9.26}$$

Expressions for P_F and P_M can be found once $f_X(x\,|\,H_1)$ and $f_X(x\,|\,H_2)$ are known. Because X is obtained from Z by scaling by σ_n, we see from (9.1) and (9.2) that

$$f_X(x\,|\,H_1) = \frac{e^{-x^2/2}}{\sqrt{2\pi}} \qquad \text{and} \qquad f_X(x\,|\,H_2) = \frac{e^{-(x-d)^2/2}}{\sqrt{2\pi}} \tag{9.27}$$

That is, under either hypothesis H_1 or hypothesis H_2, X is a unity variance Gaussian random variable. These two conditional pdf's are shown in Figure 9.2. A false alarm occurs if, given H_1,

$$X > d^{-1} \ln \eta + \tfrac{1}{2}d \tag{9.28}$$

FIGURE 9.2   Conditional probability density functions and decision regions for the problem of detecting a constant signal in zero-mean Gaussian noise

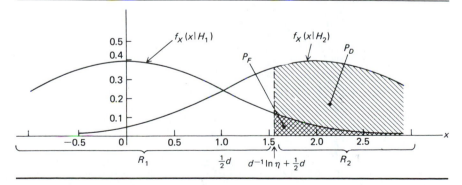

The probability of this happening is

$$P_F = \int_{d^{-1} \ln \eta + d/2}^{\infty} f_X(x \mid H_1) \, dx$$

$$= \int_{d^{-1} \ln \eta + d/2}^{\infty} \frac{e^{-x^2/2}}{\sqrt{2\pi}} \, dx = Q(d^{-1} \ln \eta + d/2) \tag{9.29}$$

which is the area under $f_X(x \mid H_1)$ to the right of $d^{-1} \ln \eta + \frac{1}{2}d$ in Figure 9.2. Detection occurs if, given H_2,

$$X > d^{-1} \ln \eta + \tfrac{1}{2}d \tag{9.30}$$

The probability of this happening is

$$P_D = \int_{d^{-1} \ln \eta + d/2}^{\infty} f_X(x \mid H_2) \, dx$$

$$= \int_{d^{-1} \ln \eta + d/2}^{\infty} \frac{e^{-(x-d)^2/2}}{\sqrt{2\pi}} \, dx = Q(d^{-1} \ln \eta - d/2) \tag{9.31}$$

Thus P_D is the area under $f_X(x \mid H_2)$ to the right of $d^{-1} \ln \eta + \frac{1}{2}d$ in Figure 9.2.

The ROC is obtained by plotting P_D versus P_F for various values of d, as shown in Figure 9.3. The curves are obtained by varying η from 0 to ∞. For

FIGURE 9.3 Receiver operating characteristic for detecting a constant signal in zero-mean Gaussian noise

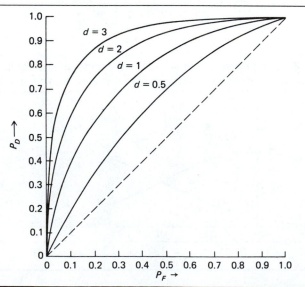

$\eta = 0$, $\ln \eta = -\infty$ and the detector always chooses H_2 ($P_F = 1$). For $\eta = \infty$, $\ln \eta = \infty$ and the detector always chooses H_1 ($P_D = P_F = 0$).

The Neyman-Pearson Detector

The design of a Bayes detector requires knowledge of the costs and *a priori* probabilities. If these are unavailable, a simple optimization procedure is to fix P_F at some tolerable level, say α, and maximize P_D (or minimize P_M) subject to the constraint $P_F \leq \alpha$. The resulting detector is known as the *Neyman-Pearson detector*. It can be shown that the Neyman-Pearson criterion leads to a likelihood ratio test identical to that of (9.12), except that the threshold η is determined by the allowed value of probability of false alarm α. This value of η can be obtained from the ROC for a given value of P_F, for it can be shown that the slope of a curve of an ROC at a particular point is equal to the value of the threshold η required to achieve the P_D and P_F of that point.*

Minimum Probability of Error Detectors

From (9.11) it follows that if $c_{11} = c_{22} = 0$ (zero cost for making right decision) and $c_{12} = c_{21} = 1$ (equal cost for making either type of wrong decision), then the risk reduces to

$$C(D) = p_0 \left[1 - \int_{R_1} f_Z(z \,|\, H_1) \, dz \right] + q_0 \int_{R_1} f_Z(z \,|\, H_2) \, dz$$

$$= p_0 \int_{R_2} f_Z(z \,|\, H_1) \, dz + q_0 \int_{R_1} f_Z(z \,|\, H_2) \, dz$$

$$= p_0 P_F + q_0 P_M \tag{9.32}$$

where we have used (9.6), (9.18), and (9.19). However, (9.32) is the probability of making a wrong decision, averaged over both hypotheses, which is the same as the probability of error used as the optimization criterion in Chapter 7. Thus Bayes receivers with this special cost assignment are minimum-probability-of-error receivers.

The Maximum a Posteriori (MAP) Detector

Letting $c_{11} = c_{22} = 0$ and $c_{21} = c_{12}$ in (9.12), we can rearrange the equation in the form

$$\frac{f_Z(Z \,|\, H_2) P(H_2)}{f_Z(Z)} \underset{H_1}{\overset{H_2}{\gtrless}} \frac{f_Z(Z \,|\, H_1) P(H_1)}{f_Z(Z)} \tag{9.33}$$

---

* Van Trees (1968), Vol. 1.

where the definitions of p_0 and q_0 have been substituted, both sides of (9.12) have been multiplied by $P(H_2)$, and both sides have been divided by

$$f_Z(Z) \triangleq f_Z(Z|H_1)P(H_1) + f_Z(Z|H_2)P(H_2) \qquad (9.34)$$

Using Bayes' rule, as given by Equation (4.7), (9.33) becomes

$$P(H_2|Z) \underset{H_1}{\overset{H_2}{\gtrless}} P(H_1|Z) \qquad (c_{11} = c_{22} = 0; c_{12} = c_{21}) \qquad (9.35)$$

Equation (9.35) states that the most probable hypothesis, given a particular observation Z, is to be chosen in order to minimize the risk, which, for the special cost assignment assumed, is equal to the probability of error. The probabilities $P(H_1|Z)$ and $P(H_2|Z)$ are called *a posteriori* probabilities, for they give the probability of a particular hypothesis *after* the observation of Z, in contrast to $P(H_1)$ and $P(H_2)$, which give us the probabilities of the same events *before* observation of Z. Because the hypothesis corresponding to the maximum *a posteriori* probability is chosen, such detectors are referred to as *maximum a posteriori* (MAP) *detectors*. Minimum-probability-of-error detectors and MAP detectors are therefore equivalent.

Minimax Detectors

The minimax decision rule corresponds to the Bayes decision rule, where the *a priori* probabilities have been chosen to make the Bayes risk a maximum. For further discussion of this decision rule, see Weber (1987).

The M-ary Hypothesis Case

The generalization of the Bayes decision criterion to M hypotheses, where $M > 2$, is straightforward but unwieldy. For the M-ary case, M^2 costs and M *a priori* probabilities must be given. In effect, M likelihood ratio tests must be carried out in making a decision. If attention is restricted to the special cost assignment used to obtain the MAP detector for the binary case (that is, right decisions cost zero and wrong decisions are all equally costly), then a MAP decision rule results that is easy to visualize for the M-hypothesis case. Generalizing from (9.35), we have the MAP decision rule for the M-hypothesis case:

Compute the M posterior probabilities $P(H_i|Z)$, $i = 1, 2, \ldots, M$, and choose as the correct hypothesis the one corresponding to the largest posterior probability.

This decision criterion will be used when *M*-ary signal detection is considered.

Decisions Based on Vector Observations

If, instead of a single observation Z, we have N observations $Z \triangleq (Z_1, Z_2, \ldots, Z_N)$, all of the preceding results hold with the exception that the N-fold joint pdf's of Z, given H_1 and H_2, are to be used. If $Z_1, Z_2, \ldots, Z_N$ are conditionally independent, these joint pdf's are easily written down, since they are simply the N-fold products of the marginal pdf's of $Z_1, Z_2, \ldots, Z_N$, given H_1 and H_2. We will make use of this generalization when the detection of arbitrary finite energy signals in white Gaussian noise is discussed. We will find the optimum Bayes detectors for such problems by resolving the possible transmitted signals into a finite dimensional *signal space*. In the next section, therefore, we continue the discussion of vector space representation of signals begun in Chapter 2, Section 2.3.

9.2   VECTOR SPACE REPRESENTATION OF SIGNALS

We recalled, in Section 2.3, that any vector in three-dimensional space can be expressed as a linear combination of any three linearly independent vectors. Recall that such a set of three linearly independent vectors is said to *span* three-dimensional vector space and is referred to as a *basis-vector set* for the space. A basis set of unit-magnitude, mutually perpendicular vectors is called an *orthonormal basis set*.

Two geometrical concepts associated with vectors are magnitude of a vector and angle between two vectors. Both are described by the scalar (or dot) product of any two vectors **A** and **B** having magnitudes A and B, defined as

$$\mathbf{A} \cdot \mathbf{B} = AB \cos \theta \tag{9.36}$$

where θ is the angle between **A** and **B**. Thus

$$A = \sqrt{\mathbf{A} \cdot \mathbf{A}} \quad \text{and} \quad \cos \theta = \frac{\mathbf{A} \cdot \mathbf{B}}{AB} \tag{9.37}$$

Generalizing these concepts to signals in Section 2.3, we expressed a signal $x(t)$ with finite energy in an interval $(t_0, t_0 + T)$ in terms of a complete set of orthonormal basis functions $\phi_1(t), \phi_2(t), \ldots$ as the series

$$x(t) = \sum_{n=1}^{\infty} X_n \phi_n(t) \tag{9.38}$$

where

$$X_n = \int_{t_0}^{t_0+T} x(t)\phi_n^*(t) \, dt \tag{9.39}$$

which is a special case of Equation (2.32) with $c_n = 1$ because the $\phi_n(t)$'s are assumed orthonormal on the interval $(t_0, t_0 + T)$. This representation provided the alternative representation for $x(t)$ as the infinite dimensional vector $(X_1, X_2, \ldots)$.

To set up a geometric structure on such a vector space, which will be referred to as *signal space*, we must first establish the linearity of the space by listing a consistent set of properties involving the members of the space and the operations between them. Second, we must establish the geometric structure of the space by generalizing the concept of scalar product, thus providing generalizations for the concepts of magnitude and angle.

Structure of Signal Space

We begin with the first task. Specifically, a collection of signals composes a linear signal space $\mathcal{S}$ if, for any pair of signals $x(t)$ and $y(t)$ in $\mathcal{S}$, the operations of addition (commutative and associative) of two signals and multiplication of a signal by a scalar are defined and obey the following axioms:

1. The signal $\alpha_1 x(t) + \alpha_2 y(t)$ is in the space for any two scalars α_1 and α_2 (establishes $\mathcal{S}$ as linear).
2. $\alpha[x(t) + y(t)] = \alpha x(t) + \alpha y(t)$ for any scalar α.
3. $\alpha_1[\alpha_2 x(t)] = (\alpha_1 \alpha_2)x(t)$.
4. The product of $x(t)$ and the scalar 1 reproduces $x(t)$.
5. The space contains a unique zero element such that $x(t) + 0 = x(t)$.
6. To each $x(t)$ there corresponds a unique element $-x(t)$ such that $x(t) + [-x(t)] = 0$.

In writing relations such as the preceding, it is convenient to suppress the independent variable t, and this will be done from now on.

Scalar Product

The second task, that of establishing the geometric structure, is accomplished by defining the scalar product, denoted (x, y), as a scalar-valued function of two signals $x(t)$ and $y(t)$ (in general complex functions), with the following properties:

1. $(x, y) = (y, x)^*$
2. $(\alpha x, y) = \alpha(x, y)$

3. $(x + y, z) = (x, z) + (y, z)$
4. $(x, x) > 0$ unless $x \equiv 0$, in which case $(x, x) = 0$

The particular definition used for the scalar product depends on the application and the type of signals involved. Because we wish to include both energy and power signals in our future considerations, at least two definitions of scalar product are required. If $x(t)$ and $y(t)$ are both of the same class, a convenient choice is

$$(x, y) \triangleq \begin{cases} \displaystyle\lim_{T' \to \infty} \int_{-T'}^{T'} x(t)y^*(t)\, dt & \text{(energy signals)} & (9.40a) \\[2mm] \displaystyle\lim_{T' \to \infty} \frac{1}{2T'} \int_{-T'}^{T'} x(t)y^*(t)\, dt & \text{(power signals)} & (9.40b) \end{cases}$$

where T' has been used to avoid confusion with the signal observation interval T. In particular, for $x(t) = y(t)$, we see that (9.40a) is the total energy contained in $x(t)$ and (9.40b) corresponds to the average power. We note that the coefficients in the series of (9.38) can be written as

$$X_n = (x, \phi_n) \qquad (9.41)$$

If the scalar product of two signals $x(t)$ and $y(t)$ is zero, they are said to be *orthogonal,* just as two ordinary vectors are said to be orthogonal if their dot product is zero.

Norm

The next step in establishing the structure of a linear signal space is to define the length, or norm $\|x\|$, of a signal. A particularly suitable choice, in view of the preceding discussion, is

$$\|x\| = (x, x)^{1/2} \qquad (9.42)$$

More generally, the norm of a signal is any nonnegative real number satisfying the following properties:

1. $\|x\| = 0$ if and only if $x \equiv 0$
2. $\|x + y\| \leq \|x\| + \|y\|$ (known as the *triangle inequality*)
3. $\|\alpha x\| = |\alpha| \, \|x\|$, where α is a scalar

Clearly, the choice $\|x\| = (x, x)^{1/2}$ satisfies these properties, and we will employ it from now on.

A measure of the distance between, or dissimilarity of, two signals x and y is provided by the norm of their difference $\|x - y\|$.

Schwarz's Inequality

An important relationship between the scalar product of two signals and their norms is Schwarz's inequality, which was used in Chapter 7 without proof. For two signals $x(t)$ and $y(t)$, it can be written as

$$|(x, y)| \leq \|x\| \, \|y\| \qquad (9.43)$$

with equality if and only if x or y is zero or if $x(t) = \alpha y(t)$ where α is a scalar.

To prove (9.43), we consider the nonnegative quantity $\|x + \alpha y\|^2$ where α is as yet an unspecified constant. Expanding it by using the properties of the scalar product, we obtain

$$
\begin{aligned}
\|x + \alpha y\|^2 &= (x + \alpha y, x + \alpha y) \\
&= (x, x) + \alpha^*(x, y) + \alpha(x, y)^* + |\alpha|^2(y, y) \\
&= \|x\|^2 + \alpha^*(x, y) + \alpha(x, y)^* + |\alpha|^2\|y\|^2 \qquad (9.44)
\end{aligned}
$$

Choosing $\alpha = -(x, y)/\|y\|^2$, which is permissible since α is arbitrary, we find that the last two terms of (9.44) cancel, yielding

$$\|x + \alpha y\|^2 = \|x\|^2 - \frac{|(x, y)|^2}{\|y\|^2} \qquad (9.45)$$

Since $\|x + \alpha y\|^2$ is nonnegative, rearranging (9.45) gives Schwarz's inequality. Furthermore, noting that $\|x + \alpha y\| = 0$ if and only if $x + \alpha y = 0$, we establish a condition under which equality holds in (9.43). Equality also holds, of course, if one or both signals are identically zero.

▼ **EXAMPLE 9.3**   A familiar example of a space that satisfies the preceding properties is ordinary two-dimensional vector space. Consider two vectors with real components,

$$\mathbf{A}_1 = a_1\hat{\imath} + b_1\hat{\jmath} \quad\text{and}\quad \mathbf{A}_2 = a_2\hat{\imath} + b_2\hat{\jmath} \qquad (9.46)$$

where $\hat{\imath}$ and $\hat{\jmath}$ are the usual orthogonal unit vectors. The scalar product is taken as the vector dot product

$$(\mathbf{A}_1, \mathbf{A}_2) = a_1a_2 + b_1b_2 = \mathbf{A}_1 \cdot \mathbf{A}_2 \qquad (9.47)$$

and the norm is taken as

$$\|\mathbf{A}_1\| = (\mathbf{A}_1, \mathbf{A}_1)^{1/2} = \sqrt{a_1^2 + b_1^2} \qquad (9.48)$$

which is just the length of the vector. Addition is defined as vector addition,

$$\mathbf{A}_1 + \mathbf{A}_2 = (a_1 + a_2)\hat{\imath} + (b_1 + b_2)\hat{\jmath} \qquad (9.49)$$

which is commutative and associative. The vector $\mathbf{C} \triangleq \alpha_1 \mathbf{A}_1 + \alpha_2 \mathbf{A}_2$, where α_1 and α_2 are real constants, is also a vector in two-space (axiom 1, page 617). The remaining axioms follow as well, with the zero element being $0\hat{i} + 0\hat{j}$.

The properties of the scalar product (pages 617 and 618) are satisfied by the vector dot product.

The properties of the norm also follow, with property 2 taking the form

$$\sqrt{(a_1 + a_2)^2 + (b_1 + b_2)^2} \le \sqrt{a_1^2 + b_1^2} + \sqrt{a_2^2 + b_2^2} \quad (9.50)$$

which is simply a statement that the length of the hypotenuse of a triangle is shorter than the sum of the lengths of the other two sides—hence the name *triangle inequality*. Schwarz's inequality squared is

$$(a_1 a_2 + b_1 b_2)^2 \le (a_1^2 + b_1^2)(a_2^2 + b_2^2) \quad (9.51)$$

which simply states that $|\mathbf{A}_1 \cdot \mathbf{A}_2|^2$ is less than or equal to the length squared of $\mathbf{A}_1$ times the length squared of $\mathbf{A}_2$.

Scalar Product of Two Signals in Terms of Fourier Coefficients

Expressing two energy or power signals $x(t)$ and $y(t)$ in the form given in (9.38), we may show that

$$(x, y) = \sum_{m=1}^{\infty} X_m Y_m^* \quad (9.52a)$$

Letting $y = x$ results in Parseval's theorem, which is

$$\|x\|^2 = \sum_{m=1}^{\infty} |X_m|^2 \quad (9.52b)$$

To indicate the usefulness of the shorthand vector notation just introduced, we will carry out the proof of (9.52) using it. Let $x(t)$ and $y(t)$ be written in terms of their respective orthonormal expansions:

$$x(t) = \sum_{m=1}^{\infty} X_m \phi_m(t) \quad \text{and} \quad y(t) = \sum_{n=1}^{\infty} Y_n \phi_n(t) \quad (9.53)$$

where, in terms of the scalar product,

$$X_m = (x, \phi_m) \quad \text{and} \quad Y_n = (y, \phi_n) \quad (9.54)$$

Thus

$$(x, y) = \left(\sum_m X_m \phi_m, \sum_n Y_n \phi_n \right) = \sum_m X_m \left(\phi_m, \sum_n Y_n \phi_n \right) \quad (9.55)$$

by virtue of properties 2 and 3 of the scalar product. Applying property 1, we obtain

$$(x, y) = \sum_m X_m \left(\sum_n Y_n \phi_n, \phi_m \right)^* = \sum_m X_m \left[\sum_n Y_n^* (\phi_n, \phi_m)^* \right] \quad (9.56)$$

the last step of which follows by virtue of another application of properties 2 and 3. But the ϕ_n's are orthonormal; that is, $(\phi_n, \phi_m) = \delta_{nm}$, where δ_{nm} is the Kronecker delta. Thus

$$(x, y) = \sum_m X_m \left[\sum_n Y_n^* \delta_{nm} \right] = \sum_m X_m Y_m^* \quad (9.57)$$

which proves (9.52a).

▼ EXAMPLE 9.4   Consider a signal $x(t)$ and the approximation to it $x_a(t)$ of Example 2.5. Both x and x_a are in the signal space consisting of all finite energy signals. All the addition and multiplication properties for signal space hold for x and x_a. Because we are considering finite energy signals, the scalar product defined by (9.40) applies. The scalar product of x and x_a is

$$(x, x_a) = \int_0^2 \sin \pi t \left[\frac{2}{\pi} \phi_1(t) - \frac{2}{\pi} \phi_2(t) \right] dt$$

$$= \left(\frac{2}{\pi} \right)^2 - \left(\frac{2}{\pi} \right) \left(-\frac{2}{\pi} \right) = 2 \left(\frac{2}{\pi} \right)^2 \quad (9.58)$$

The norm of their difference squared is

$$\| x - x_a \|^2 = (x - x_a, x - x_a) = \int_0^2 \left[\sin \pi t - \frac{2}{\pi} \phi_1(t) + \frac{2}{\pi} \phi_2(t) \right]^2 dt$$

$$= 1 - \frac{8}{\pi^2} \quad (9.59)$$

which is just the minimum integral-squared error between x and x_a.

The norm squared of x is

$$\|x\|^2 = \int_0^2 \sin^2 \pi t \, dt = 1 \tag{9.60}$$

and the norm squared of x_a is

$$\|x_a\|^2 = \int_0^2 \left[\frac{2}{\pi} \phi_1(t) - \frac{2}{\pi} \phi_2(t) \right]^2 dt = 2 \left(\frac{2}{\pi} \right)^2 \tag{9.61}$$

which follows from the orthonormality of ϕ_1 and ϕ_2. Thus Schwarz's inequality for this case is

$$2 \left(\frac{2}{\pi} \right)^2 < 1 \cdot \sqrt{2} \left(\frac{2}{\pi} \right) \tag{9.62a}$$

which is equivalent to

$$\sqrt{2} < \tfrac{1}{2}\pi \tag{9.62b}$$

▲ Since x is not a scalar multiple of x_a, we must have strict inequality.

Choice of Basis Function Sets—The Gram-Schmidt Procedure

The question naturally arises as to how we obtain suitable basis sets. For energy or power signals, with no further restrictions imposed, we require infinite sets of functions. Suffice it to say that many suitable choices exist, depending on the particular problem and the interval of interest. These include not only the sines and cosines, or complex exponentials, of harmonically related frequencies, but also the Legendre functions, Hermite functions, and Bessel functions, to name only a few. All these are complete.

A technique referred to as the *Gram-Schmidt procedure* is often useful for obtaining basis sets, especially in the consideration of *M*-ary signal detection. This procedure will now be described.

Consider the situation in which we are given a finite set of signals $s_1(t)$, $s_2(t), \ldots, s_M(t)$ defined on some interval $(t_0, t_0 + T)$, and our interest is in all signals that may be written as linear combinations of these signals:

$$x(t) = \sum_{n=1}^{M} X_n s_n(t), \qquad t_0 \leq t \leq t_0 + T \tag{9.63}$$

The set of all such signals forms an M-dimensional signal space if the $s_n(t)$'s are linearly independent [that is, no $s_n(t)$ can be written as a linear combination of the rest]. If the $s_n(t)$'s are not linearly independent, the dimension of the space is less than M. An orthonormal basis for the space can be obtained by using the Gram-Schmidt procedure, which consists of the following steps:

1. Set $v_1(t) = s_1(t)$ and $\phi_1(t) = v_1(t)/\|v_1\|$.
2. Set $v_2(t) = s_2(t) - (s_2, \phi_1)\phi_1$ and $\phi_2(t) = v_2(t)/\|v_2\|$ [$v_2(t)$ is the component of $s_2(t)$ that is linearly independent of $s_1(t)$].
3. Set $v_3(t) = s_3(t) - (s_3, \phi_2)\phi_2(t) - (s_3, \phi_1)\phi_1(t)$ and $\phi_3(t) = v_3(t)/\|v_3\|$ [$v_3(t)$ is the component of $s_3(t)$ linearly independent of $s_1(t)$ and $s_2(t)$].
4. Continue until all the $s_n(t)$'s have been used. If the $s_n(t)$'s are not linearly independent, then one or more steps will yield $v_n(t)$'s for which $\|v_n\| = 0$. These signals are omitted whenever they occur so that a set of K orthonormal functions is finally obtained where $K \leq M$.

The resulting set forms an orthonormal basis set for the space since, at each step of the procedure, we ensure that

$$(\phi_n, \phi_m) = \delta_{nm} \tag{9.64}$$

where δ_{nm} is the Kronecker delta defined in Chapter 2, and we use all signals in forming the orthonormal set.

▼| **EXAMPLE 9.5**   Consider the set of three finite-energy signals

$$s_1(t) = 1, \qquad\qquad 0 \leq t \leq 1$$
$$s_2(t) = \cos 2\pi t, \qquad 0 \leq t \leq 1 \tag{9.65}$$
$$s_3(t) = \cos^2 \pi t, \qquad 0 \leq t \leq 1$$

We desire an orthonormal basis for the signal space spanned by these three signals.

SOLUTION   We let $v_1(t) = s_1(t)$ and compute

$$\phi_1(t) = \frac{v_1(t)}{\|v_1\|} = 1, \qquad 0 \leq t \leq 1 \tag{9.66}$$

Next, we compute

$$(s_2, \phi_1) = \int_0^1 1 \cos 2\pi t \, dt = 0 \tag{9.67}$$

and we set

$$v_2(t) = s_2(t) - (s_2, \phi_1)\phi_1 = \cos 2\pi t, \qquad 0 \leq t \leq 1 \tag{9.68}$$

The second orthonormal function is found from

$$\phi_2(t) = \frac{v_2}{\|v_2\|} = \sqrt{2} \cos 2\pi t, \qquad 0 \le t \le 1 \qquad (9.69)$$

To check for another orthonormal function, we require the scalar products

$$(s_3, \phi_2) = \int_0^1 \sqrt{2} \cos 2\pi t \cos^2 \pi t \, dt = \tfrac{1}{4}\sqrt{2} \qquad (9.70a)$$

and

$$(s_3, \phi_1) = \int_0^1 \cos^2 \pi t \, dt = \tfrac{1}{2} \qquad (9.70b)$$

Thus

$$
\begin{aligned}
v_3(t) &= s_3(t) - (s_3\,\phi_2)\phi_2 - (s_3, \phi_1)\phi_1 \\
&= \cos^2 \pi t - (\tfrac{1}{4}\sqrt{2})\sqrt{2} \cos 2\pi t - \tfrac{1}{2} = 0 \qquad (9.70c)
\end{aligned}
$$

▲ so that the space is two-dimensional.

Signal Dimensionality as a Function of Signal Duration

The sampling theorem, proved in Chapter 2, provides a means of representing strictly bandlimited signals, with bandwidth W, in terms of the infinite basis function set $\mathrm{sinc}(f_s t - n)$, $n = 0, \pm 1, \pm 2, \ldots$, from Equation (2.246). Because $\mathrm{sinc}(f_s t - n)$ is not time-limited, we suspect that a strictly bandlimited signal cannot also be of finite duration (that is, time-limited). However, practically speaking, a time-bandwidth dimensionality can be associated with a signal provided the definition of bandlimited is relaxed. The following theorem, given without proof, provides an upper bound for the dimensionality of time-limited and bandwidth-limited signals.*

▷ DIMENSIONALITY THEOREM

Let $\{\phi_k(t)\}$ denote a set of orthogonal waveforms, all of which satisfy the following requirements:

1. They are identically zero outside a time interval of duration T, for example, $|t| \le \tfrac{1}{2}T$.

---

* This theorem is taken from Wozencraft and Jacobs (1965), p. 294, where it is also given without proof. However, a discussion of the equivalence of this theorem to the original ones due to Shannon and to Landau and Pollak is also given.

2. None has more than $\frac{1}{12}$ of its energy outside the frequency interval $-W < f < W$.

Then the number of waveforms in the set $\{\phi_k(t)\}$ is conservatively overbounded by $2.4TW$ when TW is large.

▼ EXAMPLE 9.6   Consider the orthogonal set of waveforms

$$
\phi_k(t) = \Pi\left[\frac{t - k\tau}{\tau}\right]
$$

$$
= \begin{cases} 1, & \frac{1}{2}(2k - 1)\tau \le t \le \frac{1}{2}(2k + 1)\tau, \quad k = 0, \pm 1, \pm 2, \pm K \\ 0, & \text{otherwise} \end{cases}
$$

(9.71)

where $(2K + 1)\tau = T$. The Fourier transform of $\phi_k(t)$ is

$$
\Phi_k(f) = \tau \operatorname{sinc} \tau f \, e^{-j2\pi k\tau f}
$$

(9.72)

The total energy in $\phi_k(t)$ is τ, and the energy for $|f| \le W$ is

$$
E_W = \int_{-W}^{W} \tau^2 \operatorname{sinc}^2 \tau f \, df = \int_{-W}^{W} \tau^2 \frac{\sin^2 \pi \tau f}{(\pi \tau f)^2} \, df
$$

$$
= \frac{2\tau}{\pi} \int_{0}^{\pi\tau W} \frac{\sin^2 u}{u^2} \, du
$$

(9.73)

where the substitution $u = \pi \tau f$ has been made in the integral and the integration is carried out only over positive values of u, owing to the evenness of the integrand. Integrating by parts once, we obtain

$$
E_W = -\frac{2\tau}{\pi} \frac{\sin^2 u}{u} \Bigg|_{0}^{\tau\pi W} + \frac{2\tau}{\pi} \int_{0}^{2\pi\tau W} \frac{\sin \alpha}{\alpha} \, d\alpha
$$

(9.74)

The second term can be written in terms of the sine-integral function Si (z), defined as

$$
\text{Si}(z) = \int_{0}^{z} \frac{\sin \alpha}{\alpha} \, d\alpha
$$

(9.75)

which is a tabulated function.* Letting $z = 2\pi\tau W$, we obtain

$$
\frac{E_W}{E} = -2\tau W \operatorname{sinc}^2(\tau W) + \frac{2}{\pi} \text{Si}(2\pi\tau W)
$$

(9.76)

---

* M. Abramowitz and I. Stegun, *Handbook of Mathematical Functions* (New York: Dover, 1972).

We want to choose τW such that $E_W/E \geq \frac{11}{12} = 0.9167$. Using tables for the sinc and Si functions, we can compute the following values for E_W/E:

τW	E_W/E
1.0	0.9027
1.1	0.9033
1.2	0.9064
1.3	0.9130
1.4	0.9218
1.5	0.9312

Thus $\tau W = 1.4$ will ensure that none of the $\phi_k(t)$'s has more than $\frac{1}{12}$ of its energy outside the frequency interval $-W < f < W$.

Now $N = [T/\tau]$ orthogonal waveforms occupy the interval $(-\frac{1}{2}T, \frac{1}{2}T)$ where [ ] signifies the integer part of T/τ. Letting $\tau = 1.4W^{-1}$, we obtain

$$N = \left[\frac{TW}{1.4}\right] = [0.714TW] \tag{9.77}$$

which satisfies the bound given by the theorem.

9.3 MAP RECEIVERS FOR DIGITAL DATA TRANSMISSION

We now apply the detection theory and signal-space concepts just developed to digital data transmission. We will consider examples of coherent and noncoherent systems.

Decision Criteria for Coherent Systems in Terms of Signal Space

In the analysis of QPSK systems in Chapter 8, the received signal plus noise was resolved into two components by the correlators comprising the receiver. This made simple the calculation of the probability of error. The QPSK receiver essentially computes the coordinates of the received signal plus noise in a signal space. The basis functions for this signal space are $\cos \omega_c t$ and $\sin \omega_c t$, $0 \leq t \leq T$, with the scalar product defined by

$$(x_1, x_2) = \int_0^T x_1(t)x_2(t)\, dt \tag{9.78}$$

which is a special case of (9.40a). These basis functions are orthogonal if $\omega_c T$ is an integer multiple of 2π, but are not normalized.

Recalling the Gram-Schmidt procedure, we see how this viewpoint might be generalized to M signals $s_1(t)$, $s_2(t)$, ..., $s_M(t)$ that have finite energy but are otherwise arbitrary. Thus, consider an M-ary communication system, depicted in Figure 9.4, wherein one of M possible signals of known form $s_i(t)$ associated with a message m_i is transmitted each T seconds. The receiver is to be constructed such that the probability of error in deciding which message was transmitted is minimized; that is, it is a MAP receiver. For simplicity, we assume that the messages are produced by the information source with equal *a priori* probability.

Ignoring the noise for the moment, we note that the ith signal can be expressed as

$$s_i(t) = \sum_{j=1}^{K} A_{ij}\phi_j(t), \qquad i = 1, 2, \ldots, M, \quad K \le M \tag{9.79}$$

where the $\phi_j(t)$'s are orthonormal basis functions chosen according to the Gram-Schmidt procedure. Thus

$$A_{ij} = \int_0^T s_i(t)\phi_j(t)\, dt = (s_i, \phi_j) \tag{9.80}$$

and we see that the receiver structure shown in Figure 9.5, which consists of a bank of correlators, can be used to compute the generalized Fourier coefficients for $s_i(t)$. Thus we can represent each possible signal as a point in a K-dimensional signal space with coordinates $(A_{i1}, A_{i2}, \ldots, A_{iK})$, for $i = 1, 2, \ldots, M$.

FIGURE 9.4 *M-ary communication system*

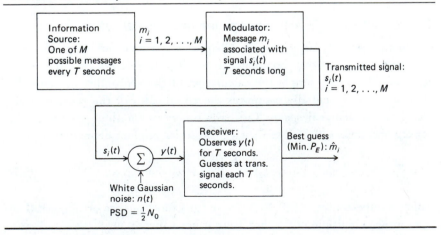

FIGURE 9.5   Receiver structure for resolving signals into K-dimensional signal space

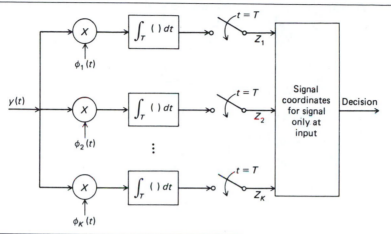

Knowing the coordinates of $s_i(t)$ is as good as knowing $s_i(t)$, since it is uniquely specified through (9.79). The difficulty is, of course, that we receive the signals in the presence of noise. Thus, instead of the receiver providing us with the actual signal coordinates, it provides us with noisy coordinates $(A_{i1} + N_1, A_{i2} + N_2, \ldots, A_{iK} + N_K)$, where

$$N_i \triangleq \int_0^T n(t)\phi_j(t) \; dt = (n, \phi_j) \tag{9.81}$$

We refer to the vector $\mathbf{Z}$ having components

$$Z_j \triangleq A_{ij} + N_j, \qquad j = 1, 2, \ldots, K \tag{9.82}$$

as the *data vector*, and the space of all possible data vectors as the *observation space*. Figure 9.6 illustrates a typical situation for $K = 3$.

The decision-making problem we are therefore faced with is one of associating sets of noisy signal points with each possible transmitted signal point in a manner that will minimize the average error probability. That is, the observation space must be partitioned into M regions R_i, one associated with each transmitted signal, such that if a received data point falls into region R_ℓ, the decision "$s_\ell(t)$ transmitted" is made with minimum probability of error.

In Section 9.1, the minimum-probability-of-error detector was shown to

FIGURE 9.6   A three-dimensional observation space

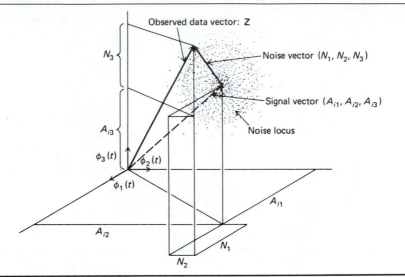

correspond to a MAP decision rule. Thus, letting hypothesis H_ℓ be "signal $s_\ell(t)$ transmitted," we want to implement a receiver that computes

$$P(H_\ell | Z_1, Z_2, \ldots, Z_K), \qquad \ell = 1, 2, \ldots, M \tag{9.83}$$

and chooses the largest.* To compute the posterior probabilities of (9.83), we use Bayes' rule and assume that

$$P(H_1) = P(H_2) = \cdots = P(H_M) \tag{9.84}$$

Application of Bayes' rule results in

$$P(H_\ell | z_1, \ldots, z_K) = \frac{f_Z(z_1, \ldots, z_K | H_\ell) P(H_\ell)}{f_Z(z_1, \ldots, z_K)} \tag{9.85}$$

However, since the factors $P(H_\ell)$ and $f_Z(z_1, \ldots, z_K)$ do not depend on ℓ, the detector can compute $f_Z(z_1, \ldots, z_K | H_\ell)$ and choose the H_ℓ corresponding to the largest. The Z_j's given by (9.82) are the results of linear operations on a Gaussian process and are therefore Gaussian random variables. All that

---

* Capital letters are used to denote components of data vectors because they represent coordinates of an observation that is *random*.

is required to write their joint pdf, given H_ℓ, are their means, variances, and covariances. Their means, given hypothesis H_ℓ, are

$$E\{Z_j | H_\ell\} = E\{A_{\ell j} + N_j\} = A_{\ell j} + \int_0^T E\{n(t)\}\phi_j(t)\, dt$$

$$= A_{\ell j}, \qquad j = 1, 2, \ldots, K \tag{9.86}$$

Their variances, given hypothesis H_ℓ, are

$$\text{var}\,\{Z_j | H_\ell\} = E\{[(A_{\ell j} + N_j) - A_{\ell j}]^2\} = E\{N_j^2\}$$

$$= E\left\{\int_0^T n(t)\phi_j(t)\, dt \int_0^T n(t')\phi_j(t')\, dt'\right\}$$

$$= \int_0^T \int_0^T E\{n(t)n(t')\}\phi_j(t)\phi_j(t')\, dt\, dt'$$

$$= \int_0^T \int_0^T \frac{N_0}{2}\delta(t - t')\phi_j(t)\phi_j(t')\, dt\, dt'$$

$$= \int_0^T \frac{N_0}{2}\phi_j^2(t)\, dt = \tfrac{1}{2}N_0, \qquad j = 1, 2, \ldots, K \tag{9.87}$$

where the orthonormality of the ϕ_j's has been used. In a similar manner, it can be shown that the covariance of Z_j and Z_k, for $j \neq k$, is zero. Thus Z_1, $Z_2, \ldots, Z_K$ are uncorrelated Gaussian random variables and, hence, are statistically independent. Thus

$$f_Z(z_1, \ldots, z_K | H_\ell) = \prod_{j=1}^K \frac{\exp\,[-(z_j - A_{\ell j})^2 / N_0]}{\sqrt{\pi N_0}}$$

$$= \frac{\exp\,[-\sum_{j=1}^K (z_j - A_{\ell j})^2 / N_0]}{(\pi N_0)^{K/2}}$$

$$= \frac{\exp\,\{-\|z - s_\ell\|^2 / N_0\}}{(\pi N_0)^{K/2}} \tag{9.88}$$

where

$$z = z(t) = \sum_{j=1}^K z_j \phi_j(t) \tag{9.89}$$

and

$$s_\ell = s_\ell(t) = \sum_{j=1}^K A_{\ell j}\phi_j(t) \tag{9.90}$$

Except for a factor independent of ℓ (9.88) is the posterior probability $P(H_\ell | z_1, \ldots, z_K)$ as obtained by applying Bayes' rule. Hence, choosing H_ℓ corresponding to the maximum posterior probability is the same as choosing the signal with coordinates $A_{\ell 1}, A_{\ell 2}, \ldots, A_{\ell K}$ so as to maximize (9.88) or, equivalently, so as to minimize the exponent. But $\|z - s_\ell\|$ is the distance between $z(t)$ and $s_\ell(t)$. Thus it has been shown that the decision criterion that minimizes the average probability of error is to choose as the transmitted signal the one whose signal point is closest to the received data point in observation space, distance being defined as the square root of the sum of the squares of the differences of the data and signal vector components. That is, choose H_ℓ such that*

$$(\text{Distance})^2 = d^2 = \sum_{j=1}^{K} (Z_j - A_{\ell j})^2$$

$$= \|z - s_\ell\|^2 = \text{minimum}, \qquad \ell = 1, 2, \ldots, M \tag{9.91}$$

which is exactly the operation to be performed by the receiver structure of Figure 9.5. We illustrate this procedure with the following example.

▼ **EXAMPLE 9.7**   In this example we consider M-ary coherent FSK in terms of signal space. The transmitted signal set is

$$s_i(t) = A \cos \{2\pi[f_c + (i - 1)\Delta f]t\}, \qquad 0 \le t \le T_s \tag{9.92}$$

where

$$\Delta f = \frac{m}{2T_s}, \qquad m \text{ an integer}, \; i = 1, 2, \ldots, M$$

For mathematical simplicity, we assume that $f_c T_s$ is an integer. The ortho-normal basis set can be obtained by applying the Gram-Schmidt procedure. Choosing

$$v_1(t) = s_1(t) = A \cos (2\pi f_c t), \qquad 0 \le t \le T_s \tag{9.93}$$

we have

$$\|v_1\|^2 = \int_0^{T_s} A^2 \cos^2 (2\pi f_c t) \, dt = \frac{A^2 T_s}{2} \tag{9.94}$$

---

* Again, Z_j is the jth coordinate of an observation $z(t)$ that is random; (9.91) is referred to as a *decision rule*.

so that

$$\phi_1(t) = \frac{v_1}{\|v_1\|} = \sqrt{\frac{2}{T_s}} \cos (2\pi f_c t), \qquad 0 \leq t \leq T_s \qquad (9.95)$$

It can be shown in a straightforward fashion that $(s_2, \phi_1) = 0$ if $\Delta f = m/(2T_s)$, so that the second orthonormal function is

$$\phi_2(t) = \sqrt{\frac{2}{T_s}} \cos [2\pi(f_c + \Delta f)t], \qquad 0 \leq t \leq T_s \qquad (9.96)$$

and similarly for $M - 2$ other orthonormal functions up to $\phi_M(t)$. Thus the number of orthonormal functions is the same as the number of possible signals; the ith signal can be written in terms of the ith orthonormal function as

$$s_i(t) = \sqrt{E_s}\,\phi_i(t) \qquad (9.97)$$

We let the received signal plus noise waveform be represented as $y(t)$. When projected into the observation space, $y(t)$ has M coordinates, the ith one of which is given by

$$Z_i = \int_0^{T_s} y(t)\phi_i(t)\,dt \qquad (9.98)$$

where $y(t) = s_i(t) + n(t)$. If $s_\ell(t)$ is transmitted, the decision rule (9.91) becomes

$$d^2 = \sum_{j=1}^{M} (Z_j - \sqrt{E_s}\,\delta_{\ell j})^2 = \text{minimum over } \ell = 1, 2, \ldots, M \qquad (9.99)$$

Taking the square root and writing the sum out, this can be expressed as

$$d = \sqrt{Z_1^2 + Z_2^2 + \cdots + (Z_\ell - \sqrt{E_s})^2 + \cdots + Z_M^2}$$
$$= \text{minimum} \qquad (9.100)$$

For two dimensions (binary FSK), the signal points lie on the two orthogonal axes at a distance $\sqrt{E_s}$ out from the origin. The decision space consists of the first quadrant, and the optimum (minimum error probability) partition is a line at 45° bisecting the right angle made by the two coordinate axes.

For *M*-ary FSK transmission, an alternative way of viewing the decision rule can be obtained by squaring the ℓth term in (9.99) so that we have

$$d^2 = \sum_{j=1}^{M} Z_j^2 + E_s - 2\sqrt{E_s}\, Z_\ell = \text{minimum} \qquad (9.101)$$

Since the sums over *j* and E_s are independent of ℓ (a constant), d^2 can be minimized with respect to ℓ by choosing as the possible transmitted signal the one that will maximize the last term; that is, the decision rule becomes: Choose the possible transmitted signal $s_\ell(t)$ such that

$$\sqrt{E_s}\, Z_\ell = \text{maximum} \quad \text{or} \quad Z_\ell = \int_0^{T_s} y(t)\phi_\ell(t)\, dt$$

$$= \text{maximum with respect to } \ell \qquad (9.102)$$

In other words, we look at the output of the bank of correlators shown in Figure 9.5 at time $t = T_s$ and choose the one with the largest output as corresponding to the most probable transmitted signal.

Sufficient Statistics

To show that (9.91) is indeed the decision rule corresponding to a MAP criterion, we must clarify one point. In particular, the decision is based on the noisy signal

$$z(t) = \sum_{j=1}^{K} Z_j \phi_j(t) \qquad (9.103)$$

Because of the noise component $n(t)$, this is *not* the same as $y(t)$, since an infinite set of basis functions would be required to represent all possible $y(t)$'s. However, we may show that only *K* coordinates, where *K* is the signal-space dimension, are required to provide all the information that is relevant to making a decision.

Assuming a complete orthonormal set of basis functions, $y(t)$ can be expressed as

$$y(t) = \sum_{j=1}^{\infty} Y_j \phi_j(t) \qquad (9.104)$$

where the first K of the ϕ_j's are chosen using the Gram-Schmidt procedure for the given signal set. Given that hypothesis H_ℓ is true, the Y_j's are given by

$$
Y_j = \begin{cases} Z_j = A_{\ell j} + N_j, & j = 1, 2, \ldots, K \\ N_j, & j = K + 1, K + 2, \ldots \end{cases} \tag{9.105}
$$

where Z_j, $A_{\ell j}$, and N_j are as defined previously. Using a procedure identical to the one used in obtaining (9.86) and (9.87), we can show that

$$
E\{Y_j\} = \begin{cases} A_{ij}, & j = 1, 2, \ldots, K \\ 0, & j > K \end{cases} \tag{9.106a}
$$

$$
\text{var}\,\{Y_j\} = \tfrac{1}{2}N_0, \qquad \text{all } j \tag{9.106b}
$$

with $\text{cov}\,\{Y_j Y_k\} = 0, j \neq k$. Thus the joint pdf of $Y_1, Y_2, \ldots$, given H_ℓ, is of the form

$$
f_Y(y_1, y_2, \ldots, y_K, \ldots | H_\ell)
$$

$$
= C \exp\left\{-\frac{1}{N_0}\left[\sum_{j=1}^{K}(y_j - A_{\ell j})^2 + \sum_{j=K+1}^{\infty} y_j^2\right]\right\}
$$

$$
= C^1 \exp\left(-\frac{1}{N_0}\sum_{j=K+1}^{\infty} y_j^2\right) f_Z(y_1, \ldots, y_K | H_\ell)
$$

$$
\tag{9.107}
$$

where C and C^1 are constants. Since this pdf factors, $Y_{K+1}, Y_{K+2}, \ldots$ are independent of $Y_1, Y_2, \ldots, Y_K$ and the former provide no information for making a decision. Thus d^2 given by (9.101) is known as a *sufficient statistic*.

Detection of M-ary Orthogonal Signals

As a more complex example of the use of signal-space techniques, let us consider an M-ary signaling scheme for which the signal waveforms have equal energies and are orthogonal over the signaling interval. Thus

$$
\int_0^{T_s} s_i(t)s_j(t)\,dt = \begin{cases} E_s, & i = j \\ 0, & i \neq j, i = 1, 2, \ldots, M \end{cases} \tag{9.108}
$$

where E_s is the energy of each signal in $(0, T_s)$.

A practical example of such a signaling scheme is the signal set for M-ary coherent FSK given by (9.92). The decision rule for this signaling scheme was considered in Example 9.7. It was found that $K = M$ orthonormal functions are required, and the receiver shown in Figure 9.5 involves M correlators. The output of the jth correlator at time T_s is given by (9.82). The decision criterion is to choose the signal point $i = 1, 2, \ldots, M$ such that d^2 given by (9.91) is minimized or, as shown in Example 9.7, such that

$$Z_\ell = \int_0^{T_s} y(t)\phi_\ell(t)\, dt = \text{maximum with respect to } \ell \qquad (9.109)$$

That is, the signal is chosen that has the maximum correlation with the received signal plus noise. To compute the probability of symbol error, we note that

$$P_E = \sum_{i=1}^{M} P[E \mid s_i(t) \text{ sent}]P[s_i(t) \text{ sent}]$$

$$= \frac{1}{M} \sum_{i=1}^{M} P[E \mid s_i(t) \text{ sent}] \qquad (9.110)$$

where each signal is assumed *a priori* equally probable. We may write

$$P[E \mid s_i(t) \text{ sent}] = 1 - P_{ci} \qquad (9.111)$$

where P_{ci} is the probability of a correct decision given that $s_i(t)$ was sent. Since a correct decision results only if

$$Z_j = \int_0^{T_s} y(t)s_j(t)\, dt < \int_0^{T_s} y(t)s_i(t)\, dt = Z_i \qquad (9.112)$$

for all $j \neq i$, we may write P_{ci} as

$$P_{ci} = P \text{ (all } Z_j < Z_i, \quad j \neq i) \qquad (9.113)$$

If $s_i(t)$ is transmitted, then

$$Z_i = \int_0^{T_s} [\sqrt{E_s}\, \phi_i(t) + n(t)]\phi_i(t)\, dt$$

$$= \sqrt{E_s} + N_i \qquad (9.114)$$

where

$$N_i = \int_0^{T_s} n(t)\phi_i(t)\, dt \qquad (9.115)$$

Since $Z_j = N_j$, $j \neq i$, given $s_i(t)$ was sent, it follows that (9.113) becomes

$$P_{ci} = P(\text{all } N_j < \sqrt{E_s} + N_i, \quad j \neq i) \qquad (9.116)$$

Now N_i is a Gaussian random variable (a linear operation on a Gaussian process) with zero mean and variance

$$\text{var}\,[N_i] = E\left\{\left[\int_0^{T_s} n(t)\phi_j(t)\,dt\right]^2\right\} = \frac{N_0}{2} \tag{9.117}$$

Furthermore, N_i and N_j, for $i \neq j$, are independent, since

$$E[N_iN_j] = 0 \tag{9.118}$$

Given a particular value of N_i, (9.116) becomes

$$P_{ci}(N_i) = \prod_{\substack{j=1 \\ j\neq i}}^{M} P[N_j < \sqrt{E_s} + N_i]$$

$$= \left(\int_{-\infty}^{\sqrt{E_s}+n_i} \frac{e^{-n_j^2/N_0}}{\sqrt{\pi N_0}}\,dn_j\right)^{M-1} \tag{9.119}$$

which follows because the pdf of N_j is $n(0, \sqrt{N_0/2})$. Averaged over all possible values of N_i, (9.119) gives

$$P_{ci} = \int_{-\infty}^{\infty} \frac{e^{-n_i^2/N_0}}{\sqrt{\pi N_0}}\left(\int_{-\infty}^{\sqrt{E_s}+n_i} \frac{e^{-n_j^2/N_0}}{\sqrt{\pi N_0}}\,dn_j\right)^{M-1}\,dn_i$$

$$= (\pi)^{-M/2}\int_{-\infty}^{\infty} e^{-y^2}\left(\int_{-\infty}^{\sqrt{E_s/N_0}+y} e^{-x^2}\,dx\right)^{M-1}\,dy \tag{9.120}$$

where the substitutions $x = n_j/\sqrt{N_0}$ and $y = n_i/\sqrt{N_0}$ have been made.

Since P_{ci} is independent of i, it follows that the probability of error is

$$P_E = 1 - P_{ci} \tag{9.121}$$

With (9.120) substituted in (9.121), a nonintegrable M-fold integral for P_E results, and one must resort to numerical integration to evaluate it.* Curves showing P_E versus $E_s/(N_0 \log_2 M)$ are given in Figure 9.7 for several values of M. We note a rather surprising behavior: As $M \to \infty$, error-free transmission can be achieved as long as $E_s/(N_0 \log_2 M) > \ln 2 = -1.59$ dB. This error-free transmission is achieved at the expense of infinite bandwidth, however, since $M \to \infty$ means that an infinite number or orthonormal functions are required. We will discuss this behavior further in Chapter 10.

---

* See Lindsey and Simon (1973), pp. 199ff, for tables giving P_E.

FIGURE 9.7 Probability of symbol error for coherent detection of *M*-ary orthogonal signals

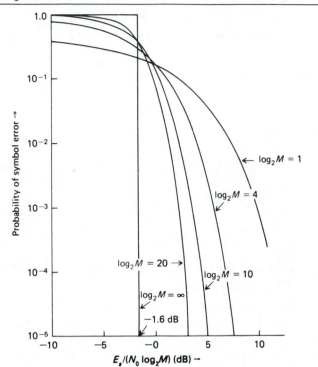

A Noncoherent Case

To illustrate the application of signal-space techniques to noncoherent digital signaling, let us consider the following binary hypothesis situation:

$$H_1: y(t) = G\sqrt{2E/T} \cos(\omega_1 t + \theta) + n(t)$$
$$H_2: y(t) = G\sqrt{2E/T} \cos(\omega_2 t + \theta) + n(t), \qquad 0 \le t \le T \qquad (9.122a)$$

where E is the energy of the transmitted signal in one bit period and $n(t)$ is white Gaussian noise with double-sided power spectral density $\frac{1}{2}N_0$. It is assumed that $|\omega_1 - \omega_2|/2\pi \gg T^{-1}$ so that the signals are orthogonal. Except for G and θ, which are assumed to be random variables, this problem would be a special case of the *M*-ary orthogonal signaling case just considered. (Recall also the consideration of coherent and noncoherent FSK in Chapter 7.)

The random variables G and θ represent random gain and phase perturbations introduced by a fading channel. The channel is modeled as introducing a random gain and phase shift during each bit interval. Because the gain and phase shift are assumed to remain constant throughout a bit interval, this channel model is called *slowly fading*. We assume that G is Rayleigh and θ is uniform in $(0, 2\pi)$ and that G and θ are independent.

Expanding (9.122a), we obtain

$$H_1: y(t) = \sqrt{2E/T} \, (G_1 \cos \omega_1 t + G_2 \sin \omega_1 t) + n(t)$$
$$H_2: y(t) = \sqrt{2E/T} \, (G_1 \cos \omega_2 t + G_2 \sin \omega_2 t) + n(t), \qquad 0 \le t \le T$$

$$(9.122b)$$

where $G_1 = G \cos \theta$ and $G_2 = -G \sin \theta$ are independent, zero-mean, Gaussian random variables (recall Example 4.15). We denote their variances by σ^2. Choosing the orthonormal basis set

$$
\left.
\begin{aligned}
\phi_1(t) &= \sqrt{2/T} \cos \omega_1 t \\
\phi_2(t) &= \sqrt{2/T} \sin \omega_1 t \\
\phi_3(t) &= \sqrt{2/T} \cos \omega_2 t \\
\phi_4(t) &= \sqrt{2/T} \sin \omega_2 t
\end{aligned}
\right\} \quad 0 \le t \le T
$$

$$
\begin{aligned}
&(9.123a)\\
&(9.123b)\\
&(9.123c)\\
&(9.123d)
\end{aligned}
$$

the term $y(t)$ can be resolved into a four-dimensional signal space, and decisions may be based on the data vector

$$\mathbf{Z} = (Z_1, Z_2, Z_3, Z_4) \tag{9.124}$$

where

$$Z_i = (y, \phi_i) = \int_0^T y(t)\phi_i(t) \, dt \tag{9.125}$$

Given hypothesis H_1, we obtain

$$Z_i = \begin{cases} \sqrt{E}G_i + N_i, & i = 1, 2 \\ N_i, & i = 3, 4 \end{cases} \tag{9.126}$$

and given hypothesis H_2, we obtain

$$Z_i = \begin{cases} N_i, & i = 1, 2 \\ \sqrt{E}G_{i-2} + N_i & i = 3, 4 \end{cases} \tag{9.127}$$

where

$$N_i = (n, \phi_i) = \int_0^T n(t)\phi_i(t) \, dt, \qquad i = 1, 2, 3, 4 \tag{9.128}$$

are independent Gaussian random variables with zero mean and variance $\frac{1}{2}N_0$. Since G_1 and G_2 are also independent Gaussian random variables with zero mean and variance σ^2, the joint conditional pdf's of Z, given H_1 and H_2, are the products of the respective marginal pdf's. It follows that

$$f_Z(z_1, z_2, z_3, z_4 | H_1)$$
$$= \frac{\exp\left[-(z_1^2 + z_2^2)/(2E\sigma^2 + N_0)\right] \exp\left[-(z_3^2 + z_4^2)/N_0\right]}{\pi^2(2E\sigma^2 + N_0)N_0} \qquad (9.129a)$$

and

$$f_Z(z_1, z_2, z_3, z_4 | H_2)$$
$$= \frac{\exp\left[-(z_1^2 + z_2^2)/N_0\right] \exp\left[-(z_3^2 + z_4^2)/(2E\sigma^2 + N_0)\right]}{\pi^2(2E\sigma^2 + N_0)N_0} \qquad (9.129b)$$

The decision rule that minimizes the probability of error is to choose the hypothesis H_ℓ corresponding to the largest posterior probability $P(H_\ell | z_1, z_2, z_3, z_4)$. But these probabilities differ from (9.129) only by a constant that is independent of ℓ. For a particular observation $\mathbf{Z} = (Z_1, Z_2, Z_3, Z_4)$, the decision rule is

$$f_Z(Z_1, Z_2, Z_3, Z_4 | H_1) \underset{H_2}{\overset{H_1}{\gtrless}} f_Z(Z_1, Z_2, Z_3, Z_4 | H_2) \qquad (9.130)$$

which, after substitution from (9.129) and simplification, reduces to

$$R_2^2 \triangleq Z_3^2 + Z_4^2 \underset{H_2}{\overset{H_1}{\gtrless}} Z_1^2 + Z_2^2 \triangleq R_1^2 \qquad (9.131)$$

The optimum receiver corresponding to this decision rule is shown in Figure 9.8.

To find the probability of error, we note that both $R_1 \triangleq \sqrt{Z_1^2 + Z_2^2}$ and $R_2 \triangleq \sqrt{Z_3^2 + Z_4^2}$ are Rayleigh random variables under either hypothesis. Given H_1 is true, an error results if $R_2 > R_1$, where the positive square root of (9.131) has been taken. From Example 4.15, it follows that

$$f_{R_1}(r_1 | H_1) = \frac{r_1 e^{-r_1^2/(2E\sigma^2 + N_0)}}{E\sigma^2 + \frac{1}{2}N_0}, \qquad r_1 > 0 \qquad (9.132a)$$

and

$$f_{R_2}(r_2 | H_1) = \frac{2r_2 e^{-r_2^2/N_0}}{N_0}, \qquad r_2 > 0 \qquad (9.132b)$$

FIGURE 9.8 Optimum receiver structures for detection of binary orthogonal signals in Rayleigh fading. (a) Implementation by correlator and squarer. (b) Implementation by matched filter and envelope detector.

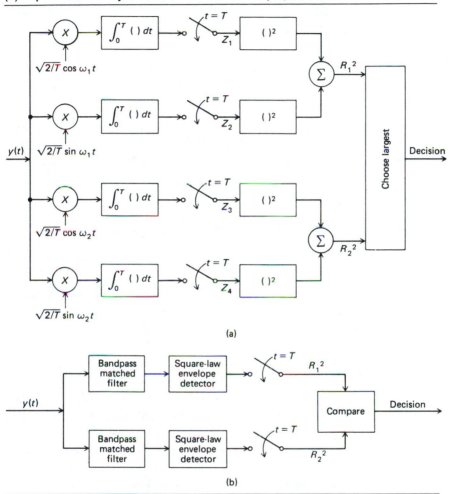

(a)

(b)

The probability that $R_2 > R_1$, averaged over R_1, is

$$P(E \mid H_1) = \int_0^\infty \left[\int_{r_1}^\infty f_{R_2}(r_2 \mid H_1)\, dr_2 \right] f_{R_1}(r_1 \mid H_1)\, dr_1$$

$$= \frac{1}{2} \frac{1}{1 + \frac{1}{2}(2\sigma^2 E/N_0)} \tag{9.133}$$

FIGURE 9.9 Comparison of P_E versus SNR for Rayleigh and fixed channels with noncoherent FSK signaling

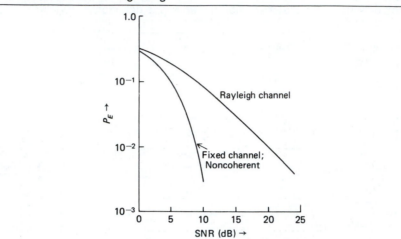

where $2\sigma^2 E$ is the average received signal energy. Because of the symmetry involved, it follows that $P(E|H_1) = P(E|H_2)$ and that

$$P_E = P(E|H_1) = P(E|H_2) \qquad (9.134)$$

The probability of error is plotted in Figure 9.9, along with the result from Chapter 7 for constant-amplitude noncoherent FSK (Figure 7.20). Whereas the error probability for nonfading, noncoherent FSK signaling decreases exponentially with the signal-to-noise ratio, the fading channel results in an error probability that decreases only inversely with signal-to-noise ratio.

One way to combat this degradation due to fading is to employ *diversity transmission;* that is, the transmitted signal power is divided among several independently fading transmission paths with the hope that not all of them will fade simultaneously. Several ways of achieving diversity were mentioned in Chapter 7. (See also Problem 9.17.)

9.4 ESTIMATION THEORY

We now consider the second type of optimization problem discussed in the introduction to this chapter—the estimation of parameters from random data. After introducing some background theory here, we will consider several applications of estimation theory to communication systems in Section 9.5.

In introducing the basic ideas of estimation theory, we will exploit several parallels with detection theory. As in the case of signal detection, we have available a noisy observation Z that depends probabilistically on a parameter of interest A.* For example, Z could be the sum of an unknown dc voltage A and an independent noise component N: $Z = A + N$.

Two different estimation procedures will be considered. These are Bayes estimation and maximum-likelihood (ML) estimation. For Bayes estimation, A is considered to be random with a known *a priori* pdf $f_A(a)$, and a suitable cost function is minimized to find the optimum estimate of A. Maximum-likelihood estimation can be used for the estimation of nonrandom parameters or a random parameter with an unknown *a priori* pdf.

Bayes Estimation

Bayes estimation involves the minimization of a cost function, as in the case of Bayes detection. Given an observation Z, we seek the *estimation rule* (or *estimator*) $\hat{a}(Z)$ that assigns a value $\hat{A}$ to A such that the cost function $C[A, \hat{a}(Z)]$ is minimized. Note that C is a function of the unknown parameter A and the observation Z. Clearly, as the absolute error $|A - \hat{a}(Z)|$ increases, $C[A, \hat{a}(Z)]$ should increase, or at least not decrease; that is, large errors should be more costly than small errors. Two useful cost functions are

$$C[A, \hat{a}(Z)] = [A - \hat{a}(Z)]^2 \quad \text{(squared-error cost)} \tag{9.135a}$$

$$C[A, \hat{a}(Z)] = \begin{cases} 1, & |A - \hat{a}(Z)| > \Delta > 0 \\ 0, & \text{otherwise} \end{cases} \quad \text{(uniform cost)} \tag{9.135b}$$

where Δ is a suitably chosen constant. For each of these cost functions, we wish to find the decision rule $\hat{a}(Z)$ that minimizes the average cost $E\{C[A, \hat{a}(Z)]\} = \overline{C[A, \hat{a}(Z)]}$. Because both A and Z are random variables, the average cost, or risk, is given by

$$\overline{C[A, \hat{a}(Z)]} = \int_{-\infty}^{\infty} \int_{-\infty}^{\infty} C[a, \hat{a}(z)] f_{AZ}(a, z) \, da \, dz$$

$$= \int_{-\infty}^{\infty} \int_{-\infty}^{\infty} C[a, \hat{a}(z)] f_{Z|A}(z \,|\, a) f_A(a) \, dz \, da \tag{9.136}$$

where $f_{AZ}(a, z)$ is the joint pdf of A and Z and $f_{Z|A}(z \,|\, a)$ is the conditional pdf of Z given A. The latter can be found if the probabilistic mechanism that

---

* For simplicity, we consider the single-observation case first and generalize to vector observations later.

produces Z from A is known. For example, if $Z = A + N$, where N is a zero-mean Gaussian random variable with variance σ_n^2, then

$$f_{Z|A}(z|a) = \frac{e^{-(z-a)^2/2\sigma_n^2}}{\sqrt{2\pi\sigma_n^2}} \qquad (9.137)$$

Returning to the minimization of the risk, we find it more advantageous to express (9.136) in terms of the conditional pdf $f_{A|Z}(a|z)$, which can be done by means of Bayes' rule, to obtain

$$\overline{C[A, \hat{a}(Z)]} = \int_{-\infty}^{\infty} f_Z(z) \left\{ \int_{-\infty}^{\infty} C[a, \hat{a}(z)] f_{A|Z}(a|z)\, da \right\} dz \qquad (9.138a)$$

where

$$f_Z(z) = \int_{-\infty}^{\infty} f_{Z|A}(z|a) f_A(a)\, da \qquad (9.138b)$$

is the pdf of Z. Since $f_Z(z)$ and the inner integral in (9.138a) are nonnegative, the risk can be minimized by minimizing the inner integral for each z. The inner integral in (9.138a) is called the *conditional risk*.

This minimization is accomplished for the squared-error cost function, (9.135a), by differentiating the conditional risk with respect to $\hat{a}$ for a particular observation Z and setting the result equal to zero. The resulting differentiation yields

$$\frac{\partial}{\partial \hat{a}} \int_{-\infty}^{\infty} [a - \hat{a}(Z)]^2 f_{A|Z}(a|Z)\, da$$

$$= -2 \int_{-\infty}^{\infty} a f_{A|Z}(a|Z)\, da + 2\hat{a}(Z) \int_{-\infty}^{\infty} f_{A|Z}(a|Z)\, da \qquad (9.139)$$

which, when set to zero, results in

$$\hat{a}_{se}(Z) = \int_{-\infty}^{\infty} a f_{A|Z}(a|Z)\, da \qquad (9.140)$$

where the fact that $\int_{-\infty}^{\infty} f_{A|Z}(a|Z)\, da = 1$ has been used. A second differentiation shows that this is a minimum. Note that $\hat{a}_{se}(Z)$, the estimator for a squared-error cost function, is the mean of the pdf of A given the observation Z, or the conditional mean. The values that $\hat{a}_{se}(Z)$ assume, $\hat{A}$, are random since the estimator is a function of the random variable Z.

In a similar manner, we can show that the uniform cost function results in the condition

$$f_{A|Z}(A|Z) \Big|_{A=\hat{a}_{\text{MAP}}(Z)} = \text{maximum} \tag{9.141}$$

for Δ in (9.135b) infinitesimally small. That is, the estimation rule, or estimator, that minimizes the uniform cost function is the maximum of the conditional pdf of A given Z, or the *a posteriori* pdf. Thus this estimator will be referred to as the *maximum a posteriori* (MAP) estimate. Necessary, but not sufficient, conditions that the MAP estimate must satisfy are

$$\frac{\partial}{\partial A} f_{A|Z}(A|Z) \Big|_{A=\hat{a}_{\text{MAP}}(Z)} = 0 \tag{9.142a}$$

and

$$\frac{\partial}{\partial A} \ln f_{A|Z}(A|Z) \Big|_{A=\hat{a}_{\text{MAP}}(Z)} = 0 \tag{9.142b}$$

where the latter condition is especially convenient for *a posteriori* pdf's of exponential type, such as Gaussian.

Often the MAP estimate is employed because it is easier to obtain than other estimates, even though the conditional-mean estimate, given by (9.140), is more general, as the following theorem indicates.

▷ THEOREM
If, as a function of a, the *a posteriori* pdf $f_{A|Z}(a|Z)$ has a single peak, about which it is symmetrical, and the cost function has the properties

$$C(A, \hat{a}) = C(A - \hat{a}) \tag{9.143a}$$
$$C(x) = C(-x) \geq 0 \qquad \text{(symmetrical)} \tag{9.143b}$$
$$C(x_1) \geq C(x_2) \text{ for } |x_1| \geq |x_2| \qquad \text{(convex)} \tag{9.143c}$$

then the conditional-mean estimator is the Bayes estimate.*

Maximum-Likelihood Estimation

We now seek an estimation procedure that does not require *a priori* information about the parameter of interest. Such a procedure is maximum-likelihood (ML) estimation. To explain this procedure, consider the MAP estimation of a random parameter A about which little is known. This lack

---

* Van Trees (1968), pp. 60–61.

of information about A is expressed probabilistically by assuming the prior pdf of A, $f_A(a)$, to be broad compared with the posterior pdf, $f_{A|Z}(a|Z)$. If this were not the case, the observation Z would be of little use in estimating A. Since the joint pdf of A and Z is given by

$$f_{AZ}(a, z) = f_{A|Z}(a|z)f_Z(z) \tag{9.144}$$

the joint pdf, regarded as a function of a, must be peaked for at least one value of a. By the definition of conditional probability, we also may write (9.144) as

$$f_{ZA}(z, a) = f_{Z|A}(z|a)f_A(a)$$
$$\cong f_{Z|A}(z|a) \quad \text{(times a constant)} \tag{9.145}$$

where the approximation follows by virtue of the assumption that little is known about A, thus implying that $f_A(a)$ is essentially constant. The ML estimate of A is defined as

$$f_{Z|A}(Z|A) \Big|_{A=\hat{a}_{ML}(Z)} = \text{maximum} \tag{9.146}$$

But, from (9.144) and (9.145), the ML estimate of a parameter corresponds to the MAP estimate if little *a priori* information about the parameter is available. From (9.146), it follows that the ML estimate of a parameter A is that value of A which is most likely to have resulted in the observaton Z; hence the name "maximum likelihood." Since the prior pdf of A is not required to obtain an ML estimate, it is a suitable estimation procedure for random parameters whose prior pdf is unknown. If a deterministic parameter is to be estimated, $f_{Z|A}(z|A)$ is regarded as the pdf of Z with A as a parameter.

From (9.146) it follows that the ML estimate can be found from the necessary, but not sufficient, conditions

$$\frac{\partial f_{Z|A}(Z|A)}{\partial A} \Big|_{A=\hat{a}_{ML}(Z)} = 0 \tag{9.147a}$$

and

$$l(A) \triangleq \frac{\partial \ln f_{Z|A}(Z|A)}{\partial A} \Big|_{A=\hat{a}_{ML}(Z)} = 0 \tag{9.147b}$$

When viewed as a function of A, $f_{Z|A}(z|A)$ is referred to as the *likelihood function*. Both (9.147a) and (9.147b) will be referred to as *likelihood equations*.

From (9.142) and Bayes' rule, it follows that the MAP estimate of a random parameter satisfies

$$\left[l(A) + \frac{\partial}{\partial A} \ln f_A(A) \right] \Bigg|_{A = \hat{a}_{MAP}(Z)} = 0 \tag{9.148}$$

which is useful when finding both the ML and MAP estimates of a parameter.

Estimates Based on Multiple Observations

If a multiple number of observations are available, say $\mathbf{Z} \triangleq (Z_1, Z_2, \ldots, Z_K)$, on which to base the estimate of a parameter, we simply substitute the K-fold joint conditional pdf $f_{\mathbf{Z}|A}(\mathbf{z}|A)$ in (9.146) and (9.147) to find the ML estimate of A. If the observations are independent, when conditioned on A, then

$$f_{\mathbf{Z}|A}(\mathbf{z} \mid A) = \prod_{k=1}^{K} f_{Z_k|A}(z_k|A) \tag{9.149}$$

where $f_{Z_k|A}(z_k|A)$ is the pdf of the kth observation Z_k given the parameter A. To find $f_{A|\mathbf{Z}}(a|\mathbf{z})$ for MAP estimation, we use Bayes' rule.

▼| **EXAMPLE 9.8**   To illustrate the estimation concepts just discussed, let us consider the estimation of a constant-level random signal A embedded in Gaussian noise $n(t)$ with zero mean and variance σ_n^2:

$$z(t) = A + n(t) \tag{9.150}$$

We assume $z(t)$ is sampled at time intervals sufficiently spaced so that the samples are independent. Let these samples be represented as

$$Z_k = A + N_k, \quad k = 1, 2, \ldots, K \tag{9.151}$$

Thus, *given* A, the Z_k's are independent, each haviang mean A and variance σ_n^2. Hence the conditional pdf of $\mathbf{Z} \triangleq (Z_1, Z_2, \ldots, Z_K)$ given A is

$$f_{\mathbf{z}|A}(\mathbf{z}|A) = \prod_{k=1}^{K} \frac{\exp\left[-(z_k - A)^2/2\sigma_n^2\right]}{\sqrt{2\pi\sigma}^{2_n}}$$

$$= \frac{\exp\left[-\sum_{k=1}^{K} (z_k - A^2/2\sigma_n^2\right]}{(2\pi\sigma_n^2)^{K/2}} \tag{9.152}$$

We will assume two possibilities for A:

1. It is Gaussian with mean m_A and variance σ_A^2.
2. Its pdf is unknown.

In the first case, we will find the conditional-mean and the MAP estimates for A. In the second case, we will compute the ML estimate.

Case 1

If the pdf of A is

$$f_A(a) = \frac{e^{-(a-m_A)^2/2\sigma_A^2}}{\sqrt{2\pi\sigma_A^2}} \tag{9.153}$$

its posterior pdf is, by Bayes' rule,

$$f_{A|Z}(a|z) = \frac{f_{Z|A}(z|a)f_A(a)}{f_Z(z)} \tag{9.154}$$

After some algebra, it can be shown that

$$f_{A|Z}(a|z) = (2\pi\sigma_p^2)^{-1/2} \exp\left(\frac{-\{a - \sigma_p^2[(Km_s/\sigma_n^2) + (m_A/\sigma_A^2)]\}^2}{2\sigma_p^2}\right) \tag{9.155}$$

where

$$\frac{1}{\sigma_p^2} = \frac{K}{\sigma_n^2} + \frac{1}{\sigma_A^2} \tag{9.156}$$

and the sample mean is

$$m_s = \frac{1}{K}\sum_{k=1}^{K} Z_k \tag{9.157}$$

Clearly, $f_{A|Z}(a|z)$ is a Gaussian pdf with variance σ_p^2 and mean

$$E\{A|Z\} = \sigma_p^2\left(\frac{Km_s}{\sigma_n^2} + \frac{m_A}{\sigma_A^2}\right)$$

$$= \frac{K\sigma_A^2/\sigma_n^2}{1 + K\sigma_A^2/\sigma_n^2}m_s + \frac{1}{1 + K\sigma_A^2/\sigma_n^2}m_A \tag{9.158}$$

Since the maximum value of a Gaussian pdf is at the mean, this is both the conditional-mean estimate (squared-error cost function, among other convex cost functions) and the MAP estimate (square-well cost function). The conditional variance var $\{A\,|\,\mathbf{Z}\}$ is σ_p^2. Because it is not a function of $\mathbf{Z}$, it follows that the average cost, or risk, which is

$$\overline{C[A,\hat{a}(Z)]} = \int_{-\infty}^{\infty} \text{var}\,\{A\,|\,\mathbf{z}\}f_{\mathbf{z}}(\mathbf{z})\,d\mathbf{z} \tag{9.159}$$

is just σ_p^2.

From the expression for $E(A\,|\,\mathbf{Z})$, we note an interesting behavior for the estimate of A, or $\hat{a}(Z)$. As $K\sigma_A^2/\sigma_n^2 \rightarrow \infty$,

$$\hat{a}(Z) \rightarrow m_s = \frac{1}{K}\sum_{k=1}^{K} Z_k \tag{9.160}$$

which says that as the ratio of signal variance to noise variance becomes large, the optimum estimate for A approaches the sample mean. On the other hand, as $K\sigma_A^2/\sigma_n^2 \rightarrow 0$ (small signal variance and/or large noise variance), $\hat{a}(\mathbf{Z}) \rightarrow m_A$, the *a priori* mean of A. In the first case, the estimate is weighted in favor of the observations; in the latter, it is weighted in favor of the known signal statistics. From the form of σ_p^2, we note that, in either case, the quality of the estimate increases as the number of independent samples of $z(t)$ increases.

Case 2

The ML estimate is found by differentiating $\ln f_{\mathbf{Z}|A}(\mathbf{z}\,|\,A)$ with respect to A and setting the result equal to zero. Performing the steps, the ML estimate is found to be

$$\hat{a}_{\text{ML}}(\mathbf{Z}) = \frac{1}{K}\sum_{k=1}^{K} Z_k \tag{9.161}$$

We note that this corresponds to the MAP estimate if $K\sigma_A^2/\sigma_n^2 \rightarrow \infty$ (that is, if the *a priori* pdf of A is broad compared with the *a posteriori* pdf).

The variance of $\hat{a}_{\text{ML}}(\mathbf{Z})$ is found by recalling that the variance of a sum of independent random variables is the sum of the variances. The result is

$$\sigma_{\text{ML}}^2 = \frac{\sigma_n^2}{K} > \sigma_p^2 \tag{9.162}$$

Thus the prior knowlege about A, available through $f_A(a)$, manifests itself as a smaller variance for the Bayes estimates (conditional-mean and MAP) than for the ML estimate.

Other Properties of ML Estimates

Unbiased Estimates   An estimate $\hat{a}(\mathbf{Z})$ is said to be *unbiased* if

$$E\{\hat{a}(\mathbf{Z})|A\} = A \tag{9.163}$$

This is clearly a desirable property of any estimation rule. If $E\{\hat{a}(\mathbf{Z})|A\} - A = B \neq 0$, B is referred to as the *bias of the estimate*.

The Cramer-Rao Inequality   In many cases it may be difficult to compute the variance of an estimate for a nonrandom parameter. A lower bound for the variance of an unbiased ML estimate is provided by the following inequality:

$$\text{var}\,\{\hat{a}(\mathbf{Z})\} \geq \left(E\left\{ \left[\frac{\partial \ln f_{\mathbf{Z}|A}(\mathbf{Z}|a)}{\partial a} \right]^2 \right\} \right)^{-1} \tag{9.164a}$$

or, equivalently,

$$\text{var}\,\{\hat{a}(\mathbf{Z})\} \geq \left(-E\left\{ \frac{\partial^2 \ln f_{\mathbf{Z}|A}(\mathbf{Z}|a)}{\partial a^2} \right\} \right)^{-1} \tag{9.164b}$$

where the expectation is only over $\mathbf{Z}$. These inequalities hold under the assumption that $\partial f_{\mathbf{Z}|A}/\partial a$ and $\partial^2 f_{\mathbf{Z}|A}/\partial a^2$ exist and are absolutely integrable. A proof is furnished by Van Trees (1968). Any estimate satisfying (9.164) with equality is said to be *efficient*.

A sufficient condition for equality in (9.164) is that

$$\frac{\partial \ln f_{\mathbf{Z}|A}(\mathbf{Z}|a)}{\partial a} = [\hat{a}(\mathbf{Z}) - A]\, g(a) \tag{9.165}$$

where $g(\cdot)$ is a function only of a. If an efficient estimate of a parameter exists, it is the maximum-likelihood estimate.

Asymptotic Qualities of ML Estimates   In the limit, as the number of independent observations becomes large, ML estimates can be shown to be *Gaussian, unbiased,* and *efficient*. In addition, the probability that the ML estimate for K observations differs by a fixed amount ϵ from the true value approaches zero as $K \to \infty$; an estimate with such behavior is referred to as *consistent*.

▼ **EXAMPLE 9.9**   Returning to Example 9.8, we can show that $\hat{a}_{\text{ML}}(\mathbf{Z})$ is an efficient estimate. We have already shown that $\sigma_{\text{ML}}^2 = \sigma_n^2/K$. Using (9.164b), we differentiate in $f_{\mathbf{Z}|A}$ once to obtain

$$\frac{\partial \ln f_{\mathbf{Z}|A}}{\partial a} = \frac{1}{\sigma_n^2} \sum_{k=1}^{K} (Z_k - a) \tag{9.166}$$

A second differentiation gives

$$\left.\frac{\partial^2 \ln f_{Z|A}}{\partial a^2}\right| = -\frac{K}{\sigma_n^2} \qquad (9.167)$$

▲| and (9.164b) is seen to be satisfied with equality.

9.5   APPLICATIONS OF ESTIMATION THEORY TO COMMUNICATIONS

We now consider two applications of estimation theory to the transmission of analog data. The sampling theorem introduced in Chapter 2 was applied in Chapter 3 in the discussion of several systems for the transmission of continuous-waveform messages via their sample values. One such technique is PAM, in which the sample values of the message are used to amplitude-modulate a pulse-type carrier. We will apply the results of Example 9.8 to find the performance of the optimum demodulator for PAM. This is a *linear* estimator because the observations are linearly dependent on the message sample values. For such a system, the only way to decrease the effect of noise on the demodulator output is to increase the SNR of the received signal, since output and input SNR are linearly related.

Following the consideration of PAM, we will derive the optimum ML estimator for the phase of a signal in additive Gaussian noise. This will result in a phase-lock loop structure. The variance of the estimate in this case will be obtained for high input SNR by applying the Cramer-Rao inequality. For low SNRs, the variance is difficult to obtain because this is a problem in *nonlinear* estimation; that is, the observations are nonlinearly dependent on the parameter being estimated.

The transmission of analog samples by pulse-position modulation (PPM) or some other modulation scheme could also be considered. An approximate analysis of its performance for low input SNRs would show the threshold effect of nonlinear modulation schemes and the implications of the tradeoff that is possible between bandwidth and output SNR. This effect was seen previously in Chapter 6 when the performance of PCM in noise was considered.

Pulse-Amplitude Modulation (PAM)

In PAM, the message $m(t)$ of bandwidth W is sampled at T-second intervals, where $T \leq 1/2W$, and the sample values $m_k = m(t_k)$ are used to amplitude-modulate a pulse train composed of time-translates of the basic pulse shape

$p(t)$, which is assumed zero for $t \leq 0$ and $t \geq T_0 < T$. The received signal plus noise is represented as

$$y(t) = \sum_{k=-\infty}^{\infty} m_k p(t - kT) + n(t) \qquad (9.168)$$

where $n(t)$ is white Gaussian noise with double-sided power spectral density $\frac{1}{2}N_0$.

Considering the estimation of a single sample at the receiver, we observe

$$y(t) = m_0 p(t) + n(t), \qquad 0 \leq t \leq T \qquad (9.169)$$

For convenience, if we assume that $\int_0^{T_0} p^2(t) \, dt = 1$, it follows that a sufficient statistic is

$$Z_0 = \int_0^{T_0} y(t) p(t) \, dt$$

$$= m_0 + N \qquad (9.170)$$

where the noise component is

$$N = \int_0^{T_0} n(t) p(t) \, dt \qquad (9.171)$$

Having no prior information about m_0, we apply ML estimation. Following procedures used many times before, we can show that N is a zero-mean Gaussian random variable with variance $\frac{1}{2}N_0$. The ML estimation of m_0 is therefore identical to the single-observation case of Example 9.8, and the best estimate is simply Z_0. As in the case of digital data transmission, this estimator could be implemented by passing $y(t)$ through a filter matched to $p(t)$, observing the output amplitude prior to the next pulse, and then setting the filter initial conditions to zero. Note that the estimator is *linearly* dependent on $y(t)$.

The variance of the estimate is equal to the variance of N, or $\frac{1}{2}N_0$. Thus the SNR at the output of the estimator is

$$(\text{SNR})_0 = \frac{2m_0^2}{N_0}$$

$$= \frac{2E}{N_0} \qquad (9.172)$$

where $E = \int_0^{T_0} m_0^2 p^2(t) \, dt$ is the average energy of the received signal sample. Thus the only way to increase $(\text{SNR})_0$ is by increasing the energy per sample or by decreasing N_0.

Estimation of Signal Phase: The PLL Revisited

We now consider the problem of estimating the phase of a sinusoidal signal $A \cos (\omega_c t + \theta)$ in white Gaussian noise $n(t)$ of double-sided power spectral density $\frac{1}{2}N_0$. Thus the observed data are

$$y(t) = A \cos (\omega_c t + \theta) + n(t), \qquad 0 \le t \le T \qquad (9.173)$$

where T is the observation interval. Expanding $A \cos (\omega_c t + \theta)$ as

$$A \cos \omega_c t \cos \theta - A \sin \omega_c t \sin \theta$$

we see that a suitable set of orthonormal basis functions for representing the data is

$$\phi_1(t) = \sqrt{\frac{2}{T}} \cos \omega_c t, \qquad 0 \le t \le T \qquad (9.174a)$$

and

$$\phi_2(t) = \sqrt{\frac{2}{T}} \sin \omega_c t, \qquad 0 \le t \le T \qquad (9.174b)$$

Thus we base our decision on

$$z(t) = \sqrt{\frac{T}{2}} A \cos \theta \, \phi_1(t) - \sqrt{\frac{T}{2}} A \sin \theta \, \phi_2(t)$$
$$+ N_1 \phi_1(t) + N_2 \phi_2(t) \qquad (9.175)$$

where

$$N_i = \int_0^T n(t)\phi_i(t) \, dt, \qquad i = 1, 2 \qquad (9.176)$$

Because $y(t) - z(t)$ involves only noise, which is independent of $z(t)$, it is not relevant to making the estimate. Thus we may base the estimate on the vector

$$\mathbf{Z} \triangleq (Z_1, Z_2) = \left(\sqrt{\frac{T}{2}} A \cos \theta + N_1, \, - \sqrt{\frac{T}{2}} A \sin \theta + N_2 \right) \qquad (9.177)$$

where

$$Z_i = (y(t), \phi_i(t)) = \int_0^T y(t)\phi_i(t) \, dt \qquad (9.178)$$

The likelihood function $f_{\mathbf{Z}|\theta}(z_1, z_2|\theta)$ is obtained by noting that the variance of Z_1 and Z_2 is simply $\frac{1}{2}N_0$, as in the PAM example. Thus the likelihood function is

$$f_{\mathbf{Z}|\theta}(z_1, z_2|\theta)$$

$$= \frac{\exp\left\{-\frac{1}{N_0}\left[\left(z_1 - \sqrt{\frac{T}{2}}A\cos\theta\right)^2 + \left(z_2 + \sqrt{\frac{T}{2}}A\sin\theta\right)^2\right]\right\}}{\pi N_0}$$

$$= C\exp\left[2\sqrt{\frac{T}{2}}\frac{A}{N_0}(z_1\cos\theta - z_2\sin\theta)\right] \qquad (9.179)$$

where the coefficient C contains all factors that are independent of θ. The logarithm of the likelihood function is

$$\ln f_{\mathbf{Z}|\theta}(z_1, z_2|\theta) = \ln C + \sqrt{2T}\frac{A}{N_0}(z_1\cos\theta - z_2\sin\theta) \qquad (9.180)$$

which, when differentiated and set to zero, yields a necessary condition for the maximum-likelihood estimate of θ in accordance with (9.147b). The result is

$$-Z_1\sin\theta - Z_2\cos\theta\Big|_{\theta=\hat{\theta}_{\mathrm{ML}}} = 0 \qquad (9.181)$$

where Z_1 and Z_2 signify that we are considering a particular (random) observation. But

$$Z_1 = (y, \phi_1) = \sqrt{\frac{2}{T}}\int_0^T y(t)\cos\omega_c t\, dt \qquad (9.182a)$$

and

$$Z_2 = (y, \phi_2) = \sqrt{\frac{2}{T}}\int_0^T y(t)\sin\omega_c t\, dt \qquad (9.182b)$$

Therefore, (9.181) can be put in the form

$$-\sin\hat{\theta}_{\mathrm{ML}}\int_0^T y(t)\cos\omega_c t\, dt - \cos\hat{\theta}_{\mathrm{ML}}\int_0^T y(t)\sin\omega_c t\, dt = 0$$

or

$$\int_0^T y(t)\sin(\omega_c t + \hat{\theta}_{\mathrm{ML}})\, dt = 0 \qquad (9.183)$$

FIGURE 9.10   ML estimator for phase

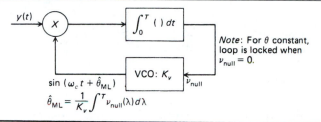

This equation can be interpreted as the feedback structure shown in Figure 9.10. Except for the integrator replacing a loop filter, this is identical to the phase-locked loop discussed in Chapter 3.

A lower bound for the variance of $\hat{\theta}_{\mathrm{ML}}$ is obtained from the Cramer-Rao inequality. Applying (9.164b), we have for the first differentiation, from (9.180),

$$\frac{\partial \ln f_{\mathbf{z}|\theta}}{\partial \theta} = \sqrt{2T}\,\frac{A}{N_0}\,(-Z_1 \sin \theta - Z_2 \cos \theta) \qquad (9.184)$$

and for the second,

$$\frac{\partial^2 \ln f_{\mathbf{z}|\theta}}{\partial \theta^2} = \sqrt{2T}\,\frac{A}{N_0}\,(-Z_1 \cos \theta + Z_2 \sin \theta) \qquad (9.185)$$

Substituting into (9.164b), we have

$$\mathrm{var}\,\{\hat{\theta}_{\mathrm{ML}}(Z)\} \geq \frac{1}{\sqrt{2T}}\,\frac{N_0}{A}\,(E\{Z_1\} \cos \theta - E\{Z_2\} \sin \theta)^{-1} \qquad (9.186)$$

The expectations of Z_1 and Z_2 are

$$E\{Z_i\} = \int_0^T E\{y(t)\}\phi_i(t)\,dt$$

$$= \int_0^T \sqrt{\frac{T}{2}}\,A[\cos \theta\,\phi_1(t) - \sin \theta\,\phi_2(t)]\phi_i(t)\,dt$$

$$= \begin{cases} \sqrt{\dfrac{T}{2}}\,A \cos \theta, & i = 1 \\[4mm] -\sqrt{\dfrac{T}{2}}\,A \sin \theta, & i = 2 \end{cases} \qquad (9.187)$$

where we used (9.175). Substitution of these results into (9.186) results in

$$\text{var}\,\{\hat{\theta}_{ML}(Z)\} \geq \frac{1}{\sqrt{2T}}\frac{N_0}{A}\left[\sqrt{\frac{T}{2}}A\,(\cos^2\theta + \sin^2\theta)\right]^{-1}$$

$$= \frac{N_0}{A^2 T} \tag{9.188}$$

Noting that the average signal power is $P_s = \frac{1}{2}A^2$ and defining $B_L = (2T)^{-1}$ as the equivalent noise bandwidth* of the estimator structure, we may write (9.188) as

$$\text{var}\,\{\hat{\theta}_{ML}\} \geq \frac{N_0 B_L}{P_s} \tag{9.189}$$

which is identical to the result given without proof in Table 8.5. As a result of the nonlinearity of the estimator, we can obtain only a lower bound for the variance. However, the bound becomes better as the SNR increases. Furthermore, because ML estimators are asymptotically Gaussian, we can approximate the conditional pdf of $\hat{\theta}_{ML}$, $f_{\hat{\theta}_{ML}|\theta}(\alpha|\theta)$, as Gaussian with mean θ ($\hat{\theta}_{ML}$ is unbiased) and variance given by (9.188).

SUMMARY

1. Two general classes of optimization problems are signal detection and parameter estimation. Although both detection and estimation are often involved simultaneously in signal reception, from an analysis standpoint, it is easiest to consider them as separate problems.

2. Bayes detectors are designed to minimize the average cost of making a decision. They involve testing a likelihood ratio, which is the ratio of the *a posteriori* (posterior) probabilities of the observations, against a threshold, which depends on the *a priori* (prior) probabilities of the two possible hypotheses and costs of the various decision/hypothesis combinations. The performance of a Bayes detector is characterized by the average cost, or risk, of making a decision. More useful in many cases, however, are the probabilities of detection and false alarm P_D and P_F in terms of which the risk can be expressed, provided the *a priori* probabilities and costs are available. A plot of P_D versus P_F is referred to as the receiver operating characteristic (ROC).

---

* The equivalent noise bandwidth of an ideal integrator of integration duration T is $(2T)^{-1}$ Hz.

3. If the costs and prior probabilities are not available, a useful decision strategy is the Neyman-Pearson detector, which maximizes P_D while holding P_F below some tolerable level. This type of receiver also can be reduced to a likelihood ratio test in which the threshold is determined by the allowed false-alarm level.

4. It was shown that a minimum-probability-of-error detector (that is, the type of detector considered in Chapter 7) is really a Bayes detector with zero costs for making right decisions and equal costs for making either type of wrong decision. Such a receiver is also referred to as a maximum *a posteriori* (MAP) detector, since the decision rule amounts to choosing as the correct hypothesis the one corresponding to the largest *a posteriori* probability for a given observation.

5. The introduction of signal-space concepts allowed the MAP criterion to be expressed as a receiver structure that chooses as the transmitted signal the signal whose location in signal space is closest to the observed data point. Two examples considered were coherent detection of *M*-ary orthogonal signals and noncoherent detection of binary FSK in a Rayleigh fading channel.

6. For *M*-ary orthogonal signal detection, zero probability of error can be achieved as $M \rightarrow \infty$ provided the ratio of energy per bit to noise spectral density is greater than -1.6 dB. This perfect performance is achieved at the expense of infinite transmission bandwidth, however.

7. For the Rayleigh fading channel, the probability of error decreases only inversely with signal-to-noise ratio rather than exponentially, as for the nonfading case. A way to improve performance is by using diversity.

8. Bayes estimation involves the minimization of a cost function, as for signal detection. The squared-error cost function results in the *a posteriori* conditional mean of the parameter as the optimum estimate, and a square-well cost function with infinitely narrow well results in the maximum of the *a posteriori* pdf of the data, given the parameter, as the optimum estimate (MAP estimate). Because of its ease of implementation, the MAP estimate is often employed even though the conditional-mean estimate is more general, in that it minimizes any symmetrical, convex-upward cost function as long as the posterior pdf is symmetrical about a single peak.

9. A maximum-likelihood (ML) estimate of a parameter A is that value for the parameter, $\hat{A}$, which is most likely to have resulted in the observed data A and is the value of A corresponding to the absolute maximum of

the conditional pdf of Z given A. The ML and MAP estimates of a parameter are identical if the *a priori* pdf of A is uniform. Since the *a priori* pdf of A is not needed to obtain an ML estimate, this is a useful procedure for estimation of parameters whose prior statistics are unknown or for estimation of nonrandom parameters.

10. The Cramer-Rao inequality gives a lower bound for the variance of an ML estimate. In the limit, ML estimates have many useful asymptotic properties as the number of independent observations becomes large. In particular, they are asymptotically Gaussian, unbiased, and efficient (satisfy the Cramer-Rao inequality with equality).

FURTHER READING

Two, by now, classic textbooks on detection and estimation theory at the graduate level are Van Trees (1968) and Helstrom (1968). Both are excellent in their own way, Van Trees being somewhat wordier and containing more examples than Helstrom, which is closely written but nevertheless very readable. Helstrom introduces complex envelope notation at the first and uses it throughout to very effectively include in one volume what Van Trees takes two volumes to accomplish (Volumes 1 and 3; Volume 2 treats nonlinear modulation theory, which Helstrom omits). A concise treatment of detection theory and signal design is the text by Weber (1987), recently reprinted. Shanmugam and Breipohl (1988) and Mendel (1987) are recommended.

At about the same (or somewhat lower) level as Van Trees and Helstrom is the book by Wozencraft and Jacobs (1965), which was the first book in the United States to use the signal-space concepts exploited by Kotel'nikov (1959) in his doctoral dissertation in 1947 to treat digital signaling and optimal analog demodulation. Many other books treat special applications of detection and estimation theory; several were referred to in Chapters 7 and 8 in connection with digital data transmission.

PROBLEMS

Section 9.1

9.1 Consider the hypotheses

$$H_1: Z = N \qquad \text{and} \qquad H_2: Z = S + N$$

where S and N are independent random variables with the pdf's

$$f_S(x) = 2e^{-2x}u(x) \qquad \text{and} \qquad f_N(x) = 10e^{-10x}u(x)$$

(a) Show that

$$f_Z(z|H_1) = 10e^{-10z}u(z)$$

and

$$f_Z(z|H_2) = 2.5(e^{-2z} - e^{-10z})u(z)$$

(b) Find the likelihood ratio $\Lambda(Z)$.
(c) If $P(H_1) = \frac{1}{4}$, $P(H_2) = \frac{3}{4}$, $c_{12} = c_{21} = 5$, and $c_{11} = c_{22} = 0$, find the threshold for a Bayes test.
(d) Show that the likelihood ratio test for part (c) can be reduced to

$$z \underset{H_1}{\overset{H_2}{\gtrless}} \gamma$$

Find the numerical value of γ for the Bayes test of part (c).
(e) Find the risk for the Bayes test of part (c).
(f) Find the threshold for a Neyman-Pearson test with P_F less than or equal to 10^{-4}. Find P_D for this threshold.
(g) Reducing the Neyman-Pearson test of part (f) to the form

$$z \underset{H_1}{\overset{H_2}{\gtrless}} \gamma$$

find P_F and P_D for arbitrary γ. Plot the ROC.

9.2   Consider a two-hypothesis decision problem where

$$f_Z(z|H_1) = \frac{\exp\left(-\frac{1}{2}z^2\right)}{\sqrt{2\pi}} \quad\text{and}\quad f_Z(z|H_2) = \frac{1}{2}\exp\left(-|z|\right)$$

(a) Find the likelihood ratio $\Lambda(Z)$.
(b) Letting the threshold η be arbitrary, find the decision regions R_1 and R_2 illustrated in Figure 9.1. Note that both R_1 and R_2 cannot be connected regions for this problem; that is, one of them will involve a multiplicity of line segments.

Section 9.2

9.3   Show that ordinary three-dimensional vector space satisfies the properties listed in the subsection entitled "Structure of Signal Space" in Section 9.2, where $x(t)$ and $y(t)$ are replaced by vectors **A** and **B**.

9.4   Show that the scalar-product definitions given by Equations (9.40a) and (9.40b) satisfy the properties listed in the subsection entitled "Scalar Product" in Section 9.2.

9.5   Using the appropriate definition, calculate (x_1, x_2) for each of the following pairs of signals:

(a) $e^{-|t|}$, $2e^{-5t}u(t)$ (b) $e^{-(2+j3)t}u(t)$, $e^{-(3+j2)t}u(t)$
(c) $\cos 2\pi t$, $\sin^2 2\pi t$ (d) $\cos 2\pi t$, $10u(t)$

9.6   Let $x_1(t)$ and $x_2(t)$ be two real-valued signals. Show that the square of the norm of the signal $x_1(t) + x_2(t)$ is the sum of the square of the norm of $x_1(t)$ and the square of the norm of $x_2(t)$ if and only if x_1 and x_2 are orthogonal; that is, $\|x_1 + x_2\|^2 = \|x_1\|^2 + \|x_2\|^2$ if and only if $(x_1, x_2) = 0$. Note the analogy to vectors in three-dimensional space: the Pythagorean theorem applies only to vectors that are orthogonal or perpendicular (zero dot product).

9.7   Evaluate $\|x_1\|$, $\|x_2\|$, $\|x_3\|$, (x_2, x_1), and (x_3, x_1) for the signals in Figure 9.11. Use these numbers to construct a vector diagram and graphically verify that $x_3 = x_1 + x_2$.

9.8   Verify Schwarz's inequality for

$$x_1(t) = \sum_{n=1}^{N} a_n\phi_n(t) \qquad \text{and} \qquad x_2(t) = \sum_{n=1}^{N} b_n\phi_n(t)$$

where the $\phi_n(t)$'s are orthonormal.

9.9   (a) Use the Gram-Schmidt procedure to find a set of orthonormal basis functions corresponding to the signals given in Figure 9.12.
(b) Express s_1, s_2, and s_3 in terms of the orthonormal basis set found in part (a).

FIGURE 9.11

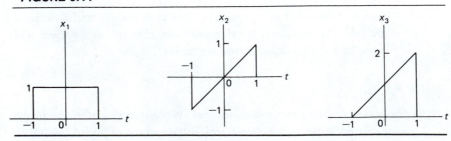

FIGURE 9.12

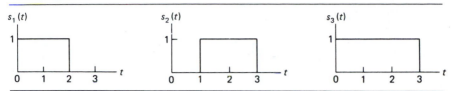

9.10   Consider the set of signals

$$s_i(t) = \begin{cases} \sqrt{2}\, A \cos(2\pi f_c t + \frac{1}{2}i\pi), & 0 \le f_c t \le N \\ 0, & \text{otherwise} \end{cases}$$

where N is an integer and $i = 1, 2, 3, 4$.

(a) Find an orthonormal basis set for the space spanned by this set of signals.

(b) Draw a set of coordinate axes, and plot the locations of $s_i(t)$, $i = 1, 2, 3, 4$, after expressing each one as a generalized Fourier series in terms of the basis set found in part (a).

Section 9.3

9.11   For M-ary PSK, the transmitted signal is of the form

$$s_i(t) = A \cos(2\pi f_c t + i2\pi/M), \quad i = 1, 2, \dots, M, \text{ for } 0 \le t \le T_s$$

(a) Find a set of basis functions for this signaling scheme. What is the dimension of the signal space? Express $s_i(t)$ in terms of these basis functions and the signal energy, $E = A^2 T_s/2$.

(b) Show the optimum partitioning of the observation space according to Equation (9.91).

(c) Sketch a block diagram of the optimum (minimum P_E) receiver.

(d) Write down an expression for the probability of error. Do not attempt to integrate it.

(e) By choosing appropriate half planes in the observation space, find upper and lower bounds for P_E derived in part (d). Sketch them versus SNR for $M = 4$, 8, and 16.

9.12   Consider Equation (9.120) for $M = 2$. Express P_E as a single Q-function. Show that the result is identical to binary, coherent FSK.

9.13 Consider *vertices-of-a-hypercube signaling*, for which the *i*th signal is of
      the form

$$s(t) = \sqrt{\frac{E_s}{n}} \sum_{k=1}^{n} \alpha_{ik} \phi_k(t), \qquad 0 \leq t \leq T$$

in which the coefficients α_{ik} are permuted through the values $+1$
and -1, E_s is the signal energy, and the ϕ_k's are orthonormal. Thus
$M = 2^n$, where *n* is an integer. For $M = 8$, $n = 3$, the signal points
in signal space lie on the vertices of a cube in three-space.

(a) Sketch the optimum partitioning of the observation space for *M*
    $= 8$.
(b) Show that the probability of symbol error is

$$P_E = 1 - P(C)$$

where

$$P(C) = \left[1 - Q\left(\sqrt{\frac{2E_s}{nN_0}} \right) \right]^n$$

(c) Plot P_E versus E_s/N_0 for $n = 1, 2, 3, 4$. Compare with Figure 9.7
    for $n = 1$ and 2.

9.14 (a) Referring to the signal set defined by Equation (9.92), show that
          the minimum possible $\Delta f = \Delta\omega/2\pi$ such that $(s_i, s_j) = 0$ is
          $\Delta f = 1/(2T_s)$.
      (b) Using the result of part (a), show that for a given time-bandwidth
          product WT_s the maximum number of signals for *M*-ary FSK sig-
          naling is given by $M = 2WT_s$, where *W* is the transmission band-
          width and T_s is the signal duration. Use null-to-null bandwidth.
      (c) For vertices-of-a-hypercube signaling, described in Problem 9.13,
          show that the number of signals grows with WT_s as $M = 2^{2WT_s}$.

9.15 Go through the steps in deriving Equation (9.133).

9.16 Generalize the fading problem of binary noncoherent FSK signaling
      to the *M*-ary case. Let the *i*th hypothesis be of the form

$$H_i: y(t) = G_i \sqrt{\frac{2E_i}{T_s}} \cos(\omega_i t + \theta_i) + n(t)$$

$$i = 1, 2, \ldots, M; 0 \leq t \leq T_s$$

where G_i is Rayleigh, θ_i is uniform in $(0, 2\pi)$, E_i is the energy of the unperturbed ith signal of duration T_s, and $|\omega_i - \omega_j| \gg T_s^{-1}$, for $i \neq j$, so that the signals are orthogonal. Note that $G_i \cos \theta_i$ and $-G_i \sin \theta_i$ are Gaussian with mean zero; assume their variances to be σ^2.

(a) Find the likelihood ratio test and show that the optimum correlation receiver is identical to the one shown in Figure 9.8(a) with $2M$ correlators, $2M$ squarers, and M summers where the summer with the largest output is chosen as the best guess (minimum P_E) for the transmitted signal if all E_i's are equal. How is the receiver structure modified if the E_i's are not equal?

(b) Write down an expression for the probability of symbol error.

9.17 Investigate the use of diversity to improve the performance of binary FSK signaling over the flat fading Rayleigh channel. Assume that the signal energy E_s is divided equally among N subpaths, all of which fade independently. For equal SNRs in all paths, the optimum receiver is shown in Figure 9.13.

(a) Referring to Problem 4.45 of Chapter 4, show that Y_1 and Y_2 are chi-squared random variables under either hypothesis.

(b) Show that the probability of error is of the form

$$P_E = \alpha^N \sum_{j=0}^{N-1} \binom{N+j-1}{j} (1-\alpha)^j$$

where

$$\alpha = \frac{\frac{1}{2}N_0}{\sigma^2 E' + N_0} = \frac{1}{2} \frac{1}{1 + \frac{1}{2}(2\sigma^2 E'/N_0)}, \qquad E' = \frac{E_s}{N}$$

FIGURE 9.13

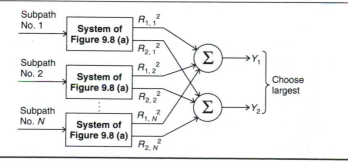

   (c) Plot P_E versus SNR $\triangleq 2\sigma^2 E_s/N_0$ for $N = 1, 2, 3, \ldots$, and show that an optimum value of N exists that minimizes P_E for a given SNR.

Section 9.4

9.18 Let an observed random variable Z depend on a parameter λ according to the conditional pdf

$$f_{Z|\Lambda}(z|\lambda) = \begin{cases} \lambda e^{-\lambda z}, & z \geq 0, \lambda > 0 \\ 0, & z < 0 \end{cases}$$

The *a priori* pdf of λ is

$$f_\Lambda(\lambda) = \begin{cases} \dfrac{\beta^m}{\Gamma(m)} e^{-\beta\lambda}\lambda^{m-1}, & \lambda \geq 0 \\ 0, & \lambda < 0 \end{cases}$$

where β and m are parameters and $\Gamma(m)$ is the gamma function. Assume that m is a positive integer.

   (a) Find $E\{\lambda\}$ and var $\{\lambda\}$ before any observations are made; that is, find the mean and variance of λ using $f_\Lambda(\lambda)$.

   (b) Assume one observation is made. Find $f_{\Lambda|Z}(\lambda|z)$ and hence the minimum mean-square error (conditional-mean) estimate of λ and the variance of the estimate. Compare with part (a). Comment on the similarity of $f_\Lambda(\lambda)$ and $f_{\Lambda|Z}(\lambda|z)$.

   (c) Making use of part (b), find the posterior pdf of λ given two observations $f_{\Lambda|Z}(\lambda|z_1, z_2)$. Find the minimum mean-square error estimate of λ based on two observations and its variance. Compare with parts (a) and (b), and comment.

   (d) Generalize the preceding to the case in which K observations are used to estimate λ.

   (e) Does the MAP estimate equal the minimum mean-square error estimate?

9.19 For which of the cost functions and posterior pdf's shown in Figure 9.14 will the conditional mean be the Bayes estimate? Tell why in each case.

9.20 Given K independent measurements $(Z_1, Z_2, \ldots, Z_K)$ of a noise voltage $Z(t)$ at the RF filter output of a receiver:

   (a) If $Z(t)$ is Gaussian with mean zero and variance σ_n^2, what is the ML estimate of the variance of the noise?

FIGURE 9.14

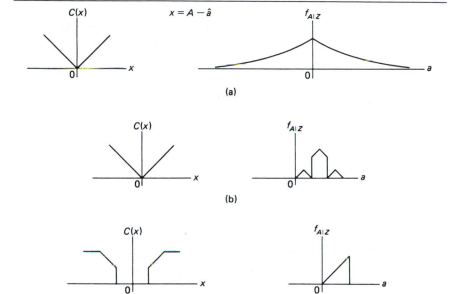

(a)

(b)

(c)

(d)

(b) Calculate the expected value and variance of this estimate as functions of the true variance.

(c) Is this an unbiased estimator?

(d) Give a sufficient statistic for estimating the variance of Z.

9.21 Generalize the estimation of a sample of a PAM signal, expressed by Equation (9.169), to the case where the sample value m_0 is a zero-mean Gaussian random variable with variance σ_m^2.

9.22 Consider the reception of a BPSK signal in noise with unknown phase, θ, to be estimated. The two hypotheses may be expressed as

$$H_1: y(t) = \cos(\omega_c t + \theta) + n(t), \qquad 0 \le t \le T_s$$
$$H_2: y(t) = -\cos(\omega_c t + \theta) + n(t), \qquad 0 \le t \le T_s$$

FIGURE 9.15

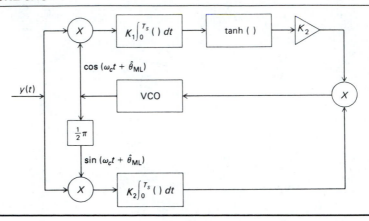

where $n(t)$ is white Gaussian noise with single-sided power spectral density N_0, and the hypotheses are equally probable $[P(H_1) = P(H_2)]$.

(a) Using ϕ_1 and ϕ_2 as given by Equation (9.174) as basis functions, write expressions for

$$f_{Z|\theta,H_i}(z_1, z_2|\theta, H_i), \qquad i = 1, 2$$

(b) Noting that

$$f_{Z|\theta}(z_1, z_2|\theta) = \sum_{i=1}^{2} P(H_i)f_{Z|\theta,H_i}(z_1, z_2|\theta, H_i)$$

show that the ML estimator can be realized as the structure shown in Figure 9.15 by employing Equation (9.147). Under what condition(s) is this structure approximated by a Costas loop? (See Chapter 3, Figure 3.57.)

(c) Apply the Cramer-Rao inequality to find an expression for var $\{\hat{\theta}_{ML}\}$. Compare with the result in Table 8.1.

9.23   Assume a biphase modulated signal in white Gaussian noise of the form

$$y(t) = \sqrt{2P} \sin (\omega_c t \pm \cos^{-1} m + \theta) + n(t), \qquad 0 \le t \le T_s$$

where the $\pm$ signs are equally probable and θ is to be estimated by a maximum-likelihood procedure. In the preceding equation,

$$T_s = \text{signaling interval}$$
$$P = \text{average signal power}$$
$$\omega_c = \text{carrier frequency (rad/s)}$$
$$m = \text{modulation constant}$$
$$\theta = \text{RF phase (rad)}$$

Let the double-sided power spectral density of $n(t)$ be $\frac{1}{2}N_0$.

(a) Show that the signal portion of $y(t)$ can be written as

$$S(t) = \sqrt{2P}\, m \sin(\omega_c t + \theta) \pm \sqrt{2P}\sqrt{1 - m^2}\cos(\omega_c t + \theta)$$

Write in terms of the orthonormal functions ϕ_1 and ϕ_2, given by Equation (9.174).

(b) Show that the likelihood function can be written as

$$L(\theta) = \frac{2m\sqrt{2P}}{N_0} \int_0^{T_s} y(t) \sin(\omega_c t + \theta)dt$$

$$+ \ln \cosh\left[\frac{2\sqrt{2P(1 - m^2)}}{N_0} \int_0^{T_s} y(t) \cos(\omega_c t + \theta)\, dt \right]$$

(c) Draw a block diagram of the ML estimator for θ and compare with the block diagram shown in Figure 9.15.

COMPUTER EXERCISES

9.1 (a) Write a program to plot the receiver operating characteristics for Example 9.2. Use the rational approximation given in Appendix C.6 to evaluate the Q-function.
   (b) Repeat for Problem 9.1.

9.2 (a) Write a program to plot the curves asked for in Problem 9.13.
   (b) Provide a justification for choosing either (8.49) or (8.51) to convert from symbol error probability to bit error probability for vertices-of-a-hypercube modulation. Use (8.52) to convert from symbol energy to bit energy, and plot curves of bit error probability versus E_b/N_0 for this modulation scheme.
   (c) Use a solve routine to obtain the value of E_b/N_0, in decibels, required for a desired probability of bit error. Do this for several values of M and note that bit error probability for a given value of E_b/N_0 appears to be independent of M. Provide an explanation for this phenomenon.

9.3 Write a computer program to justify the conclusion drawn in Example 9.6 that a value of τW of 1.4 will ensure that none of the $\phi_k(t)$'s has more than 1/12 of its energy outside the frequency band $|f| < W$.

9.4 Write a computer program to make plots of σ_p^2 versus K, the number of observations, for fixed ratios of σ_A^2/σ_n^2 thus verifying the conclusions drawn at the end of Example 9.8.

9.5 Write a computer simulation of the PLL estimation problem. Do this by generating two independent Gaussian random variables to form Z_1 and Z_2 given by (9.177). Thus for a given θ form the left-hand side of (9.181). Call the first value θ_0. Estimate the next value of θ, call it θ_1, from the algorithm

$$\theta_1 = \theta_0 + \epsilon \tan^{-1}\left(\frac{Z_{2,0}}{Z_{1,0}}\right)$$

where $Z_{1,0}$ and $Z_{2,0}$ are the first values of Z_1 and Z_2 generated and ϵ is a parameter to be varied (choose the first value to be 0.01). Generate two new values of Z_1 and Z_2 (call them $Z_{1,1}$ and $Z_{2,1}$) and form the next estimate according to

$$\theta_2 = \theta_1 + \epsilon \tan^{-1}\left(\frac{Z_{2,1}}{Z_{1,1}}\right)$$

Continue in this fashion, generating several values of θ_i. Plot the θ_i's versus i, the sequence index, to determine if they seem to converge toward zero phase. Increase the value of ϵ by a factor of 10 and repeat. Can you relate the parameter ϵ to a phase-lock loop parameter (see Chapter 3)? This is an example of Monte Carlo simulation.

INFORMATION THEORY AND CODING 10

Information theory provides a different perspective for evaluating the performance of communication systems, and significant insight into the performance characteristics of communication systems can often be gained through the study of information theory. More explicitly, information theory provides a quantitative measure of the information contained in message signals and allows us to determine the capability of a system to transfer this information from source to destination. Coding, a major topic of information theory, will be examined in this chapter in some detail. Through the use of coding, unsystematic redundancy can be removed from message signals so that channels can be used with maximum efficiency. In addition, through the use of coding, systematic redundancy can be induced into the transmitted signal so that errors caused by nonperfect practical channels can be corrected.

Information theory also provides us with the performance characteristics of an *ideal,* or optimum, communication system. The performance of an ideal system provides a meaningful basis against which to compare the performance of the realizable systems studied in previous chapters. Performance characteristics of ideal systems illustrate the gain in performance that can be obtained by implementing more complicated transmission and detection schemes.

Motivation for the study of information theory is provided by *Shannon's coding theorem,* sometimes referred to as *Shannon's second theorem,* which can be stated as follows: If a source has an information rate less than the channel capacity, there exists a coding procedure such that the source output can be transmitted over the channel with an arbitrarily small probability of error. This is a truly surprising statement. We are being led to believe that transmission and reception can be accomplished with *negligible* error, even in the presence of noise. An understanding of this process called *coding,* and an understanding of its impact on the design and performance of communication systems, requires an understanding of several basic concepts of information theory.

10.1 BASIC CONCEPTS

Consider a hypothetical classroom situation occurring early in a course at the end of a class period. The professor makes one of the following statements to the class:

A. I shall see you next period.
B. My colleague will lecture next period.
C. Everyone gets an A in the course, and there will be no more class meetings.

What is the relative information conveyed to the students by each of these statements, assuming that there had been no previous discussion on the subject? Obviously, there is little information conveyed by statement (A), since the class would normally assume that their regular professor would lecture; that is, the probability, $P(A)$, of the regular professor lecturing is nearly unity. Intuitively, we know that statement (B) contains more information, and the probability of a colleague lecturing $P(B)$ is relatively low. Statement (C) contains a vast amount of information for the entire class, and most would agree that such a statement has a very low probability of occurrence in a typical classroom situation. It appears that the lower the probability of a statement, the greater is the information conveyed by that statement. Stated another way, the students' surprise on hearing a statement seems to be a good measure of the information contained in that statement.

We next define information mathematically in such a way that the definition is consistent with the preceding intuitive example.

Information

Let x_j be an event that occurs with probability $p(x_j)$. If we are told that event x_j has occurred, we say that we have received

$$I(x_j) = \log_a \frac{1}{p(x_j)} = -\log_a p(x_j) \tag{10.1}$$

units of information. This definition is consistent with the previous example, since $I(x_j)$ increases as $p(x_j)$ decreases.

The base of the logarithm in (10.1) is quite arbitrary and determines the units by which we measure information. R. V. Hartley,* who first suggested the logarithmic measure of information in 1928, used logarithms to the base 10, and the measure of information was the hartley. Today it is standard to use logarithms to the base 2, and the unit of information is the binary unit,

---

* Hartley (1928).

or bit. Logarithms to the base e are sometimes utilized, and the corresponding unit is the nat, which stands for natural unit.

There are several reasons for us to be consistent in using the base 2 logarithm to measure information. The simplest random experiment that one can imagine is an experiment with two equally likely outcomes, such as the flipping of an unbiased coin. Knowledge of each outcome has associated with it one bit of information. Also, since the digital computer is a binary machine, each logical 0 and each logical 1 has associated with it one bit of information, assuming that each of these logical states are equally likely.

▼| **EXAMPLE 10.1**   Consider a random experiment with 16 equally likely outcomes. The information associated with each outcome is

$$I(x_j) = -\log_2 \tfrac{1}{16} = \log_2 16 = 4 \text{ bits} \qquad (10.2)$$

where j ranges from 1 to 16. The information is greater than one bit, since
▲| the probability of each outcome is much less than one-half.

Entropy

In general, the *average information* associated with the outcome of an experiment is of interest rather than the information associated with each particular event. The average information associated with a discrete random variable X is defined as the entropy $H(X)$. Thus

$$H(X) = E\{I(x_j)\} = -\sum_{j=1}^{n} p(x_j) \log_2 p(x_j) \qquad (10.3)$$

where n is the total number of possible outcomes. Entropy can be regarded as average uncertainty and therefore should be maximum when each outcome is equally likely.

▼| **EXAMPLE 10.2**   For a binary source, $p(1) = \alpha$ and $p(0) = 1 - \alpha = \beta$. Derive the entropy of the source as a function of α, and sketch $H(\alpha)$ as α varies from zero to 1.

SOLUTION   From (10.3),

$$H(\alpha) = -\alpha \log_2 \alpha - (1 - \alpha) \log_2 (1 - \alpha) \qquad (10.4)$$

This is sketched in Figure 10.1. We should note the maximum. If $\alpha = \tfrac{1}{2}$, each symbol is equally likely, and our uncertainty is a maximum. If $\alpha \neq \tfrac{1}{2}$,

FIGURE 10.1   Entropy of a binary source

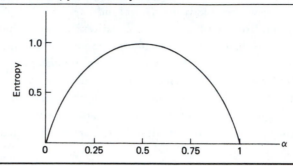

one symbol is more likely to occur than the other, and we are less uncertain as to which symbol appears on the source output. If α or β is equal to zero, our uncertainty is zero, since we know exactly which symbol will occur.

From Example 10.2 we conclude, at least intuitively, that the entropy function has a maximum, and the maximum occurs when all probabilities are equal. This fact is of sufficient importance to warrant a more complete derivation.

Assume that a chance experiment has n possible outcomes and that p_n is a dependent variable depending on the other probabilities. Thus

$$p_n = 1 - (p_1 + p_2 + \cdots + p_k + \cdots + p_{n-1}) \tag{10.5}$$

where p_j is concise notation for $p(x_j)$. The entropy associated with the chance experiment is

$$H = -\sum_{i=1}^{n} p_i \log_2 p_i \tag{10.6}$$

In order to find the maximum value of entropy, the entropy is differentiated with respect to p_k, holding all probabilities constant except p_k and p_n. This gives a relationship between p_k and p_n that yields the maximum value of H. Since all derivatives are zero except the ones involving p_k and p_n,

$$\frac{dH}{dp_k} = \frac{d}{dp_k} (-p_k \log_2 p_k - p_n \log_2 p_n) \tag{10.7}$$

by using (10.5) and since

$$\frac{d}{dx} \log_a u = \frac{1}{u} \log_a e \frac{du}{dx} \tag{10.8}$$

we obtain

$$\frac{dH}{dp_k} = -p_k \frac{1}{p_k} \log_2 e - \log_2 p_k + p_n \frac{1}{p_n} \log_2 e + \log_2 p_n \quad (10.9)$$

or

$$\frac{dH}{dp_k} = \log_2 \frac{p_n}{p_k} \quad (10.10)$$

which is zero if $p_k = p_n$. Since p_k was arbitrarily chosen,

$$p_1 = p_2 = \cdots = p_n = \frac{1}{n} \quad (10.11)$$

To show that the preceding condition yields a maximum and not a minimum, note that when $p_1 = 1$ and all other probabilities are zero, entropy is zero. From (10.6), the case where all probabilities are equal yields $H = \log_2 n$.

Channel Representations

Throughout most of this chapter we will assume the communications channel to be *memoryless*. For such channels, the channel output at a given time is a function of the channel input *at that time* and is not a function of previous channel inputs. Memoryless discrete channels are completely specified by the set of conditional probabilities that relate the probability of each output state to the input probabilities. An example illustrates the technique. A diagram of a channel with two inputs and three outputs is illustrated in Figure 10.2. Each possible input-to-output path is indicated along with a conditional probability p_{ij}, which is concise notation for $p(y_j|x_i)$. Thus p_{ij} is the conditional probability of obtaining output y_j given that the input is x_i and is called a *channel transition probability*.

We can see from Figure 10.2 that the channel is completely specified by

FIGURE 10.2 Channel diagram

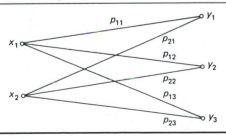

the complete set of transition probabilities. Accordingly, the channel is often specified by the matrix of transition probabilities $[P(Y|X)]$, where, for the channel of Figure 10.2,

$$[P(Y|X)] = \begin{bmatrix} p(y_1|x_1) & p(y_2|x_1) & p(y_3|x_1) \\ p(y_1|x_2) & p(y_2|x_2) & p(y_3|x_2) \end{bmatrix} \tag{10.12}$$

Since each input to the channel results in some output, each new row of the channel matrix must sum to unity.

The channel matrix is useful in deriving the output probabilities given the input probabilities. For example, if the input probabilites $P(X)$ are represented by the row matrix

$$[P(X)] = [p(x_1) \quad p(x_2)] \tag{10.13}$$

then

$$[P(Y)] = [p(y_1) \quad p(y_2) \quad p(y_3)] \tag{10.14}$$

which is computed by

$$[P(Y)] = [P(X)][P(Y|X)] \tag{10.15}$$

If $[P(X)]$ is written as a diagonal matrix, (10.15) yields a matrix $[P(X, Y)]$. Each element in the matrix has the form $p(x_i)p(y_j|x_i)$ or $p(x_j, y_j)$. This matrix is known as the *joint probability matrix*, and the term $p(x_i, y_j)$ is the joint probability of transmitting x_i and receiving y_j.

▼| **EXAMPLE 10.3**   Consider the binary input–output channel shown in Figure 10.3. The matrix of transition probabilities is

$$[P(Y|X)] = \begin{bmatrix} 0.7 & 0.3 \\ 0.4 & 0.6 \end{bmatrix} \tag{10.16}$$

FIGURE 10.3   Binary channel

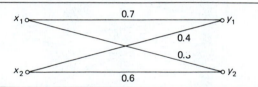

If the input probabilities are $P(x_1) = 0.5$ and $P(x_2) = 0.5$, the output probabilities are

$$[P(Y)] = [0.5 \quad 0.5] \begin{bmatrix} 0.7 & 0.3 \\ 0.4 & 0.6 \end{bmatrix} = [0.55 \quad 0.45] \tag{10.17}$$

and the joint probability matrix is

$$[P(X, Y)] = \begin{bmatrix} 0.5 & 0 \\ 0 & 0.5 \end{bmatrix} \begin{bmatrix} 0.7 & 0.3 \\ 0.4 & 0.6 \end{bmatrix} = \begin{bmatrix} 0.35 & 0.15 \\ 0.2 & 0.3 \end{bmatrix} \tag{10.18}$$

As first seen in Chapter 8, a binary satellite communication system can often be represented by the cascade combination of two binary channels. This is illustrated in Figure 10.4(a), in which the first binary channel represents the uplink and the second binary channel represents the downlink. These channels can be combined as shown in Figure 10.4(b).

FIGURE 10.4 Two-hop satellite system. (a) Binary satellite channel. (b) Composite satellite channel.

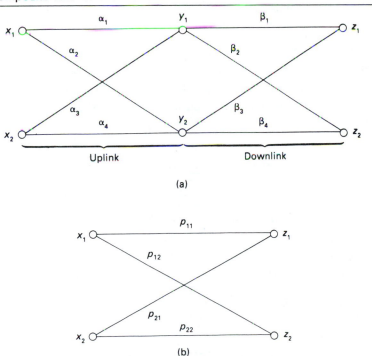

(a)

(b)

By determining all possible paths from x_i to z_j, it is clear that the following probabilities define the overall channel illustrated in Figure 10.4(b):

$$P_{11} = \alpha_1 \beta_1 + \alpha_2 \beta_3 \tag{10.19a}$$

$$P_{12} = \alpha_1 \beta_2 + \alpha_2 \beta_4 \tag{10.19b}$$

$$P_{21} = \alpha_3 \beta_1 + \alpha_4 \beta_3 \tag{10.19c}$$

$$P_{22} = \alpha_3 \beta_2 + \alpha_4 \beta_4 \tag{10.19d}$$

Thus the overall channel matrix

$$[P(Z|X)] = \begin{bmatrix} P_{11} & P_{12} \\ P_{21} & P_{22} \end{bmatrix} \tag{10.20}$$

can be represented by the matrix multiplication

$$[P(Z|X)] = \begin{bmatrix} \alpha_1 & \alpha_2 \\ \alpha_3 & \alpha_4 \end{bmatrix} \begin{bmatrix} \beta_1 & \beta_2 \\ \beta_3 & \beta_4 \end{bmatrix} \tag{10.21}$$

The right-hand side of the preceding expression is simply the uplink channel matrix multiplied by the downlink channel matrix.

Joint and Conditional Entropy

If we use the input probabilities $p(x_i)$, the output probabilities $p(y_j)$, the transition probabilities $p(y_j|x_i)$, and the joint probabilities $p(x_i, y_j)$, we can define several different entropy functions for a channel with n inputs and m outputs. These are

$$H(X) = -\sum_{i=1}^{n} p(x_i) \log_2 p(x_i) \tag{10.22}$$

$$H(Y) = -\sum_{j=1}^{m} p(y_j) \log_2 p(y_j) \tag{10.23}$$

$$H(Y|X) = -\sum_{i=1}^{n} \sum_{j=1}^{m} p(x_i, y_j) \log_2 p(y_j|x_i) \tag{10.24}$$

and

$$H(X, Y) = -\sum_{i=1}^{n} \sum_{j=1}^{m} p(x_i, y_j) \log_2 p(x_i, y_j) \tag{10.25}$$

Another useful entropy, $H(X|Y)$, which is sometimes called *equivocation,* is defined as

$$H(X|Y) = -\sum_{i=1}^{n}\sum_{j=1}^{m} p(x_i, y_j) \log_2 p(x_i|y_j) \qquad (10.26)$$

These entropies are easily interpreted. $H(X)$ is the average uncertainty of the source, whereas $H(Y)$ is the average uncertainty of the received symbol. Similarly, $H(X|Y)$ is a measure of our average uncertainty of the transmitted symbol after we have received a symbol. The function $H(Y|X)$ is the average uncertainty of the received symbol given that X was transmitted. The joint entropy $H(X, Y)$ is the average uncertainty of the communication system as a whole.

Two important and useful relationships, which can be obtained directly from the definitions of the various entropies, are

$$H(X, Y) = H(X|Y) + H(Y) \qquad (10.27)$$

and

$$H(X, Y) = H(Y|X) + H(X) \qquad (10.28)$$

These are developed in Problem 10.12.

Channel Capacity

Consider for a moment an observer at the channel output. The observer's average uncertainty concerning the channel input will have some value before the reception of an output, and this average uncertainty of the input will usually decrease when the output is received. In other words, $H(X|Y) \le H(X)$. The decrease in the observer's average uncertainty of the transmitted signal when the output is received is a measure of the average transmitted information. This is defined as *transinformation,* or *mutual information* $I(X; Y)$. Thus

$$I(X; Y) = H(X) - H(X|Y) \qquad (10.29)$$

It follows from (10.27) and (10.28) that we can also write (10.29) as

$$I(X; Y) = H(Y) - H(Y|X) \qquad (10.30)$$

It should be observed that transinformation is a function of the source probabilities as well as of the channel transition probabilities.

It is easy to show mathematically that

$$H(X) \ge H(X|Y) \qquad (10.31)$$

by showing that

$$H(X|Y) - H(X) = -I(X; Y) \le 0 \tag{10.32}$$

Substitution of (10.26) for $H(X|Y)$ and (10.22) for $H(X)$ allows us to write $-I(X; Y)$ as

$$-I(X; Y) = \sum_{i=1}^{n} \sum_{j=1}^{m} p(x_i, y_j) \log_2 \frac{p(x_i)}{p(x_i|y_j)} \tag{10.33}$$

Since

$$\log_2 x = \frac{\ln x}{\ln 2} \tag{10.34}$$

and

$$\frac{p(x_j)}{p(x_i|y_j)} = \frac{p(x_i)p(y_j)}{p(x_i, y_j)} \tag{10.35}$$

we can write $-I(X; Y)$ as

$$-I(X; Y) = \frac{1}{\ln 2} \sum_{i=1}^{n} \sum_{j=1}^{m} p(x_i, y_j) \ln \frac{p(x_i)p(y_j)}{p(x_j, y_j)} \tag{10.36}$$

In order to carry the derivation further, we need the often-used inequality

$$\ln (x) \le x - 1 \tag{10.37}$$

which we can easily prove by considering the function

$$f(x) = \ln (x) - (x - 1) \tag{10.38}$$

The derivative of $f(x)$

$$\frac{df}{dx} = \frac{1}{x} - 1 \tag{10.39}$$

is equal to zero at $x = 1$. It follows that $f(1) = 0$ is the maximum value of $f(x)$, since we can make $f(x)$ as small as we wish by choosing x sufficiently large.

Using the inequality (10.37) in (10.36) results in

$$-I(X; Y) \le \frac{1}{\ln 2} \sum_{i=1}^{n} \sum_{j=1}^{m} p(x_i, y_j) \left[\frac{p(x_i)p(y_j)}{p(x_i, y_j)} - 1 \right] \tag{10.40}$$

which yields

$$-I(X; Y) \leq \frac{1}{\ln 2} \left[\sum_{i=1}^{n} \sum_{j=1}^{m} p(x_i)p(y_j) - \sum_{i=1}^{n} \sum_{j=1}^{m} p(x_i, y_j) \right] \quad (10.41)$$

Since both the double sums equal 1, we have the desired result

$$-I(X; Y) \leq 0 \quad (10.42)$$

Thus we have shown that mutual information is always nonnegative and, consequently, that $H(X) \geq H(X|Y)$.

The *channel capacity* C is defined as the maximum value of transinformation, which is the maximum average information *per symbol* that can be transmitted through the channel. Thus

$$C = \max [I(X; Y)] \quad (10.43)$$

The maximization is with respect to the source probabilities, since the transition probabilities are fixed by the channel. However, the channel capacity is a function of only the channel transition probabilities, since the maximization process eliminates the dependence on the source probabilities. The following examples illustrate the method.

▼ EXAMPLE 10.4   Find the channel capacity of the noiseless discrete channel illustrated in Figure 10.5.

SOLUTION   We start with

$$I(X; Y) = H(X) - H(X|Y)$$

and write

$$H(X|Y) = -\sum_{i=1}^{n} \sum_{j=1}^{n} p(x_i, y_j) \log_2 p(x_i|y_j) \quad (10.44)$$

FIGURE 10.5 Noiseless channel

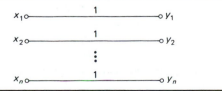

For the noiseless channel, all $p(x_i, y_j)$ and $p(x_i|y_j)$ are zero unless $i = j$. For $i = j$, $p(x_i|y_j)$ is unity. Thus $H(X|Y)$ is zero for the noiseless channel, and

$$I(X; Y) = H(X) \qquad (10.45)$$

We have seen that the entropy of a source is maximum if all source symbols are equally likely. Thus

$$C = \sum_{i=1}^{n} \frac{1}{n} \log_2 n = \log_2 n \qquad (10.46)$$

▼ **EXAMPLE 10.5**   Find the channel capacity of the *binary symmetric channel* illustrated in Figure 10.6.

SOLUTION   This problem has considerable practical importance in the area of binary digital communications. We will determine the capacity by maximizing

$$I(X; Y) = H(Y) - H(Y|X)$$

where

$$H(Y|X) = -\sum_{i=1}^{2} \sum_{j=1}^{2} p(x_i, y_j) \log_2 p(y_j|x_i) \qquad (10.47)$$

Using the probabilities defined in Figure 10.6, we obtain

$$H(Y|X) = -\alpha p \log_2 p - (1 - \alpha)p \log_2 p$$
$$- \alpha q \log_2 q - (1 - \alpha)q \log_2 q \qquad (10.48)$$

or

$$H(Y|X) = -p \log_2 p - q \log_2 q \qquad (10.49)$$

Thus

$$I(X; Y) = H(Y) + p \log_2 p + q \log_2 q \qquad (10.50)$$

FIGURE 10.6   Binary symmetrical channel

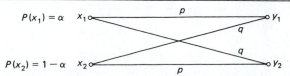

FIGURE 10.7   Capacity of a binary symmetric channel

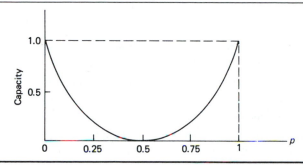

which is maximum when $H(Y)$ is maximum. Since the system output is binary, $H(Y)$ is maximum when each output has a probability of $\frac{1}{2}$ and is achieved for *equally likely inputs*. For this case, $H(Y)$ is unity, and the channel capacity is

$$C = 1 + p \log_2 p + q \log_2 q = 1 - H(p) \qquad (10.51)$$

where $H(p)$ is as defined in (10.4).

The capacity of a binary symmetric channel is sketched in Figure 10.7. As expected, if $p = 0$ or 1, the channel output is completely determined by the channel input, and the capacity is 1 bit per symbol. If p is equal to $\frac{1}{2}$, an input symbol yields either output with equal probability, and the capacity is zero.

It is worth noting that the capacity of the channel illustrated in Figure 10.5 is most easily found by starting with (10.29), while the capacity of the channel illustrated in Figure 10.6 is most easily found starting with (10.30). Choosing the appropriate expression for $I(X; Y)$ can often save considerable effort. It sometimes takes insight and careful study of a problem to choose the expression for $I(X; Y)$ that yields the capacity with minimum computational effort.

The error probability P_E of a binary symmetric channel is easily computed. From

$$P_E = \sum_{i=1}^{2} p(e \mid x_i) p(x_i) \qquad (10.52)$$

where $p(e \mid x_i)$ is the error probability given input x_i, we have

$$P_E = q p(x_1) + q p(x_2) \qquad (10.53)$$

Thus

$$P_E = q \tag{10.54}$$

which states that the *unconditional error probability* P_E is equal to the conditional error probability $p(y_j|x_i)$, $i \neq j$.

In Chapter 7 we showed that P_E is a decreasing function of the energy of the received symbols. Since the symbol energy is the received power multiplied by the symbol period, it follows that *if the transmitter power is fixed, the error probability can be reduced by decreasing the source rate.* This can be accomplished by removing the redundancy at the source through a process called *source encoding.*

▼ **EXAMPLE 10.6**   In Chapter 7 we showed that for binary coherent FSK systems, the probability of symbol error is the same for each transmitted symbol. Thus a binary symmetrical channel model is a suitable model for FSK transmission. Assume that the transmitter power is 1000 W and that the attenuation in the channel from transmitter to detector input is 30 dB. Also assume that the source rate r is 10,000 symbols per second and that the noise power spectral density N_0 is 2×10^{-5} W/Hz. Determine the channel matrix.

SOLUTION   Since the attenuation is 30 dB, the signal power P_R at the input to the detector is

$$P_R = (1000)(10^{-3}) = 1 \text{ W} \tag{10.55}$$

This corresponds to a received energy per symbol of

$$E_S = P_R T = \frac{1}{10,000} = 10^{-4} \text{ J} \tag{10.56}$$

From Chapter 7, Equation (7.83), the error probability for an FSK receiver is

$$P_E = Q\left(\sqrt{\frac{E_s}{N_0}}\right) \tag{10.57}$$

which, with the given values, is $P_E = 0.0127$. Thus the channel matrix is

$$[P(Y|X)] = \begin{bmatrix} 0.9873 & 0.0127 \\ 0.0127 & 0.9873 \end{bmatrix} \tag{10.58}$$

It is interesting to compute the change in the channel matrix resulting from moderate reduction in source symbol rate with all other parameters held constant. If the source symbol rate is reduced 25% to 7500 symbols per second, the received energy per symbol is

$$E_s = \frac{1}{7500} = 1.333 \times 10^{-4} \, \text{J} \tag{10.59}$$

With the other given parameters, the symbol error probability becomes $P_E = 0.0049$, which yields the channel matrix

$$[P(Y|X)] = \begin{bmatrix} 0.9951 & 0.0049 \\ 0.0049 & 0.9951 \end{bmatrix} \tag{10.60}$$

Thus the 25% reduction in source symbol rate results in an improvement of the system symbol error probability by a factor of almost 3.

In Section 10.2 we will investigate a technique that sometimes allows the source symbol rate to be reduced without reducing the source information rate.

10.2   SOURCE ENCODING

We determined in the preceding section that the information from a source that produced different symbols according to some probability scheme could be described by the entropy $H(X)$. Since entropy has units of bits per symbol, we also must know the symbol rate in order to specify the source information rate in bits per second. In other words, the source information rate R_s is given by

$$R_s = rH(X) \text{ bits/sec} \tag{10.61}$$

where $H(X)$ is the source entropy in bits per symbol and r is the symbol rate in symbols per second.

Let us assume that this source is the input to a channel with capacity C bits per symbol or SC bits per second, where S is the available symbol rate for the channel. An important theorem of information theory, the *noiseless coding theorem*, sometimes referred to as *Shannon's first theorem*, is stated: *Given a channel and a source that generates information at a rate less than the channel capacity, it is possible to encode the source output in such a manner that it can be transmitted through the channel.* A proof of this theorem is beyond the scope of this introductory treatment of information theory and can be found in

FIGURE 10.8 Transmission scheme

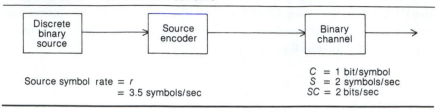

Source symbol rate = r
= 3.5 symbols/sec

C = 1 bit/symbol
S = 2 symbols/sec
SC = 2 bits/sec

any of the standard information theory textbooks.* However, we demonstrate the theorem by a simple example.

An Example of Source Encoding

Let us consider a discrete binary source that has two possible outputs A and B that have probabilities 0.9 and 0.1, respectively. Assume also that the source rate r is 3.5 symbols per second. The source output is connected to a binary channel that can transmit a binary 0 or 1 at a rate of 2 symbols per second with negligible error, as shown in Figure 10.8. Thus, from Example 10.5 with $p = 1$, the channel capacity is 1 bit per symbol, which, in this case, is an information rate of 2 bits per second.

It is clear that the source *symbol* rate is greater than the channel capacity, so the source symbols cannot be placed directly into the channel. However, the source entropy is

$$H(X) = -0.1 \log_2 0.1 - 0.9 \log_2 0.9 = 0.469 \text{ bits/symbol} \quad (10.62)$$

which corresponds to a source *information* rate of

$$rH(X) = 3.5(0.469) = 1.642 \text{ bits/s} \quad (10.63)$$

Thus, the *information* rate is less than the channel capacity, so transmission is possible.

Transmission is accomplished by the process called *source encoding*, whereby codewords are assigned to n-symbol groups of source symbols. The shortest codeword is assigned to the most probable group of source symbols, and the longest codeword is assigned to the least probable group of source symbols. Thus source encoding decreases the average symbol rate, which allows the source to be matched to the channel. The n-symbol groups of source symbols are known as the nth-order extension of the original source.

---

* See, for example, Gallagher (1968).

TABLE 10.1   First-Order Extension

Source Symbol	$P(\ )$	Codeword	$[P(\ )] \cdot$ [Number of Code Symbols]
A	0.9	0	0.9
B	0.1	1	0.1
			$\bar{L} = 1.0$

Table 10.1 illustrates the first-order extension of the original source. Clearly, the symbol rate of the encoder is equal to the symbol rate of the source. Thus the symbol rate at the channel input is still larger than the channel can accommodate.

The second-order extension of the original source is formed by taking the source symbols two at a time, as illustrated in Table 10.2. The average word length $\bar{L}$ for this case is

$$\bar{L} = \sum_{i=1}^{2^n} p(x_i)l_i = 1.29 \tag{10.64}$$

where $p(x_i)$ is the probability of the ith symbol of the extended source and l_i is the length of the codeword corresponding to the ith symbol. Since the source is binary, there are 2^n symbols in the extended source output, each of length n. Thus, for the second-order extension,

$$\frac{\bar{L}}{n} = \frac{1.29}{2} = 0.645 \text{ code symbols/source symbol} \tag{10.65}$$

and the symbol rate at the encoder output is

$$r\frac{\bar{L}}{n} = 3.5(0.645) = 2.258 \text{ code symbols/second} \tag{10.66}$$

TABLE 10.2   Second-Order Extension

Source Symbol	$P(\ )$	Codeword	$[P(\ )] \cdot$ [Number of Code Symbols]
AA	0.81	0	0.81
AB	0.09	1 0	0.18
BA	0.09	1 1 0	0.27
BB	0.01	1 1 1	0.03
			$\bar{L} = 1.29$

which is still greater than the 2 symbols per second that the channel can accept. It is clear that the symbol rate has been reduced, and this provides motivation to try again.

Table 10.3 shows the third-order extension. For this case, the source symbols are grouped three at a time. The average word length $\overline{L}$ is 1.598, and

$$\frac{\overline{L}}{n} = \frac{1.598}{3} = 0.533 \text{ code symbols/source symbol} \qquad (10.67)$$

The symbol rate at the encoder output is

$$r\frac{\overline{L}}{n} = 3.5(0.533) = 1.864 \text{ code symbols/second} \qquad (10.68)$$

This rate can be accepted by the channel.

It is worth noting in passing that if the source symbols appear at a constant rate, the code symbols at the encoder output do not appear at a constant rate. As is apparent in Table 10.3, the source output *AAA* results in a single symbol at the encoder output, whereas the source output *BBB* results in five symbols at the encoder output. Thus symbol buffering must be provided at the encoder output if the symbol rate into the channel is to be constant.

Figure 10.9 shows the behavior of $\overline{L}/n$ as a function of n. One can see that $\overline{L}/n$ always exceeds the source entropy and converges to the source entropy for large N. This is a fundamental result of information theory.

To illustrate the method used to select the codewords in this example, let us look at the general problem of source encoding.

Several Definitions

Before we discuss in detail the method of deriving codewords, we pause to make a few definitions that will clarify our work.

TABLE 10.3 Third-Order Extension

Source Symbol	$P(\ )$	Codeword	$[P(\ )] \cdot$ [Number of Code Symbols]
AAA	0.729	0	0.729
AAB	0.081	1 0 0	0.243
ABA	0.081	1 0 1	0.243
BAA	0.081	1 1 0	0.243
ABB	0.009	1 1 1 0 0	0.045
BAB	0.009	1 1 1 0 1	0.045
BBA	0.009	1 1 1 1 0	0.045
BBB	0.001	1 1 1 1 1	0.005
			$\overline{L} = 1.598$

FIGURE 10.9   Behavior of $\bar{L}/n$

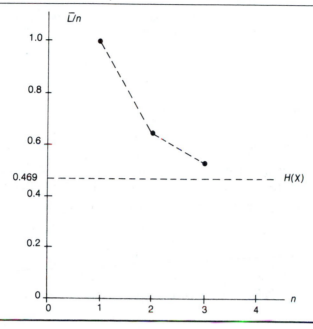

Each codeword is constructed from an *alphabet* that is a collection of symbols used for communication through a channel. For example, a binary codeword is constructed from a two-symbol alphabet, wherein the two symbols are usually taken as 0 and 1. The *word length* of a codeword is the number of symbols in the codeword.

There are several major subdivisions of codes. For example, a code can be either *block* or *nonblock*. A block code is one in which each block of source symbols is encoded into a fixed-length sequence of code symbols. A *uniquely decipherable* code is a block code in which the codewords may be deciphered without using spaces. These codes can be further classified as *instantaneous* or *noninstantaneous,* according to whether or not it is possible to decode each word in sequence without reference to succeeding code symbols. Alternatively, noninstantaneous codes require reference to succeeding code symbols, as illustrated in Figure 10.10. It should always be remembered that a noninstantaneous code can be uniquely decipherable.

A useful measure of goodness of a source code is the *efficiency,* which is defined as the ratio of the minimum average word length of the codewords $\bar{L}_{\min}$ to the average word length of the codeword $\bar{L}$. Thus

$$\text{Efficiency} = \frac{\bar{L}_{\min}}{\bar{L}} = \frac{\bar{L}_{\min}}{\sum_{i=1}^{n} p(x_i)l_i} \tag{10.69}$$

FIGURE 10.10 Instantaneous and noninstantaneous codes

Source symbols	Code 1 (noninstantaneous)	Code 2 (instantaneous)
x_1	0	0
x_2	01	10
x_3	011	110
x_4	0111	1110

where $p(x_i)$ is the probability of the ith source symbol and l_i is the length of the codeword corresponding to the ith source symbol. It can be shown that the minimum average word length is given by

$$\bar{L}_{min} = \frac{H(X)}{\log_2 D} \tag{10.70}$$

where $H(X)$ is the entropy of the message ensemble being encoded and D is the number of symbols in the encoding alphabet. This yields.

$$\text{Efficiency} = \frac{H(X)}{\bar{L} \log_2 D} \tag{10.71}$$

or

$$\text{Efficiency} = \frac{H(X)}{\bar{L}} \tag{10.72}$$

for a *binary* alphabet. It is worth noting that if the efficiency of a code is 100%, the average word length $\bar{L}$ is equal to the entropy, $H(X)$.

Entropy of an Extended Binary Source

In many problems of practical interest, the efficiency is improved by coding the nth-order extension of a source. This is exactly the scheme used in the example of source encoding in a previous subsection. Computation of the efficiency of each of the three schemes used involves calculating the efficiency of the extended source. The efficiency can, of course, be calculated directly, using the symbol probabilities of the extended source, but there is an easier method.

The entropy of the nth-order extension of a discrete memoryless source $H(X^n)$ is given by

$$H(X^n) = nH(X) \tag{10.73}$$

This is easily shown by representing a message sequence from the output of the nth source extension as $(i_1, i_2, \ldots, i_n)$, where i_k can take on one of two states with probability p_{i_k}. The entropy of the nth-order extension of the source is

$$H(X^n) = -\sum_{i_1=1}^{2}\sum_{i_2=1}^{2}\cdots\sum_{i_n=1}^{2}(p_{i_1}p_{i_2}\cdots p_{i_n})\log_2(p_{i_1}p_{i_2}\cdots p_{i_n}) \quad (10.74)$$

or

$$H(X^n) = -\sum_{i_1=1}^{2}\sum_{i_2=1}^{2}\cdots\sum_{i_n=1}^{2}(p_{i_1}p_{i_2}\cdots p_{i_n})(\log_2 p_{i_1}$$

$$+ \log_2 p_{i_2} + \cdots + \log_2 p_{i_n}) \quad (10.75)$$

We can write the preceding expression as

$$H(X^n) = -\sum_{i_1=1}^{2}p_{i_1}\log_2 p_{i_1}\left[\sum_{i_2=1}^{2}p_{i_2}\sum_{i_3=1}^{2}p_{i_3}\cdots\sum_{i_n=1}^{2}p_{i_n}\right]$$

$$-\left[\sum_{i_1=1}^{2}p_{i_1}\right]\sum_{i_2=1}^{2}p_{i_2}\log_2 p_{i_2}\left[\sum_{i_3=1}^{2}p_{i_3}\cdots\sum_{i_n=1}^{2}p_{i_n}\right]\cdots$$

$$-\left[\sum_{i_1=1}^{2}p_{i_1}\sum_{i_2=1}^{2}p_{i_2}\cdots\sum_{i_{n-1}=1}^{2}p_{i_{n-1}}\right]\sum_{i_n=1}^{2}p_{i_n}\log_2 p_{i_n} \quad (10.76)$$

Each term in brackets is equal to 1. Thus

$$H(X^n) = -\sum_{k=1}^{n}\sum_{i_k=1}^{2}p_{i_k}\log_2 p_{i_k} = \sum_{k=1}^{n}H(X) \quad (10.77)$$

which yields

$$H(X^n) = nH(X)$$

The efficiency of the extended source is therefore given by

$$\text{Efficiency} = \frac{nH(X)}{\bar{L}} \quad (10.78)$$

If efficiency tends to 100% as n approaches infinity, it follows that $\bar{L}/n$ tends to the entropy of the extended source. This is exactly the observation made from Figure 10.9.

FIGURE 10.11   Shannon-Fano encoding

Source words	Probability	Codeword	(Length) · (Probability)
X_1	0.2500	00	2 (0.25) = 0.50
X_2	0.2500	01	2 (0.25) = 0.50
		A – – – – *A'*	
X_3	0.1250	100	3 (0.125) = 0.375
X_4	0.1250	101	3 (0.125) = 0.375
X_5	0.0625	1100	4 (0.0625) = 0.25
X_6	0.0625	1101	4 (0.0625) = 0.25
X_7	0.0625	1110	4 (0.0625) = 0.25
X_8	0.0625	1111	4 (0.0625) = 0.25
			Average wordlength = 2.75

Shannon-Fano Encoding

There are several methods of encoding a source output so that an instantaneous code results. We consider two such methods here. First, we consider the Shannon-Fano method, which is very easy to apply and usually yields source codes having reasonably high efficiency. In the next subsection we consider the Huffman encoding technique, which yields the source code having the shortest average word length for a given source entropy.

Assume that we are given a set of source outputs that are to be encoded. These source outputs are first ranked in order of nonincreasing probability of occurrence, as illustrated in Figure 10.11. The set is then partitioned into two sets (indicated by line *A–A'*) that are as close to equiprobable as possible, and 0's are assigned to the upper set and 1's to the lower set, as seen in the first column of the codewords. This process is continued, each time partitioning the sets with as nearly equal probabilities as possible, until further partitioning is not possible. This scheme will give a 100% efficient code if the partitioning always results in equiprobable sets; otherwise, the code will have an efficiency of less than 100%. For this particular example,

$$\text{Efficiency} = \frac{H(X)}{\bar{L}} = \frac{2.75}{2.75} = 1 \qquad (10.79)$$

since equiprobable partitioning is possible.

Huffman Encoding

Huffman encoding results in an optimum code in the sense that the Huffman code has the minimum average word length for a source of given entropy. Thus it is the code that has the highest efficiency. We shall illustrate the

Huffman coding procedure using the same source output of eight messages used to illustrate the Shannon-Fano encoding procedure.

Figure 10.12 illustrates the Huffman encoding procedure. The source output consists of messages X_1, X_2, X_3, X_4, X_5, X_6, X_7, and X_8. They are listed in order of nonincreasing probability, as was done for Shannon-Fano encoding. The first step of the Huffman procedure is to combine the two source messages having the lowest probability, X_7 and X_8. The upper message, X_7, is assigned a binary 0 as the last symbol in the codeword, and the lower message, X_8, is assigned a binary 1 as the last symbol in the codeword. The combination of X_7 and X_8 can be viewed as a composite message having a probability equal to the sum of the probabilities of X_7 and X_8, which in this case is 0.1250, as shown. This composite message is denoted X'_4. After this initial step, the new set of messages, denoted X_1, X_2, X_3, X_4, X_5, X_6, and X'_4 are arranged in order of nonincreasing probability. Note that X'_4 could be placed at any point between X_2 and X_5, although it was given the name X'_4 because it was placed after X_4. The same procedure is then applied once again. The messages X_5 and X_6 are combined. The resulting composite message is combined with X'_4. This procedure is continued as far as possible. The resulting tree structure is then traced in reverse to determine the codewords. The resulting codewords are shown in Figure 10.12.

FIGURE 10.12   Example of Huffman encoding

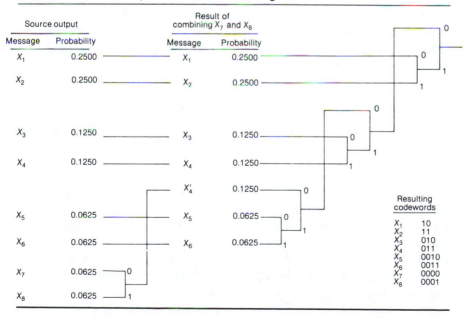

		Resulting codewords
X_1	10
X_2	11
X_3	010
X_4	011
X_5	0010
X_6	0011
X_7	0000
X_8	0001

The codewords resulting from the Huffman procedure are different from the codewords resulting from the Shannon-Fano procedure because at several points the placement of composite messages resulting from previous combinations was arbitrary. The assignment of binary 0's or binary 1's to the upper or lower messages was arbitrary. Note, however, that the average word length is the same for both procedures. This must be the case for the example chosen because the Shannon-Fano procedure yielded 100% efficiency and the Huffman procedure can be no worse. There are cases in which the two procedures do not result in equal average wordlengths.

10.3 RELIABLE COMMUNICATION IN THE PRESENCE OF NOISE

We now turn our attention to methods for achieving reliable communication in the presence of noise by combating the effects of that noise. We undertake our study with a promise from Claude Shannon of considerable success. *Shannon's theorem*, sometimes referred to as the *fundamental theorem of information theory*, is stated: *Given a discrete memoryless channel (each symbol is perturbed by noise independently of all other symbols) with capacity C and a source with positive rate R, where R < C, there exists a code such that the output of the source can be transmitted over the channel with an arbitrarily small probability of error.*

Thus Shannon's theorem predicts essentially error-free transmission in the presence of noise. Unfortunately, the theorem tells us only of the existence of codes and tells nothing of how to construct these codes.

Before we start our study of constructing codes for noisy channels, we will take a minute to discuss the continuous channel. This detour will yield considerable insight that will prove useful.

The Capacity of a White Additive Gaussian Noise Channel

The capacity, in bits per second, of a continuous channel with additive white Gaussian noise is given by

$$C_c = B \log_2 \left(1 + \frac{S}{N} \right) \tag{10.80}$$

where B is the channel bandwidth in hertz and S/N is the signal-to-noise power ratio. This particular formulation is known as the *Shannon-Hartley law*. The subscript is used to distinguish (10.80) from (10.43). Capacity, as expressed by (10.43) has units of bits per symbol, while (10.80) has units of bits per second.

The tradeoff between bandwidth and signal-to-noise ratio can be seen from the Shannon-Hartley law. For infinite signal-to-noise ratio, which is the noiseless case, the capacity is infinite for any nonzero bandwidth. We will show, however, that the capacity cannot be made arbitrarily large by increasing bandwidth if noise is present.

In order to understand the behavior of the Shannon-Hartley law for the large-bandwidth case, it is desirable to place (10.80) in a slightly different form. The energy per bit E_b is equal to the bit time T_b multiplied by the signal power S. At capacity, the bit rate R_b is equal to the capacity. Thus $T_b = 1/C_c$ seconds per bit. This yields, at capacity,

$$E_b = ST_b = \frac{S}{C_c} \tag{10.81}$$

The total noise power in bandwidth B is given by

$$N = N_0 B \tag{10.82}$$

where N_0 is the single-sided noise power spectral density in watts per hertz. The signal-to-noise ratio can therefore be expressed as

$$\frac{S}{N} = \frac{E_b}{N_0} \frac{C_c}{B} \tag{10.83}$$

This allows the Shannon-Hartley law to be written in the equivalent form

$$\frac{C_c}{B} = \log_2 \left(1 + \frac{E_b}{N_0} \frac{C_c}{B} \right) \tag{10.84}$$

Solving for E_b/N_0 yields

$$\frac{E_b}{N_0} = \frac{B}{C_c} (2^{C_c/B} - 1) \tag{10.85}$$

This expression establishes performance of the ideal system. For the case in which $B \gg C_c$,

$$2^{C_c/B} = e^{(C_c/B)\ln 2} \cong 1 + \frac{C_c}{B} \ln 2 \tag{10.86}$$

where the approximation $e^x \cong 1 + x, x \ll 1$, has been used. Substitution of (10.86) into (10.85) gives

$$\frac{E_b}{N_0} \cong \ln 2 = -1.6 \text{ dB}, \qquad B \gg C_c \tag{10.87}$$

Thus, for the ideal system, in which $R_b = C_c$, E_b/N_0 approaches the limiting value of -1.6 dB as the bandwidth grows without bound.

A plot of E_b/N_0, expressed in decibels, as a function of R_b/B is illustrated in Figure 10.13. The ideal system is defined by $R_b = C_c$ and corresponds to (10.85). There are two regions of interest. The first region, for which $R_b < C_c$, is the region in which arbitrarily small error probabilities can be obtained. Clearly this is the region in which we wish to operate. The other region, for which $R_b > C_c$, does not allow the error probability to be made arbitrarily small.

An important tradeoff can be deduced from Figure 10.13. If the bandwidth factor R_b/B is large so that the bit rate is much greater than the bandwidth, then a significantly larger value of E_b/N_0 is necessary to ensure operation in the $R_b < C_c$ region than is the case if R_b/B is small. Stated another way, assume that the source bit rate is fixed at R_b bits per second and the available bandwidth is large so that $B \gg R_b$. For this case, operation in the $R_b < C_c$ region requires only that E_b/N_0 is slightly greater than -1.6 dB. The required signal power is

$$S \cong R_b(\ln 2)N_0 \text{ W} \qquad (10.88)$$

This is the minimum signal power for operation in the $R_b < C_c$ region. Therefore, operation in this region is desired for *power-limited operation*.

FIGURE 10.13 $R_b = C_c$ relationship for additive white Gaussian noise channel

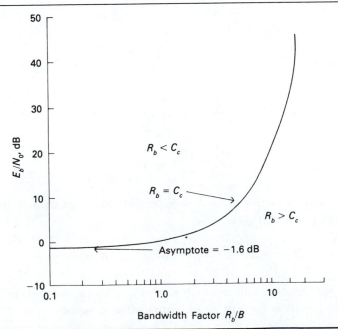

Now assume that bandwidth is limited so that $R_b \gg B$. Figure 10.13 shows that a much larger value of E_b/N_0 is necessary for operation in the $R_b < C_c$ region. Thus the required signal power is much greater than that given by (10.88). This is referred to as *bandwidth-limited operation*.

The preceding paragraphs illustrate that, at least in the additive white Gaussian noise channel where the Shannon-Hartley law applies, a tradeoff exists between power and bandwidth. This tradeoff is of fundamental importance in the design of communication systems.

Reliable Communication Using Orthogonal Signals

Now that we realize that we can theoretically achieve perfect system performance even in the presence of noise, we start our search for system configurations that yield the performance promised by Shannon's theorem. Actually one such system was analyzed in Chapter 9. Orthogonal signals were chosen for transmission through the channel, and a correlation receiver structure was chosen for demodulation. The system performance is illustrated in Figure 9.7. Shannon's bound is clearly illustrated.

Block Coding for Binary Systems

An alternative technique for approaching Shannon's limit is to use coding schemes that allow for detection and correction of errors. One class of codes for this purpose is known as *block codes*. In order to understand the meaning of a block code, let us consider a source that produces a serial stream of binary symbols at a rate of R symbols per second. Assume that these symbols are grouped into blocks T seconds long, so that each block contains $RT = k$ source or information symbols. To each of these k-symbol blocks is added a group of redundant check symbols to produce a codeword n symbols long. The $n - k$ check symbols should supply sufficient information to the decoder to allow for the correction of most errors that might occur in the channel. An encoder that operates in this manner is said to produce an (n, k) block code.

Codes can either correct or merely detect errors, depending on the amount of redundancy contained in the check symbols. Codes that can correct errors are known as *error-correcting codes*. Codes that can only detect errors are also useful. A feedback channel can be used to request a retransmission of the codeword found to be in error, and often the error can thus be corrected. We will discuss error-detection feedback later. Often the codewords with detected errors are simply discarded by the decoder. This is practical when errors are more serious than a lost codeword.

An understanding of how codes can detect and correct errors can be gained from a geometric point of view. A binary codeword is a sequence of

1's and 0's that is n symbols in length. The *Hamming weight* $w(s_j)$ of codeword s_j is defined as the number if 1's in that codeword. The *Hamming distance* $d(s_i, s_j)$ or d_{ij} between codewords s_i and s_j is defined as the number of positions in which s_i and s_j differ. It follows that Hamming distance can be written in terms of Hamming weight as

$$d_{ij} = w(s_i \oplus s_j) \tag{10.89}$$

where the symbol $\oplus$ denotes modulo-2 addition, which is binary addition without a carry.

▼ **EXAMPLE 10.7**   Compute the Hamming distance between s_1 and 101101 and $s_2 = 001100$.

SOLUTION   Since

$$101101 \oplus 001100 = 100001$$

we have

$$d_{12} = w(100001) = 2$$

▲ which simply means that s_1 and s_2 differ in 2 positions.

A geometric representation of two codewords is shown in Figure 10.14. The C's represent two codewords that are distance 5 apart. The codeword on the left is the reference codeword. The first "x" to the right of the reference codeword represents a binary sequence distance 1 from the reference codeword, where distance is understood to denote the Hamming distance. The second "x" to the right of the reference codeword is distance 2

FIGURE 10.14   Geometric representation of two codewords

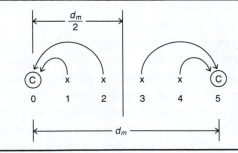

from the reference, and so on. Assuming that the two codewords shown are the closest in Hamming distance of all the codewords for a given code, the code is then a distance 5 code. Figure 10.14 illustrates the concept of a minimum-distance decoding, in which a given received sequence is assigned to the codeword closest, in Hamming distance, to the received sequence. A minimum-distance decoder will therefore assign the x's to the left of the vertical line to the codeword on the left and the x's to the right of the vertical line to the codeword on the right, as shown.

We can deduce that a minimum-distance decoder can always correct as many as e errors, where e is the largest integer not to exceed

$$\tfrac{1}{2}(d_m - 1)$$

where d_m is the minimum distance between codewords. It follows that if d_m is odd, all received words can be assigned to a codeword. However, if d_m is even, a received word can lie halfway between two codewords. For this case, errors are detected that cannot be corrected.

▼ **EXAMPLE 10.8** A code consists of codewords [0001011, 1110000, 1000110, 1111011, 0110110, 1001101, 0111101, 0000000]. If 1101011 is received, what is the decoded codeword?

SOLUTION The decoded codeword is the codeword closest in Hamming distance to 1101011. The calculations are

$$w(0001011 \oplus 1101011) = 2 \quad w(0110110 \oplus 1101011) = 5$$
$$w(1110000 \oplus 1101011) = 4 \quad w(1001101 \oplus 1101011) = 3$$
$$w(1000110 \oplus 1101011) = 4 \quad w(0111101 \oplus 1101011) = 4$$
$$w(1111011 \oplus 1101011) = 1 \quad w(0000000 \oplus 1101011) = 5$$

▲ Thus the decoded codeword is 1111011.

We will now consider several different codes.

Single-Parity-Check Codes

A simple code that can detect single errors but does not have any error-correcting capability is formed by adding one check symbol to each block of k information symbols. This yields a $(k + 1, k)$ code. The added symbol is called a *parity-check symbol,* and it is added so that the Hamming weight of each codeword is always either odd or even. If the received word contains an *even* number of errors, the decoder will not detect the errors. If the number of errors is *odd,* the decoder will detect that an error has been made.

Repetition Codes

The simplest code that allows for correction of errors consists of transmitting each symbol n times, which results in $n-1$ check symbols. This technique produces an $(n, 1)$ code having two codewords; one of all 0's and one of all 1's. A received word is decoded as a 0 if the majority of the received symbols are 0's and as a 1 if the majority are 1's. This is equivalent to minimum-distance decoding, wherein $\frac{1}{2}(n-1)$ errors can be corrected. Repetition codes have great error-correcting capability if the symbol error probability is low but have the disadvantage of transmitting many redundant symbols. For example, if the information rate of the source is R bits per symbol, the rate R_c out of the encoder is

$$R_c = \left(\frac{k}{n}\right) R = \frac{1}{n} R \text{ bits/symbol} \tag{10.90}$$

The factor k/n is called the *code rate.*

The process of repetition coding for a rate $\frac{1}{3}$ repetition code is illustrated in detail in Figure 10.15. The encoder maps the data symbols 0 and 1 into the corresponding codewords 000 and 111. There are eight possible received sequences, as shown. The mapping from the transmitted sequence

FIGURE 10.15 Example of rate $\frac{1}{3}$ repetition code

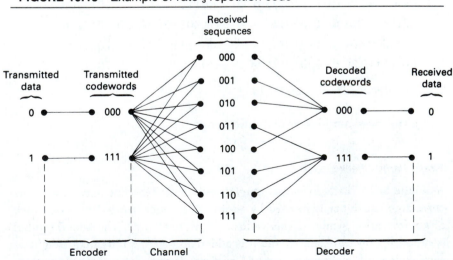

to the received sequence is random, and the statistics of the mapping are determined by the channel characteristics derived in Chapters 7 and 8. The decoder maps the received sequence into one of the two codewords by a minimum Hamming distance decoding rule. Each decoded codeword corresponds to a data symbol, as shown.

▼ **EXAMPLE 10.9**   Investigate the error-correcting capability of a repetition code having a code rate of $\frac{1}{3}$.

SOLUTION   Assume that the code is used with a binary symmetric channel with a conditional error probability equal to $(1 - p)$, that is,

$$p(y_j \mid x_i) = 1 - p, \quad i \neq j \tag{10.91}$$

Each source 0 is encoded as 000, and each source 1 is encoded as 111. An error is made if two or three symbols undergo a change in passing through the channel. Assuming that the source outputs are equally likely, the error probability P_e becomes

$$P_e = 3(1 - p)^2 p + (1 - p)^3 \tag{10.92}$$

A few simple calculations illustrate the value of the code. For $(1 - p) = 0.1$, $P_e = 0.028$, yielding an improvement factor of slightly less than 4. For $(1 - p) = 0.01$, the improvement factor is approximately 33. Thus the code performs best when $(1 - p)$ is small. Of course, performance increases as n becomes larger, but the code rate decreases. In many cases, the *information rate* must be maintained constant. Thus an increase in redundancy results in an increase in symbol rate, which changes the channel matrix. We will consider this problem in Example 10.13.   ▲

Parity-Check Codes for Correction of Single Errors

Repetition codes and single-parity-check codes are examples of codes that have either high error-correction capability or high information rate, but not both. Only codes that have a reasonable combination of these characteristics are practical for use in digital communication systems. We now examine a class of parity-check codes that satisfy these requirements.

A general codeword can be written in the form

$$a_1 a_2 \cdots a_k c_1 c_2 \cdots c_r$$

where a_i is the ith information symbol and c_j is the jth check symbol. Codes for which the first k symbols of the codeword are the information symbols are called *systematic codes*. It should be remembered that $r = n - k$. The parity check symbols are chosen to satisfy the linear equations

$$0 = h_{11}a_1 \oplus h_{12}a_2 \oplus \cdots \oplus h_{1k}a_k \oplus c_1$$
$$0 = h_{21}a_1 \oplus h_{22}a_2 \oplus \cdots \oplus h_{2k}a_k \oplus c_2$$
$$\vdots$$
$$0 = h_{r1}a_1 \oplus h_{r2}a_2 \oplus \cdots \oplus h_{rk}a_k \oplus c_r \tag{10.93}$$

Equation (10.93) can be written as

$$[H][T] = [0] \tag{10.94}$$

where $[H]$ is the parity-check matrix

$$[H] = \begin{bmatrix} h_{11} & h_{12} & \cdots & h_{1k} & 1 & 0 & \cdots & 0 \\ h_{21} & h_{22} & \cdots & h_{2k} & 0 & 1 & \cdots & 0 \\ \vdots & & & & & & & \\ h_{r1} & h_{r2} & \cdots & h_{rk} & 0 & 0 & \cdots & 1 \end{bmatrix} \tag{10.95}$$

and $[T]$ is the codeword vector

$$[T] = \begin{bmatrix} a_1 \\ a_2 \\ \vdots \\ a_k \\ c_1 \\ \vdots \\ c_r \end{bmatrix} \tag{10.96}$$

Now let the received word be $[R]$. If

$$[H][R] \neq [0] \tag{10.97}$$

we know that $[R]$ is not a codeword and at least one error has been made. If

$$[H][R] = [0] \tag{10.98}$$

we know that $[R]$ is a codeword and that it is *most likely* the transmitted codeword.

Since $[R]$ is the received codeword, it can be written

$$[R] = [T] \oplus [E] \tag{10.99}$$

where $[E]$ represents the error pattern induced by the channel. The decoding problem essentially reduces to determining $[E]$, since the codeword can be reconstructed from $[R]$ and $[E]$.

As the first step in computing [E], we multiply the received word [R] by the parity-check matrix [H]. This yields

$$[S] = [H][R] = [H][T] \oplus [H][E] \qquad (10.100)$$

or

$$[S] = [H][E] \qquad (10.101)$$

The matrix [S] is known as the *syndrome*.

Assuming that a single error has taken place, the error vector will be of the form

$$[E] = \begin{bmatrix} 0 \\ 0 \\ \vdots \\ 1 \\ \vdots \\ 0 \end{bmatrix}$$

Multiplying [E] by [H] on the left-hand side shows that the syndrome is the *i*th column of the matrix [H], where the error is in the *i*th position. The following example illustrates the method.

▼ EXAMPLE 10.10   A code has the parity-check matrix

$$[H] = \begin{bmatrix} 1 & 1 & 0 & 1 & 0 & 0 \\ 0 & 1 & 1 & 0 & 1 & 0 \\ 1 & 0 & 1 & 0 & 0 & 1 \end{bmatrix} \qquad (10.102)$$

Assuming that 1 1 1 0 1 1 is received, determine if an error has been made, and if so, determine the decoded codeword.

SOLUTION   First, we compute the syndrome, remembering that all operations are modulo-2:

$$[S] = [H][R] = \begin{bmatrix} 1 & 1 & 0 & 1 & 0 & 0 \\ 0 & 1 & 1 & 0 & 1 & 0 \\ 1 & 0 & 1 & 0 & 0 & 1 \end{bmatrix} \begin{bmatrix} 1 \\ 1 \\ 1 \\ 0 \\ 1 \\ 1 \end{bmatrix} = \begin{bmatrix} 0 \\ 1 \\ 1 \end{bmatrix} \qquad (10.103)$$

Since the syndrome is the third column of the parity-check matrix, the third symbol of the received word is assumed to be in error. Thus the decoded codeword is 110011. This can be proved by showing that 110011 has a zero syndrome.

We now pause to examine the parity-check code in more detail. It follows from (10.93) and (10.95) that the parity checks can be written as

$$
\begin{bmatrix} c_1 \\ c_2 \\ \vdots \\ c_r \end{bmatrix} =
\begin{bmatrix}
h_{11} & h_{12} & \cdots & h_{1k} \\
h_{21} & h_{22} & \cdots & h_{2k} \\
\vdots & \vdots & & \vdots \\
h_{r1} & h_{r2} & \cdots & h_{rk}
\end{bmatrix}
\begin{bmatrix} a_1 \\ a_2 \\ \vdots \\ a_k \end{bmatrix}
\tag{10.104}
$$

Thus the codeword vector $[T]$ can be written

$$
[T] =
\begin{bmatrix} a_1 \\ a_2 \\ \vdots \\ a_k \\ c_1 \\ \vdots \\ c_r \end{bmatrix} =
\begin{bmatrix}
1 & 0 & \cdots & 0 \\
0 & 1 & \cdots & 0 \\
\vdots & \vdots & & \vdots \\
0 & 0 & \cdots & 1 \\
h_{11} & h_{12} & \cdots & h_{1k} \\
\vdots & \vdots & & \vdots \\
h_{r1} & h_{r2} & \cdots & h_{rk}
\end{bmatrix}
\begin{bmatrix} a_1 \\ a_2 \\ \vdots \\ a_k \end{bmatrix}
\tag{10.105}
$$

or

$$
[T] = [G][A]
\tag{10.106}
$$

where $[A]$ is the vector of k information symbols,

$$
[A] =
\begin{bmatrix} a_1 \\ a_2 \\ \vdots \\ a_k \end{bmatrix}
\tag{10.107}
$$

and $[G]$, which is called the *generator matrix*, is

$$
[G] =
\begin{bmatrix}
1 & 0 & \cdots & 0 \\
0 & 1 & \cdots & 0 \\
\vdots & \vdots & & \vdots \\
0 & 0 & \cdots & 1 \\
h_{11} & h_{12} & \cdots & h_{1k} \\
\vdots & \vdots & & \vdots \\
h_{r1} & h_{r2} & \cdots & h_{rk}
\end{bmatrix}
\tag{10.108}
$$

The relationship between the generator matrix $[G]$ and the parity-check matrix $[H]$ is apparent if we compare (10.95) and (10.108). If the m by m identity matrix is identified by $[I_m]$ and the matrix $[H_p]$ is defined by

$$[H_p] = \begin{bmatrix} h_{11} & h_{12} & \cdots & h_{1k} \\ h_{21} & h_{22} & \cdots & h_{2k} \\ \vdots & \vdots & & \vdots \\ h_{r1} & h_{r2} & \cdots & h_{rk} \end{bmatrix} \qquad (10.109)$$

it follows that the generator matrix is given by

$$[G] = \left[\begin{array}{c} I_k \\ \hline H_p \end{array} \right] \qquad (10.110)$$

and that the parity-check matrix is given by

$$[H] = [H_p \mid I_r] \qquad (10.111)$$

which establishes the relationship between the generator and parity-check matrices for systematic codes.

Codes defined by (10.108) are referred to as *linear codes,* since the $k + r$ codeword symbols are formed as a linear combination of the k information symbols. It is also worthwhile to note that if two different information sequences are summed to give a third sequence, then the codeword for the third sequence is the sum of the two codewords corresponding to the original two information sequences. This is easily shown. If two information sequences are summed, the resulting vector of information symbols is

$$[A_3] = [A_1] \oplus [A_2] \qquad (10.112)$$

The codeword corresponding to $[A_3]$ is

$$[T_3] = [G][A_3] = [G]\{[A_1] \oplus [A_2]\} = [G][A_1] \oplus [G][A_2] \qquad (10.113)$$

Since

$$[T_1] = [G][A_1] \qquad (10.114a)$$

and

$$[T_2] = [G][A_2] \qquad (10.114b)$$

it follows that

$$[T_3] = [T_1] \oplus [T_2] \tag{10.115}$$

Codes that satisfy this property are known as *group codes*.

Hamming Codes

A Hamming code is a particular parity-check code having distance 3. There-fore, all single errors can be corrected. The parity-check matrix for the code has dimensions $2^{n-k} - 1$ by $n - k$ and is very easy to construct. The ith column of the matrix $[H]$ is the binary representation of the number i. The code has the interesting property that, for a single error, the syndrome is the binary representation of the position in error.

▼ EXAMPLE 10.11   Determine the parity-check matrix for a (7, 4) code and the decoded codeword if the received word is 1110001.

SOLUTION   Since the ith column of the matrix $[H]$ is the binary represen-tation of i, we have

$$[H] = \begin{bmatrix} 0 & 0 & 0 & 1 & 1 & 1 & 1 \\ 0 & 1 & 1 & 0 & 0 & 1 & 1 \\ 1 & 0 & 1 & 0 & 1 & 0 & 1 \end{bmatrix} \tag{10.116}$$

For the received word 1110001, the syndrome is

$$[S] = [H][R] = \begin{bmatrix} 0 & 0 & 0 & 1 & 1 & 1 & 1 \\ 0 & 1 & 1 & 0 & 0 & 1 & 1 \\ 1 & 0 & 1 & 0 & 1 & 0 & 1 \end{bmatrix} \begin{bmatrix} 1 \\ 1 \\ 1 \\ 0 \\ 0 \\ 0 \\ 1 \end{bmatrix} = \begin{bmatrix} 1 \\ 1 \\ 1 \end{bmatrix} \tag{10.117}$$

Thus the error is in the seventh position, and the decoded codeword is ▲ 1110000.

We must note in passing that for the (7, 4) Hamming code, the parity checks are in the first, second, and fourth positions in the codewords, since these are the only columns of the parity-check matrix containing only one nonzero element.

Cyclic Codes

The preceding subsections dealt primarily with the mathematical properties of parity-check codes. We have avoided any discussion of the implementation of parity-check encoders or decoders. Indeed, if we were to examine the implementation of these devices, we would find that fairly complex hardware configurations are required. However, there is a class of parity-check codes, known as *cyclic codes*, that are easily implemented using feedback shift registers. A cyclic code derives its name from the fact that a cyclic permutation of any codeword produces another codeword. For example, if $x_1 x_2 \cdots x_{n-1} x_n$ is a codeword, so is $x_n x_1 x_2 \cdots x_{n-1}$. In this section we examine not the underlying theory of cyclic codes but the implementation of encoders and decoders. We will accomplish this by means of an example.

An (n, k) cyclic code can easily be generated with an $n - k$ stage shift register with appropriate feedback. The register illustrated in Figure 10.16 produces a $(7, 4)$ cyclic code. The switch is initially in position A, and the shift register stages initially contain all zeros. The $k = 4$ information symbols are then shifted into the encoder. As each information symbol arrives, it is routed to the output and added to the value of $S_2 \oplus S_3$. The resulting sum is then placed into the first stage of the shift register. Simultaneously,

FIGURE 10.16   Coder for (7, 4) cyclic code

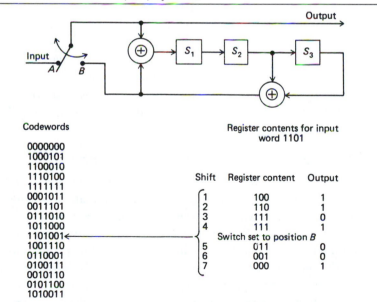

Codewords			
0000000			
1000101			
1100010		Register contents for input	
1110100		word 1101	
1111111			
0001011	Shift	Register content	Output
0011101	1	100	1
0111010	2	110	1
1011000	3	111	0
1101001←	4	111	1
1001110		Switch set to position *B*	
0110001	5	011	0
0100111	6	001	0
0010110	7	000	1
0101100			
1010011			

the contents of S_1 and S_2 are shifted to S_2 and S_3, respectively. After all information symbols have arrived, the switch is moved to position B, and the shift register is shifted $n - k = 3$ times to clear it. On each shift, the sum of S_2 and S_3 appears at the output. This sum added to itself produces a 0, which is fed into S_1. After $n - k$ shifts, a codeword has been generated that contains $k = 4$ information symbols and $n - k = 3$ parity-check symbols. It also should be noted that at this time the register contains all 0's so that the encoder is ready to receive the next $k = 4$ information symbols.

All $2^k = 16$ codewords that can be generated with the example encoder are also illustrated in Figure 10.16. The $k = 4$ information symbols, which are the first four symbols of each codeword, were shifted into the encoder beginning with the left-hand symbol. Also shown in Figure 10.16 are the contents of the register and the output symbol after each shift for the codeword 1101.

The decoder for the (7, 4) cyclic code is illustrated in Figure 10.17. The upper register is used for storage, and the lower register and feedback arrangement are identical to the feedback shift register used in the encoder. Initially, switch A is closed and switch B is open. The n received symbols are shifted into the two registers. If there are no errors, the lower register will contain all 0's when the upper register is full. The switch positions are then reversed, and the codeword that is stored in the upper register is shifted out. This operation is illustrated in Figure 10.17 for the received word 1101001.

If, after the received word is shifted into the decoder, the lower register does not contain all 0's, an error has been made. The error is corrected automatically by the decoder, since, when the incorrect symbol appears at the output of the shift register, a 1 appears at the output of the AND gate. This 1 inverts the upper register output and is produced by the sequence 100 in the lower register. The operation is illustrated in Figure 10.17.

Comparison of Errors in Block-Coded and Uncoded Systems

We will now compare the relative performance of coded and uncoded systems for block codes. The basic assumption will be that the *information rate* is the same for both systems. Since the coded and uncoded words contain the same information, the word duration T_w will be the same under the equal-information-rate assumption. Since the codeword contains more symbols than the uncoded word, as a result of the addition of parity symbols, the symbol rate will be higher for the coded system than for the uncoded system. If constant transmitter power is assumed, it follows that the energy per symbol is decreased by the use of coding, resulting in a higher proba-

FIGURE 10.17   Decoder for (7, 4) cyclic code

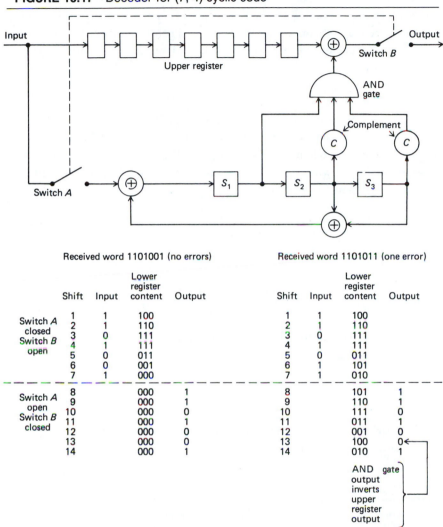

	Received word 1101001 (no errors)				Received word 1101011 (one error)			
	Shift	Input	Lower register content	Output	Shift	Input	Lower register content	Output
Switch *A* closed Switch *B* open	1	1	100		1	1	100	
	2	1	110		2	1	110	
	3	0	111		3	0	111	
	4	1	111		4	1	111	
	5	0	011		5	0	011	
	6	0	001		6	1	101	
	7	1	000		7	1	010	
Switch *A* open Switch *B* closed	8		000	1	8		101	1
	9		000	1	9		110	1
	10		000	0	10		111	0
	11		000	1	11		011	1
	12		000	0	12		001	0
	13		000	0	13		100	0
	14		000	1	14		010	1

AND gate output inverts upper register output

bility of symbol error. We must determine whether or not coding can overcome this increase in symbol error probability to the extent that a significant decrease in word error probability can be obtained.

Assume that q_u and q_c represent the probability of *symbol* error for the uncoded and coded systems, respectively. Also assume that P_{eu} and P_{ec} are the *word* error probabilities for the uncoded and coded systems. The word error probability for the uncoded system is relatively easy to compute. An uncoded word is in error if any of the symbols in that word are in error. The probability that a symbol will be received correctly is $(1 - q_u)$, and since all symbol errors are assumed independent, the probability that all k symbols in a word are received correctly is $(1 - q_u)^k$. Thus the uncoded word error probability is

$$P_{eu} = 1 - (1 - q_u)^k \tag{10.118}$$

The probability of word error for the coded system is more difficult to compute because symbol errors can possibly be corrected by the decoder, depending on the code used. If a code is capable of correcting up to e errors, the probability of word error P_{ec} is equal to the probability that more than e errors are present in the received codeword. Thus

$$P_{ec} = \sum_{i=e+1}^{n} \binom{n}{i} (1 - q_c)^{n-i} q_c^i \tag{10.119}$$

where

$$\binom{n}{i} = \frac{n!}{i!(n-i)!} \tag{10.120}$$

is the number of combinations of n symbols taken i at a time.

For single-error-correcting codes $e = 1$, and P_{ec} becomes

$$P_{ec} = \sum_{i=2}^{n} \binom{n}{i} (1 - q_c)^{n-i} q_c^i \tag{10.121}$$

By comparing (10.118) and (10.121), we determine the improvement gained through the use of coding.

▼ **EXAMPLE 10.12**   Let us investigate the effectiveness of a (7, 4) single-error-correcting code. Assume that the code is used with a BPSK transmission system, so the symbol error probability is

$$q_\mu = Q[\sqrt{2z}] \tag{10.122}$$

which is Equation (7.72) with $m = 0$. The symbol energy E is the transmitter power S times the word time T_w divided by k, since the total energy in each word is divided by k. Thus the symbol error probability without coding is

$$q_u = Q\left[\sqrt{\frac{2ST_w}{kN_0}}\right] \tag{10.123}$$

Assuming equal *word rates* for both the coded and uncoded system gives

$$q_c = Q\left[\sqrt{\frac{2ST_w}{nN_0}}\right] \tag{10.124}$$

for the coded symbol error probability, since the energy available for k information symbols must be spread over $n > k$ symbols when coding is used. Thus it follows that the *symbol* error probability is increased by the use of coding. However, we shall show that the error-correcting capability of the code can overcome the increased symbol error probability and indeed yield a net gain in word error probability for certain ranges of the signal-to-noise ratio.

From (10.121), the word error probability for the coded case is

$$P_{ec} = \sum_{i=2}^{7}\binom{7}{i}(1-q_c)^{7-i}q_c^i \tag{10.125}$$

In typical situations, the probability of two errors is much greater than the probability of three or more errors. Thus (10.125) becomes

$$P_{ec} = 21\left\{Q\left[\sqrt{\frac{2ST_w}{7N_0}}\right]\right\}^2 \tag{10.126}$$

and from (10.118), the word error probability for the uncoded system is

$$P_{eu} = 1 - \left\{1 - Q\left[\sqrt{\frac{2ST_w}{4N_0}}\right]\right\}^4 \tag{10.127}$$

The word error probabilities for the coded and uncoded systems are illustrated in Figure 10.18. The curves are plotted as a function of ST_w/N_0, which is word energy divided by the noise power spectral density.

You should note that coding has little effect on system performance unless the value of ST_w/N_0 is in the neighborhood of 11 dB or above. Also, the improvement afforded by a (7, 4) code is quite modest unless ST_w/N_0 is large, in which case system performance may be satisfactory without coding. However, in many systems, even small improvements are very important. Also illustrated in Figure 10.18 are the uncoded and coded symbol error probabilities q_u and q_c, respectively. The effect of spreading the available energy per word over a larger number of symbols is evident.

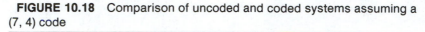

FIGURE 10.18 Comparison of uncoded and coded systems assuming a (7, 4) code

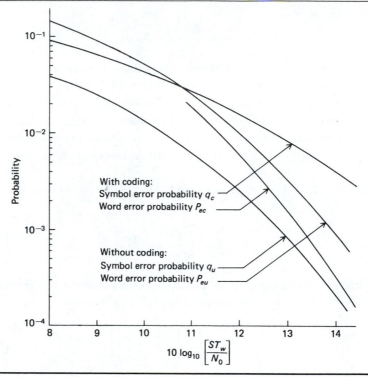

▼ EXAMPLE 10.13   In this example we examine the performance of repetition codes in two different channels. Both cases utilize FSK modulation and a noncoherent receiver structure. In the first case, an *additive white Gaussian noise* (AWGN) channel is assumed. The second case assumes a Rayleigh fading channel. Distinctly different results will be obtained.

Case 1

As was shown in Chapter 7, Equation (7.110), the error probability for a noncoherent FSK system in an additive white Gaussian noise (AWGN) channel is given by

$$q_u = \tfrac{1}{2}e^{-0.5z} \tag{10.128}$$

where z is the ratio of signal power to noise power at the output of the receiver bandpass filter. Equivalently, if the receiver filter is matched to the received signal envelope, z is the ratio of the energy per bit to the noise power spectral density at the receiver input. Without assuming a matched filter receiver, we can write z as

$$z = \frac{A^2}{2N_0B_T} \tag{10.129}$$

where N_0B_T is the noise power in the signal bandwidth B_T. The performance of the system is illustrated by the $n = 1$ curve in Figure 10.19. When an n-symbol repetition code is used with this system, the symbol error probability is given by

$$q_c = \tfrac{1}{2}e^{-z/2n} \tag{10.130}$$

This result occurs since encoding a single information symbol (bit) as n code symbols requires that the symbol duration with coding be $1/n$ times the information symbol duration without coding or, equivalently, that the signal bandwidth be increased by a factor of n with coding. Thus, with coding B_T in q_u is replaced by nB_T to give q_c. The word error probability is given by (10.119) with

$$e = \tfrac{1}{2}(n - 1) \tag{10.131}$$

Since each codeword carries one bit of information, the word error probability is equal to the bit error probability for the repetition code.

The performance of a noncoherent FSK system with an AWGN channel with rate $\tfrac{1}{3}$ and $\tfrac{1}{7}$ repetition codes is illustrated in Figure 10.19. It should be noted that system performance is degraded through the use of repetition coding. This result occurs because the increase in symbol error probability with coding is greater than can be overcome by the error-correction capability of the code. This same result occurs with coherent FSK and BPSK as well as with ASK, illustrating that the low rate of the repetition code prohibits its effective use in systems in which the dependence on symbol error probability and signal-to-noise ratio is essentially exponential.

Case 2

An example of a system in which repetition coding can be used effectively is an FSK system operating in a Rayleigh fading environment. Such a system was analyzed in Chapter 9. We showed that the symbol error probability is given by Equation (9.133), which can be written as

$$q_u = \frac{1}{2}\frac{1}{1 + \dfrac{E_a}{2N_0}} \tag{10.132}$$

FIGURE 10.19 Performance of repetition codes on AWGN and Rayleigh fading channels

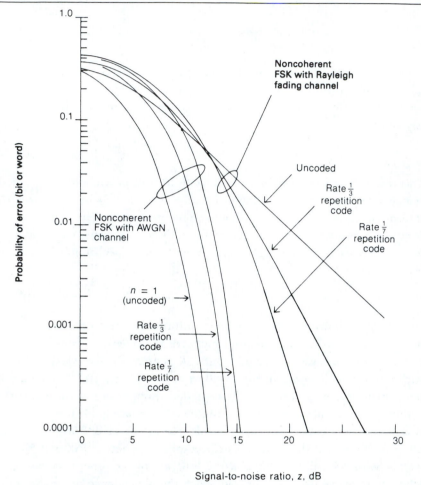

in which E_a is the average received energy per symbol (or bit). The use of a repetition code spreads the energy E_a over the n symbols in a codeword. Thus, with coding,

$$q_c = \frac{1}{2} \frac{1}{1 + \dfrac{E_a}{2nN_0}} \qquad (10.133)$$

As in Case 1, the decoded bit error probability is given by (10.119) with e given by (10.131). The Rayleigh fading results are also shown in Figure 10.19 for rate 1, $\frac{1}{3}$, and $\frac{1}{7}$ repetition codes, where, for this case, the signal-to-noise ratio z is E_a/N_0. Thus the repetition code can be used to advantage in a Rayleigh fading environment if E_a/N_0 is sufficiently large.

Repetition coding is somewhat related to diversity transmission, in which the available signal energy is divided equally among n subpaths. In Problem 9.17, it was shown that the optimal combining of the receiver outputs prior to making a decision on the transmitted information bit is as shown in Figure 10.20(a). The model for the repetition code considered in this example is shown in Figure 10.20(b). The essential difference is that a "hard decision"

FIGURE 10.20 Comparison of optimum and suboptimum systems.
(a) Optimum system. (b) Repetition code model.

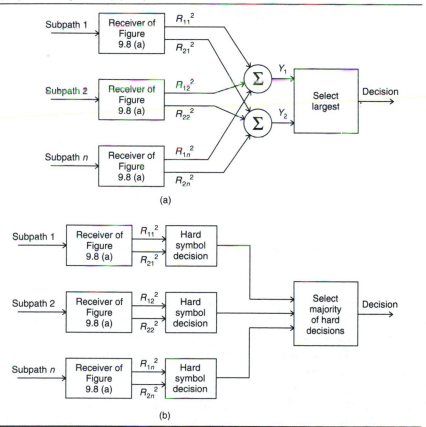

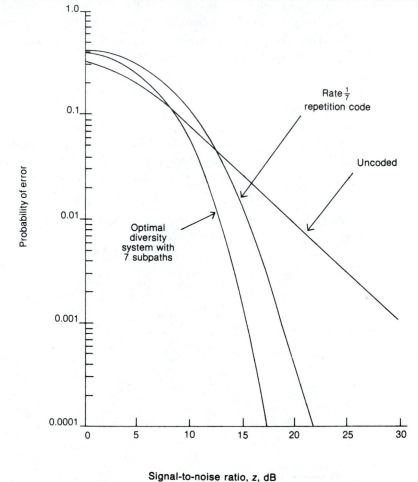

FIGURE 10.21 Performance of noncoherent FSK in Rayleigh fading channel

on each symbol of the n-symbol codeword is made at the output of the n receivers. The decoded information bit is then in favor of the majority of the n symbols of the codeword.

When a hard decision is made at the receiver output, information is clearly lost, and the result is a degradation of performance. This can be seen in Figure 10.21, which illustrates the performance of the $n = 7$ optimal system of Figure 10.20(a) and that of the rate $\frac{1}{7}$ repetition code of Figure 10.20(b). Also shown for reference is the performance of the uncoded system.

Convolutional Codes

The convolutional code is an example of a nonblock code. Rather than the parity-check symbols being calculated for a block of code symbols, the parity checks are calculated over a span of information symbols. This span, which is referred to as the *constraint span,* is shifted one information symbol each time an information symbol is input to the encoder.

   A general convolutional encoder is illustrated in Figure 10.22. The encoder is rather simple and consists of three component parts. The heart of the encoder is a shift register that holds k information symbols, where k is the constraint span of the code. The shift register stages are connected to v modulo-2 adders as indicated. Not all stages are connected to all adders. In fact, the connections are "somewhat random" and can have considerable impact on the performance of the code generated. Each time a new information symbol is shifted into the encoder, the adder outputs are sampled by the commutator. Thus v output symbols are generated for each input symbol yielding a code of rate $1/v$.

   A rate $\frac{1}{3}$ convolutional encoder is illustrated in Figure 10.23. For each input, the output of the encoder is the sequence $v_1 v_2 v_3$. For the encoder of Figure 10.23,

$$v_1 = S_1 \oplus S_2 \oplus S_3 \qquad\qquad (10.134a)$$

$$v_2 = S_1 \qquad\qquad (10.134b)$$

$$v_3 = S_1 \oplus S_2 \qquad\qquad (10.134c)$$

Thus we see that the input sequence 101001 results in the output sequence 111101011101100111.

   Convolutional codes can be decoded by tree-searching techniques. A portion of the code tree for the encoder of Figure 10.23 is illustrated in Figure 10.24. In the latter figure, the single binary symbols are inputs to the

FIGURE 10.22   General convolutional encoder

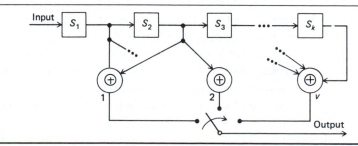

FIGURE 10.23 Example of convolutional encoder

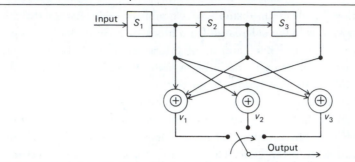

encoder, and the three binary symbols in parentheses are the output symbols corresponding to each input symbol. For example, if 1010 is fed into the encoder, the output is 111101011101 or path *A*.

The decoding procedure also follows from Figure 10.24. To decode a received sequence, we search the code tree for the path closest in Hamming distance to the input sequence. For example, the input sequence 110101011111 is decoded as 1010, indicating an error in the third and eleventh positions of the input sequence.

The exact implementation of tree-searching techniques requires considerable hardware and is not practical for many applications. For example, *N* information symbols generate 2^N branches of the code tree. Storage of the entire tree is therefore impractical for large *N*. Several algorithms have been developed that yield excellent performance with reasonable hardware requirements. Prior to taking a brief look at the most popular of these techniques, the Viterbi algorithm, we look at the trellis diagram, which is essentially a code tree in compact form.

The key to construction of the trellis diagram is recognition that the code tree is repetitive after *k* branches, where *k* is the constraint span of the encoder. This is easily recognized from the code tree shown in Figure 10.24. After the fourth input of an information symbol, 16 branches have been generated in the code tree. The encoder outputs for the first 8 branches match exactly the encoder outputs for the second 8 branches, except for the first symbol. The encoder output therefore depends only on the latest *k* inputs. In this case, the constraint span, *k*, is 3. Thus the output corresponding to the fourth information symbol depends only on the second, third, and fourth encoder inputs. It makes no difference whether the first information symbol was a binary 0 or a binary 1. (This should clarify the meaning of a constraint span.)

The analysis is simplified by defining the state of the encoder as S_1S_2, or the contents of the input register prior to inputting the current information symbol. When the current information symbol is input to S_1, S_1 is shifted to S_2 and S_2 is shifted to S_3. The state and current input then determine S_1, S_2, and S_3, which in turn determine the output. This information is summarized in Table 10.4. The outputs corresponding to given state transitions are shown in parentheses, consistent with Figure 10.24.

FIGURE 10.24   Code tree

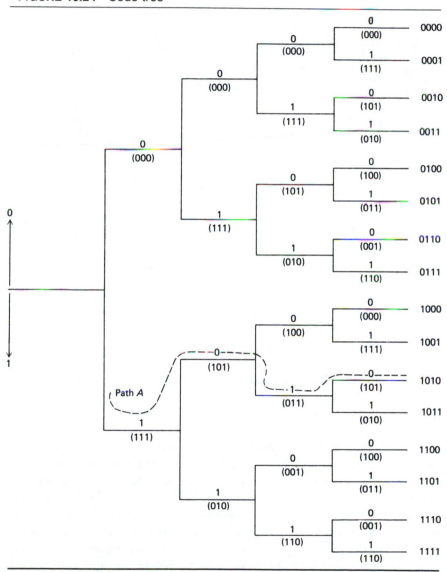

TABLE 10.4   States, Transitions, and Outputs for Convolutional Encoder Shown in Figure 10.23

(a) Definition of States

State	S_1	S_2
A	0	0
B	0	1
C	1	0
D	1	1

(b) State Transitions

Previous State	S_1	S_2	Input	S_1	S_2	S_3	Current State	Output
A	0	0	0	0	0	0	A	(000)
			1	1	0	0	C	(111)
B	0	1	0	0	0	1	A	(100)
			1	1	0	1	C	(011)
C	1	0	0	0	1	0	B	(101)
			1	1	1	0	D	(010)
D	1	1	0	0	1	1	B	(001)
			1	1	1	1	D	(110)

(c) Encoder Output for State Transition x → y

Transition	Output
$A \rightarrow A$	(000)
$A \rightarrow C$	(111)
$B \rightarrow A$	(100)
$B \rightarrow C$	(011)
$C \rightarrow B$	(101)
$C \rightarrow D$	(010)
$D \rightarrow B$	(001)
$D \rightarrow D$	(110)

It should be noted that states A and C can only be reached from states A and B. Also, states B and D can only be reached from states C and D. The information in Table 10.4 is often shown in a state diagram, such as in Figure 10.25. In the state diagram, an input of binary 0 results in the transition denoted by a dashed line, and an input of binary 1 results in the transition designated by a solid line. The resulting encoder output is denoted by the three symbols in parentheses. For any given sequence of inputs, the resulting state transitions and encoder outputs can be traced on the state diagram.

FIGURE 10.25   State diagram for example of convolutional encoder

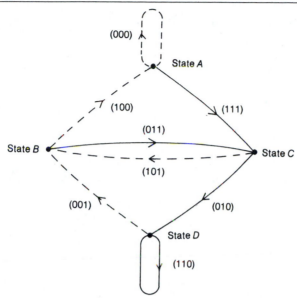

This is a very convenient method for determining the outputs resulting from a given sequence of inputs.

The trellis diagram (Figure 10.26) results directly from the state diagram. Initially, the encoder is assumed to be in state *A* (all contents are 0's). An input of binary 0 results in the encoder remaining in state *A*, as indicated by the dashed line, and an input of binary 1 results in a transition to state *C*, as indicated by the solid line. Any of the four states can be reached by a sequence of two inputs. The third input results in the possible transitions shown. The fourth input results in exactly the same set of possible transitions. Therefore, after the second input, the trellis becomes completely repetitive, and the possible transitions are those labeled steady-state transitions. The encoder can always be returned to state *A* by inputting two binary 0's as shown in Figure 10.27. As before, the output sequence resulting from any transition is shown by the sequence in parentheses.

In order to illustrate the Viterbi algorithm, we consider the received sequence that we previously considered to illustrate decoding using a code tree—namely, the sequence 110101011111. The first step is to compute the Hamming distances between the initial node (state *A*) and each of the four states three levels deep into the trellis. We must look three levels deep into the trellis because the constraint span of the example encoder is 3.

FIGURE 10.26 Trellis diagram

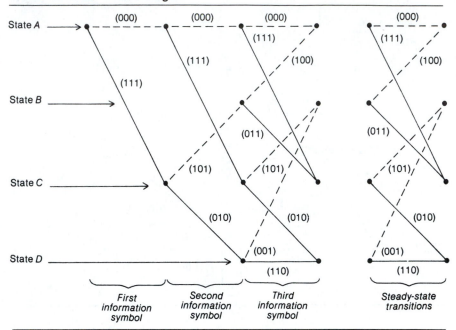

FIGURE 10.27 Termination of trellis diagram

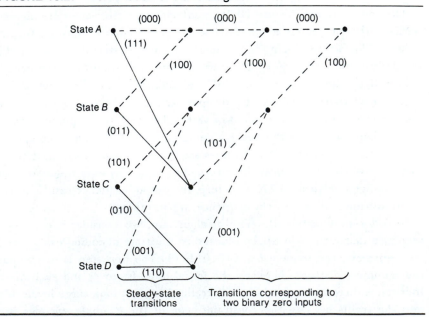

Since each of the four nodes can be reached from only two preceding nodes, eight paths must be identified, and the Hamming distance must be computed for each path. We therefore initially look three levels deep into the trellis, and since the example encoder has rate $\frac{1}{3}$, the first nine received symbols are initially considered. Thus the Hamming distances between the input sequence 110101011 and the eight paths terminating three levels deep into the trellis are computed. These calculations are summarized in Table 10.5. After the eight Hamming distances are computed, the path having the minimum Hamming distance to *each* of the four nodes is retained. These four retained paths are known as *survivors*. The other four paths are discarded from further consideration. The four survivors are identified in Table 10.5.

The next step in the application of the Viterbi algorithm is to consider the next three received symbols, which are 111 in the example being considered. The scheme is to compute once again the Hamming distance to the four states, this time four levels deep in the trellis. As before, each of the four states can be reached from only two previous states. Thus, once again, eight Hamming distances must be computed. Each of the four previous survivors, along with their respective Hamming distances, is extended to the two states reached by each surviving path. The Hamming distance of each new segment is computed by comparing the encoder output, corresponding to each of the new segments, with 111. The calculations are summarized in Table 10.6. The path having the smallest new distance is path *ACBCB*. This corresponds to information sequence 1010 and is in agreement with the previous tree search.

For a general received sequence, the process identified in Table 10.6 is

TABLE 10.5   Calculations for Viterbi Algorithm: Step One (Received Sequence = 110101011)

Path[a]	Corresponding Symbols	Hamming Distance	Survivor?
AAAA	000000000	6	No
ACBA	111101100	4	Yes
ACDB	111010001	5	Yes[b]
AACB	000111101	5	No[b]
AAAC	000000111	5	No
ACBC	111101011	1	Yes
ACDD	111010110	6	No
AACD	000111010	4	Yes

[a] The initial and terminal states are identified by the first and fourth letters, respectively. The second and third letters correspond to intermediate states.
[b] If two or more paths have the same Hamming distance, it makes no difference which is retained as a survivor.

TABLE 10.6 Calculations for the Viterbi Algorithm: Step Two
(Received Sequence 110101011111)

Path[a]	Previous Survivor's Distance	New Segment	Added Distance	New Distance	Survivor?
AC<u>BA</u>A	4	AA	3	7	Yes
AC<u>DB</u>A	5	BA	2	7	No
ACB<u>CB</u>	1	CB	1	2	Yes
AAC<u>DB</u>	4	DB	2	6	No
AC<u>BA</u>C	4	AC	0	4	Yes
AC<u>DB</u>C	5	BC	1	6	No
ACB<u>CD</u>	1	CD	2	3	Yes
AAC<u>DD</u>	4	DD	1	5	No

[a] An underscore indicates the previous survivor.

continued. After each new set of calculations, involving the next three received symbols, only the four surviving paths and the accumulated Hamming distances need be retained. At the end of the process, it is necessary to reduce the number of surviving paths from four to one. This is accomplished by inserting two dummy 0's at the end of the information sequence, corresponding to the transmission of six code symbols. As shown in Figure 10.27, this forces the trellis to terminate at state A.

The Viterbi algorithm has found widespread application in practice, especially in satellite communications. It can be shown that the Viterbi algorithm is a maximum-likelihood decoder, and in that sense, it is optimal. Viterbi and Omura (1979) give an excellent analysis of the Viterbi algorithm. A paper by Heller and Jacobs (1971) summarizes a number of performance characteristics of the Viterbi algorithm.

Burst-Error-Correcting Codes

In many practical communication channels, such as in fading channels, errors tend to group together and occur in bursts. Much attention has been devoted to code development for improving the performance of these types of channels. Most of these codes tend to be rather complex. A code for correction of a single burst, however, is rather simple to understand.

As an example, assume that the output of a source is encoded using an (n, k) block code. The ith codeword will be of the form

$$\lambda_{i1} \quad \lambda_{i2} \quad \lambda_{i3} \quad \cdot \cdot \cdot \quad \lambda_{in}$$

Assume that l of these codewords are read into a table to yield the array

$$\begin{array}{ccccc}
\lambda_{11} & \lambda_{12} & \lambda_{13} & \cdots & \lambda_{1n} \\
\lambda_{21} & \lambda_{22} & \lambda_{23} & \cdots & \lambda_{2n} \\
\lambda_{31} & \lambda_{32} & \lambda_{33} & \cdots & \lambda_{3n} \\
\vdots & \vdots & \vdots & & \vdots \\
\lambda_{l1} & \lambda_{l2} & \lambda_{l3} & \cdots & \lambda_{ln}
\end{array}$$

If transmission is accomplished by reading out of this table by columns, the transmitted stream of symbols will be

$$\lambda_{11} \quad \lambda_{21} \quad \lambda_{31} \quad \cdots \quad \lambda_{l1} \quad \lambda_{12} \quad \lambda_{22} \quad \lambda_{32} \quad \cdots$$
$$\lambda_{l2} \quad \lambda_{13} \quad \lambda_{23} \quad \lambda_{33} \quad \cdots \quad \lambda_{ln}$$

If a burst of errors affects l consecutive symbols, then each codeword will have one error, and a single error-correcting code will correct the burst *if* there are no other errors in the stream of ln symbols. Likewise, a double-error-correcting code can be used to correct a single burst spanning $2l$ symbols. These codes are known as *interleaved codes* since l codewords are interleaved to form the sequence of length ln.

Feedback Channels

In many practical systems, a feedback channel is available from receiver to transmitter. When available, this channel can be utilized to achieve a specified performance with decreased complexity of the encoding scheme. Many such schemes are possible: decision feedback, error-detection feedback, and information feedback. In a decision-feedback scheme, a null-zone receiver is used, and the feedback channel is utilized to inform the transmitter either that no decision was possible on the previous symbol and to retransmit or that a decision was made and to transmit the next symbol. The null-zone receiver is usually modeled as a binary-erasure channel.

Error-detection feedback involves the combination of coding and a feedback channel. With this scheme, retransmission of codewords is requested when errors are detected.

In general, feedback schemes tend to be rather difficult to analyze. Thus only the simplest scheme, the decision-feedback channel, will be treated here. Assume a binary transmission scheme with matched-filter detection. The signaling waveforms are $s_1(t)$ and $s_2(t)$. The conditional probability density functions of the matched filter output, conditioned on $s_1(t)$ and $s_2(t)$, were derived in Chapter 7 and are illustrated in Figure 7.7, which is redrawn for our application in Figure 10.28. We shall assume that both $s_1(t)$ and $s_2(t)$

FIGURE 10.28   Decision regions for null-zone receiver

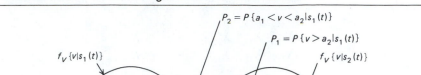

$$P_2 = P\{a_1 < v < a_2 | s_1(t)\}$$

$$P_1 = P\{v > a_2 | s_1(t)\}$$

$f_V\{v | s_1(t)\}$

$f_V\{v | s_2(t)\}$

a_1 a_2 v

have equal *a priori* probabilities. For the null-zone receiver, two thresholds, a_1 and a_2, are established. If the matched-filter output, sampled at time T, lies between a_1 and a_2, no decision is made, and the feedback channel is used to request a retransmission. The probability of error, given $s_1(t)$ was transmitted, is denoted P_1. The probability of no decision (null-zone), given $s_1(t)$ was transmitted, is denoted P_2. By symmetry, these probabilities are the same for $s_2(t)$ transmitted.

The probability of error on the jth transmission is

$$P_{Ej} = P_2^{j-1} P_1 \qquad (10.135)$$

Thus the overall probability of error P_E is

$$P_E = \sum_{j=1}^{\infty} P_2^{j-1} P_1 \qquad (10.136)$$

which is

$$P_E = \frac{P_1}{1 - P_2} \qquad (10.137)$$

The expected number of transmissions N is also easily derived. The result is

$$N = \frac{1}{1 - P_2} \qquad (10.138)$$

which is typically only slightly greater than one.

It follows from these results that the error probability can be reduced considerably without significantly increasing N. Thus performance is improved without a great sacrifice in information rate.

10.4 MODULATION SYSTEMS

In Chapter 6, signal-to-noise ratios were computed at various points in a communication system. Of particular interest were the signal-to-noise ratio at the input to the demodulator and the signal-to-noise ratio of the demodulated output. These were referred to as the *predetection SNR*, $(SNR)_T$, and the *postdetection SNR*, $(SNR)_D$, respectively. The ratio of these parameters, the detection gain, has been widely used as a figure of merit for the system. In this section we will compare the behavior of $(SNR)_D$ as a function of $(SNR)_T$ for several systems. First, however, we investigate the behavior of an *optimum*, but *unrealizable* system. This study will provide a firm basis for comparison and also provide additional insight into the concept of the tradeoff of bandwidth for signal-to-noise ratio.

Optimum Modulation

The block diagram of a communication system is illustrated in Figure 10.29. We will focus on the receiver portion of the system. The SNR at the output of the predetection filter, $(SNR)_T$ yields the *maximum* rate at which information may arrive at the receiver. From the Shannon-Hartley law, this rate, C_T, is

$$C_T = B_T \log [1 + (SNR)_T] \tag{10.139}$$

where B_T, the predetection bandwidth, is typically the bandwidth of the modulated signal. Since (10.139) is based on the Shannon-Hartley law, it is valid only for additive *white* Gaussian noise cases. The SNR of the demodu-

FIGURE 10.29 Block diagram of a communication system

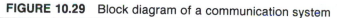

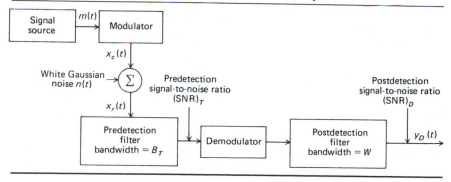

lated output, $(\text{SNR})_D$, yields the maximum rate at which information may leave the receiver. This rate, denoted C_D, is given by

$$C_D = W \log [1 + (\text{SNR})_D] \tag{10.140}$$

where W is the bandwidth of the message signal.

An optimum modulation system is defined as one for which $C_D = C_T$. For this system, demodulation is accomplished, in the presence of noise, without loss of information. Equating C_D to C_T yields

$$(\text{SNR})_D = [1 + (\text{SNR})_T]^{B_T/W} - 1 \tag{10.141}$$

which shows that the optimum exchange of bandwidth for SNR is *exponential*.

The ratio of transmission bandwidth B_T to the message bandwidth W is referred to as the *bandwidth expansion factor* γ. To fully understand the role of this parameter, we write the predetection SNR as

$$(\text{SNR})_T = \frac{P_T}{N_0 B_T} = \frac{W}{B_T} \frac{P_T}{N_0 W} = \frac{1}{\gamma} \frac{P_T}{N_0 W} \tag{10.142}$$

Thus (10.141) can be expressed as

$$(\text{SNR})_D = \left[1 + \frac{1}{\gamma} \left(\frac{P_T}{N_0 W} \right) \right]^{\gamma} - 1 \tag{10.143}$$

The relationship between $(\text{SNR})_D$ and $P_T/N_0 W$ is illustrated in Figure 10.30.

Comparison of Modulation Systems

The concept of an optimal modulation system provides a basis for comparing system performance. For example, an ideal single-sideband system has a bandwidth expansion factor of unity, since the transmission bandwidth is ideally equal to the message bandwidth. Thus the postdetection SNR of the optimal modulation system is, from (10.143) with γ equal to 1,

$$(\text{SNR})_D = \frac{P_T}{N_0 W} \tag{10.144}$$

This is exactly the same result as that obtained in Chapter 6 for an SSB system using coherent demodulation with a perfect phase reference. Therefore, if the transmission bandwidth B_T of an SSB system is *exactly* equal to the message bandwidth W, SSB is optimal, assuming that there are no other error sources. Of course, this can never be achieved in practice, since *ideal* filters

FIGURE 10.30   Performance of optimum modulation system

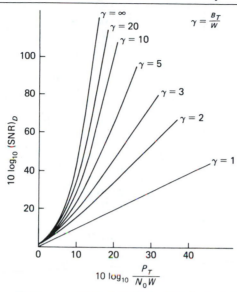

would be required in addition to *perfect* phase coherence of the demodulation carrier.

The story is quite different with DSB, AM, and QDSB. For these systems, $\gamma = 2$. In Chapter 6 we saw that the postdetection SNR for DSB and QDSB, assuming perfect coherent demodulation, is

$$(\text{SNR})_D = \frac{P_T}{N_0 W} \tag{10.145}$$

whereas for the optimal system it is given by (10.143) with $\gamma = 2$.

These results are shown in Figure 10.31, along with the result for AM with square-law demodulation. It can be seen that these systems are far from optimal, especially for large values of $P_T/N_0 W$.

Also shown in Figure 10.31 is the result for FM without preemphasis, with sinusoidal modulation, assuming a modulation index of 10. With this modulation index, the bandwidth expansion factor is

$$\gamma = \frac{2(\beta + 1)W}{W} = 22 \tag{10.146}$$

The realizable performance of the FM system is taken from Figure 6.18. It can be seen that realizable systems fall far short of optimal if γ and $P_T/N_0 W$ are large.

FIGURE 10.31   Performance comparison of analog systems

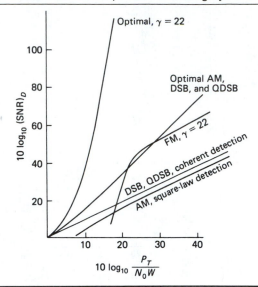

10.5   BANDWIDTH AND POWER-EFFICIENT MODULATION

A desirable characteristic of any modulation scheme is the simultaneous conservation of bandwidth and power. Since the late 1970's, the approach to this challenge has been to combine coding and modulation. There have been two approaches: (1) continuous phase modulation (CPM)[1] with memory extended over several modulation symbols by cyclical use of a set of modulation indices; and (2) combining coding with an *M*-ary modulation scheme, referred to as *trellis-coded modulation* (TCM).[2] We briefly explore the latter approach in this section. For an introductory discussion of the former

---

[1] CPM has been explored by many investigators. For introductory treatments see C.-E. Sundberg, "Continuous Phase Modulation" *IEEE Communications Magazine,* Vol. 24, Apr. 1986, pp. 25–38, and J. B. Anderson and C.-E. Sundberg, "Advances in Constant Envelope Coded Modulation," *IEEE Communications Magazine,* Vol. 29, Dec. 1991, pp. 36–45.

[2] Three introductory treatments of TCM can be found in G. Ungerboeck, "Channel Coding with Multilevel/Phase Signals," *IEEE Transactions on Information Theory,* Vol. IT-28, Jan. 1982, pp. 55–66, G. Ungerboeck, "Trellis-Coded Modulation with Redundant Signal Sets, Part I: Introduction," *IEEE Communications Magazine,* Vol. 25, Feb. 1987, pp. 5–11, and G. Ungerboeck, "Trellis-Coded Modulation with Redundant Signal Sets, Part II: State of the Art," *IEEE Communications Magazine,* Vol. 25, Feb. 1987, pp. 12–21.

approach, see Ziemer and Peterson (1985), Chapter 4. Sklar (1988) is a well-written reference with more examples on TCM than given in this short section.

In Chapter 8 it was illustrated through the use of signal-space diagrams that the most probable errors in an M-ary modulation scheme result from mistaking a signal point closest in Euclidian distance to the transmitted signal point as corresponding to the actual transmitted signal. Ungerboeck's solution to this problem was to use coding in conjunction with M-ary modulation to increase the minimum Euclidian distance between those signal points most likely to be confused without increasing the average power or bandwidth over an uncoded scheme transmitting the same number of bits per second. We illustrate the procedure with a specific example.

We wish to compare a TCM system and a QPSK system operating at the same data rates. Since the QPSK system transmits 2 bits per signal phase (signal-space point), we can keep that same data rate with the TCM system by employing an 8-PSK modulator, which carries 3 bits per signal phase, in conjunction with a convolutional encoder that produces three encoded symbols for every two input data bits, i.e., a rate $\frac{2}{3}$ encoder. Figure 10.32(a) shows an encoder for accomplishing this and Figure 10.32(b) shows the corresponding trellis diagram. The encoder operates by taking the first data bit as the input to a rate $\frac{1}{2}$ convolutional encoder that produces the first and second encoded symbols, and the second data bit directly as the third encoded symbol. These are then used to select the particular signal phase to be transmitted according to the following rules:

1. All parallel transitions in the trellis are assigned the maximum possible Euclidian distance. Since these transitions differ by one code symbol (the one corresponding to the uncoded bit in this example), an error in decoding these transitions amounts to a single bit error, which is minimized by this procedure.
2. All transitions emanating or converging into a trellis state are assigned the next to largest possible Euclidian distance separation.

The application of these rules to assigning the encoded symbols to a signal phase in an 8-PSK system can be done with a technique known as set partitioning, which is illustrated in Figure 10.33. If the encoded symbol c_1 is a 0, the left branch is chosen in the first tier of the tree, whereas if c_1 is a 1, the right branch is chosen. A similar procedure is followed for tiers 2 and 3 of the tree, with the result being that a unique signal phase is chosen for each possible encoded output.

To decode the TCM signal, the received signal plus noise in each signaling

FIGURE 10.32   (a) Convolutional encoder and (b) trellis diagram corresponding to 4-state, 8-PSK TCM.

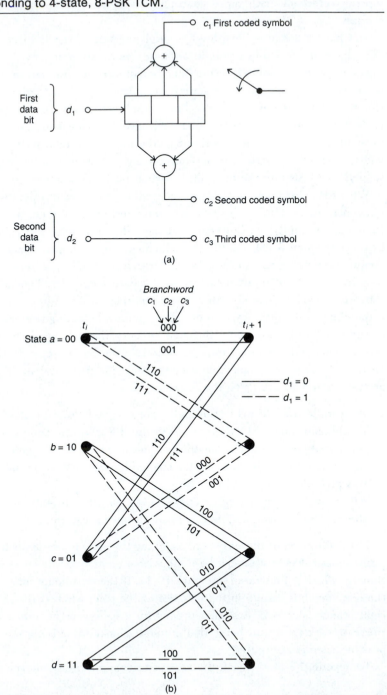

(a)

(b)

FIGURE 10.33   Set partitioning for assigning a rate ⅔ encoder output to 8-PSK signal points while obeying the rules for maximizing free distance. (From G. Ungerboeck, "Channel Coding with Multilevel/Phase Signals," *IEEE Transactions on Information Theory,* Vol. IT-28, Jan. 1982, pp. 55–66).

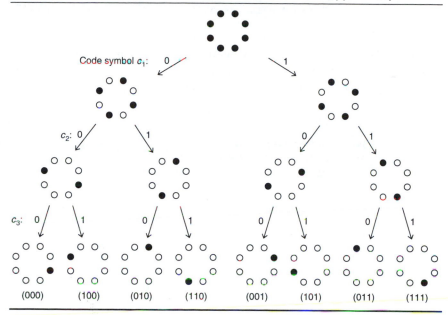

interval is correlated with each possible transition in the trellis and a search is made through the trellis by means of a Viterbi algorithm using the sum of these cross-correlations as metrics rather than Hamming distance as discussed in conjunction with Figure 10.26 (this is called the use of a soft decision metric). Also note that the decoding procedure is twice as complicated since two branches correspond to a path from one trellis state to the next due to the uncoded bit becoming the third symbol in the code. In choosing the two decoded bits for a surviving branch, the first decoded bit of the pair corresponds to the input bit b_1 that produced the state transition of the branch being decoded. The second decoded bit of the pair is the same as the third symbol c_3 of that branch word, since c_3 is the same as the uncoded bit b_2.

Ungerboeck has characterized the event error probability performance of a signaling method in terms of the free distance of the signal set. For high signal-to-noise ratios, the probability of an error event (i.e., the probability that at any given time the VA makes a wrong decision among the signals associated with parallel transitions, or starts to make a sequence of wrong decisions along some path diverging from more than one transition from the correct path) is well approximated by

$$P(\text{error event}) = N_{\text{free}} \, Q\!\left(\frac{d_{\text{free}}}{2\sigma}\right) \qquad (10.147)$$

where N_{free} denotes the number of nearest-neighbor signal sequences with distance d_{free} that diverge at any state from a transmitted signal sequence, and reemerge with it after one or more transitions. (The free distance is

FIGURE 10.34 Performance for a 4-state, 8-PSK TCM signaling scheme. (From G. Ungerboeck, "Trellis-Coded Modulation with Redundant Signal Set, Part I: Introduction," *IEEE Communications Magazine,* Feb. 1987, Vol. 25, pp. 5–11).

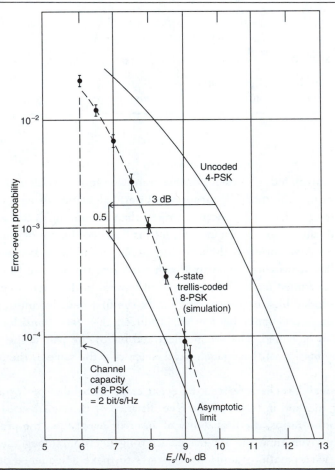

often calculated by assuming the signal energy has been normalized to unity, and that the noise standard deviation σ accounts for this normalization.)

For uncoded QPSK, we have $d_{\text{free}} = 2^{1/2}$ and $N_{\text{free}} = 2$ (there are two adjacent signal points at distance $d_{\text{free}} = 2^{1/2}$), whereas for 4-state-coded 8-PSK we have $d_{\text{free}} = 2$ and $N_{\text{free}} = 1$. Ignoring the factor N_{free}, we have an asymptotic gain due to TCM over uncoded QPSK of $2^2/(2^{1/2})^2 = 2 = 3$ dB. Figure 10.34, also from Ungerboeck, compares the asymptotic lower bound for the error event probability with simulation results.

It should be clear that the TCM encoding/modulation procedure can be generalized to higher-level *M*-ary schemes. Ungerboeck shows that this observation can be generalized as follows: (1) Of the m bits to be transmitted per encoder/modulator operation, $k \leq m$ bits are expanded to $k + 1$ coded symbols by a binary rate $k/(k + 1)$ convolutional encoder; (2) the $k + 1$ coded symbols select one of 2^{k+1} subsets of a redundant 2^{m+1}-ary signal set; (3) the remaining $m - k$ symbols determine one of 2^{m-k} signals within the selected subset. It should also be stated that one may use block codes or other modulation schemes, such as *M*-ary ASK or QASK, to implement a TCM system.

Another parameter that influences the performance of a TCM system is the constraint length of the code, ν, which is equivalent to saying that the encoder has 2^{ν} states. Ungerboeck has published asymptotic gains for TCM systems with various constraint lengths. These are given in Table 10.7.

TABLE 10.7 Asymptotic Coding Gains for TCM Systems

No. of States, 2^{ν}	k	Asymptotic Coding Gain (dB)	
		$G_{\text{8PSK/QPSK}}: m = 2$	$G_{\text{16PSK/8PSK}}: m = 3$
4	1	3.01	—
8	2	3.60	—
16	2	4.13	—
32	2	4.59	—
64	2	5.01	—
128	2	5.17	—
256	2	5.75	—
4	1	—	3.54
8	1	—	4.01
16	1	—	4.44
32	1	—	5.13
64	1	—	5.33
128	1	—	5.33
256	2	—	5.51

SOURCE: Adapted from G. Ungerboeck, "Trellis-Coded Modulation with Redundant Signal Sets. Part II: State of the Art," *IEEE Communications Magazine*, Vol. 25, Feb. 1987, pp. 12–21.

Finally, the paper by Viterbi et al. (1989) gives a simplified scheme for M-ary PSK that uses a single rate $\frac{1}{2}$, 64-state binary convolutional code for which VLSI circuit implementations are plentiful. A technique known as puncturing converts it to rate $(n - 1)/n$.

SUMMARY

1. The information associated with the occurrence of an event is defined as the logarithm of the probability of the event. If a base 2 logarithm is used, the measure of information is the bit.

2. The average information associated with a set of source outputs is known as the entropy of the source. The entropy function has a maximum, and the maximum occurs when all source states are equally likely. Entropy is average uncertainty.

3. A channel with n inputs and m outputs is represented by the nm transition probabilities of the form $p(y_j | x_i)$. The channel model can be a diagram showing the transition probabilities or a matrix of the transition probabilities.

4. A number of entropies can be defined for a system. The entropies $H(X)$ and $H(Y)$ denote the average uncertainty of the channel input and output, respectively. The quantity $H(X | Y)$ is the average uncertainty of the channel input given the output, and $H(Y | X)$ is the average uncertainty of the channel output given the input. The quantity $H(X, Y)$ is the average uncertainty of the communication system as a whole.

5. The mutual information between the input and output of a channel is given by

$$I(X; Y) = H(X) - H(X | Y)$$

or

$$I(X; Y) = H(Y) - H(Y | X)$$

The maximum value of mutual information, where the maximization is with respect to the source probabilities, is known as the channel capacity.

6. Source encoding is used to remove redundancy from a source output so that the information per transmitted symbol can be maximized. If the source rate is less than the channel capacity, it is possible to encode the source output so that the information can be transmitted through the channel. This is accomplished by forming source extensions and encod-

ing the symbols of the extended source into codewords having minimum average word length. The minimum average word length $\bar{L}$ approaches $H(X^n) = nH(X)$, where $H(X^n)$ is the entropy of the nth order extension of a source having entropy $H(X)$, as n increases.

7. Two source encoding schemes are the Shannon-Fano technique and the Huffman technique. The Huffman technique yields an optimum source code, which is the source code having minimum average word length.

8. Error-free transmission on a noisy channel can be accomplished if the source rate is less than the channel capacity. This is accomplished using channel codes.

9. The capacity of an additive white Gaussian noise (AWGN) channel is

$$C_c = B \log_2 \left(1 + \frac{S}{N} \right)$$

where B is the channel bandwidth and S/N is the signal-to-noise ratio. This is known as the Shannon-Hartley law.

10. An (n, k) block code is generated by appending $r = n - k$ parity symbols to a k-symbol source sequence. This yields an n-symbol codeword.

11. Decoding is typically accomplished by computing the Hamming distance from the received n-symbol sequence to each of the possible transmitted codewords. The codeword closest in Hamming distance to the received sequence is the most likely transmitted codeword. The two codewords closest in Hamming distance determine the minimum distance of the code d_m. The code can correct $\frac{1}{2}(d_m - 1)$ errors.

12. A single-parity-check code is formed by adding a single-parity symbol to the information sequence. This code can detect single errors but provides no error-correcting capability.

13. The rate of a block code is k/n. The best codes provide powerful error-correction capabilities in combination with high rate.

14. Repetition codes are formed by transmitting each source symbol an odd number of times. Repetition codes provide powerful error-correction capabilities but have very low rate.

15. The parity-check matrix $[H]$ is defined such that $[H][T] = [0]$, where $[T]$ is the transmitted codeword written as a column vector. If the received sequence is denoted by the column vector $[R]$, the syndrome $[S]$ is determined from $[S] = [H][R]$. This can be shown to be equivalent to $[S] = [H][E]$, where $[E]$ is the error sequence. If a single error occurs

in the transmission of a codeword, the syndrome is the column of $[H]$ corresponding to the error position.

16. The generator matrix $[G]$ of a parity-check code is determined such that $[T] = [G][A]$, where $[T]$ is the n-symbol transmitted sequence and $[A]$ is the k-symbol information sequence. Both $[T]$ and $[A]$ are written as column vectors.

17. For a group code, the modulo-2 sum of any two codewords is another codeword.

18. A Hamming code is a single-error-correcting code such that the columns of the parity-check matrix correspond to the binary representation of the column index.

19. Cyclic codes are a class of block codes such that a cyclic shift of the codeword symbols always yields another codeword. These codes are very useful because implementation is simple. Implementation of both the encoder and decoder is accomplished using shift registers and basic logic components.

20. The channel symbol error probability of a coded system is greater than the symbol error probability of an uncoded system because the available energy for transmission of k information symbols must be spread over the n-symbol codeword rather than just the k-information symbols. The error-correcting capability of the code often allows a net performance gain to be realized. The performance gain depends on the choice of code and the channel characteristics.

21. Convolutional codes are easily generated using simple shift registers and modulo-2 adders. Decoding is accomplished using a tree-search technique, which is often implemented using the Viterbi algorithm. The constraint span is the code parameter having the most significant impact on performance.

22. Interleaved codes are useful for burst noise environments.

23. The feedback channel system makes use of a null-zone receiver, and a retransmission is requested if the receiver decision falls within the null zone. If a feedback channel is available, the error probability can be significantly reduced with only a slight increase in the required number of transmissions.

24. Use of the Shannon-Hartley law yields the concept of optimum modulation for a system operating in an AWGN environment. The result is the performance of an optimum system in terms of predetection and

postdetection bandwidth. The tradeoff between bandwidth and signal-to-noise ratio is easily seen.

25. Trellis-coded modulation modulation is a scheme for combining M-ary modulation with coding in a way that increases the Euclidian distance between those signal points for which errors are most likely without increasing the average power or bandwidth over an uncoded scheme having the same bit rate. Decoding is accomplished using a Viterbi decoder that accumulates decision metrics (soft decisions) rather than Hamming distances (hard decisions).

FURTHER READING

An exposition of information theory and coding that was anywhere near complete would, of necessity, be presented at a level beyond that intended for this text. The purpose in the present chapter is to present some of the basic ideas of information theory at a level consistent with the rest of this book. You should be motivated by this to further study.

The original paper by Shannon (1948) is stimulating reading at about the same level as this chapter. This paper is available as a paperback with an interesting postscript by W. Weaver (Shannon and Weaver, 1963). A number of very brief and interesting papers have traced the development of information theory through the years since Shannon's work. Examples are those by Slepian (1973) and Viterbi (1973). Many survey papers on coding theory are also available. The paper by Forney (1970), which stresses performance characteristics, and the paper by Wolf (1973) are particularly recommended.

Several standard texts on information theory are available. Those by Blahut (1987) and Mansuripur (1987) are written at a slightly higher level than this text. The volume by Gallager (1968) is a very complete exposition of the subject and is written at the graduate level. Another standard text, which is very complete, is the one by Blahut (1984). This volume is also written at the beginning graduate level. The text by Lin and Costello (1970) is less rigorous than the other two and contains much information on the implementation of decoders. An interesting perspective on coding theory is provided by Arazi (1988). Many other well-written books and review articles are available.

Several other books deserve special mention. The book by Clark and Cain (1981) contains a wealth of practical information concerning coder and decoder design, in addition to the usual theoretical background material.

The last chapter, entitled "System Applications," contains a large number of performance curves for different channel models. The book by Viterbi and Omura (1979) is an excellent basic treatment of information theory and coding, and it contains an excellent analysis of the Viterbi algorithm. The book edited by Yuen (1983) has a very good section on coding for application to deep-space telemetry.

As mentioned in the last section of this chapter, the subject of bandwidth and power-efficient communications is very important to the implementation of modern systems. Continuous phase modulation is treated in the text by Ziemer and Peterson (1985). An introductory treatment of trellis-coded modulation, (TCM), including a discussion of coding gain, is contained in the book by Sklar (1988). The book by Biglieri, Divsalar, McLane, and Simon (1991) is a complete treatment of TCM theory, performance, and implementation.

PROBLEMS

Section 10.1

10.1 A message occurs with a probability of 0.6. Determine the information associated with the message in bits, nats, and hartleys.

10.2 Assume that you have a standard deck of 52 cards (jokers have been removed).

(a) What is the information associated with the drawing of a single card from the deck?

(b) What is the information associated with the drawing of a pair of cards, assuming that the first card drawn is replaced in the deck prior to drawing the second card?

(c) What is the information associated with the drawing of a pair of cards assuming that the first card drawn is *not* replaced in the deck prior to drawing the second card?

10.3 A source has five outputs denoted $[m_1, m_2, m_3, m_4, m_5]$ with respective probabilities $[0.3, 0.3, 0.2, 0.1, 0.1]$. Determine the entropy of the source. What is the maximum entropy of a source with five outputs?

10.4 A source consists of six outputs denoted $[A, B, C, D, E, F]$ with respective probabilities $[0.3, 0.2, 0.15, 0.15, 0.1, 0.1]$. Determine the entropy of the source.

10.5 A channel has the following transition matrix:

$$\begin{bmatrix} 0.5 & 0.3 & 0.2 \\ 0.2 & 0.6 & 0.2 \\ 0.1 & 0.2 & 0.7 \end{bmatrix}$$

(a) Sketch the channel diagram showing all transition probabilities.
(b) Determine the channel output probabilities assuming that the input probabilities are equal.
(c) Determine the channel input probabilities that result in equally likely channel outputs.
(d) Determine the joint probability matrix using (c) above.

10.6 Describe the channel transition probability matrix and joint probability matrix for a noiseless channel.

10.7 Show that $H(Y) \geq H(Y|X)$.

10.8 A channel is described by the transition probability matrix

$$[P(Y|X)] = \begin{bmatrix} \frac{2}{3} & \frac{1}{3} & 0 \\ 0 & 0 & 1 \end{bmatrix}$$

Determine the channel capacity and the source probabilities that yield capacity.

10.9 A channel is described by the matrix

$$[P(Y|X)] = \begin{bmatrix} 1 & 0 \\ 1 & 0 \\ 0 & 1 \end{bmatrix}$$

Determine the channel capacity and the source probabilities that yield capacity.

10.10 A channel has two inputs, (0, 1), and three outputs, (0, e, 1), where e indicates an erasure; that is, there is no output for the corresponding input. The channel matrix is

$$\begin{bmatrix} p & 1-p & 0 \\ 0 & 1-p & p \end{bmatrix}$$

Compute the channel capacity.

10.11   Determine the capacity of the channel described by the channel matrix shown below. Sketch your result as a function of p and give an intuitive argument that supports your sketch. (Note: $q = 1 - p$.)

$$\begin{bmatrix} p & q & 0 & 0 \\ q & p & 0 & 0 \\ 0 & 0 & p & q \\ 0 & 0 & q & p \end{bmatrix}$$

10.12   From the entropy definition given in Equations (10.22) through (10.26), derive Equations (10.27) and (10.28).

10.13   The input to a quantizer is a random signal having an amplitude probability density function

$$f_X(x) = \begin{cases} ae^{-ax}, & x \geq 0 \\ 0, & x < 0 \end{cases}$$

The signal is to be quantized using four quantizing levels x_i as shown in Figure 10.35. Determine the values x_i, $i = 1, 2, 3$, in terms of a so that the entropy at the quantizer output is maximized.

10.14   Repeat the preceeding problem assuming that the input to the quantizer has the Rayleigh probability density function

$$f_X(x) = \begin{cases} \dfrac{x}{a^2} e^{-x^2/2a^2} & x \geq 0 \\ 0, & x < 0 \end{cases}$$

10.15   A signal has a Gaussian amplitude-density function with zero mean and variance σ^2. The signal is sampled at 750 samples per second. The samples are quantized according to the following table. Deter-

FIGURE 10.35

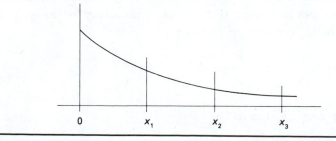

mine the entropy at the quantizer output and the information rate in samples per second.

Quantizer Input	Quantizer Output
$-\infty < x_i < -2\sigma$	m_0
$-2\sigma < x_i < -\sigma$	m_1
$-\sigma < x_i < 0$	m_2
$0 < x_i < \sigma$	m_3
$\sigma < x_i < 2\sigma$	m_4
$2\sigma < x_i < \infty$	m_5

10.16 Determine the quantizing levels, in terms of σ, so that the entropy at the output of a quantizer is maximized. Assume that there are six quantizing levels and that the quantizer is a zero-mean Gaussian process as in the previous problem.

10.17 Two binary symmetrical channels are in cascade, as shown in Figure 10.36. Determine the capacity of each channel. The overall system

FIGURE 10.36

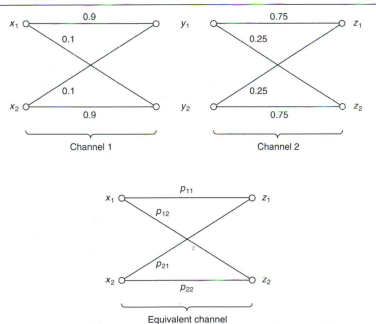

with inputs x_1 and x_2 and outputs z_1 and z_2 can be represented as shown with p_{11}, p_{12}, p_{21}, and p_{22} properly chosen. Determine these four probabilities and the capacity of the overall system. Comment on the results.

10.18   A two-hop satellite communications channel uses BPSK signaling. The uplink signal-to-noise ratio is 8 dB and the downlink signal-to-noise ratio is 5 dB, where the signal-to-noise ratio is the signal power divided by the noise power in the bit-rate bandwidth. Determine the overall error probability.

Section 10.2

10.19   A source has two outputs [A, B] with respective probabilities $[\frac{2}{3}, \frac{1}{3}]$. Determine the entropy of the fourth-order extension of this source using two different methods.

10.20   Calculate the entropy of the fourth-order extension of the source defined in Table 10.1. Determine $\overline{L}/n$ for $n = 4$ and add this result to those shown in Figure 10.9. Determine the efficiency of the resulting codes for $n = 1, 2, 3,$ and 4.

10.21   A source has five equally likely output messages. Determine a Shannon-Fano code for the source and determine the efficiency of the resulting code. Repeat for the Huffman code and compare the results.

10.22   A source has five outputs denoted [m_1, m_2, m_3, m_4, m_5] with respective probabilities [0.40, 0.19, 0.16, 0.15, 0.10]. Determine the codewords to represent the source outputs using both the Shannon-Fano and the Huffman techniques.

10.23   A binary source has output probabilities [0.85, 0.15]. The channel can transmit 300 binary symbols per second at the capacity of 1 bit/symbol. Determine the maximum source symbol rate if transmission is to be accomplished.

10.24   A source output consists of nine equally likely messages. Encode the source output using both binary Shannon-Fano and Huffman codes. Compute the efficiency of both of the resulting codes and compare the results.

10.25   Repeat the preceding problem assuming that the source has 12 equally likely outputs.

10.26 An analog source has an output described by the probability density function

$$f_x(x) = \begin{cases} 2x, & 0 \le x \le 1 \\ 0, & \text{otherwise} \end{cases}$$

The output of the source is quantized into ten measages using the eleven quantizing levels

$$x_i = 0.1k, \qquad k = 0, 1, \ldots, 10$$

The resulting messages are encoded using a binary Huffman code. Assuming that 250 samples of the source are transmitted each second, determine the resulting binary symbol rate in symbols per second. Also determine the information rate in bits per second.

10.27 A source output consists of four messages $[m_1, m_2, m_3, m_4]$ with respective probabilities $[0.4, 0.3, 0.2, 0.1]$. The second-order extension of the source is to be encoded using a code with a three-symbol alphabet. Determine the codewords using the Shannon-Fano technique and determine the efficiency of the resulting code.

10.28 It can be shown that a necessary and sufficient condition for the existence of an instantaneous binary code with wordlengths l_i, $1 \le i \le N$, is that

$$\sum_{i=1}^{N} 2^{-l_i} \le 1$$

This is known as the Kraft inequality. Show that the Kraft inequality is satisfied by the codewords given in Table 10.3. (*Note:* The inequality given above must also be satisfied for uniquely decipherable codes.)

Section 10.3

10.29 A continuous bandpass channel can be modeled as illustrated in Figure 10.37. Assuming a signal power of 20 W and a noise power spectral density of 10^{-4} W/Hz, plot the capacity of the channel as a function of the channel bandwidth.

10.30 For the preceding problem determine the capacity in the limit as $B \to \infty$.

FIGURE 10.37

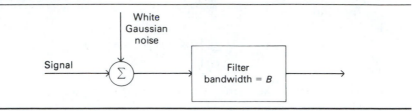

10.31   Derive an equation, similar to (10.92), that gives the error probability for a rate $\frac{1}{5}$ repetition code. Plot together, on the same set of
axes, the error probability of both a rate of $\frac{1}{3}$ and rate $\frac{1}{5}$ repetition
code as a function of $q = 1 - p$.

10.32   Show that a (7, 4) Hamming code is a distance 3 code.

10.33   Write the parity-check matrix and the generator matrix for a
(15, 11) single error-correcting code. Assume that the code is systematic. Calculate the codeword corresponding to the all one's
information sequence. Calculate the syndrome corresponding to an
error in the third position assuming the codeword corresponding to
the all one's input sequence.

10.34   A parity-check code has the parity-check matrix

$$[H] = \begin{bmatrix} 1 & 1 & 0 & 1 & 1 & 0 & 0 \\ 1 & 1 & 1 & 0 & 0 & 1 & 0 \\ 0 & 1 & 1 & 1 & 0 & 0 & 1 \end{bmatrix}$$

Determine the generator matrix and find all possible codewords.

10.35   For the code described in the preceding problem, find the codewords $[T_1]$ and $[T_2]$ corresponding to the information sequences

$$[A_1] = \begin{bmatrix} 0 \\ 1 \\ 1 \\ 0 \end{bmatrix} \qquad [A_2] = \begin{bmatrix} 1 \\ 1 \\ 1 \\ 0 \end{bmatrix}$$

Using these two codewords, illustrate the group property.

10.36   Determine the generator matrix for a rate $\frac{1}{3}$ and a rate $\frac{1}{5}$ repetition
code. Describe the generator matrix for a rate $1/n$ repetition code.

10.37   Determine the relationship between n and k for a Hamming code.
Use this result to show that the code rate approaches 1 for large n.

10.38 Determine the generator matrix for the encoder illustrated in Figure 10.16 Use the generator matrix to generate the complete set of codewords and use the results to check the codewords shown in Figure 10.16. Show that these codewords constitute a cyclic code.

10.39 Use the result of the preceding problem to determine the parity-check matrix for the encoder shown in Figure 10.16. Use the parity- check matrix to decode the received sequences 1101001 and 1101011. Compare your result with that shown in Figure 10.17.

10.40 Modify the encoder illustrated in Figure 10.16 by using feedback taps from stages S_1 and S_3 rather than from stages S_2 and S_3. Determine the resulting generator matrix and the complete set of 16 codewords. Is the resulting code a cyclic code?

10.41 Construct a set of performance curves similar to those of Figure 10.18 for a (15, 11) Hamming code.

10.42 Consider the coded system examined in Example 10.12. Show that the probability of three symbol errors in a codeword is negligible compared to the probability of two symbol errors in a codeword for signal-to-noise ratios above a certain level.

10.43 The Hamming code was defined as a code for which the ith column of the parity-check matrix is the binary representation of the number i. With a little thought it becomes clear that the columns of the parity-check matrix can be permuted without changing the distance properties, and therefore the error-correcting capabilities, of the code. Using this fact, determine a generator matrix and the corresponding parity-check matrix of the systematic code equivalent to a (7, 4) Hamming code. How many different generator matrices can be found?

10.44 In order to compare the performance of digital communications systems using block codes of different word lengths n, the probability of error for each code must be computed. One technique for accomplishing this task is to use the approximation[1]

$$P_b = \frac{d}{n} \sum_{i=e+1}^{d} \binom{n}{i} P_{sc}^i (1 - P_{sc})^{n-i} + \sum_{i=d+1}^{n} \binom{n-1}{i-1} P_{sc}^i (1 - P_{sc})^{n-i}$$

[1] D. J. Torrieri, *Principles of Secure Communication Systems*, 2nd ed. Artech House, Norwood, Mass., 1992, p. 47.

where d is the minimum distance of the code, e is the number of correctable errors per codeword, P_{sc} is the symbol error probability for the coded system and P_b is the decoded bit-error probability. Using this approximation, compare the performance of (7, 4) and (15, 11) single-error-correcting Hamming codes in an AWGN system with BPSK modulation.

10.45 Repeat the previous problem for a (7, 4) Hamming code and a (23, 12) triple-error-correcting Golay code.

10.46 What is the constraint span of the encoder shown in Figure 10.38? How many states are required to define the state diagram and the trellis diagram? Draw the state diagram, giving the output for each state transition.

10.47 Determine the state diagram for the convolutional encoder shown in Figure 10.39. Draw the trellis diagram through the first set of steady-state transitions. On a second trellis diagram, show the termination of the trellis to the all-zero state.

10.48 A source produces binary symbols at a rate of 2000 symbols per second. The channel is subjected to error bursts lasting 0.15 sec. Devise an encoding scheme using an interleaved (n, k) Hamming code which allows full correction of the error burst. Assume that the information rate out of the encoder is equal to the information rate into the encoder. What is the minimum time between bursts if your system is to operate properly?

10.49 A very popular class of codes is known as BCH (Bose-Chaudhuri-Hocquenghem) codes. These allow correction of multiple errors. For

FIGURE 10.38

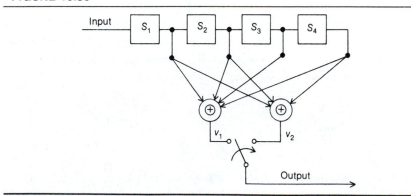

FIGURE 10.39

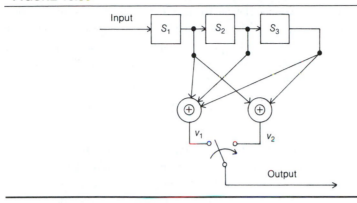

example, a (15, 7) BCH code can correct two errors per word. Analyze this code by deriving a set of performance curves, similar to those of Figure 10.18 for the (15, 7) BCH code.

Section 10.4

10.50 Compare FM *with preemphasis* to an optimal modulation system for $\beta = 1$, 5, and 10. Consider only operation above threshold, and assume 20 dB as the value of P_T/N_0W at threshold.

 COMPUTER EXERCISES

10.1 Develop a computer program that allows you to plot the entropy of a source with variable output probabilities. We wish to observe that the maximum source entropy does indeed occur when the source outputs are equally likely. Start with a simple two-output source $[m_1, m_2]$ with respective probabilities $[a, 1-a]$ and plot the entropy as a function of the parameter a. Then consider more complex cases such as a three output source $[m_1, m_2, m_3]$ with respective probabilities $[a, b, 1-a-b]$. Be creative with the manner in which the results are displayed.

10.2 In this computer exercise we wish to develop a program that allows the performance of an error-correcting code to be investigated by evaluating (10.119). Use the resulting program to verify the system performance characteristics illustrated in Examples 10.12 and 10.13.

10.3 Using your favorite high-level language, develop a program that allows you to investigate the performance of a digital communication system that utilizes a null-zone receiver and a noiseless feedback channel. Use the resulting program to evaluate the performance of a baseband system with an integrate-and-dump detector. The output of the integrate-and-dump detector is given by

$$V = \begin{cases} +AT + N, & \text{if } + A \text{ is sent} \\ -AT + N, & \text{if } - A \text{ is sent} \end{cases}$$

where N is a random variable representing the noise at the detector output at the sampling instant. The detector uses two thresholds, a_1 and a_2, where $a_1 = -\gamma AT$ and $a_2 = \gamma AT$. A retransmission occurs if $a_1 < V < a_2$. Choose several values of γ, such as $\gamma = 0.05$, 0.10 and 0.15, and plot the probability of retransmission, and the expected number of retransmissions, as a function of A^2T/N_0 with γ a parameter of the family of curves. Also plot the probability of error and compare the results to the probability of error for a standard baseband transmission system. Comment on your results.

APPENDIX A

PHYSICAL NOISE SOURCES AND NOISE CALCULATIONS IN COMMUNICATION SYSTEMS

As discussed in Chapter 1, noise originates in a communication system from two broad classes of sources: those external to the system, such as atmospheric, solar, cosmic, or manmade sources, and those internal to the system. The degree to which external noise sources influence system performance depends heavily upon system location and configuration. Consequently, the reliable analysis of their effect on system performance is difficult and depends largely on empirical formulas and on-site measurements. Their importance in the analysis and design of communication systems depends on their intensity relative to the internal noise sources. In this Appendix, we are concerned with techniques of characterization and analysis of internal noise sources.

Noise internal to the subsystems that compose a communication system arises as a result of the random motion of charge carriers within the devices composing those subsystems. We now discuss several mechanisms that give rise to internal noise, and suitable models for these mechanisms.

A.1 PHYSICAL NOISE SOURCES

Thermal Noise

Thermal noise is the noise arising from the random motion of charge carriers in a conducting, or semiconducting, medium. Such random agitation at the atomic level is a universal characteristic of matter at temperatures other than absolute zero. Nyquist was one of the first to have studied thermal noise. *Nyquist's theorem* states that the mean-square noise voltage appearing across the terminals of a resistor of R ohms at temperature T Kelvin in a frequency band B hertz is given by

$$v_{\text{rms}}^2 = \langle v_n^2(t) \rangle = 4kTRB \text{ V}^2 \tag{A.1}$$

where

$$k = \text{Boltzmann's constant} = 1.38 \times 10^{-23} \text{ J/K}$$

747

FIGURE A.1   Equivalent circuits for a noisy resistor. (a) Thevenin.
(b) Norton.

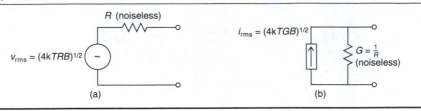

(a)   (b)

Thus a *noisy* resistor can be represented by an equivalent circuit consisting
of a *noiseless* resistor in series with a noise generator of rms voltage v_{rms} as
shown in Figure A.1(a). Short circuiting the terminals of Figure A.1(a) results
in a short-circuit noise current of mean-square value

$$i_{rms}^{\ 2} = \langle i_n^{\ 2}(t) \rangle = \frac{\langle v_n^{\ 2}(t) \rangle}{R^2} = \frac{4kTB}{R} = 4kTGB \ \text{A}^2 \qquad (A.2)$$

where $G = 1/R$ is the conductance of the resistor. The Thevenin equivalent
in Figure A.1(a) can therefore be transformed to the Norton equivalent in
Figure A.1(b).

▼ **EXAMPLE A.1**   Consider the resistor network shown in Figure A.2.
Assuming room temperature of $T = 290$ K, find the rms noise voltage
appearing at the output terminals in a 100-kHz bandwidth.

SOLUTION   We use voltage division to find the noise voltage due to each
resistor across the output terminals. Then, since powers due to independent

FIGURE A.2   Circuits for noise calculation. (a) Resistor network. (b) Noise
equivalent circuit.

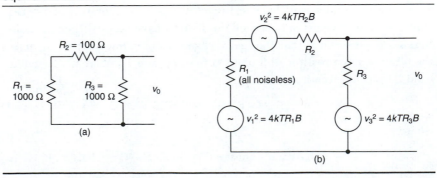

(a)   (b)

sources add, we find the rms output voltage v_0 by summing the square of the voltages due to each resistor, which gives the total mean-square voltage, and take the square root to get rms voltage. The calculation yields

$$v_0^2 = v_{01}^2 + v_{02}^2 + v_{03}^2 \qquad (A.3)$$

where

$$v_{01} = \underbrace{(4kTR_1B)^{1/2}}_{v_1 \text{ (rms)}} \frac{R_3}{R_1 + R_2 + R_3} \qquad (A.4a)$$

$$v_{02} = \underbrace{(4kTR_2B)^{1/2}}_{v_2 \text{ (rms)}} \frac{R_3}{R_1 + R_2 + R_3} \qquad (A.4b)$$

$$v_{03} = \underbrace{(4kTR_3B)^{1/2}}_{v_3 \text{ (rms)}} \frac{R_1 + R_2}{R_1 + R_2 + R_3} \qquad (A.4c)$$

Thus

$$
\begin{aligned}
v_0^2 &= (4kTB) \left[\frac{(R_1 + R_2)R_3^2}{(R_1 + R_2 + R_3)^2} + \frac{(R_1 + R_2)^2 R_3}{(R_1 + R_2 + R_3)^2} \right] \\
&= (4 \times 1.38 \times 10^{-23} \times 290 \times 10^5) \\
&\quad \times \left[\frac{(1100)(1000)^2}{(2100)^2} + \frac{(1100)^2(1000)}{(2100)^2} \right] \\
&\cong 8.39 \times 10^{-13} \ V^2 \qquad (A.5)
\end{aligned}
$$

Therefore,

$$v_0 = 9.16 \times 10^{-7} \ V \text{ (rms)} \qquad (A.6)$$

Nyquist's Formula

Although Example A.1 is instructive from the standpoint of illustrating noise computations involving several noisy resistors, it also illustrates that such computations can be exceedingly long if many resistors are involved. *Nyquist's formula,* which can be proved from thermodynamic arguments, simplifies such computations considerably. It states: The mean-square noise voltage produced at the output terminals of any one-port network containing only resistors, capacitors, and inductors is given by

$$\langle v_n^2(t) \rangle = 2kT \int_{-\infty}^{\infty} R(f) \, df \qquad (A.7a)$$

where $R(f)$ is real part of the complex impedance seen looking back into the terminals. If the network contains only resistors, the mean-square noise voltage in a bandwidth B is

$$\langle v_n^2 \rangle = 4kTR_{eq}B \text{ V}^2 \tag{A.7b}$$

where R_{eq} is the Thevenin equivalent resistance of the network.

▼ EXAMPLE A.1 (continued)   If we look back into the terminals of the network shown in Figure A.2, the equivalent resistance is

$$R_{eq} = R_3 \parallel (R_1 + R_2) = \frac{R_3(R_1 + R_2)}{R_1 + R_2 + R_3} \tag{A.8}$$

Thus

$$v_0^2 = \frac{4kTBR_3(R_1 + R_2)}{(R_1 + R_2 + R_3)} \tag{A.9}$$

▲ which can be shown to be equivalent to the result obtained previously.

Shot Noise

Shot noise arises from the discrete nature of current flow in electronic devices. For example, the *electron* flow in a saturated thermionic diode is due to the sum total of electrons emitted from the cathode, which arrive randomly at the anode, thu· providing an average *current* flow I_d (from anode to cathode when taken as positive), plus a randomly fluctuating component of mean-square value

$$i_{rms}^2 = \langle i_n^2(t) \rangle = 2eI_dB \text{ A}^2 \tag{A.10}$$

where e = charge of the electron = 1.6×10^{-19} C. Equation (A.10) is known as *Schottky's theorem*.

Since powers from independent sources add, it follows that the squares of noise voltages or noise currents from independent sources, such as two resistors or two currents originating from independent sources, add. Thus, when applying Schottky's theorem to a p-n junction, we recall that the current flowing in a p-n junction diode

$$I = I_s \left[\exp\left(\frac{eV}{kT}\right) - 1 \right] \tag{A.11}$$

where V is the voltage across the diode and I_s is the reverse saturation current, can be considered as being caused by two independent currents

$I_s \exp (eV/kT)$ and $-I_s$. Both currents fluctuate independently, producing a mean-square shot noise current given by

$$i_{\text{rms,tot}}^2 = \left[2eI_s \exp \left(\frac{eV}{kT} \right) + 2eI_s \right] B$$

$$= 2e(I + 2I_s)B \tag{A.12}$$

For normal operation, $I \gg I_s$ and the differential conductance is $g_0 = dI/dV = eI/kT$, so that (A.12) may be approximated as

$$i_{\text{rms,tot}}^2 \cong 2eIB = 2kT \left(\frac{eI}{kT} \right) B = 2kTg_0B \tag{A.13}$$

which can be viewed as *half-thermal noise* of the differential conductance g_0 since there is a factor of 2 rather than a factor of 4 as in (A.2).

Other Noise Sources

In addition to thermal and shot noise, there are three other noise mechanisms that contribute to internally generated noise in electronic devices. We summarize them briefly here. A fuller treatment of their inclusion in the noise analysis of electronic devices is given by Van der Ziel (1970).

Generation-Recombination Noise   Generation-recombination noise is the result of free carriers being generated and recombining in semiconductor material. One can consider these generation and recombination events to be random. Therefore, this noise process can be treated as a shot noise process.

Temperature-Fluctuation Noise   Temperature-fluctuation noise is the result of the fluctuating heat exchange between a small body, such as a transistor, and its environment due to the fluctuations in the radiation and heat-conduction processes. If a liquid or gas is flowing past the small body, fluctuation in heat convection also occurs.

Flicker Noise   Flicker noise is due to various causes. It is characterized by a spectral density that increases with decreasing frequency. The dependence of the spectral density on frequency is often found to be proportional to the inverse first power of the frequency. Therefore, flicker noise is sometimes referred to as *one-over-f noise*. More generally, flicker noise phenomena are characterized by power spectra that are of the form constant/f^{α}, where α is close to unity. The physical mechanism that gives rise to flicker noise is not well understood.

Available Power

Since calculations involving noise involve transfer of power, the concept of maximum power available from a source of fixed internal resistance is useful. Figure A.3 illustrates the familiar theorem regarding maximum power transfer, which states that a source of internal resistance R delivers maximum power to a resistive load R_L if $R = R_L$ and that, under these conditions, the power P produced by the source is evenly split between source and load resistances. If $R = R_L$, the load is said to be *matched* to the source, and the power delivered to the load is referred to as the *available power* P_a. Thus $P_a = \frac{1}{2} P$, which is delivered to the load only if $R = R_L$. Consulting Figure A.3(a), in which v_{rms} is the rms voltage of the source, we see that the voltage across $R_L = R$ is $\frac{1}{2}v_{rms}$, giving

$$P_a = \frac{(\frac{1}{2}v_{rms})^2}{R} = \frac{v_{rms}^2}{4R} \tag{A.14}$$

Similarly, when dealing with a Norton equivalent circuit as shown in Figure A.3(b), we can write the available power as

$$P_a = (\tfrac{1}{2}i_{rms})^2 R = \frac{i_{rms}^2}{4G} \tag{A.15}$$

where $i_{rms} = v_{rms}/R$ is the rms noise current.

Returning to (A.1) or (A.2) and using (A.14) or (A.15), we see that a noisy resistor produces the available power

$$P_{a,R} = \frac{4kTRB}{4R} = kTB \text{ W} \tag{A.16}$$

Similarly, from (A.13), a diode with load resistance matched to its differential conductance produces the available power

$$P_{a,D} = \tfrac{1}{2}kTB \text{ W}, \qquad I \gg I_s \tag{A.17}$$

FIGURE A.3 Circuits pertinent to maximum power transfer theorem. (a) Thevenin equivalent for a source with load resistance R_L. (b) Norton equivalent for a source with load conductance G_L.

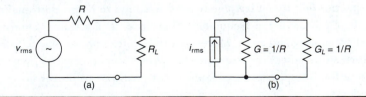

▼ **EXAMPLE A.2**   Calculate the available power per hertz of bandwidth for a resistance at room temperature, taken to be $T_0 = 290$ K. Express in decibels referenced to 1 watt (dBW) and decibels referenced to 1 milliwatt (dBm).

SOLUTION

Power/hertz $= \dfrac{P_{a,R}}{B} = (1.38 \times 10^{-23})(290) = 4.002 \times 10^{-21}$ W/Hz

Power/hertz in dBW $= 10 \log_{10} (4.002 \times 10^{-21}/1) \cong -204$ dBW

Power/hertz in dBM $= 10 \log_{10} (4.002 \times 10^{-21}/10^{-3}) \cong -174$ dBm

$$\text{(A.18)}$$

▲

Frequency Dependence

In Example A.2, available power per hertz for a noisy resistor at $T_0 = 290$ K was computed and found, to good approximation, to be -174 dBm/Hz, independent of the frequency of interest. Actually, Nyquist's theorem, as stated by (A.1), is a simplification of a more general result. The proper quantum mechanical expression for available power per hertz, or available *power spectral density $S_a(f)$*, is

$$S_a(f) \triangleq \frac{P_a}{B} = \frac{hf}{\exp(hf/kT) - 1} \quad \text{W/Hz} \qquad \text{(A.19)}$$

where h = Planck's constant = 6.6254×10^{-34} J-s.

This expression is plotted in Figure A.4, where it is seen that for all but very low temperatures and very high frequencies, the approximation is good that $S_a(f)$ is constant (that is, P_a is proportional to bandwidth B).

Quantum Noise

Taken by itself, (A.19) might lead to the false assumption that for very high frequencies where $hf \gg kT$, such as those used in optical communication, the noise would be negligible. However, it can be shown that a quantum noise term equal to hf must be added to (A.19) in order to account for the discrete nature of the electron energy. This is shown in Figure A.4 as the straight line, which permits the transition frequency between the thermal noise and quantum noise regions to be estimated. This transition frequency is seen to be above 20 GHz even for $T = 2.9$ K.

FIGURE A.4 Noise power spectral density versus frequency for thermal resistors

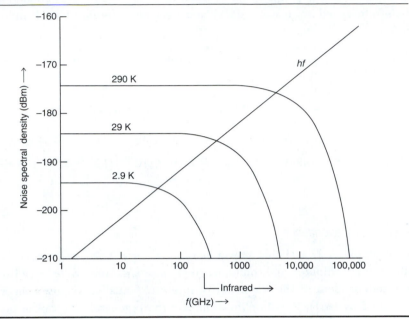

A.2 CHARACTERIZATION OF NOISE IN SYSTEMS—NOISE FIGURE AND NOISE TEMPERATURE

Having considered several possible sources of internal noise in communication systems, we now wish to discuss convenient methods for characterization of the noisiness of the subsystems that make up a system, as well as overall noisiness of the system. Figure A.5 illustrates a cascade of N stages or subsystems that make up a system. For example, if this block diagram represents a superheterodyne receiver, subsystem 1 would be the RF amplifier, subsystem 2 the mixer, subsystem 3 the IF amplifier, and subsystem 4 the detector. At the output of each stage, we wish to be able to relate the signal-to-noise power ratio to the signal-to-noise power ratio at the input. This will allow us to pinpoint those subsystems that contribute significantly to the output noise of the overall system, thereby enabling us to implement designs that minimize the noise.

Noise Figure of a System

One useful measure of system noisiness is the so-called *noise figure F,* defined as the ratio of the SNR at the system input to the SNR at the system output.

FIGURE A.5   Cascade of subsystems making up a system. (a) *N*-subsystem cascade with definition of SNR's at each point. (b) The *l*th subsystem in the cascade.

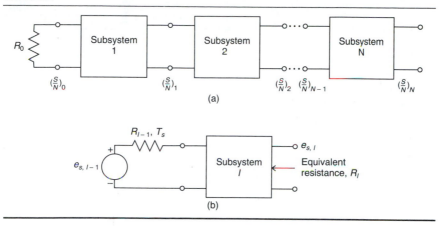

(a)

(b)

In particular, for the *l*th subsystem in Figure A.5, the noise figure F_l is defined by the relation

$$\left(\frac{S}{N}\right)_l = \frac{1}{F_l}\left(\frac{S}{N}\right)_{l-1} \tag{A.20}$$

For an ideal, noiseless subsystem, $F_l = 1$; that is, the subsystem introduces no additional noise. For physical devices, $F_l > 1$.

Noise figures for devices and systems are often stated in terms of decibels (dB). Specifically,

$$F_{dB} = 10 \log_{10} F_{ratio} \tag{A.21}$$

Table A.1 lists several devices along with typical noise figures for them.* For extremely low-noise devices, such as a maser amplifier, it is more convenient to work with noise temperature, to be discussed shortly.

The definition of noise figure given by (A.20) requires the calculation of both signal and noise powers at each point of the system. An alternative definition, equivalent to (A.20), involves the calculation of noise powers only. Although signal and noise powers at any point in the system depend on the loading of a subsystem on the preceding one, SNRs are independent of load, since both signal and noise appear across the same load. Hence any

---

* No temperature has yet been specified for the equivalent input resistance of the device. To standardize the noise figures of devices, $T_0 = 290$ K is specified as a standard temperature by standards of the Institute of Electrical and Electronic Engineers.

TABLE A.1   Typical Noise Figures and Temperatures for Several Devices[a]

Device	F (Ratio)	F (dB)	Noise Temp. (°K)	Gain (dB)	Operating Frequency (GHz)
Parametric amp.	1.26	1.0	75	10–20	0.5
(uncooled)	1.45	1.61	130		9
(liq. N_2 cooled)	1.17	0.69	50		3 & 6
(liq. He cooled)	1.03	0.13	9		4
Traveling wave	1.59	2.00	170	20–30	2.6
tube amp.	1.86	2.70	250		3
	2.69	4.30	490		9
Tunnel diode amp.					
ger.	2.38	3.77	400	20–40	
gal. ars.	1.69	2.28	200		
Low-noise superhet. rec'r (ant. noise and image response excl.)	2.38	3.77	400		
Integrated circuit IF ampl.	5.01	7	1163	50	≤11 MHz

[a] Information in column 4 from W. W. Mumford and E. H. Scheibe, *Noise: Performance Factors in Communication Systems,* Horizon House—Microwave, Inc., Dedham, Mass. (1968), pages 38, 39. Also Motorola Semiconductor Products Division, *Linear Integrated Circuits Data Book,* Motorola, Inc. (1974).

convenient load impedance may be used in making signal and noise calculations. In particular, we will use load impedances matched to the output impedance, thereby working with available signal and noise powers.

Consider the *l*th subsystem in the cascade of the system shown in Figure A.5. If we represent its input by a Thevenin equivalent circuit with rms signal voltage $e_{s,l-1}$ and equivalent resistant R_{l-1}, the available signal power is

$$P_{sa,l-1} = \frac{e_{s,l-1}^2}{4R_{l-1}} \tag{A.22}$$

If we assume that only thermal noise is present, the available noise power for a source temperature of T_s is

$$P_{na,l-1} = kT_sB \tag{A.23}$$

giving an input SNR of

$$\left(\frac{S}{N}\right)_{l-1} = \frac{e_{s,l-1}^2}{4kT_sR_{l-1}B} \tag{A.24}$$

The available output signal power, from Figure A.5(b), is

$$P_{sa,l} = \frac{e_{s,l}^2}{4R_l} \tag{A.25}$$

We can relate $P_{sa,l}$ to $P_{sa,l-1}$ by the *available power gain* G_a of subsystem l, defined to be

$$P_{sa,l} = G_a P_{sa,l-1} \tag{A.26}$$

which is obtained if all resistances are matched. The output SNR is

$$\left(\frac{S}{N}\right)_l = \frac{P_{sa,l}}{P_{na,l}} = \frac{1}{F_l} \frac{P_{sa,l-1}}{P_{na,l-1}} \tag{A.27}$$

or

$$F_l = \frac{P_{sa,l-1}}{P_{sa,l}} \frac{P_{na,l}}{P_{na,l-1}} = \frac{P_{sa,l-1}}{G_a P_{sa,l-1}} \frac{P_{na,l}}{P_{na,l-1}}$$

$$= \frac{P_{na,l}}{G_a P_{na,l-1}} \tag{A.28}$$

where any mismatches may be ignored, since they affect signal and noise the same.* Thus the noise figure is the ratio of the output noise power to the noise power that would have resulted had the system been noiseless. Noting that $P_{na,l} = G_a P_{na,l-1} + P_{int,l}$, where $P_{int,l}$ is the available internally generated noise power of subsystem l, and that $P_{na,l-1} = kT_sB$, we may write (A.28) as

$$F_l = 1 + \frac{P_{int,l}}{G_a kT_sB} \tag{A.29a}$$

or, setting $T_s = T_0 = 290$ K to standardize the noise figure,† we obtain

$$F_l = 1 + \frac{P_{int,l}}{G_a kT_0B} \tag{A.29b}$$

For $G_a \gg 1$, $F \cong 1$, which shows that the internally generated noise becomes inconsequential for systems with large gains. Conversely, systems with low gain enhance the importance of internal noise.

---

* This assumes that the gains for noise power and signal power are the same. If gain varies with frequency, then a spot noise figure can be defined, where signal power and noise power are measured in a small bandwidth Δf.

† If this were not done, the manufacturer of a receiver could claim superior noise performance of its product over that of a competitor simply by choosing T_s larger than the competitor. See Mumford and Scheibe (1968), pages 53–56, for a summary of the various definitions of noise figure used in the past.

Measurement of Noise Figure

Using (A.28), with the available noise power at the output $P_{na,out}$ referred to the device input, and representing this noise by a current generator $\overline{i_n^2}$ in parallel with the source resistance R_s or a voltage generator $\overline{e_n^2}$ in series with it, we can determine the noise figure by changing the input noise a known amount and measuring the change in noise power at the device output. In particular, if we assume a current source represented by a saturated thermionic diode, so that

$$\overline{i_n^2} = 2eI_dB \ A^2 \tag{A.30}$$

and sufficient current is passed through the diode so the noise power at the output is double the amount that appeared without the diode, then the noise figure is

$$F = \frac{eI_dR_s}{2kT_0} \tag{A.31}$$

where e is the charge of the electron in coulombs, I_d is the diode current in amperes, R_s is the input resistance, k is Boltzmann's constant, and T_0 is the standard temperature in degrees Kelvin.

A variation of the preceding method is the Y-factor method, which is illustrated in Figure A.6. Assume that two calibrated noise sources are available, one at effective temperature T_{hot} and the other at T_{cold}. With the first at the input of the unknown system with unknown temperature T_e, the available output noise power is

$$P_h = k(T_{hot} + T_e)BG \tag{A.32}$$

FIGURE A.6   *Y-factor method for measuring effective noise temperature*

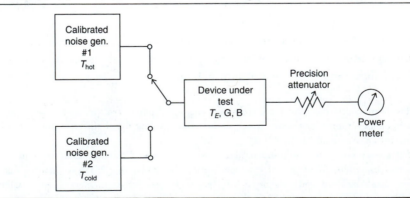

where B the noise bandwidth of the device under test and G is its available power gain. With the cold noise source present, the available output noise power is

$$P_c = k(T_{\text{cold}} + T_e)BG \qquad (A.33)$$

The two unknowns in these two equations are T_e and BG. Dividing the first by the second, we obtain

$$\frac{P_h}{P_c} = Y = \frac{T_{\text{hot}} + T_e}{T_{\text{cold}} + T_e} \qquad (A.34)$$

When solved for T_e, this equation becomes

$$T_e = \frac{T_{\text{hot}} - YT_{\text{cold}}}{Y - 1} \qquad (A.35)$$

which involves the two known source temperatures and the measured Y factor. The Y factor can be measured with the aid of the precision attenuator shown in Figure A.6 as follows: (1) connect the hot noise source to the system under test and adjust the attenuator for a convenient meter reading, (2) switch to the cold noise source and adjust the attenuator for the same meter reading as before, (3) noting the change in attenuator setting ΔA in dB, calculate $Y = 10^{-\Delta A/10}$, and (4) calculate the effective noise temperature using (A.35).

Noise Temperature

Equation (A.16) states that the available noise power of a resistor at temperature T is kTB W, independent of the value of R. We may use this result to define the *equivalent noise temperature* T_n of any noise source:

$$T_n = \frac{P_{n,\text{max}}}{kB} \qquad (A.36)$$

where $P_{n,\text{max}}$ is the maximum noise power the source can deliver in bandwidth B.

▼| **EXAMPLE A.3**   Two resistors R_1 and R_2 at temperatures T_1 and T_2 are connected in series to form a white-noise source. Find the equivalent noise temperature of the combination.

SOLUTION   The mean-square voltage generated by the combination is

$$\langle v_n^2 \rangle = 4kBR_1T_1 + 4kBR_2T_2 \qquad (A.37)$$

Since the equivalent resistance is $R_1 + R_2$, the available noise power is

$$P_{na} = \frac{\langle v_n^2 \rangle}{4(R_1 + R_2)} = \frac{4k(T_1R_1 + T_2R_2)B}{4(R_1 + R_2)} \tag{A.38}$$

The equivalent noise temperature is therefore

$$T_n = \frac{P_{na}}{kB} = \frac{R_1T_1 + R_2T_2}{R_1 + R_2} \tag{A.39}$$

Note that T_n is *not* a physical temperature unless both resistors are at the same temperature.

Effective Noise Temperature

Returning to Equation (A.29), we note that the second term, $P_{\text{int},l}/G_a k T_0 B$, which is dimensionless, is due solely to the internal noise of the system. Noting that $P_{\text{int},l}/G_a kB$ has the dimensions of temperature, we may write the noise figure as

$$F_l = 1 + \frac{T_e}{T_0} \tag{A.40}$$

where

$$T_e = \frac{P_{\text{int},l}}{G_a kB} \tag{A.41a}$$

Thus,

$$T_e = (F_l - 1)T_0 \tag{A.41b}$$

T_e is the *effective noise temperature* of the system and depends only on the parameters of the system. It is a measure of noisiness of the system referred to the input, since it is the temperature required of a thermal resistance, placed at the input of a noiseless system, in order to produce the same available noise power at the output as is produced by the internal noise sources of the system. Recalling that $P_{na,l} = G_a P_{na,l-1} + P_{\text{int},l}$ and that $P_{na,l-1} = kT_s B$, we may write the available noise power at the subsystem output as

$$P_{na,l} = G_a k T_s B + G_a k T_e B$$
$$= G_a k (T_s + T_e) B \tag{A.42}$$

where the actual temperature of the source T_s is used. Thus the available noise power at the output of a system can be found by adding the effective noise temperature of the system to the temperature of the source and multiplying by $G_a kB$, where the term G_a appears because all noise is referred to the input.

Noise Temperature and Noise Figure of a Cascade of Subsystems

Considering the first two stages in Figure A.5, we see that noise appears at the output due to the following sources:

1. Amplified source noise, $G_{a_1} G_{a_2} kT_s B$
2. Internal noise from the first stage amplified by the second stage, $G_{a_2} P_{a,\text{int}_1} = G_{a_2}(G_{a_1} kT_{e_1} B)$
3. Internal noise from the second stage, $P_{a,\text{int}_2} = G_{a_2} kT_{e_2} B$

Thus the total available noise power at the output of the cascade is

$$P_{na,2} = G_{a_1} G_{a_2} k \left(T_s + T_{e_1} + \frac{T_{e_2}}{G_{a_1}} \right) B \qquad (A.43)$$

Noting that the available gain for the cascade is $G_{a_1} G_{a_2}$ and comparing with (A.42), we see that the effective temperature of the cascade is

$$T_e = T_{e_1} + \frac{T_{e_2}}{G_{a_1}} \qquad (A.44)$$

From (A.40), the overall noise figure is

$$F = 1 + \frac{T_e}{T_0} = 1 + \frac{T_{e_1}}{T_0} + \frac{1}{G_{a_1}} \frac{T_{e_2}}{T_0}$$

$$= F_1 + \frac{F_2 - 1}{G_{a_1}} \qquad (A.45)$$

where F_1 is the noise figure of stage 1 and F_2 is the noise figure of stage 2. The generalization of this result to an arbitrary number of stages is known as *Friis's formula* and is given by

$$F = F_1 + \frac{F_2 - 1}{G_{a_1}} + \frac{F_3 - 1}{G_{a_1} G_{a_2}} + \cdots \qquad (A.46)$$

whereas the generalization of (A.44) is

$$T_e = T_{e_1} + \frac{T_{e_2}}{G_{a_1}} + \frac{T_{e_3}}{G_{a_1} G_{a_2}} + \cdots \qquad (A.47)$$

▼ EXAMPLE A.4   A parabolic dish antenna is pointed up into the sky not directly at the sun. Noise due to atmospheric radiation is equivalent to a source temperature of 70 K. A low-noise preamplifier with noise figure of 2 dB and an available power gain of 20 dB over a bandwidth of 20 MHz is mounted at the antenna feed (focus of the parabolic reflector). (a) Find the effective noise temperature at the input of the preamplifier. (b) Find the available noise power at the preamplifier output.

SOLUTION   (a) From (A.42), we have

$$T_{\text{eff,in}} = T_s + T_{e,\text{preamp}} \tag{A.48}$$

But (A.41b) gives

$$\begin{aligned} T_{e,\text{preamp}} &= T_0(F_{\text{preamp}} - 1) \\ &= 290(10^{2/10} - 1) \\ &= 169.6 \text{ K} \end{aligned} \tag{A.49}$$

(b) From (A.42), the available output noise power is

$$\begin{aligned} P_{na,\text{out}} &= G_a k(T_s + T_e)B \\ &= 10^{20/10}(1.38 \times 10^{-23})(239.6)(20 \times 10^6) \\ &= 6.61 \times 10^{-12} \text{ W} \end{aligned} \tag{A.50}$$

▲

▼ EXAMPLE A.5   A preamplifier with power gain to be found and a noise figure of 2.5 dB is cascaded with a mixer with a gain of 2 dB and a noise figure of 8 dB. Find the preamplifier gain such that the overall noise figure of the cascade is at most 4 dB.

SOLUTION   From Friis's formula

$$F = F_1 + \frac{F_2 - 1}{G_1} \tag{A.51}$$

Solving for G_1,

$$G_1 = \frac{F_2 - 1}{F - F_1} = \frac{10^{8/10} - 1}{10^{4/10} - 10^{2.5/10}} = 7.24 \text{ (ratio)} = 8.6 \text{ dB} \tag{A.52}$$

▲ Note that the gain of the mixer is immaterial.

Attenuator Noise Temperature and Noise Figure

Consider a purely resistive attenuator that imposes a loss of a factor of L in available power between input and output; thus the available power at its output $P_{a,out}$ is related to the available power at its input $P_{a,in}$ by

$$P_{a,out} = \frac{1}{L} P_{a,in} = G_a P_{a,in} \tag{A.53}$$

However, since the attenuator is resistive and assumed to be at the *same temperature* T_s as the equivalent resistance at its input, the available output power is

$$P_{na,out} = kT_s B \tag{A.54}$$

Characterizing the attenuator by an effective temperature T_e and employing (A.42), we may also write $P_{na,out}$ as

$$P_{na,out} = G_a k(T_s + T_e)B$$

$$= \frac{1}{L} k(T_s + T_e)B \tag{A.55}$$

Equating (A.54) and (A.55) and solving for T_e, we obtain

$$T_e = (L - 1)T_s \tag{A.56}$$

for the effective temperature of a noise resistance at temperature T_s, followed by an attenuator. From Equation (A.40), the noise figure of the cascade of source resistance and attenuator is

$$F = 1 + \frac{(L - 1)T_s}{T_0} \tag{A.57a}$$

or

$$F = 1 + \frac{(L - 1)T_0}{T_0} = L \tag{A.57b}$$

for an attenuator at *room temperature,* T_0.

▼ EXAMPLE A.6   Consider a receiver system consisting of an antenna with lead-in cable having a loss factor of $L = 1.5$ dB $= F_1$, an RF preamplifier with a noise figure of $F_2 = 7$ dB and a gain of 20 dB, followed by a mixer with a noise figure of $F_3 = 10$ dB and a conversion gain of 8 dB, and finally an integrated-circuit IF amplifier with a noise figure of $F_4 = 6$ dB and a gain of 60 dB.

(a) Find the overall noise figure and noise temperature of the system.
(b) Find the noise figure and noise temperature of the system with preamplifier and cable interchanged.

SOLUTION   (a) Converting decibel values to ratios and employing (A.46), we obtain

$$F = 1.41 + \frac{5.01 - 1}{1/1.41} + \frac{10 - 1}{100/1.41} + \frac{3.98 - 1}{(100)(6.3)/1.41}$$

$$= 1.41 + 5.65 + 0.13 + 6.7 \times 10^{-3} = 7.19 = 8.57 \text{ dB} \quad \text{(A.58)}$$

Note that the cable and RF amplifier essentially determine the noise figure of the system and that the noise figure of the system is enhanced because of the loss of the cable.

If we solve (A.40) for T_e, we have the effective noise temperature

$$T_e = T_0(F - 1) = 290(7.19 - 1) = 1796 \text{ K} \quad \text{(A.59)}$$

(b) Interchanging the cable and RF preamplifier, we obtain the noise figure

$$F = 5.01 + \frac{1.41 - 1}{100} + \frac{10 - 1}{100/1.41} + \frac{3.98 - 1}{(100)(6.3)/1.41}$$

$$= 5.01 + (4.1 \times 10^{-3}) + 0.127 + (6.67 \times 10^{-3})$$

$$= 5.15 = 7.12 \text{ dB} \quad \text{(A.60)}$$

The noise temperature is

$$T_e = 290(4.15) = 1203 \text{ K} \quad \text{(A.61)}$$

Now the noise figure and noise temperature are essentially determined by the noise level of the RF preamplifier. ▲

We have omitted one possibly important source of noise—namely, the antenna. If the antenna is directive and pointed at an intense source of thermal noise, such as the sun, its equivalent temperature may also be of importance in the calculation. This is particularly true when a low-noise preamplifier is employed (see Table A.1).

A.3 FREE-SPACE PROPAGATION EXAMPLE

As a final example of noise calculation, we consider a free-space electromagnetic-wave propagation channel. For the sake of illustration, suppose the

FIGURE A.7 A satellite-relay communication link

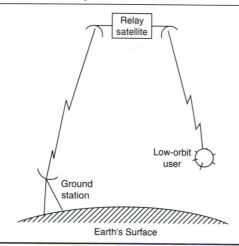

communication link of interest is between a synchronous-orbit relay satellite and a low-orbit satellite or aircraft, as shown in Figure A.7. This might represent part of a relay link between a ground station and a small scientific satellite or an aircraft. Since the ground station is high power, we assume the ground-station-to-relay-satellite link is noiseless and focus our attention on the link between the two satellites.

Assume a relay satellite transmitted signal power of P_T W. If radiated isotropically, the power density at a distance d from the satellite is given by

$$p_t = \frac{P_T}{4\pi d^2} \text{ W/m}^2 \tag{A.62}$$

If the satellite antenna has directivity, with the radiated power being directed toward the low-orbit vehicle, the antenna can be described by an antenna power gain G_T over the isotropic radiation level. For aperture-type antennas with aperture area A_T large compared with the square of the transmitted wavelength λ^2, it can be shown that the maximum gain is given by $G_T = 4\pi A_T/\lambda^2$. The power P_R intercepted by the receiving antenna is given by the product of the receiving aperture area A_R and the power density at the aperture:

$$P_R = p_t A_R = \frac{P_T G_T}{4\pi d^2} A_R \tag{A.63}$$

However, we may relate the receiving aperture antenna to its maximum gain by the expression $G_R = 4\pi A_R/\lambda_2$, giving

$$P_R = \frac{P_T G_T G_R \lambda^2}{(4\pi d)^2} \tag{A.64}$$

Equation (A.64) includes only the loss in power from isotropic spreading of the transmitted wave. If other losses, such as atmospheric absorption, are important, they may be included as a loss factor L_0 in (A.64) to yield

$$P_R = \left(\frac{\lambda}{4\pi d}\right)^2 \frac{P_T G_T G_R}{L_0} \tag{A.65}$$

The factor $(4\pi d/\lambda)^2$ is sometimes referred to as the *free-space loss*.*

In the calculation of received power, it is convenient to work in terms of decibels. Taking $10 \log_{10} P_R$, we obtain

$$10 \log_{10} P_R = 20 \log_{10} (\lambda/4\pi d) + 10 \log_{10} P_T$$
$$+ 10 \log_{10} G_T + 10 \log_{10} G_R - 10 \log_{10} L_0 \tag{A.66}$$

Now $10 \log_{10} P_R$ can be interpreted as the received power in decibels referenced to 1 W; it is commonly referred to as power in dBW. Similarly, $10 \log_{10} P_T$ is commonly referred to as the transmitted signal power in dBW. The terms $10 \log_{10} G_T$ and $10 \log_{10} G_R$ are the transmitter and receiver antenna gains (above isotropic) in decibels, while the term $10 \log_{10} L_0$ is the loss factor in decibels. When $10 \log_{10} P_T$ and $10 \log_{10} G_T$ are taken together, this sum is referred to as the *effective radiated power* in decibel watts (ERP, or sometimes EIRP, for effective radiated power referenced to isotropic). The negative of the first term is the free-space loss in decibels. For $d = 10^6$ mi $(1.6 \times 10^9$ m) and a frequency of 500 MHz $(\lambda = 0.6$ m),

$$20 \log_{10} \left(\frac{\lambda}{4\pi d}\right) = -210 \text{ dB} \tag{A.67}$$

If λ or d change by a factor of 10, this value changes by 20 dB. We now make use of (A.66) and the results obtained for noise figure and temperature to compute the signal-to-noise ratio for a typical satellite link.

▼ **EXAMPLE A.7**   We are given the following parameters for a relay-satellite-to-user link:

Relay satellite effective radiated power $(G_T = 35$ dB; $P_T = 100$ W): 55 dBW
Transmit frequency: 2 GHz $(\lambda = 0.15$ m)

---

* We take the convention here that a *loss* is a factor in the denominator of P_R; i.e., a *loss* in decibels is a positive quantity (a negative gain).

Receiver noise temperature of user (includes noise figure of receiver and
   background temperature of antenna): 700 K
User satellite antenna gain: 0 dB
Total system losses: 3 dB
Relay-user separation: 41,000 km

Find the signal-to-noise power ratio in a 50 kHz bandwidth at the user satellite receiver IF amplifier output.

SOLUTION   The received signal power is computed using (A.65) as follows
($+$ and $-$ signs in parentheses indicate whether the quantity is added or
subtracted):

Free-space loss: $-20 \log_{10} (0.15/4\pi \times 41 \times 10^6)$: 190.7 dB ($-$)
Effective radiated power: 55 dBW ($+$)
Receiver antenna gain: 0 dB ($+$)
System losses: 3 dB ($-$)
Received signal power: -138.7 dBW

The noise power level is calculated from (A.41a), written as

$$P_{int} = G_a k T_e B \tag{A.68}$$

where P_{int} is the receiver output noise power due to internal sources. Since
we are calculating the signal-to-noise ratio, the available gain of the receiver
does not enter the calculation because both signal and noise are multiplied
by the same gain. Hence we may set G_a to unity, and the noise level is

$$
\begin{aligned}
P_{int|dBW} &= 10 \log_{10} \left[kT_0 \left(\frac{T_e}{T_0} \right) B \right] \\
&= 10 \log_{10}(kT_0) + 10 \log_{10} \left(\frac{T_e}{T_0} \right) + 10 \log_{10} B \\
&= -204 + 10 \log_{10}(700/290) + 10 \log_{10}(50,000) \\
&= -153.2 \text{ dBW} \tag{A.69}
\end{aligned}
$$

Hence the signal-to-noise ratio at the receiver output is

$$\text{SNR}_0 = -138.7 + 153.2 = 14.5 \text{ dB} \tag{A.70}$$

▼| EXAMPLE A.8   To interpret the result obtained in the previous example
in terms of the performance of a digital communication system, we must
convert the signal-to-noise ratio obtained to energy-per-bit-to-noise-spec-
tral density ratio E_b/N_0 (see Chapter 7). By definition of SNR$_0$, we have

$$\text{SNR}_0 = \frac{P_R}{kT_eB} \tag{A.71}$$

Multiplying numerator and denominator by the duration of a data bit T_b, we obtain

$$\text{SNR}_0 = \frac{P_RT_b}{kT_eBT_b} = \frac{E_b}{N_0BT_b} \tag{A.72}$$

where $P_RT_b = E_b$ and $kT_e = N_0$ are the signal energy per bit and the noise power spectral density, respectively. Thus, to obtain E_b/N_0 from SNR_0, we calculate

$$E_b/N_0|_{\text{dB}} = (\text{SNR}_0)_{\text{dB}} + 10 \log_{10}(BT_b) \tag{A.73}$$

For example, from Chapter 8, we recall that the null-to-null bandwidth of a phase-shift keyed carrier is $2/T_b$ Hz. Suppose the data rate in Example A.7 is 2,500 bps. Therefore, BT_b for BPSK is 3 dB, and

$$E_b/N_0|_{\text{dB}} = 14.5 + 3 = 17.5 \text{ dB} \tag{A.74}$$

The probability of error for a binary BPSK digital communication system was derived in Chapter 7 as

$$P_E = Q(\sqrt{2E_b/N_0}) \cong \frac{\exp(-E_b/N_0)}{2\sqrt{\pi E_b/N_0}}$$

$$\cong 1.4 \times 10^{-26} \text{ for } E_b/N_0 = 17.5 \text{ dB} \tag{A.75}$$

which is a very small probability of error indeed. It appears that the system may have been overdesigned. However, no margin has been included as a safety factor. Components degrade, or the system may be operated in an environment for which it was not intended. With only 6 dB allowed for margin, the performance in terms of error probability becomes 5.5×10^{-8}.

FURTHER READING

Treatments of internal noise sources and calculations oriented toward communication systems comparable to the scope and level of the presentation here may be found in most of the books on communications referenced in Chapters 2 and 3, for instance, Carlson (1986), Schwartz (1990), and Taub and Schilling (1986). A concise, but thorough, treatment at an elementary level is available in Mumford and Scheibe (1968). An in-depth treatment of noise in solid-state devices is available in Van der Ziel (1970). A complete treatment of noise theory and calculation principles is found in Van der Ziel

(1954), which is the classical reference on noise, although noise character-istics of some of the newer devices are not included since the book was published before the invention of many of them. For discussion of satellite-link power budgets, see Ziemer and Peterson (1992).

PROBLEMS

Section A.1

A.1 A true rms voltmeter (assumed noiseless) with an effective noise bandwidth of 30 MHz is used to measure the noise voltage produced by the following devices. Calculate the meter reading in each case.

(a) A 10-kΩ resistor at room temperature, $T_0 = 290$ K
(b) A 10-kΩ resistor at 29 K
(c) A 10-kΩ resistor at 2.9 K
(d) What happens to all of the above results if the bandwidth is decreased by a factor of 4? a factor of 10? a factor of 100?

A.2 Given a junction diode with reverse saturation current $I_s = 15$ μA.

(a) At room temperature (290 K), find V such that $I > 20I_s$, thus allowing Equation (A.12) to be approximated by Equation (A.13). Find the rms noise current.
(b) Repeat part (a) for $T = 29$ K.

A.3 (a) Obtain an expression for the mean-square noise voltage appearing across R_3 in Figure A.8.
(b) If $R_1 = 2000$ Ω, $R_2 = R_L = 300$ Ω, and $R_3 = 500$ Ω, find the mean-square noise voltage per hertz.

A.4. Referring to the circuit of Figure A.8, consider R_L to be a load resis-tance, and find it in terms of R_1, R_2, and R_3 so that the maximum available noise power available from R_1, R_2, and R_3 is delivered to it.

FIGURE A.8

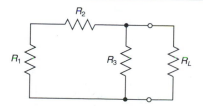

FIGURE A.9

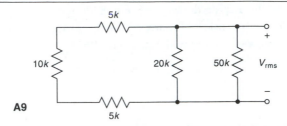

A9

A.5. Assuming a bandwidth of 2 Mhz, find the rms noise voltage across the output terminals of the circuit shown in Figure A.9 if it is at a temperature of 400 K.

Section A.2

A.6 Obtain an expression for F and T_e for the two-port resistive matching network shown in Figure A.10, assuming a source at $T_0 = 290$ K.

A.7 A source with equivalent noise temperature $T_s = 1000$ K is followed by a cascade of three amplifiers having the specifications shown in Table A.2. Assume a bandwidth of 50 kHz.

(a) Find the noise figure of the cascade.
(b) Suppose amplifiers 1 and 2 are interchanged. Find the noise figure of the cascade.
(c) Find the noise temperature of the systems of parts (a) and (b).
(d) Assuming the configuration of part (a), find the required input signal power to give an output signal-to-noise ratio of 40 dB. Perform the same calculation for the system of part (b).

FIGURE A.10

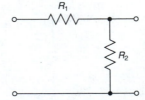

TABLE A.2

Amplifier No.	F	T_e	Gain
1		300 K	10 dB
2	6 dB		30 dB
3	11 dB		30 dB

A.8 (a) An attenuator with loss $L \gg 1$ is followed by an amplifier with noise figure F and gain $G_a = L$. Find the noise figure of the cascade.

(b) Consider the cascade of two identical attenuator-amplifier stages as in part (a). Consider the noise figure of the cascade.

(c) Generalize these results to N identical attenuators and amplifiers. How many decibels does the noise figure increase as a result of doubling the number of attenuators and amplifiers?

A.9 Given a cascade of a preamplifier, mixer, and amplifier with the specifications shown in Table A.3.

(a) Find the minimum gain of the preamplifier such that the overall noise figure of the cascade is 5 dB or greater.

(b) The preamplifier is fed by an antenna with noise temperature of 300 K (this is the temperature of the earth viewed from space). Find the temperature of the overall system using a preamplifier gain of 15 dB; for the preamplifier gain found in part (a).

(c) Find the noise power at the amplifier output for the two cases of part (b).

(d) Repeat part (b) except now assume that a transmission line with loss of 2 dB connects the antenna to the preamplifier.

A.10 An antenna with temperature of 300 K is fed into a receiver with total gain of 80 dB, $T_e = 1500$ K, and a bandwidth of 3 MHz.

(a) Find the available noise power at the output of the receiver.

(b) Find the necessary signal power P_r in dBm at the antenna terminals such that the output signal-to-noise ratio is 50 dB.

A.11 Referring to (A.35) and the accompanying discussion, suppose that two calibrated noise sources have effective temperatures of 600 K and 300 K. (a) Obtain the noise temperature of an amplifier with these noise sources used as inputs in turn if the difference in attenuator settings to get the same power meter reading at the amplifier's

TABLE A.3

	Noise Figure, dB	Gain, dB	Bandwidth
Preamplifier	2	?	*
Mixer	8	1.5	*
Amplifier	5	30	10 MHz

* The bandwidth of this stage is much greater than the amplifier bandwidth.

output is 1 dB; 1.5 dB; 2 dB. (b) Obtain the corresponding noise figures.

Section A.3

A.12 Given a relay-user link as described in Section A.3 with the following parameters:

Average transmit power of relay satellite: 35 dBW
Transmit frequency: 7.7 GHz
Effective antenna aperture of relay satellite: 1 m^2
Noise temperature of user receiver (including antenna): 1000 K
Antenna gain of user: 6 dB
Total system losses: 5 dB
System bandwidth: 1 MHz
Relay-user separation: 41,000 km

(a) Find the received signal power level at the user in dBW.
(b) Find the receiver noise level in dBW.
(c) Compute the signal-to-noise ratio at the receiver in decibels.
(d)*Find the average probability of error for the following digital signaling methods: (1) BPSK, (2) binary DPSK, (3) binary noncoherent FSK, (4) QPSK.

---

* This part of the problem requires results from Chapters 7 and 8.

APPENDIX B ▼

JOINTLY GAUSSIAN RANDOM VARIABLES

B.1 JOINT pdf AND CHARACTERISTIC FUNCTION

The joint probability density function of N jointly Gaussian random variables is*

$$f_X(\mathbf{x}) = (2\pi)^{-N/2} |\det \mathbf{C}|^{-1/2} \exp\left[-\tfrac{1}{2}(\mathbf{x} - \mathbf{m})^t \mathbf{C}^{-1}(\mathbf{x} - \mathbf{m})\right] \quad \text{(B.1)}$$

where $\mathbf{x}$ and $\mathbf{m}$ are column matrices defined as

$$\mathbf{x} = \begin{bmatrix} x_1 \\ x_2 \\ \vdots \\ x_N \end{bmatrix} \quad \text{and} \quad \mathbf{m} = \begin{bmatrix} m_1 \\ m_2 \\ \vdots \\ m_N \end{bmatrix} \quad \text{(B.2)}$$

respectively, and $\mathbf{C}$ is the positive definite matrix of correlation coefficients with elements

$$C_{ij} = \frac{E[(X_i - m_i)(X_j - m_j)]}{\sigma_{X_i}\sigma_{X_j}} \quad \text{(B.3)}$$

The joint characteristic function of the Gaussian random variables X_1, X_2, ..., X_N is

$$M_X(\mathbf{v}) = \exp\left[j\mathbf{m}^t\mathbf{v} - \tfrac{1}{2}\mathbf{v}^t\mathbf{C}\mathbf{v}\right] \quad \text{(B.4)}$$

From the power-series expansion of (B.4), it follows that for any four zero-mean Gaussian random variables,

$$E(X_1X_2X_3X_4) = E(X_1X_2)E(X_3X_4) + E(X_1X_3)E(X_2X_4)$$
$$+ E(X_1X_4)E(X_2X_3) \quad \text{(B.5)}$$

This is a rule that is useful enough to be worth memorizing.

---

* See C. W. Helstrom, *Probability and Stochastic Processes for Engineers* (New York: Macmillan, 1984), p. 208.

B.2 LINEAR TRANSFORMATIONS OF GAUSSIAN RANDOM VARIABLES

If a set of jointly Gaussian random variables is transformed to a new set of random variables by a linear transformation, the resulting random variables are jointly Gaussian. To show this, consider the linear transformation

$$\mathbf{y} = \mathbf{A}\mathbf{x} \tag{B.6}$$

where $\mathbf{y}$ and $\mathbf{x}$ are column matrices of dimension N and $\mathbf{A}$ is a nonsingular $N \times N$ square matrix with elements $[a_{ij}]$. From (B.6), the Jacobian is

$$J\begin{pmatrix} x_1, x_2, \cdots, x_N \\ y_1, y_2, \cdots, y_N \end{pmatrix} = \det(\mathbf{A}^{-1}) \tag{B.7}$$

where $\mathbf{A}^{-1}$ is the inverse matrix of $\mathbf{A}$. But $\det(\mathbf{A}^{-1}) = 1/\det(\mathbf{A})$. Using this in (B.1), along with

$$\mathbf{x} = \mathbf{A}^{-1}\mathbf{y} \tag{B.8}$$

gives

$$f_{\mathbf{Y}}(\mathbf{y}) = (2\pi)^{-N/2} |\det \mathbf{C}|^{-1/2} |\det \mathbf{A}|^{-1}$$
$$\times \exp[-\tfrac{1}{2}(\mathbf{A}^{-1}\mathbf{y} - \mathbf{m})^t \mathbf{C}^{-1}(\mathbf{A}^{-1}\mathbf{y} - \mathbf{m})] \tag{B.9}$$

Now $\det \mathbf{A} = \det \mathbf{A}^t$ and $\mathbf{A}\mathbf{A}^{-1} = \mathbf{I}$, the identity matrix, so that (B.9) can be written as

$$f_{\mathbf{Y}}(\mathbf{y}) = (2\pi)^{-N/2} |\det \mathbf{A}\mathbf{C}\mathbf{A}^t|^{-1/2}$$
$$\times \exp\left\{-\tfrac{1}{2}[\mathbf{A}^{-1}(\mathbf{y} - \mathbf{A}\mathbf{m})]^t \mathbf{C}^{-1}[\mathbf{A}^{-1}(\mathbf{y} - \mathbf{A}\mathbf{m})]\right\} \tag{B.10}$$

But the equalities $(\mathbf{A}\mathbf{B})^t = \mathbf{B}^t\mathbf{A}^t$ and $(\mathbf{A}^{-1})^t = (\mathbf{A}^t)^{-1}$ allow the term inside the braces in (B.10) to be written as

$$-\tfrac{1}{2}[(\mathbf{y} - \mathbf{A}\mathbf{m})^t(\mathbf{A}^t)^{-1}\mathbf{C}^{-1}\mathbf{A}^{-1}(\mathbf{y} - \mathbf{A}\mathbf{m})]$$

Finally, the equality $(\mathbf{A}\mathbf{B})^{-1} = \mathbf{B}^{-1}\mathbf{A}^{-1}$ allows the above term to be rearranged to

$$-\tfrac{1}{2}[(\mathbf{y} - \mathbf{A}\mathbf{m})^t(\mathbf{A}\mathbf{C}\mathbf{A}^t)^{-1}(\mathbf{y} - \mathbf{A}\mathbf{m})]$$

Thus (B.10) becomes

$$f_{\mathbf{Y}}(\mathbf{y}) = (2\pi)^{-N/2} |\det \mathbf{A}\mathbf{C}\mathbf{A}^t| \times \exp\left\{-\tfrac{1}{2}(\mathbf{y} - \mathbf{A}\mathbf{m})^t(\mathbf{A}\mathbf{C}\mathbf{A}^t)^{-1}(\mathbf{y} - \mathbf{A}\mathbf{m})\right\} \tag{B.11}$$

We recognize this as a joint Gaussian density function for a random vector $\mathbf{Y}$ with mean vector $E[\mathbf{Y}] = \mathbf{A}\mathbf{m}$ and covariance matrix $\mathbf{A}\mathbf{C}\mathbf{A}^t$.

APPENDIX C

▼

MATHEMATICAL AND NUMERICAL TABLES

This Appendix contains several tables pertinent to the material contained in this book. The tables are as follows:

C.1 Fourier Transform Pairs

Signal	Transform
1. 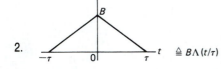 $\triangleq A\Pi(t/\tau)$	$A\tau \dfrac{\sin \pi f\tau}{\pi f\tau} \triangleq A\tau \operatorname{sinc} f\tau$
2. $\triangleq B\Lambda(t/\tau)$	$B\tau \dfrac{\sin^2 \pi f\tau}{(\pi f\tau)^2} \triangleq B\tau \operatorname{sinc}^2 f\tau$
3. $e^{-\alpha t}u(t)$	$\dfrac{1}{\alpha + j2\pi f}$
4. $\exp(-|t|/\tau)$	$\dfrac{2\tau}{1 + (2\pi f\tau)^2}$
5. $\exp[-\pi(t/\tau)^2]$	$\tau \exp[-\pi(f\tau)^2]$
6. $\dfrac{\sin 2\pi Wt}{2\pi Wt} \triangleq \operatorname{sinc} 2Wt$	$\triangleq \dfrac{\Pi(f/2W)}{2W}$
7. $\exp[j(\omega_c t + \phi)]$	$\exp(j\phi)\delta(f - f_c),\ \omega_c = 2\pi f_c$
8. $\cos(\omega_c t + \phi)$	$\tfrac{1}{2}\delta(f - f_c)\exp(j\phi) + \tfrac{1}{2}\delta(f + f_c)\exp(-j\phi)$
9. $\delta(t - t_0)$	$\exp(-j2\pi ft_0)$
10. $\displaystyle\sum_{m=-\infty}^{\infty} \delta(t - mT_s)$	$\dfrac{1}{T_s} \displaystyle\sum_{n=-\infty}^{\infty} \delta\!\left(f - \dfrac{n}{T_s}\right)$
11. $\operatorname{sgn} t = \begin{cases} +1, & t > 0 \\ -1, & t < 0 \end{cases}$	$-\dfrac{j}{\pi f}$
12. $u(t) = \begin{cases} 1, & t > 0 \\ 0, & t < 0 \end{cases}$	$\tfrac{1}{2}\delta(f) + \dfrac{1}{j2\pi f}$
13. $\dot{x}(t)$	$-j\operatorname{sgn}(f)X(f)$

C.2 Fourier Transform Theorems

Name of Theorem	Signal*	Fourier Transform
1. Superposition	$a_1 x_1(t) + a_2 x_2(t)$	$a_1 X_1(f) + a_2 X_2(f)$
2. Time delay	$x(t - t_0)$	$X(f)e^{-j\omega t_0}$
3a. Scale change	$x(at)$	$|a|^{-1} X(f/a)$
b. Time reversal	$x(-t)$	$X(-f) = X^*(f)$
4. Duality	$X(t)$	$x(-f)$
5a. Frequency translation	$x(t)e^{j\omega_0 t}$	$X(f - f_0),\ \omega_0 = 2\pi f_0$
b. Modulation	$x(t) \cos \omega_0 t$	$\frac{1}{2}X(f - f_0) + \frac{1}{2}X(f + f_0)$
6. Differentiation	$\dfrac{d^n x(t)}{dt^n}$	$(j2\pi f)^n X(f)$
7. Integration	$\displaystyle\int_{-\infty}^{t} x(t')dt'$	$(j2\pi f)^{-1} X(f) + \frac{1}{2}X(0)\delta(f)$
8. Convolution	$\displaystyle\int_{-\infty}^{\infty} x_1(t - t')x_2(t')dt'$ $= \displaystyle\int_{-\infty}^{\infty} x_1(t')x_2(t - t')dt'$	$X_1(f)X_2(f)$
9. Multiplication	$x_1(t)x_2(t)$	$\displaystyle\int_{-\infty}^{\infty} X_1(f - f')X_2(f')\,df'$ $= \displaystyle\int_{-\infty}^{\infty} X_1(f')X_2(f - f')\,df'$

*All signals are assumed to be real.

C.3   Trigonometric Identities

Euler's theorem: $e^{\pm ju} = \cos u \pm j \sin u$

$\cos u = \frac{1}{2}(e^{ju} + e^{-ju})$

$\sin u = (e^{ju} - e^{-ju})/2j$

$\sin^2 u + \cos^2 u = 1$

$\cos^2 u - \sin^2 u = \cos 2u$

$2 \sin u \cos u = \sin 2u$

$\cos^2 u = \frac{1}{2}(1 + \cos 2u)$

$\sin^2 u = \frac{1}{2}(1 - \cos 2u)$

$$\cos^{2n} u = \left[\sum_{k=0}^{n-1} 2 \binom{2n}{k} \cos 2(n-k)u + \binom{2n}{n} \right] \Big/ 2^{2n}$$

$$\cos^{2n-1} u = \left[\sum_{k=0}^{n-1} \binom{2n-1}{k} \cos (2n - 2k - 1)u \right] \Big/ 2^{2n-2}$$

$$\sin^{2n} u = \left[\sum_{k=0}^{n-1} (-1)^{n-k} 2 \binom{2n}{k} \cos 2(n-k)u + \binom{2n}{n} \right] \Big/ 2^{2n}$$

$$\sin^{2n-1} u = \left[\sum_{k=0}^{n-1} (-1)^{n+k-1} \binom{2n-1}{k} \sin (2n - 2k - 1)u \right] \Big/ 2^{2n-2}$$

where

$$\binom{n}{k} = \frac{n!}{(n-k)!k!}$$

$$\sin (u \pm v) = \sin u \cos v \pm \cos u \sin v$$
$$\cos (u \pm v) = \cos u \cos v \mp \sin u \sin v$$

$\sin u \sin v = \frac{1}{2} [\cos (u - v) - \cos (u + v)]$

$\cos u \cos v = \frac{1}{2} [\cos (u - v) + \cos (u + v)]$

$\sin u \cos v = \frac{1}{2} [\sin (u - v) + \sin (u + v)]$

C.4 Series Expansions

$$(u + v)^n = \sum_{k=0}^{n} \binom{n}{k} u^{n-k} v^k, \quad \text{where} \quad \binom{n}{k} = \frac{n!}{(n-k)!k!}$$

Letting $u = 1$ and $v = x$, where $|x| \ll 1$, results in the following approximations:

$$(1 + x)^n \cong 1 + nx$$
$$(1 + x)^{1/2} \cong 1 + \tfrac{1}{2}x$$
$$(1 + x)^{-n} \cong 1 - nx$$

$$\ln(1 + u) = \sum_{k=1}^{\infty} (-1)^{k+1} u^k / k$$

$$\log_a u = \log_e u \log_a e; \quad \log_e u = \ln u = \log_a u \log_e a$$

$$e^u = \sum_{k=0}^{\infty} u^k / k! \cong 1 + u, \quad |u| \ll 1$$
$$a^u = e^{u \ln a}$$

$$\sin u = \sum_{k=0}^{\infty} (-1)^k u^{2k+1} / (2k+1)! \cong u - u^3/3!, \quad |u| \ll 1$$

$$\cos u = \sum_{k=0}^{\infty} (-1)^k u^{2k} / (2k)! \cong 1 - u^2/2!, \quad |u| \ll 1$$

$$\tan u = u + \tfrac{1}{3} u^3 + \tfrac{2}{15} u^5 + \cdots$$
$$\sin^{-1} u = u + \tfrac{1}{6} u^3 + \tfrac{3}{40} u^5 + \cdots$$
$$\tan^{-1} u = u - \tfrac{1}{3} u^3 + \tfrac{1}{5} u^5 - \cdots$$
$$\operatorname{sinc} u = 1 - (\pi u)^2/3! + (\pi u)^4/5! - \cdots$$

$$J_n(u) \cong \begin{cases} \dfrac{u^n}{2^n n!} \left[1 - \dfrac{u^2}{2^2(n+1)} + \dfrac{u^4}{2 \cdot 2^4 (n+1)(n+2)} - \cdots \right], \\[2em] \sqrt{\dfrac{2}{\pi u}} \cos\left(u - \dfrac{n\pi}{2} - \dfrac{\pi}{2} \right), \quad u \gg 1 \end{cases}$$

$$I_0(u) \cong \begin{cases} 1 + \dfrac{u^2}{2^2} + \dfrac{u^4}{2^2 4^2} + \cdots \cong e^{u^2/4}, \quad 0 \le u \ll 1 \\[2em] \dfrac{e^u}{\sqrt{2\pi u}}, \quad u \gg 1 \end{cases}$$

C.5 Definite Integrals

$$\int_0^\infty \frac{x^{m-1}}{1+x^n}\,dx = \frac{\pi/n}{\sin(m\pi/n)}, \qquad n > m > 0$$

$$\int_0^\infty x^{\alpha-1}e^{-x}\,dx = \Gamma(\alpha), \qquad \alpha > 0$$

where

$$\Gamma(\alpha + 1) = \alpha\Gamma(\alpha)$$
$$\Gamma(1) = 1; \; \Gamma(\tfrac{1}{2}) = \sqrt{\pi}$$
$$\Gamma(n) = (n-1)! \quad \text{if } n \text{ is an integer}$$

$$\int_0^\infty x^{2n}e^{-ax^2}\,dx = \frac{1\cdot 3\cdot 5\ldots(2n-1)}{2^{n+1}a^n}\sqrt{\frac{\pi}{a}}$$

$$\int_0^\infty e^{-ax}\cos bx\,dx = \frac{a}{a^2+b^2}, \qquad a > 0$$

$$\int_0^\infty e^{-ax}\sin bx\,dx = \frac{b}{a^2+b^2}, \qquad a > 0$$

$$\int_0^\infty e^{-a^2x^2}\cos bx\,dx = \frac{\sqrt{\pi}\,e^{-b^2/4a^2}}{2a}, \qquad a > 0$$

$$\int_0^\infty x^{\alpha-1}\cos bx\,dx = \frac{\Gamma(\alpha)}{b^\alpha}\cos\tfrac{1}{2}\pi\alpha, \qquad 0 < a < 1, \; b > 0$$

$$\int_0^\infty x^{\alpha-1}\sin bx\,dx = \frac{\Gamma(\alpha)}{b^\alpha}\sin\tfrac{1}{2}\pi\alpha, \qquad 0 < |a| < 1, \; b > 0$$

$$\int_0^\infty xe^{-ax^2}I_k(bx)\,dx = \frac{1}{2a}e^{-b^2/4a}$$

$$\int_0^\infty \operatorname{sinc} x\,dx = \int_0^\infty \operatorname{sinc}^2 x\,dx = \tfrac{1}{2}$$

$$\int_0^\infty \frac{\cos ax}{b^2+x^2}\,dx = \frac{\pi}{2b}e^{-ab}, \qquad a > 0, b > 0$$

$$\int_0^\infty \frac{x\sin ax}{b^2+x^2}\,dx = \frac{\pi}{2}e^{-ab}, \qquad a > 0, b > 0$$

C.6 The Q-Function*

The Gaussian probability density function of unit variance and zero mean is

$$Z(x) = \frac{e^{-x^2/2}}{\sqrt{2\pi}} \tag{C.1}$$

and the corresponding cumulative distribution function is

$$P(x) = \int_{-\infty}^{x} Z(t)dt \tag{C.2}$$

The Q-function is defined as†

$$Q(x) = 1 - P(x) = \int_{x}^{\infty} Z(t)dt, \ x \geq 0 \tag{C.3}$$

An asymptotic expansion for $Q(x)$ valid for large x is

$$Q(x) = \frac{Z(x)}{x} \left[1 - \frac{1}{x^2} + \frac{1 \cdot 3}{x^4} - \cdots \right.$$

$$\left. + \frac{(-1)^n \cdot 3 \cdots (2n-1)}{x^{2n}} \right] + R_n \tag{C.4}$$

where the remainder is given by

$$R_n = (-1)^{n+1} \, 1 \cdot 3 \cdots (2n+1) \int_{x}^{\infty} \frac{Z(t)}{t^{2n+2}} \, dt \tag{C.5}$$

which is less in absolute value than the first neglected term. For $x \geq 3$, less than 10% error results if only the first term in (C.4) is used to approximate the Q-function.

For moderate values of x, several rational approximations are available. One such approximation is

$$Q(x) = Z(x)(b_1 t + b_2 t^2 + b_3 t^3 + b_4 t^4 + b_5 t^5) + \epsilon(x) \tag{C.6}$$

where

$$t = \frac{1}{1 + px} \tag{C.7}$$

* The information given in this appendix is extracted from M. Abramowitz and I. Stegun, *Handbook of Mathematical Functions*, New York: Dover, 1972 (Originally published in 1964 as part of the National Bureau of Standards Applied Mathematics Series 55).

† For $x < 0$, $Q(x) = 1 - Q(|x|)$.

The error term is bounded by $|\epsilon(x)| < 7.5 \times 10^{-8}$, and the constants are given by

$p = 0.2316419$

$b_1 = 0.319381530$ $\qquad$ $b_2 = -0.356563782$ $\qquad$ $b_3 = 1.781477937$

$b_4 = -1.821255978$ $\qquad$ $b_5 = 1.330274429$

The error function can be related to the Q-function by

$$\text{erf}(x) \triangleq \frac{2}{\sqrt{\pi}} \int_0^x e^{-t^2} \, dt = 1 - 2Q(\sqrt{2}x) \tag{C.8}$$

The complementary error function is $\text{erfc}(x) = 1 - \text{erf}(x)$, and can be approximated similarly to the Q-function as in (C.4) and (C.6). These approximations will not be given here.

A short table of values for $Q(x)$ and $Z(x)$ follows. Extensive tables of $P(x)$, $Q(x)$, and $Z(x)$ and its derivatives can be found in Abramowitz and Stegun (1972).

x	$Q(x)$	$Z(x)$	x	$Q(x)$	$Z(x)$
0.0	0.50000	0.39894	2.2	0.01390	0.03547
0.1	0.46017	0.39695	2.3	0.01072	0.02833
0.2	0.42074	0.39104	2.4	0.00820	0.02239
0.3	0.38209	0.38138	2.5	0.00621	0.01753
0.4	0.34458	0.36827	2.6	0.00466	0.01358
0.5	0.30854	0.35206	2.7	0.00347	0.01042
0.6	0.27425	0.33322	2.8	0.00256	0.00792
0.7	0.24196	0.31225	2.9	0.00187	0.00595
0.8	0.21186	0.28969	3.0	0.00135	0.00443
0.9	0.18406	0.26608	3.1	0.00097	0.00327
1.0	0.15866	0.24197	3.2	0.00069	0.00238
1.1	0.13567	0.21785	3.3	0.00048	0.00172
1.2	0.11507	0.19419	3.4	0.00035	0.00123
1.3	0.09680	0.17137	3.5	0.00023	0.00087
1.4	0.08076	0.14973	3.6	0.00016	0.00061
1.5	0.06681	0.12952	3.7	0.00011	0.00042
1.6	0.05480	0.11092	3.8	7.24×10^{-5}	0.00029
1.7	0.04457	0.09405			
1.8	0.03593	0.07895	3.9	4.81×10^{-5}	0.00020
1.9	0.02872	0.06562			
2.0	0.02275	0.05399	4.0	3.17×10^{-5}	0.00013
2.1	0.01786	0.04398			

BIBLIOGRAPHY

Historical References

Middleton, D. *An Introduction to Statistical Communication Theory*. McGraw-Hill, New York, 1960. Reprinted by Peninsula Publishing, 1989.

North, D. O. "An Analysis of the Factors Which Determine Signal/Noise Discrimination in Pulsed-Carrier Systems," *RCA Tech. Rept.* PTR-6-C, June 1943, reprinted in *Proc. IEEE,* **51**, 1016–1027 (July 1963).

Rice, S. O. "Mathematical Analysis of Random Noise," *Bell System Technical Journal,* **23**, 282–332 (July 1944); **24**, 46–156 (January 1945).

Shannon, C. E. "A Mathematical Theory of Communications," *Bell System Technical Journal,* **27**, 379–423, 623–656 (July 1948).

Wiener, Norbert. *Extrapolation, Interpolation, and Smoothing of Stationary Time Series with Engineering Applications*. MIT Press, Cambridge, 1949.

Woodward, P. M. *Probability and Information Theory with Applications to Radar*. Pergamon Press, New York, 1953.

Woodward, P. M., and J. L. Davies. "Information Theory and Inverse Probability in Telecommunication," *Proc. Inst. Elect. Engr.,* **99**, Pt. III, pp. 37–44 (March 1952).

Books

Arazi, B., *A Commonsense Approach to the Theory of Error Correcting Codes*. MIT Press, Cambridge, Mass., 1988.

Ash, C. *The Probability Tutoring Book*. IEEE Press, New York, 1992.

Barr, D. R., and P. W. Zhena. *Probability: Modeling and Uncertainty*. Addison-Wesley, Reading, Mass., 1983.

Beckmann, P. *Probability in Communication Engineering*. Harcourt, Brace, and World, New York, 1967.

Benedetto, S., E. Biglieri, and V. Castellani. *Digital Transmission Theory*. Prentice-Hall, Englewood Cliffs, N.J., 1987.

Biglieri, B., D. Divsalar, P. J. McLane, and M. K. Simon. *Introduction to Trellis-Coded Modulation with Applications,* Macmillan, New York, 1991.

Blahut, R. E. *Theory and Practice of Error Control Codes*. Addison-Wesley, Reading, Mass., 1984.

Blahut, R. E. *Principles and Practice of Information Theory*. Addison-Wesley, Reading, Mass., 1987.

Bracewell, R. *The Fourier Transform and Its Applications,* 2nd ed. McGraw-Hill, New York, 1986.

Breipohl, A. M. *Probabilistic Systems Analysis*. Wiley, New York, 1970.

Brigham, E. O. *Fast Fourier Transform and Its Applications*. Prentice-Hall, Englewood Cliffs, N.J., 1988.

Carlson, A. B. *Communication Systems,* 3rd ed. McGraw-Hill, New York, 1986.

Clark, G. C., Jr., and J. B. Cain. *Error-Correction Coding for Digital Communications*. Plenum Press, New York, 1981.

Couch, L. W., II. *Digital and Analog Communication Systems,* 3rd ed. Macmillan, New York, 1990.

Cunningham, D. R., and J. A. Stuller. *Basic Circuit Analysis,* Houghton Mifflin, Boston, Mass., 1991.

Downing, J. J. *Modulation Systems and Noise*. Prentice-Hall, Englewood Cliffs, N.J., 1964.

Frederick, D. K., and A. B. Carlson. *Linear Systems in Communication and Control,* 2nd ed. Wiley, New York, 1988.

Gallager, R. G. *Information Theory and Reliable Communication*. Wiley, New York, 1968.

Gardner, F. M. *Phaselock Techniques,* 2nd ed. Wiley, New York, 1979.

Haykin, S. *Communication Systems,* 3rd ed. Wiley, New York, 1989.

Hayt, W. H., Jr., and J. E. Kemmerly. *Engineering Circuit Analysis,* 5th ed. McGraw-Hill, New York, 1993.

Helstrom, C. W. *Statistical Theory of Signal Detection,* 2nd ed. Pergamon Press, New York, 1968.

Helstrom, C. W. *Probability and Stochastic Processes for Engineers,* 2nd ed. Macmillan, New York, 1990.

Kotel'nikov, V. A. *The Theory of Optimum Noise Immunity*. McGraw-Hill, New York, 1959. Doctoral dissertation presented in January 1947 before Academic Council of the Molatov Energy Institute in Moscow.

Lathi, B. P. *Modern Digital and Analog Communication Systems*. Holt, Rinehart and Winston, New York, 1983.

Lee, E. A., and D. G. Messerschmitt, *Digital Communication*. Kluwer, Boston, 1988.

Leon-Garcia, A. *Probability and Random Processes for Electrical Engineering,* Addison-Wesley, Reading, Mass., 1994.

Lin, S., and D. J. Costello, Jr. *Error Control Coding: Fundamentals and Applications*. Prentice-Hall, Englewood Cliffs, N. J., 1983.

Lindsey, W. C., and M. K. Simon. *Telecommunication Systems Engineering*. Prentice-Hall, Englewood Cliffs, N.J., 1973.

Mansuripur, M. *Introduction to Information Theory*. Prentice-Hall, Englewood Cliffs, N.J., 1987.

Mendel, J. M. *Lessons in Digital Estimation Theory*. Prentice Hall, Englewood Cliffs, N.J., 1987.

Mumford, W. W., and E. H. Scheibe. *Noise Performance Factors in Communication Systems*. Horizon House-Microwave, Dedham, Mass., 1968.

Papoulis, A. *Signal Analysis*. McGraw-Hill, New York, 1977.

Papoulis, A. *Probability, Random Variables, and Stochastic Processes,* 2nd ed. McGraw-Hill, New York, 1984.

Papoulis, A. *Probability and Statistics*. Prentice-Hall, Englewood Cliffs, N.J., 1990.

Peebles, P. Z. *Probability, Random Variables, and Ranaom Signal Principles,* 2nd ed. McGraw-Hill, New York, 1987.

Peterson, R. L., R. E. Ziemer, and D. A. Borth, *Introduction to Spread Spectrum Communications*. Prentice-Hall, Englewood Cliffs, N.J., 1995.

Proakis, J. G. *Digital Communications.* McGraw-Hill, New York, 1989.

Roberts, R. A. *An Introduction to Applied Probability.* Addison-Wesley, Reading, Mass., 1992.

Roden, M. S. *Digital and Data Communication Systems,* 2nd ed. Prentice-Hall, Englewood Cliffs, N.J., 1985.

Schwartz, M. *Information Transmission, Modulation, and Noise,* 4th ed. McGraw-Hill, New York, 1990.

Schwartz, M., W. R. Bennett, and S. Stein. *Communication Systems and Techniques.* McGraw-Hill, New York, 1966.

Schwartz, R. J., and B. Friedland. *Linear Systems.* McGraw-Hill, New York, 1965.

Shanmugam, K. S. *Digital and Analog Communication Systems.* Wiley, New York, 1979.

Shanmugam, K. S., and A. M. Breipohl. *Random Signals: Detection, Estimation and Data Analysis.* Wiley, New York, 1988.

Shannon, C. E., and W. Weaver. *The Mathematical Theory of Communication.* University of Illinois Press, Urbana, Ill., 1963.

Sklar, B. *Digital Communications: Fundamentals and Applications.* Prentice-Hall, Englewood Cliffs, N.J., 1988.

Skolnik, M. I. (editor). *Radar Handbook,* 2nd ed. McGraw-Hill, New York, 1970.

Smith, R. J., and R. C. Dorf. *Circuits, Devices, and Systems: A First Course in Electrical Engineering,* 5th ed. Wiley, New York, 1991.

Stiffler, J. J. *Theory of Synchronous Communications.* Prentice-Hall, Englewood Cliffs, N.J., 1971.

Stremler, F. G. *Introduction to Communication Systems,* 2nd ed. Addison-Wesley, Reading Mass., 1982.

Taub, H., and D. L. Schilling. *Principles of Communication Systems,* 2nd ed. McGraw-Hill, New York, 1986.

Van der Ziel, A. *Noise.* Prentice-Hall, Englewood Cliffs, N.J., 1954.

Van der Ziel, A. *Noise: Sources Characterization, Measurement.* Prentice-Hall, Englewood Cliffs, N.J., 1970.

Van Trees, H. L. *Detection, Estimation, and Modulation Theory,* Vols I–III. Wiley, New York, 1968 (Vol. 1), 1971 (Vol. II), 1970 (Vol. III).

Viterbi, A. J. *Principles of Coherent Communication.* McGraw-Hill, New York, 1966.

Viterbi, A. J., and J. K. Omura. *Principles of Digital Communication and Coding.* McGraw-Hill, New York, 1979.

Weber, C. L. *Elements of Detection and Signal Design.* Springer-Verlag, New York, 1987.

Williams A. B., and F. J. Taylor. *Electronic Filter Design Handbooks,* 2nd ed. McGraw-Hill, New York, 1988.

Williams, R. H. *Electrical Engineering Probability.* West St. Paul, Minn., 1991.

Wozencraft, J. M., and I. M. Jacobs. *Principles of Communication Engineering.* Wiley, New York, 1965.

Yuen, J. H. (editor). *Deep Space Telecommunications Engineering.* Plenum Press, New York, 1983.

Ziemer, R. E., and R. L. Peterson. *Digital Communications and Spread Spectrum Systems.* Macmillan, New York, 1985.

Ziemer, R. E., and R. L. Peterson. *Introduction to Digital Communication,* Macmillan, New York, 1992.

Ziemer, R. E., W. H. Tranter, and D. R. Fannin. *Signals and Systems,* 3rd ed. Macmillan, New York, 1993.

Articles

Bennett, W. R. "Envelope Detection of a Unit-Index Amplitude Modulated Carrier Accompanied by Noise," *IEEE Trans. Information Theory*, **IT-20**, 723–728 (November 1974).

Forney, G. D., Jr. "Coding and Its Application in Space Communications," *IEEE Spectrum*, pp. 46–58 (June 1970).

Hartley, R. V. L. "Transmission of Information," *Bell System Tech. J.*, **7**, 535–563 (1928).

Heller, J. A., and I. M. Jacobs. "Viterbi Decoding for Satellite and Space Communication," *IEEE Trans. Commun. Technol.*, **COM-19**, 835–848 (October 1971, Part II).

Houts, R. C., and R. S. Simpson. "Analysis of Waveform Distortion in Linear Systems," *IEEE Trans. Education*, pp. 122–125 (June 1968).

Slepian, D. "Information Theory in the Fifties," *IEEE Trans. Information Theory*, **IT-19**, 145–148 (March 1973).

Viterbi, A. J. "Information Theory in the Sixties," *IEEE Trans. Information Theory*, **IT-19**, 257–262 (May 1973).

Viterbi, A. J., J. K. Wolf, E. Zehavi and R. Padovani. "A Pragmatic Approach to Trellis-Coded Modulation," *IEEE Communications Magazine*, **27**, 11–19 (July 1989).

Wolf, J. K. "A Survey of Coding Theory: 1967–1972," *IEEE Trans. Information Theory*, **IT-19**, 381–389 (July 1973).

AUTHOR INDEX

SUBJECT INDEX